LIVERPOOL JMU LIBRARY
3 1111 01444 6809

Steroid Dimers

Steroid Dimers

Chemistry and Applications in Drug Design and Delivery

LUTFUN NAHAR

Leicester School of Pharmacy, De Montfort University, Leicester, UK

SATYAJIT D. SARKER

Department of Pharmacy, University of Wolverhampton, Wolverhampton, UK

A John Wiley & Sons, Ltd., Publication

This edition first published 2012

Registered office
John Wiley & Sons Ltd, The Atrium, Southern Gate, Chichester, West Sussex, PO19 8SQ, United Kingdom

For details of our global editorial offices, for customer services and for information about how to apply for permission to reuse the copyright material in this book please see our website at www.wiley.com.

Library of Congress Cataloging-in-Publication Data

Nahar, Lutfun, Ph. D.
Steroid dimers [electronic resource] : chemistry and applications in drug design and delivery / Lutfun Nahar, Satyajit D. Sarker.
1 online resource.
Includes bibliographical references and index.
Description based on print version record and CIP data provided by publisher; resource not viewed.
ISBN 978-1-119-97285-3 (MobiPocket) – ISBN 978-1-119-97284-6 (ePub) – ISBN 978-1-119-97094-1 (Adobe PDF) – ISBN 978-0-470-74657-8 (cloth) (print)
I. Sarker, Satyajit D. II. Title.
[DNLM: 1. Steroids–chemistry. 2. Dimerization. 3. Drug Delivery Systems. 4. Drug Design. QU 85]
572′.579—dc23

2011052550

A catalogue record for this book is available from the British Library.

Print ISBN: 9780470746578

Set in 10/12 Times Roman by Thomson Digital, Noida, India
Printed and bound in Malaysia by Vivar Printing Sdn Bhd

1 2012

Dedicated to our parents

Mariam Sattar *Sadhan Sarker*
Abdus Sattar *Madhuri Sarker*

Contents

Preface *ix*
List of Abbreviations *xi*

1 Introduction **1**
1.1 Steroids and Steroid Dimers 1
1.2 General Physical and Spectroscopic Properties of Steroid Dimers 2
1.3 Chromatographic Behaviour of Steroid Dimers 5
1.4 Applications of Steroid Dimers 6
References 6

2 Synthesis of Acyclic Steroid Dimers **7**
2.1 Dimers *via* Ring A–Ring A Connection 7
2.1.1 Direct Connection 7
2.1.2 Through Spacer Groups 21
2.2 Dimers *via* Ring B–Ring B Connection 68
2.2.1 Direct Connection 68
2.2.2 Through Spacer Groups 74
2.3 Dimers *via* Ring C–Ring C Connection 84
2.3.1 Through Spacer Groups 84
2.4 Dimers *via* Ring D–Ring D Connection 87
2.4.1 Direct Connection 87
2.4.2 Through Spacer Groups 89
2.4.3 Through Side Chain and Spacer Groups 100
2.5 Dimers *via* Ring A–Ring D Connection 151
2.5.1 Direct Connection 151
2.6 Dimers *via* Connection of C-19 169
2.7 Molecular Umbrellas 170
2.8 Miscellaneous 174
References 182

3 Synthesis of Cyclic Steroid Dimers **187**
3.1 With Spacer Groups: Cholaphanes 187
3.2 Without Spacer Groups: Cyclocholates 232
References 238

4 Naturally Occurring Steroid Dimers **241**
4.1 Cephalostatins 242
4.2 Crellastatins 254

4.3 Ritterazines 262
4.4 Others 277
References 284

5 Synthesis of Cephalostatin and Ritterazine Analogues 287
5.1 Introduction 287
5.2 Synthesis of Cephalostatin and Ritterazine Analogues 288
5.3 Total Synthesis of Naturally Occurring Cephalostatin 1 371
References 376

6 Applications of Steroid Dimers 379
6.1 Application of Steroid Dimers as '*Molecular Umbrellas*': Drug Delivery 379
6.2 Biological and Pharmacological Functions of Steroid Dimers: Drug Discovery and Design 382
6.2.1 Antimalarial Activity 383
6.2.2 Cytotoxicity and Anticancer Potential 386
6.2.3 Effect on Micellar Concentrations of Bile Salts and Serum Cholesterol Level 401
6.2.4 Effect on Bilayer Lipid Membranes 402
6.2.5 Supramolecular Transmembrane Ion Channels, and Artificial Receptors and Ionophores 402
6.2.6 Other Properties 404
References 405

***Index* 409**

Preface

Steroid dimers form an important group of pharmacologically active compounds that are predominantly biosynthesized by various marine organisms, and also synthesized in laboratories. These dimers can also be used to create *'molecular umbrella'* for drug delivery. While there are hundreds of such compounds and numerous research papers on these compounds available to date, there is no book documenting the chemistry and applications of these compounds. However, there are two reviews, one by Li and Dias, and the other one by us (Nahar, Sarker and Turner) published, respectively, in 1997 and 2007. We believe that this is the right time to publish a book on steroid dimers covering their chemistry and applications. This book will be a handy reference for the organic synthetic, medicinal and natural-products chemists working in the area of steroids, and drug design, discovery and development in general.

The primary readership of this book is expected to be the postgraduate synthetic organic, medicinal and natural-product chemists working either in academia or industries, especially in the area of drug design, discovery and delivery. This book will also be suitable for the postgraduate students (and undergraduate students to some extent) within the subject areas of Chemistry, Pharmacy, Biochemistry, Food Sciences, Health Sciences, Environmental Sciences and Life Sciences.

This book comprises six chapters. *Chapter 1* introduces the topic, 'steroid dimers', and builds the foundation of the subsequent chapters. *Chapters 2* and *3* deal with the synthesis and the chemistry of various classes of steroid dimers, including cyclic and acyclic dimers, placing particular emphasis on the types of connectivities. *Chapter 4* presents an overview on the naturally occurring steroidal dimers, e.g., cephalostatins, crellastatins and ritterazines. *Chapter 5* discusses the synthesis of cephalostatin and ritterazine analogues, as well as the total synthesis of the naturally occurring extremely cytotoxic steroidal dimer cephalostatin 1. *Chapter 6* looks into the applications of both synthetic and natural steroid dimers, and evaluates the importance of these dimeric compounds in drug design, discovery and delivery. It also elaborates the concept of *'molecular umbrella'* in the context of steroid dimers.

The major features of this book include easy-to-follow synthetic protocols for various classes of important dimeric steroids, source details, valuable spectroscopic data and depiction of unique structural features of natural steroidal dimers, applications of steroidal dimers, especially in relation to drug design, development and delivery, and the Structure-Activity-Relationships (SARs) of some pharmacologically active dimeric steroids.

Dr Lutfun Nahar
De Montfort University, Leicester

Professor Satyajit D. Sarker
University of Wolverhampton, Wolverhampton

October 2011

List of Abbreviations

Å	Angstrom
AcCl	Acetyl chloride
Ac_2O	Acetic anhydride
AcOH	Acetic acid
$AgBF_4$	Silver tetrafluoroborate
$AgNO_3$	Silver nitrate
Ag_2O	Silver oxide
AIBN	Azobisisobutyronitrile
Al(O-*t*-Bu)$_3$	Aluminum *tert*-butoxide
Ar	Argon
$BaCO_3$	Barium carbonate
BF_3	Boron trifluoride
$BF_3.Et_2O$	Boron trifluoride etherate
BH_3	Borane
$BH_3 \cdot SMe_2$	Borane-dimethyl sulphide
BnBr	Benzylbromide
BnOH	Benzyl alcohol
$(Boc)_2O$	Di-*tert*-butyl dicarbonate
$BrCH_2COBr$	Bromoacetyl bromide
n-BuLi	Butyllithium
Bu_4NOAc	Tetrabutylammonium acetate
2-BuOH	2-Butanol
t-BuOH	*tert*-Butanol or *tert*-butyl alcohol
t-BuOK	Potassium *tert*-butoxide
Bu_3Sb	Tributylstibine
Bu_2SnCl_2	Dibutyltin dichloride or dichlorodibutylstannane
Bu_3SnH	Tributyltin hydride or tributylstannane
^{13}C NMR	Carbon Nuclear Magnetic Resonance
$CaCO_3$	Calcium carbonate
CaH_2	Calcium hydride
$CaSO_4$	Calcium sulphate
m-CBPA	*meta*-Chloroperoxybenzoic acid
CC	Column chromatography
CCl_4	Carbon tetrachloride
C_5D_5N	Deuterated pyridine
$CeCl_3$	Cerium trichloride
$CeCl_3.7H_2O$	Cerium trichloride heptahydrate

$(CF_3CO)_2O$	Trifluorocacetic anhydride
C_6F_5OH	Pentafluorophenol
C_6F_5SH	Pentafluorothiophenol
C_6H_6	Benzene
C_6H_{12}	Cyclohexane
$CHCl_3$	Chloroform
CH_2I_2	Diiodomethane
CH_2N_2	Diazomethane
C_5H_5N	Pyridine
CH_3NO_2	Nitromethane
$C_2H_6O_2$	Ethane-1,2-diol or ethylene glycol
$CH(OMe)_3$	Trimethoxymethane or trimethyl orthoformate
$(C_3H_5O)_2O$	Propionic anhydride
CH_3ReO_3	Methyltrioxorhenium
CIMS	Chemical Ionization Mass Spectroscopy
$ClCH_2CO_2H$	Chloroethanoic acid
$(Cl_3CO)_2CO$	Triphosgene
$Cl_2P(O)OEt$	Dichloroethylphosphate
CME	Chloromethyl ethyl ether
CO_2	Carbon dioxide
$(COCl)_2$	Oxalyl chloride
COSY	Correlation Spectroscopy
CrO_3	Chromium trioxide
CSA	Camphorsulfonic acid
$Cu(AcO)_2$	Copper acetate
CuCN	Copper cyanide
CuI	Copper Iodide
$Cu(OTf)_2$	Copperbistrifluoromethanesulfonate or copper triflate
d	Day (s)
DABCO	1,4-diazabicyclo[2.2.2]octane
DBE	Di-*n*-butyl ether
DBU	1,8-Diazabicyclo[5.4.0]undec-7-ene
DCBC	2,6-Dichlorobenzoyl chloride
DCC	Dicyclohexyl carbodiimide
DCCC	Droplet counter current chromatography
DCE	Dichloroethene or dichloroethylene
DCM	Dichloromethane
DEAD	Diethyl azodicarboxylate
DEG	Diethylene glycol
DEPC	Diethylphosphoryl cyanide
DEPT	Distortionless Enhancement by Polarisation Transfer
DHEA	Dehydro-*epi*-androsterone
DHP	2,3-Dihydropyran
DHT	Dihydrotestosterone
$(DHQ)_2PHAL$	Hydroquinine 1,4-phthalazinediyl diether
DIAD	Diisopropyl azodicarboxylate

DIB	Diacetoxyiodobenzene
Dibromo-PEG	Dibromo-polyethyleneglycol
DIEA	*N,N*-diisopropylethylamine
DIPA	Diisopropylamine
DIPEA	Diisopropylethylamine
DMAP	4-Dimethylaminopyridine
DMDO	Dimethyldioxirane
DME	Dimethyl ether
DMF	Dimethylformamide
DMSO	Dimethyl sulphoxide
2,4-DNP	2,4-Dinitrophenylhydrazine
D_2O	Deuterated water or heavy water
DTBB	4,4′-Di-*tert*-butylbiphenyl
EDC	1-Ethyl-3-(3-dimethyl aminopropyl)-carbodiimide hydrochloride
EDTA	Ethylenediaminetetraacetic acid
EEDQ	*N*-Ethoxycarbonyl-2-ethoxy-1,2-dihydroquinoline
EIMS	Electron Ionisation Mass Spectroscopy
ESIMS	Electron Spray Ionisation Mass Spectroscopy
$EtCO_2Cl$	Ethyl chloroformate
Et_3N	Triethyl amine
$EtNO_2$	Nitroethane
Et_2O	Ether
EtOAc	Ethyl acetate
$EtO_2CC(N_2)PO(OEt)_2$	Ethyldiazophosphonate
$(EtO)_3CH$	Triethyl orthoformate
EtOH	Ethanol
$EtPPh_3Br$	Ethyltriphenylphosphonium bromide
FABMS	Fast Atom Bombardment Mass Spectroscopy
FCC	Flash column chromatography
$FeCl_3.Et_2O$	Ferric chloride etherate
$Fe(ClO_4)_3$	Ferric perchlorate
FTIR	Fourier-Transfer Infrared Spectroscopy
h	Hour (s)
H_2	Hydrogen
H_3BO_3	Boric acid
HBr	Hydrobromic acid or hydrogen bromide
HCl	Hydrochloric acid or hydrogen chloride
$HClO_4$	Perchloric acid
HCO_2H	Formic acid
H_2CrO_4	Chromic Acid
HDTC	Ethylene glycol
HF	Hydrofluoric acid or Hydrogen fluoride
HgO	Mercury oxide
HIO_4	Periodic acid
HMBC	Heteronuclear Multiple Bond Correlation

$(HMe_2Si)_2O$	Tris(dimethylsilyl)methane
$H_2N(CH_2)_2NH_2$	Ethylenediamine
HN_3	Hydrogen azide or hydrazoic acid
^1H NMR	Proton Nuclear Magnetic Resonance
H_2O	Water
H_2O_2	Hydrogen peroxide
HPLC	High Performance Liquid Chromatography
HREIMS	High Resolution Electron Impact Mass Spectroscopy
HRFABMS	High Resolution Fast Atom Bombardment Mass Spectroscopy
H_2SO_4	Sulphuric acid
HSQC	Heteronuclear Single Quantum Coherence
Hz	Hertz
I_2	Iodine
IR	Infrared Spectroscopy
KBr	Potassium bromide
K_2CO_3	Potassium carbonate
K_2CrO_4	Potassium chromate
$K_3Fe(CN)_6$	Potassium ferricyanide
KH	Potassium hydride
$KHCO_3$	Potassium bicarbonate or potassium hydrogen carbonate
KHMDS	Potassium hexamethyldisilazane
KI	Potassium iodide
$KMnO_4$	Potassium permanganate
KOAc	Potassium acetate
K_2OsO_4	Potassium osmate
$K_2S_2O_5$	Potassium metabisulphite
LDA	Lithium diisopropylamide
Li	Lithium
$LiAlH_4$	Lithium aluminium hydride
$LiBH_4$	Lithium borohydride
LiBr	Lithium bromide
$LiClO_4$	Lithium perchlorate
Li_2CO_3	Lithium carbonate
LiOH	Lithium hydroxide
LSIMS	Liquid secondary ion mass spectrometry
MALDI-TOF	Matrix-assisted laser desorption ionisation-time of flight
MeCN	Acetonitrile
MeCO	Acetyl or Ac
Me_2CO	Acetone
MeI	Methyl iodide
MeLi	Methyllithium
MeMgBr	Methyl magnesium bromide
Me_2NEt	*N*,*N*-Dimethylethylamine
MeOH	Methanol
$MeONH_2.HCl$	*O*-Methylhydroxylamine hydrochloride
MsCl	Methanesulfonyl chloride

Me_3SI	Trimethylsulfonium iodide
$MeSO_2NH_2$	Methane sulfonamide
Mg	Magnesium
$MgSO_4$	Magnesium sulphate
MHz	Megahertz
Mp	Melting point
MS	Mass Spectroscopy
MsCl	Methanesulfonyl chloride or mesyl chloride
MTBE	Methyl tertiary butyl ether
MTO	Methyltrioxorhenium
MWAM	Microwave assisted metathesis
m/z	Mass to charge ratio
N_2	Nitrogen
$NaBH_4$	Sodium borohydride
$NaBH(OAc)_3$	Sodium triacetoxyborohydride
$NaClO_2$	Sodium chlorite
$NaClO_4$	Sodium perchlorate
$NaCNBH_3$	Sodium cyanoborohydride
Na_2CO_3	Sodium carbonate
$Na_2Cr_2O_7$	Sodium dichromate
NaH	Sodium hydride
$NaHCO_3$	Sodium bicarbonate or sodium hydrogen carbonate
Na-Hg	Sodium-mercury or sodium amalgam
NaI	Sodium iodide
$NaIO_4$	Sodium periodate or sodium metaperiodate
NaN_3	Sodium azide
NaOAc	Sodium acetate
$NaOAc.3H_2O$	Sodium acetate trihydrate
NaOH	Sodium hydroxide
NaOMe	Sodium methoxide
Na_2SO_4	Sodium sulphate
$Na_2S_2O_3$	Sodium thiosulphate
$Na_2S_2O_5$	Sodium metabisulphite
NaTeH	Sodium hydrogen telluride
NBS	*N*-Bromosuccinimide
NH_3	Ammonia
N_2H_4	Hydrazine
NH_4Cl	Ammonium chloride
$N_2H_4.H_2O$	Hydrazine hydrate
NH_4OAc	Ammonium acetate
NH_4OH	Ammonium hydroxide
$NH_2OH.HCl$	Hydroxylamine hydrochloride
$NH_2OMe\cdot HCl$	Methoxyamine hydrochloride
$(NH_4)_2SO_4$	Ammonium Sulphate
NH_2SO_3H	Sulfamic acid
NMO	*N*-Methylmorpholine-*N*-oxide

NMR	Nuclear Magnetic Resonance
nOe	Nuclear Overhauser Effect
NOESY	Nuclear Overhauser Effect Spectroscopy
O_3	Ozone
OsO_4	Osmium tetroxide
OTs	Tosylate
$Pb(OAc)_4$	Lead tetraacetate
PCC	Pyridinium chlorochromate
Pd-C	Palladium on carbon
PDC	Pyridinium dichromate
$Pd(MeCN)_2Cl_2$	Bis(acetonitrile)dichloropalladium
Petroleum ether	Petroleum ether (40–60 °C)
PFPOH	Pentafluorophenol
Ph	Phenyl
$PhI(OAc)_2$	Diacetoxyiodobenzene
PhLi	Phenyllithium
$PhMe_3NBr_3$	Phenyltrimethylammonium tribromide
Ph_3P	Triphenylphosphine
Ph_3PAuCl	Chloro(triphenylphosphine)gold
$Ph_3P{=}CH_2Br$	Triphenyl(methyl)phosphonium bromide
PPh_3Cl	Dichlorotriphenylphosphane
Ph_3PO	Triphenylphosphine oxide
PPh_3Cl	Dichlorotriphenylphosphane
$(Ph_3P)_4Pd$	Tetrakis(triphenylphosphine)palladium
PhSeBr	Phenylselenyl bromide
$(PhSeO)_2O$	Phenylseleninic anhydride
$PhSO_2Cl$	Phenylsulphonyl chloride
PMP	*p*-Methoxyphenol
PPTS	Pyridinium *p*-toluenesulfonate
PTAB	Phenyltrimethylammoniumbromide tribromide
PTAP	Phenyltrimethylammonium perbromide
PTLC	Preparative Thin Layer Chromatography
PtO_2	Platinum dioxide or Adams' catalyst
P_2O_5	Phosphorus pentoxide
POPC	1-Palmitoyl-2-oleoyl-*sn*-glycero-3-phosphocholine
$i\text{-}Pr_2NEt$	*N,N*-Diisopropylethylamine
$n\text{-}PrNH_2$	*n*-propylamine
$i\text{-}Pr_2O$	Diisopropyl ether
i-PrOH	*iso*-Propanol or propan-2-ol
PVP	Polyvinylpyridine
RCM	Ring-closing metathesis
$Rh_2(OAc)_4$	Rhodium acetate
ROSEY	Rotating-frame Overhauser Effect SpectroscopY
r.t.	Room temperature
$RuCl_3$	Ruthenium trichloride
$RuCl_3.H_2O$	Ruthenium trichloride hydrate

L-Selectride	Lithium tris(*sec*-butyl)hydroborate
SiO_2	Silica or silicon dioxide
SO_2	Sulphur dioxide
SO_3	Sulphur trioxide
$SOCl_2$	Thionyl chloride
SPHRSIMS	Super Probe High Resolution Secondary Ion Mass Spectrometry
TBAF	Tetrabutylammonium fluoride
TBAHS	Tetrabutylammonium hydrogen sulphate
TBDMSCl	*tert*-butyldimethylsilyl chloride
TBDPS	*tert*-Butyldiphenylsilyl
TBDPSCl	*tert*-Butylchlorodiphenylsilane
TBMS	*tert*-Butyldimethylsilyl
TBDMSOTf	*tert*-Butyldimethylsilyl trifluoromethanesulfonate
TBSOTf	*tert*-Butyldimethylsilyl triflate
TEA	Triethanolamine
TEG	Triethylene glycol
TFAT	Trifluoroacetyl trifluoromethanesulfonate
THF	Tetrahydrofuran
THP	Tetrahydropyran
$TiCl_3$	Titanium trichloride
$TiCl_4$	Titanium tetrachloride
TIPS	Triisopropylsilyl
TIPSCl	Triisopropylsilyl chloridc
TLC	Thin Layer Chromatography
TMGA	Tetramethylguanidinium azide
TMS	Trimethylsilyl
TMSCN	Trimethylsilyl cyanide
$(TMS)_2O_2$	Bis(trimethylsilyl) peroxide
TMSOTf	Trimethylsilyl trifluoromethanesulfonate
TPAP	Tetrapropylammonium perruthenate
TPSCl	*tert*-Butylchloro diphenylsilane
TrCl	Triphenylmethylchloride
p-TsCl	*p*-Toluenesulfonyl chloride or tosyl chloride
p-TsOH	*p*-Toluenesulfonic acid
p-TsOH.H_2O	*p*-Toluenesulfonic acid hydrate
UV	Ultra violet
VLC	Vacuum Liquid Chromatography
Zn	Zinc
$ZnCl_2$	Zinc chloride

1

Introduction

1.1 Steroids and Steroid Dimers

Steroids are a family of biologically active lipophilic molecules that include cholesterol, steroidal hormones, bile acids and plant sterols (also known as phytosterols). These metabolic derivatives of terpenes are biosynthesized by plants as well as animals including humans, and play an important role in biological systems (Li and Dias, 1997; Nahar *et al.*, 2007a). Structurally, a steroid is a lipid molecule having a carbon skeleton with four fused rings; three fused cyclohexane rings, known as phenanthrene, are fused with a cyclopentane ring (Sarker and Nahar, 2007). The basic tetracyclic seventeen carbon steroidal ring system is known as *1,2-cyclopentano-perhydrophenanthrene* or simply *cyclopentaphenanthrene* (*Figure 1.1.1*). All steroids are derived from the acetyl CoA biosynthetic pathway. The four rings are lettered A, B, C, and D, and the carbon atoms are numbered beginning in the A ring. In steroids, the B, C, and D rings always are *trans*-fused, and in most natural steroids, rings A and B also are *trans*-fused. Each member of the steroid family has a structure that differs from the basic cyclopentaphenanthrene skeleton in the degrees of unsaturation within the rings and the identities of the hydrocarbon side chain substituents, *e.g.*, alkyl, alcohol, aldehyde, ketone or carboxylic acid functional groups, attached to the rings.

Even minor changes in the functionalities attached to the steroid skeleton can lead to significant changes in their biological and pharmacological activities (Nahar *et al.*, 2007a). That is why synthetic chemists have always been keen to carry out structural modifications of steroids to optimize their biological and pharmacological properties or to discover new properties. Steroid dimers are one of such group of modified steroids that are well known for their rigid, predictable and inherently asymmetric architecture.

Steroid dimer formation was first noticed during photochemical studies on steroids. During the investigation of the effect of sensitized light on the activation of ergosterol (**1**) in the absence of oxygen, it was discovered that in an alcoholic solution containing sensitizer, ergosterol on exposure to sunlight had undergone dehydrogenation to form a strongly

Steroid Dimers: Chemistry and Applications in Drug Design and Delivery, First Edition. Lutfun Nahar and Satyajit D. Sarker. Published 2012 by John Wiley & Sons, Ltd.

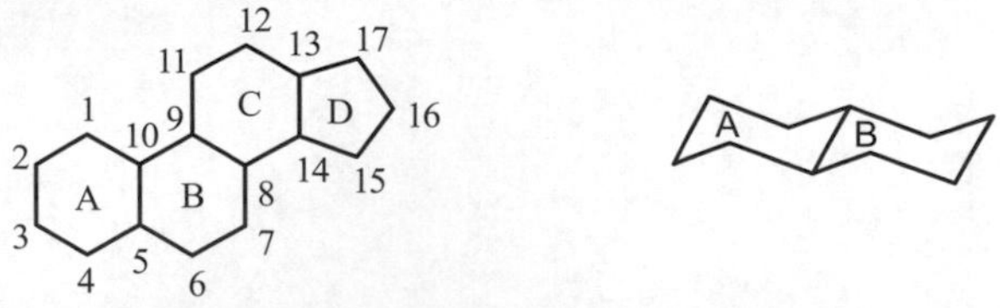

Figure 1.1.1 Cyclopentaphenanthrene skeleton (left) and trans-fused rings (right)

levorotatory substance ($[\alpha]_D$: −209°, mp: 205 °C) having double the original molecular weight and two hydroxyl groups. This bimolecular product was named bisergostatrienol (**2**) (*Scheme 1.1.1*) (Windaus and Borgeaud, 1928). Since this discovery, several dimeric steroids have been found in nature, particularly from marine sponges, and also have been synthesized in the laboratory (Nahar *et al.*, 2007a).

Steroid dimers can be classified broadly into acyclic dimers (also known as *'linear dimers'*) and cyclic dimers (*Figure 1.1.2*). Acyclic dimers involving connections between A, B, C or D rings, or *via* C-19, direct or through spacers, form the major group of steroid dimers (see *Chapter 2*). In the cyclic steroid dimers, dimerization of steroids, direct or through spacers, leads to formation of new ring systems or macrocyclic structures, *e.g.*, cyclocholates or cholaphanes, respectively (see *Chapter 3*). Steroid dimers can also be classified as symmetrical and unsymmetrical dimers; when a dimer is composed of two identical steroid monomeric units, it is called a symmetrical dimer, and when two different monomeric steroid units are involved or two identical monomeric steroid units are joined in a way that there is no symmetry in the resulting dimer, the dimer is known as an unsymmetrical dimer (*Figure 1.1.2*). One other way of classifying steroid dimers is to divide them into natural and synthetic dimers (*Figure 1.1.2*).

1.2 General Physical and Spectroscopic Properties of Steroid Dimers

In general, like most monomeric steroids, steroid dimers are lipophilic in nature and are not water soluble. However, depending on the monomeric steroid, spacer group or other

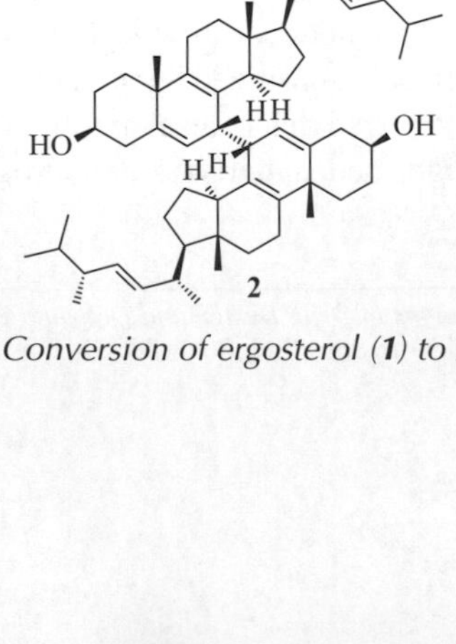

Scheme 1.1.1 Conversion of ergosterol (**1**) to bisergostatrienol (**2**)

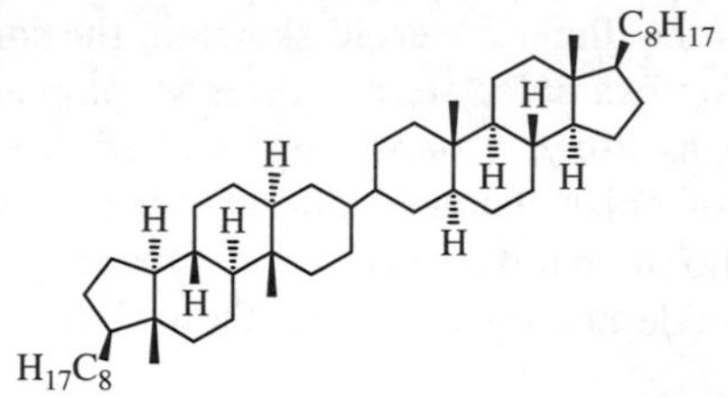

Bischolestane, a synthetic symmetrical acyclic (linear) steroid dimer

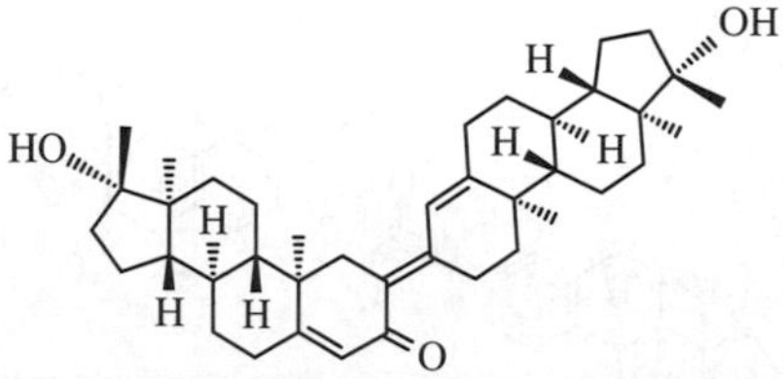

17α-methyltestosterone dimers, a synthetic acyclic unsymmetrical steroid dimer

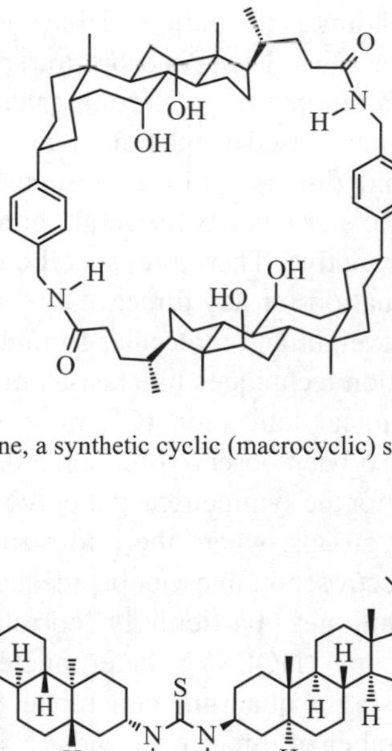

Cholaphane, a synthetic cyclic (macrocyclic) steroid dimer

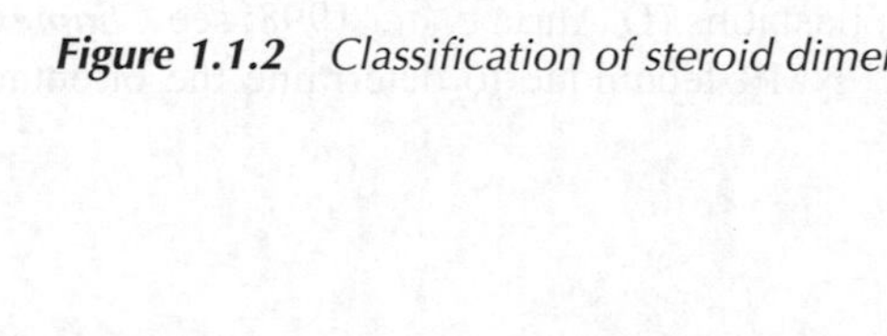

Japindine, a natural symmetrical steroid dimer

Figure 1.1.2 *Classification of steroid dimers*

functionalities present on the dimeric steroid skeleton, the solubility of such molecules can be quite variable. For example, steroid dimers composed of two sterol (steroid alcohol) units where the hydroxyl groups are not altered, as in bisergostatrienol (**2**), will retain some degree of polar character due to the hydroxyl groups, while keeping its nonpolar or hydrophobic nature because of the ring systems and other alkyl substituents or aliphatic side chains, and thus, these dimers will have properties like amphipathic lipids.

Most dimeric steroids are solids and can be transformed into well-formed crystals from various solvents (see *Chapters 2* and *3*), *e.g.*, bis[estra-1,3,5(10)-trien-17-on-3-yl]oxalate (**3**) was crystallized from $CHCl_3$-EtOAc (2:1) (Nahar, 2003).

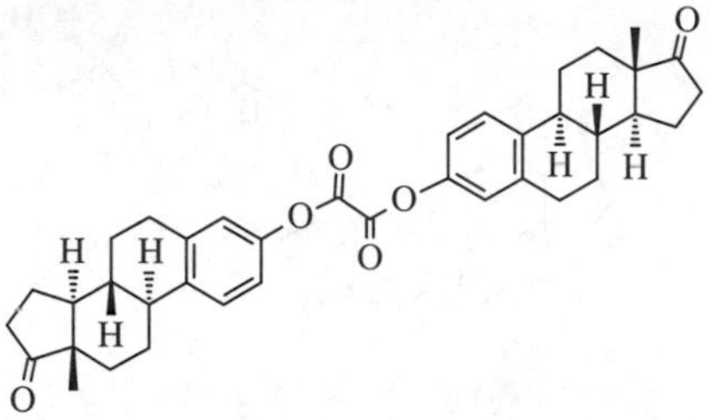

Bis[estra-1,3,5(10)-trien-17-on-3-yl]oxalate (**3**)

The melting points of steroid dimers are quite variable and depend on the monomer, the spacer groups and other functionalities. The UV absorption spectra of steroid dimers depend on the presence or absence of chromophores, *e.g.*, conjugated double bonds. The IR spectra can be different from dimer to dimer based on the functional groups present. Details on these spectral data of various steroid dimers will be presented in *Chapters 2–5*. Like the monomeric steroids, the dimeric steroids have several chiral centres in the molecule that make these molecules optically active. Therefore, specific rotation $[\alpha]_D$ data can provide additional characteristic information for any dimer.

To determine the molecular weight and molecular formula of steroid dimers, it is often essential to employ soft ionization techniques like fast-atom bombardment (FAB), electrospray ionization (ESI) or chemical ionization (CI) mass spectroscopy. The use of the MALDI–TOF technique has also been observed for some dimers very recently. MS information is particularly important for the symmetrical dimers composed of two identical steroid momoners without any spacer groups, where the information obtained from the nuclear magnetic resonance (NMR) spectroscopy may not be adequate to confirm the structure.

A range of 2D NMR techniques, particularly, correlation spectroscopy (COSY), nuclear Overhauser spectroscopy (NOESY), heteronuclear multiple bond coherence (HMBC) and heteronuclear single quantum coherence (HSQC), could be useful to confirm the structures of a number of dimeric steroids (Nahar, 2003; Nahar and Turner, 2003; Nahar *et al.*, 2006, 2007b). Sometimes, the use of the rotating frame Overhauser effect spectroscopy (ROESY) could be useful in establishing the relative stereochemistry, as in the case of crellastatins (D'Auria *et al.*, 1998; see *Chapter 4*). Fuzukawa *et al.* (1996) used ^{15}N-HMBC NMR technique to determine the orientation of the steroidal units about the pyrazine ring in ritterazine A (**4**). However, the use of the ^{15}N-HMBC NMR technique is rather limited.

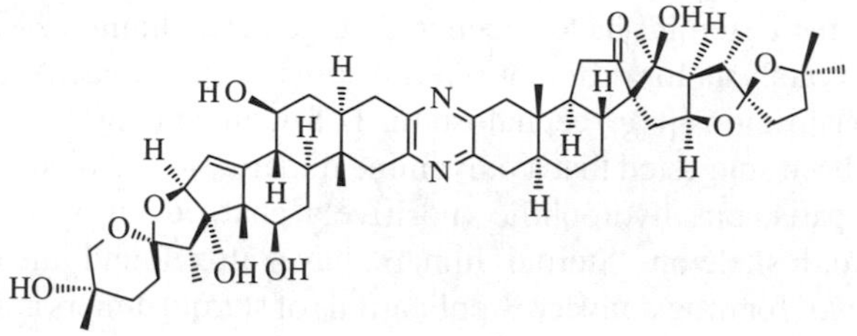

Ritterazine A (**4**)

1.3 Chromatographic Behaviour of Steroid Dimers

Most steroid dimers are nonpolar in nature and can be separated by normal-phase column, flash or thin layer chromatography (FCC or TLC) on silica gel (SiO_2) as the stationary phase and using various solvent mixtures, *e.g.*, *n*-hexane-EtOAc or $CHCl_3$-MeOH, as the mobile phase or eluent (Nahar, 2003). However, alumina or celite as the stationary phase has also been utilized for the separation of several steroid dimers.

On the TLC plates, steroid dimers can be detected by I_2 vapour, or using various sprays reagents, *e.g.*, vanillin-H_2SO_4 and Liebermann–Burchard reagents. For the detection of steroidal alkaloid dimers, *e.g.*, cephalostatin 1 (**5**), any alkaloid-detecting reagents, *e.g.*, Dragendorff's reagent, may be used.

The use of the reversed-phase high-performance liquid chromatography (HPLC) can equally be useful, and generally, MeOH-H_2O or MeCN-H_2O as the mobile phase, and a C_{18} reversed-phase column as the stationary phase can be used (Nahar, 2003). However, for the purification of some cephalostatins and ritterazines, a C_8 reversed-phase column was reported to be used (see *Chapter 4*).

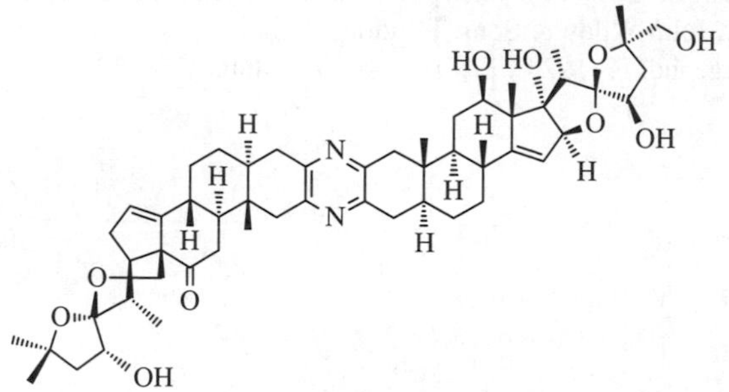

Cephalostatin 1 (**5**)

In some cases, for the initial separation of naturally occurring cytotoxic steroid dimers, *e.g.*, cephalostatins or ritterazines, solvent partitioning methods and droplet countercurrent chromatography (DCCC) have been regularly employed (see *Chapter 4*).

1.4 Applications of Steroid Dimers

Dimerization of steroid skeleton renders some unique characteristics that are applicable to different areas. Dimeric steroids have miceller, detergent, and liquid-crystal properties, and have been used as catalysts for different types of organic reactions. A number of dimeric steroids, *e.g.*, cephalostatins [*e.g.*, cephalostatin 1 (**5**)], are among the most potent natural cytotoxins. It has been suggested that a polyamine dimeric steroid binds to DNA due to the presence of two parts, one hydrophilic (positively charged nitrogen) and the other is hydrophobic steroid skeleton. Steroid dimers have also found their applications as *'molecular umbrella'* for drug delivery. Applications of steroid dimers are discussed further in *Chapter 6*.

References

D'Auria, M. V., Giannini, C., Zampella, A., Minale, L., Debitus, C. and Roussakis, C. (1998). Crellastatin A: a cytotoxic bis-steroid sulfate from the Vanuatu marine sponge *Crella* sp. *Journal of Organic Chemistry* **63**, 7382–7388.

Fukuzawa, S., Matsunaga, S. and Fusetani, N. (1996). Use of ^{15}N-HMBC techniques to determine the orientation of the steroidal units in ritterazine A. *Tetrahedron Letters* **37**, 1447–1448.

Li, Y. X. and Dias, J. R. (1997). Dimeric and oligomeric steroids. *Chemical Review* **97**, 283–304.

Nahar, L. (2003). PhD thesis: *Synthesis and some reactions of steroid dimers*, University of Aberdeen, UK.

Nahar, L. and Turner, A. B. (2003). Synthesis of ester-linked lithocholic acid dimers. *Steroids* **68**, 1157–1161.

Nahar, L., Sarker, S. D. and Turner, A. B. (2007a). A review on synthetic and natural steroid dimers: 1997–2006. *Current Medicinal Chemistry* **14**, 1349–1370.

Nahar, L., Sarker, S. D. and Turner, A. B. (2007b). Synthesis of 17β-hydroxy-steroidal oxalate dimers from naturally occurring steroids. *Acta Chimica Slovenica* **54**, 903–906.

Nahar, L., Sarker, S. D. and Turner, A. B. (2006). Facile synthesis of oxalate dimers of naturally occurring 3-hydroxysteroids. *Chemistry of Natural Compounds* **42**, 549–552.

Sarker, S. D. and Nahar, L. (2007). *Chemistry for Pharmacy Students: General, Organic and Natural Product Chemistry*, John Wiley & Sons, London.

Windaus, A. and Borgeaud, P. (1928). *Liebig's Annuals* **460**, 235–237.

2

Synthesis of Acyclic Steroid Dimers

Several steroid dimers have been synthesized over the years (Li and Dias, 1997a; Nahar *et al.*, 2007a), and acyclic dimers involving connections between A, B, C or D rings, direct or through spacers, form the major group of such molecules. These dimers are also referred to as *'linear dimers'*. In this chapter, dimers connected *via* ring A–ring A, ring B–ring B, ring C–ring C, ring D–ring D, ring A–ring D, dimers *via* C-19 and *'molecular umbrellas'* are discussed.

2.1 Dimers *via* Ring A–Ring A Connection

2.1.1 Direct Connection

Direct ring A–ring A connection between two steroid units can be achieved by using an active metal reduction or a photochemical condensation. Several dimers were synthesized from steroidal 4-en-3-one by photochemical, electrolytic, and metal reduction (Squire, 1951; Lund, 1957; Bladon *et al.*, 1958). Squire (1951) reported the synthesis of bicholestane (**5**) in three steps by chlorination of cholesterol (**1**, cholest-5-en-3β-ol), hydrogenation of cholesteryl chloride (**2**, 3β-chlorocholest-5-ene) and finally the coupling between cholestanyl chloride (**3**, 3β-chloro-5α-cholestane) and cholestenyl magnesium chloride (**4**, 3β-magnesium chloro-5α-cholestane) (prepared *in situ*) (*Scheme 2.1.1*).

Cholesteryl chloride (**2**) was first obtained from cholesterol (**1**) using the method described in the literature (Daughenbaugh, 1929). To a stirred solution of **1** (5.0 g, 12.93 mmol) in dry C_5H_5N (1 mL), $SOCl_2$ (10 mL) was added and the reaction mixture was refluxed for 60 min (*Scheme 2.1.1*). After complete liberation of SO_2, the reaction mixture was cooled, H_2O was added, and extracted with Et_2O. The etheral solution was dried (Na_2CO_3) and evaporated under pressure to a semicrystalline residue, which was recrystallized from EtOH and identified as cholesteryl chloride (**2**, 3.3 g, 63%, mp: 95–96 °C) (Daughenbaugh, 1929). Compound **2** was then utilized for the synthesis of cholestanyl chloride (**3**) (Squire, 1951).

To a stirred solution of **2** (630 mg, 1.56 mmol) in Et_2O (15 mL) and EtOH (15 mL), PtO_2 (38 mg) was added (*Scheme 2.1.1*). The reaction mixture was kept under 2 atm of H_2 for

Steroid Dimers: Chemistry and Applications in Drug Design and Delivery, First Edition. Lutfun Nahar and Satyajit D. Sarker.

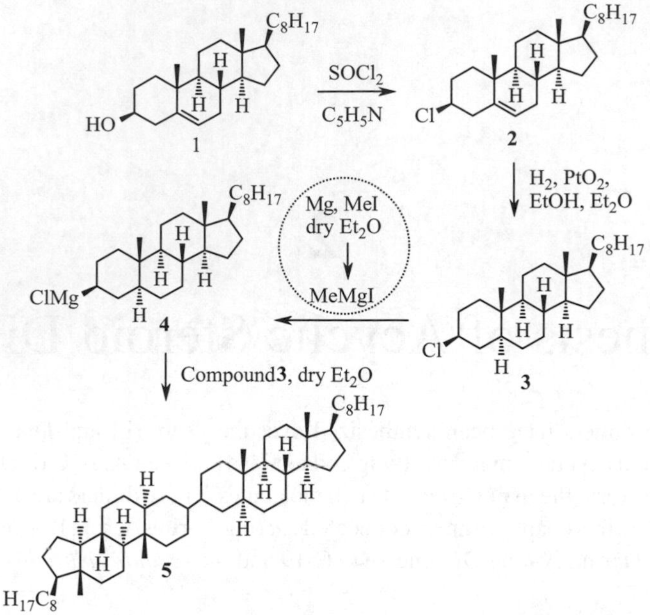

Scheme 2.1.1 *Synthesis of 3β-chlorocholest-5-ene (**2**), 3β-chloro-5α-cholestane (**3**) and bicholestane (**5**)*

60 min with stirring. On completion of the hydrogenation, the reaction mixture was filtered and rotary evaporated to yield cholestanyl chloride (**3**, 510 mg, 80%, mp: 115–115.5 °C), which was employed for the synthesis of bicholestane (**5**) (Squire, 1951).

Compound **3** (5.0 g), Mg powder (10 g, 41.0 mmol) and MeI (10 mL) in dry Et_2O (5 mL) were refluxed under N_2 (*Scheme 2.1.1*). After cholestenyl magnesium chloride (**4**) had formed, a solution of **3** (820 mg, 2.01 mmol) in dry Et_2O (15 mL) was added carefully, and the reaction mixture was stirred for 22 h. The reaction mixture was cooled (0 °C) and treated with CO_2 gas at 1.5 atm pressure for another 18 h, after which, a mixture of ice (100 g) in conc. HCl (20 mL) was added and left for a further 3 h. The ice cooled solution was extracted with Et_2O (100 mL), the Et_2O layer was successively washed with 0.1N NaOH (5 × 20 mL) and H_2O (5 × 10 mL), dried (Na_2SO_4), filtered and evaporated under vacuum. Recrystallization of the residue from Et_2O afforded the ring A-ring A dimer, bischolestane (**5**, 37 mg, 5%, mp: 265 °C) (Squire, 1951).

Cholestenone pinacol, bis(3β-hydroxycholest-4-en-3′α-yl) (**7**), was produced with high yield (85%) from the reaction between cholestenone (**6**, cholest-4-en-3-one) and Na-Hg in AcOH (*Scheme 2.1.2*) (Squire, 1951). Later, the same dimer **7** was also synthesized by electrolytic reduction of **6** (Bladon *et al.*, 1958). Cholestenone (**6**) was produced from a stirred solution of **1** (20 g, 51.73 mmol) in Me_2CO by the treatment with $Al(O\text{-}t\text{-}Bu)_3$ according to procedure described in the literature (Oppenauer, 1937) (*Scheme 2.1.2*). The crude dimer was purified initially by TLC on Al_2O_3, and recrystallized from MeOH-petroleum ether to yield **6** (17.9 g, 90%, mp: 78–79 °C), which was employed to prepare cholestenone pinacol **7**.

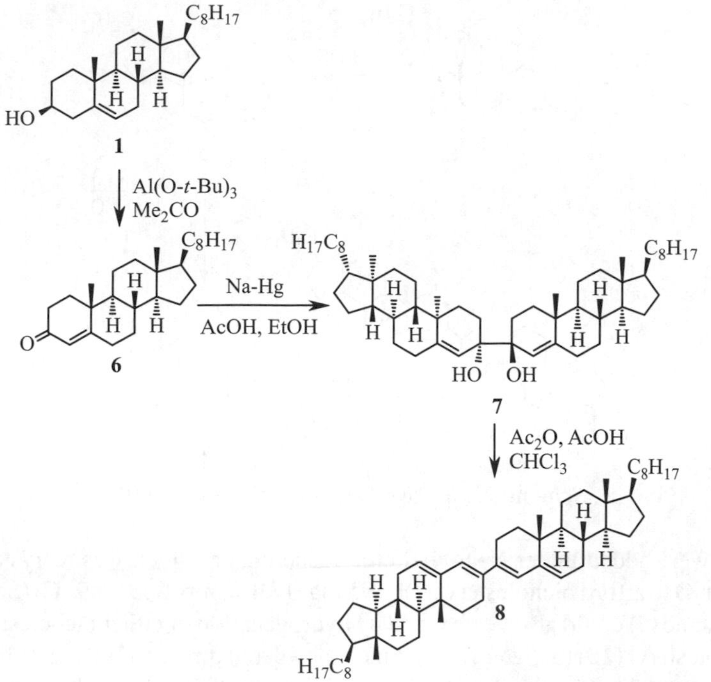

*Scheme 2.1.2 Synthesis of cholest-4-en-3-one (**6**), bis(3β-hydroxycholest-4-en-3′α-yl) (7), bicholesta-3,5-dienyl (**8**)*

To a stirred solution of **6** (2.0 g, 5.19 mmol) in AcOH (75 mL) and EtOH (75 mL), 2% Na-Hg (300 g) was added slowly over 30 min under moderate heat (*Scheme 2.1.2*). The reaction mixture was boiled for 5 min, cooled and poured into H_2O (600 mL), resulting in the formation of precipitates, which was filtered under vacuum, washed with H_2O and dried (Na_2SO_4). The crude dimer was dissolved in hot C_6H_6 (100 mL), filtered and concentrated down to 25 mL, to which, hot Me_2CO (100 mL) was added. The resulting mixture was cooled in an ice bath to generate colourless crystals upon standing overnight. The crystals were filtered and dried *in vacuo* to give cholestenone pinacol **7** (1.7 g, 85%, mp: 200–205 °C) (Squire, 1951).

Cholestenone pinacol **7** was employed for the synthesis of bicholesta-3,5-dienyl (**8**) (*Scheme 2.1.2*). Glacial AcOH (5 mL) and Ac_2O (5 mL) were added to a stirred solution of **7** (3.0 g, 3.89 mmol) in $CHCl_3$ (250 mL) under reflux. After 4 h, $CHCl_3$ was slowly evaporated over another 4 h, MeOH (100 mL) was added to the reaction mixture to form precipitation, which was filtered off and dried *in vacuo* to obtain a light yellow solid. Recrystallization of this solid from Me_2CO-$CHCl_3$ yielded **8** (2.6 g, 83%, mp: 244–246 °C) (Squire, 1951; Bladon, 1958).

The synthesis of bi(cholest-5-ene) (**10**, bicholesteryl) was accomplished in two steps: by chlorination of **1** forming cholesteryl chloride (**2**), and coupling of **2** and cholesteryl magnesium chloride (**9**) (prepared *in situ*) (*Scheme 2.1.3*). The synthetic protocol for **10** was similar to that described earlier for the dimer **5**. To a solution of **9** (1.30 g, 3.04 mmol) in dry Et_2O, a solution of **2** (1.23 g, 3.04 mmol) in dry 2-BuOH (50 mL) and conc. H_2SO_4

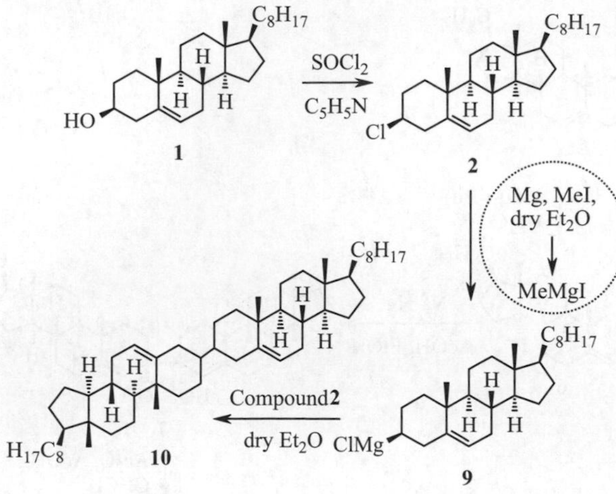

Scheme 2.1.3 *Synthesis of bicholesteryl (10)*

(20 drops) were added under N_2 to yield the crude dimer, which was recrystallized from 2-BuOH-Et_2O to afford bicholesteryl (**10**, 195 mg, 17.4%, mp: 266–269 °C) (Squire, 1951).

Bicholestane (**5**) could also be obtained by hydrogenation of either bicholesta-3,5-dienyl (**8**) or bicholesteryl (**10**) (*Scheme 2.1.4*). Finely powdered dimer **8** (100 mg, 0.14 mmol) and PtO_2 catalyst (100 mg) were dissolved in cyclohexane (250 mL) under constant stirring

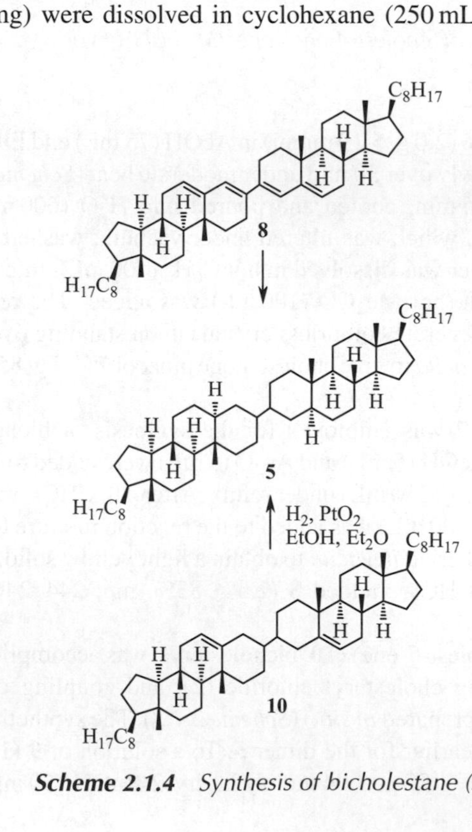

Scheme 2.1.4 *Synthesis of bicholestane (5)*

(3 atm of H_2 for 18 h). The disappearance of the yellow colour in the solution indicated completion of the reaction. After filtration, the solution was evaporated to dryness to produce a colourless solid, which was recrystallized from $CHCl_3$-cyclohexane to give bicholestanyl (**5**, 60 mg, 57.7%, mp: 265–270 °C). The hydrogenation of the dimer **10** (100 mg, 0.14 mmol) was also carried out following the same method to yield fine needles (97 mg), which were further purified by recrystallization as outlined above to yield the dimer **5** (70 mg, 67.3%) (Squire, 1951).

Pinacol-like synthetic approach was also adopted to synthesize cholestenone pinacol **7** from cholestenone (**6**) by reductive condensation with Na-Hg in presence of $Pb(OAc)_4$ (Dulou *et al.*, 1951). To a stirred ethanolic (600 mL) solution of cholestenone (**6**, 20 g, 51.73 mmol) and $Pb(OAc)_4$, 4% Na-Hg (600 g) was added over 30 min under moderate heat (*Scheme 2.1.5*). The reaction mixture was left overnight to form white precipitates, which were filtered under vacuum, washed with H_2O and dried under vacuum. The precipitates were dissolved in hot C_6H_6 (100 mL), filtered, concentrated down to 25 mL, to which, hot Me_2CO (100 mL) was added and subsequently cooled in an ice bath to induce crystal formation upon standing overnight. The colourless crystals were filtered and dried *in vacuo* to obtain cholestenone pinacol **7** (1.7 g, 8.9%, mp: 200–205 °C) (Dulou *et al.*, 1951).

Instead of Na-Hg, Zn dust was also used in the synthesis of a symmetrical C-3 linked testosterone dimer, bis(17β-acetyloxy-4-chloro-3β-hydroxyandrost-4-en-3′α-yl) (**13**), which was, in fact, a major reaction byproduct (Templeton *et al.*, 1990). Dimer **13** was prepared in two steps from testosterone acetate (**11**, 17β-acetyloxyandrost-4-en-3-one): by chlorination of **11** to afford 4-chlorotestosterone acetate (**12**, 17β-acetyloxy-4-chloroandrost-4-en-3-one, mp: 228–231 °C), following the method described in the literature (Mori, 1962), and then by Zn in AcOH reduction of **12**.

To a stirred solution of **12** (5.84 g, 17.98 mmol) in AcOH (300 mL), Zn dust (120 g) was added and the stirring was continued for 5.5 h at room temperature (*Scheme 2.1.6*). After completion of the reaction, the solution was filtered and evaporated under pressure. The resulting residue was dissolved in H_2O and extracted with Et_2O. The Et_2O solution was washed with H_2O followed by saturated aqueous $NaHCO_3$ and dried (Na_2SO_4). After evaporation of Et_2O, the residue was redissolved in DCM and purified by FCC, eluted with 1% Me_2CO in DCM, to produce the crude dimer (885 mg), which was recrystallized from MeOH-DCM yielding bis(17β-acetyloxy-4-chloro-3β-hydroxyandrost-4-en-3′α-yl) (**13**, 233 mg, 4%, mp: 218–222 °C) (Templeton *et al.*, 1990).

Similarly, Zn was utilized in the synthesis of symmetrical C-3 linked isomeric dimer, bis (methyl-3α-hydroxyandrost-4-en-17-on-19-oate-3′α-yl) (**17**) from 19-hydroxyandrost-4-ene-3,17-dione (**14**) in three steps (Templeton *et al.*, 2000). The use of Li in NH_3 almost doubled (12%) the yield of androstenedione dimer **17**. The same group also synthesized bis

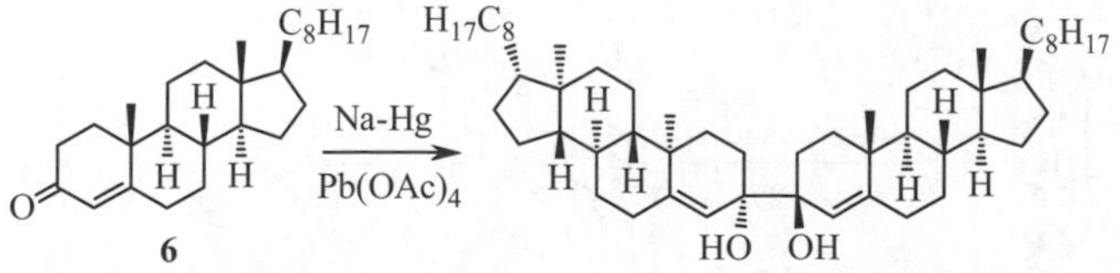

Scheme 2.1.5 *Synthesis of bis(3β-hydroxycholest-4-en-3′α-yl) (7)*

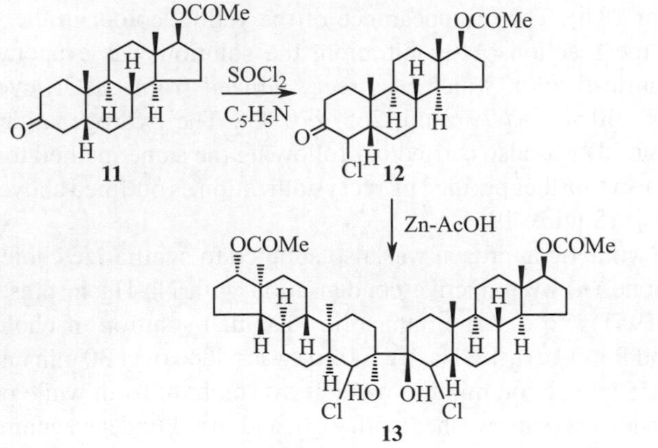

Scheme 2.1.6 *Synthesis of 17β-acetyloxy-4-chloroandrost-4-en-3-one (**12**) and bis(17β-acetyloxy-4-chloro-3β-hydroxyandrost-4-en-3′α-yl) (**13**)*

(methyl 3α-trimcthylsilyloxyandrost-4-en-17-on-19-oate-3′α-yl) (**18**) and bis(methyl 3α-trimethylsilyloxyandrost-4-en-17-on-19-oate-3′β-yl) (**19**) in 6% and 2.5% yields, respectively (Templeton *et al*., 2000).

To a stirred solution of 19-hydroxyandrost-4-ene-3,17-dione (**14**, 10 g, 33 mmol) in Me_2CO (250 mL) in an ice bath, Jones' reagent (30 mL, 80 mmol) was added over 40 min (*Scheme 2.1.7*). The resulting solid was crystallized from DCM-petroleum ether to give **15** (5.0 g, 48%, mp: 146–148 °C). ^{1}H NMR (300 MHz, $CDCl_3$): δ 0.91 (s, 3H, 19-Me), 5.95 (d, 1H, $J = 1.2$ Hz, 4-CH). ^{13}C NMR (75 MHz, $CDCl_3$): δ 33.7 (C-1), 34.8 (C-2), 198.7 (C-3), 127.2 (C-4), 161.6 (C-5), 32.6 (C-6), 31.4 (C-7), 35.6 (C-8), 53.7 (C-9), 50.5 (C-10), 21.6 (C-11), 30.0 (C-12), 47.6 (C-13), 50.9 (C-14), 22.0 (C-15), 35.7 (C-16), 220.0 (C-17), 13.9 (C-18), 175.5 (C-19) (Templeton *et al*., 1990).

The treatment of androst-4-ene-3,17-dion-19-oic acid (**15**, 2.26 g, 7.14 mmol) with CH_2N_2 in Et_2O produced **16** (2.1 g, 89%, mp: 142–145 °C) (*Scheme 2.1.7*). ^{1}H NMR (300 MHz, $CDCl_3$): δ 0.90 (s, 3H, 18-Me), 3.76 (s, 3H, 19-OMe), 5.90 (d, 1H, $J = 1.5$ Hz, 4-CH). ^{13}C NMR (75 MHz, $CDCl_3$): δ 33.8 (C-1), 34.9 (C-2), 198.6 (C-3), 126.8 (C-4), 162.0 (C-5), 32.6 (C-6), 31.4 (C-7), 35.6 (C-8), 53.8 (C-9), 50.9 (C-10), 21.6 (C-11), 30.1 (C-12), 47.5 (C-13), 50.9 (C-14), 21.9 (C-15), 35.7 (C-16), 220.0 (C-17), 13.8 (C-18), 171.6 (C-19), 50.9 (19-OMe) (Templeton *et al*., 2000).

Zn powder (10 g) was added to a stirred solution of **16** (526 mg, 1.6 mmol) in aqueous AcOH (20 mL) (*Scheme 2.1.8*). After continuous stirring of the mixture for 6 h at room temperature, it was filtered, and the filtrate was extracted with DCM. The organic solvent

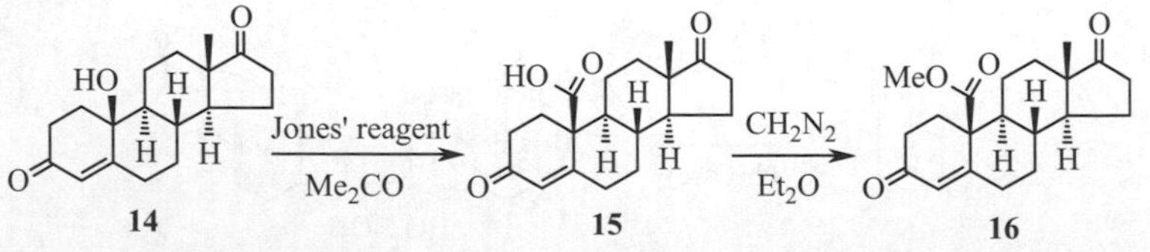

Scheme 2.1.7 *Synthesis of androst-4-ene-3,17-dion-19-oic acid (**15**) and methyl androst-4-ene-3,17-dion-19-oate (**16**)*

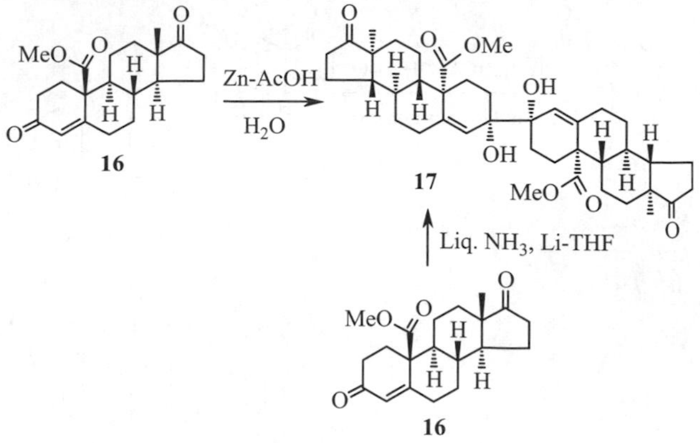

Scheme 2.1.8 *Synthesis of bis(methyl-3α-hydroxyandrost-4-en-17-on-19-oate-3′α-yl) (**17**)*

was removed by evaporation and purified by FCC (mobile phase: 3–25% Me_2CO in petroleum ether), followed by crystallization from EtOAc-DCM to afford bis(methyl-3α-hydroxyandrost-4-en-17-on-19-oate-3α′-yl) (**17**, 67 mg, 6.5%, mp: 217–220 °C). IR (DCM): ν_{max} cm^{-1} 3595br (O–H), 3553br (O–H), 1735s (C=O). ^1H NMR (300 MHz, $CDCl_3$): δ 0.93 (s, 3H, 18-Me), 2.38 (m, 1H, 6β-CH), 2.45 (dd, 1H, J = 9.3, 19.3 Hz, 16β-CH), 3.69 (s, 3H, 19-OMe), 5.69 (s, 1H, 4-CH). ^{13}C NMR (75 MHz, $CDCl_3$): δ 29.7 (2 × C-1), 28.6 (2 × C-2), 73.6 (2 × C-3), 124.6 (2 × C-4), 143.1 (2 × C-5), 33.9 (2 × C-6), 31.6 (2 × C-7), 36.2 (2 × C-8), 52.7 (2 × C-9), 50.2 (2 × C-10), 21.7 (2 × C-11), 31.7 (2 × C-12), 47.8 (2 × C-13), 51.5 (2 × C-14), 21.7 (2 × C-15), 35.7 (2 × C-16), 221.0 (2 × C-17), 13.8 (2 × C-18), 174.3 (2 × C-19), 51.4 (2 × 19-OMe). EIMS: *m/z* 644 $[M - H_2O]^+$, 626 $[M - 2H_2O]^+$. FABMS *m/z*: 685 $[M + Na]^+$, 663 $[M + H]^+$ (Templeton *et al*, 2000).

A solution of methyl androst-4-ene-3,17-dion-19-oate (**16**, 709 mg, 2.15 mmol) in THF (20 mL) was added dropwise to a stirred solution of liquid NH_3 (150 mL) and THF (10 mL) containing dissolved Li metal (700 mg, 0.1 mol) over 20 min (*Scheme 2.1.8*). After 4 h, solid NH_4Cl (6.0 g, 11.0 mmol) and DCM (150 mL) was added to the reaction mixture. The mixture was rotary evaporated and purified by FCC, eluted with 20% EtOAc in petroleum ether, followed by crystallization from EtOAc-DCM to yield the dimer **17** (170 mg, 12%). The mp, ^1H and ^{13}C NMR data were in good agreement with the product above (Templeton *et al*, 2000).

Bis(methyl 3α-trimethylsilyloxyandrost-4-en-17-on-19-oate-3′α-yl) (**18**) and bis(methyl 3α-trimethylsilyloxyandrost-4-en-17-on-19-oate-3′β-yl) (**19**) were prepared from methyl androst-4-ene-3,17-dion-19-oate (**16**) using Zn and TMS-imidazole (*Scheme 2.1.9*). A solution of **16** (2.22 g, 6.72 mmol) in aqueous AcOH (40 mL), Zn powder (40 g) was added in one portion and stirred for 2.5 h. The resulting residue was treated with TMS-imidazole (2.24 mL, 15.3 mmol) in DCM (5 mL) for 60 min. The reaction mixture was evaporated to dryness and purified by FCC to afford silylated dimers **18** and **19**. After crystallization from EtOAc-DCM, dimer **18** (270 mg, 6%, mp: 258–260 °C) could be obtained. IR (DCM): ν_{max} cm^{-1} 1735s (C=O). ^1H NMR (300 MHz, $CDCl_3$): δ 0.04 (s, 6H, 2 × $SiMe_3$), 0.90 (s, 6H, 2 × 18-Me), 2.46 (dd, 2H, J = 8.8, 19.2 Hz, 2 × 16β-CH), 2.62, (td, 2H, J = 3.3, 13.7 Hz, 2 × 6β-CH), 3.67 (s, 6H, 2 × 19-OMe), 5.63 (s, 2H, 2 × 4-CH). ^{13}C NMR (75 MHz, $CDCl_3$): δ 31.0 (2 × C-1), 30.5 (2 × C-2), 77.9 (2 × C-3), 127.3 (2 × C-4),

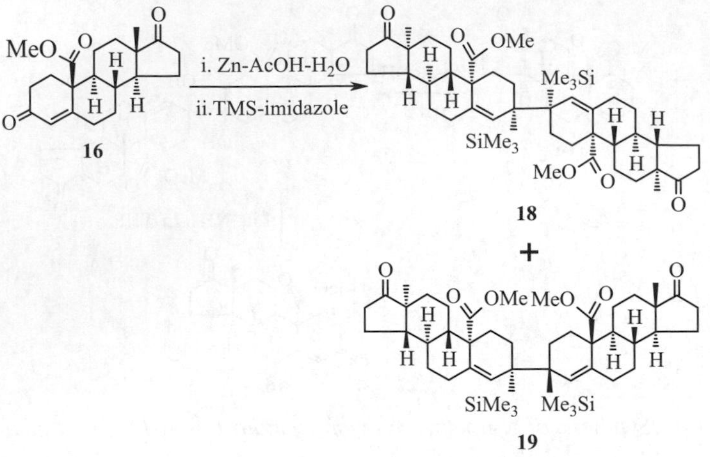

Scheme 2.1.9 *Synthesis of bis(methyl 3α-trimethylsilyloxyandrost-4-en-17-on-19-oate-3′α-yl) (**18**) and bis(methyl 3α-trimethylsilyloxyandrost-4-en-17-on-19-oate-3′β-yl) (**19**)*

140.9 (2 × C-5), 34.0 (2 × C-6), 31.6 (2 × C-7), 36.0 (2 × C-8), 53.1 (2 × C-9), 50.0 (2 × C-10), 21.6 (2 × C-11), 31.4 (2 × C-12), 47.8 (2 × C-13), 51.3 (2 × C-14), 21.7 (2 × C-15), 35.8 (2 × C-16), 220.9 (2 × C-17), 13.9 (2 × C-18), 174.2 (2 × C-19), 51.4 (2 × 19-OMe), 2.6 (2 × 3-$SiMe_3$). EIMS: *m/z* 644 $[M - H_2O]^+$, 626 $[M - 2H_2O]^+$. FABMS: *m/z* 807 $[M + H]^+$, 829 $[M + Na]^+$ (Templeton *et al.*, 2000).

Crystallization from MeOH-EtOAc provided dimer **19** (130 mg, 2.5%, mp: 181–182 °C) (*Scheme 2.1.9*). IR (DCM): ν_{max} cm^{-1} 1735s (C=O). ^1H NMR (300 MHz, $CDCl_3$): δ 0.02 and 0.06 (s, 6H, 2 × 3-$SiMe_3$), 0.84 and 0.86 (s, 6H, 2 × 18-Me), 2.42 (dd, 2H, $J = 8.8$, 18.8 Hz, 2 × 16β-CH), 2.52 and 2.62 Hz (td, 2H, $J = 3.3$, 13.7 Hz, 6β-CH), 3.62 and 3.65 (s, 6H, 2 × 19-OMe), 5.35 and 5.69 (s, 2H, 2 × 4-CH). ^{13}C NMR (75 MHz, $CDCl_3$): δ 29.7 and 30.3 (2 × C-1), 29.0 and 29.7 (2 × C-2), 77.9 and 78.1 (2 × C-3), 127.3 and 127.6 (2 × C-4), 140.1 and 140.8 (2 × C-5), 33.8 and 34.0 (2 × C-6), 31.4 and 31.5 (2 × C-7), 35.9 and 36.0 (2 × C-8), 52.3 and 52.9 (2 × C-9), 49.9 and 50.0 (2 × C-10), 21.6 and 21.8 (2 × C-11), 30.7 and 30.9 (2 × C-12), 47.6 and 47.7 (2 × C-13), 51.1 and 51.14 (2 × C-14), 22.1 and 22.6 (2 × C-15), 35.71 (2 × C-16), 220.5 and 220.6 (2 × C-17), 13.76 and 13.81 (2 × C-18), 173.5 and 173.8 (2 × C-19), 51.2 and 51.3 (2 × 19-OMe), 2.7 and 3.2 (2 × 3-Me_3Si). EIMS: *m/z* 644 $[M - H_2O]^+$, 626 $[M - 2H_2O]^+$. FABMS: *m/z* 829 $[M + Na]^+$, 807 $[M + H]^+$ (Templeton *et al.*, 2000).

Titanium trichloride ($TiCl_3$) along with $LiAlH_4$ was used in the synthesis of cholestenone-based dimers **20** (*anti*) and **21** (*syn*) from cholestenone (**6**) (*Scheme 2.1.10*) (McMurry, 1983). Another dimer, bis(cholesta-3,5-diene) (**22**, bischolesterienyl) as the oxidized product of cholesterol (**1**) with $Fe(ClO_4)_3$, has been reported (*Scheme 2.1.11*) (Martin and Hartney, 1978).

A simple metal reduction of androstadienedione (**23**, androsta-1,4-diene-3,17-dione) using Zn produced a mixture of *E/Z* androstadienedione dimers **24** and **25** (Schmidt *et al.*, 1992). To a stirred solution of Zn dust (19 g, 154 mmol) in THF (60 mL), conc. HCl (1 mL) was added (*Scheme 2.1.12*). When the gas formation was complete, androstadienedione (**23**, 1.0 g, 3.52 mmol) was added and the mixture was refluxed for 10 h.

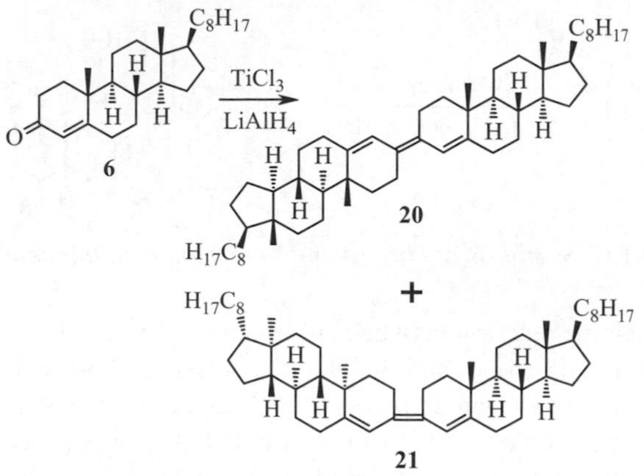

Scheme 2.1.10 Synthesis of cholestanone-based dimers **20** and **21**

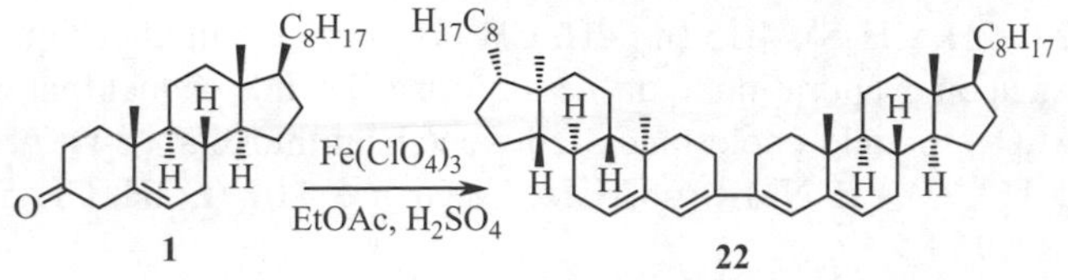

Scheme 2.1.11 Synthesis of bis(cholesta-3,5-diene) (**22**)

The reaction mixture was cooled, filtered and the solvent was evaporated under pressure. Crystallization of the residue from MeOH-H_2O afforded a mixture of *E/Z* androstadienedione dimers **24** and **25** (910 mg, 97%, mp: (*E/Z*-mixture) = 230–240 °C). IR (KBr): ν_{max} cm^{-1} 1735s (C=O). UV ($CHCl_3$): λ_{max} nm 327, 340, 358. ^{1}H NMR (200 MHz, $CDCl_3$): δ 0.86 (s, 6H, 2 × 18-Me), 1.15 (s, 6H, 2 × 19-Me), 5.84 (d, 2H, *J* = 10.2 Hz, 2 × CH), 6.35 (m, 2H, 2 × CH), 6.65 (s, 2H, *J* = 10.3 Hz, 2 × CH) (Schmidt *et al.*, 1992).

Cholestanone dimer, 2α-(2′-cholesten-3′-yl)-3-cholestanone (**30**) could be synthesized in three steps from cholestanone (**26**) and the dimerization could be carried out without the use

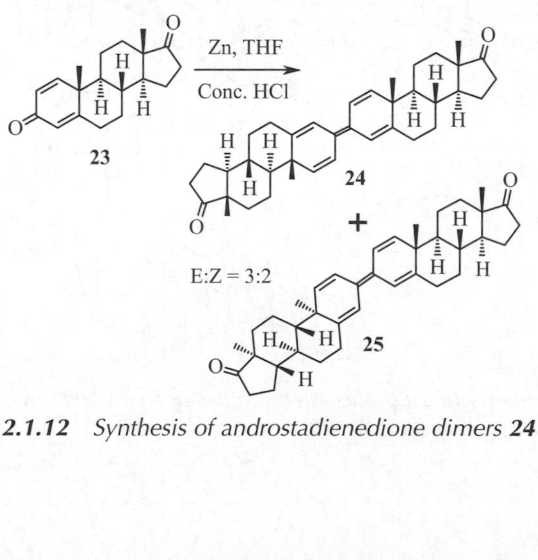

Scheme 2.1.12 Synthesis of androstadienedione dimers **24** and **25**

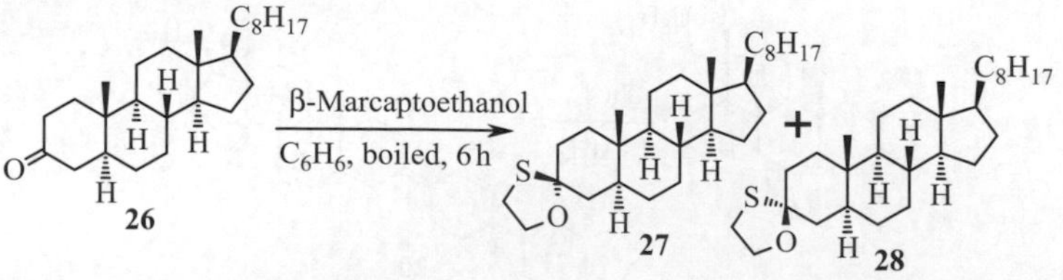

Scheme 2.1.13** Synthesis of cholestane 3-ethylene hemithioketals **27** and **28

of any metals through the reaction between cholestane 3-ethylene hemithioketal (α-O,β-S) (**27**) of saturated steroidal ketones and Ac_2O in the presence of *p*-TsOH catalyst. However, the same dimer **30** was also prepared in one step *via* the Aldol condensation (Karmas, 1968). In order to produce, the hemiketals **27** (α-O,β-S) and **28** (β-O,α-S), cholestanone (**26**, 5.0 g) and β-marcaptoethanol (2.24 g) were boiled in C_6H_6 (100 mL) for 6 h to give a mixture of hemiketals (*Scheme 2.1.13*). Recrystallization from Me_2CO-MeOH provided cholestane 3-ethylene hemithioketal (α-O,β-S) (**27**, 2.19 g, 38%, mp: 135–136 °C). ^1H NMR (60 MHz, $CDCl_3$): δ 3.02 (t, 2H, CH_2S), 4.15 (m, 2H, CH_2O). From the mother liquor of the above recrystallization, upon evaporation to dryness furnished solid product that was purified by CC (on basic Al_2O_3) to yield cholestane 3-ethylene hemithioketal (β-O,α-S) (**28**, 500 mg, 8.7%, mp: 112–113 °C). ^1H NMR (60 MHz, $CDCl_3$): δ 3.01 (t, 2H, CH_2S), 4.17 (t, 2H, CH_2O) (Karmas, 1968).

In the next step, cholestane 3-ethylene hemithioketal (α-O,β-S) (**27**) was treated with Ac_2O in presence of *p*-TsOH to obtain the acetylthioethoxy dimer **29**, which was acid hydrolyzed to afford the cholestanone dimer **30** (Karmas, 1968). A mixture of *p*-TsOH (1.0 g) in Ac_2O (11 mL) and cholestane 3-ethylene hemithioketal (α-O,β-S) (**27**, 1.0 g, 2.25 mmol) in DCM (4 mL) was refluxed for 45 min, after which, the suspension was cooled (0 °C) and Ac_2O (20 mL) and C_5H_5N (2 mL) were added to form precipitates (*Scheme 2.1.14*). The solid was filtered off, washed with MeOH-C_5H_5N and air-dried to provide the crude dimer

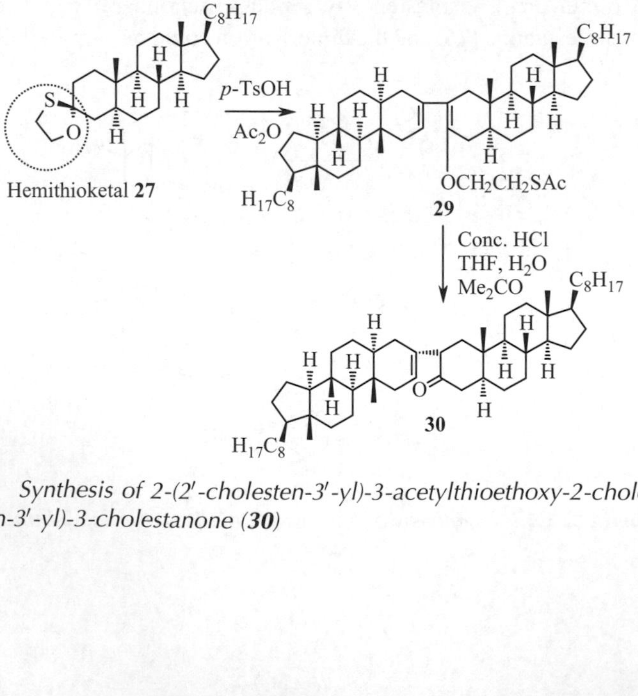

***Scheme 2.1.14** Synthesis of 2-(2′-cholesten-3′-yl)-3-acetylthioethoxy-2-cholestene (**29**) and 2α-(2′-cholesten-3′-yl)-3-cholestanone (**30**)*

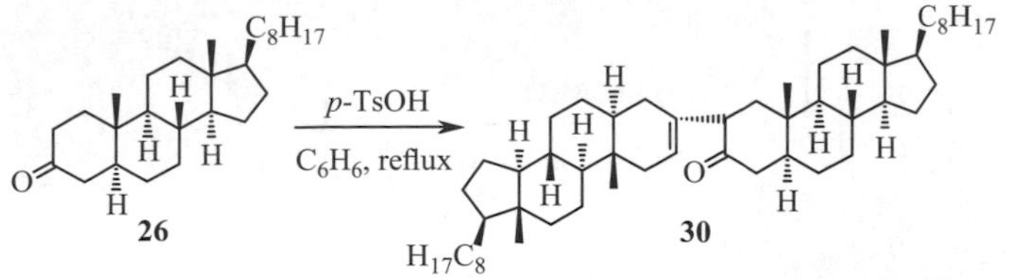

Scheme 2.1.15 *Synthesis of 2α-(2′-cholesten-3′-yl)-3-cholestanone (**30**)*

(900 mg), which was recrystallized from Me_2CO to afford 2-(2′-cholesten-3′-yl)-3-acetylthioethoxy-2-cholestene (**29**, mp: 140–143 °C). ^{1}H NMR (60 MHz, $CDCl_3$): δ 2.31 (s, 3H, MeCO-S), 3.05 (t, 2H, CH_2S), 3.72 (t, 2H, CH_2O), 5.31 (m, 1H, 2′-CH) (Karmas, 1968).

Conc. HCl (300 μL) in THF (20 mL) was added to a stirred solution of 2-(2′-cholesten-3′-yl)-3-acetylthioethoxy-2-cholestene (**29**, 1.0 g, 1.33 mmol) in Me_2CO (10 mL), and the mixture was left at 25 °C (*Scheme 2.1.14*). After 60 min, H_2O (100 mL) was added, resulting in the formation of solids, which were filtered off, dried *in vacuo* and recrystallized from EtOAc-DCM to obtain 2α-(2′-cholesten-3′-yl)-3-cholestanone (**30**, 750 mg, mp: 206–208 °C). ^{1}H NMR (60 MHz, $CDCl_3$): δ 1.06 (s, 3H, 19-Me), 2.92 (m, 1H, 2-CH), 5.30 (m, 1H, 2′-CH) (Karmas, 1968).

The cholestanone dimer **30** could also be obtained *via* the Aldol condensation in presence of dry *p*-TsOH as follows (*Scheme 2.1.15*). Dry *p*-TsOH (10 mL) was prepared from boiling a stirred solution of *p*-TsOH (800 mg) in C_6H_6 (50 mL). A solution of cholestanone (**26**, 1.5 g, 3.86 mmol) in C_6H_6 (10 mL) was added to the dry *p*-TsOH (10 mL) and refluxed. After 3 h, the reaction mixture was cooled, washed with H_2O, and the organic solvent was removed under vacuum. Repeated recrystallization of the crude dimer from EtOAc yielded 2α-(2′-Cholesten-3′-yl)-3-cholestanone (**30**, 150 mg, mp: 200–204 °C) (Karmas, 1968).

The cholestanone dimer **30** was also synthesized from cholestanone (**26**) in one step by the catalytic hydrogenation over Pt in di-*n*-butylether and 3.3% HBr solution (Corey and Young, 1955). A stirred solution of **26** (5.0 g, 12.86 mmol) in di-*n*-butylether (125 mL) was treated with 3.3% HBr in AcOH (15 mL) and the mixture was heated at 60 °C for 4 h (*Scheme 2.1.16*). The catalyst was removed by filtration and the filtrate was washed with 10% NaOH followed by H_2O. The solvent was removed and the crude dimer was recrystallized twice from Me_2CO-DCM to yield 2α-(2′-Cholesten-3′-yl)-3-cholestanone (**30**, 2.0 g, 41%). The dimer **30** could also be prepared easily by shaking the above reaction mixture with H_2 and pre-reduced Pt catalyst (Corey and Young, 1955).

17β-Acetyloxy-5α-androstan-3-one (**31**, 5α-dihydrotestosterone acetate) was employed for the synthesis of 5α-dihydrotestosterone acetate dimers **34–36**, following the similar protocol as outlined for dimer **30** (Karmas, 1968). First, 5α-dihydrotestosterone acetate

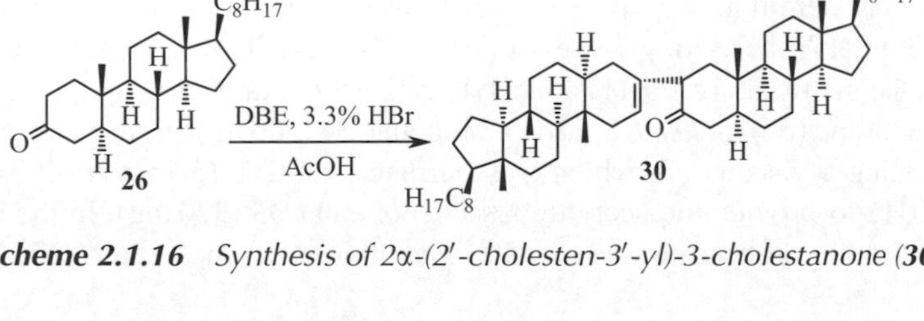

Scheme 2.1.16 *Synthesis of 2α-(2′-cholesten-3′-yl)-3-cholestanone (**30**)*

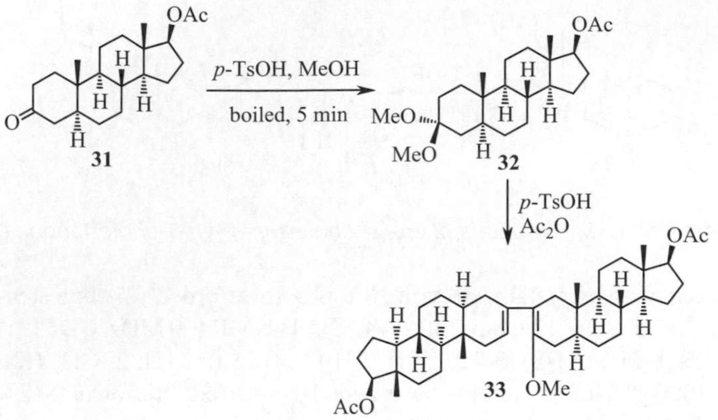

Scheme 2.1.17 *Synthesis of 2-[17′β-Acetyloxy-2′-(5α-androsten)-3′-yl]-17β-acetyloxy-3-methoxy-2-(5α-androstane) (**33**)*

3α,3β-dimethyl ketal (**32**) was obtained from a mixture of 5α-dihydrotestosterone acetate (**31**, 1.5 g) in dry MeOH (20 mL) and *p*-TsOH (50 mg) (*Scheme 2.1.17*). The reaction mixture was boiled for 5 min, solid NaOMe was added, and finally H_2O (500 mL) was added. The solution was extracted twice with Et_2O, the ethereal layers were washed twice with H_2O, dried ($MgSO_4$), evaporated *in vacuo*, and the residue was recrystallized from MeOH-C_5H_5N to afford **32** (1.35 g, mp: 144–146 °C). ^{1}H NMR (60 MHz, $CDCl_3$): δ 3.18 and 3.22 (3α-OMe and 3β-OMe), 2.04 (s, 3H, 17β-CH_3CO), 0.82 (s, 6H, 18-Me and 19-Me).

The synthesis of 2-[17′β-acetyloxy-2′-(5α-androsten)-3′-yl]-17β-acetyloxy-3-methoxy-2-(5α-androstane) (**33**) was achieved from 3,3-dimethyl ketal **32** (3.1 g) in Ac_2O (25 mL), treated with *p*-TsOH (750 mg) at 25 °C (*Scheme 2.1.17*). The reaction mixture was stirred for 20 min, and ice-H_2O containing C_5H_5N (78 mL) was added to form precipitates. The solid was filtered off, dissolved in MeOH (50 mL), and refiltered to give crude dimer (2.5 g), which was recrystallized from C_5H_5N-Me_2CO to obtain methyl enol ether dimer **33** (2.0 g, mp: 192–199 °C). ^{1}H NMR (60 MHz, $CDCl_3$): δ 5.30 (m, 2′-CH), 3.41 (s, 3H, 3-OMe), 2.01 (s, 6H, 2 × 17β-MeCO), 0.80 (s, 12H, 2 × 18-Me and 2 × 19-Me) (Karmas, 1968).

5α-Dihydrotestosterone acetate 3-ethylene hemithioketal (β-O,α-S) (**34**) was prepared from a mixture of 5α-dihydrotestosterone acetate (**31**, 10 g), β-mercaptoethanol (10 g) and *p*-TsOH (500 mg) in C_6H_6 (500 mL) (*Scheme 2.1.18*). The mixture was refluxed for 3 h and H_2O was separated with the Dean–Stark trap. The reaction mixture was washed several times with H_2O and the solvent was removed under vacuum to yield a mixed hemithiokeial. Successive crystallization from EtOAc-Et_2O and EtOAc produced 3-ethylene hemithioketal **34** (1.1 g, mp: 185–186 °C). ^{1}H NMR (60 MHz, $CDCl_3$): δ 4.12 (t, CH_2O), 3.03 (t, CH_2S), 2.01 (s, 3H, 17β-MeCO), 0.82 (s, 3H, 19-Me), 0.79 (s, 3H, 18-Me) (Karmas, 1968).

The 3-ethylene hemithioketal (β-O,α-S) (**34**) was converted to 2-[17′β-acetyloxy-2′-(5α-androsten)-3′-yl]-17β-acetyloxy-3-acetylthioethyl-2-(5α-androstane) (**35**), which was then utilized for the synthesis of 2α-[17′β-acetyloxy-2′-(5α-androsten)-3′-yl]-17β-acetyloxy-5α-androstan-3-one (**36**) (*Scheme 2.1.18*). Compound **34** (500 mg) in Ac_2O was treated with *p*-TsOH yielding a viscous oil, which was purified by CC on acidic Al_2O_3, eluted with *n*-hexane-C_6H_6, to provide the acetylthioethyl enol ether **35** (320 mg). In the next step, a solution of 5α-dihydrotestosterone acetate acetylthioethyl enol ether (**35**, 320 mg) in

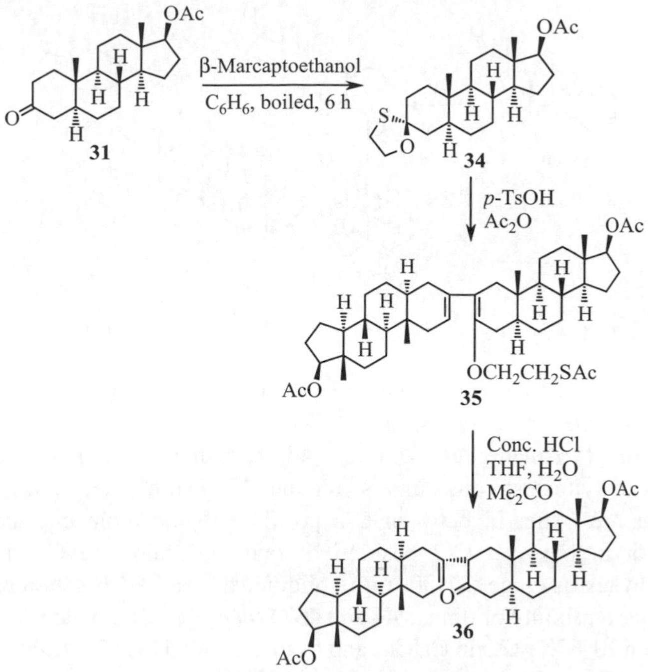

Scheme 2.1.18 *Synthesis of 5α-dihydrotestosterone acetate dimers* ***35*** *and* ***36***

Me_2CO (10 mL) was treated with conc. HCl (300 μL), the mixture was kept for 30 min at 5 °C and filtered to get crude dimer (250 mg), which was recrystallized from Me_2CO to afford dimer **36** (200 mg, mp: 258–261 °C). ^{1}H NMR (60 MHz, $CDCl_3$): δ 5.23 (m, 1H, 2′-CH), 2.90 (q, 1H, $J = 6.0$, 12.0 Hz, 2-CH), 2.01 (s, 6H, 2 × 17β-MeCO), 1.07 (s, 3H, 19-Me), 0.80 (s, 6H, 18′-Me and 19′-Me) (Karmas, 1968).

Dimerization of steroidal molecules could also be accomplished by photochemical reactions, *e.g.*, the synthesis of cholestenone dimers **38** and **39** from cholest-4,6-dien-3-one (**37**) (*Scheme 2.1.19*) (Devaquet and Salem, 1969).

Similar photochemical (*hv*) dimerization of androsta-3,5-diene-7,17-dione (**40**) afforded isomeric androsta-3,5-diene-7,17-dione dimers **41** and **42** (*Scheme 2.1.20*). In cholestenone dimers **38** and **39**, the A ring from two monomers are connected *via* 4,4′ and 5,5′ links

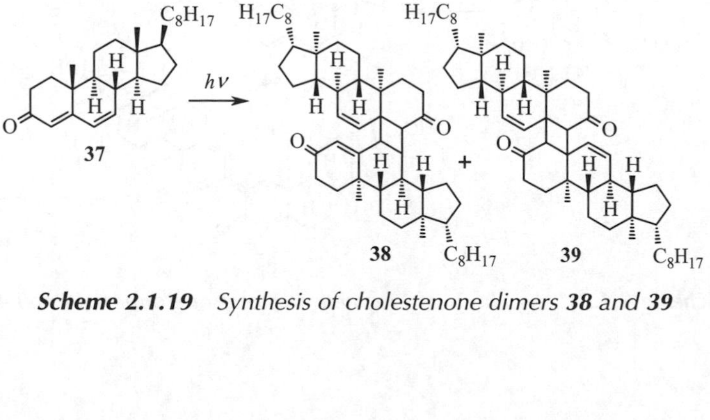

Scheme 2.1.19 *Synthesis of cholestenone dimers* ***38*** *and* ***39***

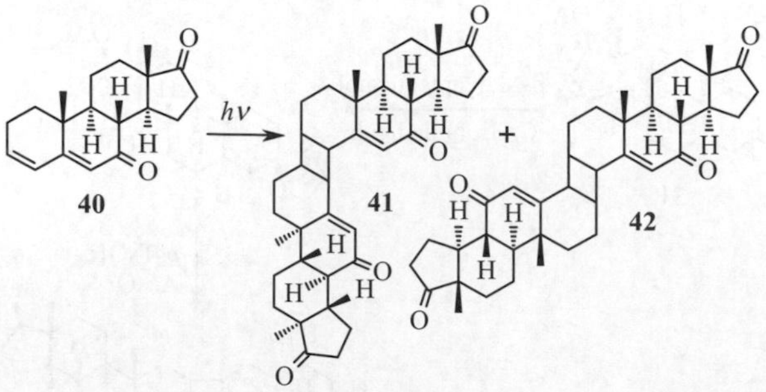

Scheme 2.1.20 *Synthesis of androsta-3,5-diene-7,17-dione dimers* ***41*** *and* ***42***

forming a fused cyclobutane ring system, and similarly, in **41** and **42**, the connections are *via* 3,3′ and 4,4′ links (Devaquet and Salem, 1969).

Irradiation of 17α-methyltestosterone (**43**, 17α-methyl-androst-4-en-3-on-17β-ol) provided 17α-methyltestosterone dimers **44** and **45** (DellaGreca *et al.*, 2002), where the connections are either in between C-2 position of one molecule and C-3 position of the other, or in between C-2 position of one molecule and C-2 position of the other. 17α-Methyltestosterone (**37**, 300 mg), solidified from a $CHCl_3$ solution, was irradiated for 4 h to produce a mixture of dimers **44** and **45** (*Scheme 2.1.21*), which were separated by FCC, eluted with 20% Me_2CO in $CHCl_3$, and further purified by TLC (mobile phase: C_6H_6:Me_2CO:hexane = 7.5:1.5:1) as dimers **44** (15 mg, 5%) and **45** (5.0 mg, 1%).

Dimer **44**: IR ($CHCl_3$): ν_{max} cm^{-1} 1654s (C=C). ^{1}H NMR (500 MHz, $CDCl_3$): δ 5.85 (s, 1H, 4′-CH), 5.73 (s, 1H, 4-CH), 2.71 (d, 1H, J = 9.2 Hz, 1-CH), 1.23 (s, 3H, 20-Me), 1.21 (s, 3H, 20′-Me), 1.06 (s, 6H, 2 × 19-Me), 0.88 (s, 6H, 2 × 18-Me). ^{13}C NMR (125 MHz, $CDCl_3$): δ 37.2 (C-1), 35.0 (C-1′), 134.6 (C-2), 25.6 (C-2′), 199.6 (C-3), 129.3 (C-3′), 126.6 (C-4), 118.6 (C-4′), 167.2 (C-5), 151.3 (C-5′), 35.1 (C-6), 33.9 (C-6′), 32.6 (2 × C-7), 37.5 (2 × C-8), 54.4 (C-9), 54.1 (C-9′), 37.2 (C-10), 40.1 (C-10′), 20.9 (C-11), 20.7 (C-11′), 31.5 (C-12), 31.3 (C-12′), 45.6 (C-13), 45.4 (C-13′), 50.5 (C-14), 50.4 (C-14′), 23.2 (2 × C-15), 38.9 (2 × C-16), 81.7 (C-17), 81.5 (C-17′), 13.9 (2 × C-18), 16.4 (C-19), 18.2 (C-19′), 25.7 (C-20), 25.8 (C-20′). MS (MALDI–TOF): m/z 586 $[M]^+$, 568 $[M - H_2O]^+$ (DellaGreca *et al.*, 2002).

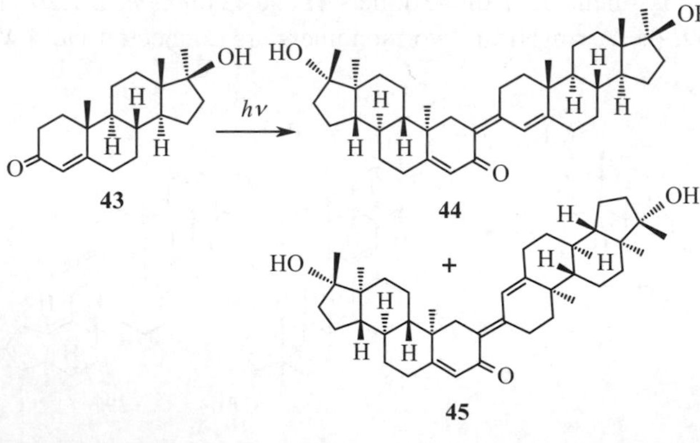

Scheme 2.1.21 *Synthesis of 17α-methyltestosterone dimers* ***44*** *and* ***45***

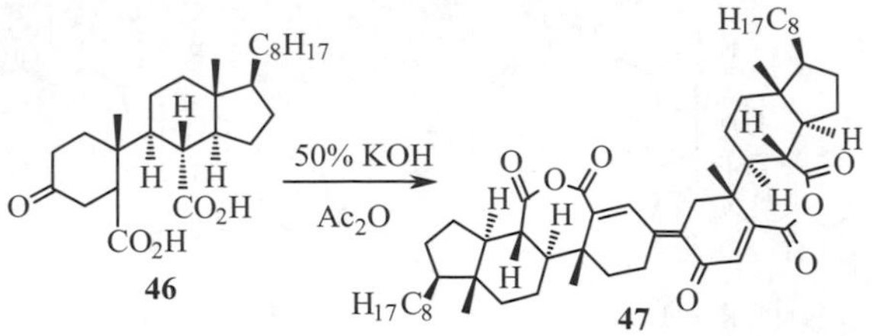

Scheme 2.1.22 *Synthesis of 3-oxo-2(3′)-en-bis(6,7-seco-*cholest-4-en-6,7-dioic) anhydride (**47**)

Dimer **45**: IR ($CHCl_3$): ν_{max} cm^{-1} 1652s (C=C). ^1H NMR (500 MHz, $CDCl_3$): δ 6.08 (s, 1H, 4′-CH), 5.65 (s, 1H, 4-CH), 2.88 (d, 1H, J = 12.5 Hz, 1-CH), 1.24 (s, 3H, 20′-Me), 1.19 (s, 3H, 20-Me), 1.06 (s, 3H, 2 × 19-Me), 0.90 (s, 3H, 18-Me), 0.89 (s, 3H, 18′-Me). ^{13}C NMR (125 MHz, $CDCl_3$): δ 37.9 (C-1), 35.0 (C-1′), 133.8 (C-2), 25.6 (C-2′), 199.5 (C-3), 129.3 (C-3′), 125.4 (C-4), 118.6 (C-4′), 167.5 (C-5), 151.3 (C-5′), 34.3 (C-6), 33.7 (C-6′), 33.0 (C-7), 32.6 (C-7′), 37.9 (C-8), 37.0 (C-8′), 54.2 (C-9), 54.2 (C-9′), 40.0 (C-10), 36.6 (C-10′), 20.9 (2 × C-11), 31.5 (2 × C-12), 45.6 (C-13), 45.4 (C-13′), 50.5 (2 × C-14), 23.3 (2 × C-15), 38.9 (2 × C-16), 81.7 (C-17), 81.7 (C-17′), 13.9 (C-18), 14.0 (C-18′), 16.4 (C-19), 18.1 (C-19′), 25.7 (C-20), 25.7 (C-20′). MS (MALDI–TOF): m/z 586 $[M]^+$, 568 $[M - H_2O]^+$ (DellaGreca *et al.*, 2002).

3-Oxo-2(3′)-en-bis(6,7-*seco*-cholest-4-en-6,7-dioic)anhydride (**47**, 15%, mp: 259–260 °C) was obtained (Knoll *et al.*, 1986), when butenandt acid (**46**) reacted with 50% KOH (hot), followed by the Ac_2O treatment (*Scheme 2.1.22*). ^1H NMR (400 MHz, $CDCl_3$): δ 3.20 and 2.50 (d, 2H, J = 15.0 Hz, 1-CH_2), 2.00 (ddd, 2H, J = 4.0, 14.0 Hz, 1′-CH_2), 3.70 and 2.45 (ddd, 2H, J = 3.0, 19.0 Hz, 2′-CH_2), 6.20 (s, 1H, 4-CH), 6.85 (s, 1H, 4′-CH), 2.70 and 2.65 (dd, 4H, J = 11.0 Hz, 2 × 8-CH_2), 2.18 and 2.08 (ddd, 4H, J = 3.0, 14.0 Hz, 2 × 12-CH_2), 2.18 and 2.08 (ddd, 4H, J = 3.0, 14.0 Hz, 2 × 12-CH_2), 0.70 and 0.65 (s, 6H, 2 × 18-Me), 1.40 and 1.30 (s, 6H, 2 × 19-Me). ^{13}C NMR (100 MHz, $CDCl_3$): δ 38.8 and 33.9 (2 × C-1), 129.8 and 23.5 (2 × C-2), 188.6 and 154.8 (2 × C-3), 128.7 and 131.9 (2 × C-4), 143.1 and 148.6 (2 × C-5), 161.7 and 163.6 (2 × C-6), 171.3 and 171.8 (2 × C-7), 47.1 and 47.4 (2 × C-8), 49.6 or 50.6 (2 × C-9), 42.0 and 38.9 (2 × C-10), 23.4 and 22.8 (2 × C-11), 39.0 and 39.8 (2 × C-12), 42.5 and 42.2 (2 × C-13), 50.8 and 51.2 (2 × C-14), 24.6 and 24.7 (2 × C-15), 28.1 and 28.2 (2 × C-16), 56.0 and 56.3 (2 × C-17), 12.2 and 12.1 (2 × C-18), 20.1 and 19.6 (2 × C-19), 35.8 (2 × C-20), 18.7 (2 × C-21), 36.2 (2 × C-22), 23.8 (2 × C-23), 39.5 (2 × C-24), 28.0 (2 × C-25), 22.6 (2 × C-26), 22.8 (2 × C-27) (Knoll *et al.*, 1986).

2.1.2 Through Spacer Groups

Synthesis of dimeric steroids involving A rings of the steroid monomers and connecting through spacer groups can be achieved by using suitable spacers. Most of the ring A–ring A steroid dimers with a spacer group were synthesized through simple esterification or etherification reactions. For the esterification, most often, diacids (*e.g.*, oxalic acid) were used to create the spacer. Lithocholic acid (**48**, 3α-hydroxy-5β-cholanic acid or 3α-hydroxy-5β-cholan-24-oic acid) and cholic acid (**49**, 3α,7α,12α-trihydroxy-5β-cholanic

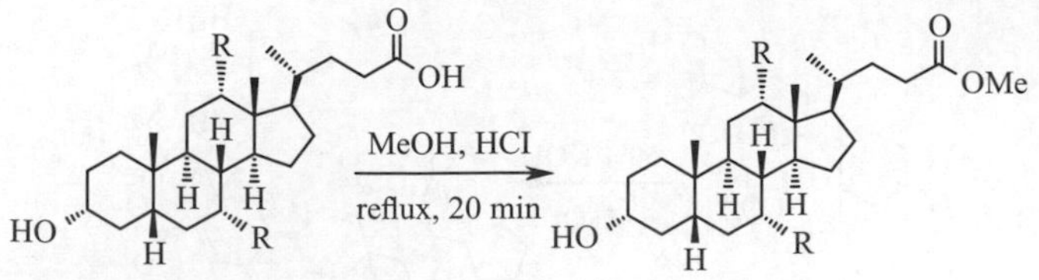

MeOH, HCl
reflux, 20 min

48: R = H
49: R = OH

50: R = H
51: R = OH

Scheme 2.1.23 *Synthesis of methyl lithocholate (**50**) and methyl cholate (**51**)*

acid or 3α,7α,12α-trihydroxy-5β-cholan-24-oic acid) were employed to synthesize 3α- and 3β-dimers *via* ring A–ring A by the formation of ester or ether linkages (Gouin and Zhu, 1996). A solution of lithocholic acid (**48**, 67.0 mmol) in MeOH (125 mL) was acidified with conc. HCl (1 mL), the mixture was stirred and refluxed for 20 min (*Scheme 2.1.23*). The solution was cooled to 0 °C until crystallization occurred. The crystalline product was filtered and washed twice with cold MeOH to give methyl lithocholate (methyl 3α-hydroxy-5β-cholanate or methyl 3α-hydroxy-5β-cholan-24 oate) (**50**, 95%, mp: 120–121 °C) (Fieser and Rajagopalan, 1950).

Similarly, methyl cholate (**51**, methyl 3α,7α,12α-trihydroxy-5β-cholanate or methyl 3α, 7α,12α-trihydroxy-5β-cholan-24-oate) was prepared by esterification of cholic acid (**49**) (*Scheme 2.1.23*). A solution of **49** (27.38 g, 67.0 mmol) in MeOH (125 mL) was acidified with conc. HCl (1 mL), the mixture was stirred and refluxed for 20 min. The solution was cooled to 0 °C until crystallization took place. The crystalline product was filtered and washed twice with cold MeOH to obtain methyl cholate (**51**, 26.96 g, 95%, mp: 154–155 °C). ^{1}H NMR (250 MHz, $CDCl_3$): δ 3.94 (m, 1H, 12β-CH), 3.82 (m, 1H, 7β-CH), 3.66 (s, 3H, 24-OMe), 3.43 (br s, 1H, 3β-CH), 0.99 (d, 3H, J = 6.1 Hz, 21-Me), 0.88 (s, 3H, 19-Me), 0.67 (s, 3H, 18-Me) (Fieser and Rajagopalan, 1950).

The general method for the synthesis of 3α-lithocholic acid-based dimers (**52** and **53**) and 3α-cholic acid-based dimers (**54** and **55**) can be outlined as follows (*Scheme 2.1.24*). A stirred mixture of bile acid (3.0 mmol) in distilled DCM (5 mL) containing Et_3N (500 mg, 5.0 mmol) was cooled under N_2, and a solution of the dicarbonyl acid chloride (1.8 mmol) in dry DCM (1 mL) was added over 10 min. The mixture was allowed to warm to room temperature and stirred for 20 h. The reaction mixture was filtered off, and the solvent was evaporated under pressure, the solid was dissolved in THF and refiltered. The solvent was rotary evaporated, the crude product was redissolved in $CHCl_3$ and purified by CC (mobile phase: EtOAc-light petroleum ether), to yield 3α-lithocholic acid-based dimers (**52** and **53**) and 3α-cholic acid-based dimers (**54** and **55**) from corresponding bile acids (Gouin and Zhu, 1996).

Bis(methyl 5β-cholan-24-oate-3α-yl)suberate (**52**, 72%, mp: 122–124 °C). IR (KBr): ν_{max} cm^{-1} 2932m (C–H), 2864m (C–H), 1743s (C=O), 1738s (C=O). ^{1}H NMR (300 MHz, $CDCl_3$): δ 4.70 (m, 2H, 2 × 3β-CH), 3.66 (s, 6H, 2 × 24-OMe), 2.23 (t, 4H, 2 × CH_2 of suberate), 1.30 (m, 8H, 4 × CH_2 of suberate), 0.96 (d, 6H, J = 6.0 Hz, 2 × 21-Me), 0.92 (s, 6H, 2 × 19-Me), 0.72 (s, 6H, 2 × 18-Me). ^{13}C NMR (75 MHz, $CDCl_3$): δ 34.8 (2 × C-1), 26.7 (2 × C-2), 74.3 (2 × C-3), 35.0 (2 × C-4), 42.1 (2 × C-5), 27.2 (2 × C-6), 26.4 (2 × C-7), 36.0 (2 × C-8), 40.6 (2 × C-9), 34.8 (2 × C-10), 21.0 (2 × C-11), 40.3 (2 × C-12), 42.9 (2 × C-13), 56.2 (2 × C-14), 24.3 (2 × C-15), 28.4 (2 × C-16), 56.7 (2 × C-17), 12.3

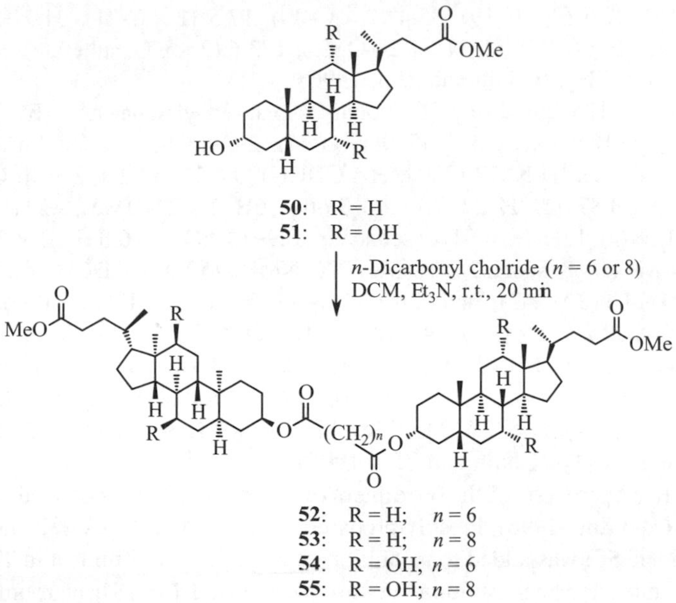

50: R = H
51: R = OH

n-Dicarbonyl cholride (*n* = 6 or 8)
DCM, Et_3N, r.t., 20 min

52: R = H; *n* = 6
53: R = H; *n* = 8
54: R = OH; *n* = 6
55: R = OH; *n* = 8

Scheme 2.1.24 *Synthesis of 3α-lithocholic acid-based dimers (**52** and **53**) and 3α-cholic acid-based dimers (**54** and **55**)*

(2 × C-18), 23.5 (2 × C-19), 35.5 (2 × C-20), 18.5 (2 × C-21), 31.1 (2 × C-22), 31.2 (2 × C-23), 174.8 (2 × C-24), 51.5 (2 × 24-OMe), 173.5 (2 × CO, suberate), 31.1, 28.8 and 25.0 (suberate 6 × CH_2) (Gouin and Zhu, 1996).

Bis(methyl 5β-cholan-24-oate-3α-yl)sebacate (**53**, 79%, mp: 124–125 °C). IR (KBr): ν_{max} cm^{-1} 2930m (C–H), 2864m (C–H), 1738s (C=O), 1733s (C=O). ^{1}H NMR (300 MHz, $CDCl_3$): δ 4.71 (m, 2H, 2 × 3β-CH), 3.67 (s, 6H, 2 × 24-OMe), 2.25 (t, 4H, 2 × CH_2 of sebacate), 1.31 (m, 12H, 6 × CH_2 of sebacate), 0.96 (d, 6H, *J* = 6.0 Hz, 2 × 21-Me), 0.92 (s, 6H, 2 × 19-Me), 0.71 (s, 6H, 2 × 18-Me). ^{13}C NMR (75 MHz, $CDCl_3$): δ 34.9 (2 × C-1), 26.9 (2 × C-2), 74.2 (2 × C-3), 35.3 (2 × C-4), 42.1 (2 × C-5), 27.2 (2 × C-6), 26.5 (2 × C-7), 36.0 (2 × C-8), 40.6 (2 × C-9), 34.8 (2 × C-10), 21.0 (2 × C-11), 40.4 (2 × C-12), 42.9 (2 × C-13), 56.2 (2 × C-14), 24.4 (2 × C-15), 28.5 (2 × C-16), 56.7 (2 × C-17), 12.2 (2 × C-18), 23.5 (2 × C-19), 35.6 (2 × C-20), 18.5 (2 × C-21), 31.1 (2 × C-22), 31.3 (2 × C-23), 174.9 (2 × C-24), 51.5 (2 × 24-OMe), 173.5 (2 × CO, sebacate), 32.5, 29.3, 28.3 and 25.2 (sebacate 8 × CH_2) (Gouin and Zhu, 1996).

Bis(methyl 7α,12α-dihydroxy-5β-cholan-24-oate-3α-yl)suberate (**54**, 61%, mp: 97–99 °C). IR (KBr): ν_{max} cm^{-1} 3511br (O–H), 2941m (C–H), 2868m (C–H), 1738s (C=O), 1734s (C=O). ^{1}H NMR (300 MHz, $CDCl_3$): δ 4.55 (m, 2H, 2 × 3β-CH), 3.99 (m, 2H, 2 × 12β-CH), 3.87 (m, 2H, 2 × 7β-CH), 3.67 (s, 6H, 2 × 24-OMe), 2.22 (t, 4H, 2 × CH_2 of suberate), 1.31 (m, 8H, 4 × CH_2 of suberate), 0.98 (d, 6H, *J* = 6.0 Hz, 2 × 21-Me), 0.92 (s, 6H, 2 × 19-Me), 0.68 (s, 6H, 2 × 18-Me). ^{13}C NMR (75 MHz, $CDCl_3$): δ 35.1 (2 × C-1), 26.9 (2 × C-2), 74.4 (2 × C-3), 35.0 (2 × C-4), 42.1 (2 × C-5), 34.9 (2 × C-6), 68.4 (2 × C-7), 39.6 (2 × C-8), 26.7 (2 × C-9), 34.8 (2 × C-10), 28.4 (2 × C-11), 73.1 (2 × C-12), 46.7 (2 × C-13), 41.4 (2 × C-14), 23.4 (2 × C-15), 27.6 (2 × C-16), 47.3 (2 × C-17),

12.7 (2 × C-18), 22.6 (2 × C-19), 35.4 (2 × C-20), 17.5 (2 × C-21), 31.2 (2 × C-22 and 2 × C-23), 174.9 (2 × C-24), 51.5 (2 × 24-OMe), 173.6 (2 × CO, suberate), 31.1, 28.7 and 25.0 (suberate 6 × CH_2) (Gouin and Zhu, 1996).

Bis(methyl 7α,12α-dihydroxy-5β-cholan-24-oate-3α-yl)sebacate (**55**, 68%, mp: 99–101 °C). IR (KBr): ν_{max} cm^{-1} 3520br (O–H), 2939m (C–H), 2867m (C–H), 1731s (C=O), 1727s (C=O). ^{1}H NMR (300 MHz, $CDCl_3$): δ 4.52 (m, 2H, 2 × 3β-CH), 4.00 (m, 2H, 2 × 12β-CH), 3.87 (m, 2H, 2 × 7β-CH), 3.66 (s, 6H, 2 × 24-OMe), 2.22 (t, 4H, 2 × CH_2 of sebacate), 1.28 (m, 12H, 6 × CH_2 of sebacate), 0.99 (d, 6H, J = 6.0 Hz, 2 × 21-Me), 0.932 (s, 6H, 2 × 19-Me), 0.68 (s, 6H, 2 × 18-Me). ^{13}C NMR (75 MHz, $CDCl_3$): δ 35.1 (2 × C-1), 26.9 (2 × C-2), 74.3 (2 × C-3), 34.9 (2 × C-4), 42.1 (2 × C-5), 34.9 (2 × C-6), 68.4 (2 × C-7), 39.6 (2 × C-8), 26.7 (2 × C-9), 34.8 (2 × C-10), 28.4 (2 × C-11), 73.1 (2 × C-12), 46.7 (2 × C-13), 41.4 (2 × C-14), 23.4 (2 × C-15), 27.6 (2 × C-16), 47.3 (2 × C-17), 12.6 (2 × C-18), 22.6 (2 × C-19), 35.4 (2 × C-20), 17.5 (2 × C-21), 31.2 (2 × C-22 and 2 × C-23), 174.9 (2 × C-24), 51.5 (2 × 24-OMe), 173.6 (2 × CO, sebacate), 31.0, 29.1, 29.0 and 25.2 (sebacate 8 × CH_2) (Gouin and Zhu, 1996).

Regioselective hydrolysis of the 3α-dimers of bile acids **52–55** produced further dimers **56–59** (Gouin and Zhu, 1996). Freshly prepared 2N LiOH (1 mL) (for **52** and **53**) or 0.5N LiOH (for **54** and **55**) was added to a solution of 3α-dimer (0.11 mmol) in THF (2 mL) at room temperature, and the resulting mixture was stirred for 18 h at room temperature (*Scheme 2.1.25*). THF was removed by rotary evaporation, and the remaining solution was acidified with 10% HCl to neutral pH. The precipitate was filtered off, washed with cold H_2O, and dried *in vacuo* to afford respective dimers **56–59**.

Bis(5β-cholan-24-oic acid 3α-yl)suberate (**56**, mp: 139–141 °C). IR (KBr): ν_{max} cm^{-1} 3440br (O–H), 2928m (C–H), 2866m (C–H), 1735s (C=O), 1708s (C=O). ^{1}H NMR (300 MHz, $CDCl_3$): δ 4.70 (m, 2H, 2 × 3β-CH), 2.22 (t, 4H, 2 × CH_2 of suberate), 1.30 (m, 8H, 4 × CH_2 of suberate), 0.96 (d, 6H, J = 6.0 Hz, 2 × 21-Me), 0.92 (s, 6H, 2 × 19-Me), 0.72 (s, 6H, 2 × 18-Me) (Gouin and Zhu, 1996).

Bis(5β-cholan-24-oic acid 3α-yl)sebacate (**57**, mp: 142–144 °C). IR (KBr): ν_{max} cm^{-1} 3443br (OH), 2927m (C–H), 2865m (C–H), 1734s (C=O), 1707s (C=O). ^{1}H NMR (300 MHz, $CDCl_3$): δ 4.72 (m, 2H, 2 × 3β-CH), 2.23 (t, 4H, 2 × CH_2 of sebacate), 1.32 (m, 12H, 6 × CH_2 of sebacate), 0.96 (d, 6H, J = 6.0 Hz, 2 × 21-Me), 0.91 (s, 6H, 2 × 19-Me), 0.71 (s, 6H, 2 × 18-Me) (Gouin and Zhu, 1996).

Compounds **52-55**
LiOH, THF
18 h, r.t.

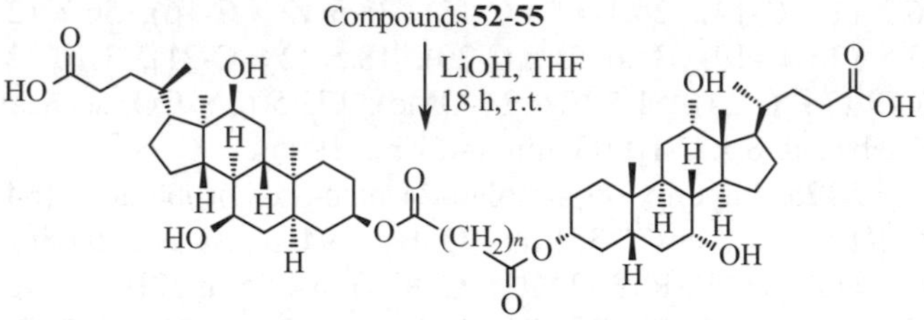

56: R = H; n = 6
57: R = H; n = 8
58: R = OH; n = 6
59: R = OH; n = 8

Scheme 2.1.25 *Synthesis of 3α-lithocholic acid-based dimers (**56** and **57**) and 3α-cholic acid-based dimers (**58** and **59**)*

Bis(7α,12α-dihydroxy-5β-cholan-24-oic acid 3α-yl)suberate (**58**, mp: 134–136 °C). IR (KBr): ν_{max} cm^{-1} 3495br (O–H), 2945m (C–H), 2871m (C–H), 1736s (C=O), 1705s (C=O). ^{1}H NMR (300 MHz, $CDCl_3$): δ 4.55 (m, 2H, 2 × 3β-CH), 3.98 (m, 2H, 2 × 12β-CH), 3.87 (m, 2H, 2 × 7β-CH), 2.23 (t, 4H, 2 × CH_2 of suberate), 1.32 (m, 8H, 4 × CH_2 of suberate), 0.98 (d, 6H, J = 6.0 Hz, 2 × 21-Me), 0.92 (s, 6H, 2 × 19-Me), 0.66 (s, 6H, 2 × 18-Me) (Gouin and Zhu, 1996).

Bis (7α,12α-dihydroxy-5β-cholan-24-oic acid 3α-yl)sebacate (**59**, mp: 135–137 °C). IR (KBr): ν_{max} cm^{-1} 3500br (O–H), 2943m (C–H), 2870m (C–H), 1734s (C=O), 1705s (C=O). ^{1}H NMR (300 MHz, $CDCl_3$): δ 4.56 (m, 2H, 2 × 3β-CH), 4.01 (m, 2H, 2 × 12β-CH), 3.88 (m, 2H, 2 × 7β-CH), 2.23 (t, 4H, 2 × CH_2 of sebacate), 1.30 (m, 12H, 6 × CH_2 of sebacate), 0.99 (d, 6H, J = 6.0 Hz, 2 × 21-Me), 0.92 (s, 6H, 2 × 19-Me), 0.68 (s, 6H, 2 × 18-Me) (Gouin and Zhu, 1996).

3β-Lithocholic acid-based dimers **62** and **63** were obtained by ether bond formation (Gouin and Zhu, 1996). In this process, methyl 3α-tosyloxy-5β-cholan-24-oate (**60**) was first prepared from methyl lithocholate (**50**), using the procedure described in the literature (Kramer and Kurz, 1983). To a cooled solution (0 °C in ice-water bath under a dry inert gas) of **50** (2.4 g) in dry C_5H_5N (10 mL), a solution of *p*-TsCl (1.28 g, 6.75 mmol) in C_5H_5N (5 mL) was added dropwise over 10 min, the reaction mixture was allowed to warm gradually to room temperature, and stirred for 24 h (*Scheme 2.1.26*). The reaction mixture was filtered and the solvent was removed by evaporation *in vacuo*. The residue was dissolved in THF, filtered off again and the solvent was evaporated under pressure. Recrystallization of the crude solid from EtOAc-petroleum ether produced **60** (3.0 g, 90%, mp: 114–115 °C). IR (KBr): ν_{max} cm^{-1} 3120m (C–H), 2940m (C–H), 2870m (C–H), 1730s (C=O), 1250s (C–O). ^{1}H NMR (300 MHz, $CDCl_3$): δ 7.76 (d, 2H, J = 8.0 Hz, Ph-CH), 7.32 (d, 2H, J = 8.0 Hz, Ph-CH), 4.44 (m, 1H, 3β-CH), 3.66 (s, 3H, 24-OMe), 2.42 (s, 3H, Ts-Me), 0.65 (s, 3H, 18-Me), 0.88 (s, 3H, C-19), 0.93 (d, 3H, J = 6.0 Hz, 21-Me) (Gouin and Zhu, 1996).

The next step involved the transformation of methyl 3α-tosyloxy-5β-cholan-24-oate (**60**) to methyl 3β-(2-hydroxyethoxy)-5β-cholan-24-oate (**61**) (*Scheme 2.1.26*). A mixture of **60**

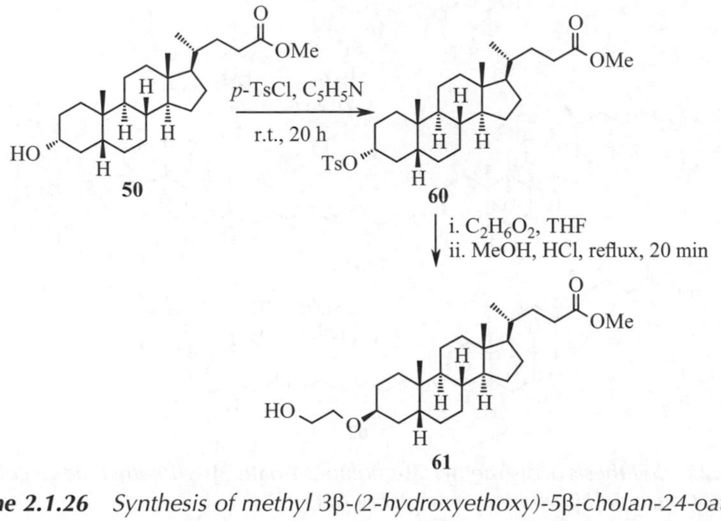

p-TsCl, C_5H_5N, r.t., 20 h

i. $C_2H_6O_2$, THF
ii. MeOH, HCl, reflux, 20 min

Scheme 2.1.26 *Synthesis of methyl 3β-(2-hydroxyethoxy)-5β-cholan-24-oate (**61**)*

(1.0 g, 1.84 mmol) in dry $C_2H_6O_2$ (25 mL) and dry THF (5 mL) was heated to 120 °C for 2 h under N_2, followed by cooling to room temperature. The reaction mixture was diluted with distilled H_2O (15 mL) and extracted with $CHCl_3$ (3 × 15 mL), and the pooled $CHCl_3$ layers were dried ($MgSO_4$), and the solvent was removed by rotary evaporation. As some of the methyl ester groups might have been replaced by hydroxyethoxy ester groups because of the excess use of $C_2H_6O_2$ the crude product needed to be treated with MeOH to convert it back to methyl ester. The solid was dissolved in MeOH (25 mL), acidified with conc. H_2SO_4 (100 μL) and refluxed for 60 min. The reaction mixture was cooled to 0 °C resulting in the formation of precipitates, which were filtered off under vacuum. The crude residue was dissolved in $CHCl_3$ and purified by CC, eluted with 40% DCM in EtOAc, to provide **61** (400 mg, 50%, mp: 123–125 °C). IR (KBr): ν_{max} cm^{-1} 3480br (H–O), 2936m (C–H), 2864m (C–H), 1729s (C=O), 1250m (C–O). ^{1}H NMR (300 MHz, $CDCl_3$): δ 3.72 and 3.48 (s, 4H, 2 × OCH_2 of ethane-1,2-diol), 3.67 (s, 3H, 24-OMe), 3.65 (m, 1H, 3β-CH), 0.95 (d, 3H, J = 6.0 Hz, 21-Me), 0.87 (s, 3H, 19-Me), 0.71 (s, 3H, 18-Me). ^{13}C NMR (75 MHz, $CDCl_3$): δ 30.8 (C-1), 24.6 (C-2), 74.8 (C-3), 30.3 (C-4), 36.9 (C-5), 26.7 (C-6), 26.2 (C-7), 35.6 (C-8), 40.0 (C-9), 34.9 (C-10), 21.0 (C-11), 40.2 (C-12), 42.7 (C-13), 55.9 (C-14), 24.1 (C-15), 28.1 (C-16), 56.5 (C-17), 12.0 (C-18), 23.8 (C-19), 35.3 (C-20), 18.2 (C-21), 30.9 (C-22), 31.0 (C-23), 174.7 (C-24), 51.4 (24-OMe), 68.5 and 62.0 (2 × OCH_2 of ethane-1,2-diol) (Gouin and Zhu, 1996).

Bis(methyl 5β-cholan-24-oate-3β-yl)diethylene glycol (**62**) was synthesized from **61** (*Scheme 2.1.27*). A stirred solution of **61** (500 mg, 1.15 mmol) in dry C_5H_5N (10 mL) was cooled under N_2, and MsCl (7.0 mg, 0.61 mmol) in dry C_5H_5N (5 mL) was added dropwise over 15 min. The reaction mixture was warmed to room temperature while stirring. After 48 h, the mixture was evaporated to dryness and purified by CC (mobile phase:

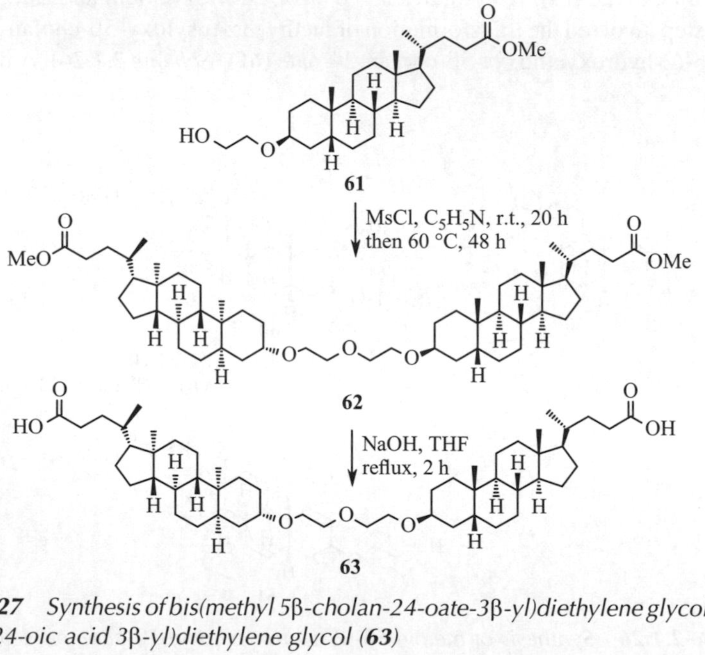

Scheme 2.1.27 *Synthesis of bis(methyl 5β-cholan-24-oate-3β-yl)diethylene glycol (**62**) and bis (5β-cholan-24-oic acid 3β-yl)diethylene glycol (**63**)*

toluene-$CHCl_3$) to produce **62** (200 mg, 40%, mp: 117–119 °C). IR (KBr): ν_{max} cm^{-1} 2932m (C–H), 2865m (C–H). 1732s (C=O), 1245s (C–O). ^{1}H NMR (300 MHz, $CDCl_3$): δ 3.67 (s, 6H, 2 × 24-OMe), 3.61 (m, 2H, 2 × 3β-CH), 3.59 and 3.58 (t, 8H, CH_2 of ethane-1,2-diol), 0.96 (d, 6H, J = 6.0 Hz, 2 × 21-Me), 0.85 (s, 6H, 2 × 19-Me), 0.70 (s, 6H, 2 × 18-Me). ^{13}C NMR (75 MHz, $CDCl_3$): δ 30.7 (2 × C-1), 24.7 (2 × C-2), 75.0 (2 × C-3), 30.3 (2 × C-4), 36.9 (2 × C-5), 26.7 (2 × C-6), 26.3 (2 × C-7), 35.7 (2 × C-8), 40.0 (2 × C-9), 34.9 (2 × C-10), 21.0 (2 × C-11), 40.2 (2 × C-12), 42.7 (2 × C-13), 55.9 (2 × C-14), 24.1 (2 × C-15), 28.1 (2 × C-16), 56.6 (2 × C-17), 12.0 (2 × C-18), 23.8 (2 × C-19), 35.3 (2 × C-20), 18.2 (2 × C-21), 30.9 (2 × C-22), 31.0 (2 × C-23), 174.7 (2 × C-24), 51.4 (2 × OMe), 67.9 and 63.2 (4 × OCH_2 of ethane-1,2-diol) (Gouin and Zhu, 1996).

Another glycol dimer, bis(5β-cholan-24-oic acid 3β-yl)diethylene glycol (**63**), was produced from dimer **62** (*Scheme 2.1.27*). A mixture of **62** (1.0 g) in dioxane (10 mL) and 2.5N NaOH (10 mL) was refluxed for 2 h, and the solution was allowed to stand at room temperature to cool. The mixture was then acidified with 15% HCl to neutral pH resulting in the formation of precipitates, which were filtered under vacuum, washed three times with distilled H_2O, and dried in a vacuum oven at 40 °C, yielding **63** (950 mg, 97%, mp: 121–122 °C). IR (KBr): ν_{max} cm^{-1} 3430br (O–H), 2934m (C–H), 2865m (C–H), 1720s (C=O). ^{1}H NMR (300 MHz, $CDCl_3$): δ 3.62 (m, 2H, 2 × 3β-CH) 3.58 (t, 8H, CH_2 of ethane-1,2-diol) (Gouin and Zhu, 1996).

A high-yielding estrone oxalate dimer, bis[estra-1,3,5(10)trien-17-on-3-yl]oxalate (**65**) *via* ring A–ring A connection through an oxalate spacer group, was prepared from 3-hydroxyestra-1,3,5(10)trien-17-one (**64**, estrone) using dry C_5H_5N and $(COCl)_2$ (Nahar *et al.*, 2006). A stirred solution of estrone (**64**, 300 mg, 1.11 mmol) in C_5H_5N (5 mL) was treated dropwise with $(COCl)_2$ (71 mg, 0.56 mmol) at room temperature under N_2 (strongly exothermic reaction and white fumes evolved) (*Scheme 2.1.28*). After 18 h, H_2O (10 mL) was added dropwise to the reaction mixture, precipitation occurred, and the precipitate was filtered off. It was washed several times with H_2O to remove pyridine hydrochloride. Finally, the solid was dissolved in Et_2O (10 mL) and rotary evaporated at 35 °C. Recrystallization from $CHCl_3$-EtOAc (2:1) yielded estrone oxalate dimer **65** (198 mg, 59%, mp: 254–256 °C). IR ($CHCl_3$): ν_{max} cm^{-1} 3026w (C–H), 2954s (C–H), 2923s (C–H), 2856s (C–H), 1760s (C=O), 1740s (C=O), 1703s (C=O), 1608m (C=C), 1598m (C=C), 1492s (C–H), 1453m (C–H), 1310m (C–H), 1260m (C–O), 1158s (C–O), 1084m (C–O), 1008m (C–O), 911m, 732m. ^{1}H NMR (400 MHz, $CDCl_3$): δ 7.31 (d, 2H, J = 8.2 Hz, 2 × 1-CH), 6.98 (dd, 2H, J = 1.8, 8.2 Hz, 2 × 2-CH), 6.95 (d, 2H, J = 1.8 Hz, 2 × 4-CH), 0.89 (s, 6H, 2 × 18-Me). ^{13}C NMR (100 MHz, $CDCl_3$): δ 126.7 (2 × C-1), 118.0 (2 × C-2), 147.9

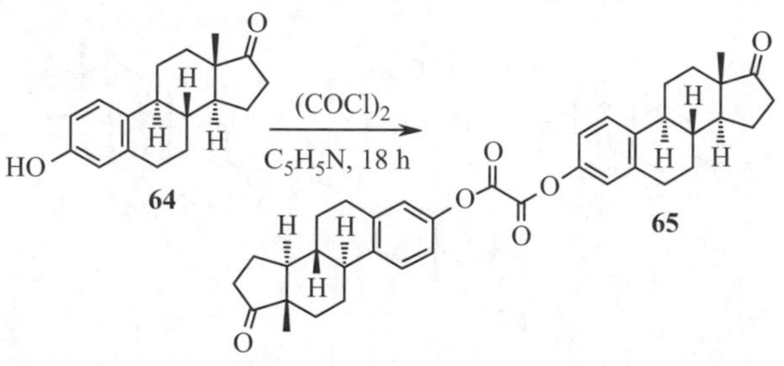

Scheme 2.1.28 *Synthesis of bis[estra-1,3,5(10)trien-17-on-3-yl]oxalate (**65**)*

(2 × C-3), 120.9 (2 × C-4), 138.6 (2 × C-5), 29.4 (2 × C-6), 26.2 (2 × C-7), 37.9 (2 × C-8), 44.1 (2 × C-9), 138.5 (2 × C-10), 25.7 (2 × C-11), 31.5 (2 × C-12), 47.9 (2 × C-13), 50.4 (2 × C-14), 21.6 (2 × C-15), 35.8 (2 × C-16), 220.6 (2 × C-17), 13.8 (2 × C-18), 156.1 (2 × oxalate **C**O). FABMS *m/z*: 595 $[M + H]^+$, 617 $[M + Na]^+$. HRFABMS: *m/z* 595.3059 calculated for $C_{38}H_{43}O_6$, found: 595.3058 (Nahar *et al.*, 2006).

Similarly, several oxalate dimers **70–75** *via* ring A–ring A connection through an oxalate spacer group were synthesized, respectively, from DHEA (**66**, dehydro-*epi*-androsterone or 3β-hydroxyandrost-5-en-17-one), pregnenolone (**67**, 3β-hydroxypregn-5-en-20-one or pregn-5-en-3β-ol-20-one), methyl lithocholate (**50**), cholesterol (**1**), cholestanol (**68**, 3β-hydroxy-5α-cholestane or 5α-cholestan-3β-ol), and stigmasterol (**69**, stigmasta-5,22*t*-dien-3β-ol) (Nahar and Turner, 2003; Nahar *et al.*, 2006). These oxalate ester-linked dimers **70–75** were obtained by a simple esterification reaction using dry C_5H_5N and $(COCl)_2$ as reagents as described for the synthesis of estrone oxalate dimer **65**.

A stirred solution of DHEA (**66**, 500 mg, 1.73 mmol) in dry C_5H_5N (6 mL) was treated with $(COCl)_2$ (110 mg, 0.87 mmol) under N_2 to give bis(androst-5-en-17-on-3β-yl)oxalate (**70**, 306 mg, 56%, mp: 281–282 °C) (*Scheme 2.1.29*). IR ($CHCl_3$): ν_{max} cm^{-1} 2954s (C–H), 2889s (C–H), 2858s (C–H), 1764s (C=O), 1740s (C=O), 1669m (C=O), 1637m (C=C), 1471m (C–H), 1372m (C–O), 1250m (C–O), 1190s (C–O), 1059m (C–O), 966w, 772w. ^1H NMR (400 MHz, $CDCl_3$): δ 5.39 (d, 2H, $J = 5.1$ Hz, 2 × 6-CH), 4.73 (m, 2H, 2 × 3α-CH–O), 1.02 (s, 6H, 2 × 19-Me), 0.84 (s, 6H, 2 × 18-Me). ^{13}C NMR (100 MHz, $CDCl_3$): δ 36.8 (2 × C-1), 27.3 (2 × C-2), 76.9 (2 × C-3), 31.4 (2 × C-4), 139.3 (2 × C-5), 122.5 (2 × C-6), 30.8 (2 × C-7), 31.4 (2 × C-8), 51.7 (2 × C-9), 36.7 (2 × C-10), 20.3 (2 × C-11), 37.6 (2 × C-12), 47.5 (2 × C-13), 50.1 (2 × C-14), 21.8 (2 × C-15), 35.8 (2 × C-16), 220.8 (2 × C-17), 13.5 (2 × C-18), 19.3 (2 × C-19), 157.5 (2 × oxalate CO). FABMS: *m/z* 631 $[M + H]^+$, 653 $[M + Na]^+$. HRFABMS: *m/z* 631.3998 calculated for $C_{40}H_{55}O_6$, found: 631.3999 (Nahar *et al.*, 2006).

A stirred solution of pregnenolone (**67**, 550 mg, 1.74 mmol) in dry C_5H_5N (6 mL) was treated dropwise with $(COCl)_2$ (110 mg, 0.87 mmol) under N_2 to yield bis(pregn-5-en-20-on-3β-yl)oxalate (**71**, 269 mg, 45%, mp: 246–247 °C) (*Scheme 2.1.30*). IR ($CHCl_3$): ν_{max} cm^{-1} 2954s (C–H), 2889s (C–H), 2858s (C–H), 1762s (C=O), 1740s (C=O), 1669m (C=O), 1637m (C=C), 1471m (C–H), 1439m (C–H), 1372m (C–H), 1250s (C–O), 1190s (C–O), 1138m (C–O), 1059m (C–O), 1021m (C–O), 966m, 927m, 772w. ^1H NMR (400 MHz, $CDCl_3$): δ 5.36 (t, 2H, $J = 2.1$ Hz, 2 × 6-CH), 4.72 (m, 2H, 2 × 3α-CH–O), 2.08 (s, 6H, $J = 6.1$ Hz, 2 × 21-Me), 0.99 (s, 6H, 2 × 19-Me), 0.58 (s, 6H, 2 × 18-Me). ^{13}C NMR (100 MHz, $CDCl_3$): δ 36.8 (2 × C-1), 27.3 (2 × C-2), 77.1 (2 × C-3), 37.6 (2 × C-4),

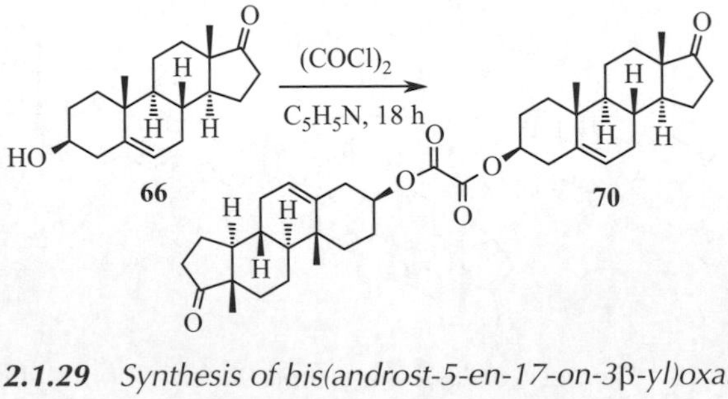

Scheme 2.1.29 *Synthesis of bis(androst-5-en-17-on-3β-yl)oxalate (**70**)*

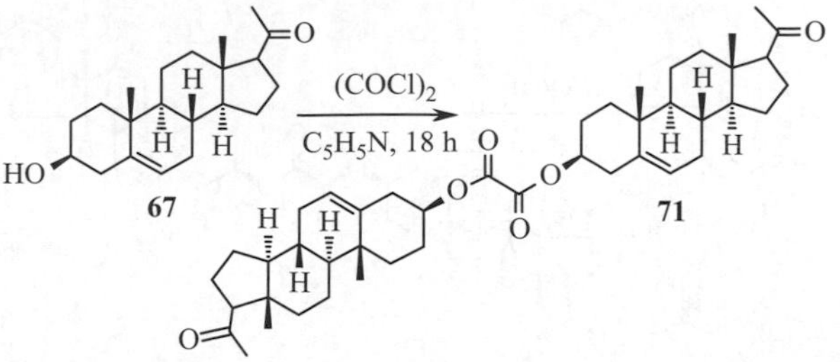

Scheme 2.1.30 *Synthesis of bis(androst-5-en-17-on-3β-yl)oxalate (**71**)*

139.0 (2 × C-5), 123.0 (2 × C-6), 31.8 (2 × C-7), 31.7 (2 × C-8), 49.8 (2 × C-9), 36.6 (2 × C-10), 21.0 (2 × C-11), 38.7 (2 × C-12), 43.9 (2 × C-13), 56.8 (2 × C-14), 24.5 (2 × C-15), 22.8 (2 × C-16), 63.6 (2 × C-17), 13.2 (2 × C-18), 19.2 (2 × C-19), 209.4 (2 × C-20), 31.5 (2 × C-21), 157.6 (2 × oxalate CO). FABMS: *m/z* 687 $[M + H]^+$, 709 $[M + Na]^+$. HRFABMS: *m/z* 687.4624 calculated for $C_{44}H_{63}O_6$, found: 687.4624 (Nahar *et al.*, 2006).

The treatment of a stirred solution of methyl lithocholate (**50**, 1.0 g, 2.56 mmol) in dry C_5H_5N (10 mL) dropwise with $(COCl)_2$ (2.04 g, 16.02 mmol) under N_2 afforded bis(methyl 5β-cholan-24-oate-3α-yl)oxalate (**72**, 1.2 g, 56%, mp: 195–196 °C) (*Scheme 2.1.31*). IR ($CHCl_3$): ν_{max} cm^{-1} 2934s (C–H), 2859s (C–H), 1762s (C=O), 1739s (C=O), 1449m (C–H), 1376m (C–H), 1175s (C–O), 758w. ^{1}H NMR (400 MHz, $CDCl_3$): δ 4.85 (m, 2H, 2 × 3β-CH–O), 3.62 (s, 6H, 2 × 24-OMe), 0.89 (s, 6H, 2 × 19-Me), 0.86 (d, 6H, J = 6.2 Hz, 2 × 21-Me), 0.60 (s, 6H, 2 × 18-Me). ^{13}C NMR (100 MHz, $CDCl_3$): δ 36.9 (2 × C-1), 27.4 (2 × C-2), 77.8 (2 × C-3), 31.8 (2 × C-4), 41.9 (2 × C-5), 27.0 (2 × C-6), 26.3 (2 × C-7), 35.8 (2 × C-8), 40.3 (2 × C-9), 34.6 (2 × C-10), 20.8 (2 × C-11), 40.1 (2 × C-12), 42.7 (2 × C-13), 56.5 (2 × C-14), 24.2 (2 × C-15), 28.2 (2 × C-16), 56.0 (2 × C-17), 12.0 (2 × C-18), 23.2 (2 × C-19), 35.4 (2 × C-20), 18.3 (2 × C-21), 31.1 (2 × C-22), 31.0 (2 × C-23), 174.7 (2 × C-24), 51.5 (2 × 24-OMe), 157.8 (2 × oxalate CO). HRFABMS: *m/z* 852.6358 calculated for $C_{52}H_{86}NO_8$, found: 852.6355 (Nahar *et al.*, 2006).

Similarly, dropwise addition of $(COCl)_2$ (164 mg, 1.29 mmol) under N_2 to a stirred solution of cholesterol (**1**, 1.0 g, 2.59 mmol) in dry C_5H_5N (10 mL) produced the bis (cholest-5-en-3β-yl)oxalate (**73**, 438 mg, 41%, mp: 220–221 °C) (*Scheme 2.1.32*). IR ($CHCl_3$): ν_{max} cm^{-1} 2943s (C–H), 2867s (C–H), 1764s (C=O), 1742s (C=O), 1665m (C=C), 1466m (C–H), 1375m (C–H), 1174s (C–O), 912w, 733m. ^{1}H NMR (400 MHz,

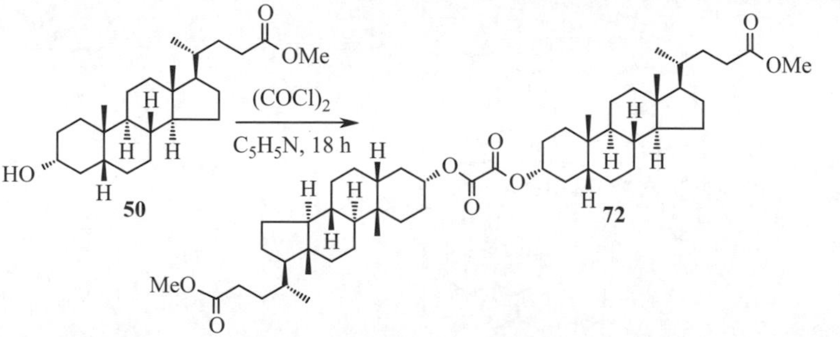

Scheme 2.1.31 *Synthesis of bis(methyl 5β-cholan-24-oate-3α-yl)oxalate (**72**)*

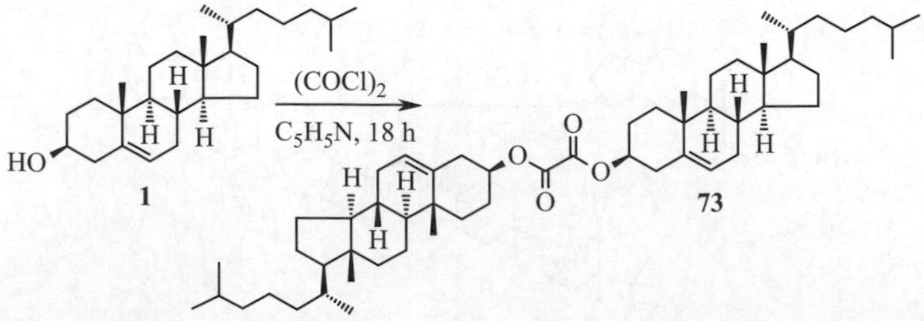

Scheme 2.1.32 *Synthesis of bis(cholest-5-en-3β-yl)oxalate (**73**)*

$CDCl_3$): δ 5.36 (t, 2H, $J = 2.7$ Hz, 2 × 6-CH), 4.71 (m, 2H, 2 × 3α-CH–O), 0.99 (s, 6H, 2 × 19-Me), 0.88 (d, 6H, $J = 6.5$ Hz, 2 × 21-Me), 0.83 (d, 6H, $J = 6.8$ Hz, 2 × 27-Me), 0.82 (d, 6H, $J = 6.8$ Hz, 2 × 26-Me), 0.64 (s, 6H, 2 × 18-Me). ^{13}C NMR (100 MHz, $CDCl_3$): δ 36.9 (2 × C-1), 27.4 (2 × C-2), 77.2 (2 × C-3), 37.6 (2 × C-4), 139.0 (2 × C-5), 123.3 (2 × C-6), 31.9 (2 × C-7), 31.8 (2 × C-8), 50.0 (2 × C-9), 36.6 (2 × C-10), 21.0 (2 × C-11), 39.7 (2 × C-12), 42.3 (2 × C-13), 56.7 (2 × C-14), 24.3 (2 × C-15), 28.2 (2 × C-16), 56.1 (2 × C-17), 11.8 (2 × C-18), 19.3 (2 × C-19), 35.8 (2 × C-20), 18.7 (2 × C-21), 36.2 (2 × C-22), 23.8 (2 × C-23), 39.5 (2 × C-24), 28.0 (2 × C-25), 22.6 (2 × C-26), 22.8 (2 × C-27), 157.7 (2 × oxalate CO). FABMS: *m/z* 827 $[M + H]^+$, 849 $[M + Na]^+$. HRFABMS: *m/z* 827.6917 calculated for $C_{56}H_{91}O_4$, found: 827.6919 (Nahar *et al.*, 2006).

When a stirred solution of cholestanol (**68**, 1.0 g, 2.57 mmol) in dry C_5H_5N (10 mL) was treated dropwise with $(COCl)_2$ (164 mg, 1.29 mmol) under N_2, it formed bis(5α-cholestan-3β-yl)oxalate (**74**, 451 mg, 42%, mp: 209–210 °C) (*Scheme 2.1.33*). IR ($CHCl_3$): v_{max} cm^{-1} 2930s (C–H), 2863s (C–H), 1764s (C=O), 1745s (C=O), 1471m (C–H), 1373m (C–H), 1183s (C–O), 997m, 927m, 755m, 607w. ^1H NMR (400 MHz, $CDCl_3$): δ 4.80 (m, 2H, 2 × 3α-CH–O), 0.85 (d, 6H, $J = 6.5$ Hz, 2 × 27-Me), 0.82 (d, 6H, $J = 6.8$ Hz, 2 × 21-Me), 0.81 (d, 6H, $J = 6.5$ Hz, 2 × 26-Me), 0.79 (s, 6H, 2 × 19-Me), 0.60 (s, 6H, 2 × 18-Me). ^{13}C NMR (100 MHz, $CDCl_3$): δ 36.7 (2 × C-1), 27.1 (2 × C-2), 77.0 (2 × C-3), 37.6 (2 × C-4), 44.7 (2 × C-5), 28.6 (2 × C-6), 31.9 (2 × C-7), 35.4 (2 × C-8), 54.2 (2 × C-9), 33.5 (2 × C-10), 21.2 (2 × C-11), 40.0 (2 × C-12), 42.6 (2 × C-13), 56.4 (2 × C-14), 24.2 (2 × C-15), 28.2 (2 × C-16), 56.3 (2 × C-17), 12.1 (2 × C-18), 18.9 (2 × C-19), 35.8 (2 × C-20), 12.2 (2 × C-21), 36.2 (2 × C-22), 23.8 (2 × C-23), 39.5 (2 × C-24), 28.0 (2 × C-25), 22.6 (2 × C-26), 22.8 (2 × C-27), 157.9 (2 × oxalate CO). FABMS: *m/z* 831 $[M + H]^+$, 853 $[M + Na]^+$. HRFABMS: *m/z* 831.7229 calculated for $C_{56}H_{95}O_4$, found: 831.7228 (Nahar *et al.*, 2006).

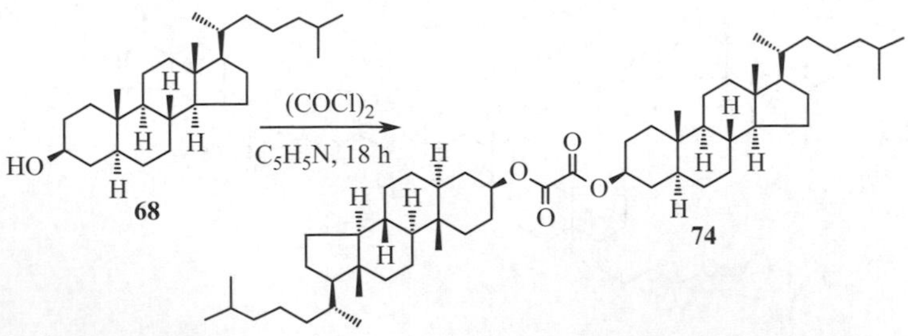

Scheme 2.1.33 *Synthesis of bis(5α-cholestan-3β-yl)oxalate (**74**)*

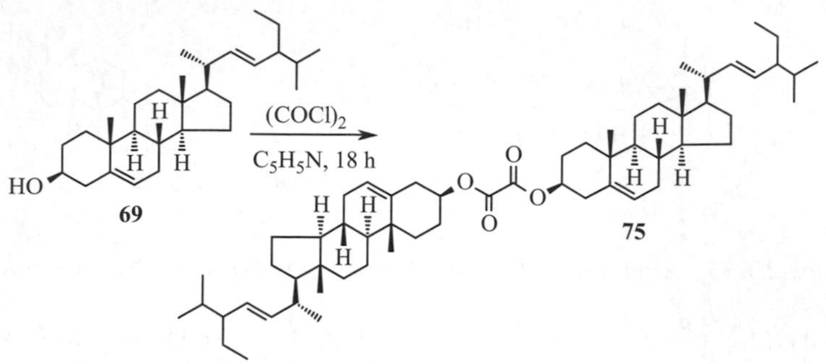

Scheme 2.1.34 *Synthesis of bis(stigmasta-5,22t*-dien-3β-yl)oxalate (**75**)

A stirred solution of stigmasterol (**69**, 1.0 g, 2.42 mmol) in dry C_5H_5N (10 mL) was treated dropwise with $(COCl)_2$ (154 mg, 1.21 mmol) under N_2 to afford bis(stigmasta-5,22*t*-dien-3β-yl)oxalate (**75**, 491 mg, 46%, mp: 184–185 °C) (*Scheme 2.1.34*). IR ($CHCl_3$): ν_{max} cm^{-1} 2947s (C–H), 2872s (C–H), 1762s (C=O), 1739s (C=O), 1665m (C=C), 1457m (C–H), 1368m (C–H), 1195s (C–O), 970m and 758w. ^{1}H NMR (400 MHz, $CDCl_3$): δ 5.36 (d, 2H, *J* = 5.1 Hz, 2 × 6-CH), 5.12 (dd, 2H, *J* = 8.5, 15.1 Hz, 2 × 22-CH), 4.98 (dd, 2H, *J* = 8.5, 15.1 Hz, 2 × 23-CH), 4.73 (m, 2H, 2 × 3α-CH–O), 1.49 (m, 2H, 2 × 25-CH), 1.00 (s, 6H, 2 × 19-Me), 0.97 (s, 6H, *J* = 6.1 Hz, 2 × 21-Me), 0.81 (d, 6H, *J* = 6.5 Hz, 2 × 27-Me), 0.77 (t, 6H, *J* = 7.5 Hz, 2 × 29-Me), 0.75 (d, 6H, *J* = 7.2 Hz, 3 × 26-Me), 0.66 (s, 6H, 2 × 18-Me). ^{13}C NMR (100 MHz, $CDCl_3$): δ 36.9 (2 × C-1), 27.4 (2 × C-2), 77.2 (2 × C-3), 37.6 (2 × C-4), 139.0 (2 × C-5), 123.3 (2 × C-6), 31.9 (2 × C-7), 31.8 (2 × C-8), 51.2 (2 × C-9), 36.6 (2 × C-10), 21.0 (2 × C-11), 39.6 (2 × C-12), 42.2 (2 × C-13), 56.8 (2 × C-14), 24.3 (2 × C-15), 28.9 (2 × C-16), 55.9 (2 × C-17), 12.0 (2 × C-18), 19.3 (2 × C-19), 40.5 (2 × C-20), 21.2 (2 × C-21), 128.3 (2 × C-22), 129.3 (2 × C-23), 50.0 (2 × C-24), 31.8 (2 × C-25), 19.0 (2 × C-26), 21.1 (2 × C-27), 25.4 (2 × C-28), 12.2 (2 × C-29), 157.7 (2 × oxalate CO). FABMS: *m/z* 879 $[M + H]^+$, 901 $[M + Na]^+$. HRFABMS: *m/z* 879.7229 calculated for $C_{60}H_{95}O_4$, found: 879.7229 (Nahar *et al.*, 2006).

Cholestane-derived *gem*-dihydroperoxide dimer, bis(3-hydroperoxy-5β-cholestan-3′-yl) peroxide (**78**) and a disatereomeric mixture (1:5) of tetraoxane dimer, bis(3-dioxy-5α-cholestane) (**79**) were prepared, respectively, from 5β-cholestan-3-one (**76**) and 5α-cholestan-3-one (**77**) by acid-catalyzed addition of H_2O_2 to the ketone (Todorovic *et al.*, 1996). In the case of **78**, the spacer was a dioxane, whereas in **79**, it was a 1,2,4,5-tetraoxane system.

To synthesize **78**, a stirred solution of **76** (400 mg, 1.0 mmol) in cyclohexane (4 mL) and 30% H_2O_2 (250 μL, 2.2 mmol) was treated with a stirred mixture of H_2O (1.4 mL), EtOH (1.5 mL), and H_2SO_4 (2.70 mL) at 0 °C (*Scheme 2.1.35*). After 4 h, C_6H_6 (20 mL) and H_2O (10 mL) were added to the reaction mixture and sequentially washed with H_2O (5 × 15 mL), saturated aqueous $NaHCO_3$ (3 × 15 mL), and brine (2 × 15 mL), dried (Na_2SO_4) and evaporated under vacuum. The crude dimer was purified by CC, eluted with 5% EtOAc in toluene, produced dihydroperoxy-peroxide **78** (50 mg, 12%). IR (KBr): ν_{max} cm^{-1} 3421m (O–H), 2942s (C–H), 2867s (C–H), 1467m (C–H), 1449m (C–H), 1382m (C–H). IR (CCl_4): ν_{max} cm^{-1} 3420m (O–H), 2937s (C–H), 2868s(C–H). ^{1}H NMR (600 MHz, $CDCl_3$): δ 9.57 (s, 2H, 2 × 3-CO_2H, exchangeable with D_2O), 0.96 (s, 6H, 2 × 19-Me), 0.90 (d, 6H, *J* = 6.4 Hz, 2 × 21-Me), 0.88 and 0.86 (d, 12H, *J* = 6.1 Hz, 2 × 26-Me and 2 × 27-Me), 0.64

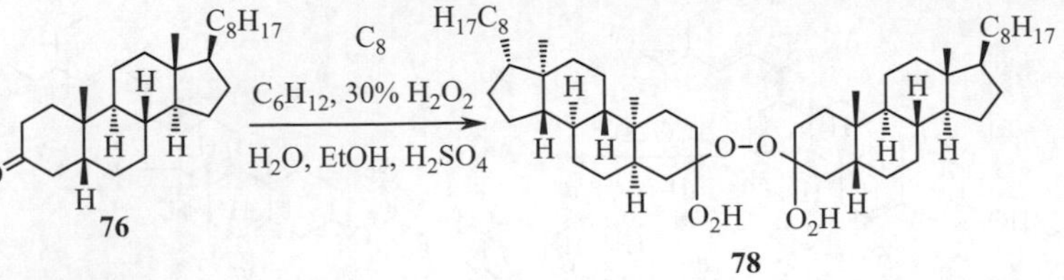

Scheme 2.1.35 *Synthesis of bis(3-hydroperoxy-5β-cholestan-3′-yl)peroxide (**78**)*

(s, 6H, 2 × 18-Me). CIMS: *m*/*z* 805 [(M + H) - H_2O_2]$^+$. FABMS: *m*/*z* 861 [M + Na]$^+$, 827 [(M + Na) – H_2O_2]$^+$ (Todorovic *et al.*, 1996).

Bis(3-dioxy-5α-cholestane) (**79**) was also synthesized in the same manner (*Scheme 2.1.36*). To a stirred mixture of H_2O (4.2 mL), EtOH (4.5 mL), and H_2SO_4 (8.1 mL) at 0 °C, a solution of 5α-cholestan-3-one (**77**, 1.50 g, 3.9 mmol) in C_6H_6 (20 mL) and 30% H_2O_2 (1.30 mL, 11.5 mmol) was added. After 3 h, H_2O (30 mL) was added to the reaction mixture and extracted with C_6H_6 (3 × 20 mL). The pooled organic extracts were successively washed with H_2O (5 × 20 mL), saturated aqueous $NaHCO_3$ (3 × 20 mL) and brine (2 × 20 mL), and dried (Na_2SO_4). The solvent was evaporated under pressure, the crude dimer was purified by CC (mobile phase: 1% petroleum ether in toluene) and crystallized from *n*-hexane-DCM yielded a diastereomeric mixture (1:5) of tetraoxane dimer **79** (800 mg, 51%, mp: 187–190 °C). IR (KBr): ν_{max} cm^{-1} 3436w (O–H), 2932s(C–H), 2868s(C–H), 1468m (C–H), 1377m (C–H), 1037m (C–O). ^{1}H NMR (600 MHz, $CDCl_3$): δ 0.90 (d, 6H, *J* = 6.3 Hz, 2 × 21-Me), 0.87 and 0.86 (d, 12H, *J* = 6.5 Hz, 2 × 26-Me and 2 × 27-Me), 0.80 (s, 6H, 2 × 19-Me), 0.64 (s, 6H, 2 × 18-Me). CIMS: *m*/*z* 805 [M + H]$^+$, 773 [(M + H) – O_2]$^+$ (Todorovic *et al.*, 1996).

These dioxane or tetraoxane systems were considered to be of great importance because of their potential contribution to any antimalarial property of the prepared dimers. This consideration prompted the synthesis of a tetraoxane spacer group containing cholic acid-based dimers with H_2O_2. A series of tetraoxane dimers **87–94** were synthesized starting from methyl cholate (**51**), which was converted to its derivatives **80–86** in the following manner (Opsenica *et al.*, 2000).

Methyl cholate (**51**, 1.0 g, 2.37 mmol) was dissolved in freshly prepared solution of Ac_2O (1 mL) and TMSOTf (26 μL, 0.14 mmol), and stirred for 5 min at room temperature (*Scheme 2.1.37*). Saturated aqueous $NaHCO_3$ was added to the reaction mixture to form solids, which after crystallization from *n*-hexane-Me_2CO, produced methyl 3α,7α,12α-triacetoxy-5β-cholan-24-oate (**80**, mp: 93–96 °C). IR (film): ν_{max} cm^{-1} 3021m, 2948s (C–H), 2871s(C–H), 1734s (C=O), 1468m (C–H), 1438s (C–H), 1378s (C–H), 1245s (C–O), 1063s, 1022s, 756s. ^{1}H NMR (200 MHz, $CDCl_3$): δ 5.09 (br s, 1H, 12β-CH), 4.91

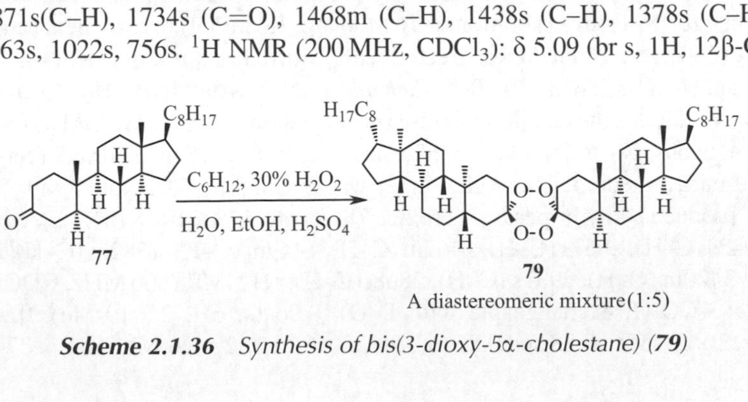

A diastereomeric mixture(1:5)

Scheme 2.1.36 *Synthesis of bis(3-dioxy-5α-cholestane) (**79**)*

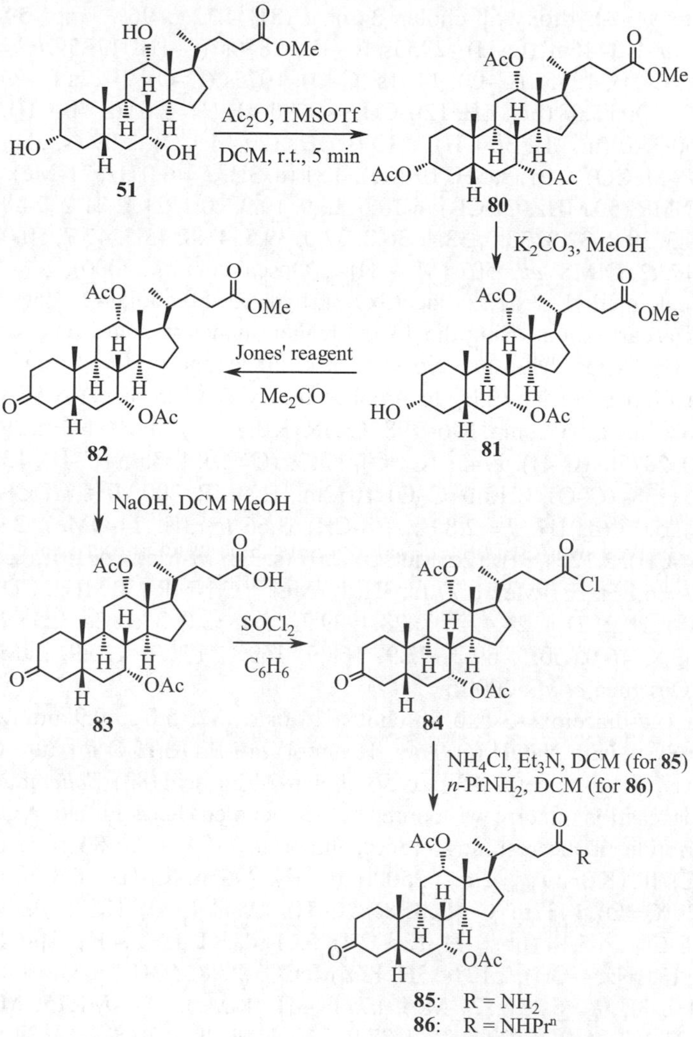

Scheme 2.1.37 *Synthesis of cholic acid derivatives* ***80–86***

(d, 1H, $J = 2.6$ Hz, 7β-CH), 4.65-4.50 (m, 1H, 3β-CH), 3.66 (s, 3H, 24-OMe), 2.14 (s, 3H, 12α-MeCO), 2.09 (s, 3H, 7α-MeCO), 2.05 (s, 3H, 3α-MeCO), 0.92 (s, 3H, 19-Me), 0.81 (d, 3H, $J = 6.2$ Hz, 21-Me), 0.73 (s, 3H, 18-Me). ^{13}C NMR (50 MHz, $CDCl_3$): δ 12.1, 17.4, 21.5, 21.6, 22.5, 22.7, 25.5, 26.8, 27.1, 28.8, 30.7, 30.8, 31.2, 34.3, 34.6, 37.7, 40.9, 43.4, 45.0, 47.3, 51.5, 70.6, 74.05, 75.36, 170.4, 170.5, 174.6. CIMS: *m*/z 549 $[M + H]^+$ (Opsenica *et al.*, 2000).

To a stirred solution of methyl 3α,7α,12α-triacetoxy-5β-cholan-24-oate (**80**, 5.30 g, 9.66 mmol) in dry MeOH (80 mL), anhydrous K_2CO_3 (2.4 g, 17.4 mmol) was added, and the mixture was stirred at room temperature for 2 h (*Scheme 2.1.37*). AcOH (3 mL) was added, and once CO_2 had liberated, the remaining solution was rotary evaporated. The residue was dissolved in EtOAc and washed with brine, dried (Na_2SO_4) and evaporated under vacuum. Crystallization of the crude product from EtOAc-*n*-hexane provided methyl

7α,12α-diacetoxy-3α-hydroxy-5β-cholan-24-oate (**81**, 4.72 g, 96%, mp: 59–62 °C). IR (KBr): ν_{max} cm^{-1} 3446m (O–H), 2953s (C–H), 2870s (C–H), 1735s (C=O), 1467m (C–H), 1379s (C–H), 1245s (C–O), 1101s (C–O), 1075s (C–O), 1022s (C–O). ^{1}H NMR (200 MHz, $CDCl_3$): δ 5.08 (br s, 1H, 12β-CH), 4.90 (d, 1H, $J = 2.6$ Hz, 7β-CH), 3.66 (s, 3H, 24-OMe), 3.60-3.40 (m, 1H, 3β-CH), 2.13 (s, 3H, 12α-MeCO), 2.09 (s, 3H, 7α-MeCO), 2.05 (s, 3H, 3α-MeCO), 0.91 (s, 3H, 19-Me), 0.81 (d, 3H, $J = 6.0$ Hz, 21-Me), 0.73 (s, 3H, 18-Me). ^{13}C NMR (50 MHz, $CDCl_3$): δ 10.6, 15.9, 19.9, 20.1, 21.1, 21.2, 24.0, 25.6, 27.4, 28.8, 29.2, 29.3, 29.8, 32.8, 33.0, 33.4, 36.2, 37.0, 39.5, 41.8, 43.5, 47.7, 50.0, 69.4, 69.9, 74.0, 169.4, 173.2. CIMS: *m/z* 507 $[M + H]^+$ (Opsenica *et al.*, 2000).

Oxidation of methyl 7α,12α-diacetoxy-3α-hydroxy-5β-cholan-24-oate (**81**, 2.0 g, 3.95 mmol) was carried out using the Jones' reagent under standard reaction conditions (*Scheme 2.1.37*). After the usual workup as outlined above, the crude product was recrystallized from *n*-hexane-C_6H_6 to produce methyl 7α,12α-diacetoxy-3-oxo-5β-cholan-24-oate (**82**, 1.61 g, 81%, mp: 196–198 °C). IR (KBr): ν_{max} cm^{-1} 3441m, 2965m (C–H), 2923m (C–H), 2875m (C–H), 1745s (C=O), 1713s (C=O), 1435m (C–H), 1382m (C–H), 1258s (C–O), 1238s (C–O), 1215m (C–O), 1072m. ^{1}H NMR (200 MHz, $CDCl_3$): 5.13 (br s, 1H, 12β-CH), 5.00 (d, 1H, $J = 2.8$ Hz, 7β-CH), 3.66 (s, 3H, 24-OMe), 2.99 (dd, CH, $J = 13.2$ Hz, 4-CH), 2.12 (s, 3H, 12α-MeCO), 2.07 (s, 3H, 7α-MeCO), 1.02 (s, 3H, 19-Me), 0.82 (d, 3H, $J = 6.2$ Hz, 21-Me), 0.77 (s, 3H, 18-Me). ^{13}C NMR (50 MHz, $CDCl_3$): δ 10.8, 16.1, 19.9, 20.1, 20.2, 21.4, 24.4, 25.7, 28.4, 29.3, 29.4, 33.0, 33.2, 34.7, 35.2, 36.3, 40.8, 41.9, 43.2, 43.8, 46.0, 50.1, 69.2, 73.9, 168.9, 169.1, 173.2, 210.9. CIMS: *m/z* 506 $[M + H]^+$ (Opsenica *et al.*, 2000).

Methyl 7α,12α-diacetoxy-3-oxo-5β-cholan-24-oate (**82**, 5.0 g, 9.9 mmol) underwent alkaline hydrolysis with NaOH (600 mg, 15 mmol) in *i*-PrOH/H_2O mixture (80 mL, 3:1) at 80 °C to give **7**α,12α-diacetoxy-3-oxo-5β-cholan-24-oic acid (**83**) (*Scheme 2.1.37*). After hydrolysis, the reaction mixture was poured onto ice-cooled HCl acidified H_2O, and filtered off. Crystallization of the resulting crude solid from EtOH gave **83** (4.53 g, 93%, mp: 200–201.5 °C). IR (KBr): ν_{max} cm^{-1} 3500br (O–H), 2965m (C–H), 2878m (C–H), 1734s (C=O), 1713s (C=O), 1437m (C–H), 1381s (C–H), 1254s (C–O), 1032m (C–O). ^{1}H NMR (200 MHz, $CDCl_3$): δ 5.14 (br s, 1H, 12β-CH), 5.01 (d, 1H, $J = 2.4$ Hz, 7β-CH), 3.00 (dd, 1H, $J = 13.8$, 15.0 Hz, 4-CH), 2.12 (s, 3H, 12α-MeCO), 2.08 (s, 3H, 7α-MeCO), 1.02 (s, 3H, 19-Me), 0.84 (d, 3H, $J = 6.2$ Hz, 21-Me), 0.77 (s, 3H, 18-Me). ^{13}C NMR (50 MHz, $CDCl_3$): δ 12.2, 17.4, 21.3, 21.4, 21.6, 22.7, 25.7, 27.0, 29.7, 30.4, 30.7, 30.8, 34.3, 34.5, 36.0, 36.5, 37.6, 42.0, 43.2, 44.5, 45.0, 47.3, 70.6, 75.2, 170.3, 170.5, 179.6, 212.5. CIMS: *m/z* 491 $[M + H]^+$ (Opsenica *et al.*, 2000).

7α,12α-Diacetoxy-3-oxo-5β-cholan-24-oic acid (**83**) was converted to its amide, 7α,12α-diacetoxy-3-oxo-5β-cholan-24-amide (**85**), and also to *N*-(*n*-propyl)-7α,12α-diacetoxy-3-oxo-5β-cholan-24-amide (**86**) through the formation of **7**α,12α-diacetoxy-3-oxo-5β-cholan-24-oyl chloride (**84**) (*Scheme 2.1.37*). To a cooled (0 °C) solution of **77** (3.0 g, 6.11 mmol) in dry C_6H_6 (50 mL), $SOCl_2$ (534 μL, 7.35 mmol) was added under N_2, and the mixture was refluxed for 3 h. The solution was cooled to room temperature, and the solvent was evaporated to yield the crude acid chloride **84**. The resulting acid chloride was dissolved in dry DCM (25 mL), cooled to 0 °C, a suspension of NH_4Cl (3.27 g, 61.1 mmol) and Et_3N (8.47 mL, 61.1 mmol) in dry DCM (50 mL) was added, stirred for 15 min at 0 °C and then at room temperature overnight. The reaction mixture was poured onto H_2O, extracted with DCM (3 × 30 mL), the pooled DCM extracts were washed with H_2O

followed by brine, and dried (Na_2SO_4). Crude product was purified by dry-flash chromatography and crystallized from *n*-hexane-C_6H_6 produced 24-amide **85** (1.97 g, 61%, mp: 214 °C. IR (KBr): ν_{max} cm^{-1} 3426m (N–H), 2964m (C–H), 2878m (C–H), 1714s (C=O), 1688s (C=O), 1626m (N–H), 1380m (C–N), 1254s (C–O), 1032m. ^{1}H NMR (200 MHz, $CDCl_3$): δ 5.46 (br s, NH_2, exchangeable with D_2O), 5.14 (br s, 1H, 12α-CH), 5.00 (d, 1H, *J* = 2.6 Hz, 7β-CH), 2.99 (dd, 1H, *J* = 13.4, 13.4 Hz, 4-CH), 2.12 (s, 3H, 12α-MeCO), 2.07 (s, 3H, 7α-MeCO), 1.02 (s, 3H, 19-Me), 0.84 (d, 3H, *J* = 6.2 Hz, 21-Me), 0.77 (s, 3H, 18-Me). ^{13}C NMR (50 MHz, $CDCl_3$): δ 12.2, 17.6, 21.3, 21.5, 21.6, 22.7, 25.8, 27.1, 29.7, 30.9, 31.3, 32.7, 34.3, 34.7, 36.1, 36.6, 37.7, 42.1, 43.2, 44.5, 45.1, 47.5, 70.5, 75.2, 170.2, 170.5, 175.6, 212.2. CIMS: *m/z* 490 [M + H]$^+$ (Opsenica *et al*., 2000).

The conversion of 7α,12α-diacetoxy-3-oxo-5β-cholan-24-oyl chloride (**84**) to *N*-(*n*-propyl)-3-oxo-7α,12α-diacetoxy-5β-cholan-24-amide (**86**) was performed using the same procedure as outlined above for **85**, with the only difference that **84** (2.95 g, 6.01 mmol) reacted with *n*-$PrNH_2$ (2 equiv.) instead of NH_4Cl/Et_3N mixture (*Scheme 2.1.37*). Crystallization of the crude product, as described above, furnished **86** (2.17 g, 68%, mp: 216 °C). IR (KBr): ν_{max} cm^{-1} 3407m (N–H), 2954m (C–H), 2935m(C–H), 2871m (C–H), 1735s (C=O), 1718s (C=O), 1664s (C=O), 1534m (N–H), 1376m (C–N), 1258s (C–O), 1244s (C–O), 1028m. ^{1}H NMR (200 MHz, $CDCl_3$): δ 5.59–5.43 (m, H–N, exchangeable with D_2O), 5.13 (br s, 1H, 12β-CH), 5.00 (d, 1H, *J* = 2.8 Hz, 7β-CH), 3.30–3.17 (m, 2H, $CH_3CH_2\mathbf{CH_2}N$), 2.99 (dd, 1H, *J* = 13.4 Hz, 4-CH), 2.11 (s, 3H, 12α-MeCO), 2.07 (s, 3H, 7α-MeCO), 1.57–1.43 (m, 2H, $CH_3\mathbf{CH_2}CH_2N$), 1.02 (s, 3H, 19-Me), 0.92 (t, 3H, *J* = 7.4 Hz, $\mathbf{CH_3}CH_2CH_2N$), 0.83 (d, 3H, *J* = 6.2 Hz, 21-Me), 0.77 (s, 3H, 18-Me). ^{13}C NMR (50 MHz, $CDCl_3$): δ 12.3, 17.6, 21.3, 21.6, 22.7, 25.8, 27.1, 29.7, 30.9, 31.3, 32.7, 34.3, 34.7, 36.1, 36.6, 37.7, 42.1, 43.2, 44.5, 45.1, 47.5, 70.5, 75.2, 170.2, 170.5, 175.6, 212.2. CIMS: *m/z* 532 [M + H]$^+$ (Opsenica *et al*., 2000).

7α,12α-Diacetoxy-3-oxo-5β-cholan-24-oic acid (**83**) was utilized for the synthesis of bis(methyl 3-dioxy-7α,12α-diacetoxy-5β-cholan-24-oate) (**87** and **88**) (*Scheme 2.1.38*). To a cooled (0 °C) solution of EtOH (1.20 mL), H_2O (1.1 mL) and H_2SO_4 (2.16 mL), a solution of **75** (400 mg, 0.79 mmol) in toluene (8.3 mL) was added. After 15 min, 32% H_2O_2 (200 μL) was added at 0 °C and stirring was continued for 2 h at the same temperature. The reaction mixture was diluted with H_2O (20 mL) and toluene (30 mL), the organic layer was separated and successively washed with H_2O (2 × 10 mL), saturated aqueous $NaHCO_3$ (2 × 10 mL) and brine, dried (Na_2SO_4) and rotary evaporated. Purification on CC (mobile phase: 30% EtOAc in heptanes), of the resulting mixture provided two main fractions, which after crystallization from *n*-hexane-Me_2CO yielded dimers **87** (104 mg, 25%, mp: 251–252 °C) and **88** (116 mg, 28%, mp: 167–170 °C) (Opsenica *et al*., 2000).

Dimer **87**: IR (KBr): ν_{max} cm^{-1} 2995m (C–H), 1737s (C=O), 1440m (C–H), 1378m (C–H), 1250s (C–O), 1027m. ^{1}H NMR (200 MHz, $CDCl_3$): δ 5.09 (br s, 2H, 2 × 12β-CH), 4.92 (br s, 2H, 2 × 7β-CH), 3.66 (s, 6H, 2 × 24-OMe), 2.10 (br s, 12H, 2 × 12α-MeCO and 2 × 7α-MeCO), 0.94 (s, 6H, 2 × 19-Me), 0.81 (d, 6H, *J* = 5.8 Hz, 2 × 21-Me), 0.73 (s, 6H, 2 × 18-Me). ^{13}C NMR (50 MHz, $CDCl_3$): δ 12.1, 17.4, 21.2, 21.5, 22.0, 22.7, 25.6, 27.0, 28.3, 30.6, 30.7, 34.5, 34.5, 37.6, 43.2, 44.9, 47.2, 51.4, 70.5, 75.2, 108.6, 170.5, 174.5. ESIMS: *m/z* 1041.7 [M + H]$^+$, 1063.8 [M + Na]$^+$ (Opsenica *et al*., 2000).

Dimer **88**: IR (KBr): ν_{max} cm^{-1} 2960m (C–H), 1738s (C–O), 1439m (C–H), 1378m (C–H), 1238s (C–O), 1026m. ^{1}H NMR (200 MHz, $CDCl_3$): δ 5.09 (br s, 2H, 2 × 12β-CH),

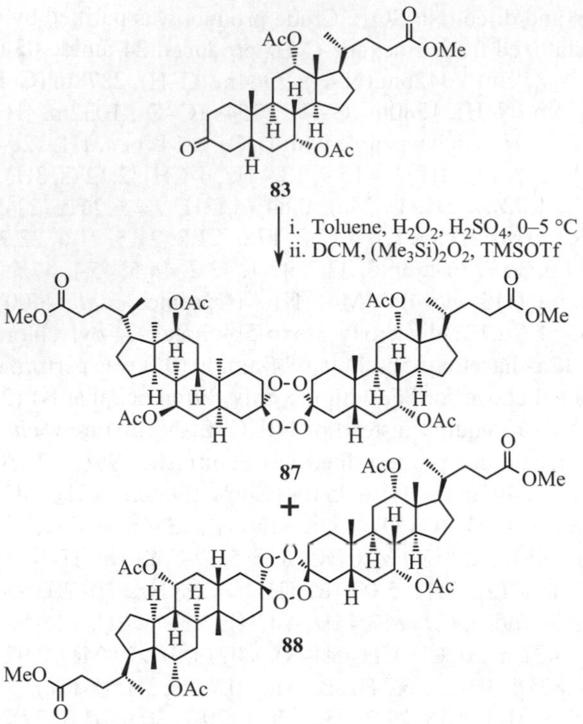

Scheme 2.1.38 *Synthesis of cholic acid-based dimers* ***87*** *and* ***88***

4.92 (br s, 2H, 2 × 7β-CH), 3.66 (s, 6H, 2 × 24-OMe), 2.12 (br s, 6H, 2 × 12α-MeCO), 2.07 (br s, 6H, 2 × 7α-MeCO), 0.94 (s, 6H, 2 × 19-Me), 0.81 (d, 6H, $J = 5.8$ Hz, 2 × 21-Me), 0.73 (s, 6H, 2 × 18-Me). ^{13}C NMR (50 MHz, $CDCl_3$): δ 12.1, 17.4, 21.3, 21.5, 22.0, 22.7, 25.6, 27.1, 28.3, 30.7, 30.8, 34.5, 34.6, 37.6, 43.2, 45.0, 47.3, 51.5, 70.6, 75.2, 108.6, 170.5, 174.5. ESIMS: *m/z* 1041.7 $[M + H]^+$, 1063.5 $[M + Na]^+$ (Opsenica *et al.*, 2000).

Another cholic acid dimer, bis(3-dioxy-7α,12α-diacetoxy-5β-cholan-24-oic acid) (**89**) was obtained from methyl ester dimer **87** by alkaline hydrolysis (*Scheme 2.1.39*). To a solution of dimer **87** (250 mg, 0.24 mmol) in DCM-MeOH mixture (100 mL), 1.25M methanolic solution of NaOH (29.95 mL) was added and stirred for 3 days at room temperature. The reaction was completed by adding glacial AcOH (until pH 5), then DCM (30 mL) and H_2O (50 mL) were added. The H_2O layer was separated and extracted with DCM (5 × 30 mL), the pooled DCM layers were washed with H_2O followed by brine, dried (Na_2SO_4) and evaporated under pressure. Crystallization from *n*-hexane-Me_2CO of the resulting crude solid afforded acid dimer **89** (192 mg, 79%, mp: 228–232 °C) IR (KBr): ν_{max} cm^{-1} 3450s (O–H), 2953s (C–H), 1737s (C=O), 1441m (C–H), 1380s (C–O), 1253s (C–O), 1126m (C–O), 1081m (C–O). ^{1}H NMR (200 MHz, DMSO-d_6): δ 11.94 (br s, 2H, 2 × 24-CO_2H), exchangeable with D_2O), 4.98 (br s, 2H, 2 × 12β-CH), 4.82 (br s, 2H, 2 × 7β-CH), 2.04 (s, 6H, 2 × 12α-MeCO), 2.00 (s, 6H, 2 × 7α-MeCO), 0.91 (s, 6H, 2 × 19-Me), 0.74 (d, 6H, $J = 6.0$ Hz, 2 × 21-Me), 0.70 (s, 6H, 2 × 18-Me). ^{13}C (50 MHz, DMSO-d_6): δ 175.0, 170.0, 169.8, 108.2, 74.7, 70.3, 47.2,

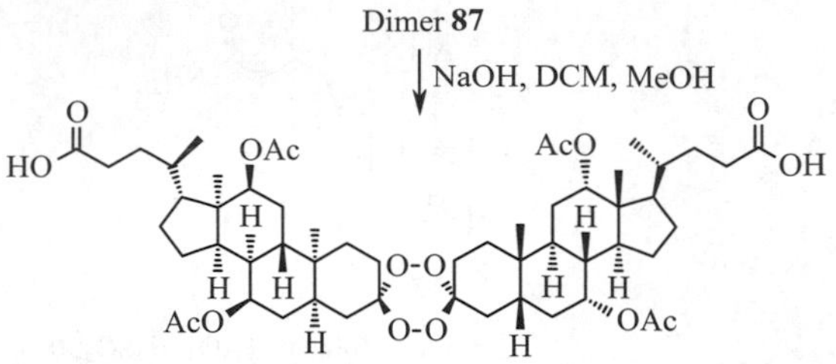

Dimer **87**

NaOH, DCM, MeOH

89

Scheme 2.1.39 *Synthesis of bis(3-dioxy-7α,12α-diacetoxy-5β-cholan-24-oic acid) (**89**)*

44.8, 43.2, 36.9, 34.4, 30.8, 30.6, 28.0, 26.8, 25.4, 22.4, 22.1, 21.4, 21.1, 17.4, 12.1. LSIMS: *m*/*z* 1035.6 $[M + Na]^+$ (Opsenica *et al.*, 2000).

Similarly, bis(3-dioxy-7α,12α-diacetoxy-5β-cholan-24-oic acid) (**90**, 91 mg, 72%, mp: 199–202 °C) was prepared from methyl ester dimer **88** (130 mg, 0.12 mmol) (*Scheme 2.1.40*), using the same method as outlined above, and the crude dimer was purified by crystallization from DCM in *i*-Pr_2O. IR (KBr): ν_{max} cm^{-1} 3473m (O–H), 3456m (O–H), 2953s (C–H), 1737s (C=O), 1441m (C–H), 1380s (C–O), 1250s (C–O), 1125w (C–O), 1081w (C–O), 1027m (C–O). ^1H NMR (200 MHz, $CDCl_3$): δ 5.09 (br s, 2H, 2 × 12β-CH), 4.93 (br s, 2H, 2 × 7β-CH), 2.12 (br s, 6H, 2 × 12α-MeCO), 2.08 (br s, 6H, 2 × 7α-MeCO), 0.94 (s, 6H, 2 × 19-Me), 0.82 (d, 6H, $J = 5.6$ Hz, 2 × 21-Me), 0.73 (s, 6H, 2 × 18-Me). ^{13}C NMR (50 MHz, $CDCl_3$): δ 12.2, 17.5, 21.4, 21.5, 22.1, 22.7, 25.7, 27.1, 28.4, 30.5, 30.7, 34.5, 34.6, 37.6, 43.3, 45.0, 47.3, 70.7, 75.3, 108.7, 170.4, 170.6, 179.5. LSIMS: *m*/*z* 1013.5 $[M + H]^+$, 1035.6 $[M + Na]^+$ (Opsenica *et al.*, 2000).

Bis(3-dioxy-7α,12α-diacetoxy-5β-cholan-24-amide) (**91** and **92**) was obtained from 24-amide **85** (500 mg, 1.02 mmol) by the treatment with TMSOTf/$(TMS)_2O_2$ solution in MeCN (*Scheme 2.1.41*). The reaction mixture was poured into a stirred solution of C_6H_6/$NaHCO_3$/ice-H_2O and extracted with C_6H_6, washed with brine, dried (Na_2SO_4) and evaporated *in vacuo*. Purification on CC (mobile phase: 3% THF in EtOAc), of the resulting mixture furnished two major fractions, which after crystallization from DCM in *i*-Pr_2O produced 24-amide dimers **91** (133 mg, 26%, mp: 211–217 °C) and **92** (124 mg, 24%, mp: 196–199 °C) (Opsenica *et al.*, 2000).

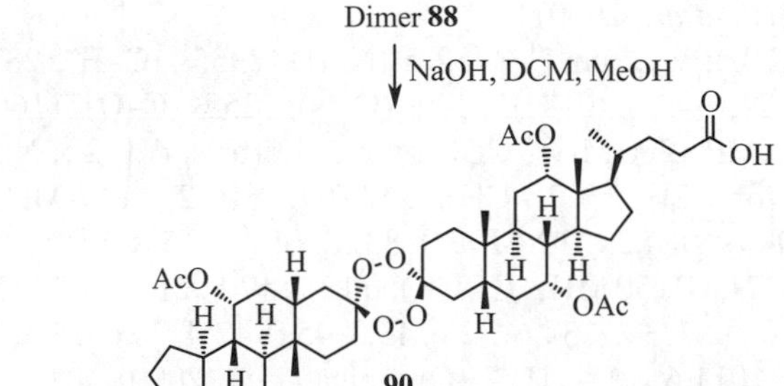

Dimer **88**

NaOH, DCM, MeOH

90

Scheme 2.1.40 *Synthesis of bis(3-dioxy-7α,12α-diacetoxy-5β-cholan-24-oic acid) (**90**)*

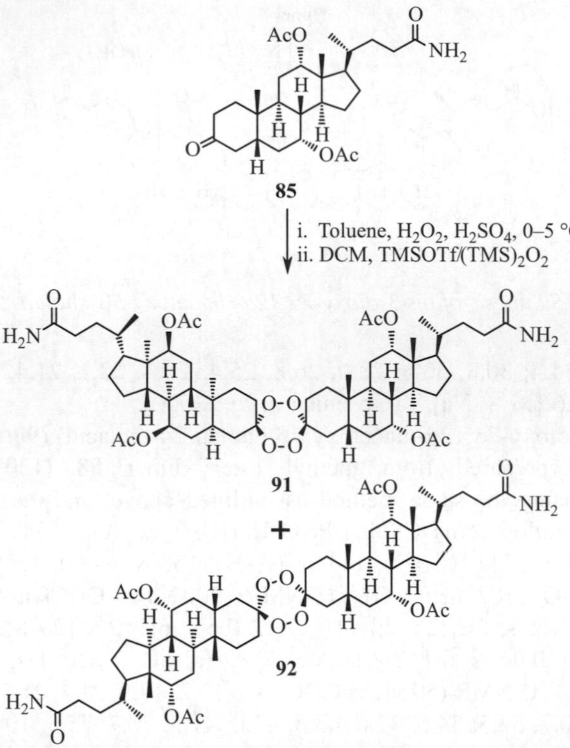

Scheme 2.1.41 *Synthesis of cholic acid-based dimers **91** and **92***

Dimer **91**: IR (KBr): ν_{max} cm^{-1} 3452m (N–H), 2955m (C–H), 2873m (C–H), 1718s (C=O), 1672s (C=O), 1443m (C–H), 1379s (C–N), 1256s (C–O), 1166w, 1126m, 1104m, 1080w, 1029w, 967w. ^{1}H NMR (200 MHz, $CDCl_3$): δ 5.45 (br s, 4H, 2 × NH_2), 5.09 (br s, 2H, 2 × 12β-CH), 4.92 (br s, 2H, 2 × 7β-CH), 2.10 (br s, 12H, 2 × 12α-MeCO and 2 × 7α-MeCO), 0.94 (s, 6H, 2 × 19-Me), 0.83 (d, 6H, J = 5.8 Hz, 2 × 21-Me), 0.73 (s, 6H, 2 × 18-Me). ^{13}C NMR (50 MHz, $CDCl_3$): δ 12.2, 17.5, 21.3, 21.6, 22.1, 22.7, 25.6, 27.2, 28.4, 30.6, 31.3, 32.7, 34.6, 37.6, 43.2, 45.0, 47.4, 70.6, 75.2, 108.6, 170.5, 175.7. LSIMS: *m/z* 1011.6 $[M + H]^+$ (Opsenica *et al.*, 2000).

Dimer **92**: IR (KBr): ν_{max} cm^{-1} 3452m (N–H), 2955s (C–H), 2877m (C–H), 1738s (C=O), 1670s (C=O), 1441m (C–H), 1379s (C–N), 1255s (C–O), 1166w, 1125m, 1081m, 1027w, 966w. ^{1}H NMR (200 MHz, $CDCl_3$): δ 5.42 (br s, 4H, 2 × NH_2), 5.10 (br s, 2H, 2 × 12β-CH), 4.92 (br s, 2H, 2 × 7β-CH), 2.12 (br s, 6H, 2 × 12α-MeCO), 2.07 (br s, 6H, 2 × 7α-MeCO), 0.94 (s, 6H, 2 × 19-Me), 0. 83 (d, 6H, J = 5.8 Hz, 2 × 21-Me), 0.73 (s, (s, 6H, 2 × 18-Me). ^{13}C NMR (50 MHz, $CDCl_3$): δ 12.2, 17.5, 21.4, 21.5, 22.0, 22.7, 25.7, 27.2, 28.4, 30.5, 31.3, 32.0, 32.7, 34.2, 34.6, 37.6, 43.3, 45.0, 47.4, 70.6, 75.3, 108.7, 170.3, 170.5, 175.6. LSIMS: *m/z* 1011.6 $[M + H]^+$ (Opsenica *et al.*, 2000).

The synthesis of two further isomeric dimers, bis[*N*-(*n*-propyl)-3-dioxy-7α,12α-diacetoxy-5β-cholan-24-amide] (**93** and **94**) (*Scheme 2.1.42*), were accomplished from 24-amide **86** (532 mg, 1.0 mmol) using the same procedure as above. The crude product was purified by CC, eluted with EtOAc, followed by crystallization from *i*-Pr_2O-DCM yielded dimers **93**

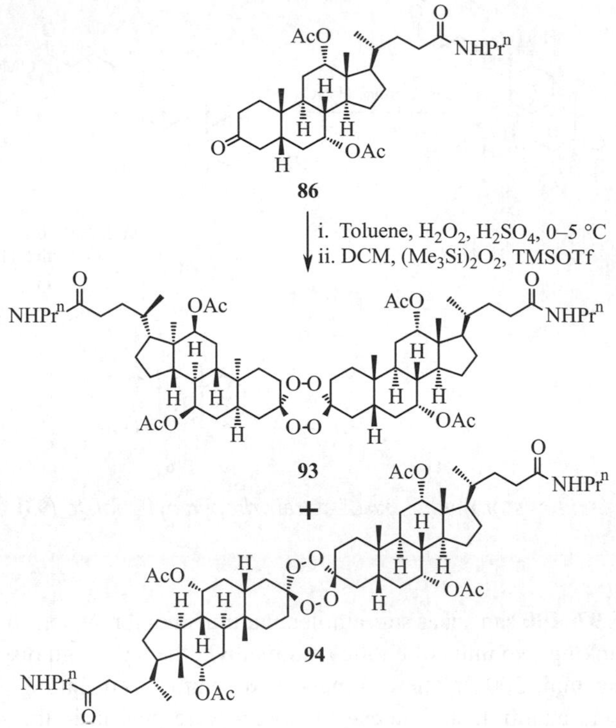

Scheme 2.1.42 *Synthesis of cholic acid-based dimers* ***93*** *and* ***94***

(74 mg, 14%, mp: 240–243 °C) and **94** (62 mg, 11%, mp: 171–174 °C) (Opsenica *et al.*, 2000).

Dimer **93**: IR (KBr): ν_{max} cm^{-1} 3422m (N–H), 2960m (C–H), 2874m (C–H), 1738s (C=O), 1650s (C=O), 1551m (N–H), 1441m (C–H), 1378s (C–N), 1250s (C–O), 1027m. ^{1}H NMR (200 MHz, $CDCl_3$): δ 5.38–5.43 (m, 2H, 2 × NH), 5.09 (br s, 2H, 2 × 12β-CH), 4.93 (br s, 2H, 2 × 7β-CH), 3.28-3.12 (m, 4H, 2 × CH_2**CH_2**N), 2.10 (br s, 12H, 2 × 12α-MeCO and 2 × 7α-MeCO), 1.60–1.40 (m, 4H, 2 × **CH_2**CH_2N), 0.96-0.87 (m, 12H, 2 × 19-Me and 2 × **CH_3**CH_2), 0.82 (d, 6H, J = 6.0 Hz, 2 × 21-Me), 0.73 (s, 6H, 2 × 18-Me). ^{13}C NMR (50 MHz, $CDCl_3$): δ 11.3, 12.2, 17.5, 21.3, 21.6, 22.0, 22.8, 25.6, 27.1, 28.4, 30.6, 31.5, 33.5, 34.6, 34.7, 37.6, 41.1, 43.2, 45.0, 47.4, 70.6, 75.2, 108.6, 170.5, 173.2. LSIMS: m/z 1095.7 $[M + H]^+$ (Opsenica *et al.*, 2000).

Dimer **94**: IR (KBr): ν_{max} cm^{-1} 3651m (N–H), 3424m (N–H), 2960m (C–H), 2875m (C–H), 1738s (C=O), 1651m (C=O), 1547m (N–H), 1441m (C–H), 1379s (C–N), 1250s (C–O), 1165w, 1125w, 1026m, 966w. ^{1}H NMR (200 MHz, $CDCl_3$): δ 5.38–5.49 (m, 2H, 2 × NH), 5.09 (br s, 2H, 2 × 12β-CH), 4.92 (br s, 2H, 2 × 7β-CH), 3.28-3.13 (m, 4H, 2 × CH_2**CH_2**N), 2.12 (br s, 6H, 2 × 12α-MeCO), 2.07 (br s, 6H, 2 × 7α-MeCO), 1.60-1.40 (m, 4H, 2 × **CH_2**CH_2N), 0.96-0.86 (m, 12H, 2 × 19-Me and 2 × **CH_3**CH_2), 0.82 (d, 6H, J = 6.0 Hz, 2 × 21-Me), 0.73 (s, 6H, 2 × 18-Me). ^{13}C NMR (50 MHz, $CDCl_3$): δ 11.3, 12.2, 17.5, 21.4, 21.5, 22.0, 22.7, 22.9, 25.7, 27.2, 28.4, 31.6, 33.6, 34.6, 34.7, 37.6, 41.1, 43.3, 45.0, 47.5, 70.6, 75.3, 108.7, 170.5, 173.2. LSIMS: m/z 1095.8 $[M + H]^+$ (Opsenica *et al.*, 2000).

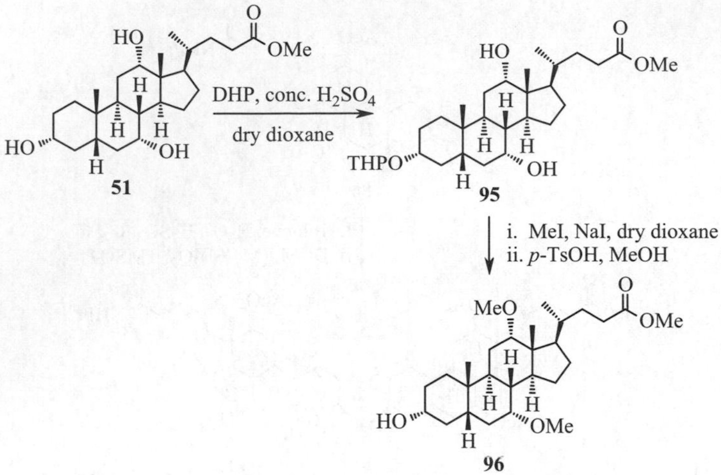

Scheme 2.1.43 *Synthesis of methyl 3α-(2′-tetrahydropyranyl)cholate (**95**) and methyl 7,12-dimethylcholate (**96**)*

Steroid dimers **97–100** can act as supramolecular transmembrane ion channels and could be prepared by linking two units of cholic acid methyl ethers through biscarbamate bonds (Kobuke and Nagatani, 2001). These dimers were synthesized linking 1,3-disubstituted benzene ring where an amide and an ester linkages were present in the spacer.

Methyl cholate (**51**) was employed to prepare the precursors for the dimer **97**, methyl 3α-(2′-tetrahydropyranyl)cholate (**95**) and methyl 7,12-dimethylcholate (**96**) (*Scheme 2.1.43*). To a stirred solution of **51** (5.0 g, 14 mmol) in dry dioxane (50 mL), freshly distilled DHP (1.22 g, 14.5 mmol) was added at room temperature, followed by the addition of a few drops of conc. H_2SO_4 while the reaction mixture was cooling in an ice-water bath. On completion of the reaction, as confirmed by the TLC, the reaction mixture was allowed to warm to room temperature and neutralized with saturated aqueous $NaHCO_3$. The solvent was removed *in vacuo*, the residue was extracted with $CHCl_3$, and purified by CC (mobile phase: EtOAc-C_6H_6), to provide protected methyl cholate **95** (3.15 g, 43%). ^{1}H NMR (270 MHz, $CDCl_3$): δ 4.72 (m, 1H, OCHO-THP), 3.95 (m, 1H, CH–O), 3.90 (m, 1H, THP-OCH), 3.82 (m, 1H, CH–O), 3.65 (s, 3H, 24-OMe), 3.45 (m, 2H, OCH_2-THP), 0.6–2.4 (m, 44H, steroidal CH and CH_2) (Kobuke and Nagatani, 2001).

To a solution of **95** (2.0 g, 4 mmol) in dry dioxane (50 mL), 60% NaH (750 mg, 19 mmol) was added and the mixture was refluxed for 60 min (*Scheme 2.1.43*). After the addition of MeI (2.3 g, 16 mmol) at room temperature, the mixture was refluxed again. Once the reaction had been completed, as indicated by TLC, NaH (3.0 g, 75 mmol) and MeI (23 g, 160 mmol) were added in portions over 46 h. Finally, the reaction mixture was neutralized with 1N HCl, concentrated, and extracted with Et_2O. Solvent was removed, and the resulting solid was dissolved in MeOH (100 mL). *p*-TsOH (20 mg) was then added to the solution to cleave off the THP protecting group. The solution was stirred at room temperature and neutralized with saturated aqueous $NaHCO_3$. After filtration, the reaction mixture was concentrated under vacuum and the residue was purified as above to afford methyl 7,12-dimethylcholate (**96**, 400 mg, 16%). ^{1}H NMR (270 MHz, $CDCl_3$): δ 3.66 (s, 3H, 24-OMe),

3.41 (m, 1H, CH–O), 3.36 (m, 1H, CH–O), 3.26 (m, 3H, OMe), 3.20 (m, 3H, OMe), 3.15 (m, 1H, CH–O), 0.6-2.4 (m, 34H, steroidal CH and CH_2) (Kobuke and Nagatani, 2001).

Methyl 7,12-dimethylcholate (**96**) was utilized to synthesize bis[(methyl 7,12-dimethyl-24-carboxylate)-3-cholanyl]-*N,N*′-xylylene dicarbamate (**97**) as follows (*Scheme 2.1.44*). To a solution of **96** (439 mg, 0.97 mmol) in dry dioxane (10 mL), xylylene diisocyanate (91 mg, 0.49 mmol) and DABCO (12 mg, 0.097 mmol) were added at room temperature. The mixture was heated at 70 °C for 7 h, and the solvent was evaporated under vacuum. 1N HCl was added, DABCO was removed and the supernatant was concentrated *in vacuo*. The crude dimer was purified by CC, eluted with 20% EtOAc in C_6H_6, to obtain the dimer **97** (240 mg, 45%). ^{1}H NMR (270 MHz, $CDCl_3$): δ 4.90 (m, 1H, NH), 4.47 (m, 1H, CHOCO), 4.32 (m, 2H, Ph-CH_2), 3.65 (s, 3H, 24-OMe), 3.37 (m, 1H, CHOC), 3.25 (m, 3H, OMe), 3.20 (m, 3H, OMe), 3.15 (m, 1H, CHOC), 0.6–2.4 (m, 34H, alkyl CH). MS (MALDI–TOF): *m/z* 1088.73 calculated for $C_{64}H_{100}N_2O_{12}$, found: 1111.21 $[M + Na]^+$, 1126.08 $[M + K]^+$ (Kobuke and Nagatani, 2001).

The synthesis of another dimer, bis(7,12-dimethyl-24-carboxy-3-cholanyl)-*N,N*′-xylylene dicarbamate (**98**) was achieved from the dimer **97** (*Scheme 2.1.44*). To a hot methanolic solution (60–70 °C, 12 mL) of dimer **97** (20 mg, 0.018 mmol), 1N NaOH (3 mL) was added and the mixture was stirred for 2 h. The solution was acidified with 1N HCl (pH 4), concentrated and extracted with $CHCl_3$. Finally, $CHCl_3$ was rotary evaporated to produce the dimer **98** (17 mg, 87%). ^{1}H NMR (270 MHz, $CDCl_3$): δ 4.98 (m, 1H, NH), 4.47 (m, 1H, CHOCO), 4.38 (m, 2H, Ph-CH_2), 3.39 (m, 1H, CHOC), 3.25 (m, 3H, OMe), 3.20 (m, 3H,

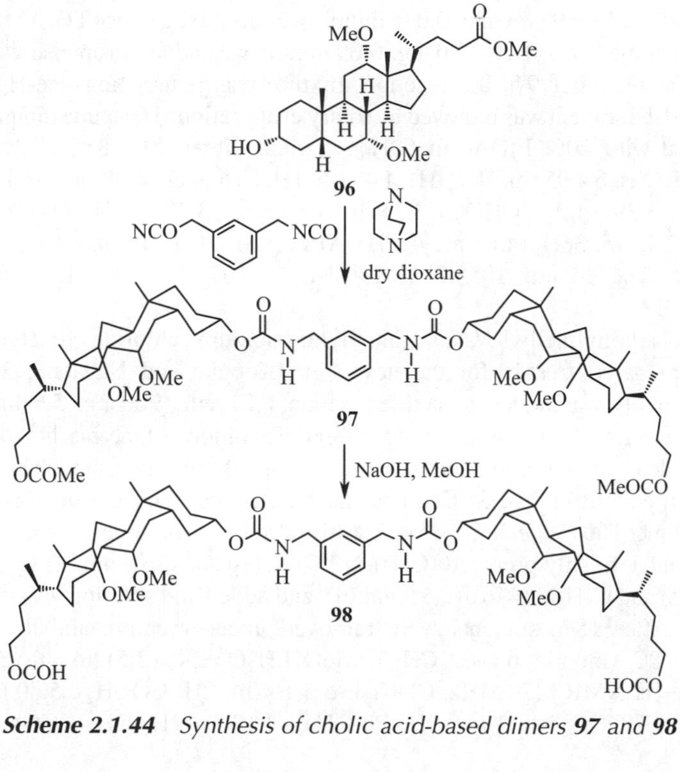

Scheme 2.1.44 *Synthesis of cholic acid-based dimers **97** and **98***

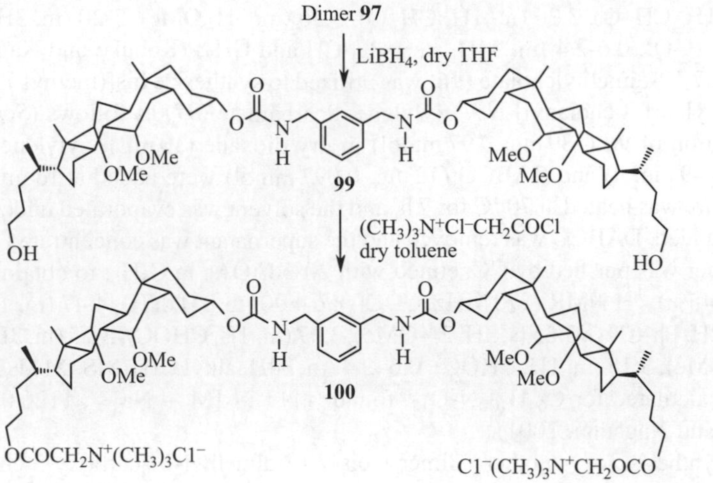

Scheme 2.1.45 *Synthesis of cholic acid-based dimers* ***99*** *and* ***100***

OMe), 3.18 (m, 1H, CHOC). MS (MALDI–TOF): *m/z* 1060.70 calculated for $C_{62}H_{96}N_2O_{12}$, found: 1083.94 $[M + Na]^+$, 1099.98 $[M + K]^+$ (Kobuke and Nagatani, 2001).

In order to synthesize dimer, bis[7,12-dimethyl-24-(*N,N,N*-trimethyl ethanaminium-2-carboxylate)-3-cholanyl]-*N,N′*-xylylenedicarbamate dichloride (**100**), dimer **97** was converted to bis(7,12-dimethyl-24-hydroxy-3-cholanyl)-*N,N′*-xylylene dicarbamate (**99**) (*Scheme 2.1.45*). Dimer **97** (50 mg, 0.046 mmol) was dissolved in dry THF (5 mL) and a dry THF solution (1 mL) of $LiBH_4$ (2.0 mg, 0.09 mmol) was added dropwise with stirring at room temperature. After 9 h, the reaction mixture was poured into ice-H_2O, extracted with Et_2O, and the solvent was removed by rotary evaporation. The crude dimer was purified by CC, eluted with 20% EtOAc in C_6H_6, to yield dimer **99** (18 mg, 38%). ^{1}H NMR (270 MHz, $CDCl_3$): δ 4.95 (m, 1H, NH), 4.47 (m, 1H, CHOCO), 4.39 (m, 2H, Ph-CH_2), 3.65 (t, 2H, CH_2O), 3.39 (m, 1H, CHOC), 3.25 (m, 3H, OMe), 3.20 (m, 3H, OMe), 3.18 (m, 1H, CHOC), 0.6-2.4, (m, 36H, other alkyl CH). MS (MALDI–TOF): *m/z* 1032.73 calculated for $C_{62}H_{100}N_2O_{10}$, found: 1055.73 $[M + Na]^+$, 1071.71 $[M + K]^+$ (Kobuke and Nagatani, 2001).

N-2-Chlorocarbonylmethyl-*N,N,N*-trimethylammonium chloride $[(CH_3)_3N^+Cl^-CH_2COCl]$ was prepared freshly for the next step (Kobuke and Nagatani, 2001). $SOCl_2$ (1.2 mL, 16 mmol) was added to a dried betain HCl salt (900 mg, 5.9 mmol) and the mixture was heated to 75 °C while stirring. After the evolution of the gas, hot toluene (80 °C, 5 mL) was added to the mixture and upper toluene phase was decanted. This method was repeated six times to remove $SOCl_2$, and the final toluene suspension was used for the synthesis of dimer **100** (*Scheme 2.1.45*). To a stirred solution of dimer **99** (40 mg, 3.9 mmol) in toluene (5 mL), freshly prepared $(CH_3)_3N^+Cl^-CH_2COCl$ (5.9 mmol) was dissolved in toluene (5 mL) and C_5H_5N (476 μL, 5.9 mmol), and added and solution was stirred at room temperature. After 18 h, solvents were removed under vacuum, and the residue was subjected to CC (mobile phase: $CHCl_3$:MeOH:H_2O = 30:12:5) to afford dimer **100** (3.0 mg, 2%). ^{1}H NMR (270 MHz, $CDCl_3$): δ 5.10 (m, 2H, $COCH_2$), 5.00 (m, 1H, NH), 4.49 (m, 1H, CHOCO), 4.32 (m, 2H, Ph-CH_2), 4.18 (s, 2H, CH_2-OCO), 3.65 (s, 9H,

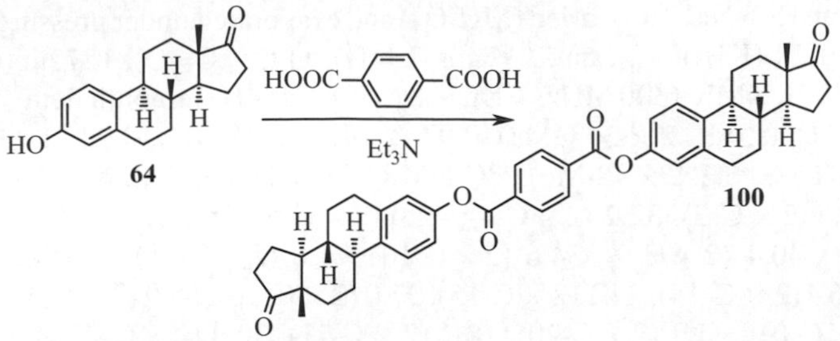

Scheme 2.1.46 Synthesis of estrone terephthaloate dimer **101**

$N^+(CH_3)_3$), 3.39 (m, 1H, CHOC), 3.27 (m, 3H, OMe), 3.19 (m, 3H, OMe), 3.15 (m, 1H, CHOC), 0.6–2.4 (m, 36H, alkyl CH). ESIMS (MALDI–TOF): *m/z* 616.445 calculated for $C_{72}H_{120}N_4O_{12}$. Found: 616.300 (Kobuke and Nagatani, 2001).

Two molecules of estrone (**64**) could be joined together through the A-rings *via* the spacer terephthalic acid to give estrone terephthaloate dimer **101** (*Scheme 2.1.46*) (Hoffmann and Kumpf, 1986).

The same strategy could also be applied for the synthesis of bile acid cholate dimers **102–104** with aromatic spacer group of ester type from bile acid cholates **50**, **51** and **81** which are potential receptors for neutral molecules and metal cations (Joachimiak and Paryzek, 2004). The precursors, methyl lithocholate (**50**), methyl cholate (**51**) and 7α,12α-diacetoxy methyl cholate (**81**), for the synthesis of their respective terephthalate dimers **102–104**, were synthesized according to literature procedures as outlined below (*Scheme 2.1.47*).

The synthesis of bis(methyl 5β-cholan-24-oate-3α-yl)terephthalate (**102**) was carried out from a solution of **50** (200 mg, 0.52 mmol) in dry C_6H_6 (5 mL) and dry C_5H_5N (2 mL), after the treatment with terephthaloyl chloride (68 mg, 0.33 mmol, 1.2 equiv.) and the solution was stirred at room temperature for 24 h (*Scheme 2.1.47*). After completion of reaction (monitored by TLC), C_6H_6-Et_2O (a few mL) mixture was added and resulting precipitate was filtered off and dissolved in $CHCl_3$. The $CHCl_3$ solution was sequentially washed with

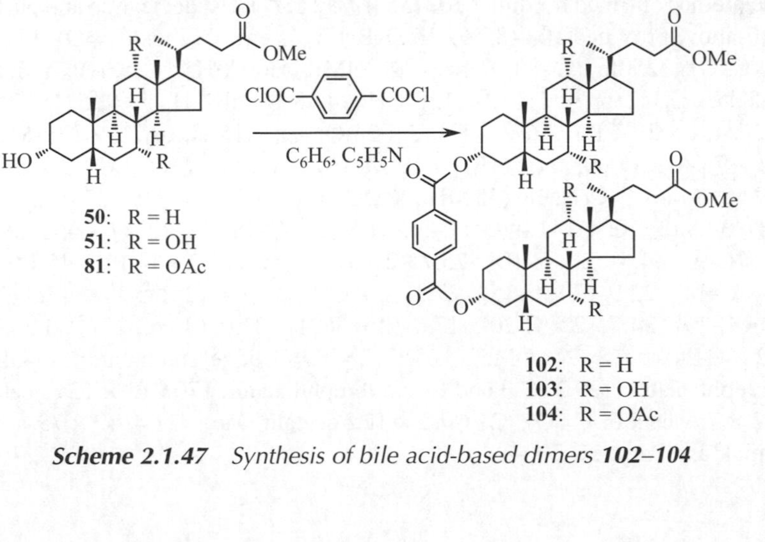

Scheme 2.1.47 Synthesis of bile acid-based dimers **102–104**

H_2O, 1N HCl and 5% $NaHCO_3$, dried ($MgSO_4$) and evaporated under pressure to obtain **102** (203 mg, 87%). IR (KBr): ν_{max} cm^{-1} 1740s (C=O), 1717s (C=O), 1272m (C–O), 1252m (C–O), 730m. ^1H NMR (300 MHz, $CDCl_3$): δ 8.08 (s, 4H, terephthalate), 4.98 (m, 2H, 2 × 3β-CH), 3.66 (s, 6H, 2 × 24-OMe), 0.97 (s, 6H, 2 × 19-Me), 0.92 (d, 6H, J = 6.3 Hz, 2 × 21-Me), 0.66 (s, 6H, 2 × 18-Me). ^{13}C NMR (75 MHz, $CDCl_3$): δ 35.0 (2 × C-1), 28.1 (2 × C-2), 75.5 (2 × C-3), 32.2 (2 × C-4), 42.0 (2 × C-5), 26.7 (2 × C-6), 26.3 (2 × C-7), 35.7 (2 × C-8), 40.4 (2 × C-9), 34.6 (2 × C-10), 21.8 (2 × C-11), 40.0 (2 × C-12), 42.7 (2 × C-13), 56.4 (2 × C-14), 24.1 (2 × C-15), 27.0 (2 × C-16), 55.9 (2 × C-17), 12.0 (2 × C-18), 23.3 (2 × C-19), 35.3 (2 × C-20), 18.2 (2 × C-21), 30.9 (2 × C-22), 31.0 (2 × C-23), 174.7 (2 × C-24), 51.4 (2 × 24-OMe), 165.3 (2 × C=O of terephthalate), 134.4 (C-2 and C-6 of terephthalate), 129.3 (C-3 and C-5 of terephthalate) (Joachimiak and Paryzek, 2004).

A solution of methyl cholate (**51**, 500 mg, 1.18 mmol) in dry C_6H_6 (10 mL) and dry C_5H_5N (500 μL) was treated with terephthaloyl chloride (180 mg, 0.72 mmol) to synthesize another terephthalate dimer, bis(methyl 7α,12α-dihydroxy-5β-cholan-24-oate-3α-yl)terephthalate (**103**) (*Scheme 2.1.47*). The mixture was stirred for 12 h at room temperature, diluted with $CHCl_3$ (10 mL), filtered off and successively washed with 1N HC1, 10% Na_2CO_3 and brine. The solvents were evaporated to dryness and the residue was purified by CC (mobile phase: $CHCl_3$:Me_2CO = 20:1) to provide the dimer **103** (200 mg, 35%). IR (KBr): ν_{max} cm^{-1} 3614br (O–H), 3530br (O–H), 1740s (C=O), 1711s (C=O), 1274s (C–O), 1251s (C–O). ^1H NMR (300 MHz, $CDCl_3$): δ 8.07 (s, 4H, terephthalate), 4.85 (m, 2H, 2 × 3β-CH), 4.02 (br s, 2H, 2 × 12β-CH), 3.88 (br s, 2H, 2 × 7β-CH), 3.67 (s, 6H, 2 × 24-OMe), 0.99 (d, 6H, J = 6 Hz, 2 × 21-Me), 0.95 (s, 6H, 2 × 19-Me), 0.72 (s, 6H, 2 × 18-Me). ^{13}C NMR (75 MHz, $CDCl_3$): δ 34.8 (2 × C-1), 26.7 (2 × C-2), 75.4 (2 × C-3), 35.1 (2 × C-4), 42.1 (2 × C-5), 34.4 (2 × C-6), 68.3 (2 × C-7), 39.5 (2 × C-8), 26.8 (2 × C-9), 34.7 (2 × C-10), 28.4 (2 × C-11), 72.9 (2 × C-12), 46.5 (2 × C-13), 41.2 (2 × C-14), 23.1 (2 × C-15), 27.4 (2 × C-16), 47.2 (2 × C-17), 12.6 (2 × C-18), 22.5 (2 × C-19), 35.1 (2 × C-20), 17.3 (2 × C-21), 31.0 (2 × C-22), 30.8 (2 × C-23), 174.7 (2 × C-24), 51.5 (2 × 24-OMe), 165.5 (2 × C=O of terephthalate), 134.4 (C-2 and C-6 of terephthalate), 129.4 (C-3 and C-5 of terephthalate) (Joachimiak and Paryzek, 2004).

Similarly, bis(methyl 7α,12α-diacetoxy-5β-cholan-24-oate-3α-yl)terephthalate (**104**) was prepared from methyl 7α,12α-diacetoxy-3α-hydroxy-5β-cholan-24-oate (**81**) according to the method outlined for dimer **102** (*Scheme 2.1.47*). Usual cleanup and purification as described above provided **104** (85%). IR (KBr): ν_{max} cm^{-1} 1736s (C=O), 1713s (C=O), 1271m (C–O), 1250m (C–O), 734m. ^1H NMR (300 MHz, $CDCl_3$): δ 8.09 (s, 4H, terephthalate), 5.11 (br s, 2H, 2 × 12β-CH), 4.93 (br d, 2H, J = 2.5 Hz, 2 × 7β-CH), 4.82 (m, 2H, 2 × 3β-CH), 3.67 (s, 6H, 2 × 24-OMe), 2.15 (s, 6H, 2 × 12α-MeCO), 2.07 (s, 6H, 2 × 7α-MeCO), 0.96 (s, 6H, 2 × 19-Me), 0.82 (d, 6H, J = 6.3 Hz, 2 × 21-Me), 0.75 (s, 6H, 2 × 18-Me). ^{13}C NMR (75 MHz, $CDCl_3$): δ 34.7 (2 × C-1), 27.0 (2 × C-2), 75.4 (2 × C-3), 34.8 (2 × C-4), 41.0 (2 × C-5), 31.3 (2 × C-6), 70.7 (2 × C-7), 37.8 (2 × C-8), 29.0 (2 × C-9), 34.4 (2 × C-10), 25.7 (2 × C-11), 75.4 (2 × C-12), 45.1 (2 × C-13), 43.5 (2 × C-14), 22.9 (2 × C-15), 27.3 (2 × C-16), 47.4 (2 × C-17), 12.3 (2 × C-18), 22.7 (2 × C-19), 34.7 (2 × C-20), 17.6 (2 × C-21), 31.0 (2 × C-22), 30.9 (2 × C-23), 174.3 (2 × C-24), 51.5 (2 × OMe), 165.1 (2 × C=O of terephthalate), 134.4 (C-2 and C-6 of terephthalate), 129.3 (C-3 and C-5 of terephthalate), 170.2 (2 × 12α-acetate C=O), 170.0 (2 × 7α-acetate C=O), 21.6 (2 × 12α-acetate Me), 21.4 (2 × 7α-acetate Me) (Joachimiak and Paryzek, 2004).

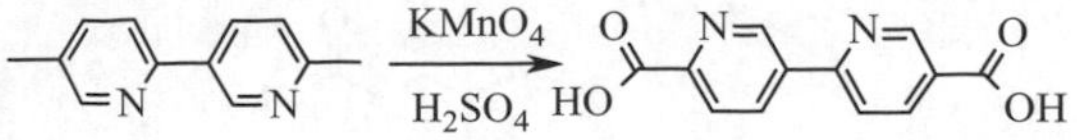

Scheme 2.1.48 *Synthesis of 2,2′-bipyridine-4,4′-dicarboxylic acid*

Several 2,2′-bipyridine-4,4′-dicarboxylates **107–110** of bile acid methyl cholates were synthesized, respectively, from methyl lithocholate (**50**), methyl cholate (**51**), methyl chenodeoxycholate (**105**, methyl 3α,7α-dihydroxy-5β-cholanate or methyl 3α,7α-dihydroxy-5β-cholan-24-oate) and methyl deoxycholate (**106**, 3α,12α-dihydroxy-5β-cholanate or methyl 3α,12α-dihydroxy-5β-cholan-24-oate) (Tamminen *et al.*, 2000a).

2,2′-Bipyridine-4,4′-dicarboxylic acid was freshly prepared for the synthesis of bile acid-based dimers **107–110** (Tamminen *et al.*, 2000a). A mixture of 4,4′-dimethyl-2,2′-bipyridine (2.5 g, 13.57 mmol) in 25% H_2SO_4 (132 mL, distilled H_2O) and $KMnO_4$ (5.0 g, 31.64 mmol) was stirred at 5 °C for 30 min (*Scheme 2.1.48*). The mixture was allowed to rise to 35 °C and stirred for another 20 min, then cooled again to 5 °C and more $KMnO_4$ (5.0 g, 31.64 mmol) was added. After a further 10 min, the mixture was refluxed for 12 h at 130 °C, the excess $KMnO_4$ was destroyed by adding $K_2S_2O_5$ (5.0 mg) and the precipitate was filtered and dried under vacuum to yield 2,2′-bipyridine-4,4′-dicarboxylic acid (1.67 g, 50%) as solid (Tamminen *et al.*, 2000a).

The synthesis of dimethyl-3α,3α′-bis(2,2′-bipyridine-4,4′-dicarboxy)-5β,5β′-dicholan-24,24′-dioate (**107**) was accomplished from methyl lithocholate (**50**) (*Scheme 2.1.49*). To a stirred mixture of **50** (2.17 g, 5.39 mmol) and 2,2′-bipyridine-4,4′-dicarboxylic acid (660 mg, 2.70 mmol) in Na-dried toluene (150 mL), DMAP (2.5 g, 20.46 mmol) was added and the mixture was heated to 100 °C. DCBC (1.2 g, 5.73 mmol) was added and the mixture was kept at 100 °C. After 90 h, the solvent was removed by rotary evaporation, the residue was dissolved in DCM (80 mL), extracted with saturated aqueous $NaHCO_3$ (2 × 60 mL), washed with H_2O (1 × 60 mL), dried ($MgSO_4$), and evaporated to dryness under pressure. The crude dimer was purified by a step gradient CC (mobile phase: 5% Me_2CO in $CHCl_3$, 100% DCM then 100% Me_2CO) to produce **107** (810 mg, 30.3%) (Tamminen *et al.*, 2000a).

^{13}C NMR (67.7 MHz, $CDCl_3$): δ 35.1 (2 × C-1), 26.7 (2 × C-2), 76.3 (2 × C-3), 32.2 (2 × C-4), 42.0 (2 × C-5), 27.0 (2 × C-6), 26.3 (2 × C-7), 35.8 (2 × C-8), 40.5 (2 × C-9), 34.7 (2 × C-10), 20.9 (2 × C-11), 40.1 (2 × C-12), 42.8 (2 × C-13), 56.5 (2 × C-14), 24.2 (2 × C-15), 28.2 (2 × C-16), 56.0 (2 × C-17), 12.0 (2 × C-18), 23.3 (2 × C-19), 35.4 (2 × C-20), 18.3 (2 × C-21), 31.1 (2 × C-22), 31.0 (2 × C-23), 174.7 (2 × C-24), 51.4 (2 × 24-OMe), 164.7 (2 × aroyl CO), aryl carbons: 134.0 (2 × C-1), 156.6 (2 × C-2), 120.6 (2 × C-3), 139.5 (2 × C-4), 123.3 (2 × C-5), 150.0 (2 × C-6). MS (MALDI–TOF): *m/z* 989.17 $[M + H]^+$ (Tamminen *et al.*, 2000a).

Dimethyl-3α,3α-bis(2,2′-bipyridine-4,4′-dicarboxy)-7α,7α′,12α,12α′-tetrahydroxy-5β,5β′-dicholan-24,24′-dioate (**108**) was obtained from methyl cholate (**51**) (*Scheme 2.1.49*), according to the procedure as described for dimer **107**. Usual workup and purification as described above furnished dimer **108** (10 mg, 0.4%). ^{13}C NMR (67.7 MHz, $CDCl_3$): δ 35.2 (2 × C-1), 26.7 (2 × C-2), 76.8 (2 × C-3), 34.9 (2 × C-4), 42.2 (2 × C-5), 34.4 (2 × C-6), 68.2 (2 × C-7), 39.7 (2 × C-8), 26.9 (2 × C-9), 34.8 (2 × C-10), 28.5 (2 × C-11), 72.9 (2 × C-12), 46.6 (2 × C-13), 41.4 (2 × C-14), 23.2 (2 × C-15), 27.4 (2 × C-16), 47.3 (2 × C-17), 12.6 (2 × C-18), 22.9 (2 × C-19), 35.1 (2 × C-20), 17.4 (2 × C-21), 31.1 (2 × C-22), 31.0 (2 × C-23), 174.7 (2 × C-24), 51.5 (2 × 24-OMe), 164.8

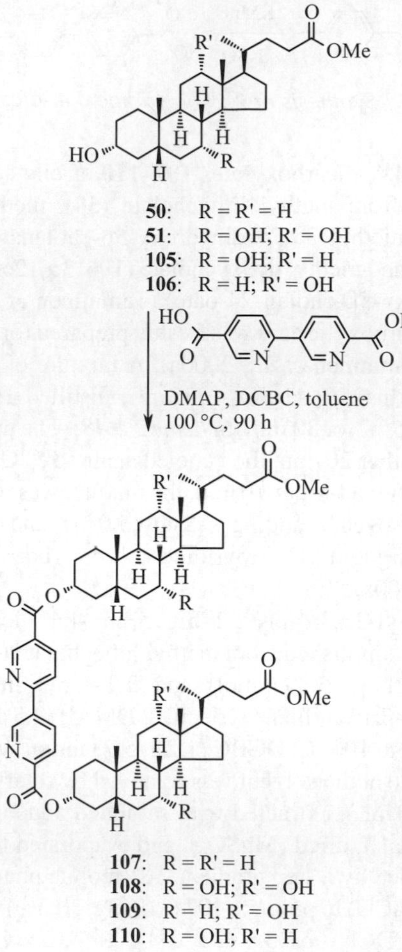

50: R = R' = H
51: R = OH; R' = OH
105: R = OH; R' = H
106: R = H; R' = OH

DMAP, DCBC, toluene
100 °C, 90 h

107: R = R' = H
108: R = OH; R' = OH
109: R = H; R' = OH
110: R = OH; R' = H

Scheme 2.1.49 *Synthesis of bile acid-based dimers* ***107–110***

(2 × aroyl CO), aryl carbons: 133.9 (2 × C-1), 156.7 (2 × C-2), 120.6 (2 × C-3), 139.4 (2 × C-4), 123.3 (2 × C-5), 150.0 (2 × C-6). MS (MALDI–TOF): *m/z* 1054.15 $[M + H]^+$ (Tamminen *et al.*, 2000a).

Dimethyl-3α,3α-bis(2,2′-bipyridine-4,4′-dicarboxy)-7α,7α′-dihydroxy-5β,5β′-dicholan-24,24′-dioate (**109**) was produced from methyl chenodeoxycholate (**105**), according to the method described for dimer **107** (*Scheme 2.1.49*). After the usual cleanup as above, the crude dimer was purified by CC (mobile phase: 4% Me_2CO in DCM, then Me_2CO in $CHCl_3$) to provide dimer **109** (200 mg, 8%). ^{13}C NMR (67.7 MHz, $CDCl_3$): δ 35.1 (2 × C-1), 26.6 (2 × C-2), 76.0 (2 × C-3), 34.3 (2 × C-4), 41.2 (2 × C-5), 34.9 (2 × C-6), 68.2 (2 × C-7), 39.3 (2 × C-8), 32.8 (2 × C-9), 35.0 (2 × C-10), 20.5 (2 × C-11), 39.4 (2 × C-12), 42.6 (2 × C-13), 50.3 (2 × C-14), 23.6 (2 × C-15), 28.0 (2 × C-16), 55.7 (2 × C-17), 11.6 (2 × C-18), 22.6 (2 × C-19), 35.2 (2 × C-20), 18.2 (2 × C-21), 30.9 (2 × C-22), 30.8 (2 × C-23), 174.5 (2 × C-24), 51.3 (2 × 24-OMe), 164.6 (2 × aroyl CO), aryl carbons: 133.9 (2 × C-1),

156.4 (2 × C-2), 120.4 (2 × C-3), 139.3 (2 × C-4), 123.1 (2 × C-5), 149.8 (2 × C-6). MS (MALDI–TOF): *m/z* 1022.22 $[M + H]^+$ (Tamminen *et al.*, 2000a).

Methyl deoxycholate (**106**) was converted to dimethyl-3α,3α-bis(2,2′-bipyridine-4,4′-dicarboxy)-12α,12α′-dihydroxy-5β,5β′-dicholan-24,24′-dioate (**110**) as follows (*Scheme 2.1.49*). Usual workup and purification as described above afforded dimer **110** (70 mg, 2.9%). ^{13}C NMR (67.7 MHz, $CDCl_3$): δ 35.0 (2 × C-1), 27.5 (2 × C-2), 76.1 (2 × C-3), 32.3 (2 × C-4), 42.0 (2 × C-5), 26.6 (2 × C-6), 26.1 (2 × C-7), 36.1 (2 × C-8), 33.8 (2 × C-9), 34.2 (2 × C-10), 28.6 (2 × C-11), 73.2 (2 × C-12), 46.6 (2 × C-13), 48.4 (2 × C-14), 23.6 (2 × C-15), 27.0 (2 × C-16), 47.5 (2 × C-17), 12.8 (2 × C-18), 23.2 (2 × C-19), 35.1 (2 × C-20), 17.4 (2 × C-21), 31.1 (2 × C-22), 30.9 (2 × C-23), 174.7 (2 × C-24), 51.5 (2 × 24-OMe), 164.8 (2 × aroyl CO), aryl carbons: 134.1 (2 × C-1), 156.7 (2 × C-2), 120.6 (2 × C-3), 139.4 (2 × C-4), 123.3 (2 × C-5), 150.0 (2 × C-6). MS (MALDI–TOF): *m/z* 1020.91 $[M + H]^+$ (Tamminen *et al.*, 2000a).

Five lithocholic acid-based dimers **124–126** were synthesized *via* ring A–ring A connection from lithocholic acid (**48**) with excellent yields (Valkonen *et al.*, 2007). The precursor 5β-cholane-3α,24-diol (**111**) was prepared from lithocholic acid (**48**) by $LiAlH_4$ reduction (*Scheme 2.1.50*). To a cooled (0 °C) solution of **48** (300 mg, 0.796 mmol) in dry THF (8 mL) under Ar, $LiAlH_4$ (90 mg, 2.39 mmol) in THF (2 mL) was added at room temperature and stirred for 30 min, then refluxed for 2 h, cooled again (0 °C) and the excess hydride was destroyed by adding 10% HCl (10 mL). The mixture was extracted with Et_2O (3 × 30 mL), the pooled extracts were dried ($MgSO_4$), concentrated *in vacuo* and the residue was recrystallized from Me_2CO to afford **111** (289 mg, 100%, mp: 172–173 °C). ^{1}H NMR (500 MHz, $CDCl_3$): δ 3.65-3.55 (m, 3H, 3β-CH and 24-CH_2), 0.92 (s, 3H, 19-Me and d, 3H, $J = 5.6$ Hz, 21-Me), 0.64 (s, 3H, 18-Me). ^{13}C NMR (125 MHz, $CDCl_3$): δ 35.4 (C-1), 30.6 (C-2), 71.9 (C-3), 36.5 (C-4), 42.1 (C-5), 27.2 (C-6), 26.4 (C-7), 35.9 (C-8), 40.5 (C-9), 34.6 (C-10), 20.8 (C-11), 40.2 (C-12), 42.7 (C-13), 56.5 (C-14), 24.2 (C-15), 28.3 (C-16), 56.2 (C-17), 12.0 (C-18), 23.4(C-19), 35.6 (C-20), 18.6 (C-21), 31.9 (C-22), 29.5 (C-23), 63.6 (C-24) (Valkonen *et al.*, 2007).

24-Triphenylmethoxy-5β-cholan-3β-ol (**112**) was obtained from diol **111** (*Scheme 2.1.50*). A mixture of **111** (90 mg, 0.25 mmol), TrCl (104 mg, 0.37 mmol), DMAP

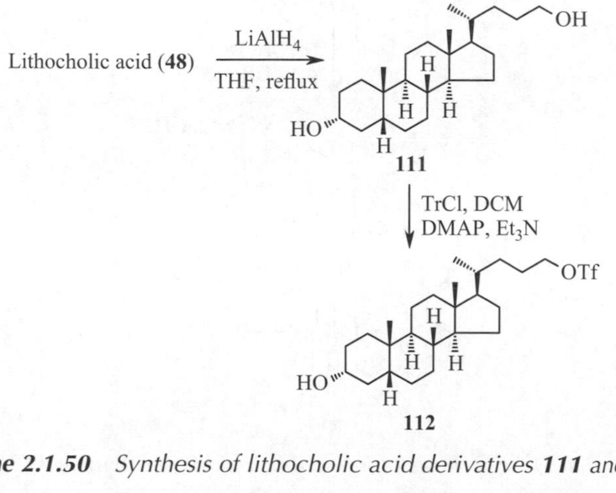

Lithocholic acid (**48**) → ($LiAlH_4$; THF, reflux) → **111** → (TrCl, DCM; DMAP, Et_3N) → **112**

Scheme 2.1.50 *Synthesis of lithocholic acid derivatives **111** and **112***

(5.0 mg, 0.04 mmol), and Et_3N (84 μL, 0.6 mmol) in DCM (1.5 mL) was heated on an oil bath (60 °C). After 15 h, the solvent was removed under vacuum and the residue was purified by CC, eluted with DCM, to give **112** (119 mg, 83%, mp: 93–94 °C). ^{1}H NMR (500 MHz, $CDCl_3$): δ 7.44–7.19 (m, 15H, OTr), 3.61 (m, 1H, 3β-CH), 3.00 (m, 2H, 24-CH_2O), 0.90 (s, 3H, 19-Me), 0.88 (d, $J = 6.1$ Hz, 3H, 21-Me), 0.60 (s, 3H, 18-Me) (Valkonen *et al.*, 2007).

24-Triphenylmethoxy **112** was employed for the synthesis of the dimer, bis(24-triphenylmethoxy)-3α,3′α-(isophthaloyloxy)-bis(5β-cholane) (**113**) (*Scheme 2.1.51*). A mixture of **112** (120 mg, 0.2 mmol), DMAP (7.0 mg, 0.06 mmol), Et_3N (44 μL, 0.3 mmol), and isophthaloyl dichloride (20 mg, 0.1 mmol) in dry toluene (2 mL) was stirred at 100 °C for 15 h. After cooling to room temperature the mixture was diluted with DCM, washed twice with saturated aqueous $NaHCO_3$, the pooled organic solvent was dried ($MgSO_4$) and the crude residue was purified by CC (mobile phase: 10% EtOAc in petroleum ether) to provide **113** (112 mg, 85%, mp: 150–152 °C). ^{1}H NMR (400 MHz, $CDCl_3$): δ 8.69 (m, 1H, 29-CH), 8.19 (m, 2H, 2 × 30-CH), 7.51–7.18 (m, 16H, 31-CH, OTr), 4.98 (m, 2H, 2 × 3β-CH), 3.00 (m, 4H, 2 × 24-CH_2), 0.95 (s, 6H, 2 × 19-Me), 0.89 (d, 6H, 2 × 21-Me), 0.62 (s, 6H, 18-Me) (Valkonen *et al.*, 2007).

3α,3′α-(Isophthaloyloxy)-bis(5β-cholan-24-ol) (**114**) was achieved from **113** as follows (*Scheme 2.1.51*). Dimer **113** (106 mg, 0.08 mmol) was dissolved in MeOH (6 mL) and DCM

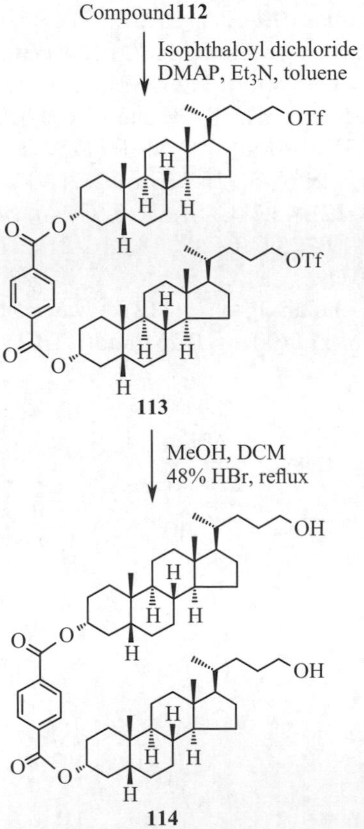

Compound **112**
Isophthaloyl dichloride
DMAP, Et_3N, toluene
OTf
OTf
113
MeOH, DCM
48% HBr, reflux
OH
OH
114

Scheme 2.1.51 *Synthesis of lithocholic acid-based dimers* ***113*** *and* ***114***

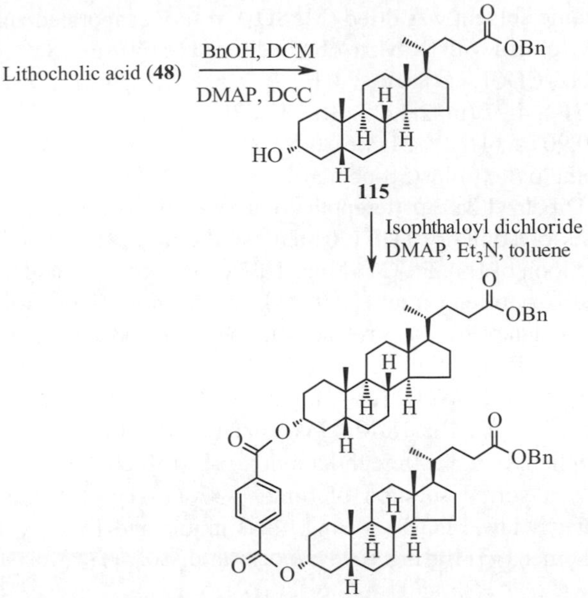

Scheme 2.1.52 *Synthesis of lithocholic acid-based dimer* ***116***

(2 mL), 48% HBr (200 μL) was added and refluxed for 3 h. The reaction mixture was extracted with DCM (2 × 20 mL), the pooled DCM extracts were washed with H_2O, dried ($MgSO_4$), concentrated and the residue was purified by CC, eluted with DCM-MeOH = 20:1, to obtain **114** (50 mg, 74%, mp: 188–190 °C). ^{1}H NMR (400 MHz, $CDCl_3$): δ 8.68 (m, 1H, 29-CH), 8.19 (m, 2H, 2 × 30-CH), 7.49 (m, 1H, 31-CH), 4.98 (m, 2H, 2 × 3β-H), 3.60 (m, 4H, 2 × 24-CH_2), 0.95 (s, 6H, 2 × 19-Me), 0.92 (d, 6H, 2 × 21-Me), 0.65 (s, 6H, 2 × 18-Me) (Valkonen *et al.*, 2007).

The synthesis of benzyl 3α-hydroxy-5β-cholan-24-oate (**115**) was accomplished from a stirred solution of lithocholic acid (**48**, 500 mg, 1.33 mmol) in dry DCM (3 mL), treated with DMAP (32 mg, 0.27 mmol), BnOH (216 mg, 2 mmol) and DCC (357 mg, 1.73 mmol) (*Scheme 2.1.52*). After stirring overnight in a closed flask, the reaction mixture was diluted with DCM and filtered. The filtrate was sequentially washed with 5% HCl, saturated aqueous $NaHCO_3$ and H_2O. The organic solvent was dried ($MgSO_4$), rotary evaporated to dryness, the residue was purified by CC (mobile phase: 20% EtOAc in petroleum ether), and recrystallized from *n*-hexane-EtOAc to yield **115** (432 mg, 74%). ^{1}H NMR (400 MHz, $CDCl_3$): δ 7.33 (m, 5H, Ph-CH), 5.09 (m, 2H, Ph-CH_2), 3.58 (m, 1H, 3β-CH), 2.42-2.21 (m, 2H, 23-CH_2), 0.88 (s and d, 6H, 19-Me and 21-Me), 0.59 (s, 3H, 18-Me) (Valkonen *et al.*, 2007).

Dibenzyl-3α,3′α-(terephthaloyloxy)-bis(5β-cholan-24-oate) (**116**) was synthesized from benzyl 3α-hydroxy-5β-cholan-24-oate (**115**) (*Scheme 2.1.52*). To a stirred solution of **115** (430 mg, 0.98 mmol), terephthaloyl dichloride (90 mg, 0.443 mmol) in dry toluene (3 mL), DMAP (36 mg, 0.29 mmol), Et_3N (500 μL) were added at room temperature, and stirred under Ar first at room temperature for 60 min and then at 100 °C for 12 h. After cooling to room temperature, the mixture was diluted with DCM, washed twice with saturated aqueous

$NaHCO_3$, the organic solvent was dried ($MgSO_4$), rotary evaporated and the crude dimer was purified by CC, eluted with DCM, to obtain dimer **116** (401 mg, 86%, mp: 132–134 °C). ^1H NMR (400 MHz, $CDCl_3$): δ 8.07 (s, 4H, 2 × 29-CH_2), 7.34 (m, 10H, 2 × Ph-CH), 5.09 (m, 4H, 2 × Ph-CH_2), 4.97 (m, 2H, 2 × 3β-CH), 2.43–2.22 (m, 4H, 2 × 23-CH_2), 0.95 (s, 6H, 2 × 19-Me), 0.90 (d, 6H, 2 × 21-Me), 0.59 (s, 6H, 2 × 18-Me) (Valkonen *et al.*, 2007).

3α,3′α-(Terephthaloyloxy)-bis(5β-cholan-24-oic acid) (**117**) was obtained from **116** (*Scheme 2.1.53*). Dibenzyl-3α,3′α-(terephthaloyloxy)-bis(5β-cholan-24-oate) (**117**, 2.05 g, 1.92 mmol) was dissolved in dry THF (20 mL) and the flask was purged with H_2 for 1-2 s (connected to a balloon of H_2), Pd-C (61 mg, 10% on carbon) was added under Ar and the mixture was stirred at room temperature. After 3 h, the reaction mixture was filtered through a short plug of SiO_2, and the filtrate was rotary evaporated to produce dimer **117** (1.64 g, 96%). ^1H NMR (400 MHz, $CDCl_3$): δ 11.56 (br s, 2H, 2 × 24-CO_2H), 7.98 (s, 4H, 2 × 29-CH_2), 4.86 (m, 2H, 2 × 3β-CH), 2.24-2.02 (m, 4H, 2 × 23-CH_2), 0.88 (s, 6H, 2 × 19-Me), 0.84 (d, 6H, 2 × 21-Me), 0.57 (s, 6H, 2 × 18-Me) (Valkonen *et al.*, 2007).

Bis(pentafluorophenyl)3α,3′α-(terephthaloyloxy)-bis(5β-cholan-24-oate) (**118**) could be produced from a stirred solution of dimer **117** (1.64 g, 1.84 mmol) in dry DCM (50 mL) was treated with PFPOH (846 mg, 4.60 mmol) and DCC (140 mg, 5.53 mmol)

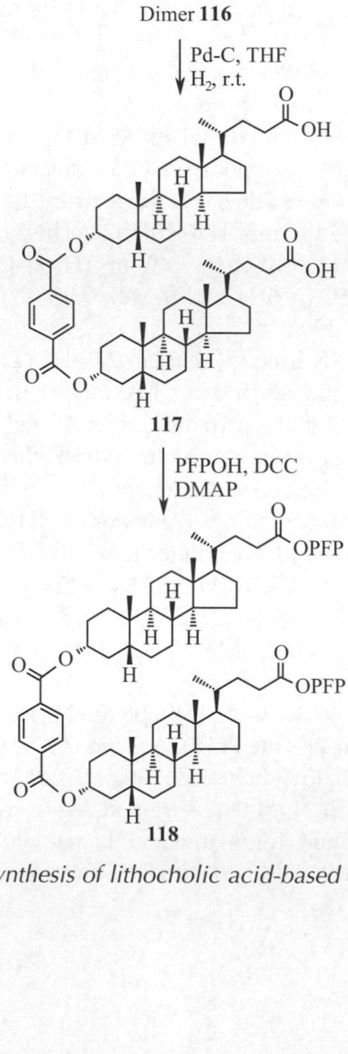

Dimer **116**

Pd-C, THF
H_2, r.t.

117

PFPOH, DCC
DMAP

118

Scheme 2.1.53 *Synthesis of lithocholic acid-based dimers* ***117*** *and* ***118***

(*Scheme 2.1.53*). The reaction mixture was stirred overnight, diluted with DCM and filtered. The filtrate was washed with saturated aqueous $NaHCO_3$ followed by H_2O and the organic solvent was dried (Na_2SO_4), concentrated under vacuum and the crude dimer was purified by CC (mobile phase: petroleum ether-DCM), to give dimer **118** (1.88 g, 84%). ^{1}H NMR (400 MHz, $CDCl_3$): δ 8.07 (s, 4H, 2 × 29-CH_2), 4.98 (m, 2H, 2 × 3β-CH), 2.73-2.53 (m, 4H, 2 × 23-CH_2), 0.97 (s, d, s and d, 12H, 2 × 19-Me and 2 × 21-Me), 0.67 (s, 6H, 2 × 18-Me). ESIMS: *m/z* 1237.7355 $[M + MeOH + Na]^+$ (Valkonen *et al.*, 2007).

Łotowski and Guzmanski (2006) reported the synthesis of cholic acid-based dimers **122–128** from methyl 7α,12α-diacetoxy-3α-hydroxy-5β-cholan-24-oate (**81**) with oxamide spacers *via* ring A–ring A connections. They also reported similar dimers connected *via* ring D side chains with oxamide and hydrazide spacers (*Schemes 2.4.27* and *2.4.28*) and *via* ring A–ring D side chain with oxamide spacers (*Schemes 2.5.10* and *2.5.11*).

Methyl 7α,12α-diacetoxy-3α-iodo-5β-cholan-24-oate (**119**) was prepared from methyl 7α,12α-diacetoxy-3α-hydroxy-5β-cholan-24-oate (**81**) (*Scheme 2.1.54*). A mixture of Ph_3P (8.40 g, 32.06 mmol) and imidazole (2.30 g, 33.82 mmol) was added to a stirred solution of **81** (2.94 g, 5.81 mmol) in a mixture of 20% MeCN in C_6H_6 (100 mL). I_2 (7.4 g, 29.13 mmol) was added after a few minutes and stirred for 15 min. H_2O containing a few drops 30% H_2O_2 was added to the reaction mixture and extracted with C_6H_6. The organic layer was washed with saturated aqueous $Na_2S_2O_5$, dried ($MgSO_4$), rotary evaporated and the crude product was purified by CC, eluted with 20% EtOAc in *n*-hexane, to provide iodide **119** (3.54 g, 99%). ^{1}H NMR (200 MHz, $CDCl_3$): δ 5.08 (m, 1H, 12β-CH), 4.91 (m, 2H, 7β-CH and 3α-CH), 3.66 (s, 3H, 24-OMe), 2.12 (s, 3H, 12α-MeCO), 2.09 (s, 3H, 7α-MeCO), 0.92 (s, 3H, 19-Me), 0.86 (d, 3H, $J = 6.1$ Hz, 21-Me), 0.73 (s, 3H, 18-Me) (Łotowski and Guzmanski, 2006).

Methyl 7α,12α-diacetoxy-3α-iodo-5β-cholan-24-oate (**119**) was then utilized for the synthesis of methyl 3α-azido-7α,12α-diacetoxy-5β-cholan-24-oate (**120**) (*Scheme 2.1.54*). NaN_3 (3.0 g, 46.2 mmol) and AcOH (3.5 mL, 61.25 mmol) were added to a stirred solution of iodide **119** (3.5 g, 5.68 mmol) in *N*-methylpyrrolidone (250 mL) at room temperature and

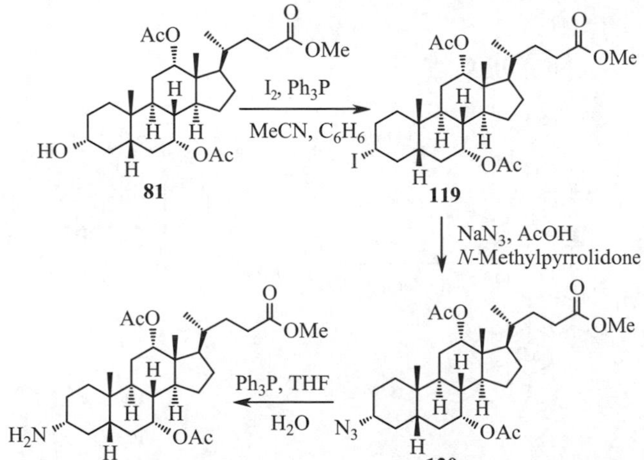

Scheme 2.1.54 *Synthesis of cholic acid derivatives* ***119–121***

stirring was continued overnight. Saturated aqueous $NaHCO_3$ was added to the reaction mixture and extracted several times with C_6H_6. The pooled C_6H_6 layers were dried ($MgSO_4$), concentrated under pressure and purification on CC (mobile phase: 5% EtOAc in C_6H_6) afforded azide **120** (2.74 g, 91%). ^{1}H NMR (200 MHz, $CDCl_3$): δ 5.08 (m, 1H, 12β-CH), 4.91 (m, 1H, 7β-CH), 3.66 (s, 3H, 24-OMe), 3.14 (m, 1H, 3β-CH), 2.12 (s, 3H, 12α-MeCO), 2.09 (s, 3H, 7α-MeCO), 0.92 (s, 3H, 19-Me), 0.86 (d, 3H, J = 6.1 Hz, 21-Me), 0.73 (s, 3H, 18-Me) (Łotowski and Guzmanski, 2006).

Methyl 3α-amino-7α,12α-diacetoxy-5β-cholan-24-oate (**121**) was obtained from a stirred solution methyl 3α-azido-7α,12α-diacetoxy-5β-cholan-24-oate (**120**, 2.3 g, 4.33 mmol) in THF (80 mL), treated with Ph_3P (3.36 g, 12.82 mmol) and H_2O (2 mL) at room temperature under Ar and allowed to stand overnight (*Scheme 2.1.54*). The solution was concentrated *in vacuo* and the residue was purified by CC, eluted with 20% MeOH in $CHCl_3$, to obtain amine **121** (2.05 g, 94%). IR ($CHCl_3$): ν_{max} cm^{-1} 3368m (N–H), 1724s (C=O), 1256m (C–O), 1024m, 666m, 542w. ^{1}H NMR (200 MHz, $CDCl_3$): δ 5.08 (m, 1H, 12β-CH), 4.89 (m, 1H, 7β-CH), 3.66 (s, 3H, 24-OMe), 2.62 (m, 1H, 3β-CH), 2.13 (s, 3H, 12α-MeCO), 2.08 (s, 3H, 7α-MeCO), 0.91 (s, 3H, 19-Me), 0.81 (d, 3H, J = 6.1 Hz, 21-Me), 0.73 (s, 3H, 18-Me) (Łotowski and Guzmanski, 2006).

Methyl 3α-amino-7α,12α-diacetoxy-5β-cholan-24-oate (**121**) was converted to *N,N′*-di (methyl 7α,12α-diacetoxy-5β-cholan-24-oate)3α,3′α-oxamide (**122**) (*Scheme 2.1.55*). To a stirred solution of amine **121** (2.0 g, 3.96 mmol) in dry C_5H_5N (80 mL) and $(COCl)_2$ (170 μL, 1.98 mmol) was added dropwise over 15 min. Acidified H_2O to the reaction mixture, extracted with DCM, the solvent was dried ($MgSO_4$), evaporated under pressure and the crude dimer was purified by CC (mobile phase: 30% EtOAc in C_6H_6) to produce

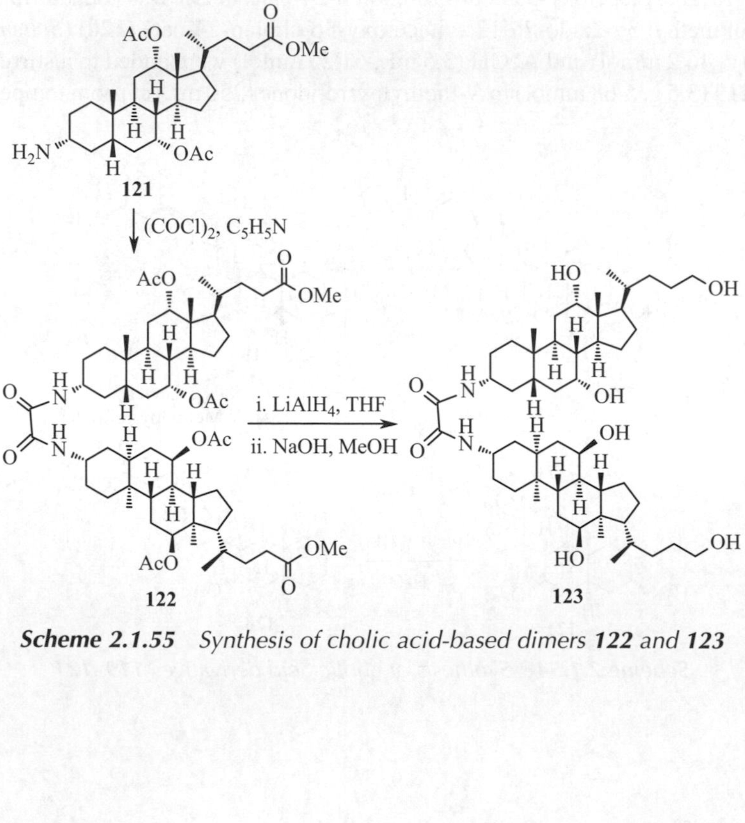

Scheme 2.1.55 *Synthesis of cholic acid-based dimers **122** and **123***

diester **122** (91.87 g, 89%). IR ($CHCl_3$): ν_{max} cm^{-1} 3528m (N–H), 3390m (N–H), 1726s (C=O), 1670s (C=O), 1508m, 1255s (C–O), 1021m, 682w. ^1H NMR (200 MHz, $CDCl_3$): δ 7.37 (d, 2H, $J = 8.5$ Hz, 2 × NH), 5.10 (m, 2H, 2 × 12β-CH), 4.91 (m, 2H, 2 × 7β-CH), 4.15 (m, 2 × 3β-CH), 3.67 (s, 6H, 2 × 24-OMe), 2.19 (s, 6H, 2 × 12α-MeCO), 2.07 (s, 6H, 2 × 7α-MeCO), 0.98 (s, 6H, 2 × 19-Me), 0.82 (d, 6H, $J = 5.9$ Hz, 2 × 21-Me), 0.74 (s, 6H, 2 × 18-Me). ^{13}C NMR (50 MHz, $CDCl_3$): δ 12.2, 17.5, 21.5, 21.7, 22.7, 22.8, 25.5, 27.1, 27.5, 28.9, 30.7, 30.8, 31.3, 34.3, 34.6, 35.3, 35.4, 37.7, 41.4, 43.3, 45.1, 47.3, 50.0, 51.5, 70.7, 75.3, 159.0, 170.3, 170.6, 174.5. FABMS: *m/z* 1087 $[M + Na]^+$ (Łotowski and Guzmanski, 2006).

N,N'-Di(methyl 7α,12α-diacetoxy-5β-cholan-24-oate)3α,3′α-oxamide (**122**) was reduced to *N,N'*-di(5β-cholane-7α,12α,24-triol)3α,3′α-oxamide (**123**) (*Scheme 2.1.55*). $LiAlH_4$ (240 mg, 6.32 mmol) was added to a stirred solution of diester **122** (1.8 g, 1.69 mmol) in dry THF (100 mL) at 15 °C under Ar. After 30 min, the excess hydride was destroyed by adding a few drops H_2O, the resulting precipitate was filtered off. The solvent was rotary evaporated and the solid was dissolved in MeOH (100 mL), H_2O (1.5 mL) and NaOH (220 mg, 5.5 mmol) were added, and the reaction mixture was kept at 40 °C for 7 days. The solvent was removed under pressure and the crude dimer was purified by CC, eluted with 5% MeOH in $CHCl_3$, and crystallization from MeOH to yield triol **123** (1.21 g, 85%, mp: 303–306 °C). IR ($CHCl_3$): ν_{max} cm^{-1} 3386br (O–H and N–H), 1681s (C=O), 1514m, 1084s (C–O), 1036s (C–O), 979m, 916w. ^1H NMR (200 MHz, $CDCl_3$): δ 7.63 (d, 2H, $J = 8.7$ Hz, 2 × N–H), 3.83 (m, 2H, 2 × 12β-CH), 3.68 (m, 2H, 2 × 7β-CH), 3.46 (m, 2H, 2 × 3β-CH), 2.28 (m, 4h, 2 × CH_2OH), 0.86 (d, 6H, $J = 5.3$ Hz, 2 × 21-Me), 0.79 (s, 6H, 2 × 19-Me), 0.55 (s, 6H, 2 × 18-Me). ^{13}C NMR (50 MHz, $CDCl_3$): δ 12.6, 17.6, 22.7, 23.6, 27.2, 27.4, 28.2, 28.8, 29.8, 32.6, 35.0, 35.3, 36.4, 36.4, 40.3, 42.4, 42.8, 46.9, 47.7, 50.9, 63.1, 68.4, 73.4, 160.1. FABMS: *m/z* 863 $[M + Na]^+$ (Łotowski and Guzmanski, 2006).

Later, similar dimers **124–128** were prepared from *N,N'*-di(methyl 7α,12α-diacetoxy-5β-cholan-24-oate)3α,3′α-oxamide (**122**) in a similar fashion (Łotowski *et al.*, 2008). To a stirred solution of diester **122** (2.87 g, 2.70 mmol) in MeOH (100 mL), sequentially, H_2O (500 μL) and LiOH (130 mg, 5.4 mmol) were added at room temperature (*Scheme 2.1.56*). After 48 h, a few drops of diluted HCl (pH 3–4) was added to the reaction mixture, the resulting solid was filtered off and the filtrate was dried ($MgSO_4$). The solvent was rotary evaporated, purified by CC and recrystallized from EtOAc to afford *N,N'*-di(7α,12α-diacetoxy-5β-cholan-24-ioc acid)3α,3′α-oxamide (**124**, 2.48 g, 94%, mp: 191-194 °C). IR ($CHCl_3$): ν_{max} cm^{-1} 3389br (O–H), 2957m (C–H), 1721s (C=O), 1670s (C=O), 1508m (N–H), 1255m (C–O). ^1H NMR (200 MHz, $CDCl_3$): δ 7.52 (d, 2H, $J = 8.5$ Hz, 2 × NH), 5.07 (m, 2H, 2 × 12β-CH), 4.88 (m, 2H, 2 × 7β-CH), 3.59 (m, 2H, 2 × 3β-CH), 2.15 (s, 6H, 2 × 12α-MeCO), 2.04 (s, 6H, 2 × 7α-MeCO), 0.92 (s, 6H, 2 × 19-Me), 0.80 (d, 6H, $J = 5.2$ Hz, 2 × 21-Me), 0.71 (s, 6H, 2 × 18-Me). ^{13}C NMR (50 MHz, $CDCl_3$): δ 12.2, 17.4, 21.4, 21.6, 22.6, 22.7, 25.5, 27.4, 28.8, 30.4, 30.8, 31.2, 34.2, 34.5, 35.2, 37.6, 41.3, 43.3, 45.0, 47.3, 50.0, 53.4, 70.6, 75.2, 159.0, 170.4, 170.6, 179.4. FABMS: *m/z* 1059.6 $[M + Na]^+$ (Łotowski *et al.*, 2008).

N,N'-Di(methyl 7α,12α-diacetoxy-5β-cholan-24-ol)3α,3′α-oxamide (**125**) was obtained from a stirred solution of diester **122** (1.60 g, 1.50 mmol) in dry THF (50 mL), treated with $LiAlH_4$ (140 mg, 3.68 mmol) at 15 °C under Ar (*Scheme 2.1.56*). After the usual cleanup as described earlier for triol **123**, the crude dimer was purified by CC (mobile phase: 2% MeOH

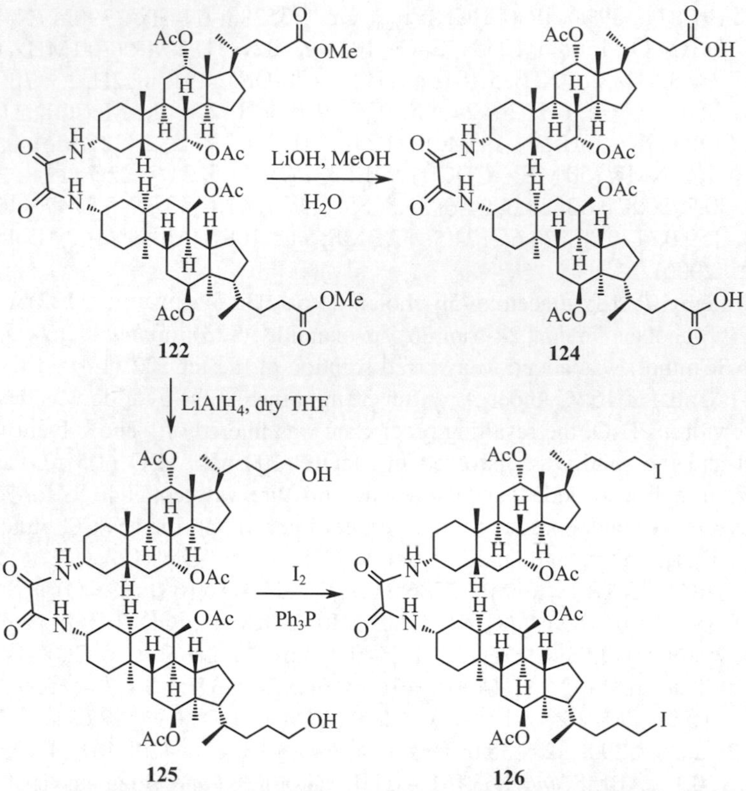

Scheme 2.1.56** Synthesis of cholic acid-based dimers **124–126

in $CHCl_3$) and crystallized from *n*-hexane-DCM to give diol **125** (920 mg, 61%, mp 177–179 °C). IR ($CHCl_3$): ν_{max} cm^{-1} 3672br (O–H), 3513br (O–H), 3390m (N–H), 1724s (C=O), 1670s (C=O), 1255m (C–O), 1024m (C–O). ^{1}H NMR (200 MHz, $CDCl_3$): δ 7.38 (d, 2H, $J = 8.6$ Hz, 2 × NH), 5.10 (m, 2H, 2 × 12β-CH), 4.91 (m, 2H, 2 × 7β-CH), 3.61 (m, 2H, 2 × 3β-CH and t, 4H, 2 × CH_2OH), 2.18 (s, 6H, 2 × 12α-MeCO), 2.07 (s, 6H, 2 × 7α-CH3CO), 0.94 (s, 6H, 2 19-Me), 0.84 (d, 6H, $J = 6.3$ Hz, 2 × 21-Me), 0.73 (s, 6H, 2 × 18-Me). ^{13}C NMR (50 MHz, $CDCl_3$): δ 12.1, 17.8, 21.5, 21.6, 22.6, 22.7, 25.5, 27.2, 27.4, 28.8, 29.1, 31.2, 31.5, 34.2, 34.7, 35.2, 35.3, 37.6, 41.3, 43.2, 44.9, 47.4, 49.9, 63.3, 70.6, 75.3, 158.9, 170.3, 170.6 (Łotowski *et al.*, 2008).

Iodination of diol **125** furnished *N,N′*-di(7α,12α-diacetoxy-24-iodo-5β-cholan)3α,3′α-oxamide (**126**) (*Scheme 2.1.56*). A mixture of I_2 (1.38 g, 5.4 mmol) in dry C_6H_6 (20 mL) and Ph_3P (1.38 g, 5.4 mmol) in dry C_5H_5N (880 μL, 11.0 mmol) was stirred at room temperature, after 15 min *N,N′*-di(methyl 7α,12α-diacetoxy-5β-cholan-24-ol)3α,3′α-oxamide (**125**, 920 mg, 0.91 mmol) was added and the mixture had been heated at 40 °C. After 3 days, the organic solvents were evaporated to dryness and the crude dimer was purified by CC, eluted with 25% EtOAc in *n*-hexane, and crystallized from *n*-hexane-DCM to obtain diiodide **126** (872 mg, 78%, mp: 170–172 °C). IR ($CHCl_3$): ν_{max} cm^{-1} 3389m (N–H), 1724s

(C=O), 1670s (C=O), 1508m (N–H), 1255m (C–O), 1023m. ^{1}H NMR (200 MHz, $CDCl_3$): δ 7.41 (d, 2H, $J=9.1$ Hz, 2 × NH), 5.10 (m, 2H, 2 × 12β-CH), 4.91 (m, 2H, 2 × 7β-CH), 3.58 (m, 2H, 2 × 3β-CH), 3.14 (m, 4H, 2 × CH_2I), 2.19 (s, 6H, 2 × 12α-MeCO), 2.07 (s, 6H, 2 × 7α-MeCO), 0.94 (s, 6H, 2 × 19-Me), 0.83 (d, 6H, $J=6.3$ Hz, 2 × 21-Me), 0.74 (s, 6H, 2 × 18-Me). ^{13}C NMR (50 MHz, $CDCl_3$): δ 7.6, 12.1, 17.9, 21.4, 21.6, 22.6, 22.7, 25.4, 27.1, 27.4, 28.8, 30.0, 31.1, 34.2, 35.2, 35.3, 36.5, 37.6, 41.3, 43.2, 44.9, 47.3, 49.8, 70.6, 75.2, 158.9, 170.2, 170.4 (Łotowski *et al.*, 2008).

N,N′-Di(24-Azido-7α,12α-diacetoxy-5β-cholan)3α,3′α-oxamide (**127**) was produced from diiodide **126** (*Scheme 2.1.57*). To a mixture of NaN_3 (640 mg, 9.8 mmol) and AcOH (570 μL, 10 mmol), a solution of *N,N′*-di(7α,12α-diacetoxy-24-iodo-5β-cholan) 3α,3′α-oxamide (**126**, 870 mg, 0.78 mmol) in *N*-methylpyrrolidone (10 mL) was added at room temperature and stirred overnight. Saturated aqueous $NaHCO_3$ was added to the reaction mixture and extracted several times with C_6H_6 and the pooled extracts were washed twice with H_2O, dried ($MgSO_4$) and evaporated under pressure. Purification on CC (mobile phase: 20% EtOAc in hexane) of the crude dimer followed by crystallization from *n*-hexane-EtOAc yielded diazide **127** (510 mg, 68%, mp: 159–161 °C). IR ($CHCl_3$): ν_{max} cm^{-1} 3390m (N–H), 2099s (N_3), 1721s (C=O), 1671s (C=O), 1256m (C–O), 1022m. ^{1}H NMR

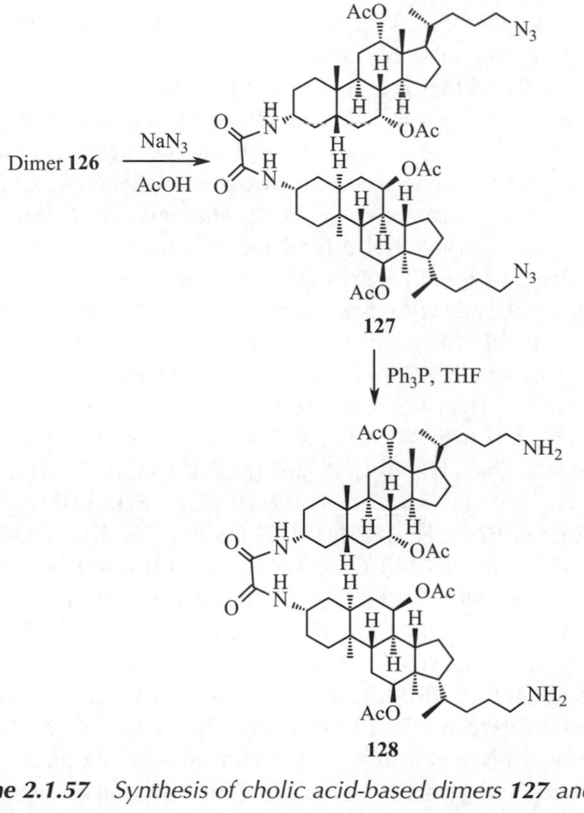

Dimer **126** $\xrightarrow[\text{AcOH}]{NaN_3}$ **127** $\xrightarrow{Ph_3P,\ THF}$ **128**

Scheme 2.1.57 *Synthesis of cholic acid-based dimers **127** and **128***

(200 MHz, $CDCl_3$): δ 7.38 (d, 2H, $J = 8.6$ Hz, 2 × NH), 5.11 (m, 2H, 2 × 12β-CH), 4.92 (m, 2H, 2 × 7β-H), 3.59 (m, 2H, 2 × 3β-CH), 3.25 (m, 4H, 2 × CH_2N_3), 2.19 (s, 6H, 2 × 12α-MeCO), 2.07 (s, 6H, 2 × 7β-MeCO), 0.95 (s, 6H, 2 × 19-Me), 0.85 (d, 6H, $J = 6.2$ Hz, 2 × 21-Me), 0.74 (s, 6H, 2 × 18-Me). ^{13}C NMR (50 MHz, $CDCl_3$): δ 12.1, 17.7, 21.4, 21.5, 22.6, 22.7, 25.3, 25.4, 27.1, 27.4, 28.7, 31.1, 32.5, 34.1, 34.5, 35.1, 35.3, 37.6, 41.2, 43.2, 44.9, 47.3, 49.8, 51.7, 70.5, 75.1, 158.9, 170.2, 170.4 (Łotowski *et al.*, 2008).

N,N′-Di(24-Amino-7α,12α-diacetoxy-5β-cholan)3α,3′α-oxamide (**128**) was prepared from diazide **127** (*Scheme 2.1.57*). Ph_3P (503 mg, 1.92 mmol) was added to a stirred solution of *N,N′*-di(24-azido-7α,12α-diacetoxy-5β-cholan)3α,3′α-oxamide (**127**, 510 mg, 0.48 mmol) in dry THF (12 mL) under Ar and stirred at 45 °C. After 15 min, H_2O (144 μL) was added dropwise and the reaction mixture was stirred at room temperature. After 24 h, the solvent was evaporated under pressure and the solid was dissolved in DCM, dried ($MgSO_4$) and rotary evaporated. Purification of the crude dimer on CC, eluted with DCM-MeOH with a few drops aqueous NH_3, and crystallization from MeOH gave diamine **128** (380 mg, 78%, mp: 174–176 °C). IR ($CHCl_3$): ν_{max} cm^{-1} 3388m (N–H), 1723s (C=O), 1670s (C=O), 1508m (N–H), 1255m (C–O). ^1H NMR (200 MHz, $CDCl_3$): δ 7.36 (d, 2H, $J = 8.6$ Hz, 2 × NH), 5.07 (m, 2H, 2 × 12β-CH), 4.87 (m, 2H, 2 × 7β-CH), 3.61 (m, 2H, 2 × 3β-CH), 2.62 (m, 4H, 2 × CH_2NH_2), 2.14 (s, 6H, 2 × 12α-MeCO), 2.03 (s, 6H, 2 × 7α-MeCO), 0.91 (s, 6H, 2 × 19-Me), 0.79 (d, 6H, $J = 6.2$ Hz, 2 × 21-Me), 0.70 (s, 6H, 2 × 18-Me). ^{13}C NMR (50 MHz, $CDCl_3$): δ 12.0, 17.7, 21.3, 21.5, 22.5, 22.6, 25.4, 27.1, 27.3, 28.7, 29.7, 31.1, 32.7, 34.1, 34.7, 35.1, 35.2, 37.5, 41.2, 42.3, 43.1, 44.8, 47.3, 49.8, 70.6, 75.2, 158.7, 170.2, 170.5 (Łotowski *et al.*, 2008).

Paryzek *et al.* (2010) synthesized cholic acid biscarbamate, N^1,N^2-bis(methyl-7α,12α-diacetoxy-5β-cholan-24-oate-3α-yl oxycarbonyl)ethylenediamine (**130**), from methyl cholate (**51**). Initially, methyl 7α,12α-diacetoxy-3α-hydroxy-5β-cholan-24-oate (**81**) was prepared from **51**, following the same procedure discussed earlier (*Scheme 2.1.37*). It was then converted to methyl 3α-chlorocarbonyloxy-7α,12α-diacetoxy-5β-cholan-24-oate (**129**). Triphosgene (514 mg, 1.7 mmol) and C_5H_5N (250 μL, 3.3 mmol) were added to a stirred solution of diacetate **79** (560 mg, 1.1 mmol) in dry C_6H_6 (6 mL) at room temperature (*Scheme 2.1.58*). After 60 min, C_6H_6 (5 mL) and H_2O (5 mL) were added, the organic layer was separated and washed with HCl (0.1 mol/L) followed by brine, dried ($MgSO_4$) and evaporated under vaccum to yield **129** (500 mg, 81%). IR (KBr): ν_{max} cm^{-1} 2951m (C–H), 2873m (C–H), 1781s (C=O), 1735s (C=O), 1469m (C–H), 1449m (C–H), 1438m (C–H), 1237s (C–O), 1166m, 1021m, 876m, 827m, 691w (C–Cl), 608w (C–Cl), 590w (C–Cl). ^1H NMR (500 MHz, $CDCl_3$): δ 5.08 (br s, 1H, 12β-CH), 4.91 (br s, 1H, 7β-CH), 4.67 (m, 1H, 3β-CH), 3.66 (s, 3H, 24-OMe), 2.15 (s, 3H, 12α-MeCO), 2.10 (s, 3H, 7α-MeCO), 0.92 (s, 3H, 19-Me), 0.82-0.79 (d, 3H, $J = 6.0$ Hz, 21-Me), 0.73 (s, 3H, 18-Me). ^{13}C NMR (125 MHz, $CDCl_3$): δ 34.3 (C-1), 26.5 (C-2), 83.3 (C-3), 149.4 (3α-CO), 34.6 (C-4), 41.0 (C-5), 31.2 (C-6), 70.4 (C-7), 170.0 (7α-CO), 21.5 (7α-MeCO), 37.8 (C-8), 29.0 (C-9), 34.1 (C-10), 25.6 (C-11), 75.2 (C-12), 170.2 (12α-CO), 21.6 (12α-MeCO), 45.1 (C-13), 43.4 (C-14), 22.9 (C-15), 27.2 (C-16), 47.4 (C-17), 12.3 (C-18), 22.4 (C-19), 34.4 (C-20), 17.6 (C-21), 30.9 (C-22), 30.8 (C-23), 174.2 (C-24), 51.5 (24-OMe). ESIMS: m/z 511.4 $[M - CH_3CO_2H]^+$ (Paryzek *et al.*, 2010).

Methyl 3α-chlorocarbonyloxy-7α,12α-diacetoxy-5β-cholan-24-oate (**129**) was employed to prepare cholic acid biscarbamate, N^1,N^2-bis(methyl-7α,12α-diacetoxy-5β-cholan-24-oate-3α-yl oxycarbonyl) ethylenediamine (**130**) in the following manner (*Scheme 2.1.58*). A solution of **129** (569 mg, 1.0 mmol) in dry THF (10 mL) was treated with a mixture of H_2N

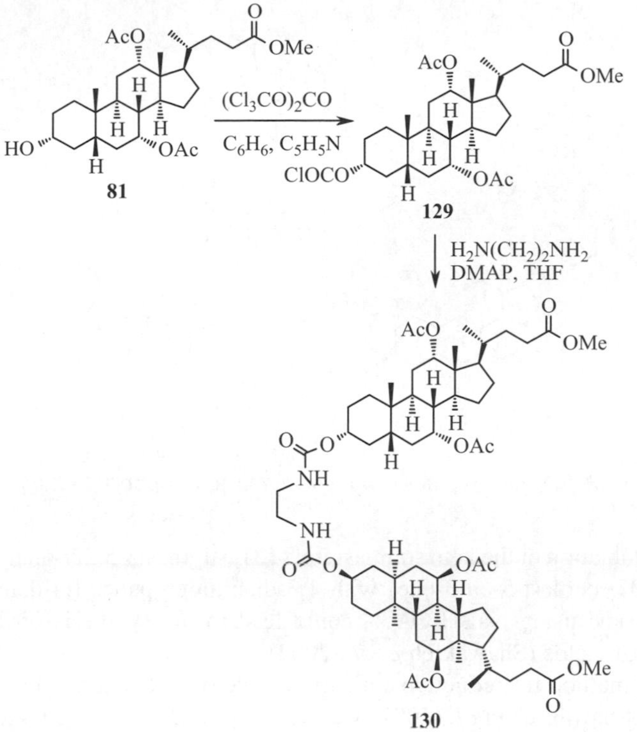

Scheme 2.1.58 *Synthesis of cholic acid biscarbamate* ***130***

$(CH_2)_2NH_2$ (33 μL, 0.5 mmol) and DMAP (111.98 mg, 1.0 mmol) in THF at room temperature. After 24 h, the solvent was evaporated to dryness and the residue was dissolved in C_6H_6-Et_2O and successively washed with 1N HCl, 5% $NaHCO_3$ and brine, dried ($MgSO_4$) and rotary evaporated. The residue was crystallized from *n*-hexane to afford cholic acid biscarbamate **130** (439 mg, 74%, mp: 124–125 °C). IR (KBr): ν_{max} cm^{-1} 3388m (N–H), 2951m (C–H), 2872m (C–H), 1736s (C=O), 1715s (C=O), 1653s (C=O), 1524m (N–H), 1467m (C–H), 1448m (C–H), 1438m (C–H), 1377m (C–N), 1248s (C–O), 1071m, 1023m, 963m, 937m, 896m, 778m, 607w, 592w. UV (MeOH): λ_{max} nm 219.5, 265, 270.5, 290. ^1H NMR (500 MHz, $CDCl_3$): δ 5.08 (br s, 2H, 2 × 12β-CH), 4.90 (br s, 2H, 2 × 7β-CH), 4.44 (m, 2H, 2 × 3β-CH), 3.66 (s, 6H, 2 × 24-OMe), 3.48, 3.30, 3.12 (4H, -NH-C_2H_4-NH-), 2.13 (s, 6H, 2 × 12α-MeCO), 2.08 (s, 6H, 2 × 7α-MeCO), 0.91 (s, 6H, 2 × 19-Me), 0.81 (d, 6H, $J = 6.3$ Hz, 2 × 21-Me), 0.73 (s, 6H, 2 × 18-Me). ^{13}C NMR (125 MHz, $CDCl_3$): δ 34.5 (2 × C-1), 27.0 (2 × C-2), 74.6 (2 × C-3), 156.6 (3α-$NHCO_2$), 41.3 (CH_2NHCO), 35.0 (2 × C-4), 40.8 (2 × C-5), 31.2 (2 × C-6), 70.6 (2 × C-7), 170.2 (7α-MeCO), 21.3 (2 × 7α-MeCO), 37.6 (2 × C-8), 28.7 (2 × C-9), 34.2 (2 × C-10), 25.4 (2 × C-11), 75.3 (C-12), 170.4 (2 × 12α-MeCO), 21.5 (2 × 12α-MeCO), 44.9 (2 × C-13), 43.3 (2 × C-14), 22.7 (2 × C-15), 27.1 (2 × C-16), 47.2 (2 × C-17), 12.1 (2 × C-18), 22.4 (2 × C-19), 34.5 (2 × C-20), 17.4 (2 × C-21), 30.8 (C-22), 30.7 (C-23), 174.4 (C-24), 51.4 (2 × 24-OMe). ESIMS: *m/z* 1148.3 $[M + Na]^+$, 1160.1 $[M + Cl]^+$. FABMS: *m/z* 1148.5 $[M + Na]^+$, 1126.6 $[M + H]^+$ (Paryzek *et al.*, 2010).

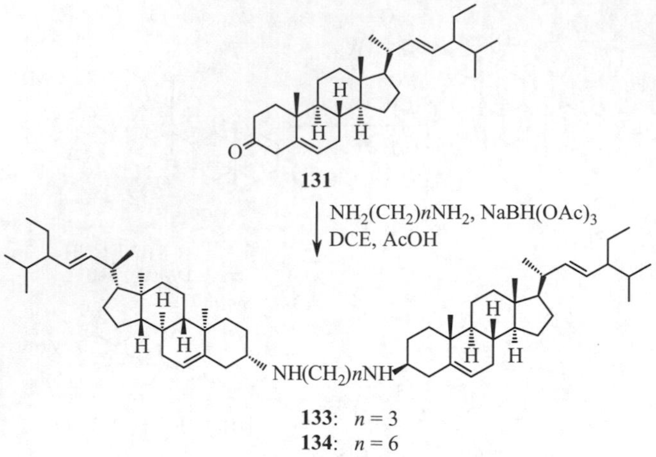

131

$NH_2(CH_2)nNH_2$, $NaBH(OAc)_3$
DCE, AcOH

$NH(CH_2)nNH$

133: $n = 3$
134: $n = 6$

Scheme 2.1.59 *Synthesis of keto stigmasterol dimers **133** and **134***

Reductive amination of the keto stigmasterol (**131**, stigmasta-5,22*t* dien-3-one) and keto cholesterol (**132**, cholest-5-en-3-one) with 1,3-diaminopropane, 1,4-diaminobutane and 1,6-diaminohexane using $NaBH(OAc)_3$ could lead to the synthesis of dimeric steroids **133–139** in high yields (Shawakfeh *et al.*, 2002).

The general method for reductive amination of keto stigmasterol (**131**) with diamines as reported by Shawakfeh *et al.* (2002) was as follows. Diamines (1.0 mmol) and NaBH $(OAc)_3$ (850 mg, 4.0 mmol) were added to a stirred solution of **131** (410 mg, 1.0 mmol) in DCE (40 mL) under Ar at room temperature (*Scheme 2.1.59*). After 48 h, glacial AcOH (1 mL) was added and the mixture was stirred for a further 88 h. The reaction was quenched with 1N NaOH and the mixture was extracted with $CHCl_3$ (4×40 mL). The organic layer was washed with brine (2×25 mL), dried ($MgSO_4$), and rotary evaporated to produce crude dimer that was recrystallized from EtOH to give dimers **133** (60%, mp: 170–172 °C) and **134** (55%, mp: 120–123 °C) (Shawakfeh *et al.*, 2002).

1,3-Diaminopropane stigmasterol dimer (**133**): IR (KBr): ν_{max} cm^{-1} 3450m (N–H), 2956m (C–H). ^1H NMR (300 MHz, $CDCl_3$): δ 5.25 (br s, 2H, $2 \times$ 6-CH), 4.90–5.10 (m, 4H, $2 \times$ 22-CH_2 and $2 \times$ 23-CH_2), 3.10 (br, 2H, $2 \times 3\alpha$-CH), 2.60-2.90 (m, 4H, $2 \times$ NH-CH_2), 1.01 (s, 6H, $2 \times$ 19-Me), 0.82 (s, 6H, $2 \times$ 18-Me). ^{13}C NMR (75 MHz, $CDCl_3$): δ 54.5 ($2 \times$ C-3), 146.0 ($2 \times$ C-5), 125.0 ($2 \times$ C-6), 56.0 ($2 \times$ C-14), 55.8 ($2 \times$ C-17), 12.4 ($2 \times$ C-18), 19.1 ($2 \times$ C-19), 138.0 ($2 \times$ C-22), 129.0 ($2 \times$ C-23), 39.67 ($2 \times$ C-30) (Shawakfeh *et al.*, 2002).

1,6-Diaminohexane stigmasterol dimer (**134**): IR (KBr): ν_{max} cm^{-1} 3459m (N–H), 2940m (C–H). ^1H NMR (300 MHz, $CDCl_3$): δ 5.30 (br s, 2H, $2 \times$ 6-CH), 4.90–5.10 (m, 4H, $2 \times$ 22-CH_2 and $2 \times$ 23-CH_2), 3.10 (br, 2H, $2 \times$ 3-CH), 2.60-2.90 (m, 4H, $2 \times$ NH-CH_2), 1.06 (s, 6H, $2 \times$ 19-Me), 0.77 (s, 6H, $2 \times$ 18-Me). ^{13}C NMR (75 MHz, $CDCl_3$): δ 54.6 ($2 \times$ C-3), 147.0 ($2 \times$ C-5), 123.0 ($2 \times$ C-6), 12.1 ($2 \times$ C-18), 19.3 ($2 \times$ C-19), 138.0 ($2 \times$ C-22), 129.0 ($2 \times$ C-23), 39.3 ($2 \times$ C-30) (Shawakfeh *et al.*, 2002).

The general procedure for reductive amination of keto cholesterol (**132**) with diamines was similar to that described for dimers **135** and **136** (Shawakfeh *et al.*, 2002). Diamines (1.35 mmol) and $NaBH(OAc)_3$ (439 mg, 2.0 mmol) were added at room temperature to a

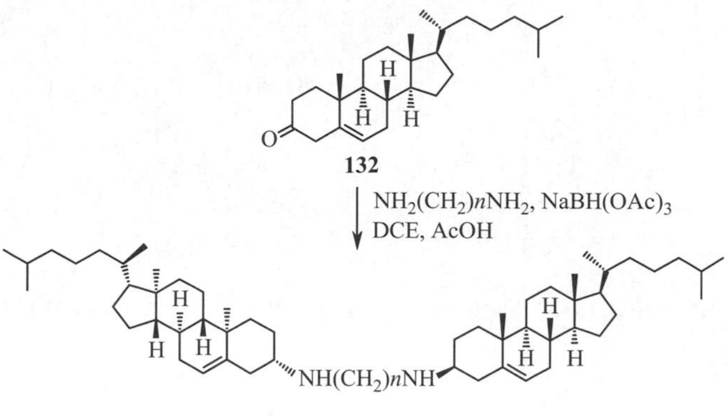

132

$NH_2(CH_2)nNH_2$, $NaBH(OAc)_3$
DCE, AcOH

$NH(CH_2)nNH$

135: $n = 3$
136: $n = 4$
137: $n = 6$

Scheme 2.1.60 *Synthesis of keto cholesterol dimers **135–137***

stirred solution of **132** (380 mg, 1.0 mmol) in DCE (40 mL) (*Scheme 2.1.60*) under Ar. After 40 h, glacial AcOH (1 mL) was added and the mixture was stirred for further 36 h. The usual workup as above produced the crude residue, which was washed with Et_2O to yield dimers **135** (60%, mp: 178–180 °C), **136** (60%, mp: 174–176 °C) and **137** (60%, mp: 134–136 °C) (Shawakfeh *et al.*, 2002).

1,3-Diaminopropane cholesterol dimer (**135**): IR (KBr): ν_{max} cm^{-1} 3450m (N–H), 2970m (C–H). ^1H NMR (300 MHz, $CDCl_3$): δ 5.25 (br s, 2H, 2 × 6-CH), 2.70-2.40 (br, 6H, 2 × 3α-CH and 2 × 28-CH_2), 2.60-2.90 (m, 4H, 2 × NH-CH_2 of 1,3-diaminopropane), 1.10 (s, 6H, 2 × 19-Me), 0.70 (s, 6H, 2 × 18-Me). ^{13}C NMR (75 MHz, $CDCl_3$): δ 54.1 (2 × C-3), 150.0 (2 × C-5), 116.0 (2 × C-6), 55.9 (2 × C-14), 54.6 (2 × C-17), 11.9 (2 × C-18), 18.8 (2 × C-19), 39.5 (2 × C-28) (Shawakfeh *et al.*, 2002).

1,4-Diaminobutane cholesterol dimer (**136**): IR (KBr): ν_{max} cm^{-1} 3465m (N–H), 2976m (C–H). ^1H NMR (300 MHz, $CDCl_3$): δ 5.30 (br s, 2H, 2 × 6-CH), 3.20 (br, 2H, 2 × 3α-CH), 2.70-2.90 (br, 4H, 2 × 28-CH_2), 2.60-2.90 (m, 4H, 2 × NH-CH_2 of 1,4-diaminobutane), 1.12 (s, 6H, 2 × 19-Me), 0.73 (s, 6H, 2 × 18-Me). ^{13}C NMR (75 MHz, $CDCl_3$): δ 54.3 (2 × C-3), 148 (2 × C-5), 120 (2 × C-6), 11.8 (2 × C-18), 18.6 (2 × C-19), 39.5 (2 × C-28) (Shawakfeh *et al.*, 2002).

1,6-Diaminohexane cholesterol dimer (**137**): IR (KBr): ν_{max} cm^{-1} 3459m (N–H), 2980m (C–H). ^1H NMR (300 MHz, $CDCl_3$): δ 5.40 (br s, 2H, 6-CH), 3.10–2.70 (br, 6H, 2 × 3α-CH and 2 × 28-CH_2), 2.60–2.90 (m, 4H, 2 × NH-CH_2 of 1,3-diaminohexane), 1.11 (s, 6H, 2 × 19-Me), 0.71 (s, 6H, 2 × 18-Me). ^{13}C NMR (75 MHz, $CDCl_3$): δ 54.1 (2 × C-3), 152.0 (2 × C-5), 114.0 (2 × C-6), 55.9 (2 × C-14), 54.9 (2 × C-17), 12.1 (2 × C-18), 19.0 (2 × C-19), 39.3 (2 × C-28) (Shawakfeh *et al.*, 2002).

The catalytic oxidation of keto stigmasterol dimer **133** and keto cholesterol dimer **135** was performed using MTO and H_2O_2, which selectively oxidized the internal double bond in both cases and produced, respectively, 1,2-diol dimers **138** and **139** (Shawakfeh *et al.*, 2002). The general method for epoxidation with MTO and H_2O_2 was as follows. Keto steroid dimer **133** or **135** (0.15 mmol) was added to a stirred mixture of MTO (10 mg, 4 μmol), 30% H_2O_2 (1 mL) and 1M $HClO_4$ (1 mL) in THF (15 mL) at room temperature

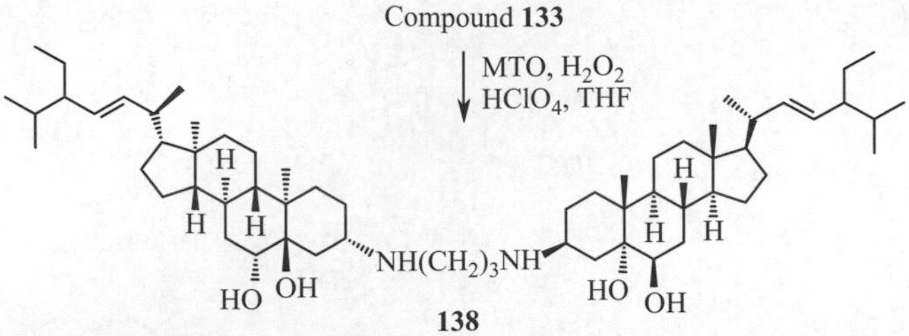

Compound **133**

MTO, H_2O_2
$HClO_4$, THF

$NH(CH_2)_3NH$

138

Scheme 2.1.61 *Synthesis of stigmasterol 1,2-diol dimer (**138**)*

Compound **135**

MTO, H_2O_2
$HClO_4$, THF

$NH(CH_2)_3NH$

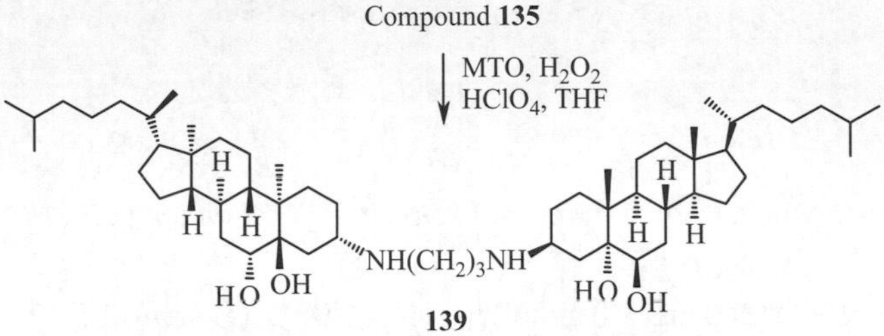

139

Scheme 2.1.62 *Synthesis of cholesterol 1,2-diol dimer **139***

(*Schemes 2.1.61* and *2.1.62*). The reaction mixture was protected from light for 24 h, and was neutralized with 1N NaOH, and extracted with $CHCl_3$. The organic layer was washed with brine, dried ($MgSO_4$), solvent was evaporated under pressure and the crude dimer was recrystallized from $CHCl_3$ to achieve the dimer **138** (60%, mp: 185–187 °C) or **139** (67%, mp: 199–201 °C) (Shawakfeh *et al.*, 2002).

Stigmasterol 1,2-diol dimer (**138**): IR (KBr): ν_{max} cm^{-1} 3489br (O–H and N–H), 2970m (C–H). ^{1}H NMR (300 MHz, $CDCl_3$): δ 3.40 (brs, 2H, 6α-CH), 2.80-3.20 (br s, 6H, 2 × 3α-CH and 2 × 28-CH_2), 1.10 (s, 6H, 2 × 19-Me), 0.75 (s, 6H, 2 × 18-Me). ^{13}C NMR (75 MHz, $CDCl_3$): δ 54.5 (2 × C-3), 67.5 (2 × C-5), 75.1 (2 × C-6), 13.1 (2 × C-18), 19.9 (2 × C-19) (Shawakfeh *et al.*, 2002).

Cholesterol 1,2-diol dimer (**139**): IR (KBr): ν_{max} cm^{-1} 3510br (O–H and N–H), 2990m (C–H). ^{1}H NMR (300 MHz, $CDCl_3$): δ 4.90–5.20 (m, 4H, 2 × 22-CH and 2 × 23-CH), 3.05-3.40 (br s, 4H, 2 × 3α-CH and 2 × 6α-CH), 1.40–2.20 (m, 4H, 2 × $NHCH_2$), 1.00 (s, 6H, 2 × 19-Me), 0.71 (s, 6H, 2 × 18-Me). ^{13}C NMR (75 MHz, $CDCl_3$): δ 55.7 (2 × C-3), 67.6 (2 × C-5), 76.0 (2 × C-6), 12.8 (2 × C-18), 19.6 (2 × C-19), 137.8 (2 × C-22), 129.5 (2 × C-23) (Shawakfeh *et al.*, 2002).

Four 6*E*-hydroximinosteroid homodimers **150–153** were synthesized from cholesterol (**1**) *via* a ruthenium-catalyzed crossmetathesis reaction (Rega *et al.*, 2007). For this synthesis several cholesterol derivatives **140–149** were prepared as follows (*Scheme 2.1.63*). The first of these cholesterol derivatives was propargyl cholesteryl ether (**140**), which was obtained from the reaction involving a mixture of propargyl alcohol (4 mL, 68 mmol), $CHCl_3$ (50 mL), cholesterol (**1**, 2.0 g, 4.91 mmol) and Montmorillonite K10 (4.5 g, activated at 120 °C overnight prior to use), and stirring at 55 °C in an oil bath. After 7 days, the mixture was filtered, and the filtrate was washed with Et_2O, the solvent was evaporated to dryness and the crude solid was purified by FCC (mobile phase: 5% Et_2O in *n*-hexane) to afford

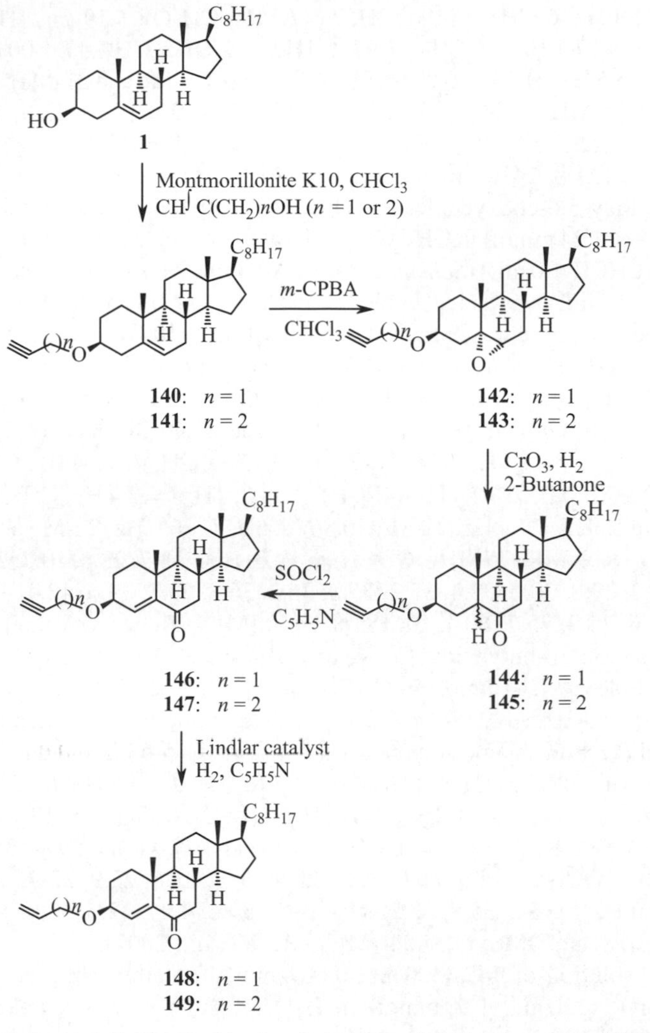

Montmorillonite K10, $CHCl_3$
$CH \equiv C(CH_2)nOH$ (n = 1 or 2)

m-CPBA
$CHCl_3$

140: $n = 1$
141: $n = 2$

142: $n = 1$
143: $n = 2$

CrO_3, H_2
2-Butanone

$SOCl_2$
C_5H_5N

146: $n = 1$
147: $n = 2$

144: $n = 1$
145: $n = 2$

Lindlar catalyst
H_2, C_5H_5N

148: $n = 1$
149: $n = 2$

Scheme 2.1.63 *Synthesis of cholesterol derivatives **140–149***

propargyl cholesteryl ether (**140**, 1.67 g, 80%, mp: 113.6–114.4 °C). ^{1}H NMR (300 MHz, $CDCl_3$): δ 5.36 (br d, 1H, $J = 5.2$ Hz, 6-CH), 4.19 (d, 2H, d, $J = 2.4$ Hz, CH_2O), 3.38 (m, 1H, 3α-CH), 2.39 (t, 1H, $J = 2.4$ Hz, CHC≡C), 1.00 (s, 3H, 19-Me), 0.91 (d, 3H, $J = 6.5$ Hz, 21-Me), 0.87 (d, 3H, $J = 6.6$ Hz, 27-Me), 0.86 (d, 3H, $J = 6.6$ Hz, 26-Me), 0.67 (s, 3H, 18-Me). ^{13}C NMR (75 MHz, $CDCl_3$): δ 12.0, 18.8, 19.4, 21.2, 22.7, 22.9, 23.9, 24.4, 28.1, 28.2, 28.3, 32.0, 32.1, 35.9, 36.3, 36.9, 37.2, 38.8, 39.6, 39.9, 42.4, 50.3, 55.2, 56.3, 56.9, 73.8, 78.3, 80.6, 122.0, 140.7. EIMS: *m/z* 424 $[M]^+$ (Rega *et al.*, 2007).

3-Butynyl cholesteryl ether (**141**) was synthesized from **1** in a similar fashion as outlined above for **140**, just by using 3-butynyl alcohol (4 mL, 53 mmol) instead of propargyl alcohol (*Scheme 2.1.63*). Usual cleanup and purification as described above produced 3-butynyl cholesteryl ether (**141**, 1.83 g, 85%, mp: 81.3–82.5 °C). ^{1}H NMR (300 MHz, $CDCl_3$): δ 5.35

(br d, 1H, $J = 5.1$ Hz, 6-CH), 3.60 (t, 2H, $J = 7.1$ Hz, CH_2O), 3.19 (m, 1H, 3α-CH), 2.45 (dt, 2H, $J = 2.6$ and 7.1 Hz, $\equiv CCH_2$), 1.97 (t, 1H, $J = 2.6$ Hz, CHC≡), 1.00 (s, 3H, 19-Me), 0.92 (d, 3H, $J = 6.5$ Hz, 21-Me), 0.87 (d, 6H, $J = 6.7$ Hz, 26-Me and 27-Me), 0.68 (s, 3H, 18-Me). ^{13}C NMR (75 MHz, $CDCl_3$): δ 12.0, 18.8, 19.5, 20.4, 21.2, 22.7, 22.9, 23.9, 24.4, 28.1, 28.3, 28.5, 32.0, 32.1, 35.9, 36.3, 37.0, 37.3, 39.2, 39.6, 39.9, 42.4, 50.3, 56.3, 56.9, 66.3, 69.3, 79.4, 81.6, 121.8, 141.0. EIMS: *m/z* 438 $[M]^+$ (Rega *et al*., 2007).

3β-Propargyloxy-5,6-epoxycholestane (**142**) was prepared by treating dropwise a solution of **140** (1.8 g, 4.24 mmol) in $CHCl_3$ (30 mL) at 0 °C, with a solution of *m*-CPBA (2.5 g, 11.1 mmol) in $CHCl_3$ (30 mL) (*Scheme 2.1.63*). After the solution had been stirred for 20 h, Na_2SO_3 (10%, 150 mL) was added, cooled in an ice-water bath and kept at this temperature for 6 h. The aqueous phase was extracted with $CHCl_3$ (3 × 100 mL) and the pooled extracts were washed with H_2O (200 mL), dried (Na_2SO_4), concentrated under pressure to obtain the crude residue, which was purified by FCC, eluted with 10% EtOAc in *n*-hexane, to produce 3β-propargyloxy-5,6-epoxycholestane (**142**) as the major isomer (1.68 g, 90%, mp: 81.4–82.9 °C). ^{1}H NMR (300 MHz, $CDCl_3$): δ 4.15 (d, 2H, $J = 2.4$ Hz, CH_2O), 3.72 (m, 1H, 3α-CH), 2.89 (d, 1H, $J = 4.4$ Hz, 6β-CH), 2.38 (t, 1H, $J = 2.4$ Hz, CHC≡), 1.05 (s, 3H, 19-Me), 0.88 (d, 3H, $J = 7.4$ Hz, 21-Me), 0.86 (d, 6H, $J = 6.6$ Hz, 26-Me and 27-Me), 0.55 (s, 3H, 18-Me). ^{13}C NMR (75 MHz, $CDCl_3$): δ 12.0, 16.0, 18.7, 22.6, 20.7, 22.9, 23.9, 24.1, 27.8, 28.1, 28.2, 29.0, 30.0, 32.4, 35.2, 35.8, 36.2, 36.8, 39.5, 39.6, 42.4, 42.7, 55.6, 56.0, 57.0, 59.5, 65.6, 74.0, 75.7, 80.4. EIMS: *m/z* 440 $[M]^+$ (Rega *et al*., 2007).

The synthesis of 3β-butynyloxy-5,6-epoxycholestane (**143**) was carried out from a solution of **141** (1.8 g, 4.10 mmol) in $CHCl_3$ (30 mL) at 0 °C, following exactly the same procedure as described for **142** (*Scheme 2.1.63*). Usual workup and purification as described above afforded **143** as the major isomer (1.6 g, 86%, mp: 51.5-53.5 °C). ^{1}H NMR (300 MHz, $CDCl_3$): δ 3.55 (m, 3H, 3α-CH and CH_2O), 2.87 (d, 1H, $J = 4.3$ Hz, 6-CH), 2.40 (dt, 2H, $J = 2.7$, 7.0 Hz, $\equiv CCH_2$), 1.94 (t, 1H, $J = 2.7$ Hz, CH≡C), 1.03 (s, 3H, 19-Me), 0.87 (d, 3H, $J = 7.4$ Hz, 21-Me), 0.85 (d, 6H, $J = 6.6$ Hz, 26-Me and 27-Me), 0.59 (s, 3H, 18-Me). ^{13}C NMR (75 MHz, $CDCl_3$): δ 11.9, 16.0, 18.7, 20.4, 20.7, 22.6, 22.9, 23.9, 24.1, 28.0, 28.1, 28.2, 29.0, 30.0, 32.4, 35.2, 35.8, 36.2, 37.1, 39.5, 39.6, 42.4, 42.7, 56.0, 56.9, 59.5, 65.5, 66.5, 69.3, 76.3, 81.4. EIMS: *m/z* 454 $[M]^+$ (Rega *et al*., 2007).

To a stirred solution of **142** (300 mg, 0.68 mmol) in methyl ethyl ketone (20 mL), a solution of CrO_3 (150 mg, 1.50 mmol) in H_2O (500 μL) was added dropwise at 0 °C (*Scheme 2.1.63*). This addition was repeated at room temperature and stirred for 2 h. *i*-PrOH (3 mL) was added, the organic solvent was rotary evaporated and the resulting solid was purified by FCC (mobile phase: 20% EtOAc in *n*-hexane) to provide 3β-propargyloxy-5α-hydroxycholestan-6-one (**144**, 200 mg, 65%, mp: 141.7–142.8 °C). ^{1}H NMR (300 MHz, $CDCl_3$): δ 4.17 (d, 2H, $J = 2.2$ Hz, CH_2O), 3.84 (m, 1H, 3α-CH), 2.69 (t, 1H, $J = 12.6$ Hz, 7-CH), 2.39 (t, 1H, $J = 2.2$ Hz, CHC≡), 0.90 (d, 3H, $J = 6.4$ Hz, 21-Me), 0.85 (d, 6H, $J = 6.6$ Hz, 26-Me and 27-Me), 0.79 (s, 3H, 19-Me), 0.63 (s, 3H, 18-Me). ^{13}C NMR (75 MHz, $CDCl_3$): δ 12.1, 14.1, 18.7, 21.5, 22.6, 22.9, 23.9, 24.0, 27.0, 28.1, 28.2, 29.7, 33.0, 35.8, 36.2, 37.3, 39.6, 39.7, 41.9, 42.7, 43.2, 44.6, 55.3, 56.2, 56.5, 73.8, 74.1, 80.4, 80.7, 212.3. EIMS: *m/z* 456 $[M]^+$ (Rega *et al*., 2007).

Following the same protocol as outlined above, from a solution of **143** (311 mg, 0.68 mmol) in methyl ethyl ketone (20 mL) at 0 °C, 3β-butynyloxy-5α-hydroxycholestan-6-one (**145**, 190 mg, 60%, mp: 140.1–141.2 °C) was obtained (*Scheme 2.1.63*). ^{1}H NMR (300 MHz, $CDCl_3$): δ 3.67 (m, 1H, 3α-CH), 3.58 (t, 2H, $J = 7.1$ Hz, CH_2O), 2.68 (t, 1H,

$J = 12.6$ Hz, 7-CH), 2.42 (dt, 2H, $J = 2.6$, 7.1 Hz, ≡CCH_2), 1.96 (t, 1H, $J = 2.6$ Hz, CHC≡), 0.89 (d, 3H, $J = 6.4$ Hz, 21-Me), 0.85 (d, 6H, $J = 6.6$ Hz, 26-Me and 27-Me), 0.78 (s, 3H, 19-Me), 0.61 (s, 3H, 18-Me). ^{13}C NMR (75 MHz, $CDCl_3$): δ 12.1, 14.1, 18.7, 20.4, 21.5, 22.6, 22.9, 23.9, 24.0, 27.2, 28.1, 28.2, 29.7, 33.2, 35.8, 36.2, 37.3, 39.5, 39.7, 41.9, 42.7, 43.2, 44.6, 56.2, 56.4, 66.4, 69.4, 74.9, 80.7, 81.6, 212.6. EIMS: *m/z* 470 $[M]^+$ (Rega *et al.*, 2007).

Another cholestrerol derivative, 3β-propargyloxycholest-4-en-6-one (**146**), was synthesized from 3β-propargyloxy-5α-hydroxycholestan-6-one (**144**) (*Scheme 2.1.63*). To a cooled (0 °C) solution of **144** (200 mg, 0.44 mmol) in C_5H_5N (10 mL), $SOCl_2$ (100 μL, 1.36 mmol) was added dropwise and stirred for 45 min, after which the mixture was poured into H_2O (50 mL). The resulting precipitate was extracted with EtOAc (2 × 20 mL), the pooled extracts were washed with 5% HCl followed by brine, dried ($MgSO_4$), evaporated under pressure. Purification of the crude residue on FCC, eluted with 5% EtOAc in *n*-hexane, produced **146** (173 mg, 90%). ^{1}H NMR (300 MHz, $CDCl_3$): δ 6.25 (t, 1H, $J = 1.7$ Hz, 4-CH), 4.24 (d, 2H, $J = 2.4$ Hz, CH_2O), 4.18 (ddd, 1H, $J = 1.7$, 6.1, 9.9 Hz, 3α-CH). 2.55 (d, 1H, $J = 11.9$ Hz, 7-CH), 2.41 (t, 1H, $J = 2.4$ Hz, CHC≡), 1.00 (s, 3H, 19-Me), 0.91 (d, 3H, $J = 6.5$ Hz, 21-Me), 0.86 (d, 6H, $J = 6.6$ Hz, 26-Me and 27-Me), 0.69 (s, 3H, 18-Me). ^{13}C NMR (75 MHz, $CDCl_3$): δ 12.1, 18.7, 19.8, 20.8, 22.6, 22.9, 23.8, 24.0, 24.8, 28.0, 28.1, 34.1, 34.7, 35.7, 36.1, 38.4, 39.4, 39.5, 42.6, 46.4, 51.3, 55.5, 56.1, 56.7, 73.1, 74.4, 79.9, 129.8, 147.1, 202.4 (Rega *et al.*, 2007).

The same method as above was followed for the synthesis of 3β-butynyloxycholest-4-en-6-one (**147**, 182 mg, 92%) from a solution of **145** (200 mg, 0.44 mmol) in C_5H_5N (10 mL) at 0 °C (*Scheme 2.1.63*). ^{1}H NMR (300 MHz, $CDCl_3$): δ 6.24 (t, 1H, $J = 1.7$ Hz, 4-CH), 3.97 (ddd, 1H, $J = 1.7$, 6.1, 10 Hz, 3α-CH), 3.65 (m, 2H, CH_2O), 2.54 (d, 1H, $J = 11.9$ Hz, 7-CH), 2.46 (dt, 2H, $J = 2.6$ and 7.1 Hz, ≡CCH_2), 1.97 (t, 1H, $J = 2.6$ Hz, CH≡C), 0.99 (s, 3H, 19-Me), 0.91 (d, 3H, $J = 6.5$ Hz, 21-Me), 0.86 (d, 6H, $J = 6.6$ Hz, 26-Me and 27-Me), 0.69 (s, 3H, 18-Me). ^{13}C NMR (75 MHz, $CDCl_3$): δ 12.1, 18.7, 19.8, 20.3, 20.8, 22.6, 22.8, 23.8, 24.0, 24.8, 28.0, 28.1, 34.1, 34.7, 35.7, 36.1, 38.4, 39.4, 39.5, 42.6, 46.4, 51.3, 56.1, 56.7, 66.4, 69.4, 74.6, 81.2, 130.4, 146.9, 202.4 (Rega *et al.*, 2007).

Cholesterol derivative **146** was used to synthesize 3β-(2-propenyloxy)-4-cholesten-6-one (**148**) in the following manner (*Scheme 2.1.63*). To a stirred solution of **146** (300 mg, 0.684 mmol) in C_5H_5N (5 mL), Lindlar's catalyst (14 mg) was added. The reaction flask was purged three times with H_2 and stirred under H_2 at room temperature. After 5 h, the mixture was filtered over Celite and rotary evaporated to dryness. Purification on FCC (mobile phase: 5% EtOAc in *n*-hexane) of the crude product furnished **148** (265 mg, 88%, mp: 68.4–70.2 °C). ^{1}H NMR (300 MHz, $CDCl_3$): δ 6.27 (br s, 1H, 4-CH), 5.91 (tdd, 1H, $J = 5.4$, 10.3, 17.2 Hz, =$CHCH_2O$), 5.28 (d, 1H, $J = 17.2$ Hz, CHa=C), 5.16 (d, 1H, $J = 10.3$ Hz, CHb=C), 4.07 (t, 2H, $J = 5.4$ Hz, CH_2O), 3.96 (ddd, 1H, $J = 1.7$, 6.2, 9.9 Hz, 3α-CH), 2.54 (d, 1H, $J = 12.1$ Hz, 7-CH), 0.99 (s, 3H, 19-Me), 0.91 (d, 3H, $J = 6.4$ Hz, 21-Me), 0.85 (d, 6H, $J = 6.6$ Hz, 26-Me and 27-Me), 0.68 (s, 3H, 18-Me). ^{13}C NMR (75 MHz, $CDCl_3$): δ 12.0, 18.7, 19.9, 20.9, 22.6, 22.9, 23.9, 24.0, 25.1, 28.1, 28.2, 34.1, 34.8, 35.8, 36.2, 38.4, 39.5, 39.6, 42.7, 46.4, 51.3, 56.1, 56.8, 69.3, 73.7, 117.1, 130.7, 135.0, 146.8, 202.8 (Rega *et al.*, 2007).

Following the same protocol as utilized for **148**, 3β-(3-butenyloxy)-4-cholesten-6-one (**149**, 277 mg, 92%, mp: 65.3–66.5 °C) was accomplished from a solution of **147** (300 mg, 0.663 mmol) in C_5H_5N (5 mL) (*Scheme 2.1.63*). ^{1}H NMR (300 MHz, $CDCl_3$): δ 6.26 (br s, 1H, 4-CH), 5.81 (tdd, 1H, $J = 6.7$, 10.2, 17.0 Hz, =$CHCH_2O$), 5.05 (m, 2H, CH_2=C), 3.91 (ddd, 1H, $J = 1.8$, 6.1, 10.0 Hz, 3α-CH), 3.56 (m, 2H, CH_2O), 2.54 (d, 1H, $J = 11.9$ Hz,

7-CH), 2.32 (dd, 2H, $J = 6.7$, 13.5 Hz, $=CHCH_2$), 0.99 (s, 3H, 19-Me), 0.91 (d, 3H, $J = 6.4$ Hz, 21-Me), 0.85 (d, 6H, $J = 6.6$ Hz, 26-Me and 27-Me), 0.68 (s, 3H, 18-Me). ^{13}C NMR (75 MHz, $CDCl_3$): δ 12.0, 18.7, 19.9, 20.9, 22.6, 22.9, 23.9, 24.0, 24.9, 28.1, 28.2, 34.1, 34.6, 34.8, 35.8, 36.2, 38.4, 39.5, 39.6, 42.7, 46.4, 51.3, 56.1, 56.8, 67.8, 74.4, 116.5, 130.9, 135.2, 146.7, 202.9 (Rega *et al.*, 2007).

3β-(2-Propenyloxy)-4-cholesten-6-one (**148**) was employed for the preparation of the dimer, 1,4-bis(4-cholesten-6-on-3β-oxy)-2*E*-butene (**150**) (*Scheme 2.1.64*). A mixture of cholesterol derivative **148** (100 mg, 0.227 mmol) and the Grubbs 1st-generation catalyst (15 mg, 0.018 mmol) were stirred overnight in DCM (4 mL) at 45 °C. MeOH was added, the organic solvent was evaporated to dryness and the residue was initially purified by FCC, eluted with 1% EtOAc in *n*-hexane, to yield a mixture of dimeric isomers ($E/Z = 7{:}1$) which were separated on FCC using SiO_2 impregnated with $AgNO_3$ as the stationary phase (mobile phase: 15% EtOAc in *n*-hexane to 10% MeOH in DCM). 1,4-Bis(4-cholesten-6-one-3β-oxy)-2*E*-butene (**150**, 79 mg, 82%) was obtained as the major isomer. ^1H NMR (500 MHz, $CDCl_3$): δ 6.24 (d, 2H, $J = 1.9$ Hz, 2 × 4-CH), 5.80 (t, 2H, $J = 3.3$ Hz, CH=CH), 4.09 (dd, 2H, $J = 3.3$, 12.7 Hz, OCH_2), 4.04 (dd, 2H, $J = 3.3$ and 12.7 Hz, CH_2O), 3.95 (ddd, 2H, $J = 1.9$, 6.0, 10.0 Hz, 2 × 3α-CH), 2.53 (dd, 2H, $J = 3.5$ and 15.5 Hz, 2 × 7-CH), 0.98 (s, 6H, 2 × 19-Me), 0.90 (d, 6H, $J = 6.4$ Hz, 2 × 21-Me), 0.85 (d, 12H, $J = 6.4$ Hz, 2 × 26-Me and 2 × 27-Me), 0.68 (s, 6H, 18-Me). ^{13}C NMR (125 MHz, $CDCl_3$): δ 12.0, 18.7, 19.9, 20.9,

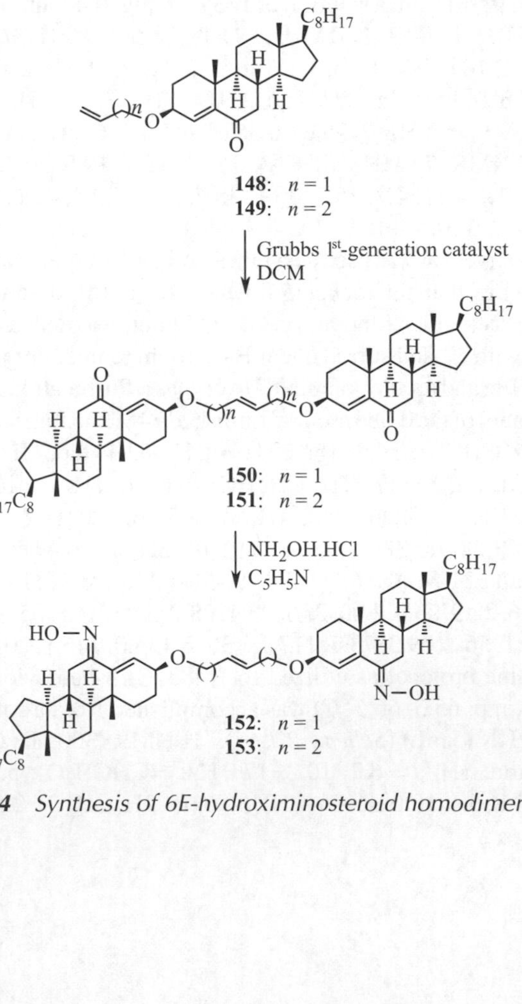

Scheme 2.1.64 *Synthesis of 6E-hydroximinosteroid homodimers* ***150–153***

22.6, 22.9, 23.9, 24.0, 25.0, 28.1, 28.2, 34.1, 34.8, 35.8, 36.2, 38.4, 39.4, 39.5, 42.6, 46.4, 51.3, 56.1, 56.8, 68.3, 73.6, 129.6, 130.6, 146.7, 202.7. HRESIMS: *m/z* 853.7074 calculated for $C_{58}H_{93}O_4$, found: 853.7064 (Rega *et al.*, 2007).

1,6-Bis(4-cholesten-6-on-3β-oxy)-3*E*-hexene (**151**) was also prepared in a similar fashion as depicted above, by utilizing 3β-(3-butenyloxy)-4-cholesten-6-one (**149**, 103 mg, 0.227 mmol) as the starting material (*Scheme 2.1.64*), and was purified by repeated FCC as outlined for **150**, to afford the major dimeric isomer, 1,6-bis(4-cholesten-6-on-3β-oxy)-3*E*-hexene (**151**, 80 mg, 80%). ^{1}H NMR (500 MHz, $CDCl_3$): δ 6.25 (t, 2H, t, $J=2.0$ Hz, 2 × 4-CH), 5.50 (m, 2H, CH=CH), 3.90 (ddd, 2H, $J=2.0$, 6.0, 10.3 Hz, 3α-CH), 3.52 (m, 4H, 2 × CH_2O), 2.54 (dd, 2H, $J=3.6$, 14.2 Hz, 2 × 7-CH), 2.27 (m, 4H, 2 × CH_2CH=), 0.99 (s, 6H, 19-Me), 0.91 (d, 6H, $J=6.5$ Hz, 2 × 21-Me), 0.86 (d, 12H, $J=6.6$ Hz, 2 × 26-Me and 2 × 27-Me), 0.69 (s, 6H, 18-Me). ^{13}C NMR (125 MHz, $CDCl_3$): δ 12.0, 18.7, 19.9, 20.9, 22.7, 22.9, 23.9, 24.0, 25.0, 28.1, 28.2, 33.5, 34.1, 34.8, 35.8, 36.2, 38.4, 39.5, 39.6, 46.4, 51.3, 56.1, 56.8, 68.2, 74.4, 128.6, 131.0, 146.6, 202.8. HRESIMS: *m/z* 881.7387 calculated for $C_{60}H_{97}O_4$, found: 881.7408 (Rega *et al.*, 2007).

For the synthesis of 1,4-bis(6*E*-hydroximino-4-cholesten-3β-oxy)-2*E*-butene (**152**), dimer **150** (31 mg, 0.036 mmol) and $NH_2OH.HCl$ (30 mg, 0.427 mmol) were dissolved in dry C_5H_5N (1 mL) and stirred at room temperature (*Scheme 2.1.64*). After 60 min, H_2O (4 mL) was added to form a precipitate. The mixture was extracted with EtOAc (3 × 5 mL) and the pooled extracts were sequentially washed with 5% HCl, saturated aqueous $NaHCO_3$ and H_2O, dried ($MgSO_4$) and evaporated to dryness. The crude dimer was subjected to FCC, eluted with 30% EtOAc in *n*-hexane, to provide dimer **152** (20 mg, 65%). ^{1}H NMR (500 MHz, $CDCl_3$): δ 8.74 (br s, 2H, 2 × NOH), 5.86 (t, 2H, $J=2.0$ Hz, 2 × 4-CH), 5.83 (m, 2H, CH=CH), 4.06 (m, 4H, 2 × CH_2O), 3.93 (ddd, 2H, $J=2.0$, 6.0, 10.2 Hz, 2 × 3α-CH), 3.31 (dd, 2H, $J=4.5$, 15.1 Hz, 2 × 7-CH), 0.96 (s, 6H, 19-Me), 0.90 (d, 6H, $J=6.4$ Hz, 21-Me), 0.86 (d, 12H, $J=6.6$ Hz, 2 × 26-Me and 2 × 27-Me), 0.66 (s, 6H, 18-Me). ^{13}C NMR (125 MHz, $CDCl_3$): δ 12.1, 18.8, 18.9, 21.0, 22.7, 22.9, 23.9, 24.2, 25.3, 28.1, 28.3, 30.0, 33.8, 34.6, 35.8, 36.2, 38.1, 39.6, 39.7, 42.8, 52.7, 56.2, 56.8, 68.2, 74.1, 125.3, 129.9, 142.5, 158.4. HRESIMS: *m/z* 883.7292 calculated for $C_{58}H_{95}N_2O_4$, found: 883.7297 (Rega *et al.*, 2007).

Following the above procedure, 1,6-bis(6*E*-hydroximino-4-cholesten-3β-oxy)-3*E*- hexene (**153**) was prepared from dimer **151** (18 mg, 0.0204 mmol) (*Scheme 2.1.64*). Purification of the crude dimer on FCC (mobile phase: 20–50% EtOAc in *n*-hexane), provided dimer **153** (12 mg, 64%). ^{1}H NMR (500 MHz, $CDCl_3$): δ 8.54 (br s, 2H, 2 × NOH), 5.84 (t, 2H, $J=2.0$ Hz, 2 × 4-CH), 5.50 (m, 2H, CH=CH), 4.06 (m, 4H, 2 × CH_2O), 3.89 (ddd, 2H, $J=2.0$, 6.0, 10.2 Hz, 2 × 3α-CH), 3.31 (dd, 2H, $J=4.7$, 15.2 Hz, 2 × 7-CH), 2.26 (m, 4H, 2 × CH_2CH=), 0.96 (s, 6H, 2 × 19-Me), 0.90 (d, 6H, $J=6.4$ Hz, 2 × 21-Me), 0.86 (d, 12H, $J=6.6$ Hz, 2 × 26-Me and 2 × 27-Me), 0.67 (s, 6H, 18-Me). ^{13}C NMR (125 MHz, $CDCl_3$): δ 12.1, 18.8, 18.9, 21.0, 22.7, 22.9, 23.9, 24.2, 25.1, 28.1, 28.3, 30.0, 33.5, 33.8, 34.6, 35.8, 36.2, 38.1, 39.6, 39.7, 42.8, 52.7, 56.2, 56.8, 67.8, 74.4, 125.4, 128.6, 142.4, 158.4. HRESIMS: *m/z* 911.7605 calculated for $C_{60}H_{99}N_2O_4$, found: 911.7609 (Rega *et al.*, 2007).

Potential antileukemic steroid dimers **156–158** were synthesized by linking two 5α-androstanolone (**154**, 17β-hydroxy-5α-androstan-3-one or 5α-androstan-3-one-17β-ol) or by two 5β-androstanolone (**155**, 17β-hydroxy-5β-androstan-3-one or 5β-androstan-3-one-17β-ol) units through a 5-fluorouracil moiety using a Fischer-type reaction with *N,N′*-dimethylhydrazine (Sucrow and Chondromatidis, 1970).

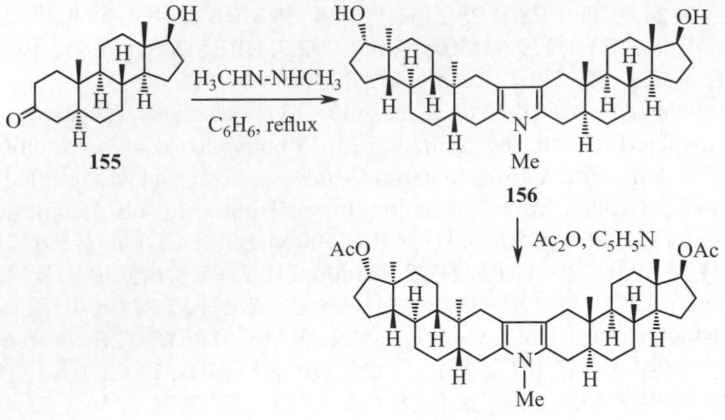

Scheme 2.1.65 *Synthesis of 5α-androstanolone diol dimer* ***156*** *and 5α-androstanolone diacetate dimer* ***157***

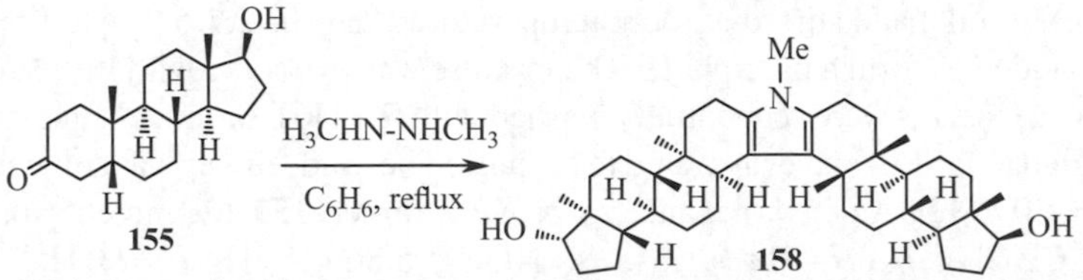

Scheme 2.1.66 *Synthesis of 5β-androstanolone diol dimer* ***158***

To synthesize 5α-androstanolone diol dimer (**156**), a mixture of 5α-androstanolone (**154**, 2.9 g), *N*,*N*′-dimethylhydrazine (420 mg) in C_6H_6 (25 mL) and glacial AcOH (40 mg) were refluxed for 3 h (*Scheme 2.1.65*). The reaction mixture was quenched with boiling H_2O, and extracted with C_6H_6. The organic layer was separated, dried ($MgSO_4$) and evaporated to dryness to give crude dimer, which was crystallized from DMF to obtain diol dimer **156** (2.1 g, 73%, mp: 258–263 °C) (Sucrow and Chondromatidis, 1970).

In the next step, dimer **156** was acetylated using Ac_2O in C_5H_5N to achieve 5α-androstanolone diacetate dimer (**157**, 39 mg, mp: 189–194 °C) (*Scheme 2.1.65*). ^{1}H NMR (100 MHz, $CDC1_3$): δ 4.61 (t, 2H, 2 × 1-CH), 3.28 (s, 3H, CH_3-N), 2.04 (s, 6H, 2 × 17β-MeCO), 0.81 (s, 6H, 2 × 19-Me), 0.76 (s, 6H, 2 × 18-Me) (Sucrow and Chondromatidis, 1970).

Following the same protocol as outlined for **156**, but utilizing 5β-androstanolone (**155**, 2.0 g), 5β-androstanolone diol dimer (**158**) was obtained (*Scheme 2.1.66*), and purified by crystallization from MeOH yielded **158** (1.45 g, 73%, mp: 209–213 °C). ^{1}H NMR (100 MHz, $CDC1_3$): δ 3.60 (t, 2H, 2 × 17α-CH), 3.28 (s, 3H, CH_3-N), 2.30–2.50 [m, 6H, 2 × CH_2 and 2 × $(CH_3)_3CH$], 1.03 (s, 6H, 2 × 19-Me), 0.76 (s, 6H, 2 × 18-Me) (Sucrow and Chondromatidis, 1970).

Valverde *et al*. (2006) reported the synthesis of pregnenolone dimer **161** *via* ring A–ring A connection through a spacer in three steps from pregnenolone (**67**). The reaction involved the preparation of pregnenolone hemisuccinate (**159**) using succinic anhydride and followed by formation of pregnenolone-3β-yl-*N*-(2-amino-ethyl)-succinamic acid (**160**), both of which were utilized to produce the dimer **161**.

Pregnenolone hemisuccinate (**159**) was accomplished from a solution of **67** (200 mg, 0.95 mmol) in C_5H_5N (3 mL), refluxing with succinic anhydride (142 mg, 1.42 mmol) and toluene (10 mL) for 8 h (*Scheme 2.1.67*). The mixture was cooled to room temperature and concentrated *in vacuo* to a smaller volume that was diluted with H_2O and extracted with $CHCl_3$. The organic solvent was evaporated to dryness and the residue was purified by crystallization (*n*-hexane:MeOH:H_2O = 1:2:1) to afford **159** (235 mg, 80%, mp: 165 °C). IR (KBr): ν_{max} cm^{-1} 3505br (O–H), 1700s (C=O). UV (MeOH): λ_{max} nm 215. ^{1}H NMR (300 MHz, CDC1$_3$): δ 10.0 (br, 1H, COOH), 5.33 (d, 1H, J = 4.5 Hz, 6-CH), 3.42 (m, 1H, 3α-CH), 2.35 (m, 5H), 2.06 (s, 3H, J = 6.1 Hz, 21-Me), 1.93–2.19 (m, 2H), 1.58-1.73 (m, 4H), 1.45–1.57 (m, 3H), 1.12-1.35 (m, 4H), 1.01 (s, 3H, 19-Me), 0.92-1.14 (m, 1H), 0.62 (s, 3H, 18-Me). ^{13}C NMR (125 MHz, $CDCl_3$): δ 37.0 (C-1), 27.8 (C-2), 74.0 (C-3), 36.5 (C-4), 139.6 (C-5), 122.7 (C-6), 32.6 (C-7), 31.8 (C-8), 49.9 (C-9), 36.7 (C-10), 21.9 (C-11), 38.8

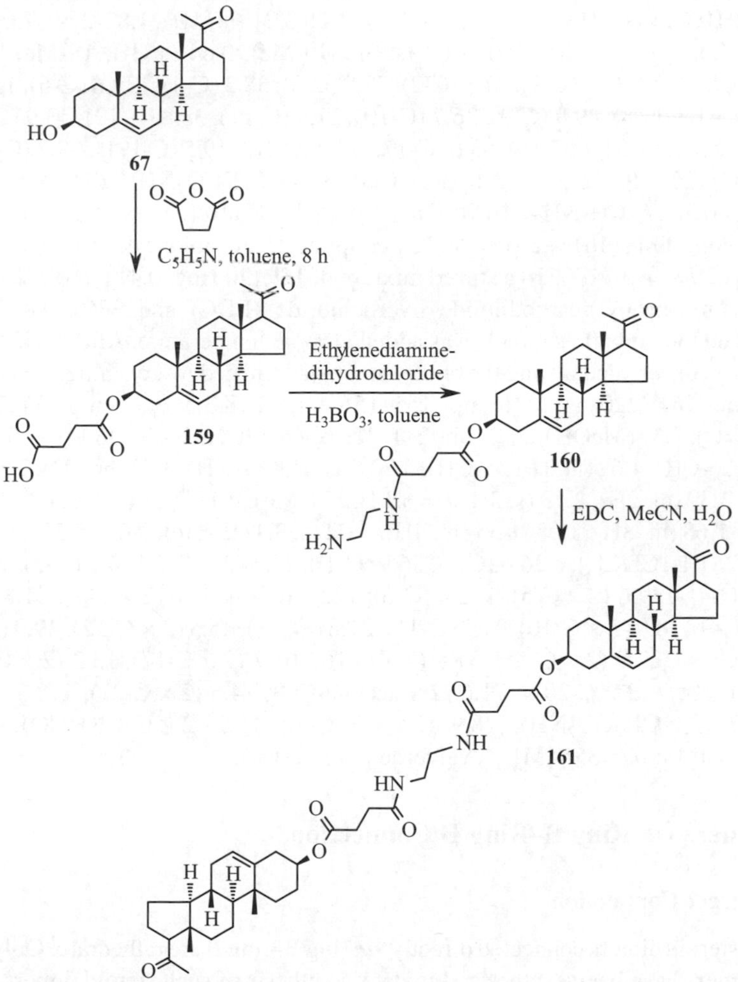

Scheme 2.1.67 *Synthesis of pregnenolone dimer* ***161***

(C-12), 43.9 (C-13), 56.8 (C-14), 22.9 (C-15), 31.4 (C-16), 63.6 (C-17), 13.4 (C-18), 19.5 (C-19), 209.7 (C-20), 173.9 (acetate CO), 23.6 (C-21), 177.3 (acid CO), 29.5 (CH_2CO). EIMS: *m/z* 416 $[M]^+$ (Valverde *et al.*, 2006).

Pregnenolone-3β-yl-*N*-(2-amino-ethyl)-succinamic acid (**160**) was synthesized from **159** (*Scheme 2.1.67*). A mixture of ethylenediamne dihydrochloride (82 mg, 0.62 mmol), H_3BO_3 (26 mg, 0.42 mmol) and **159** (200 mg, 0.41 mmol) in toluene (10 mL) was refluxed for 5 h, the reaction was allowed to cool to room temperature, and the solvent was rotary evaporated. The reduced residue was dissolved in EtOAc (10 mL) and washed with H_2O (20 mL). The organic layer was separated and the aqueous layer was extracted with EtOAc (10 mL). The pooled organic layers were washed with H_2O, dried (Na_2CO_3), evaporated under vacuum and the crude solid was purified by crystallization from 25% H_2O in MeOH to give **160** (120 mg, 22%, mp: 212–213 °C). IR (KBr): ν_{max} cm-1 3360m (N–H), 1685s (C=O). UV (MeOH): λ_{max} nm 218. ^{1}H NMR (300 MHz, $CDCl_3$): δ 5.39 (d, 1H, $J = 4.5$ Hz, 6-CH), 4.62 (m, 1H, 3α-CH), 3.72 (br, 1H, CO_2NH), 3.63 (br, 1H), 2.55-2.71 (m, 5H), 2.30-2.39 (m, 2H), 2.13 (s, 3H, J = 6.1 Hz, 21-Me), 2.11-2.21 (m, 1H), 1.82-2.09 (m, 3H), 1.39-1.68 (m, 8H), 1.08-1.35 (m, 4H), 1.02 (s, 3H, 19-Me), 0.64 (s, 3H, 18-Me). ^{13}C NMR (75 MHz, $CDCl_3$): δ 37.0 (C-1), 27.8 (C-2), 73.9 (C-3), 38.3 (C-4), 139.6 (C-5), 122.7 (C-6), 32.7 (C-7), 31.8 (C-8), 49.9 (C-9), 36.7 (C-10), 21.9 (C-11), 38.8 (C-12), 43.9 (C-13), 56.7 (C-14), 22.9 (C-15), 31.4 (C-16), 63.6 (C-17), 13.7 (C-18), 19.5 (C-19), 209.7 (C-20), 173.9 (aceate CO), 23.6 (C-21), 177.3 (amide CO), 30.4 (**C**H_2CO_2NH), 29.5 (**C**H_2CO), 41.9 (CH_2NHCO), 42.7 (CH_2NH_2). EIMS: *m/z* 458 $[M]^+$ (Valverde *et al.*, 2006).

Prenenolone dimer **161** was prepared by coupling pregnenolone derivatives **159** and **160** as follows (*Scheme 2.1.67*). To a stirred mixture of **159** (200 mg, 0.41 mmol), 1-ethyl-3-(3-dimethyl aminopropyl)-carbodiimide hydrochloride (EDC) and 33% H_2O in MeCN (10 mL), **160** (303 mg, 0.72 mmol) was added at room temperature. After 36 h, the solvent was removed under vacuum and the crude dimer was purified by crystallization as above to obtain dimer **161** (220 mg, 55%, mp: 155–156 °C). IR (KBr): ν_{max} cm^{-1} 3447m (N–H), 1685s (C=O). UV (MeOH): λ_{max} nm 211. ^{1}H NMR (300 MHz, $CDCl_3$): δ 5.39 (d, 1H, $J = 4.5$ Hz, 6-CH), 4.62 (m, 1H, 3α-CH), 3.72 (br, 1H, CO_2NH), 3.63 (br, 1H), 2.55-2.71 (m, 5H), 2.30-2.39 (m, 2H), 2.13 (s, 3H, $J = 6.1$ Hz, 21-Me), 2.11-2.21 (m, 1H), 1.82–2.09 (m, 3H), 1.39–1.68 (m, 8H), 1.08–1.35 (m, 4H), 1.02 (s, 3H, 19-Me), 0.64 (s, 3H, 18-Me). ^{13}C NMR (125 MHz, $CDCl_3$): δ 36.6 (C-1), 36.9 (C-1′), 27.6 (2 × C-2), 74.3 (C-3), 74.4 (C′-3), 38.8 (2 × C-4), 139.6 (2 × C-5), 122.4 (C-6), 122.5 (C′-6), 32.0 (2 × C-7), 31.8 (2 × C-8), 50.0 (2 × C-9), 38.0 (2 × C-10), 21.2 (C-11), 21.8 (C′-11), 45.5 (2 × C-12), 49.9 (2 × C-13), 56.8 (2 × C-14), 22.9 (2 × C-15), 31.8 (2 × C-16), 63.7 (2 × C-17), 13.2 (2 × C-18), 19.3 (2 × C-19), 209.7 (2 × C-20), 171.6 (2 × acetate CO), 24.5 (2 × C-21), 177.2 (2 × amide CO), 29.2 (2 × **C**H_2CO_2NH), 28.9 (2 × **C**H_2CO), 31.5 (2 × CH_2NHCO), 44.0 (2 × CH_2NH_2). EIMS: *m/z* 856 $[M]^+$ (Valverde *et al.*, 2006).

2.2 Dimers *via* Ring B–Ring B Connection

2.2.1 Direct Connection

Synthetic steroid dimers connected directly *via* ring B–ring B are rather rare. Only a handful of such dimers have been synthesized to date. Synthesis of such steroid dimers connected through C-7 and C-7′ could be achieved by irradiation or by a reductive process of two

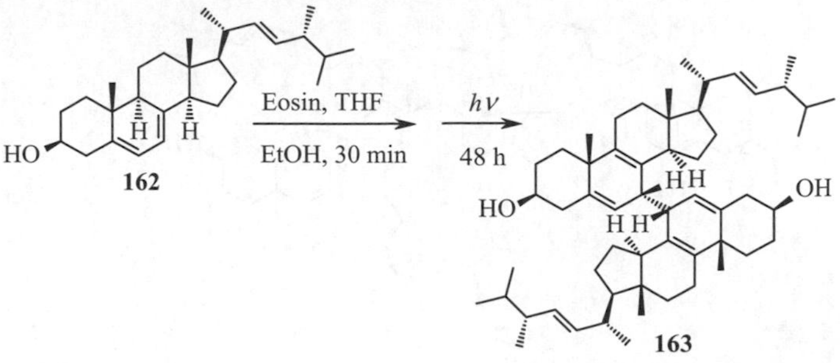

Scheme 2.2.1 Synthesis of 7α,7′α- bisergostatrienol (163)

steroidal units having unsaturated B-rings. For example, 7α,7′α-bisergostatrienol (**163**) was prepared with 46% yield by irradiation of (22*E*)-ergosta-5,7,22-trien-3β-ol (**162**, ergosterol) (Mosettig and Scheer, 1952).

To a mixture of ergosterol (**162**, 60 g) and eosin (60 g) in THF (800 mL), 95% EtOH (1.6 L) was added and the mixture was kept in the dark for 30 min under N_2, after which the reaction mixture was irradiated for 48 h (*Scheme 2.2.1*). The solution was diluted with 95% EtOH (5 L) and the crystals were filtered and successively washed with EtOH and Et_2O, dried under pressure at 50–60 °C to yield 7α,7′α-bisergostatrienol (**163**, 46 g, 77%, mp: 201–203 °C) (Mosettig and Scheer, 1952).

A mixture of isomeric dimers, 7α,7′α-bisergostatrienol (**163**), 7α,7′β-bisergostatrienol (**164**) and 7β,7′β-bisergostatrienol (**165**), were achieved by irradiation of ergosterol (**162**) (*Scheme 2.2.2*) (Crabbe and Mislow, 1968).

7α,7′α-Bisergostatrienol (**163**, 34%, mp: 189–190 °C): ^{1}H NMR (100 MHz, $CDCl_3$): δ 5.20, 3.40, 3.05, 1.19, 1.09, 1.02, 0.97, 0.89, 0.84, 0.82, 0.72 (Crabbe and Mislow, 1968).

7α,7′β-Bisergostatrienol (**164**, 4.4%, mp: 177–179 °C): ^{1}H NMR (100 MHz, $CDCl_3$): δ 6.30, 3.70, 3.20, 1.32, 1.04, 0.99, 0.92, 0.85 (Crabbe and Mislow, 1968).

7β,7′β-Bisergostatrienol (**165**, 14.5%, mp: 193–194 °C): ^{1}H NMR (100 MHz, $CDCl_3$): δ 3.50, 1.07, 1.00, 0.94, 0.87, 0.86, 0.80, 0.79, 0.70 (Crabbe and Mislow, 1968).

Mosettig and Scheer (1952) reported the synthesis of bis(methyl 3β-acetyloxybisnorchola-5,7-dien-22-oate) (**168**) and bis(3β-acetyloxy-22-isospirosta-5,7-diene) (**169**), respectively, from methyl 3β-acetyloxy-bisnorchola-5,7-dien-22-oate (**166**) and 3β-acetyloxy-22-isospirosta-5,7-diene (**167**) following the same procedure as described for dimer **163**.

Bis(methyl 3β-acetyloxybisnorchola-5,7-dien-22-oate) (**168**) was accomplished from the reaction mixture of **166** (20 g) and eosin (20 g) in THF (100 mL) and 95% EtOH (700 mL) in the dark for 30 min, followed by irradiation for 90 h (*Scheme 2.2.3*). The precipitate was filtered off, washed with EtOH, and dried under vacuum at 50–60 °C to afford dimer **168** (12.8 g, 64%, mp: 200–201 °C). IR (CS_2): ν_{max} cm^{-1} 1736s (C=O), 1238s (C–O) (Mosettig and Scheer, 1952).

Following the similar method as described for dimer **168**, bis(3β-acetyloxy-22-isospirosta-5,7-diene) (**169**) was achieved from the reaction of 3β-acetyloxy-22-isospirosta-5,7-diene (**167**, 20 g) and eosin (20 g) in THF (200 mL) and 95% EtOH (500 mL) was kept in the dark for 30 min, followed by irradiation for 7 days (*Scheme 2.2.4*). The reaction

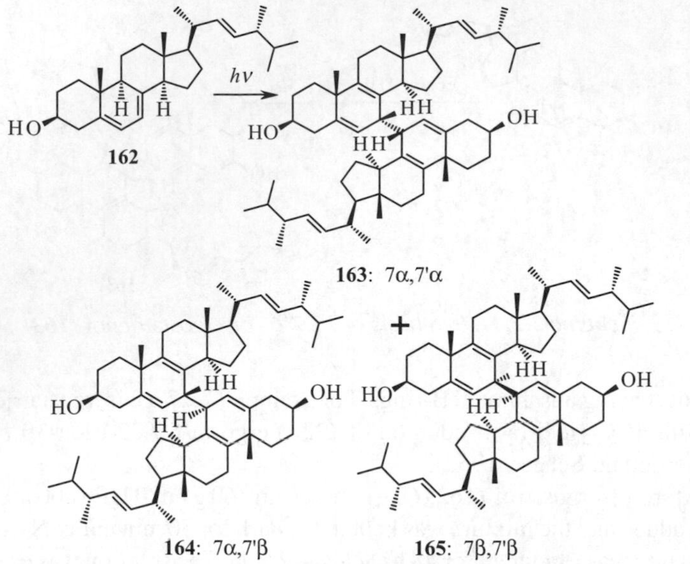

Scheme 2.2.2 *Synthesis of 7α,7′α-bisergostatrienol (**163**), 7α,7′β-bisergostatrienol (**164**) and 7β,7′β-bisergostatrienol (**165**) isomers*

mixture was poured into 95% EtOH (1.5 L), the crystalline precipitate was filtered, washed with EtOH, and recrystallized from C_6H_6-EtOH to provide **169** (14 g, 70%, mp: 212 °C). IR (CS_2): ν_{max} cm^{-1} 1736s (C=O), 1239s (C–O) (Mosettig and Scheer, 1952).

The synthesis of dimeric ketone, 6β,6β′-bicholesta-4,4′-diene-3,3′-dione (**170**) was carried out by oxidation of cholesterol (**1**) with freshly prepared Ag_2O (Stohrer, 1971). Cholesterol (**1**) was recrystallized twice from AcOH, dried over KOH *in vacuo*, and the dried **1** (20 g) and Ag_2O (80 g) were refluxed in toluene (300 mL) for 3 days (*Scheme 2.2.5*). The brown solution was separated and extracted with $CHCl_3$ and the pooled extracts were evaporated under pressure. The solid was extracted with Me_2CO leaving colloidal silver residue (1.2 g). The crude dimer was then purified by CC, eluted with 5% MeOH in heptane, to produce dimer **170** (1.8 g, mp: 214–215 °C), which was further employed to synthesize another cholesterol dimer, 3β,3′β-bisdinitrophenylhydrazone-6β,6′β-bicholesta-4,4′-diene (**171**), (Stohrer, 1971).

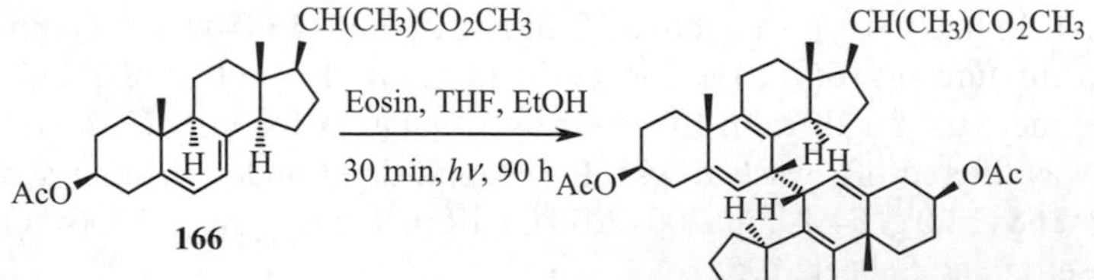

Scheme 2.2.3 *Synthesis of bis(methyl 3β-acetyloxybisnorchola-5,7-dien-22-oate) (**168**)*

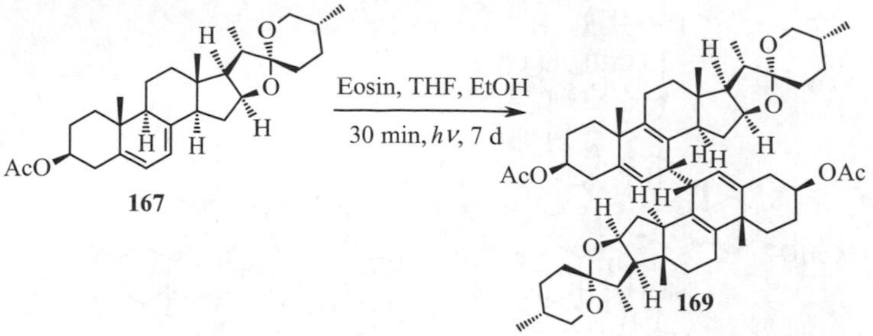

Scheme 2.2.4 *Synthesis of bis(3β-acetyloxy-22-isospirosta-5,7-diene) (**169**)*

Bicholestadienedione (**170**, 30 mg) in EtOH (50 mL) was treated with a solution of 2,4-DNP (80 mg) in EtOH (25 mL) at 100 °C (*Scheme 2.2.5*). After the addition of two drops of conc. HCl, a precipitate was formed. The product was initially purified by CC on Al_2O_3 (mobile phase: 20% $CHCl_3$ in C_6H_6), and the resulting crude dimer was further purified by crystallization from dioxane to give bisdinitrophenylhydrazone **171** (66%, mp: 247–248 °C) (Stohrer, 1971).

The dimer **170** was also utilized to synthesize 3β,3′β-dihydroxy-6β,6′β-bicholesta-4,4′-diene (**172**) (*Scheme 2.2.6*). $LiAlH_4$ (1.0 g) was added to a stirred solution of **170** (600 mg) in Et_2O (500 mL) at 25 °C. After 5 h, the excess hydride was destroyed with conc. $(NH_4)_2SO_4$, the solvent was rotary evaporated to yield the crude diol that was recrystallized twice from Et_2O to obtain diol **172** (400 mg, mp: 281–283 °C) (Stohrer, 1971), which could easily be acetylated by Ac_2O in C_5H_5N to produce 3β,3′β-diacetoxy-6β,6′β-bicholesta-4,4′-diene (**173**) (*Scheme 2.2.6*). Recrystallization of the crude dimer from $CHCl_3$-EtOH furnished acetylated dimer **173** (29 mg, mp: 209–210 °C) (Stohrer, 1971).

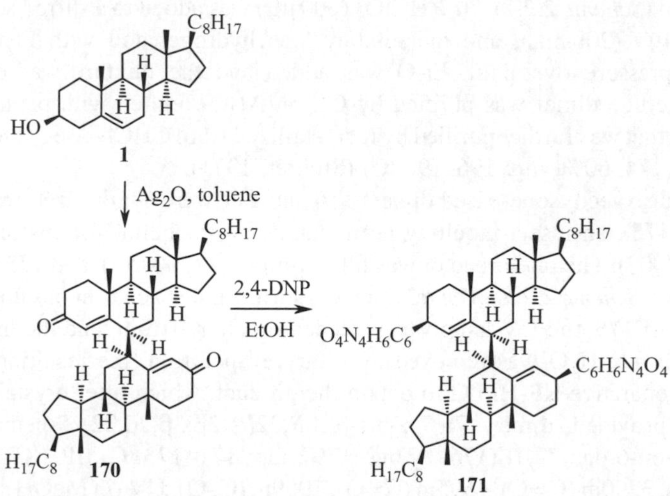

Scheme 2.2.5 *Synthesis of 6β,6′β-bicholesta-4,4′-diene-3,3′-dione (**170**) and 3β,3′β-bisdinitrophenylhydrazone-6β,6′β-bicholesta-4,4′-diene (**171**)*

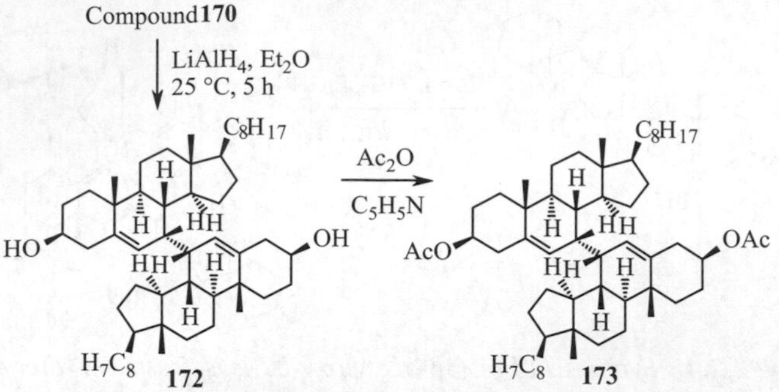

Scheme 2.2.6 *Synthesis of 3β,3′β-dihydroxy-6β,6′β-bicholesta-4,4′-diene (**172**) and 3β,3′β-diacetoxy-6β,6′β-bicholesta-4,4′-diene (**173**)*

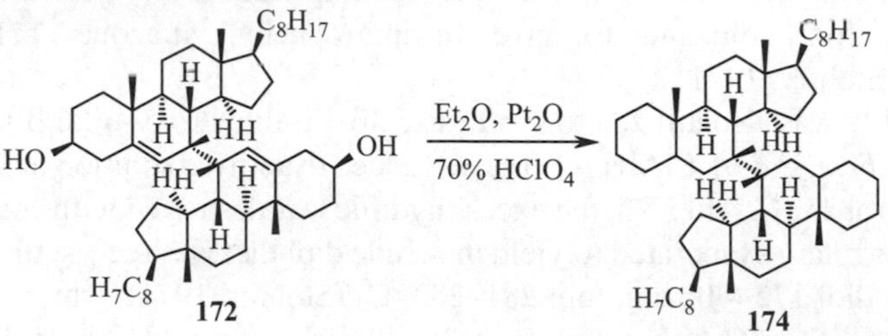

Scheme 2.2.7 *Synthesis of 6β,6′β-bicholestane (**174**)*

The synthesis of 6β,6′β-bicholestane (**174**) was accomplished from the dimer **172** by Pt_2O reduction (*Scheme 2.2.7*). 70% $HClO_4$ (200 μL) was added to a stirred solution of **172** (77 mg) in Et_2O (200 mL), and the mixture was hydrogenated with Pt_2O (40 mg) at atmospheric pressure overnight. Et_2O was added and the mixture was evaporated to dryness. The crude dimer was purified by CC on Al_2O_3, eluted with pentane, to afford crude product that was further purified by recrystallized from $CHCl_3$-MeOH to give 6β,6′β-bicholestane (**174**, 60%, mp: 196–198 °C) (Stohrer, 1971).

Two 20-hydroxyecdysone-based dimers **176** and **177** were synthesized from 20-hydroxyecdysone (**175**), an insect-moulting hormone, by photochemical transformation (Harmatha *et al.*, 2002). The photoreactor was filled with H_2O (500 mL), and **175** (500 mg) was added under Ar (*Scheme 2.2.8*). The mixture was stirred under a continuous flow of Ar. After dissolving all of **175**, the UV lamp was switched on after 10 min, and the irradiation was carried out for 4 h. H_2O was removed by rotary evaporation, the resulting residue was purified by preparative RP-HPLC to obtain the product, which after crystallization from MeOH-H_2O provided dimer, 7α,7′α-bis-[(20*R*,22*R*)-2β,3β,20,22,25-pentahydroxy-5β-cholest-8(14)-en-6-one-7-yl] (**176**, 87 mg, 17%, mp: 171–173 °C). IR (KBr): ν_{max} cm^{-1} 3485br (O–H), 3290m (C=C), 1695m (C=O), 1059m (C–O). UV (in MeOH): λ_{max} nm 210, 242, 310. ^{13}C NMR (125 MHz, C_5D_5N): δ 37.1 (2 × C-1), 67.7 (2 × C-2), 68.5 (2 × C-3), 33.7 (2 × C-4), 55.5 (2 × C-5), 214.7 (2 × C-6), 58.3 (2 × C-7), 123.8 (2 × C-8), 32.7 (2 ×

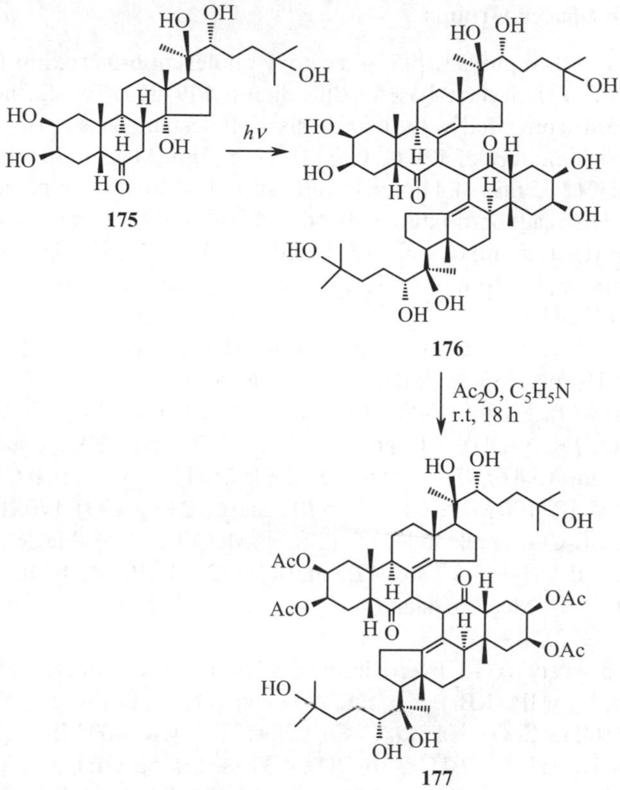

Scheme 2.2.8 *Synthesis of 20-hydroxyecdysone-based dimers **176** and **177***

C-9), 40.6 (2 × C-10), 20.8 (2 × C-11), 36.1 (2 × C-12), 43.4 (2 × C-13), 151.9 (2 × C-14), 26.0 (2 × C-15), 22.7 (2 × C-16), 57.3 (2 × C-17), 19.4 (2 × C-18), 25.3 (2 × C-19), 76.7 (2 × C-20), 21.9 (2 × C-21), 77.0 (2 × C-22), 27.2 (2 × C-23), 41.6 (2 × C-24), 69.9 (2 × C-25), 29.8 (2 × C-26), 30.3 (2 × C-27). HRFABMS: *m/z* calculated 949.6111 for $C_{54}H_{86}O_{12}Na$, found: 949.6017 (Harmatha *et al.*, 2002).

7α,7′α-Bis-[(20*R*,22*R*)-2β,3β-diacetoxy-20,22,25-trihydroxy-5β-cholest-8(14)-en-6-one-7-yl] (**177**) was obtained from dimer **176** by simple acetylation using Ac_2O in C_5H_5N and at room temperature overnight (*Scheme 2.2.8*). Crystallization of the crude acetylated dimer from MeOH-H_2O afforded dimer **177** (mp: 182–188 °C). IR (KBr): ν_{max} cm^{-1} 3430br (O–H), 1745s (C=O), 1702s (C=O), 1248m (C–O), 1231m (C–O), 1045m (C–O). ^{13}C NMR (125 MHz, CH_3OD): δ 34.2 (2 × C-1), 69.9 (2 × C-2), 69.3 (2 × C-3), 31.2 (2 × C-4), 56.7 (2 × C-5), 214.7 (2 × C-6), 58.9 (2 × C-7), 123.7 (2 × C-8), 33.5 (2 × C-9), 41.4 (2 × C-10), 21.3 (2 × C-11), 36.7 (2 × C-12), 44.0 (2 × C-13), 153.8 (2 × C-14), 26.7 (2 × C-15), 23.0 (2 × C-16), 57.5 (2 × C-17), 19.7 (2 × C-18), 25.1 (2 × C-19), 78.4 (2 × C-20), 21.3 (2 × C-21), 77.5 (2 × C-22), 27.3 (2 × C-23), 41.3 (2 × C-24), 71.6 (2 × C-25), 28.9 (2 × C-26), 29.9 (2 × C-27). HRFABMS: *m/z* calculated 1117.6439 for $C_{62}H_{94}O_{16}Na$, found: 1117.6283 (Harmatha *et al.*, 2002).

2.2.2 Through Spacer Groups

Duddeck *et al.*, (1992) employed 3β,5β-diacetoxy-cholestan-6-one oxime (**178**) to produce a complex mixture of cholestane-based oxime dimers **179–181**. Two further oxime dimers **182** and **183** were also prepared by the hydrolysis of cholestane-based oxime dimers **179** and **181**, respectively. A mixture of **178** (1.16 g, 2.23 mmol) in MeOH (30 mL), Et_2O (15 mL) and 7.5% NaOH (2.35 mL, 4.4 mmol) were stirred at room temperature for 30 min (*Scheme 2.2.9*). The reaction mixture was diluted with H_2O, saturated aqueous $NaHCO_3$ was added to the reaction mixture and extracted with Et_2O-C_6H_6. The crude product was purified by step gradient CC (mobile phase: 15–25% Et_2O in C_6H_6) to give cholestane-based oxime dimers **179–181**.

(6*E*,6′*E*)-6-(3′β-Acetyloxy-6′-hydroximino-cholestan-5′α-yl-oximino)-cholestan-3β,5β-yl diacetate (**179**, 197 mg, 18%, mp: 188–193 °C). IR (KBr): ν_{max} cm^{-1} 3480br (O–H), 1750s (C=O), 1254m (C–O), 1168m (C–O), 1053m (C–O), 1025m, 920m. ^{1}H NMR (400 MHz, $CDCl_3$): δ 7.77 (s, 1H, N-OH), 5.14 and 4.81 (m, 2H, 2 × 3α-CH), 3.24 and 3.07 (m, 2H, 2 × 7β-CH), 1.92 and 1.94 (s, 9H, 3 × MeCO), 0.89 (s, 6H, 2 × 19-Me), 0.69 and 0.62 (s, 6H, 2 × 18-Me). ^{13}C NMR (100 MHz, $CDCl_3$): δ 70.2 and 68.2 (2 × C-3), 170.1, 170.0 and 169.6 (2 × 3β-CO and 5β-CO), 22.1 and 21.2 (2 × 3β-MeCO and 5β-MeCO), 85.9 and 84.0 (2 × C-5), 161.5 and 154.9 (2 × C-6), 42.5 and 41.7 (2 × C-10), 43.0 and 42.9 (2 × C-13), 12.1 and 11.9 (2 × 18-Me), 16.4 and 14.8 (2 × 19-Me). LSIMS: *m*/*z* 998 $[M + Na]^+$ (Duddeck *et al.*, 1992).

(6*E*,6′*E*)-6-(3′β-Acetyloxy-6′-hydroximino-cholestan-5′α-yl-oximino)-cholestane-3β,5β-diol (**180**, 39 mg, 4%): IR (KBr): ν_{max} cm^{-1} 3465br (O–H), 1747s (C=O), 1250m (C–O), 1168m (C–O), 1100m (C–O), 1053m (C–O), 920w. ^{1}H NMR (400 MHz, $CDCl_3$): δ 7.70 (s, 1H, N-OH), 4.60 (d, 1H, *J* = 9.4 Hz, 3β-OH), 4.37 (s, 1H, 5β-OH), 3.99 and 4.74 (m, 2H, 2 × 3α-CH), 3.12 (m, 2H, 2 × 7-CH), 2.25 and 2.19 (dd, 2H, *J* = 4.5, 13.6 Hz/3.3, 14.5 Hz, 2 × 4-CH), 1.95 (s, 3H, MeCO), 0.90 and 0.72 (s, 6H, 2 × 19-Me), 0.69 and 0.63 (s, 6H, 2 × 18-Me). ^{13}C NMR (100 MHz, $CDCl_3$): δ 69.9 and 66.5 (2 × C-3), 170.0 (3β-CO), 21.3 (3β-MeCO), 86.1 and 77.8 (2 × C-5), 162.2 and 160.2 (2 × C-6), 42.5 and 41.9 (2 × C-10), 43.0 (2 × C-13), 12.2 and 12.1 (2 × 18-Me), 16.6 and 14.9 (2 × 19-Me). LSIMS: *m*/*z* 914 $[M + Na]^+$ (Duddeck *et al.*, 1992).

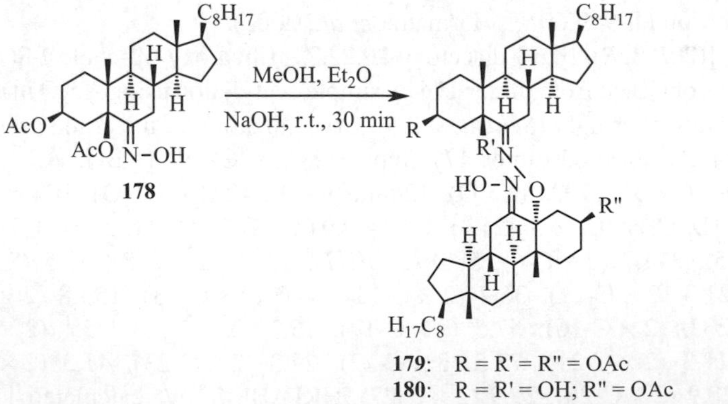

MeOH, Et_2O / NaOH, r.t., 30 min

178

179: R = R' = R'' = OAc
180: R = R' = OH; R'' = OAc
181: R = OH; R' = $\Delta^{4,5}$; R'' = OAc

Scheme 2.2.9 *Synthesis of cholestane-based oxime dimers* ***179–181***

(6*E*,6′*E*)-6-(3′β-Acetyloxy-6′-hydroximino-cholestan-5′α-yl-oximino)-cholest-4-en-3β-ol (**181**, 99 mg, 10%, mp: 197-201 °C): IR (KBr): ν_{max} cm^{-1} 3410br (O–H), 1744s (C=O), 1248m (C–O), 1170m (C–O), 1052m (C–H), 1027m, 920w. ^{1}H NMR (400 MHz, $CDCl_3$): δ 9.39 (s, 1H, N-OH), 5.60 (s, 1H, 4-CH), 4.83 and 4.06 (m, 2H, 2 × 3α-CH), 3.75 (d, 1H, J = 6.8 Hz, 3β-OH), 3.23 and 3.11 (dd, 2H, J = 4.4, 14.9 Hz/4.0, 13.1 Hz, 2 × 7-CH), 2.33 (dd, 2H, J = 3.8, 13.2 Hz, 2 × 4-CH), 1.98 (s, 3H, MeCO), 0.91 (s, 6H, 2 × 19-Me), 0.70 and 0.62 (s, 6H, 2 × 18-Me). ^{13}C NMR (100 MHz, $CDCl_3$): δ 69.9 and 67.1 (2 × C-3), 170.1 (3β-CO), 21.5 (3β-MeCO), 128.3 (C-4), 141.6 and 85.2 (2 × C-5), 161.0 and 160.5 (2 × C-6), 42.1 and 38.0 (2 × C-10), 43.1 and 42.7 (2 × C-13), 12.2 and 12.0 (2 × 18-Me), 18.7 and 14.8 (2 × 19-Me). LSIMS: m/z 896 [M + Na]$^{+}$ (Duddeck *et al.*, 1992).

The cholestane-based oxime dimer **179** (58 mg) was dissolved in 20% Et_2O in MeOH (3.5 mL) and treated with aqueous NaOH (600 μL) at room temperature for 50 min to obtain the crude diol (*Scheme 2.2.10*), which was further purified by CC, eluted with 35% C_6H_6 in Et_2O, to provide a diol dimer, (6*E*,6′*E*)-6-(3′β-acetyloxy-6′-hydroximino-cholestan-5′α-yl-oximino)-cholestane-3β,5β-diol (**182**, 52 mg). IR (KBr): ν_{max} cm^{-1} 3460br (O–H), 1167m (C–O), 1100m (C–O), 1013m, 918w. ^{1}H NMR (400 MHz, $CDCl_3$): δ 8.12 (s, 1H, NOH), 4.63 (d, 2H, J = 9 Hz, 2 × 3β-OH), 4.42 (s, 1H, 5β-OH), 3.99 and 3.66 (m, 2H, 2 × 3α-CH), 3.10 (m, 2H, 2 × 7-CH), 2.31 and 2.18 (dd, 2H, J = 3.8, 13.7 Hz/3.4, 14.4 Hz, 2 × 4-CH), 0.89 and 0.76 (s, 6H, 2 × 19-Me), 0.64 (s, 6H, 2 × 18-Me). ^{13}C NMR (100 MHz, $CDCl_3$): δ 67.2 and 66.5 (2 × C-3), 86.4 and 77.7 (2 × C-5), 161.0 and 160.5 (2 × C-6), 42.3 and 41.8 (2 × C-10), 43.0 and 42.9 (2 × C-13), 12.2 and 12.1 (2 × 18-Me), 16.9 and 15.0 (2 × 19-Me). LSIMS: m/z 872 [M + Na]$^{+}$ (Duddeck *et al.*, 1992).

A similar hydrolysis of the cholestane oxime dimer **181** (54 mg) provided crude diol **183** (*Scheme 2.2.10*), which was further purified by CC (mobile phase: 20% C_6H_6 in Et_2O) to afford another dimer, (6*E*,6′*E*)-6-(3′β-hydroxy-6′-hydroximino-cholestan-5′α-yl oximino)-cholest-4-en-3β-ol (**183**, 45 mg, mp: 220–223 °C). IR (KBr): ν_{max} cm^{-1} 3420br (O–H), 1165m (C–O), 1085m (C–O), 1000m, 920w. ^{1}H NMR (400 MHz, $CDCl_3$): δ 9.60 (s, 1H,

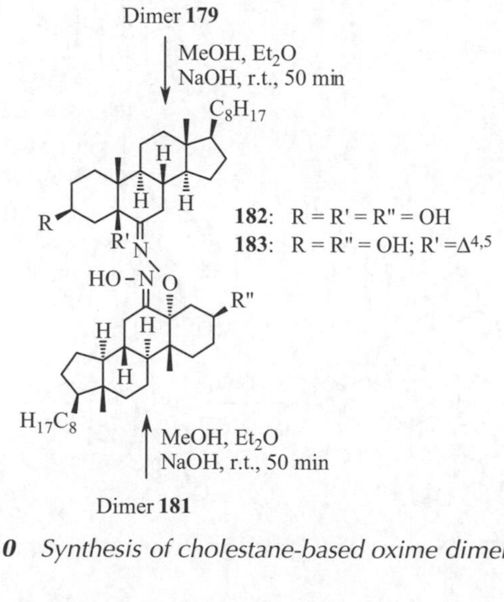

Dimer **179**
MeOH, Et_2O
NaOH, r.t., 50 min

182: R = R' = R'' = OH
183: R = R'' = OH; R' = $\Delta^{4,5}$

MeOH, Et_2O
NaOH, r.t., 50 min
Dimer **181**

Scheme 2.2.10 *Synthesis of cholestane-based oxime dimers **182** and **183***

NOH), 5.61 (s, 1H, 4-CH), 4.05 and 3.72 (m, 2H, 2 × 3α-CH), 3.81 and 2.94 (s, 2H, 2 × 3β-OH), 3.15 and 3.09 (dd, 2H, $J = 3.8$, 15.1 Hz/3.8, 13.8 Hz, 2 × 7β-CH), 0.91 and 0.88 (s, 6H, 2 × 19-Me), 0.67 and 0.62 (s, 6H, 2 × 18-Me). ^{13}C NMR (100 MHz, $CDCl_3$): δ 66.9 and 66.8 (2 × C-3), 128.4 (C-4), 141.5 and 85.6 (2 × C-5), 160.8 and 159.8 (2 × C-6), 41.9 and 37.8 (2 × C-10), 43.0 and 42.7 (2 × C-13), 12.2 and 12.1 (2 × 18-Me), 18.6 and 14.9 (2 × 19-Me). LSIMS: *m/z* 854 $[M + Na]^+$ (Duddeck *et al*., 1992).

Blunt *et al*. (1966) reported the synthesis of several cholestane-based ether dimers **185–189** from 3β-acetyloxy-5α,6α-epoxycholestane (**184**). To a stirred solution of **184** (24 g) in dry C_6H_6 (30 mL), redistilled $BF_3.Et_2O$ (24 mL) was added and the mixture was kept at 20 °C for 45 s. The reaction mixture was poured into mixture of dry Et_2O and saturated aqueous $NaHCO_3$. The organic solvent was separated, evaporated under pressure and crystallized from Et_2O to obtain dimer **185** (6.4 g, mp: 181–183 °C). IR (CS_2): ν_{max} cm^{-1} 3460br (O–H), 1740s (C=O), 1242s (C–O). ^{1}H NMR (100 MHz, $CDCl_3$): δ 5.15 (2H, 2 × 3α-CH), 2.97 (2H, 2 × 6β-CH), 2.05 (6H, 2 × MeCO), 1.15 (3H, 19′-Me), 1.01 (6H, 5β-Me and 21-Me), 0.91 (6H, 14β-Me and 21′-Me), 0.90 (6H, 27-Me and 27′-Me), 0.81 (3H, 26′-Me), 0.80 (3H, 26-Me), 0.67 (3H, 18′-Me) (Blunt *et al*., 1966).

A mixture of dimer **185** (9.4 g) and KOH (6 g) in MeOH (600 mL) was refluxed for 15 min (*Scheme 2.2.11*). On cooling, a solid mass was separated out from the solution, the crude solid was purified by recrystallization from Me_2CO to achieve dimer **186** (8.6 g, mp: 136–137 183 °C). IR (nujol): ν_{max} cm^{-1} 3390br (O–H), 1736s (C=O), 1250s (C–O). ^{1}H NMR (100 MHz, $CDCl_3$): δ 5.11 (1H, 3α-CH), 4.00 (1H, 3′α-CH), 3.13 (1H, 6′β-CH), 2.93 (1H, 6β-CH), 2.05 (3H, 2 × MeCO), 1.15 (3H, 19′-Me), 0.99 (3H, 5β-Me), 0.98 (3H,

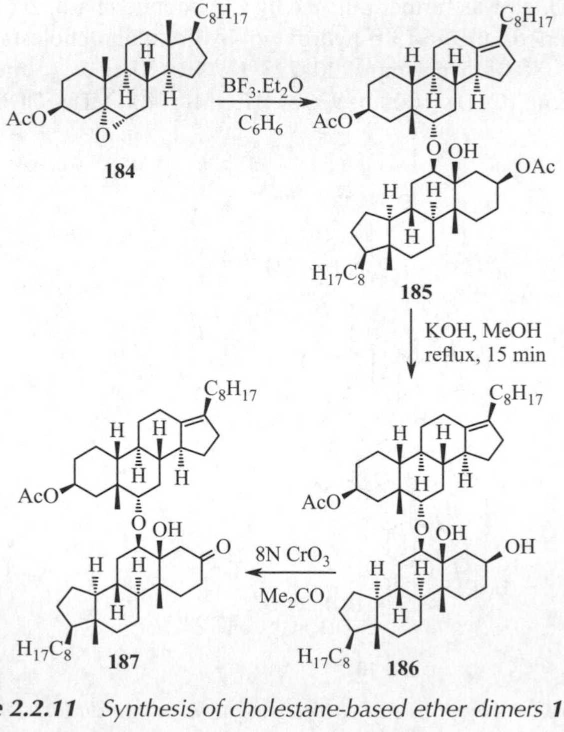

Scheme 2.2.11 *Synthesis of cholestane-based ether dimers* ***185–187***

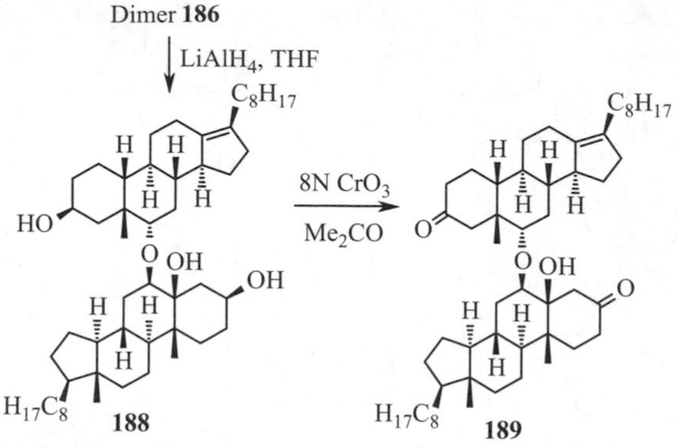

Scheme 2.2.12 *Synthesis of cholestane-based ether dimers* ***188*** *and* ***189***

21-Me), 0.91 (3H, 14β-Me), 0.90 (3H, 27′-Me), 0.88 (6H, 21′-Me and 27-Me), 0.81 (3H, 26′-Me), 0.80 (3H, 26-Me), 0.67 (18′-Me) (Blunt *et al*., 1966).

Dimer **186** could easily be oxidized to its keto product **187** by the treatment with 8N CrO_3 in Me_2CO (*Scheme 2.2.11*). The crude dimer was recrystallized from Me_2CO to produce dimer **187** (6.3 g, mp: 209–210 °C). IR (CS_2): ν_{max} cm^{-1} 3390br (O–H), 1733s (C=O), 1235s (C–O). ^{1}H NMR (100 MHz, $CDCl_3$): δ 5.10 (1H, 3α-CH), 4.00 (1H, 3′α-CH), 3.11 (1H, 6′β-CH), 2.95 (1H, 6β-CH), 2.00 (6H, 2 × 3β-MeCO), 1.31 (3H, 19′-Me), 1.00 (6H, 5β-Me and 21′-Me), 0.91 (3H, 27′-Me), 0.90 (6H, 14β-Me and 21-Me), 0.88 (3H, 27-Me), 0.81 (3H, 26′-Me), 0.80 (3H, 26-Me), 0.68 (3H, 18′-Me) (Blunt *et al*., 1966).

$LiAlH_4$ reduction of the dimer **186** afforded the triol dimer **188** (*Scheme 2.2.12*). A solution of **186** (3.78 g) in dry Et_2O (150 mL) was treated with $LiAlH_4$ (2.0 g) at 35 °C for 45 min. Extraction with DCM produced dimer **188** (23 g, mp: 233–237 °C). IR (nujol): ν_{max} cm^{-1} 3390br (O–H). ^{1}H NMR (100 MHz, $CDCl_3$): δ 4.20 (3α-CH), 4.17 (3′α-CH), 3.15 (6′β-CH), 2.93 (6β-CH), 1.11 (19′-Me), 1.00 (5β-Me and 21-Me), 0.91 (14β-Me), 0.90 (21′-Me, 26′-Me and 26-Me), 0.82 (27′-Me), 0.80 (27-Me), 0.67 (18′-Me) (Blunt *et al*., 1966). In a similar manner as described for dimer **187**, dimer **188** could be oxidized to its diketone dimer **189**, by the treatment with 8N CrO_3 in Me_2CO (*Scheme 2.2.12*). IR (nujol): ν_{max} cm^{-1} 3390br (O–H), 1715s (C=O) (Blunt *et al*., 1966).

In a similar fashion as described for the synthesis of the ether dimer **185**, the BF_3-catalyzed rearrangement of cholestane monomers, 3β-acetyloxy-5α,6α-epoxycholestane (**184**) and 10β-ethyl-5α,6α-epoxycholestane (**190**), furnished the backbone rearranged dimer **191** with a 26% yield (*Scheme 2.2.13*). ^{1}H NMR (100 MHz, $CDCl_3$): δ 5.80–7.20 (α-vinyl-2H), 4.60–5.50 (β-vinyl-4H and 2 × 3α-CH), 3.34 and 3.24 (2H, 2 × 6β-CH), 0.94 (d, 3H, $J = 6.4$ Hz, 20-Me) (Guest *et al*., 1971).

Edelsztein *et al*. (2009) reported the synthesis of pregnenolone homodimers, 1,4-bis-(6β-hydroxy-20-oxo-pregnan-6-yl)-2*E*-butene (**198**) and (*E*)-1,2-bis-(3β,20β-diacetoxypregn-5-en-19-yl)ethene (**199**) from pregnenolone acetate (**192**, 3β-acetyloxypregn-5-en-20-one) by olefin metathesis. In order to synthesize these dimers, 20β-acetyloxy-5α-pregn-2-en-6-one (**193**) was synthesized from pregnenolone acetate (**192**, 200 mg, 0.56 mmol) according to the method described in the literature (Masakazu *et al*., 1987) (*Scheme 2.2.14*). Usual cleanup and purification as described above gave **193** (164 mg, 82%). IR (KBr): ν_{max} cm^{-1} 2942m (C–H), 2906m (C–H), 1730s (C=O), 1708s (C=O), 1433m (C–H), 1373m (C–H),

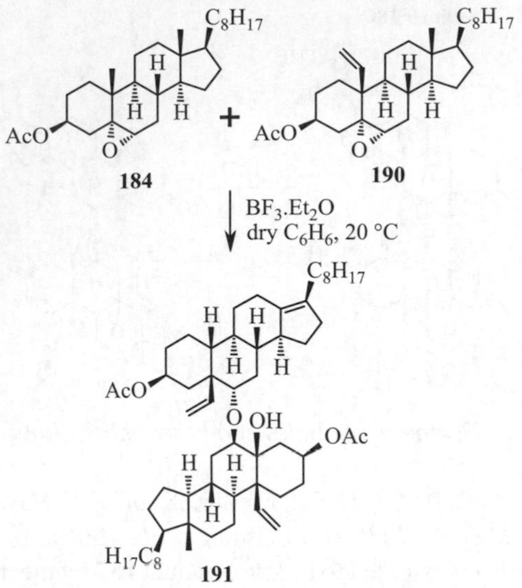

Scheme 2.2.13 *Synthesis of cholestane-based ether dimer* ***191***

1075m, 1024m. ^{1}H NMR (500 MHz, $CDCl_3$): δ 5.69 (m, 1H, 3-CH), 5.56 (m, 1H, 2-CH), 4.85 (br s, 1H, 20α-CH), 2.02 (s, 3H, 20β-MeCO), 1.16 (d, $J = 5.8$ Hz, 3H, 21-Me), 0.71 (s, 3H, 19-Me), 0.65 (s, 3H, 18-Me). ^{13}C NMR (125 MHz, $CDCl_3$): δ 39.4 (C-1), 125.0 (C-2), 124.5 (C-3), 21.6 (C-4), 53.5 (C-5), 211.6 (C-6), 46.9 (C-7), 37.5 (C-8), 53.9 (C-9), 40.0 (C-10), 21.0 (C-11), 39.0 (C-12), 42.6 (C-13), 56.1 (C-14), 24.0 (C-15), 25.2 (C-16), 54.9 (C-17), 12.5 (C-18), 13.5 (C19), 72.7 (C-20), 170.3 (20β-CO), 21.5 (20β-MeCO), 19.9 (C-21). EIMS: *m/z* 381 $[M + Na]^+$ (Edelsztein *et al.*, 2009).

20β-Acetyloxy-5α-pregnan-6-one (**194**) was obtained from 20β-acetyloxy-5α-pregn-2-en-6-one (**193**) by hydrogenation (*Scheme 2.2.14*). Pd-C (15 mg, 10% on carbon) was added to a stirred solution of **193** (150 mg, 0.42 mmol) in EtOAc (15 mL), and the solution was hydrogenated at 3 bar pressure. After 8 h, the catalyst was filtered off and the crude product was purified by CC, eluted with *n*-hexane-EtOAc, to yield **194** (148 mg, 98%, mp: 163–165 °C). IR (KBr): ν_{max} cm^{-1} 2981m (C–H), 2937m (C–H), 2873m (C–H), 1732s (C=O), 1646s (C=C), 1372m (C–H), 1244m (C–O), 1075m, 1019m, 877m. ^{1}H NMR (500 MHz, $CDCl_3$): δ 4.83 (dq, 1H, $J = 6.0$, 10.6 Hz, 20α-CH), 2.28 (dd, 1H, $J = 4.5$, 13.1 Hz, 7-CHa), 2.13 (dd, 1H, $J = 3.2$, 12.3 Hz, 5-CH), 2.02 (s, 3H, 20β-MeCO), 1.15 (d, 3H, $J = 6.1$ Hz, 21-Me), 0.72 (s, 3H, 19-Me), 0.62 (s, 3H, 18-Me). ^{13}C NMR (125 MHz, $CDCl_3$): δ 38.2 (C-1), 21.4 (C-2), 24.0 (C-3), 20.5 (C-4), 58.9 (C-5), 212.5 (C-6), 46.8 (C-7), 37.8 (C-8), 54.4 (C-9), 41.8 (C-10), 21.1 (C-11), 39.0 (C-12), 42.8 (C-13), 56.2 (C-14), 25.2 (C-15), 25.3 (C-16), 54.9 (C-17), 12.5 (C-18), 13.1 (C-19), 72.7 (C-20), 170.4 (20β-CO), 21.5 (20β-MeCO), 19.9 (C-21). EIMS: *m/z* 360 $[M]^+$ (Edelsztein *et al.*, 2009).

A mixture of isomeric diol dimers, 6β-(1-propenyl)-5α-pregnane-20β,6α-diol (**195**) and 6α-(1-propenyl)-5α-pregnane-20β,6β-diol (**196**), was prepared from 20β-acetyloxy-5α-pregnan-6-one (**194**) as follows (*Scheme 2.2.14*). A solution of **194** (200 mg, 0.56 mmol) in THF (800 μL) was added at 0 °C to a stirred solution of allylmagnesium chloride (1.2 mL,

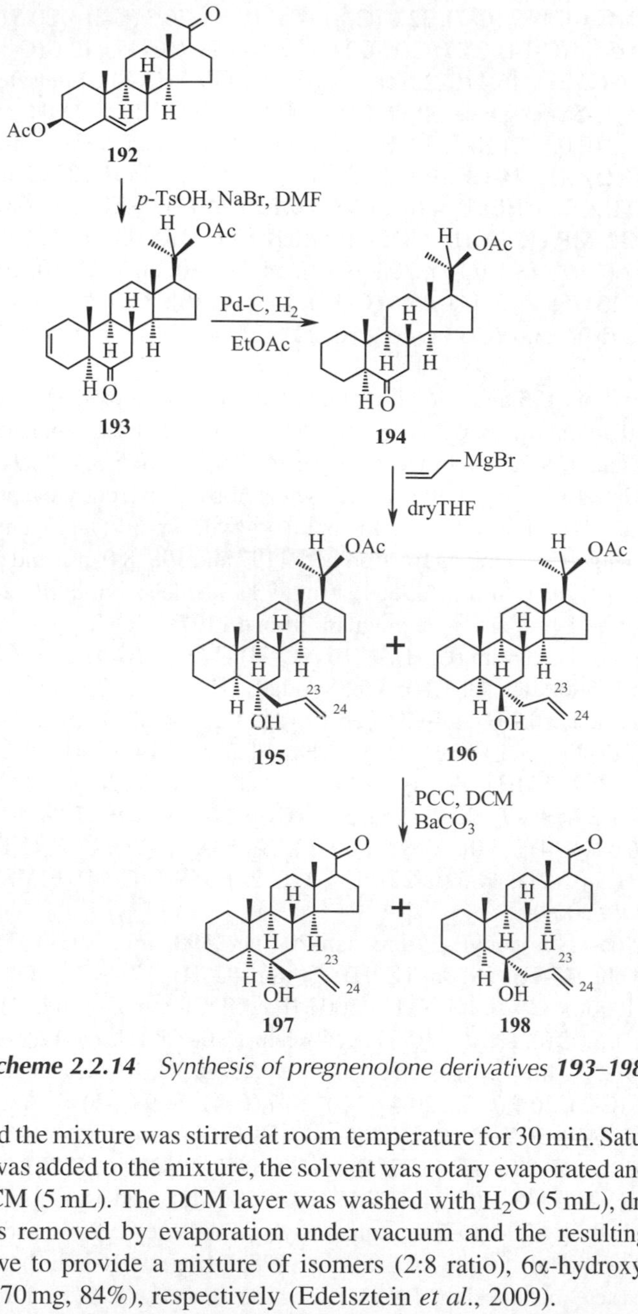

Scheme 2.2.14 *Synthesis of pregnenolone derivatives **193–198***

2M in THF), and the mixture was stirred at room temperature for 30 min. Saturated aqueous NH_4Cl (1 mL) was added to the mixture, the solvent was rotary evaporated and the solid was dissolved in DCM (5 mL). The DCM layer was washed with H_2O (5 mL), dried (Na_2SO_4), the solvent was removed by evaporation under vacuum and the resulting residue was purified as above to provide a mixture of isomers (2:8 ratio), 6α-hydroxy **195** and 6β-hydroxy **196** (170 mg, 84%), respectively (Edelsztein *et al.*, 2009).

6β-(1-Propenyl)-5α-pregnane-20β,6α-diol (**195**): ^{1}H NMR (500 MHz, $CDCl_3$): δ 5.86 (dddd, 1H, J = 6.9, 8.1, 9.9, 17.4 Hz, 23-CH), 5.19 (dd, 1H, J = 2.3, 10.2 Hz, 24-CHa), 5.14 (dd, 1H, J = 2.3, 17.1 Hz, 24-CHb), 2.29 (dd, 1H, J = 6.7, 13.7 Hz, 1H, 22-CHa), 1.97 (dd, 1H, J = 8.1, 13.7 Hz, 22-CHb), 1.13 (d, 3H, J = 6.1 Hz, 21-Me), 1.12 (s, 3H, 19-Me), 0.76 (s, 3H, 18-Me). ^{13}C NMR (125 MHz, $CDCl_3$): δ 38.8 (C-1), 20.9 (C-2), 26.9 (C-3), 25.0 (C-4),

52.2 (C-5), 75.0 (C-6), 39.7 (C-7), 32.0 (C-8), 41.1 (C-9), 36.3 (C-10), 20.5 (C-11), 40.3 (C-12), 42.7 (C-13), 55.8 (C-14), 25.7 (C-15), 24.5 (C-16), 58.7 (C-17), 12.6 (C-18), 27.7 (C-19), 70.6 (C-20), 23.6 (C-21), 46.4 (C-22), 133.4 (C-23), 119.4 (C-24) (Edelsztein *et al.*, 2009).

6α-(1-Propenyl)-5α-pregnane-20β,6β-diol (**196**): ^{1}H NMR (500 MHz, $CDCl_3$): δ 5.76 (ddt, 1H, $J = 7.5$, 10.0, 17.1 Hz, 23-CH), 5.07 (dd, 1H, $J = 2.2$, 9.9 Hz, 24-CHa), 5.05 (dd, 1H, $J = 2.5$, 16.8 Hz, 1H, 24-CHa), 2.22 (dd, 1H, $J = 7.9$ and 14 Hz, 22-CHa), 2.17 (dd, 1H, $J = 7.7$ and 13.7 Hz, 22-CHb), 1.14 (d, 3H, $J = 6.1$ Hz, 21-Me), 1.02 (s, 3H, 19-Me), 0.77 (s, 3H, 18-Me). ^{13}C NMR (125 MHz, $CDCl_3$): δ 40.8 (C-1), 21.8 (C-2), 26.9 (C-3), 20.4 (C-4), 50.9 (C-5), 73.7 (C-6), 43.1 (C-7), 30.8 (C-8), 54.3 (C-9), 36.7 (C-10), 20.5 (C-11), 40.2 (C-12), 42.5 (C-13), 55.7 (C-14), 25.6 (C-15), 24.4 (C-16), 58.6 (C-17), 12.6 (C-18), 15.5 (C-19), 70.6 (C-20), 23.6 (C-21), 46.9 (C-22), 134.4 (C-23), 118.0 (C-24) (Edelsztein *et al.*, 2009).

The isomeric diols **195** and **196** (170 mg, 0.47 mmol) were oxidized with a mixture of PCC (253 mg, 0.94 mmol), $BaCO_3$ (172 mg, 0.71 mmol), and 4 Å molecular sieves in dry DCM (15 mL) under N_2 at room temperature for 45 min (*Scheme 2.2.14*). The reaction mixture was diluted with Et_2O, the solvent was removed by rotary evaporation, and the resulting residue was purified as above, to obtain 6α-hydroxy-6β-(1-propenyl)-5α-pregnan-20-one (**197**, 15 mg, 9%), a mixed fraction both (**197** and **198**, 8.0 mg) and 6β-hydroxy-6α-(1-propenyl)-5α-pregnan-20-one (**198**, 121 mg, 72%) (Edelsztein *et al.*, 2009).

6α-Hydroxy-6β-(1-propenyl)-5α-pregnan-20-one (**197**): IR (KBr): ν_{max} cm^{-1} 3406br (O–H), 2920m (C–H), 2848m (C–H), 1716s (C–H), 1457m (C–H), 1261s (C–O), 1096m (C–O). ^{1}H NMR (500 MHz, $CDCl_3$): δ 5.88 (dddd, 1H, $J = 5.5$, 7.7, 9.9, 17.3 Hz, 23-CH), 5.22 (dd, 1H, $J = 2.1$, 10.2 Hz, 24-CHb), 5.17 (dd, 1H, $J = 2.0$, 17.1 Hz, 24-CHa), 2.56 (t, 1H, $J = 8.9$ Hz, 17-CH), 2.14 (s, 3H, $J = 6.1$ Hz, 21-Me), 1.14 (s, 3H, 19-Me), 0.65 (s, 3H, 18-Me). ^{13}C NMR (125 MHz, $CDCl_3$): δ 39.2 (C-1), 26.9 (C-2), 20.9 (C-3), 20.6 (C-4), 46.4 (C-5), 74.9 (C-6), 39.6 (C-7), 30.9 (C-8), 52.1 (C-9), 36.3 (C-10), 22.9 (C-11), 38.8 (C-12), 41.1 (C-13), 56.6 (C-14), 25.0 (C-15), 24.5 (C-16), 63.9 (C-17), 13.5 (C-18), 27.6 (C-19), 209.9 (C-20), 31.5 (C-21), 44.5 (C-22), 133.2 (C-23), 119.6 (C-24). EIMS: *m/z* 358 $[M]^+$ (Edelsztein *et al.*, 2009).

6β-Hydroxy-6α-(1-propenyl)-5α-pregnan-20-one (**198**, mp: 105–107 °C). IR (KBr): ν_{max} cm^{-1} 3406br (O–H), 2920m (C–H), 2848m (C–H), 1716s (C=O), 1457m (C–H), 1261s (C–O), 1096m (C–O). ^{1}H NMR (500 MHz, $CDCl_3$): δ 5.75 (ddt, 1H, $J = 7.5$, 10.0, 17.0 Hz, 23-CH), 5.08 (d, 1H, $J = 10.0$ Hz, 24-CHb), 5.04 (d, 1H, $J = 17.0$ Hz, 24-CH), 2.12 (s, 3H, $J = 6.1$ Hz, 21-Me), 1.01 (s, 3H, 19-Me), 0.63 (s, 3H, 18-Me). ^{13}C NMR (125 MHz, $CDCl_3$): δ 40.8 (C-1), 26.9 (C-2), 20.4 (C-3), 20.7 (C-4), 50.9 (C-5), 73.6 (C-6), 43.0 (C-7), 31.0 (C-8), 54.2 (C-9), 36.7 (C-10), 21.8 (C-11), 39.1 (C-12), 44.2 (C-13), 56.5 (C-14), 24.4 (C-15), 22.8 (C-16), 63.9 (C-17), 13.5 (C-18), 15.5 (C-19), 209.6 (C-20), 31.5 (C-21), 46.9 (C-22), 134.3 (C-23), 118.1 (C-24). EIMS: *m/z* 358 $[M]^+$ (Edelsztein *et al.*, 2009).

1,4-Bis-(6β-hydroxy-20-oxo-5α-pregnan-6-yl)-2*E*-butene (**199**) was achieved from 6β-hydroxy-6α-(1-propenyl)-5α-pregnan-20-one (**198**, 60 mg, 0.016 mmol) using the general microwave-assisted metathesis (MWAM) reaction (*Scheme 2.2.15*). Purification of the crude dimer on CC using the same solvent system afforded pregnenolone homodimer **199** (53 mg, 97%, mp: 240–245 °C). IR (KBr): ν_{max} cm^{-1} 3401br (O–H), 2931m (C–H), 2906m (C–H), 1696s (C=C), 1197m (C–O), 1144m (C–O), 750m. ^{1}H NMR (500 MHz, $CDCl_3$): δ 5.38 (dd, 2H, $J = 3.4$, 5.1 Hz, 2 × 23-CH), 2.32 (dd, 2H, $J = 4.5$ and 13.5 Hz, 2 × 22-CHa), 2.14 (s, 6H, $J = 6.1$ Hz, 2 × 21-Me), 1.98 (dd, 2H, $J = 6.3$ and 13.7 Hz, 2 × 22-CHb), 1.05 (s,

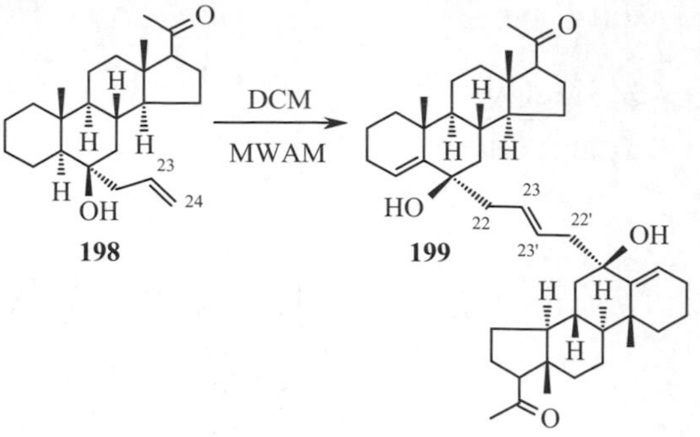

Scheme 2.2.15 *Synthesis of 1,4-bis-(6β-hydroxy-20-oxo-5α-pregnan-6-yl)-2E-*butene (**199**)

6H, 2 × 19-Me), 0.67 (s, 6H, 2 × 18-Me). ^{13}C NMR (125 MHz, $CDCl_3$): δ 40.9 (2 × C-1), 21.8 (2 × C-2), 27.0 (2 × C-3), 20.6 (2 × C-4), 50.2 (2 × C-5), 74.3 (2 × C-6), 42.7 (2 × C-7), 31.0 (2 × C-8), 54.6 (2 × C-9), 36.7 (2 × C-10), 20.7 (2 × C-11), 39.3 (2 × C-12), 44.2 (2 × C-13), 56.8 (2 × C-14), 24.4 (2 × C-15), 22.8 (2 × C-16), 64.0 (2 × C-17), 13.5 (2 × C-18), 15.5 (2 × C-19), 209.4 (2 × C-20), 31.5 (2 × C-21), 45.6 (2 × C-22), 129.5 (2 × C-23). ESIHRMS: *m/z* 711.5328 calculated for $C_{46}H_{72}O_4Na$, found: 711.5323 (Edelsztein *et al.*, 2009).

The synthesis of two easily separable isomeric testosterone homodimers, *trans*-T_2 (**204**) and *cis*-T_2 (**205**), was reported (Bastien *et al.*, 2010). These dimers were synthesized from testosterone (**200**, 17β-hydroxyandrost-4-en-3-one, a male sex hormone) in five steps with a moderate yield (36%). Later, dimers *trans*-T_2 (**204**) and *cis*-T_2 (**205**) were simply hydrolyzed to deacetyl-*trans*-T_2 (**206**) and deacetyl-*cis*-T_2 (**207**) with high yields (95%), respectively. Testosterone (**200**) was first converted to androsta-3,5-dien-3,17β-yl diacetate (**201**) (*Scheme 2.2.16*). To a stirred solution of **200** (5.57 g, 19.3 mmol) in C_5H_5N (1.82 mL, 19.3 mmol), AcCl (20.9 mL, 281.53 mmol) and Ac_2O (6.24 mL, 77.2 mmol) were added and refluxed for 4 h, and stirred for 30 min at room temperature. The reaction mixture was rotary evaporated and the crude solid was dissolved in DCM, filtered through a silica gel pad, and the solvent was evaporated under pressure to yield diacetate **201** (6.46 g, 90%). IR (NaCl): ν_{max} cm^{-1} 1736s (C=O), 1666s (C=C), 1248m (C–O). ^{1}H NMR (200 MHz, $CDCl_3$): δ 5.67 (s, 1H, 4-CH), 5.36 (m, 1H, 6-CH), 4.59 (t, 1H, J = 8.2 Hz, 17α-CH), 2.11 (s, 3H, 3-MeCO), 2.02 (s, 3H, 17β-MeCO), 0.99 (s, 3H, 19-Me), 0.81 (s, 3H, 18-Me). ^{13}C NMR (50 MHz, $CDCl_3$): δ 12.3, 19.1, 20.9, 21.3, 21.4, 23.7, 25.0, 27.7, 31.6, 31.8, 33.9, 35.2, 36.9, 42.7, 48.1, 51.4, 82.9, 117.1, 123.7, 139.7, 147.3, 169.6, 171.4. EIMS: *m/z* 372 $[M]^+$. HREIMS: *m/z* 372.2300 calculated for $C_{23}H_{32}O_4$, found: 372.2297 (Bastien *et al.*, 2010).

Androsta-3,5-dien-3,17β-yl diacetate (**201**) was converted to androsta-4,6-dien-3-on-17β-yl acetate (**202**) in the following manner (*Scheme 2.2.16*). To a stirred and cooled (0 °C) solution of **201** (6.46 g, 17.3 mmol) in DMF (70 mL) and H_2O (3 mL), NBS was added over a period of 60 min under N_2 and stirred for an additional 40 min at 0 °C. Li_2CO_3 and LiBr were added to the mixture at room temperature, and heated for 4 h at 95 °C. A mixture of ice-H_2O (150 mL) and AcOH (10 mL) was added to the reaction, resulting the formation of solid, which was filtered, washed with H_2O, dried ($MgSO_4$) and purified by FCC, eluted with 10% Me_2CO in *n*-hexane, to produce ketone **202** (4.33 g, 76%). IR (NaCl): ν_{max} cm^{-1} 1735s (C=O), 1664s (C=O), 1613s (C=C), 1252m (C–O). ^{1}H NMR (200 MHz, $CDCl_3$): δ 6.09

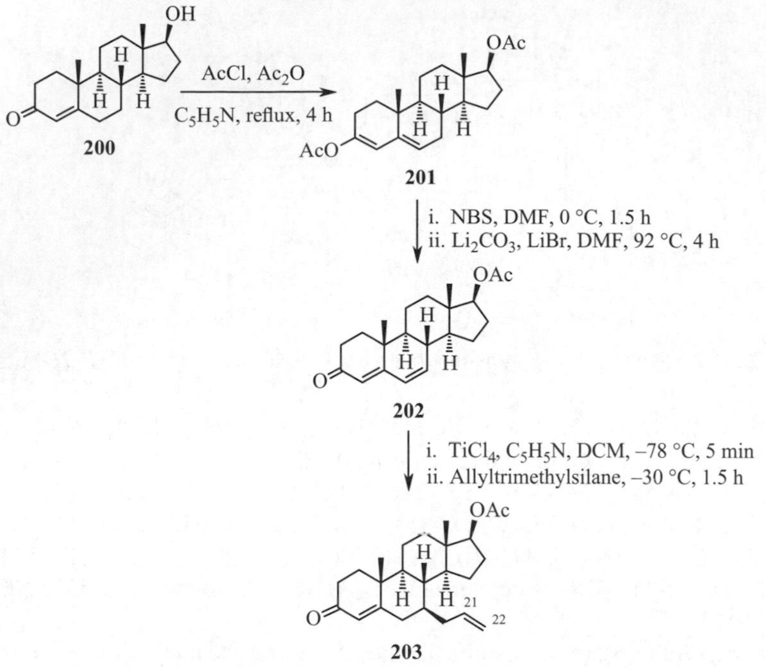

Scheme 2.2.16 *Synthesis of 7α-allyl-androst-4-en-3-on-17β-yl acetate (**203**)*

(s, 2H, 6-CH and 7-CH), 5.65 (s, 1H, 4-CH), 4.61 (t, 1H, $J = 7.8$ Hz, 17α-CH), 2.03 (s, 3H, 17β-MeCO) 1.10 (s, 3H, 19-Me), 0.86 (s, 3H, 18-Me). ^{13}C NMR (50 MHz, $CDCl_3$): δ 12.2, 16.5, 20.4, 21.3, 23.3, 27.7, 34.1, 36.3, 36.7, 36.8, 37.6, 43.6, 48.3, 50.8, 82.3, 124.0, 128.4, 140.3, 163.8, 171.3, 199.3. EIMS: *m*/*z* 328 $[M]^+$. HREIMS: 328.2038 calculated for $C_{21}H_{28}O_3$, found: 328.2032 (Bastien *et al.*, 2010).

Androsta-4,6-dien-3-on-17β-yl acetate (**202**) was transformed to 7α-allyl-androst-4-en-3-on-17β-yl acetate (**203**) (*Scheme 2.2.16*). $TiCl_4$ (3.58 mL, 32.6 mmol) and C_5H_5N (0.65 mL, 6.39 mmol) were added to a stirred solution of ketone **202** (2.38 g) in dry DCM under N_2, cooled to −78 °C, and stirred for 5 min. Allyltrimethylsilane was added to the reaction mixture and stirred for 1.5 h at −78 °C and then 1.5 h at −30 °C. Et_2O was added to the reaction mixture and sequentially washed with 2% HCl (2 × 20 mL) and H_2O (4 × 20 mL). The etheral layer was dried ($MgSO_4$), rotary evaporated and purified as above to afford **203** (1.67 g, 70%). IR (NaCl): ν_{max} cm^{-1} 1736s (C=O), 1678s (C=O), 1616s (C=C), 1243m (C–O). ^{1}H NMR (200 MHz, $CDCl_3$): δ 5.70 (s, 1H, 4-CH), 5.60 (m, 1H, 21-CH), 5.00 (m, 2H, 22-CH_2), 4.60 (t, 1H, $J = 8.4$ Hz, 17α-CH), 2.03 (s, 3H, 17β-MeCO), 1.20 (s, 3H, 19-Me), 0.84 (s, 3H, 18-Me). ^{13}C NMR (50 MHz, $CDCl_3$): δ 12.1, 18.2, 20.9, 21.4, 23.1, 27.6, 30.4, 34.2, 36.1, 36.2, 36.3, 36.7, 38.5, 38.9, 42.8, 46.2, 47.2, 82.6, 117.0, 126.4, 137.0, 169.4, 171.3, 199.3. EIMS: *m*/*z* 370 $[M]^+$. HREIMS: *m*/*z* 370.2508 calculated for $C_{24}H_{34}O_3$, found: 370.2505 (Bastien *et al.*, 2010).

7α-Allyl-androst-4-en-3-on-17β-yl acetate (**203**) was utilized for the synthesis of testosterone homodimers, *trans*-T_2 (**204**) and *cis*-T_2 (**205**) aas follows (*Scheme 2.2.17*). Grubbs 2nd-generation catalyst (90 mg, 0.15 mmol) was added to the solution of **203** (560 mg,

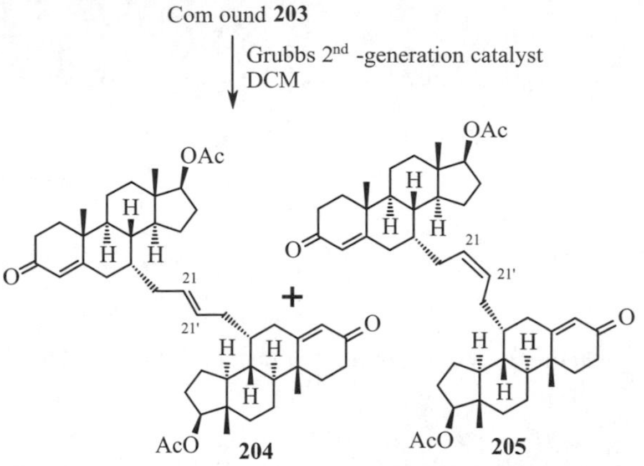

Com ound **203**

Grubbs 2nd -generation catalyst
DCM

204 **205**

Scheme 2.2.17 *Synthesis of testosterone homodimers* trans-T_2 *(**204**) and* cis-T_2 *(**205**)*

1.51 mmol) in dry DCM (8 mL), and refluxed overnight, and then stirred for 30 min at room temperature. After evaporation of the solvent, the crude product was purified by FCC (mobile phase: 5% Me_2CO in *n*-hexane) to give two separable isomeric dimers, *trans*-T_2 (**204**, 260 mg, 50%, mp: 123–126 °C) as the major product, and *cis*-T_2 (**205**, 130 mg, 25%, mp: 241–244 °C) as the minor (Bastien *et al.*, 2010).

Testosterone homodimer *trans*-T_2 (**204**): IR (NaCl): ν_{max} cm^{-1} 1734s (C=O), 1673s (C=O), 1611s (C=C), 1250m (C–O). ^{1}H NMR (200 MHz, $CDCl_3$): δ 5.63 (s, 2H, 2 × 4-CH), 5.14 (m, 2H, 2 × 21-CH), 4.59 (t, 2H, *J* = 8.2 Hz, 2 × 17α-CH), 2.03 (s, 6H, 2 × 17β-MeCO), 1.19 (s, 6H, 2 × 19-Me), 0.83 (s, 6H, 2 × 18-Me). ^{13}C NMR (50 MHz, $CDCl_3$): δ 12.1, 18.2, 20.9, 21.4, 23.1, 27.6, 29.3, 31.2, 34.2, 36.1, 36.4, 36.6, 38.3, 38.9, 42.7, 46.2, 47.1, 82.7, 131.1, 169.8, 171.4, 199.3. HRESIMS: *m*/*z* 713.4776 calculated for $C_{46}H_{65}O_6$, found: 713.4773 (Bastien *et al.*, 2010).

Testosterone homodimer *cis*-T_2 (**205**): IR (NaCl): ν_{max} cm^{-1} 1734s (C=O), 1673s (C=O), 1250m (C–O). ^{1}H NMR (200 MHz, $CDCl_3$): δ 5.61 (s, 2H, 2 × 4-CH), 5.30 (m, 2H, 2 × 21-CH), 4.65 (t, 2H, *J* = 8.2 Hz, 2 × 17α-CH), 2.03 (s, 6H, 2 × 17β-MeCO), 1.17 (s, 6H, 2 × 19-Me), 0.83 (s, 6H, 2 × 18-Me). ^{13}C NMR (50 MHz, $CDCl_3$): δ 12.1, 18.3, 21.0, 21.4, 23.2, 24.7, 27.7, 34.2, 36.1, 36.7, 36.4, 37.0, 38.5, 38.8, 42.7, 46.1, 47.0, 82.8, 126.6, 129.8, 169.8, 171.3, 198.9. HRESIMS: *m*/*z* 713.4776 calculated for $C_{46}H_{65}O_6$, found: 713.4773 (Bastien *et al.*, 2010).

Testosterone homodimers **204** and **205** could be hydrolyzed separately using a 5N HCl solution in MeOH at reflux for 4.5 h, to obtain deacetyl-*trans*-T_2 (**206**) and deacetyl-*cis*-T_2 (**207**), respectively (*Schemes 2.2.18 and 2.2.19*). The crude dimer was washed with a 5% $NaHCO_3$ followed by H_2O. The solvent was dried ($MgSO_4$) and concentrated *in vacuo* to produce dimers without any need for further purification.

Deacetyl-*trans*-T_2 (**205**, 95%, mp: 225–228 °C). IR (NaCl): ν_{max} cm^{-1} 3422br (O–H), 1656s (C=O), 1217m (C–O). ^{1}H NMR (200 MHz, $CDCl_3$): δ 5.65 (s, 2H, 2 × 4-CH), 5.17 (m, 2H, 2 × 21-CH), 3.65 (t, 2H, *J* = 5.1 Hz, 2 × 17α-CH), 1.21 (s, 6H, 2 × 19-Me), 0.80 (s, 6H, 2 × 18-Me). ^{13}C NMR (50 MHz, $CDCl_3$): δ 11.1, 18.2, 21.1, 23.0, 29.3, 29.9, 30.6, 34.2, 36.1, 36.4, 36.6, 38.6, 39.0, 43.1, 46.4, 47.3, 81.9, 126.3, 131.1, 169.9, 199.3. HRESIMS: *m*/*z* 629.4564 calculated for $C_{42}H_{61}O_4$, found: 629.4563 (Bastien *et al.*, 2010).

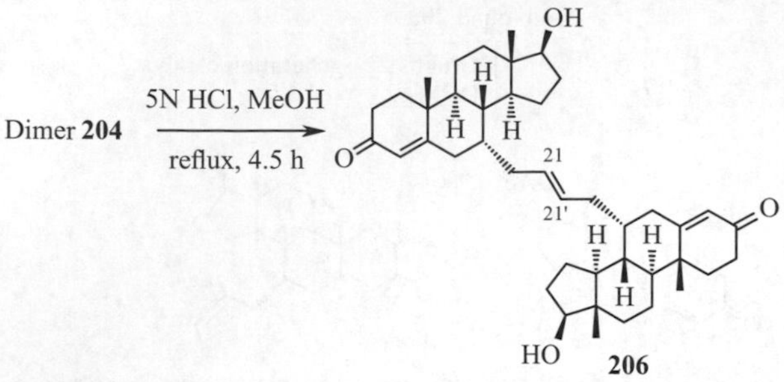
Dimer 204
5N HCl, MeOH
reflux, 4.5 h
OH
H
H
H
O
21
21'
H
H
O
H
HO
206

Scheme 2.2.18 *Synthesis of deacetyl-*trans*-T_2 (**206**)*

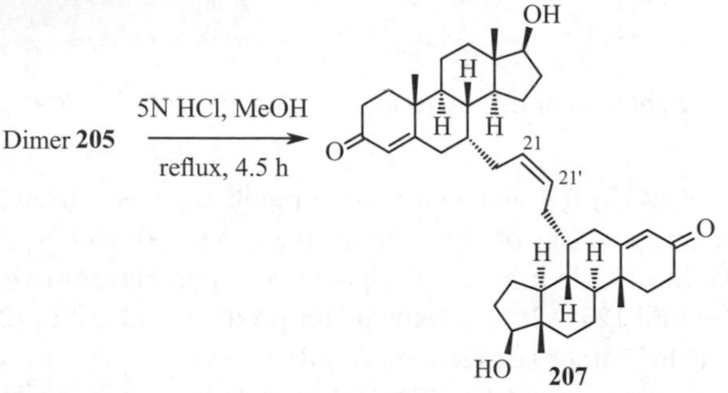
Dimer 205
5N HCl, MeOH
reflux, 4.5 h
OH
H
H
H
O
21
21'
H
H
O
H
HO
207

Scheme 2.2.19 *Synthesis of deacetyl-*cis*-T_2 (**207**)*

Deacetyl-*cis*-T_2 (**206**, 95%, mp: 127–130 °C): IR (NaCl): ν_{max} cm^{-1} 3424br (O–H), 1658m (C=O), 1217m (C–O). ^{1}H NMR (200 MHz, $CDCl_3$): δ 5.60 (s, 2H, 2 × 4-CH), 5.29 (m, 2H, 2 × 21-CH), 3.78 (t, 2H, $J = 6.9$ Hz, 2 × 17α-CH), 1.18 (s, 6H, 2 × 19-Me), 0.79 (s, 6H, 2 × 18-Me). ^{13}C NMR (50 MHz, $CDCl_3$): δ 11.2, 18.3, 21.1, 23.0, 24.9, 29.9, 30.4, 34.2, 36.2, 36.7, 36.9, 38.7, 38.8, 43.1, 46.3, 47.3, 81.6, 126.6, 129.9, 170.5, 199.0. HRESIMS: *m/z* 629.4564 calculated for $C_{42}H_{61}O_4$, found: 629.4558 (Bastien *et al.*, 2010).

2.3 Dimers *via* Ring C–Ring C Connection

2.3.1 Through Spacer Groups

A symmetrical cholic acid-based dimer, bishydrazone **210** through a hydrazone spacer group *via* ring C-ring C, was prepared in three steps starting from methyl cholate (**51**) (Gao and Dias, 1998). First, methyl ester **51** was acetylated to methyl 3α,7α-diacetoxy-12α-hydroxy-5β-cholan-24-oate (**208**, 86%, mp: 183–185 °C) (*Scheme 2.3.1*). ^{1}H NMR (250 MHz, $CDCl_3$): δ 4.89 (s, 1H, 7β-CH), 4.58 (br, 1H, 3β-CH), 3.99 (s, 1H, 12β-CH), 3.66 (s, 3H, 24-OMe), 2.30 (m, 2H, 23-CH_2), 2.03 (s, 3H, 7α-OAc), 2.02 (s, 3H, 3α-OAc), 0.97 (d, 3H, $J = 6.1$ Hz, 21-Me), 0.92 (s, 3H, 19-Me), 0.68 (s, 3H, 18-Me). EIMS *m/z*: 506 $[M]^+$ (Gao and Dias, 1998).

Compound **208** (in AcOH) was then oxidized to methyl 3α,7α-diacetoxy-12-oxo-5β-cholan-24-oate (**209**, 94%, mp: 179–181 °C), using the oxidizing agent K_2CrO_4 according to the protocol reported by Fieser and Rajagopalan (1950) (*Scheme 2.3.1*). ^{1}H NMR (250 MHz,

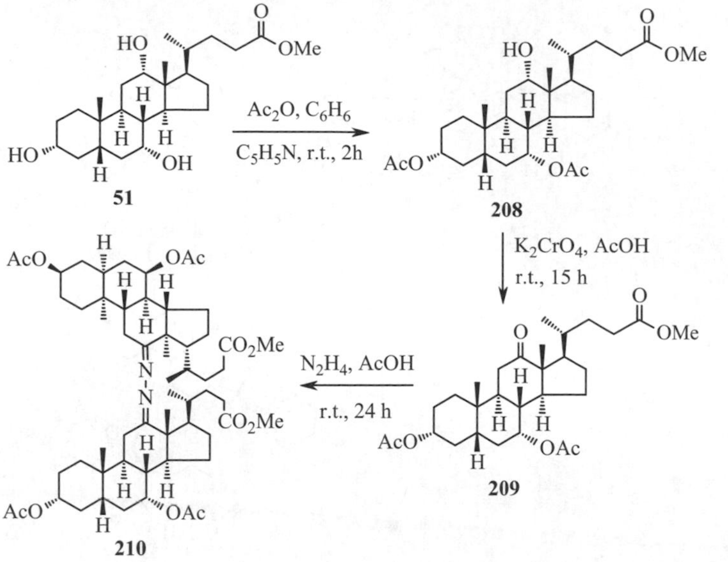

Scheme 2.3.1 *Synthesis of bishydrazone* ***210***

$CDCl_3$): δ 4.99 (s, 1H, 7β-CH), 4.58 (br, 1H, 3β-CH), 3.66 (s, 3H, 24-OMe), 2.50 (dd, 2H, 11-CH_2), 2.32 (m. 2H, 23-CH_2), 2.03 (s. 3H, 7α-MeCO), 2.02 (s. 3H, 3α-MeCO), 1.03 (s. 6H, 18-Me and 19-Me), 0.85 (d, $J = 6.1$ Hz, 3H, 21-Me). EIMS *m/z*: 504 $[M]^+$ (Gao and Dias, 1998).

Methyl 3α,7α-diacetoxy-12-oxo-5β-cholan-24-oate (**209**) was converted to bishydrazone **210** (*Scheme 2.3.1*). Anhydrous N_2H_4 (7 μL, 2.1 mmol) was added dropwise to a cooled (0 °C) solution of ketone **209** (1.0 g, 2 mmol) in AcOH (15 mL), and stirred continuously at room temperature. After 24 h, H_2O (50 mL) was added to the reaction mixture and extracted with EtOAc. The organic solvent was successively washed with saturated aqueous Na_2CO_3 and saturated aqueous NaCl, dried (Na_2SO_4) and evaporated to dryness to yield a crude dimer, which was subjected to CC, eluted with 90% EtOAc in *n*-hexane, to provide dimer **210** (680 mg, 68%, mp: 125–127 °C). ^1H NMR (250 MHz, $CDCl_3$): δ 4.99 (br s, 2H, 2 × 7β-CH), 4.62 (m, 2H, 2 × 3β-CH), 3.66 (s, 6H, 2 × 24-OMe), 2.04 (s, 6H, 2 × 7α-MeCO), 2.00 (s, 6H, 2 × 3α-MeCO), 1.04 (s, 6H, 2 × 19-Me), 0.97 (s, 6H, 2 × 18-Me), 0.96 (d, 6H, 2 × 21-Me). ^{13}C NMR (62.5 MHz, $CDCl_3$): δ 35.0 (2 × C-1), 27.3 (2 × C-2), 74.0 (2 × C-3), 169. 9 (2 × 3α-MeCO), 21.7 (2 × 3α-MeCO), 120.9 (2 × C-4), 40.9 (2 × C-5), 32.0 (2 × C-6), 71.0 (2 × C-7), 170.4 (2 × 7α-MeCO), 21.4 (2 × 7α-MeCO), 38.1 (2 × C-8), 36.2 (2 × C-9), 35.0 (2 × C-10), 36.2 (2 × C-11), 184.6 (2 × C-12), 51.7 (2 × C-13), 47.3 (2 × C-14), 23.9 (2 × C-15), 27.6 (2 × C-16), 53.8 (2 × C-17), 12.4 (2 × C-18), 22.4 (2 × C-19), 35.9 (2 × C-20), 20.0 (2 × C-21), 31.6 (2 × C-22), 30.9 (2 × C-23), 174.5 (2 × C-24), 51.6 (2 × 24-OMe). FABMS *m/z*: 1005.6 $[M + H]^+$ (Gao and Dias, 1998).

The synthesis of two symmetrical hecogenin acetate dimers **213** and **214** through ether spacer groups *via* ring C–ring C, starting from (25*R*)-3β-acetyloxy-5α-spirostan-12-one (**211**, hecogenin acetate) was achieved in two steps (Jautelat *et al.*, 1996).

Photohecogenin acetate (**212**) was prepared from a stirred solution of hecogenin acetate (**211**, 12 g, 25.4 mmol) in dry degassed dioxane (230 mL), and irradiated at room temperature under Ar (*Scheme 2.3.2*). After 24 h, the solvent was evaporated to dryness and the

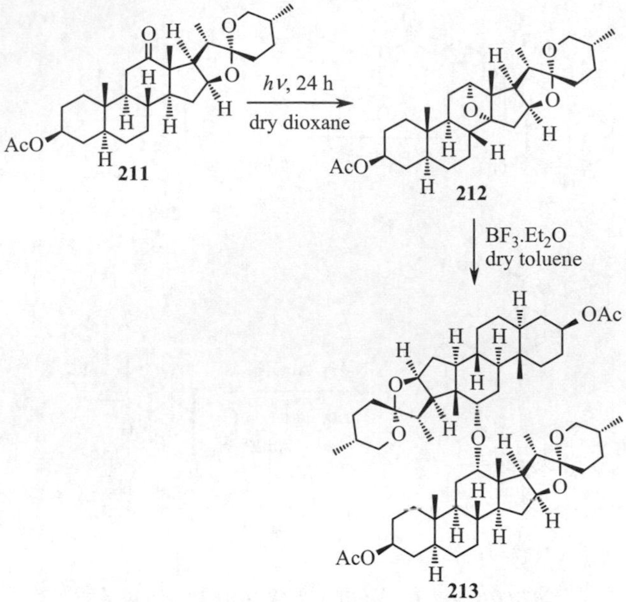

Scheme 2.3.2 Synthesis of hecogenine acetate dimer **213**

crude solid was purified by recrystallization from Et_2O-petroleum ether to produce **212** (8.52 g, 71%, mp: 205 °C). IR (KBr): ν_{max} cm^{-1} 2930s (C–H), 2859m (C–H), 1728s (C=O), 1243s (C–O), 1070s (C–O). ^1H NMR (200 MHz, $CDCl_3$): δ 4.99 (q, 1H, $J=8.0$ Hz), 4.69 (m, 1H), 4.39 (dd, 1H, $J=5.7$ Hz), 3.47 (m, 2H), 2.75 (dd, 1H, $J=8.0$, 12.0 Hz), 2.20–1.90 (m, 6H), 1.32 (s, 3H), 1.03 (d, 3H, $J=7.0$ Hz), 0.93 (s, 3H), 0.80 (d, 3H, $J=6.0$ Hz). ^{13}C NMR (50 MHz, $CDCl_3$): δ 11.2, 13.4, 17.1, 20.6, 21.3, 27.0, 28.6, 28.8, 29.1, 30.3, 32.6, 33.2, 35.1, 35.6, 37.5, 38.1, 38.8, 43.7, 44.2, 56.2, 58.2, 59.4, 67.1, 73.4, 80.9, 85.0, 91.0, 107.7, 170.3. EIMS: *m/z* 472 $[M]^+$ (Jautelat *et al.*, 1996).

Photohecogenin acetate (**212**) in dry toluene could be treated with in $BF_3.Et_2O$ following the procedure described in the literature (Bladon *et al.*, 1963) to obtain hecogenin acetate dimer **213** (*Scheme 2.3.3*). Recrystallization of crude dimer from Et_2O-petroleum ether furnished dimer **213** (mp: 290 °C). IR (KBr): ν_{max} cm^{-1} 2932s (C–H), 2861s (C–H), 1735s (C=O), 1652w (CS), 1242s (C–O), 1065s (C–O). ^1H NMR (200 MHz, $CDCl_3$): δ 5.31 (s br, 2H), 4.70 (m, 4H), 3.55–3.35 (m, 4H), 3.26 (s br, 2H), 2.47 (dd, 2H, $J=8.0$, 10.0 Hz), 2.02 (s, 6H), 1.04 (s, 6H), 0.97 (d, 6H, $J=7.0$ Hz), 0.88 (s, 6H), 0.82 (d, 6H, $J=6.0$ Hz). ^{13}C NMR (50MHz, $CDCl_3$): δ 11.8, 14.0, 17.3, 18.5, 21.5, 22.9, 27.3, 28.4, 28.8, 29.9, 30.4, 31.2, 33.9, 34.0, 35.8, 35.9, 44.0, 44.6, 49.9, 51.0, 53.9, 67.0, 73.4, 80.4, 85.6, 106.1, 120.9. FABMS: *m/z* 927 $[M + H]^+$ (Jautelat *et al.*, 1996).

Dimer **213** (200 mg, 0.216 mmol) and KOH (70 mg, 1.0 mmol) were dissolved in a mixture of MeOH (1.5 mL) and DCM (1.5 mL) and stirred at room temperature (*Scheme 2.3.3*). After 2 h, H_2O was added and the aqueous layer was extracted with DCM. The pooled organic layers were dried (Na_2SO_4), concentrated, and the crude dimer was recrystallized from Et_2O to provide dimer **214** (174 mg, 87%, mp: > 300 °C). IR (KBr): ν_{max} cm^{-1} 3476br (O–H), 2929s (C–H), 2860s (C–H), 1651w (C=C), 1243m (C–O). ^1H NMR

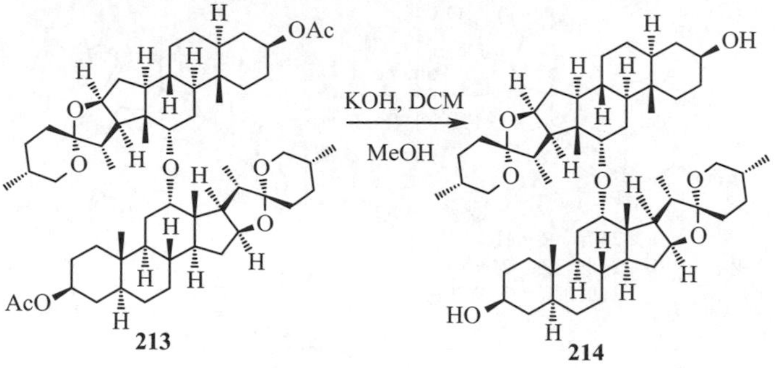

Scheme 2.3.3 Synthesis of hecogenine acetate dimer **214**

(200 MHz, $CDCl_3$): δ 5.34 (s, 2H), 4.74 (d, 2H, $J = 8.0$ Hz), 3.7–3.5 (m 2H), 3.50–3.30 (m, 4H), 3.25 (s br, 2H), 2.42 (dd, 2H, $J = 8.0$ and 10.0 Hz), 1.04 (s, 6H), 0.94 (d, 6H, $J = 7.0$ Hz), 0.87 (s, 6H), 0.81 (d, 6H, $J = 6.0$ Hz). ^{13}C NMR (50 MHz, $CDCl_3$): δ 11.3, 13.3, 16.7, 17.8, 22.6, 28.1, 28.2, 29.8, 30.0, 30.5, 30.7, 33.5, 35.6, 35.7, 37.2, 44.0, 44.3, 50.0, 50.7, 53.6, 66.6, 70.1, 80.4, 85.5, 105.9, 120.2, 152.5. FABMS: *m/z* 843 $[M + H]^+$ (Jautelat *et al.*, 1996).

2.4 Dimers *via* Ring D–Ring D Connection

2.4.1 Direct Connection

Steroid dimers *via* direct ring D–ring D connection could be obtained using the Diels–Alder reaction. For example, 16-dehydropregnenolone acetate dimer **216** was prepared by a transient Diels–Alder condensation after acid-catalyzed dienolether formation from 16-dehydropregnenolone acetate (**215**, 3β-acetyloxypregna-5,16-dien-20-one) (Morita *et al.*, 1962). A stirred solution of **215** in MeOH was treated with $CH(OMe)_3$ in the presence of an acid catalyst to produce precipitates after 3 min at room temperature (*Scheme 2.4.1*). The corresponding 20-dimethylketal was produced *in situ*, which was redissolved in the reaction mixture and stirred for another 15 min to yield the dimer **216** (mp: 301–304 °C) (Morita *et al.*, 1962).

Another example of such dimerization by the Diels–Alder reaction is the dimer **220** of 16,17-dehydrodigitoxigenin-3β-yl acetate (**217**) (Hashimoto *et al.*, 1979). The reaction occurred simply upon heating **217** (500 mg) in toluene (500 μL) at 120 °C for 180 h under N_2 (*Scheme 2.4.2*). The product was purified by crystallization from EtOAc to afford **220**

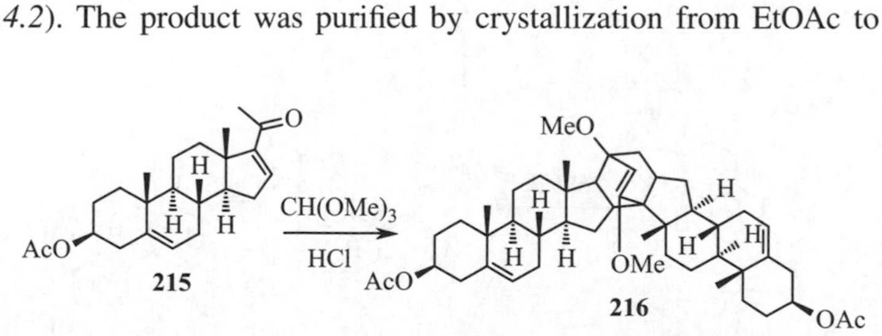

Scheme 2.4.1 Synthesis of 16-dehydropregnenolone acetate dimer **216**

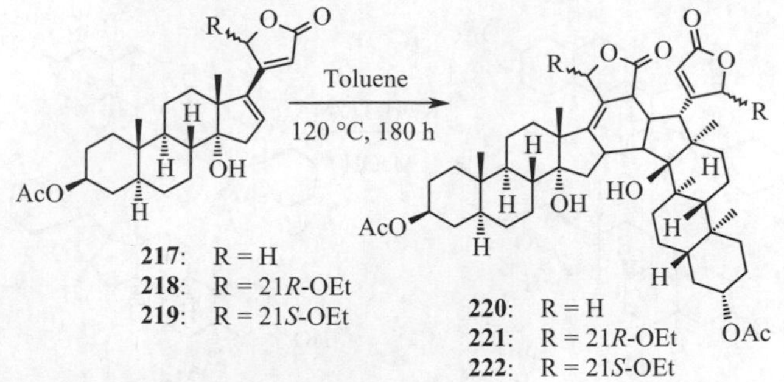

Scheme 2.4.2 *Synthesis of 16,17-dehydrodigitoxigenin-3β-yl acetate dimers* ***220–222***

(236 mg, mp: 179–182 °C). IR (KBr): ν_{max} cm^{-1} 3500br (O–H), 1760s (C=O), 1740s (C=O), 1610w (C=C). ^{1}H NMR (250 MHz, $CDCl_3$): δ 6.04 (s, 1H, 22′-CH), 4.80 (m, 2H, 2 × 21-CH), 4.06 (s, 1H, 22-CH). ^{13}C NMR (63 MHz, $CDCl_3$): δ 87.5 (C-14), 87.0 (C-14′), 123.9 (C-17), 149.0 (C-20), 173.8 (C-20′), 67.1 (C-21), 76.7 (C-21′), 65.3 (C-22), 117.7 (C-22′) (Hashimoto *et al.*, 1979).

Two other similar dimers **221** and **222** were synthesized in the same way from the 21-ethoxy derivatives **218** and **219** of 16,17-dehydrodigitoxigenin-3β-yl acetate (**217**) (*Scheme 2.4.2*). 21*R*-Ethoxy-16,17-dehydrodigitoxigenin-3β-yl acetate dimer (**221**, 174 mg, mp: 162–163 °C). IR (KBr): ν_{max} cm^{-1} 3525br (O–H), 1770s (C=O), 1740s (C=O), 1610w (C=C). ^{1}H NMR (250 MHz, $CDCl_3$): δ 6.26 (s, 1H, 22′-CH), 5.66 (m, 2H, 21-CH and 21′-CH), 4.21 (s, 1H, 22-CH). ^{13}C NMR (62.5 MHz, $CDCl_3$): δ 87.7 (C-14), 87.0 (C-14′), 125.0 (C-17), 155.3 (C-20), 170.1 (C-20′), 100.4 (C-21), 104.1 (C-21′), 64.4 (C-22), 125.6 (C-22′) (Hashimoto *et al.*, 1979).

21*S*-Ethoxy-16,17-dehydrodigitoxigenin-3β-yl acetate dimer (**222**, 281 mg, mp: 298–300 °C). IR (KBr): ν_{max} cm^{-1} 3525br (O–H), 1770s (C=O), 1740s (C=O), 1620w (C=C). ^{1}H NMR (250 MHz, $CDCl_3$): δ 6.01 (s, 1H, 22′-CH), 5.71 (m, 1H, 21′-CH), 5.67 (m, 1H, 21-CH), 3.91 (s, 1H, 22-CH). ^{13}C NMR (62.5 MHz, $CDCl_3$): δ 87.9 (C-14), 86.9 (C-14′), 125.1 (C-17), 155.5 (C-20), 171.3 (C-20′), 99.4 (C-21), 106.8 (C-21′), 66.4 (C-22), 120.7 (C-22′) (Hashimoto *et al.*, 1979).

Similarly, the estrone dimer **224**, connected directly through the D rings of the estrone 3-methyl ether (**223**), was obtained using an Aldol reaction (*Scheme 2.4.3*) (Iriarte *et al.*, 1972). Dimer **226** was reported as the byproduct of a Pd-catalyzed of 17-iodoandrosta-5,16-dien-3β-ol (**225**) (*Scheme 2.4.4*) (Iriarte *et al.*, 1972). However, the yield of this byproduct was only 5%.

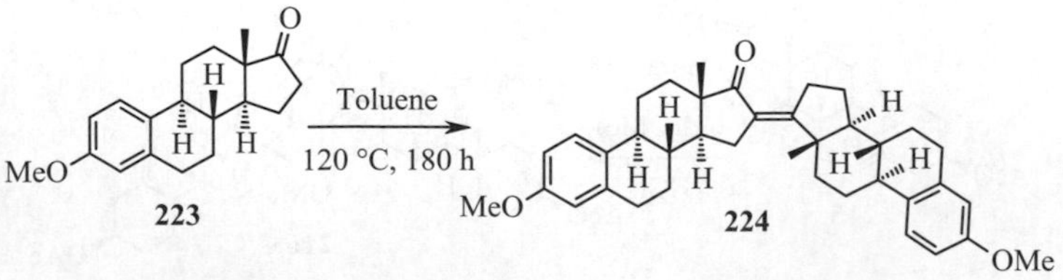

Scheme 2.4.3 *Synthesis of estrone dimer* ***224***

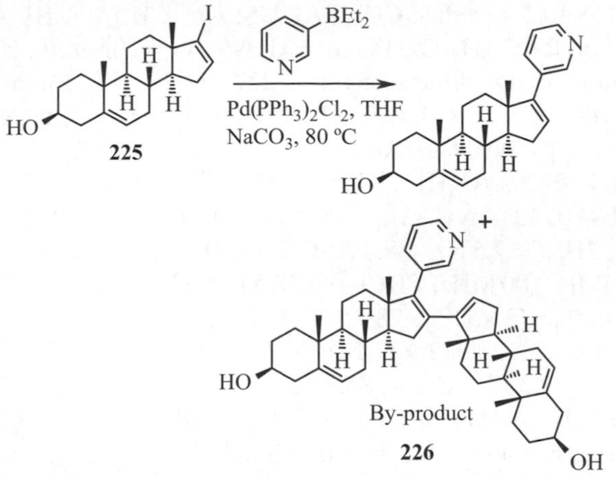

Scheme 2.4.4 Synthesis of 3β-hydroxyandrosta-5,16-diene dimer (**226**)

2.4.2 Through Spacer Groups

The synthesis of three symmetrical 17β-oxalate dimers **229-231** through oxalate spacer groups *via* ring D–ring D from naturally occurring 17β-hydroxysteroids, namely, testosterone (**200**), dihydrotestosterone (**227**, DHT, 17β-hydroxy-5α-androstan-3-one), and 5α-androst-2-en-17β-ol (**228**), using $(COCl)_2$ in the presence of C_5H_5N was reported (Nahar *et al.*, 2007b). These dimers were prepared using the same protocol as employed for the synthesis of estrone oxalate dimer **65** (*Scheme 2.1.28*).

A stirred solution of **200** (600 mg, 2.08 mmol) in dry C_5H_5N (10 mL) was treated dropwise with $(COCl)_2$ (132 mg, 1.04 mmol) under N_2 to give bis(androst-4-en-3-on-17β-yl)oxalate (**229**, 335 mg, 51%, mp: 256–257 °C) (*Scheme 2.4.5*). IR ($CHCl_3$): ν_{max} cm^{-1} 2943s (C–H), 2858s (C–H), 1763s (C=O), 1741s (C=O), 1675s (C=O), 1613m (C=C), 1433m (C–H), 1318m (C–H), 1190s (C–O), 912m, 755m. ^1H NMR (400 MHz, $CDCl_3$): δ 5.69 (m, 2H, 2 × 4-CH), 4.68 (t, 2H, J = 7.9 Hz, 2 × 17α-CH–O), 1.15 (s, 6H, 2 × 19-Me), 0.86 (s, 6H, 2 × 18-Me). ^{13}C NMR (100 MHz, $CDCl_3$): δ 35.7 (2 × C-1), 33.9 (2 × C-2), 199.4 (2 × C-3), 124.0 (2 × C-4), 170.7 (2 × C-5), 32.7 (2 × C-6), 31.4 (2 × C-7), 35.3 (2 × C-8), 53.9 (2 × C-9), 38.6 (2 × C-10), 20.5 (2 × C-11), 36.5 (2 × C-12), 42.3 (2 × C-13), 50.1 (2 × C-14), 23.5 (2 × C-15), 27.3 (2 × C-16), 85.1 (2 × C-17), 12.0 (2 × C-18),

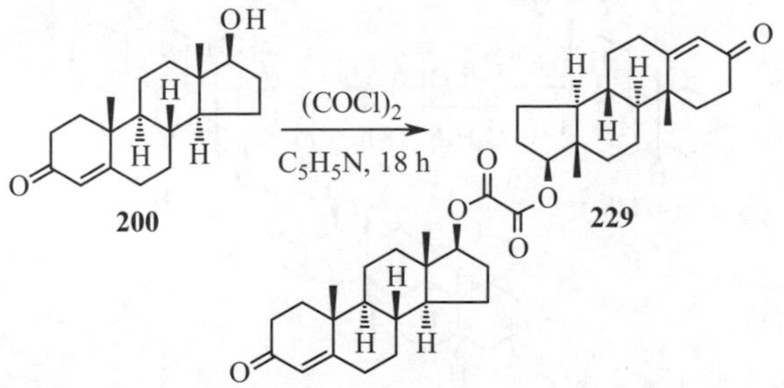

Scheme 2.4.5 Synthesis of bis(androst-4-en-3-on-17β-yl)oxalate (**229**)

17.4 (2 × C-19), 158.1 (2 × oxalate CO). FABMS: *m/z*: 631 $[M+H]^+$. HRFABMS: *m/z* 631.39984 calculated for $C_{40}H_{55}O_6$, found: 631.39986 (Nahar *et al.*, 2007b).

A stirred solution of dihydrotestosterone (**227**, 3.82 g, 13.15 mmol) in dry C_5H_5N (15 mL) was treated dropwise with $(COCl)_2$ (835 mg, 6.58 mmol) under N_2 to obtain bis (5α-androstan-3-on-17β-yl)oxalate (**230**, 1.98 g, 47%, mp: 283–284 °C) (*Scheme 2.4.6*). IR ($CHCl_3$): ν_{max} cm^{-1} 2929s (C–H), 2854s (C–H), 1762s (C=O), 1740s (C=O), 1712m (C=O), 1446m (C–H), 1314m (C–H), 1182s (C–O), 996w, 755m. ^{1}H NMR (400 MHz, $CDCl_3$): δ 4.67 (t, 2H, $J = 7.5$ Hz, 2 × 17α-CH–O), 0.97 (s, 6H, 2 × 19-Me), 0.83 (s, 6H, 2 × 18-Me). ^{13}C NMR (100 MHz, $CDCl_3$): δ 38.5 (2 × C-1), 38.1 (2 × C-2), 211.7 (2 × C-3), 44.6 (2 × C-4), 46.6 (2 × C-5), 28.7 (2 × C-6), 31.2 (2 × C-7), 35.2 (2 × C-8), 53.6 (2 × C-9), 35.7 (2 × C-10), 20.9 (2 × C-11), 36.7 (2 × C-12), 43.0 (2 × C-13), 50.4 (2 × C-14), 23.5 (2 × C-15), 27.3 (2 × C-16), 85.3 (2 × C-17), 11.4 (2 × C-18), 12.2 (2 × C-19), 157.6 (2 × oxalate CO). FABMS: *m/z* 635 $[M+H]^+$. HRFABMS: *m/z* 635.4311 calculated for $C_{40}H_{59}O_6$, found: 635.4313 (Nahar *et al.*, 2007b).

Similarly, when a stirred solution of 5α-androst-2-en-17β-ol (**228**, 200 mg, 0.73 mmol) in dry C_5H_5N (6 mL) was treated dropwise with $(COCl)_2$ (47 mg, 0.37 mmol) over 5 min under N_2, it yielded bis(5α-androst-2-en-17β-yl)oxalate (**231**, 121 mg, 54%, mp: 182–183 °C) (*Scheme 2.4.7*). IR ($CHCl_3$): ν_{max} cm^{-1} 2963s (C–H), 2848s (C–H), 1768s (C=O), 1738s (C=O), 1655m (C=C), 1445m (C–H), 1379m (C–N), 1314m (C–H), 1180s (C–O), 908m, 736m. ^{1}H NMR (400 MHz, $CDCl_3$): δ 5.56 (br m, 4H, 2 × 2-CH and 2 × 3-CH), 4.69 (t, 2H, $J = 7.9$ Hz, 2 × 17β-CH–O), 0.84 (s, 6H, 2 × 19-Me), 0.74 (s, 6H, 2 × 18-Me). ^{13}C NMR (100 MHz, $CDCl_3$): δ 39.6 (2 × C-1), 125.6 (2 × C-2), 125.6 (2 × C-3), 30.1 (2 × C-4), 41.3

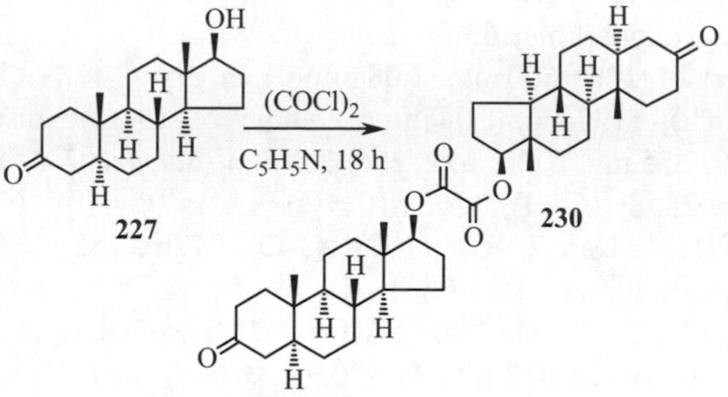

OH H O H H H H H H (COCl)$_2$ C_5H_5N, 18 h O O O O H 227 230 H H H O H

Scheme 2.4.6 *Synthesis of bis(5α-androstan-3-on-17β-yl)oxalate (**230**)*

OH H H H H H H H (COCl)$_2$ C_5H_5N, 18 h O O O O H 228 231 H H H H

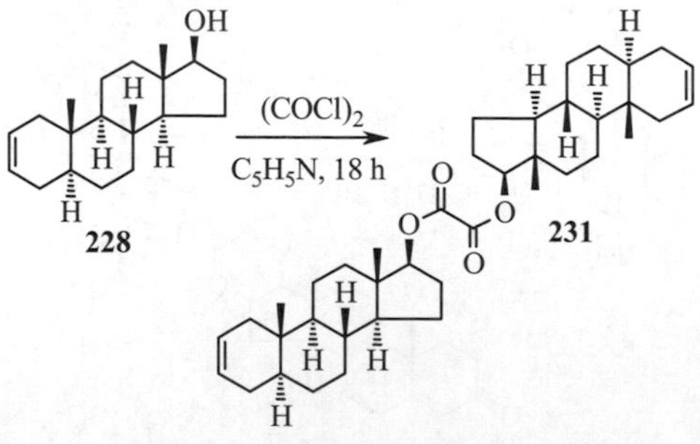

Scheme 2.4.7 *Synthesis of bis(5α-androst-2-en-17β-yl)oxalate (**231**)*

(2 × C-5), 28.4 (2 × C-6), 31.2 (2 × C-7), 35.2 (2 × C-8), 53.8 (2 × C-9), 34.5 (2 × C-10), 20.2 (2 × C-11), 36.7 (2 × C-12), 42.8 (2 × C-13), 50.5 (2 × C-14), 23.4 (2 × C-15), 27.2 (2 × C-16), 85.3 (2 × C-17), 11.6 (2 × C-18), 11.9 (2 × C-19), 158.1 (2 × oxalate CO). FABMS: *m/z* 603 $[M + H]^+$, 625 $[M + Na]^+$. HRFABMS: *m/z* 603.44130 calculated for $C_{40}H_{59}O_4$, found: 603.44131 (Nahar *et al.*, 2007b).

The same group also reported the synthesis of oxyminyl oxalate dimers **236–238** from pregnenolone derivatives through oxyminyl oxalate spacer groups *via* ring D–ring D. All those dimers contained an oxalate ester linkage between the 20-oxime positions of two molecules using dry C_5H_5N and $(COCl)_2$ as reagents (Nahar *et al.*, 2008).

Bis(3β-acetyloxypregn-5-en-20-one oximinyl) oxalate (**236**) was synthesized in three steps by acetylation of pregnenolone (**67**), oximinylation of pregnenolone acetate (**192**), and finally esterification of 3β-acetyloxypregn-5-en-20-one oxime (**233**) using dry C_5H_5N and $(COCl)_2$ as reagent, as described for the synthesis of estrone oxalate dimer **65** earlier (Nahar *et al.*, 2006). Pregnenolone (**67**, 1.0 g, 3.16 mmol) was acetylated using dry C_5H_5N (5 mL) and Ac_2O (3 mL) by stirring the mixture for 18 h at room temperature (*Scheme 2.4.8*). The reaction mixture was quenched with ice-H_2O and a white precipitate was collected, taken into Et_2O, dried ($MgSO_4$) and evaporated under pressure to yield pregnenolone acetate (**192**, 990 mg, 87%, mp: 136–137 °C) (Nahar *et al.*, 2008).

Pregnenolone acetate (**192**, 930 mg, 2.59 mmol) in dry C_5H_5N (5 mL) was treated with $HONH_2{\cdot}HCl$ (360 mg, 5.19 mmol) and Et_3N (5 mL), and refluxed for 6 h (*Scheme 2.4.8*). Finally, the reaction mixture was quenched with ice-H_2O and a white precipitate was formed. The precipitate was collected and taken into EtOAc, dried ($MgSO_4$) and rotary evaporated to afford 3β-acetyloxypregn-5-en-20-one oxime (**233**, 851 mg, 88%, mp: 183–185 °C). IR ($CHCl_3$): ν_{max} cm^{-1} 3313br (O–H), 2937s (C–H), 2854s (C–H), 1732s

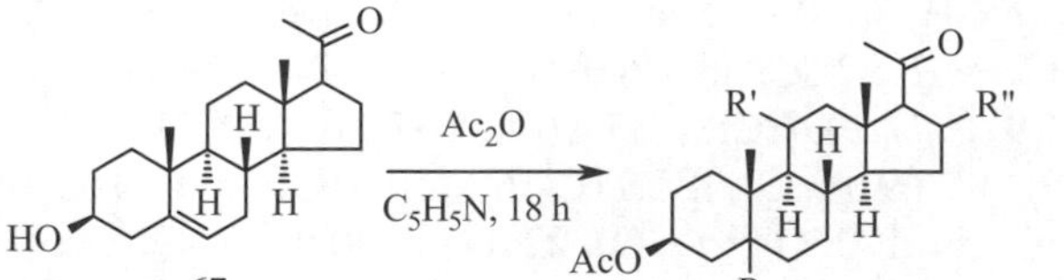

192: R = 5-dehydro, R' = H and R" = H
215: R = 5-dehydro, R' = H and R" = 16-dehydro
232: R = 5 α-H, R' = 11 α-OAc and R" = 16-dehydro

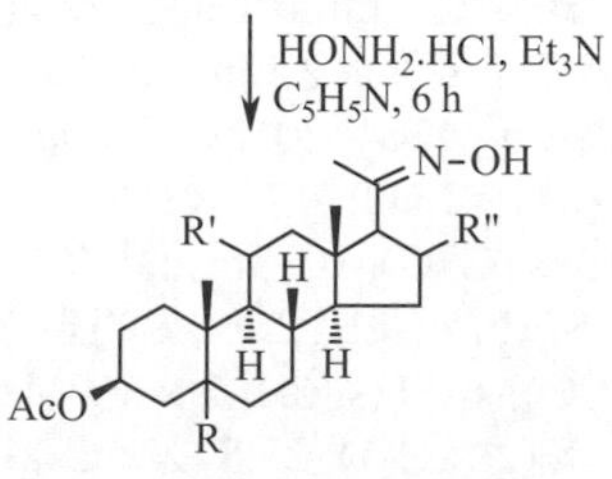

233: R = 5-dehydro, R' = H and R" = H
234: R = 5-dehydro, R' = H and R" = 16-dehydro
235: R = 5 α-H, R' = 11 α-OAc and R" = 16-dehydro

Scheme 2.4.8 *Synthesis of pregnenolone 20-one oximes* ***233–235***

(C=O), 1654s (C=C), 1439m (C–H), 1369m (C–N), 1249s (C–O), 1041m (C–O), 668m. ^{1}H NMR (400 MHz, $CDCl_3$): δ 8.88 (br s, 1H, =N-OH), 5.36 (br m, 1H, 6-CH), 4.57 (m, 1H, 3α-CH–O), 2.01 (s, 3H, $J = 6.1$ Hz, 21-Me), 1.99 (s, 3H, 3β-MeCO), 1.00 (s, 3H, 19-Me), 0.63 (s, 3H, 18-Me). ^{13}C NMR (100 MHz, $CDCl_3$): δ 37.0 (C-1), 27.8 (C-2), 73.9 (C-3), 38.1 (C-4), 139.7 (C-5), 122.4 (C-6), 31.8 (C-7), 32.0 (C-8), 50.1 (C-9), 36.6 (C-10), 21.0 (C-11), 38.6 (C-12), 43.8 (C-13), 56.8 (C-14), 24.2 (C-15), 23.1 (C-16), 56.1 (C-17), 13.1 (C-18), 15.2 (C-19), 158.7 (C-20), 19.3 (C-21), 170.6 (3β-CO), 21.5 (3β-MeCO). ESIMS: *m/z* 374 $[M + H]^+$, 396 $[M + Na]^+$ (Nahar *et al.*, 2008).

The same synthetic method, as described for 3β-acetyloxypregn-5-en-20-one oxime (**233**), was followed for the oximinylation of 16-dehydropregnenolone acetate (**215**) and 3β,11α-diacetoxy-5α-pregn-16-en-20-one (**232**) to produce 3β-acetyloxypregna-5,16-dien-20-one oxime (**234**) and 3β,11α-diacetoxy-5α-pregn-16-en-20-one oxime (**235**), respectively (*Scheme 2.4.8*).

A stirred solution of **215** (950 mg, 2.66 mmol) in dry C_5H_5N (5 mL) was treated with $HONH_2{\cdot}HCl$ (369 mg, 5.32 mmol) and Et_3N (5 mL) to give **234** (510 mg, 52%, mp: 182–183 °C) (*Scheme 2.4.8*). ^{13}C NMR (100 MHz, $CDCl_3$): δ 36.9 (C-1), 27.7 (C-2), 74.0 (C-3), 38.2 (C-4), 140.1 (C-5), 122.3 (C-6), 31.6 (C-7), 30.3 (C-8), 50.4 (C-9), 36.8 (C-10), 20.9 (C-11), 35.6 (C-12), 46.6 (C-13), 57.1 (C-14), 31.6 (C-15), 133.0 (C-16), 151.5 (C-17), 11.7 (C-18), 15.9 (C-19), 154.5 (C-20), 19.3 (C-21), 170.7 (3β-CO), 21.5 (3β-MeCO). ESIMS: *m/z* 372 $[M + H]^+$, 394 $[M + Na]^+$ (Nahar *et al.*, 2008).

Similarly, a stirred solution of **232** (1.0 g, 2.40 mmol) in dry C_5H_5N (5 mL) was treated with $HONH_2{\cdot}HCl$ (333 mg, 4.80 mmol) and Et_3N (5 mL) to yield **235** (500 mg, 48%, mp: 120–122 °C) (*Scheme 2.4.8*). IR ($CHCl_3$): ν_{max} cm^{-1} 3378 br (O–H), 2928s (C–H), 2857s (C–H), 1731s (C=O), 1590s (C=C), 1443m (C–H), 1372m (C–N), 1247s (C–O), 1021m (C–O), 962m, 753s. ^{1}H NMR (400 MHz, $CDCl_3$): δ 7.51 (br s, 1H, =N-OH), 5.96 (t, $J = 2.1$ Hz, 1H, 16-CH=C), 5.17 (m, 1H, 11β-CH–O), 4.61 (m, 1H, 3α-CH–O), 1.99 (s, 3H, J = 6.1 Hz, 21-Me), 1.91 (s, 3H, 3β-MeCO and 11α-MeCO), 0.91 (s, 3H, 19-Me), 0.89 (s, 3H, 18-Me). ^{13}C NMR (100 MHz, $CDCl_3$): δ 36.9 (C-1), 27.7 (C -2), 73.1 (C-3), 34.4 (C-4), 44.7 (C-5), 31.4 (C-6), 31.9 (C-7), 33.4 (C-8), 55.4 (C-9), 37.2 (C-10), 71.5 (C-11), 42.7 (C-12), 46.6 (C-13), 56.9 (C-14), 28.9 (C-15), 132.1 (C-16), 150.8 (C-17), 11.0 (C-18), 12.8 (C-19), 153.3 (C-20), 17.0 (C-21), 170.7 (11α-CO), 22.0 (11α-MeCO), 170.5 (3β-CO), 21.4 (3β-MeCO). FABMS: *m/z* 432 $[M + H]^+$, 454 $[M + Na]^+$. HRFABMS: *m/z* 432.274978 calculated for $C_{25}H_{38}NO_5$, found: 432.274978 (Nahar *et al.*, 2008).

A stirred solution of 20-one oxime **233** (200 mg, 0.54 mmol) in dry C_5H_5N (5 mL) was treated with $(COCl)_2$ (34 mg, 0.27 mmol), according to the procedure described for estrone oxalate dimer **65** (*Scheme 2.1.28*), to afford bis(3β-acetyloxypregn-5-en-20-one oximinyl) oxalate (**236**, 125 mg, 58%, mp: 204–205 °C) (*Scheme 2.4.9*). IR ($CHCl_3$): ν_{max} cm^{-1} 2943s (C–H), 2853s (C–H), 1764s (C=O), 1732s (C=O), 1638s (C=C), 1438m (C–O), 1374m (C–N), 1247s (C–O), 1134s (C–O), 1033m and 761m. ^{1}H NMR (400 MHz, $CDCl_3$): δ 5.35 (br m, 2H, 2 × 6-CH), 4.57 (br m, 2H, 2 × 3α-CH–O), 2.02 (s, 6H, $J = 6.1$ Hz, 2 × 21-Me), 1.99 (s, 6H, 2 × 3β-MeCO), 1.00 (s, 6H, 2 × 19-Me), 0.64 (s, 6H, 2 × 18-Me). ^{13}C NMR (100 MHz, $CDCl_3$): δ 36.8 (2 × C-1), 27.8 (2 × C-2), 73.8 (2 × C-3), 38.1 (2 × C-4), 139.7 (2 × C-5), 122.2 (2 × C-6), 31.7 (2 × C-7), 32.0 (2 × C-8), 50.0 (2 × C-9), 36.6 (2 × C-10), 21.0 (2 × C-11), 38.4 (2 × C-12), 44.0 (2 × C-13), 56.2 (2 × C-14), 24.1 (2 × C-15), 23.4 (2 × C-16), 56.6 (2 × C-17), 13.2 (2 × C-18), 19.3 (2 × C-19), 167.5 (2 × C-20), 17.1 (2 × C-21), 170.5 (2 × oxalate CO), 170.5 (3β-CO), 21.4 (3β-MeCO). FABMS: *m/z* 801

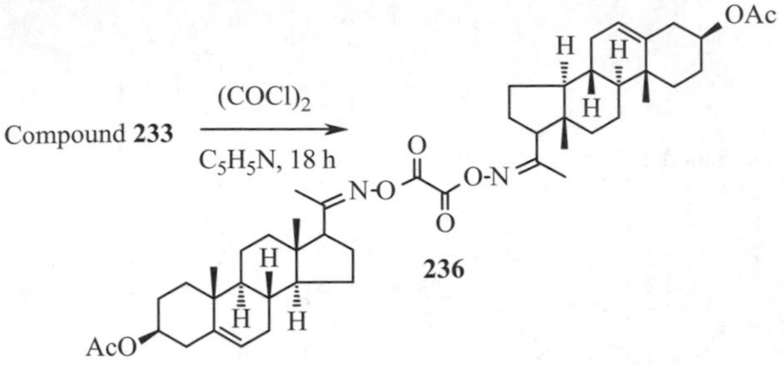

Compound **233** $\xrightarrow[C_5H_5N,\ 18\ h]{(COCl)_2}$ **236**

*Scheme 2.4.9 Synthesis of bis(3β-acetyloxypregn-5-en-20-one oximinyl)oxalate (**236**)*

$[M + H]^+$, 823 $[M + Na]^+$. HRFABMS: *m/z* 801.50536 calculated for $C_{48}H_{69}N_2O_8$, found: 801.50536 (Nahar *et al.*, 2008).

Other oxyminyl oxalate dimers, bis(3β-acetyloxypregna-5,16-dien-20-one oximinyl) oxalate (**237**) and bis(3β,11α-diacetoxy-5α-pregn-16-en-20-one oximinyl)oxalate (**238**) were synthesized, from **234** and **235**, respectively, according to the method as outlined for dimer **236** (Nahar *et al.*, 2008).

A stirred solution of 20-one oxime **234** (300 mg, 0.81 mmol) in dry C_5H_5N (5 mL) was treated with $(COCl)_2$ (51 mg, 0.40 mmol) to provide dimer **237** (150 mg, 47%, mp: 165–167 °C) (*Scheme 2.4.10*). IR ($CHCl_3$): ν_{max} cm^{-1} 3313 br (O–H) 2937s (C–H), 2854s (C–H), 1764s (C=O), 1732s (C=O), 1654m (C=C), 1439m (C–H), 1369m (C–N), 1249s (C–O), 1132s (C–O), 1041m (C–O), 668m. ^1H NMR (400 MHz, $CDCl_3$): δ 6.33 (br m, 2H, 2 × 16-CH), 5.33 (br m, 2H, 2 × 6-CH), 4.56 (m, 2H, 2 × 3α-CH–O), 2.09 (s, 6H, J = 6.1 Hz, 2 × 21-Me), 1.99 (s, 6H, 2 × 3β-MeCO), 0.98 (s, 6H, 2 × 19-Me), 0.83 (s, 6H, 2 × 18-Me). ^{13}C NMR (100 MHz, $CDCl_3$): δ 36.9 (2 × C-1), 27.7 (2 × C-2), 73.8 (2 × C-3), 34.6 (2 × C-4), 140.2 (2 × C-5), 122.0 (2 × C-6), 32.1 (2 × C-7), 32.0 (2 × C-8), 50.2 (2 × C-9), 36.7 (2 × C-10), 20.9 (2 × C-11), 38.1 (2 × C-12), 46.5 (2 × C-13), 56.7 (2 × C-14), 31.4 (2 × C-15), 139.5 (2 × C-16), 149.7 (2 × C-17), 13.4 (2 × C-18), 15.4 (2 × C-19), 159.5 (2 × C-20), 167.5 (2 × C-20), 19.2 (2 × C-21), 170.5 (2 × oxalate CO), 170.5 (3β-CO), 21.4 (3β-MeCO). FABMS: *m/z* 797 $[M + H]^+$, 819 $[M + Na]^+$. HRFABMS: *m/z* 797.47406 calculated for $C_{48}H_{65}N_2O_8$, found: 797.47406 (Nahar *et al.*, 2008).

Similarly, when a stirred solution of **235** (1.0 g, 2.42 mmol) in dry C_5H_5N (15 mL) was treated with $(COCl)_2$ (154 mg, 1.21 mmol), it produced bis(3β-acetoxpregn-5-en-20-one oximinyl)oxalate (**238**, 424 mg, 38%, mp: 229–230 °C) (*Scheme 2.4.11*). IR ($CHCl_3$): ν_{max}

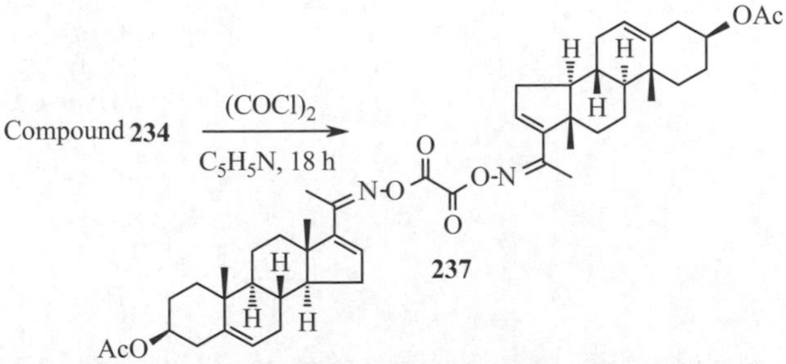

Compound **234** $\xrightarrow[C_5H_5N,\ 18\ h]{(COCl)_2}$ **237**

*Scheme 2.4.10 Synthesis of bis(3β-acetyloxypregn-5-en-20-one oximinyl)oxalate (**237**)*

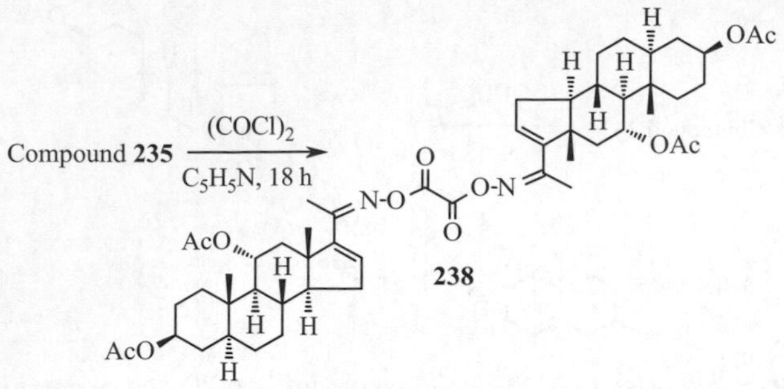

Compound **235** $\xrightarrow[C_5H_5N,\ 18\ h]{(COCl)_2}$ **238**

Scheme 2.4.11 *Synthesis of bis(3β-acetyloxpregn-5-en-20-one oximinyl)oxalate (**238**)*

cm^{-1} 3364 br (O–H) 2930s (C–H), 2860s (C–H), 1765s (C=O), 1731s (C=O), 1594s (C=C), 1449m (C–H), 1373m (C–N), 1258s (C–O), 1131s (C–O), 1027m (C–O), 916m, 733s. ^{1}H NMR (400 MHz, $CDCl_3$): δ 5.97 (t, J = 2.1 Hz, 2H, 2 × 16-CH), 5.18 (m, 2H, 2 × 11β-CH–O), 4.60 (m, 2H, 2 × 3α-CH–O), 1.94 (s, 6H, J = 6.1 Hz, 2 × 21-Me), 1.91 (s, 6H, 2 × 3β-MeCO), 1.89 (s, 6H, 2 × 11α-MeCO), 0.94 (s, 6H, 2 × 19-Me), 0.91 (s, 6H, 2 × 18-Me). ^{13}C NMR (100 MHz, $CDCl_3$): δ 37.1 (2 × C-1), 27.7 (2 × C-2), 73.1 (2 × C-3), 34.4 (2 × C-4), 44.6 (2 × C-5), 28.8 (2 × C-6), 33.2 (2 × C-7), 33.5 (2 × C-8), 55.5 (2 × C-9), 36.7 (2 × C-10), 70.5 (2 × C-11), 42.2 (2 × C-12), 46.3 (2 × C-13), 56.9 (2 × C-14), 31.9 (2 × C-15), 139.2 (2 × C-16), 149.0 (2 × C-17), 12.7 (2 × C-18), 13.4 (2 × C-19), 159.0 (2 × C-20), 16.7 (2 × C-21), 170.5 (2 × oxalate CO), 170.5 (3β-CO), 21.4 (3β-MeCO), 170.7 (11α-CO), 21.9 (11α-MeCO). FABMS: m/z 917 $[M + H]^+$, 939 $[M + Na]^+$. HRFABMS: m/z 917.51631 calculated for $C_{52}H_{73}N_2O_{12}$, found: 917.51633 (Nahar *et al*., 2008).

During the synthesis of functionalized enantiopure steroids from estrone (**64**) through organolithium intermediates, a C_2-symmetric estrone diol dimer **241** was obtained as a result of the reaction of *O*-protected estrone as the electrophilic component and the organolithium intermediate in three steps (Yus *et al*., 2001). First, *O*-ethoxymethylestrone (**239**) was prepared from **64** as follows (*Scheme 2.4.12*). *n*-BuLi (1.6M solution in *n*-hexane,

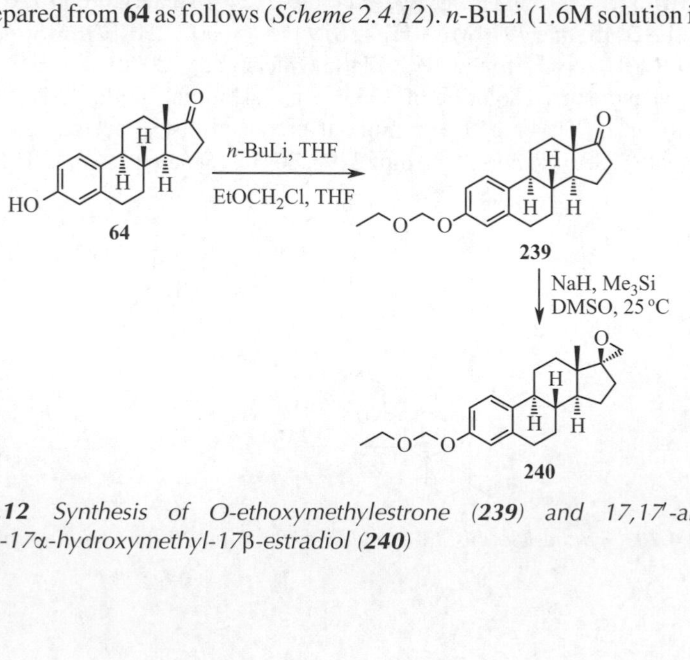

64 $\xrightarrow[EtOCH_2Cl,\ THF]{n\text{-BuLi, THF}}$ **239** $\xrightarrow[DMSO,\ 25\,^\circ C]{NaH,\ Me_3Si}$ **240**

Scheme 2.4.12 *Synthesis of O-ethoxymethylestrone (**239**) and 17,17′-anhydro-3-O-ethoxymethyl-17α-hydroxymethyl-17β-estradiol (**240**)*

1.62 mL, 2.4 mmol) was added dropwise to a THF solution (40 mL) of **64** (540 mg, 2.0 mmol) at −78 °C under N_2. After 5 min, $EtOCH_2Cl$ (0.83 mL, 4.88 mmol) was added and the reaction temperature was allowed to rise to 25 °C overnight, H_2O (20 mL) was added and the mixture was extracted with EtOAc (3 × 20 mL). The pooled organic solvents were dried (Na_2SO_4), rotary evaporated and the crude solid was purified by CC, eluted with 20% EtOAc in *n*-hexane, to obtain **239** (95%). IR (film): ν_{max} cm^{-1} 3039m (C–H), 3008m (C–H), 1736s (C=O), 1613s (C=C), 1502s (C=C), 1008s (C–O). ^{1}H NMR (300 MHz, $CDCl_3$): δ 7.19 (d, 1H, *J* = 8.5 Hz, Ph-CH), 6.83 (d, 1H, *J* = 8.5 Hz, Ph-CH), 6.79 (s, 1H, Ph-CH), 5.19 (s, 2H, OCH_2O), 3.72 (q, 2H, *J* = 7.3 Hz, CH_3CH_2), 2.87-2.90 (m, 2H, CH_2), 1.93–2.54 (m, 7H, CH), 1.28–1.66 (m, 6H, CH_2), 1.23 (t, 3H, *J* = 7.0 Hz, CH_3CH_2), 0.90 (s, 3H, Me). ^{13}C NMR (75 MHz, $CDCl_3$): δ 13.8, 15.0, 21.5, 25.8, 26.5, 29.5, 31.5, 35.8, 38.2, 44.0, 47.9, 50.3, 64.0, 93.1, 113.8, 116.2, 126.2, 133.1, 137.7, 155.4, 220.8. EIMS: *m/z* 328 [M^+] (Yus *et al.*, 2001).

O-Ethoxymethylestrone (**239**) was transformed into 7,17′-anhydro-3-*O*-ethoxymethyl-17α-hydroxymethyl-17β-estradiol (**240**) (*Scheme 2.4.12*). To a mixture of **239** (1.64 g, 5.0 mmol) in THF (5 mL) and DMSO (10 mL), a suspension of 95% NaH (630 mg, 25.0 mmol) and Me_3SI (2.1 g, 10.0 mmol) in DMSO (30 mL) was added dropwise at 25 °C, stirred for 12 h, H_2O (40 mL) was added and the mixture was extracted with EtOAc (4 × 30 mL). The combined organic layers were washed with H_2O (5 × 20 mL), dried (Na_2SO_4), evaporated to dryness and the crude product was purified by CC (mobile phase: *n*-hexane-EtOAc) to provide **240** (86%). IR (film): ν_{max} cm^{-1} 3036m (C–H), 1619s (epoxide), 1502s (C=C), 1013s (epoxide C–O). ^{1}H NMR (300 MHz, $CDCl_3$): δ 7.18 (d, 1H, *J* = 8.5 Hz, Ph-CH), 6.77 (s, 1H, Ph-CH), 6.28 (d, 1H, *J* = 8.5 Hz, Ph-CH), 5.18 (s, 2H, OCH_2O), 3.71 (q, 2H, *J* = 7.3 Hz, CH_3CH_2), 2.94 (d, 1H, *J* = 5.2 Hz, CH), 2.84-2.89 (m, 2H, CH_2), 2.62 (d, 1H, *J* = 5.2 Hz, CH), 2.22-2.34 (m, 2H, CH), 1.83-2.07 (m, 5H, CH), 1.34-1.54 (m, 6H, CH_2), 1.22 (t, 3H, *J* = 7.0 Hz, CH_3CH_2), 0.91 (s, 3H, Me). ^{13}C NMR (75 MHz, $CDCl_3$): δ 14.2, 15.0, 23.2, 25.9, 27.1, 29.0, 29.6, 33.9, 38.8, 40.3, 43.8, 51.8, 53.5, 64.0, 70.4, 93.1, 113.6, 116.1, 26.2, 133.4, 137.8, 155.3. EIMS: *m/z* 342 [M^+] (Yus *et al.*, 2001).

The synthesis of the C_2-symmetric estrone diol dimer **241** was achieved from *O*-ethoxymethylestrone (**239**) and 17,17′-anhydro-3-*O*-ethoxymethyl-17α-hydroxy methyl-17β-estradiol (**240**) (*Scheme 2.4.13*). To a cooled (–78 °C) suspension of powdered Li (100 mg, 14.0 mmol) and a catalytic amount of DTBB (4.0 mg, 0.15 mmol) in THF (5 mL)

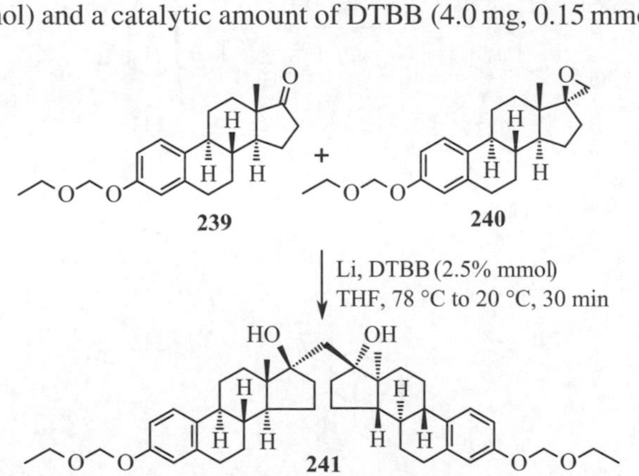

239

240

Li, DTBB (2.5% mmol)
THF, 78 °C to 20 °C, 30 min

241

Scheme 2.4.13 *Synthesis of C_2-symmetric estrone diol dimer **241***

at −78 °C, compound **240** (1.0 mmol) was added under N_2 and stirred for 2 h. Then compound **239** (1.2 mmol) was added at -78 °C. After 10 min, H_2O (20 mL) was added and the reaction temperature was allowed to rise to 20 °C overnight. The reaction mixture was extracted with EtOAc (3 × 20 mL), the organic solvent was dried (Na_2SO_4), and evaporated under pressure. The crude dimer was purified by CC as above to produce diol **241** (26%). IR (film): ν_{max} cm^{-1} 3644-3110br (O–H), 3020m (C–H), 1661s (C=C), 1499s, 1014s (C–O). ^{1}H NMR (300 MHz, $CDCl_3$): δ 8.43 (d, 2H, J = 8.5 Hz, Ph-CH), 7.18 (d, 2H, J = 8.5 Hz, Ph-CH), 6.77 (s, 2H, Ph-CH), 5.19 (s, 4H, 2 × OCH_2O), 3.72 (q, 4H, J = 7.1 Hz, 2 × OCH_2), 2.83-2.85 (m, 4H, CH), 1.29-2.32 (m, 28H, CH), 1.22 (t, 6H, J = 7.1 Hz, 2 × CH_2CH_3), 0.93 (s, 6H, 2 × Me). ^{13}C NMR (75 MHz, $CDCl_3$): δ 14.1, 15.1, 23.2, 26.2, 27.4, 29.7, 32.4, 33.9, 39.35, 43.8, 45.6, 50.3, 64.1, 67.0, 83.4, 93.2, 113.7, 116.2, 126.25, 133.6, 138.0, 155.3. EIMS: m/z 655 $[M - OH]^+$ (Yus *et al.*, 2001).

The bis-steroid of androsterone derivatives, 3α-hydroxy-5α-androstan-17-one azine (**245**) and 3α-acetyloxy-5α-androstan-17-one azinc (**246**) were synthesized, respectively, from androsterone (**242**, 3α-hydroxy-5α-androstan-17-one) in two steps and androsterone acetate (**243**, 3α-acetyloxy-5α-androstan-17-one) in one step (Suginome and Uchida, 1979).

A mixture of androsterone (**242**, 200 mg) and $N_2H_4.H_2O$ (3 mL) in EtOH (5 mL) was refluxed for 60 min, resulting crystals were filtered off, washed with EtOH and followed by H_2O, and dried ($MgSO_4$) to yield androsterone hydrazone (**244**, 183 mg, 87%, mp: 230–231 °C) (*Scheme 2.4.14*). IR (Nujol): ν_{max} cm^{-1} 3372br (O–H and N–H), 1629s (C=N), 1271s (C–O), 1007m, 761m. ^{1}H NMR (100 MHz, $CDCl_3$): δ 4.58 (br, 1H, N–H), 4.02 (br s, J = 8.1 Hz, 3-CH), 2.17 (br t, 2H, J = 7.5 Hz, 16-CH_2), 0.83 (s, 3H, 18-Me), 0.80 (s, 3H, 19-Me) (Suginome and Uchida, 1979).

A solution of androsterone hydrazone (**244**, 1.0 g) in dry dioxane (250 mL) was irradiated under N_2 (*Scheme 2.4.14*). After for 7 h, the solvent was rotary evaporated and the solid was purified by prep-TLC on SiO_2 to give the crude azine, which was recrystallized from Me_2CO to provide 3α-hydroxy-5α-androstan-17-one azine (**245**, 81 mg, mp: 247–248 °C). UV (MeOH): λ_{max} 210 and 228 nm. IR (Nujol): ν_{max} cm^{-1} 3343br (O–H), 1656s (C=N-N=C),

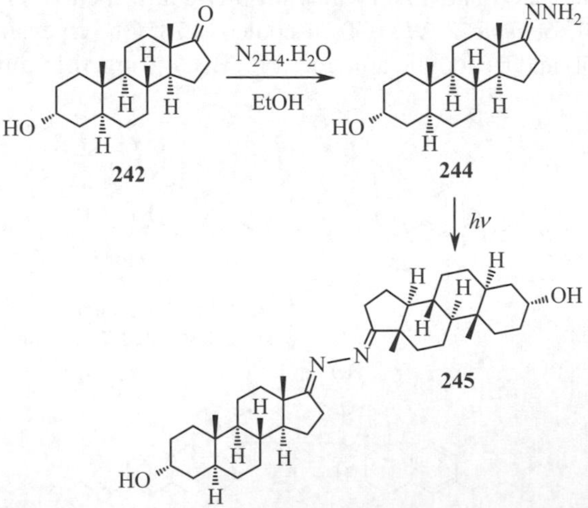

Scheme 2.4.14 *Synthesis of 3α-hydroxy-5α-androstan-17-one azine (**245**)*

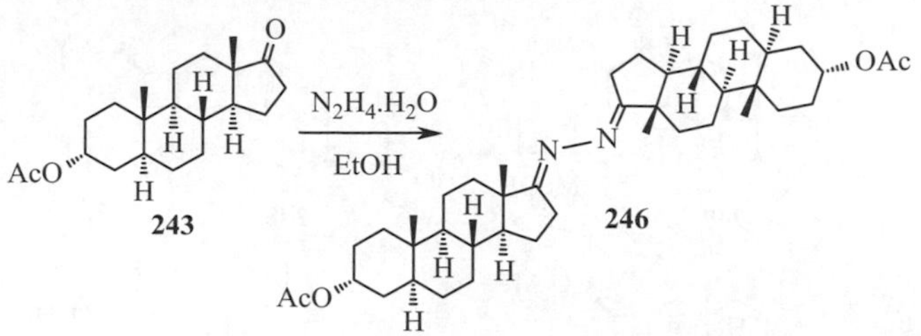

Scheme 2.4.15 *Synthesis of 3α-acetyloxy-5α-androstan-17-one azine (**246**)*

1247s (C–O), 1074m, 1007m. ^{1}H NMR (100 MHz, $CDCl_3$): δ 5.76 (br, 1H, N–H), 3.98 (br s, $J = 7.5$ Hz, 3β-CH), 0.85 (s, 3H, 18-Me), 0.78 (s, 3H, 19-Me) (Suginome and Uchida, 1979).

A mixture of androsterone acetate (**243**, 272 mg) in EtOH (4 mL), $N_2H_4.H_2O$ (600 μL) and concentrated HCl (300 μL) was stirred at room temperature (*Scheme 2.4.15*). The mixture was then heated in a water bath for 30 min, and left at room temperature for 60 min to cool. The crude azine was filtered off and recrystallized from MeOH to afford 3α-acetyloxy-5α-androstan-17-one azine (**246**, 501 mg, mp: 238–240 °C). IR (Nujol): ν_{max} cm^{-1} 1727s (C=O), 1641s (C=N), 1259s (C–O), 1243s (C–O), 1231s, 1023m, and 978m. ^{1}H NMR (100 MHz, $CDCl_3$): δ 4.90 (br s, $J = 7.5$ Hz, 2 × 3β-CH), 1.99 (s, 3H, 2 × 3α-MeCO), 0.86 (s, 3H, 2 × 18-Me), 0.80 (s, 3H, 2 × 19-Me) (Suginome and Uchida, 1979).

The synthesis of D-nor-5α-androstan-16-one azine (**257**) from D-nor-5α-androstan-16-one (**255**) or D-nor-5α-androstan-16-one hydrazine (**256**) was achieved in multisteps from 16-diazo-3β-hydroxyandrost-5-en-17-one (**247**) (Suginome *et al.*, 1980). Initially, methyl 3β-hydroxy-D-norandrost-3-en-16β-oate (**248**) was prepared from a stirred solution of **247** (18 g) in dry THF (720 mL) and MeOH (180 mL) was irradiated under N_2 After 34 h, the solvent was rotary evaporated and the crude solid was recrystallized from Me_2CO to provide methyl ester **248** (11.29 g) (*Scheme 2.4.16*).

Catalytic hydrogenation of methyl 3β-hydroxy-D-norandrost-3-en-16β-oate (**248**) provided methyl 3β-hydroxy-D-nor-5α-androstan-16β-oate (**249**) (*Scheme 2.4.16*). To a mixture of methyl ester **248** (11 g) in AcOH (100 mL), Adams' catalyst (300 mg) was added and the mixture was hydrogenated under Ar. After 24 h, the catalyst was filtered off and the solvent was evaporated under pressure, the crude residue was dissolved in DCM, sequentially washed with 5% $NaHCO_3$, saturated aqueous NaCl, and H_2O, dried (Na_2SO_4) and concentrated *in vacuo*. The resulting solid was recrystallized from Me_2CO to obtain alcohol **249** (8.35 g, mp: 141–142 °C). IR (Nujol): ν_{max} cm^{-1} 3527br (O–H), 3438br (O–H), 1736s (C=O), 1714s (C=O), 1258s (C–O), 1198s (C–O), 1049s (C–O). ^{1}H NMR (300 MHz, $CDCl_3$): δ 9.17 (s, 3H, 19-Me), 9.07 (s, 3H, 18-Me), 7.15 (dd, 1H, $J = 7.2$, 8.4 Hz, 16α-CH), 6.44 (br, 1H, 3α-CH), 6.35 (s, 3H, 3β-MeCO). EIMS: m/z 320 $[M]^+$ (Suginome *et al.*, 1980).

Oxidation of methyl 3β-hydroxy-D-nor-5α-androstan-16β-oate (**249**) afforded methyl 3-oxo-D-nor-5α-androstan-16β-oate (**250**) (*Scheme 2.4.16*). Jones' reagent (20 mL in Me_2CO) was added dropwise to a stirred cooled (0 °C) solution of alcohol **249** (8.345 g) in Me_2CO (180 mL). After 1.5 h, saturated aqueous $NaHSO_3$ was added until the colour of the solution turned blue, the solution was filtered off and the filtrate was partly rotary evaporated. The solid of ketone **250** (4.622 g) was collected by filtration and the filtrate was extracted with Et_2O, the ethereal solution was dried (Na_2SO_4) and concentrated *in vacuo*.

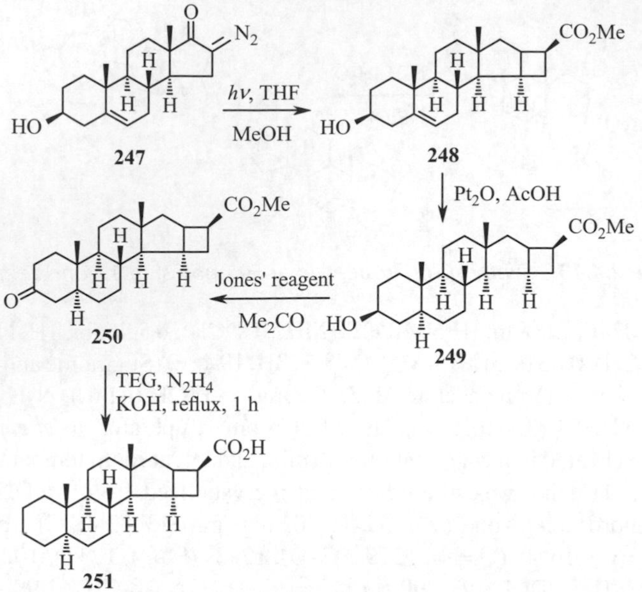

Scheme 2.4.16 *Synthesis of D-nor-5α-androstan-16β-oic acid (**251**)*

The resulting product was recrystallized from Me_2CO to produce ketone **250** (1.243 g, mp: 157–158 °C). The filtrate of this recrystallization was evaporated under pressure and purified by CC, eluted with C_6H_6, to provide more of ketone **250** (750 mg). The combined yield was 81%. IR (Nujol): ν_{max} cm^{-1} 1731s (C=O), 1710s (C=O), 1350m (C–H), 1224s (C–O), 1192m, 1172m, 1047m, 1028m. ^1H NMR (300 MHz, $CDCl_3$): δ 8.99 (s, 3H, 18-Me), 8.95 (s, 3H, 19-Me), 7.24 (dd, 1H, $J=6.8$, 8.6 Hz, 16α-CH), 6.44 (br, 1H, 3α-CH), 6.34 (s, 3H, 3β-MeCO). EIMS: *m/z* 318 $[M]^+$ (Suginome *et al.*, 1980).

The Wolff–Kishner reduction of methyl 3-oxo-D-nor-5α-androstan-16β-oate (**250**) furnished D-nor-5α-androstan-16β-oic acid (**251**) (*Scheme 2.4.16*). A mixture of ketone **250** (5.92 g) in N_2H_4 (16 mL), KOH (16 g) and TEG (160 mL) was refluxed. After 60 min, the excess of N_2H_4 and H_2O were removed by rotary evaporation, the solution was refluxed again for 1.5 h. A mixture of ice-H_2O (200 mL) was added to the cooled solution and the pH of the solution was adjusted by adding HCl (6 mol/mL) and the solution was stirred for 30 min and then left at room temperature for a further 30 min. The crude product was filtered and dissolved in DCM, washed with H_2O and dried (Na_2SO_4). The solid was recrystallized from MeOH to yield carboxylic acid **251** (3.16 g, mp: 226–227 °C). The filtrate of this recrystallization was rotary evaporated and purified as above to get more of carboxylic acid **251** (1.668 g). The combined yield was 90%. IR (Nujol): ν_{max} cm^{-1} 3200–2400br (O–H), 1692s (C=O), 1268s (C–O), 958m. ^1H NMR (300 MHz, $CDCl_3$): δ 9.20 (s, 3H, 19-Me), 8.98 (s, 3H, 18-Me), 7.23 (dd, 1H, $J=7.2$, 9.0 Hz, 16α-CH). EIMS: *m/z* 290 $[M]^+$ (Suginome *et al.*, 1980).

The synthesis of D-norpregnan-20-methyl ether (**252**) was accomplished from D-nor-5α-androstan-16β-oic acid (**251**) (*Scheme 2.4.17*). A solution of 1M MeLi (40 mL) in Et_2O was added dropwise to a stirred solution of carboxylic acid **251** (3.65 g) in C_6H_6 (300 mL) at room temperature. After 60 min, the reaction mixture was poured into ice-H_2O, extracted

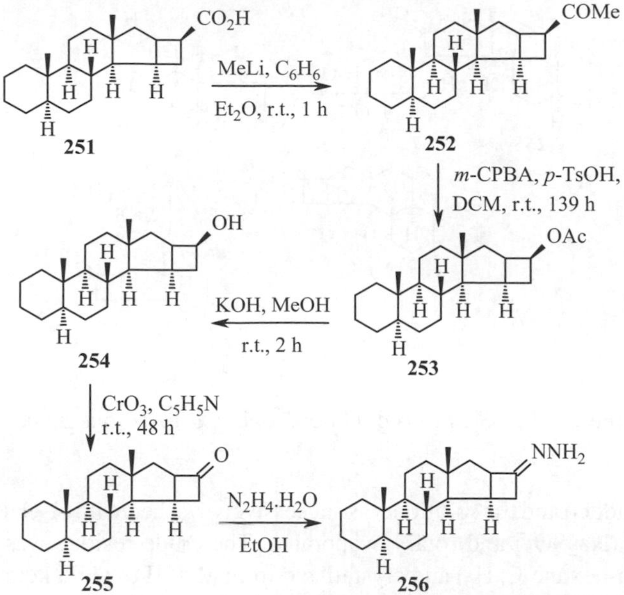

Scheme 2.4.17 *Synthesis of D-nor-5α-androstan-16-one (**255**) and D-nor-5α-androstan-16-one hydrazone (**256**)*

with Et_2O and worked up as above. The solvent was evaporated under pressure and crude product was recrystallized from MeOH to afford methyl ether **252** (3.25 g, 90%, mp: 125–127 °C). IR (Nujol): ν_{max} cm^{-1} 1715s (C=O), 1354m (C–H), 1186m (C–O), 1161m (C–O). ^{1}H NMR (300 MHz, $CDCl_3$): δ 9.20 (s, 3H, 19-Me), 9.11 (s, 3H, 18-Me), 7.98 (s, 3H, 20β-COMe), 7.18 (dd, 1H, $J=6.0$, 9.6 Hz, 16α-CH). EIMS: m/z 288 $[M]^+$ (Suginome *et al.*, 1980).

The Baeyer–Villiger oxidation of D-norpregnan-20-methyl ether (**252**) yielded D-nor-5α-androstan-16β-ol (**254**) in two steps *via* the formation of D-nor-5α-androstan-16β-yl acetate (**253**) (*Scheme 2.4.17*). A mixture of methyl ether **252** (3.41 g), *m*-CPBA (4.0 g) and *p*-TsOH (160 mg) in DCM (100 mL) was stirred at room temperature. After 139 h, 5% $Na_2S_2O_3$ was added to decompose the excess peracids, the solution was washed with 10% $NaHCO_3$ followed by H_2O, dried (Na_2SO_4) and concentrated *in vacuo* to give D-nor-5α-androstan-16β-yl acetate (**253**), which was dissolved in MeOH (100 mL) and hydrolyzed with KOH (3.0 g) at room temperature. After 2 h, the solution was neutralized with HCl (2 mol/mL), extracted with Et_2O, and worked up as usual. The crude product was recrystallized from Me_2CO to obtain alcohol **254** (1.62 g, mp: 132–135 °C). The residue from the filtrate was purified as outlined above to produce more of alcohol **254** (993 mg). The combined yield was 84%. IR (Nujol): ν_{max} cm^{-1} 3600–3200br (O–H), 1191m, 1131m, 1062s (C–O). ^{1}H NMR (300 MHz, $CDCl_3$): δ 9.19 (s, 3H, 19-Me), 9.05 (s, 3H, 18-Me), 7.96 (s, 1H, 20α-CH), 5.51 (dd, 1H, $J=6.0$, 8.0 Hz, 16α-CH). EIMS: m/z 262 $[M]^+$ (Suginome *et al.*, 1980).

Oxidation of D-nor-5α-androstan-16β-ol (**254**) provided D-nor-5α-androstan-16-one (**255**) (*Scheme 2.4.17*). A solution of CrO_3 (320 mg) in C_5H_5N (2 mL) was added dropwise to a stirred solution of alcohol **254** (320 mg) in C_5H_5N (2 mL) at room temperature. After

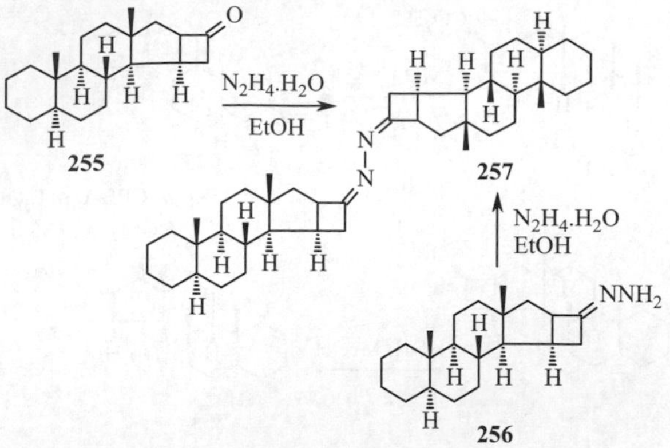

Scheme 2.4.18 *Synthesis of D-nor-5α-androstan-16-one azine (**257**)*

48 h, DCM was added and the solution was successively washed with H_2O, HCl (2 mol/mL), H_2O, and dried (Na_2SO_4) and rotary evaporated. The crude residue was purified by CC (mobile phase: *n*-hexane-C_6H_6) and crystallized from MeOH to yield ketone **255** (178 mg, mp: 128–219 °C). IR (Nujol): ν_{max} cm^{-1} 1775s (C=O), 1143m, 1019m, 987m, 973m. ^1H NMR (300 MHz, $CDCl_3$): δ 9.17 (s, 3H, 19-Me), 8.99 (s, 3H, 18-Me), 6.26 (dd, 1H, $J = 5.5$, 7.9 Hz, 16α-CH). EIMS: *m/z* 260 $[M]^+$ (Suginome *et al.*, 1980).

D-Nor-5α-androstan-16-one (**255**) transformed to D-nor-5α-androstan-16-one hydrazone (**256**) (*Scheme 2.4.17*). A stirred mixture of $N_2H_4.H_2O$ (5 mL) and ketone **255** (500 mg) in EtOH (15mL) was refluxed for 1.5 h. The solvent was rotary evaporated, H_2O was added and the crystalline solid was filtered off and dried *in vacuo* to furnish hydrazone **256** (214 mg, mp: 116 °C). IR (Nujol): ν_{max} cm^{-1} 3345m (N–H), 1162m, 1112m, 926m, 910m, 896w. ^1H NMR (300 MHz, $CDCl_3$): δ 9.17 (s, 3H, 19-Me), 8.86 (s, 3H, 18-Me), 5.82 (br, 2H, 16-NNH_2). EIMS: *m/z* 274 $[M]^+$.

The synthesis of D-nor-5α-androstan-16-one azine (**257**) was accomplished from D-nor-5α-androstan-16-one (**255**) or D-nor-5α-androstan-16-one hydrazone (**256**) (*Scheme 2.4.18*). A mixture of ketone **255** or hydrazone **256** (50 mg) in EtOH (1.5 mL) and $N_2H_4.H_2O$ (500 μL) was refluxed for 2 h, a part of the solvent was evaporated under pressure and H_2O was added, the mixture was extracted with EtOH. The etheral layer was washed with H_2O, dried (Na_2SO_4), the solvent was rotary evaporated and the azine was recrystallized from MeOH to provide azine **257** (5.0 mg, mp: 220 °C). IR (Nujol): ν_{max} cm^{-1} 1699s (C=N-N=C). ^1H NMR (300 MHz, $CDCl_3$): δ 9.17 (s, 6H, 2 × 19-Me), 8.81 (s, 6H, 2 × 18-Me). EIMS: *m/z* 516 $[M]^+$ (Suginome *et al.*, 1980).

2.4.3 Through Side Chain and Spacer Groups

The unique characteristics of the bile acids in relation to their chiral, rigid and curved framework and chemically diverse hydroxyl functionalities have made them suitable building blocks in tailoring supramolecular hosts (Nahar *et al.*, 2007a). As a consequence, bile acids have been successfully utilized to synthesize various cholaphanes and cyclocholates (See *Chapter 3*). During the synthesis of macrocyclic cholaphane molecules (See *Chapter 3*), several bile acid-based dimers **263–268** were successfully synthesized from

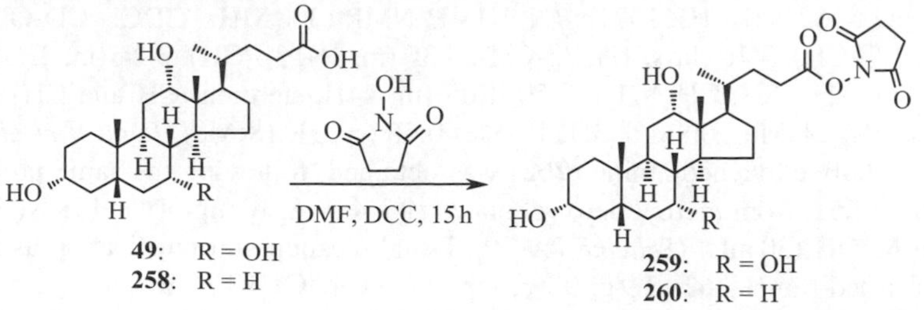

Scheme 2.4.19 *Synthesis of succinimido cholate (**259**) and succinimido deoxycholate (**260**)*

cholic acid (**49**), methyl cholate (**51**), deoxycholic acid (**258**, 3α,12α-dihydroxy-5β-cholanic acid or 3α,12α-dihydroxy-5β-cholan-24-oic acid) and methyl deoxycholate (**106**) by connecting the side chains through spacer groups (Pandey *et al*., 2002).

To a stirred mixture of cholic acid (**49**, 8.50 g, 20.8 mmol) in dry DMF (30 mL) and *N*–hydroxysuccinimide (2.87 g, 24.9 mmol), after 10 min DCC (6.0 g, 29.12 mmol) was added and stirred at room temperature for 15 h (*Scheme 2.4.19*). The reaction mixture was filtered and ice-H_2O (500 mL) was added to the filtrate, the resulting solid was filtered off, washed with ice-H_2O, and dried under vacuum. The dried crude product was purified by CC, eluted with 5% MeOH in $CHCl_3$, to afford succinimido cholate (**259**, 9.8 g, 90%, mp: 110–112 °C). IR (KBr): ν_{max} cm^{-1} 3314m (O–H), 1814m (C=O), 1784s (C=O), 1736s (C=O). ^{1}H NMR (300 MHz, $CDCl_3$): δ 3.98 (br s, 1H, 12β-CH), 3.85 (br s, 1H, 7β-CH), 3.45 (m, 1H, 3β-CH), 2.84 (s, 4H, CH_2CH_2), 2.50-1.10 (m, 24H, steroidal CH and CH_2), 1.00 (d, $J = 6$ Hz, 3H, 21-Me), 0.89 (s, 3H, 19-Me), 0.70 (s, 3H, 18-Me) (Pandey *et al*., 2002).

Following the same method as described above, succinimido deoxycholate (**260**, 7.3 g, 94%, mp: 145–146 °C) was prepared from deoxycholic acid (**258**, 6.0 g, 15.31 mmol) (*Scheme 2.4.19*). IR (KBr): ν_{max} cm^{-1} 3314m (O–H), 1814m (C=O), 1783s (C=O), 1735s (C=O). ^{1}H NMR (300 MHz, $CDCl_3$): δ 3.98 (br s, 1H, 12β-CH), 3.61 (m, 1H, 3β-CH), 2.82 (s, 4H, CH_2CH_2), 2.50-1.03 (m, 25H, steroidal CH and CH_2), 1.00 (d, $J = 6.0$ Hz, 3H, 21-Me), 0.91 (s, 3H, 19-Me), 0.69 (s, 3H, 18-Me) (Pandey *et al*., 2002).

A stirred solution of methyl cholate (**51**, 10 g, 23.66 mmol) in MeOH (50 mL) was treated with an excess amount of $H_2N(CH_2)_2NH_2$ (15 mL) at room temperature (*Scheme 2.4.20*). After 48 h, ice-cold H_2O (400 mL) was added, the resulting solid was filtered off, dried *in vacuo*, and purified by recrystallization from $CHCl_3$-MeOH to give *N*-cholylethylenedimine (**261**, 9.9 g, 98%, mp: 175–177 °C). IR (KBr): ν_{max} cm^{-1} 3390m (O–H and N–H),

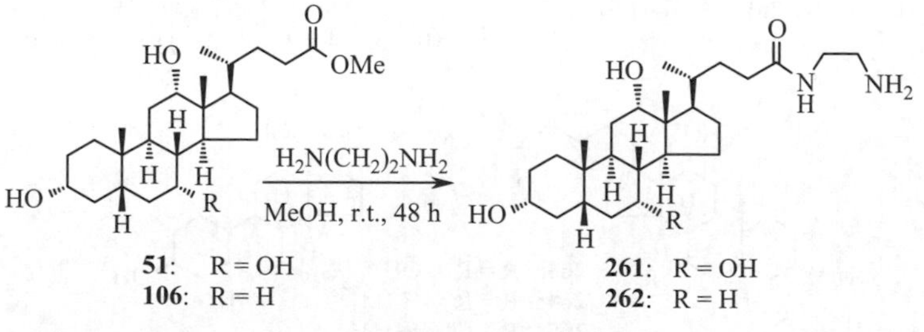

Scheme 2.4.20 *Synthesis of N-cholylethylenedimine (**261**) and N-deoxycholylethylenedimine (**262**)*

1663s (C=O), 1622s (N–H), 1571m (N–H). ^{1}H NMR (300 MHz, $CDCl_3$-CD_3OD): δ 3.90 (br s, 1H, 12β-CH), 3.76 (br s, 1H, 7β-CH), 3.35 (m, 1H, 3β-CH), 3.24 (m, 2H, $NHCH_2$), 2.77 (t, $J = 6.0$ Hz, 2H, CH_2NH_2), 2.50–1.05 (m, 24H, steroidal CH and CH_2), 1.00 (d, $J = 6.0$ Hz, 3H, 21-Me), 0.88 (s, 3H, 19-Me) 0.66 (s, 3H, 18-Me) (Pandey *et al.*, 2002).

N-Deoxycholylethylenediamine (**262**) was obtained, following the same procedure as described for **261**, from methyl deoxycholate (**106**, 4.06 g, 10 mmol) and $H_2N(CH_2)_2NH_2$ (8 mL) in MeOH (30 mL) (*Scheme 2.4.20*). Usual cleanup and purification as described above furnished dimer **262** (3.9 g, 98%, mp: 115–117 °C). IR (KBr): ν_{max} cm^{-1} 3368m (O–H and N–H), 1636s (C=O), 1558m (N–H). ^{1}H NMR (300 MHz, $CDCl_3$-CD_3OD): δ 3.95 (br s, 1H, 12β-CH), 3.56 (m, 1H, 3β-CH), 3.27 (m, 2H, $NHCH_2$), 2.80 (m, 2H, CH_2NH_2), 2.10-1.20 (m, 25H, steroidal CH and CH_2), 0.99 (br s, 3H, $J = 6.1$ Hz, 21-Me), 0.90 (s, 3H, 19-Me), 0.67 (s, 3H, 18-Me) (Pandey *et al.*, 2002).

A mixture of *N*-Cholylethylenedimine (**261**, 901 mg, 2.0 mmol) and succinimido cholate (**259**, 1.01 g, 2.0 mmol) in dry DMF (15 mL) was stirred at room temperature (*Scheme 2.4.21*). After 24 h, ice-H_2O (500 mL) and brine were added to the reaction mixture and the resulting precipitate was filtered off, washed with cold H_2O, and dried *in vacuo*. The compound was purified by CC (mobile phase: 10% MeOH in $CHCl_3$) to provide *N*,*N*′-dicholylethylenedimine (**263**, 1.62 g, 96%, mp: 160–162 °C). IR (KBr): ν_{max} cm^{-1} 3400m (O–H and N–H), 1652m (C=O). ^{1}H NMR (300 MHz, $CDCl_3$-CD_3OD): δ 3.96 (br s, 2H, 2 × 12β-CH), 3.84 (br s, 2H, 2 × 7β-CH), 3.24 (m, 2H, 2 × 3β-CH), 3.42 (m, 4H, 2 × CH_2NHCO), 2.19–1.20 (m, 48H, steroidal CH and CH_2), 0.99 (br s, 6H, $J = 6.1$ Hz, 2 × 21-Me), 0.89 (s, 6H, 2 × 19-Me), 0.67 (s, 6H, 2 × 18-Me). ^{13}C NMR (75 MHz, $CDCl_3$-CD_3OD): δ 39.3 ($NHCH_2CH_2NH$), 73.0 (2 × C-3), 68.3 (2 × C-7), 71.6 (2 × C-12), 31.7 (2 × C-22), 35.3 (2 × C-23), 175.9 (2 × C-24). FABMS: *m/z* 863 [M + Na]$^{+}$; 841 [M + H]$^{+}$ (Pandey *et al.*, 2002).

Bisdeoxycholamide (**264**) was synthesized from a mixture of *N*-deoxycholylethylenedimine (**262**, 869 mg, 2.0 mmol) and succinimido deoxycholate (**260**, 979 mg, 2.0 mmol) in DMF (10 mL) (*Scheme 2.4.21*), following the same method as described for bischolamide (**263**). Purification of the crude product was achieved by CC (mobile phase: 8% MeOH in

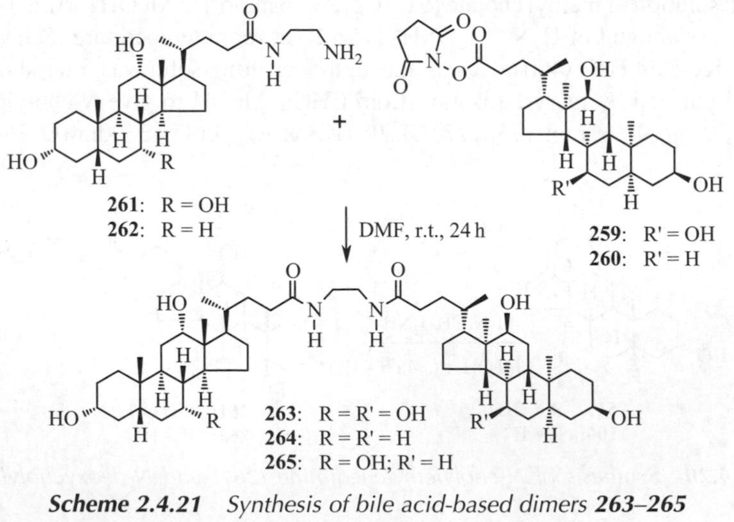

261: R = OH
262: R = H
DMF, r.t., 24 h
259: R′ = OH
260: R′ = H
263: R = R′ = OH
264: R = R′ = H
265: R = OH; R′ = H

Scheme 2.4.21 *Synthesis of bile acid-based dimers* ***263–265***

$CHCl_3$) to obtain *N,N′*-bisdeoxycholylethylenedimine (**264**, 1.5 g, 95%, mp: 152–153 °C). IR (KBr): ν_{max} cm^{-1} 3334m (O–H and N–H), 1659s (C=O). ^{1}H NMR (300 MHz, $CDCl_3$-CD_3OD): δ 3.95 (br s, 2H, 2 × 12β-CH), 3.55 (m, 2H, 2 × 3β-CH), 3.37 (m, 4H, 2 × CH_2NHCO), 2.40–1.10 (m, 50H, steroidal CH and CH_2), 0.98 (d, 6H, *J* = 6.0 Hz, 2 × 21-Me), 0.90 (s, 6H, 2 × 19-Me), 0.67 (s, 6H, 2 × 18-Me). ^{13}C NMR (75 MHz, $CDCl_3$-CD_3OD): δ 39.3 ($NHCH_2CH_2NH$), 72.9 (2 × C-3), 71.3 (2 × C-12), 31.6 (2 × C-22), 35.3 (2 × C-23), 175.9 (2 × C-24). FABMS: *m/z* 832 $[M + Na]^+$; 810 $[M + H]^+$ (Pandey *et al.*, 2002).*N*-Cholyl-*N′*-deoxycholylethylenediamine (**265**, 795 mg, 96%, mp: 151–153 °C) was prepared from a mixture of *N*-cholylethylenedimine (**261**, 450 mg, 1.0 mmol) and succinimido deoxycholate (**260**, 489 mg, 1.0 mmol) in DMF (10 mL) (*Scheme 2.4.21*), following the procedure as described earlier for bischolamide (**263**). IR (KBr): ν_{max} cm^{-1} 3391m (O–H and N–H), 1653s (C=O). ^{1}H NMR (300 MHz, $CDCl_3$-CD_3OD): δ 3.96 (br s, 2H, 2 × 12β-CH), 3.83 (br s, 1H, 2 × 7β-CH), 3.57 (m, 2H, 2 × 3β-CH), 3.38 (m, 4H, 2 × CH_2NHCO), 2.50-1.01 (m, 49H, steroidal CH and CH_2), 0.99 (br s, 2H, 2 × 21-Me), 0.90 (s, 6H, 2 × 19-Me), 0.67 (s, 6H, 2 × 18-Me). ^{13}C NMR (75 MHz, $CDCl_3$-CD_3OD): δ 39.3 (CH_2NHCO), 73.0 (2 × C-3), 68.2 (C-7), 27.1 (C-7′), 71.5 (C-12), 71.4 (C-12′), 31.6 (2 × C-22), 35.3 (2 × C-23), 175.9 (2 × C-24), 175.8 (C-24′). FABMS: *m/z* 825 $[M + H]^+$ (Pandey *et al.*, 2002).

Anhydrous K_2CO_3 (139 mg, 1.0 mmol) was added to a solution of bischolamide (**263**, 420 mg, 0.5 mmol) in dry $CHCl_3$ (20 mL) at 55–60 °C, then $BrCH_2COBr$ in $CHCl_3$ (202 mg, 1.0 mmol) was added dropwise (*Scheme 2.4.22*). After 10 min, the mixture was poured over ice-II_2O (20 mL), the organic layer was separated, dried (Na_2SO_4), rotary evaporated and the crude dimer was purified by gradient CC (mobile phase: 0–5% MeOII in $CHCl_3$) to yield *N, N′*-bis(3α-*O*-bromoacetylcholyl)ethylenediamine (**266**, 379 mg, 70%). IR (KBr): ν_{max} cm^{-1} 3420m (O–H and N–II), 1733s (C=O), 1653s (C=O). ^{1}H NMR (300 MHz, $CDCl_3$): δ 4.57 (m, 2H, 2 × 3β-CH), 3.92 (br s, 2H, 2 × 12β-CH), 3.79 (br s, 2H, 2 × 7β-CH), 3.71 (s, 4H,

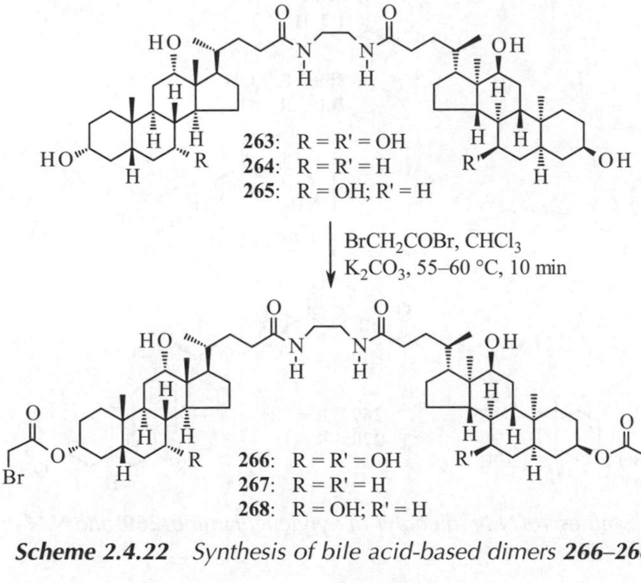

263: R = R′ = OH
264: R = R′ = H
265: R = OH; R′ = H

$BrCH_2COBr$, $CHCl_3$
K_2CO_3, 55–60 °C, 10 min

266: R = R′ = OH
267: R = R′ = H
268: R = OH; R′ = H

Scheme 2.4.22 *Synthesis of bile acid-based dimers **266–268***

2 × CH_2Br), 3.29 (m, 4H, 2 × CH_2NH), 2.30–1.00 (m, 48H, steroidal CH and CH_2), 0.92 (br s, 6H, 2 × 21-Me), 0.84 (s, 6H, 2 × 19-Me), 0.62 (s, 6H, 2 × 18-Me) (Pandey *et al.*, 2002).

N,N′-Bis(3α-*O*-bromoacetyldeoxycholyl)ethylenediamine (**267**) was prepared, following the same procedure as above, from a mixture of bisdeoxycholamide (**264**, 650 mg, 0.80 mmol), $BrCH_2COBr$ (323 mg, 1.6 mmol) and anhydrous K_2CO_3 (221 mg, 1.6 mmol) in dry $CHCl_3$ (10 mL) (*Scheme 2.4.22*). The crude dimer was purified by gradient CC, eluted with 0–3% MeOH in $CHCl_3$, to provide dimer **267** (633 mg, 75%). IR (KBr): ν_{max} cm^{-1} 3400m (O–H and N–H), 1730s (C=O), 1652s (C=O). ^1H NMR (300 MHz, $CDCl_3$): δ 4.71 (m, 2H, 2 × 3β-CH), 3.93 (br s, 2H, 2 × 12β-CH), 3.73 (s, 4H, 2 × CH_2Br), 3.31 (m, 4H, 2 × CH_2NH), 2.40-1.00 (m, 50H, steroidal CH and CH_2), 0.93 (d, 6H, *J* = 6 Hz, 2 × 21-Me), 0.86 (s, 6H, 2 × 19-Me), 0.62 (s, 6H, 2 × 18-Me) (Pandey *et al.*, 2002).

3α,3α′-*O*-Bis(bromoacetyl)cholyl(deoxycholyl)ethylenediamine (**268**, 519 mg, 73%) was obtained from the reaction of cholyldeoxycholamide (**265**, 550 mg, 0.66 mmol), $BrCH_2COBr$ (267 mg, 1.32 mmol) and anhydrous K_2CO_3 (183 mg, 1.32 mmol) in dry $CHCl_3$ (10 mL) (*Scheme 2.4.22*), following the same method as described above for dimer **266**. IR (KBr): ν_{max} cm^{-1} 3425m (O–H and N–H), 1735s (C=O), 1655s (C=O). ^1H NMR (300 MHz, $CDCl_3$): δ 4.77 and 4.68 (m, 2H, 2 × 3β-CH), 3.98 (br s, 2H, 2 × 12β-CH), 3.85 (br s, 1H, 7β-CH), 3.78 (s, 4H, 2 × CH_2Br), 3.36 (m, 4H, 2 × CH_2NH), 2.40-1.20 (m, 49H, steroidal CH and CH_2), 1.00 (br s, 6H, *J* = 6.1 Hz, 2 × 21-Me), 0.91 (s, 6H, 2 × 19-Me), 0.68 (s, 6H, 2 × 18-Me) (Pandey *et al.*, 2002).

To a stirred solution of succinimido cholate (**259**, 2.22 g, 4.3 mmol) in DMF (10 mL), *m*-xylylenediamine (299 mg, 2.15 mmol) was added at room temperature under N_2 (*Scheme 2.4.23*). After 10 h, the reaction mixture was filtered and ice-cold brine

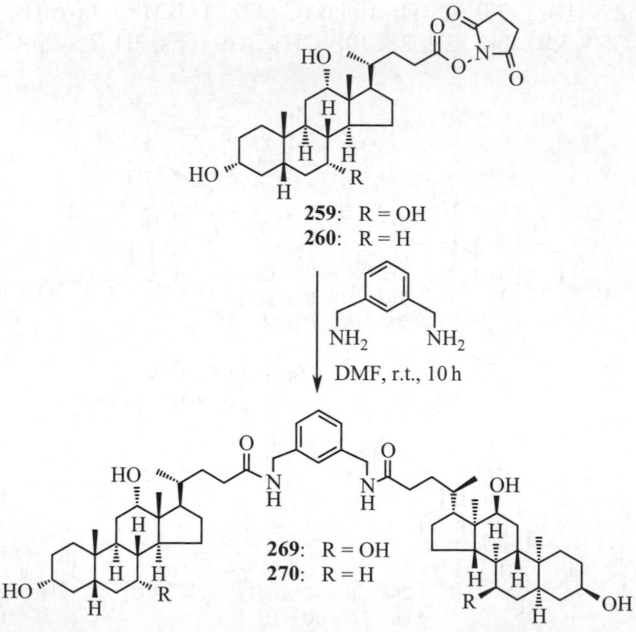

259: R = OH
260: R = H

DMF, r.t., 10 h

269: R = OH
270: R = H

Scheme 2.4.23 *Synthesis of N,N′-dicholyl-m-xylylenediamine (**269**) and N,N′-bisdeoxycholyl-m-xylylenediamine (**270**)*

(400 mL) was added to the filtrate. The resulting precipitate was filtered off, washed with cold H_2O, and dried *in vacuo*, the crude dimer was purified by CC (mobile phase: 10% MeOH in $CHCl_3$) to afford *N,N′*-dicholyl-*m*-xylylenediamine (**269**, 1.78 g, 88%, mp: 153–155 °C). IR (KBr): ν_{max} cm^{-1} 3399m (O–H and N–H), 1653s (C=O). ^1H NMR (00 MHz, $CDCl_3$-CD_3OD): δ 7.21 (br s, 4H, Ph-CH), 4.37 (m, 4H, 2 × CH_2NH), 3.91 (br s, 2H, 2 × 12β-CH), 3.80 (br s, 2H, 2 × 7β-CH), 3.21 (m, 2H, 2 × 3β-CH), 2.25-1.10 (m, 48H, steroidal CH and CH_2), 0.98 (br s, 6H, J = 6.1 Hz, 2 × 21-Me), 0.86 (s, 6H, 2 × 19-Me), 0.66 (s, 6H, 2 × 18-Me). ^{13}C NMR (75 MHz, $CDCl_3$-CD_3OD): δ 72.8 (C-3), 43.0 (CH_2NHCO), 68.1 (C-7), 71.4 (C-12), 31.5 (C-22), 32.4 (C-23), 128.5 (Ph-CH), 126.8 (Ph-CH), 138.6 (Ph-C), 174.7 (C-24). FABMS: *m/z* 940 $[M + Na]^+$; 918 $[M + H]^+$ (Pandey *et al.*, 2002).

Following the same protocol as outlined for dimer **269**, *N,N′*-bisdeoxycholyl-*m*-xylylenediamine (**270**, 1.92 g, 87%, mp: 145–146 °C) was synthesized from succinimido deoxycholate (**260**, 2.45 g, 5.02 mmol) and *m*-Xylylenediamine (341 mg, 2.45 mmol) *Scheme 2.4.23*). IR (KBr): ν_{max} cm^{-1} 3386m (O–H and N–H), 1654s (C=O). ^1H NMR (300 MHz, $CDCl_3$-CD_3OD): δ 7.21 (br s, 4H, Ph-CH), 4.36 (m, 4H, 2 × CH_2NH), 3.94 (br s, 2H, 2 × 12β-CH), 3.43 (m, 2H, 2 × 3β-CH), 2.60–1.00 (m, 50H, steroidal CH and CH_2), 0.95 (d, J = 6 Hz, 6H, 2 × 21-Me), 0.89 (s, 6H, 2 × 19-Me), 0.65 (s, 6H, 2 × 18-Me). ^{13}C NMR (75 MHz, $CDCl_3$-CD_3OD): δ 43.2 (CH_2NHCO), 72.9 (2 × C-3), 27.0 (2 × C-7), 71.3 (2 × C-12), 31.5 (2 × C-22), 32.1 (2 × C-23), 138.7 (Ph-C), 128.9 (Ph-CH), 126.8 (Ph-CH), 174.6 (2 × C-24). FABMS: *m/z* 908 $[M + Na]^+$; 886 $[M + H]^+$ (Pandey *et al.*, 2002).

Similarly following the same method as described above for dimers **266–268** (*Scheme 2.4.22*), *N,N′*-bis(3α-*O*-bromoacetylcholyl)-*m*-xylylenediamine (**271,** 718 mg, 71%) was achieved from *N,N′*-dicholyl-*m*-xylylenediamine (**269**, 800 mg, 0.87 mmol) (*Scheme 2.4.24*). IR (KBr): ν_{max} cm^{-1} 3325m (O–H and N–H) 1733s (C=O), 1653s (C=O). ^1H NMR (300 MHz, $CDCl_3$): δ 7.18 (br s, 4H, Ph-CH), 4.56 (m, 2H, 2 × 3β-CH),

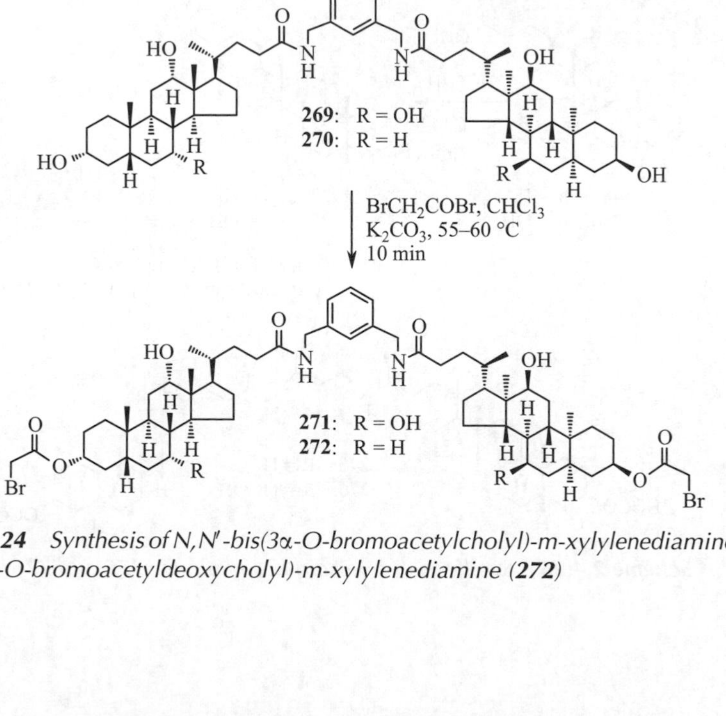

Scheme 2.4.24 *Synthesis of N,N′-bis(3α-O-bromoacetylcholyl)-m-xylylenediamine (**271**) and N,N′-bis(3α-O-bromoacetyldeoxycholyl)-m-xylylenediamine (**272**)*

4.30 (m, 4H, 2 × CH_2NH), 3.90 (br s, 2H, 2 × 12β-CH), 3.76 (br s, 2H, 2 × 7β-CH), 3.70 (s, 4H, 2 × CH_2Br), 2.40–1.00 (m, 48H, steroidal CH and CH_2), 0.89 (br s, 6H, $J = 6.1$ Hz, 2 × 21-Me), 0.84 (br s, 6H, 2 × 19-Me), 0.60 (s, 6H, 2 × 18-Me) (Pandey *et al.*, 2002).

N,N′-Bis(3α-*O*-bromoacetyldeoxycholyl)-*m*-xylylenediamine (**272**, 629 mg, 76%) was prepared, in a similar fashion as above, from bisdeoxycholyl-*m*-xylylenediamine (**270**, 650 mg, 0.73 mmol) (*Scheme 2.4.24*). IR (KBr): ν_{max} cm^{-1} 3415br (O–H and N–H), 1730s (C=O), 1652s (C=O). ^{1}H NMR (300 MHz, $CDCl_3$): δ 7.17 (m, 4H, Ph-CH), 4.72 (m, 2H, 2 × 3β-CH), 4.30 (m, 4H, 2 × CH_2NH), 3.90 (s, 2H, 2 × 12β-CH), 3.74 (s, 4H, 2 × CH_2Br), 2.30-1.03 (50H, steroidal CH and CH_2), 0.91 (d, 6H, $J = 6.0$ Hz, 2 × 21-Me), 0.86 (s, 6H, 2 × 19-Me), 0.60 (s, 6H, 2 × 18-Me) (Pandey *et al.*, 2002).

Synthesis of similar bile acid-based dimers **275**–**280** was accomplished from lithocholic acid (**48**) and deoxycholic acid (**258**) by connecting the side chains through spacer groups *via* ring D–ring D (Chattopadhyay and Pandey, 2006).

Formylation of lithocholic acid (**48**, 3.0 g, 7.96 mmol) produced 3α-*O*-formyllithocholic acid (**273**) (*Scheme 2.4.25*). HCO_2H (20 mL) was added to **48** (3.0 g, 7.96 mmol) and stirred at 60 °C for 4 h, cooled to room temperature, H_2O (100 mL) was added dropwise, the resulting solid was filtered and dried under vacuum to obtain **273** (3.15 g, 98%, mp: 127–128 °C). IR (KBr): ν_{max} cm^{-1} 3446br (O–H), 2948m (C–H), 1725s (C=O), 1450m, 1182m (C–O). ^{1}H NMR (300 MHz, $CDCl_3$): δ 8.04 (s, 1H, 3α-OCOH), 4.85 (m, 1H, 3β-CH), 1.08–2.40 (m, 28H, steroidal CH and CH_2), 0.91 (br s, 6H, 21-Me and 19-Me), 0.65 (s, 3H, 18-Me). ^{13}C NMR (75 MHz, $CDCl_3$): δ 12.0, 18.2, 20.8, 23.3, 24.1, 26.3, 26.6, 27.0, 28.1, 30.7, 31.0, 32.2, 34.6, 34.9, 35.3, 35.8, 40.1, 40.4, 41.9, 42.7, 55.9, 56.4, 74.4, 160.8, 180.4. HRMS: 427.2824 calculated for $C_{25}H_{40}O_4Na$), found: 427.2823 (Chattopadhyay and Pandey, 2006).

Deoxycholic acid (**258**, 3.0 g, 7.64 mmol) was employed for the synthesis of 3α,12α-*O*-diformyldeoxycholic acid (**274**, 3.35 g, 98%, mp: 173-175 °C), following the same

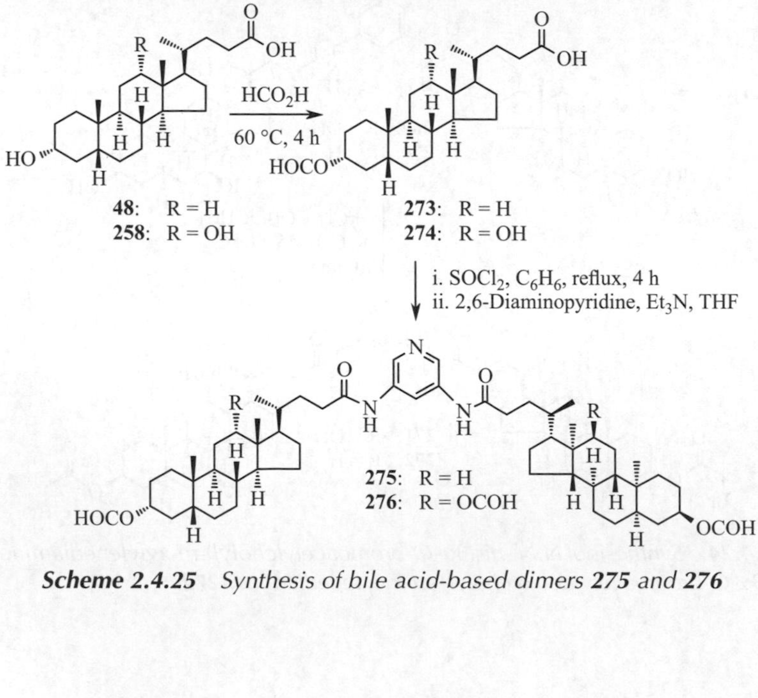

Scheme 2.4.25 *Synthesis of bile acid-based dimers **275** and **276***

procedure as described for **273** (*Scheme 2.4.25*). IR (KBr): ν_{max} cm^{-1} 3446br (O–H), 2948m (C–H), 1725s (C=O), 1721s (C=O), 1450m, 1182m (C–O). ^{1}H NMR (300 MHz, $CDCl_3$): δ 8.13 (s, 1H, 12α-OCOH), 8.03 (s, 1H, 3α-OCOH), 5.25 (m, 1H, 12β-CH), 4.85 (m, 1H, 3β-CH), 1.08-2.39 (m, 26H, steroidal CH and CH_2), 0.92 (s, 3H, 19-Me), 0.84 (d, 3H, J = 6.2 Hz, 21-Me), 0.75 (s, 3H, 18-Me). ^{13}C NMR (75 MHz, $CDCl_3$): δ 12.3, 17.4, 22.9, 23.4, 25.7, 25.0, 26.5, 26.8, 27.3, 30.5, 31.0, 32.1, 34.0, 34.2, 34.7, 34.8, 36.0, 41.7, 45.0, 47.4, 49.3, 73.2, 76.0, 160.6, 160.9, 180.1. HRMS: m/z 487.2462 calculated for $C_{26}H_{40}O_6K$, found: 487.2464 (Chattopadhyay and Pandey, 2006).

N,N′-Bis(3α-*O*-formyllithocholyl)-pyridine-2,6-diamine (**275**) was prepared from a stirred solution of **273** (4 g, 9.9 mmol) in dry C_6H_6 (20 mL), by treating it with freshly distilled $SOCl_2$ (1 mL) and DMF (few drops) at 0 °C, and stirring at 60 °C for 4 h (*Scheme 2.4.25*). The solvent was evaporated to dryness and dry C_6H_6 (10 mL) was added, rotary evaporated twice under pressure to remove excess $SOCl_2$. The dried acid chloride was dissolved in dry THF (10 mL) and added dropwise to a solution of 2,6-diaminopyridine (440 mg, 4.03 mmol) and Et_3N (1.4 mL) in dry THF (15 mL) at 0 °C. After completion of the reaction (monitored on TLC), the reaction mixture was concentrated and extracted with $CHCl_3$. The organic extract was dried (Na_2SO_4), rotary evaporated and the crude dimer was purified by FCC (mobile phase: *n*-hexane:EtOAc = 6:1) to produce dimer **275** (2.84 g, 80%, mp: 105–107 °C). IR (KBr): ν_{max} cm^{-1} 3326m (N–H), 2937m (C–H), 2863m (C–H), 1720s (C=O), 1588m (N–H), 1507m (N–H). ^{1}H NMR (300 MHz, $CDCl_3$): δ 8.04 (s, 2H, 2 × OCHO), 7.89 (d, 2H, J = 8.1 Hz, Py-3-CH and Py-5-CH), 7.67-7.72 (m, 3H, 2 × HCO and Py-4-CH), 4.85 (m, 2H, 2 × 3β-CH), 1.08–2.44 (m, 56H, steroidal CH and CH_2), 0.96 (br s, 6H, J = 6.1 Hz, 2 × 21-Me), 0.93 (s, 6H, 2 × 19-Me), 0.65 (s, 6H, 2 × 18-Me). ^{13}C NMR (75 MHz, $CDCl_3$): δ 12.1, 18.4, 20.8, 23.3, 24.2, 26.3, 26.6, 27.0, 28.2, 30.9, 31.4, 32.2, 34.8, 34.9, 35.4, 35.9, 40.1, 40.4, 41.9, 42.8, 56.0, 56.5, 74.4, 109.4, 141.0, 149.4, 160.8, 172.0. HRMS: m/z 882.6360 calculated for $C_{55}H_{84}N_3O_6$, found 882.6361 (Chattopadhyay and Pandey, 2006).

N,N′-Bis(3α,12α-*O*-diformyldeoxycholyl)-pyridine-2,6-diamine (**276**) was afforded from **274**, following the method as described for the synthesis of dimer **275** (*Scheme 4.2.25*). To a stirred solution of **274** (4.0 g, 8.91 mmol) in dry THF (10 mL), a solution of 2,6-diaminopyridine (440 mg, 4.03 mmol) in a mixture Et_3N (1.4 mL) and dry THF (15 mL) was added dropwise at 0 °C. After the usual workup and purification as above provided dimer **276** (3.02 g, 77%, mp: 154–155 °C). IR (KBr): ν_{max} cm^{-1} 3334m (N–H), 2946m (C–H), 2869m (C–H), 1722s (C=O), 1586m (N–H), 1506m (N–H). ^{1}H NMR (300 MHz, $CDCl_3$): δ 8.14 (s, 2H, 2 × OCHO), 8.03 (s, 2H, 2 × OCHO), 7.88 (d, 2H, J = 8.0 Hz, Py-3-CH and Py-5-CH), 7.69 (t, 1H, J = 8.0 Hz, Py-4-CH), 7.59 (s, 2H, 2 × NHCO), 5.26 (s, 2H, 2 × 12β-CH), 4.84 (m, 2H, 2 × 3β-CH), 1.04-2.42 (m, 52H, steroidal CH and CH_2), 0.93 (s, 6H, 2 × 19-Me), 0.87 (d, 6H, J = 6 Hz, 2 × 21-Me), 0.75 (s, 6H, 2 × 18-Me). ^{13}C NMR (75 MHz, $CDCl_3$): δ 12.4, 17.6, 22.9, 23.4, 25.7, 25.8, 26.5, 26.7, 27.4, 30.6, 31.0, 32.1, 34.0, 34.2, 34.6, 34.9, 35.6, 41.7, 45.0, 47.5, 49.2, 74.0, 76.0, 109.3, 141.1, 149.2, 160.5, 160.6, 171.8. HRMS: m/z 970.6157 calculated for $C_{57}H_{84}N_3O_{10}$, found: 970.6161 (Chattopadhyay and Pandey, 2006).

N,N′-Bis(3α-*O*-formyllithocholyl)-pyridine-2,6-diamine (**275**) was utilized for the synthesis of *N,N′*-bislithocholyl-pyridine-2,6-diamine (**277**) (*Scheme 2.4.26*). To a stirred solution of **275** (2.0 g, 2.26 mmol) in THF-H_2O (10:1, 25 mL), LiOH (200 mg, 4.65 mmol) was added at room temperature. After 12 h, the solvent was removed under pressure

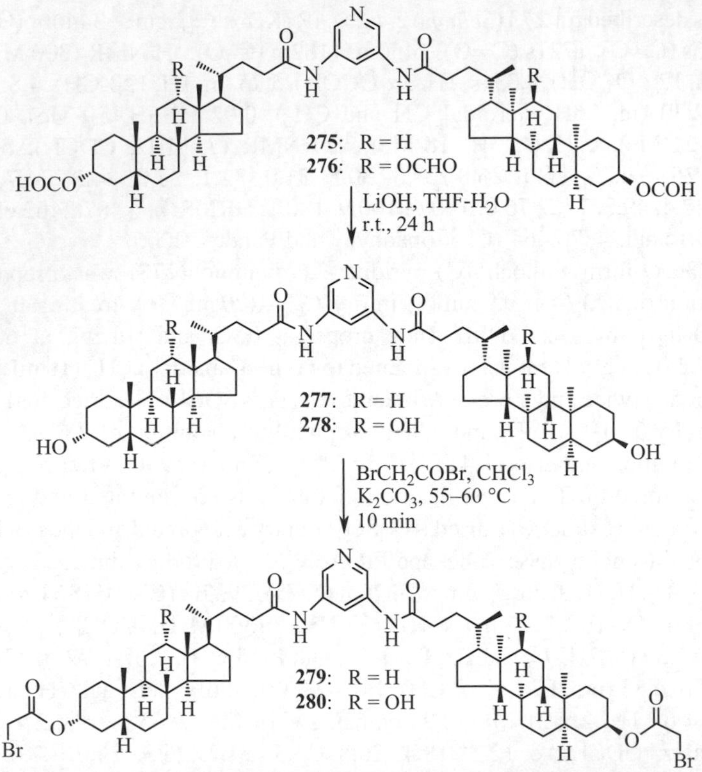

275: R = H
276: R = OCHO

LiOH, THF-H_2O
r.t., 24 h

277: R = H
278: R = OH

$BrCH_2COBr$, $CHCl_3$
K_2CO_3, 55–60 °C
10 min

279: R = H
280: R = OH

Scheme 2.4.26 *Synthesis of bile acid-based dimers **277–280***

and the residue was extracted with $CHCl_3$, dried (Na_2SO_4) and evaporated to dryness. The crude dimer was purified by FCC, eluted with *n*-hexane:EtOAc = 5:1, to give dimer **277** (1.57 g, 84%, mp:132-134 °C). IR (KBr): ν_{max} cm^{-1} 3422br (O–H and N–H), 2934m (C–H), 1683s (C=O), 1585m (N–H). ^1H NMR (300 MHz, $CDCl_3$): δ 0.58 (s, 6H, 2 × 18-Me), 0.84 (s, 6H, 2 × 19-Me), 0.88 (d, 6H, *J* = 6.1 Hz, 2 × 21-Me), 1.0-2.3 (m, 56H, steroidal CH and CH_2), 3.56 (m, 2H, 2 × 3β-CH), 7.53 (s, 2H, 2 × NHCO), 7.62 (t, 1H, *J* = 8.0 Hz, Py-4-CH), 7.82 (d, 2H, *J* = 8.0 Hz, Py-3-CH and py-5-CH). ^{13}C NMR (75 MHz, DMSO-d_6): δ 11.9, 18.3, 20.4, 23.3, 23.9, 26.2, 26.9, 27.8, 30.4, 31.2, 33.2, 34.2, 35.0, 35.4, 36.3, 38.9, 38.9, 40.1, 41.5, 42.3, 55.6, 56.1, 69.9, 108.9, 139.8, 150.4, 172.6. HRMS: *m/z* 826.6462 calculated for $C_{53}H_{84}N_3O_4$, found: 826.6434 (Chattopadhyay and Pandey, 2006).

Similarly, *N,N′*-bisdeoxycholyl-pyridine-2,6-diamine (**278**) was obtained from a stirred solution of **276** (2.0 g, 2.06 mmol) in THF-H_2O (10:1, 25 mL), treated with LiOH (400 mg, 9.3 mmol) at room temperature, following the same procedure as described for dimer **277** (*Scheme 2.4.26*). After the usual cleanup as above, the crude dimer was purified by FCC (mobile phase: 25% EtOAc in *n*-hexane) to yield dimer **278** (1.42 g, 80%, mp: 176–178 °C). IR (KBr): ν_{max} cm^{-1} 3420br (O–H and N–H), 2932m (C–H), 1680s (C=O), 1586m (N–H). ^1H NMR (300 MHz, $CDCl_3$): δ 0.63 (s, 6H, 2 × 18-Me), 0.83 (s, 6H, 2 × 19-Me), 0.94 (d, 6H, *J* = 6.0 Hz, 2 × 21-Me), 1.04–2.50 (m, 52H, steroidal CH and CH_2), 3.54 (m, 2H,

2 × 3β-CH), 3.9 (s, 2H, 2 × 12β-CH), 7.60 (t, 1H, J = 7.8 Hz, Py-4-CH), 7.72 (d, 2H, J = 7.8 Hz, Py-3-CH and Py-5-CH), 8.91 (s, 2H, 2 × NHCO). ^{13}C NMR (75 MHz, DMSO-d_6): δ 13.3, 17.9, 23.2, 23.9, 24.3, 26.6, 27.6, 27.80, 29.4, 31.0, 32.2, 33.7, 34.1, 34.6, 35.9, 36.5, 37.1, 42.4, 46.8, 48.3, 70.8, 71.9, 79.3, 108.8, 139.1, 151.2, 173.2. HRMS: *m/z* 858.6360 calculated for $C_{53}H_{84}N_3O_6$, found: 858.6346 (Chattopadhyay and Pandey, 2006).

N,N′-Bis(3α-*O*-bromoacetyllithocholyl)-pyridine-2,6-diamine (**279**) was synthesized from *N,N*′-bislithocholyl-pyridine-2,6-diamine (**277**) (*Scheme 2.4.26*). Anhydrous K_2CO_3 and $BrCH_2COBr$ (1.09 g, 5.44 mmol) in dry $CHCl_3$ (10 mL) were added, sequentially, to a stirred solution of **277** (1.5 g, 1.81 mmol) in dry $CHCl_3$ (20 mL) at 55–60 °C. After 10 min, the heating was stopped, ice-cold H_2O (20 mL) was added, the organic layer was separated, dried (Na_2SO_4), evaporated under pressure and the crude dimer was purified by FCC, using the same solvent system as mentioned above for dimer **275**, to afford dimer **279** (1.65 g, 85%, mp: 152–154 °C). IR (KBr): ν_{max} cm^{-1} 3337m (N–H), 2936m (C–H), 1734s (C=O), 1702s (C=O), 1586m, 1284m (C–O). ^{1}H NMR (300 MHz, $CDCl_3$): δ 0.65 (s, 6H, 2 × 18-Me), 0.86 (s, 6H, 2 × 19-Me), 0.93 (d, 6H, J = 6.2 Hz, 2 × 21-Me), 1.08-2.43 (m, 56H, steroidal CH and CH_2), 3.8 (s, 4H, 2 × $COCH_2Br$), 4.79 (m, 2H, 2 × 3β-CH), 7.52 (s, 2H, 2 × NHCO), 7.69 (t, 1H, J = 7.9 Hz, Py-4-CH), 7.88 (d, 2H, J = 7.9 Hz, Py-3-CH and Py-5-CH). ^{13}C NMR (75 MHz, $CDCl_3$): δ 12.0, 18.4, 20.8, 23.3, 24.2, 26.3, 26.9, 28.2, 29.7, 30.2, 31.4, 31.9, 32.0, 34.6, 34.9, 35.4, 35.7, 40.1, 40.4, 41.8, 42.7, 56.0, 56.4, 76.6, 109.3, 140.9, 149.4, 166.5, 171.9. HRMS: *m/z* 1066.4883 calculated for $C_{57}H_{86}N_3O_6Br_2$, found 1066.4932 (Chattopadhyay and Pandey, 2006).

N,N′-Bis(3α-*O*-bromoacetyldeoxycholyl)-pyridine-2,6-diamine (**280**) was achieved from *N,N*′-bisdeoxycholyl-pyridine-2,6 diamine (**278**, 1.5 g, 1.74 mmol), following the method as described for dimer **279** (*Scheme 2.4.26*). After the usual workup as above, the crude dimer was purified by FCC, using the same solvent system as mentioned for dimer **278**, to provide dimer **280** (1.33 g, 70%, mp: 162–165 °C). IR (KBr): ν_{max} cm^{-1} 3423br (O–H and N–H), 2939m (C–H), 1734s (C=O), 1585m (N–H), 1287m (C–O). ^{1}H NMR (300 MHz, $CDCl_3$): δ 7.88 (d, 2H, J = 7.9 Hz, Py-3-CH and Py-5-CH), 7.69 (t, 1H, J = 7.9 Hz, Py-4-CH), 7.61 (s, 2H, 2 × NHCO), 4.78 (m, 2H, 2 × 3β-CH), 4.00 (s, 2H, 2 × 12β-CH), 3.79 (s, 4H, 2 × $COCH_2Br$), 1.08-2.44 (m, 52H, steroidal CH and CH_2), 1.01 (d, 6H, J = 5.8 Hz, 2 × 21-Me), 0.93 (s, 6H, 2 × 19-Me), 0.69 (s, 6H, 2 × 18-Me). ^{13}C NMR (75 MHz, $CDCl_3$): δ 12.7, 17.4, 19.3, 23.0, 23.5, 25.9, 26.2, 26.8, 27.4, 28.7, 30.5, 31.1, 33.6, 34.0, 34.4, 34.7, 35.0, 35.9, 41.7, 46.4, 47.1, 48.2, 71.4, 76.5, 109.3, 141.0, 149.2, 166.7, 171.9. HRMS: *m/z* 1098.4782 calculated for $C_{57}H_{86}N_3O_8Br_2$, found: 1098.4810 (Chattopadhyay and Pandey, 2006).

Later, the same group reported the synthesis of cholic acid-based dimer **282** from cholic acid (**49**) in two steps, following the same procedure as described for bile acid-based dimers **275**–**280** (Chattopadhyay and Pandey, 2007). First, formylation of cholic acid (**49**) could be carried out, following the method described earlier for **273**, with HCO_2H at 60 °C to obtain 3α-*O*-formylcholic acid (**281**), which was employed for the synthesis of *N,N*′-bis(3α-*O*-formylcholyl)-pyridine-2,6-diamine (**282**) (*Scheme 4.2.27*), following the same protocol as utilized for dimer **275**. To a stirred solution of **281** (4.0 g, 8.12 mmol) in dry C_6H_6 (20 mL), freshly distilled $SOCl_2$ (1 mL) and a drop of DMF were added at 0 °C and stirred for 4 h at 60 °C. After the usual cleanup, the crude dimer was purified by FCC, eluted with 20% EtOAc in *n*-hexane, to produce dimer **282** (3.36 g, 79%, mp: 130 °C). IR (KBr): ν_{max} cm^{-1} 3341m (N–H and O–H), 2944m (C–H), 2872m (C–H), 1718s (C=O), 1585m, 1511m. ^{1}H

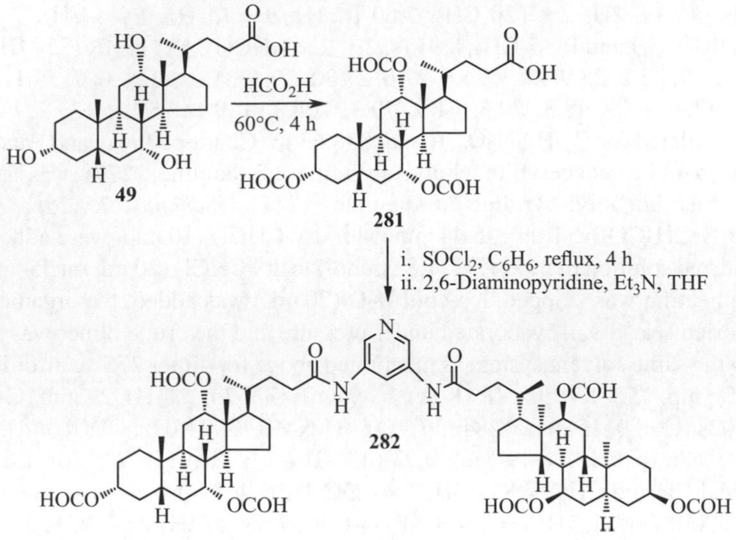

Scheme 2.4.27 *Synthesis of cholic acid-based dimer* ***282***

NMR (300 MHz, $CDCl_3$): δ 8.17 (s, 2H, 2 × 12α-OCOH), 8.10 (s, 2H, 2 × 7α-OCOH), 8.02 (s, 2H, 2 × 3α-OCOH), 7.87 (d, 2H, J = 7.9 Hz, Py-3-CH and Py-5-CH), 7.68 (t, 1H, J = 8.0 Hz, Py-4-CH), 7.55 (s, 2H, 2 × NHCO), 5.28 (br s, 2H, 2 × 12β-CH), 5.08 (br s, 2H, 2 × 7β-CH), 4.72 (m, 2H, 2 × 3β-CH), 1.05-2.45 (m, 48H, steroidal CH and CH_2), 0.95 (s, 6H, 2 × 19-Me), 0.88 (d, 6H, J = 6.0 Hz, 2 × 21-Me), 0.76 (s, 6H, 2 × 18-Me). ^{13}C NMR (75 MHz, $CDCl_3$): δ 12.2, 17.6, 22.3, 22.8, 25.6, 26.6, 27.2, 28.6, 31.0, 31.3, 34.3, 34.4, 34.5, 34.8, 37.7, 40.8, 43.0, 45.1, 47.3, 70.7, 73.7, 75.3, 77.3, 109.4, 141.0, 149.4 160.5, 160.6, 171.6. HRMS: m/z 1058.5953 calculated for $C_{59}H_{84}N_3O_{14}$, found: 1058.5944 (Chattopadhyay and Pandey, 2007).

Łotowski and Guzmanski (2006) prepared cholic acid-based dimers **287**–**290** *via* ring D side chains with oxamide and hydrazide spacers, according to the method as described earlier (Schemes *2.1.55* and *2.1.56*). In order to achieve the oxamide dimers **287** and **288**, 23-iodo-24-nor-5β-cholan-3α,7α,12α-yl triacetate (**284**) was obtained from 3α,7α,12α-triacetoxy-5β-cholan-24-oic acid (**283**) (*Scheme 2.4.28*). To a stirred solution of carboxylic acid **283** (4.0 g, 7.49 mmol) in CCl_4 (100 mL), red HgO (1.9 g, 8.76 mmol) was added and the mixture was refluxed for 2 h, a solution of I_2 (1.91 g, 7.52 mmol) in CCl_4 (50 mL) was added and refluxing continued for another 12 h. The reaction mixture was cooled, concentrated under pressure and the crude residue was purified by CC (mobile phase: 6% EtOAc in C_6H_6) to yield iodide **284** (3.0g, 65%). IR ($CHCl_3$) ν_{max} cm^{-1} 1724s (C=O), 1254s (C–O), 1025m. ^1H NMR (200MHz, $CDCl_3$): δ 5.08 (m, 1H, 12β-CH), 4.91 (m, 1H, 7β-CH), 4.56 (m, 1H, 3β-CH), 3.29 (m, 1H, 23-CH) and 3.06 (m, 1H, 23-CH), 2.12 (s, 3H, 12α-MeCO), 2.07 (s, 3H, 7α-MeCO), 2.03 (s, 3H, 3α-MeCO), 0.91 (s, 3H, 19-Me), 0.81 (d, 3H, J = 5.9 Hz, 21-Me), 0.74 (s, 3H, 18-Me). ^{13}C NMR (50 MHz, $CDCl_3$): δ 4.6, 12.2, 17.0, 21.3, 21.4, 21.5, 22.5, 22.7, 25.5, 26.8, 27.1, 28.8, 31.2, 34.3, 34.5, 34.6, 36.4, 37.7, 40.0, 40.8, 43.3, 45.1, 47.2, 70.6, 74.0, 75.3, 170.2, 170.3, 170.4. FABMS: m/z 639 $[M + Na]^+$ (Łotowski and Guzmanski, 2006).

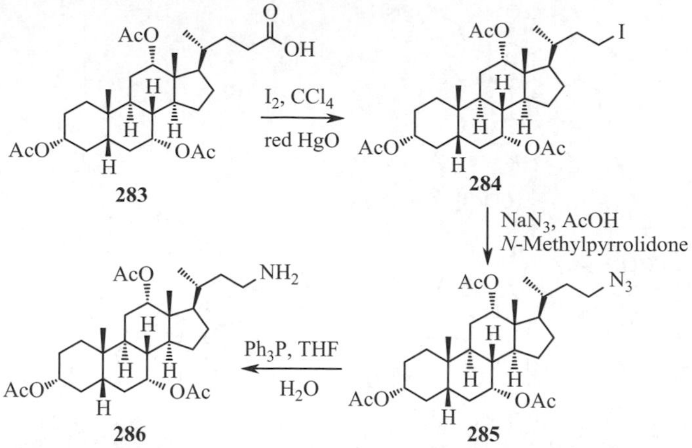

Scheme 2.4.28 Synthesis of cholic acid derivatives **284–286**

23-Iodo-24-nor-5β-cholan-3α,7α,12α-yl triacetate (**284**) was converted to 23-azido-24-nor-5β-cholan-3α,7α,12α-yl triacetate (**285**) (*Scheme 2.4.28*), following the procedure as described for **120** (*Scheme 2.1.54*). Usual workup and purification produced azide **285** (2.56 g, 99%). ^{1}H NMR (200 MHz, $CDCl_3$): δ 5.11 (m, 1H, 12β-CH), 4.91 (m, 1H, 7β-CH), 4.59 (m, 1H, 3β-CH), 3.33 (m, 1H, 23-CH) and 3.26 (m, 1H, 23-CH), 2.14 (s, 3H, 12α-MeCO), 2.09 (s, 3H, 7α-MeCO), 2.05 (s, 3H, 3β-MeCO), 0.93 (s, 3H, 19-Me), 0.84 (d, 3H, $J = 6.3$ Hz, 21-Me), 0.75 (s, 3H, 18-Me) (Łotowski and Guzmanski, 2006).

23-Amino-24-nor-5β-cholan-3α,7α,12α-yl triacetate (**286**) was produced from 23-azido-24-nor-5β-cholan-3α,7α,12α-yl triacetate (**285**) (*Scheme 2.4.28*), following the method as described for **121** (*Scheme 2.1.54*). Usual cleanup and purification as described above furnished amine **286** (1.52 g, 63%). IR ($CHCl_3$): ν_{max} cm^{-1} 1724s (C=O), 1254m (C–O), 1025m, 668w. ^{1}H NMR (200 MHz, $CDCl_3$): δ 5.10 (m, 1H, 12β-CH), 4.91 (m, 7β-CH), 4.57 (m, 3β-CH), 2.72 (m, 2H, 23-CH_2), 2.14 (s, 3H, 12α-MeCO), 2.08 (s, 3H, 7α-MeCO), 2.04 (s,3H, 3α-MeCO), 0.91 (s, 3H, 19-Me), 0.81 (d, 3H, $J = 5.9$ Hz, 21-Me), 0.73 (s, 3H, 18-Me) (Łotowski and Guzmanski, 2006).

23-Amino-24-nor-5β-cholan-3α,7α,12α-yl triacetate (**286**) was converted to triacetate **287** (*Scheme 2.4.29*), following the procedure as described for the dimer **122** (*Scheme 2.1.55*). After the usual workup and purification as described above produced, *N,N′*-di(3α,7α,12α-triacetoxy-24-nor-5β-cholan)23,23′-oxamide (**287**, 91.12 g, 82%). IR ($CHCl_3$): ν_{max} cm^{-1} 3399br (O–H and N–H), 1724s (C=O), 1675s (C=O), 1514m, 1254s (C–O), 1024s (C–O). ^{1}H NMR (200 MHz, CD_3OD): δ 7.41 (d, 2H, $J = 6.0$ Hz, 2 × NH), 5.09 (m, 2H, 2 × 12β-CH), 4.92 (m, 1H, 2 × 7β-CH), 4.58 (m, 2H, 2 × 3β-CH), 3.32 (m, 2H, 2 × 23-CH) and 3.20 (m, 2H, 2 × 23-CH), 2.14 (s, 6H, 2 × 12α-MeCO), 2.10 (s, 6H, 2 × 7α-MeCO), 2.05 (s, 6H, 2 × 3α-MeCO), 0.92 (s, 6H, 2 × 19-Me), 0.87 (d, 6H, $J = 6.0$ Hz, 2 × 21-Me), 0.73 (s, 6H, 2 × 18-Me) (Łotowski and Guzmanski, 2006).

N,N′-Di(3α,7α,12α-trihydroxy-24-nor-5β-cholan)23,23′-oxamide (**288**) was synthesized from a stirred solution of triacetate **287** (1.1 g, 1.03 mmol) in MeOH (50 mL), treated with NaOH (300 mg, 7.5 mmol) and H_2O (1 mL) at 40 °C for 4 days (*Scheme 2.4.29*). The solvent was removed, the crude dimer was purified by CC (mobile phase: 20% MeOH in

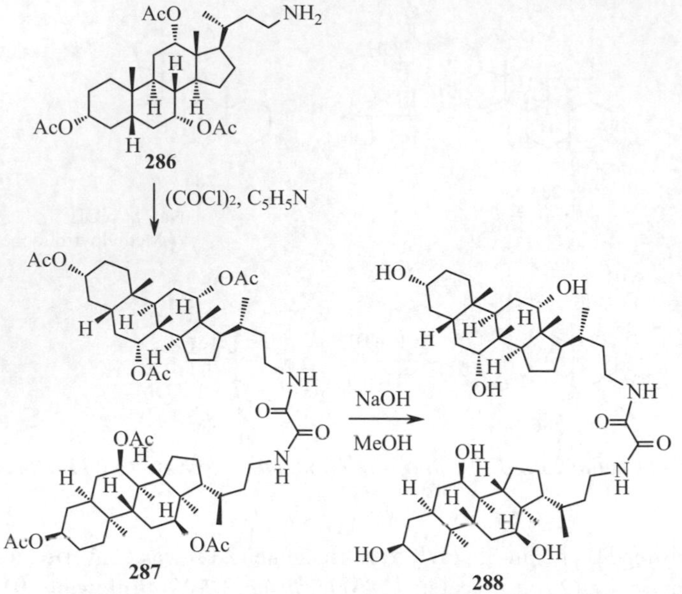

Scheme 2.4.29 *Synthesis of cholic acid-based dimers **287** and **288***

$CHCl_3$) and recrystallized from MeOH-EtOAc to afford triol **288** (650 mg, 77%, mp: 200–203 °C). IR ($CHCl_3$): ν_{max} cm^{-1} 1667s (C=O), 1516s (N–H), 1246m, 1079m, 1046m, 612w. ^{1}H NMR (200 MHz, CD_3OD): δ 8.53 (d, 2H, $J = 6.0$ Hz, 2 × NH), 3.95 (m, 2H, 2 × 12β-CH), 3.78 (m, 2H, 2 × 7β-CH), 3.60 (m, 2H, 2 × 3β-CH), 2.27 (m, 4H, 2 × 23-CH_2), 1.06 (d, 6H, $J = 6.3$ Hz, 2 × 21-Me), 0.91 (s, 6H, 2 × 19-Me), 0.71 (s, 6H, 2 × 18-Me). ^{13}C NMR (50 MHz, CD_3OD): δ 12.6, 17.6, 22.8, 23.9, 27.5, 28.4, 29.2, 30.8, 34.7, 35.1, 35.5, 36.0, 36.1, 37.8, 40.1, 40.6, 42.6, 42.8, 47.1, 47.8, 68.6, 72.5, 73.6, 161.3. FABMS: *m/z* 835 $[M + Na]^+$ (Łotowski and Guzmanski, 2006).

Applying the similar protocol as outlined above, two hydrazide dimers **289** and **290** were prepared from 3α,7α,12α-triacetoxy-5β-cholan-24-oic acid (**283**) (Łotowski and Guzmanski, 2006). 3α,7α,12α-Triacetoxy-5β-cholan-24-oic acid (**283**) was transformed to *N,N′*-di(5β-cholan-3α,7α,12α-yl triacetate)hydrazine (**289**) (*Scheme 2.4.30*). To a stirred solution of **283** (520 mg, 0.97 mmol) in dry C_6H_6 (20 mL), $(COCl)_2$ (1 mL, 11.6 mmol) was added dropwise at 60 °C. After 15 min, the reaction mixture was rotary evaporated to dryness and the residue was dissolved in dry $CHCl_3$ (EtOH free) and then N_2H_4 (1 mL, 1.0 mmol, 1M solution in THF) was added and stirred for 15 min. Acidified H_2O was added to the reaction mixture, extracted with $CHCl_3$, the organic layer was dried ($MgSO_4$) and the solvent was removed *in vacuo*. The crude dimer was purified by CC, eluted with 25% EtOAc in C_6H_6, and crystallized from hexane-DCM to give triacetate **289** (325 mg, 63%, mp: 153–156 °C). IR ($CHCl_3$): ν_{max} cm^{-1} 3411m (N–H), 1724s (C=O), 1634s (C=O), 1254m (C–O), 1025m. ^{1}H NMR (200 MHz, $CDCl_3$): δ 8.64 (s, 2H, 2 × NH), 5.07 (m, 2H, 2 × 12β-CH), 4.98 (m, 2H, 2 × 7β-CH), 4.56 (m, 2H, 2 × 3β-CH), 2.12 (s, 6H, 2 × 12β-MeCO), 2.08 (s, 6H, 2 × 7β-MeCO), 2.04 (s, 6H, 2 × 3β-MeCO), 0.90 (s, 6H, 2 × 19-Me), 0.80 (d, 6H, $J = 5.56$ Hz, 2 × 21-Me), 0.71 (s, 6H, 2 × 18-Me). ^{13}C NMR (50 MHz, $CDCl_3$): δ 12.2, 17.5, 21.3, 21.4, 21.6, 22.5, 22.7, 25.5, 26.8, 27.2, 28.8, 30.9, 31.1,

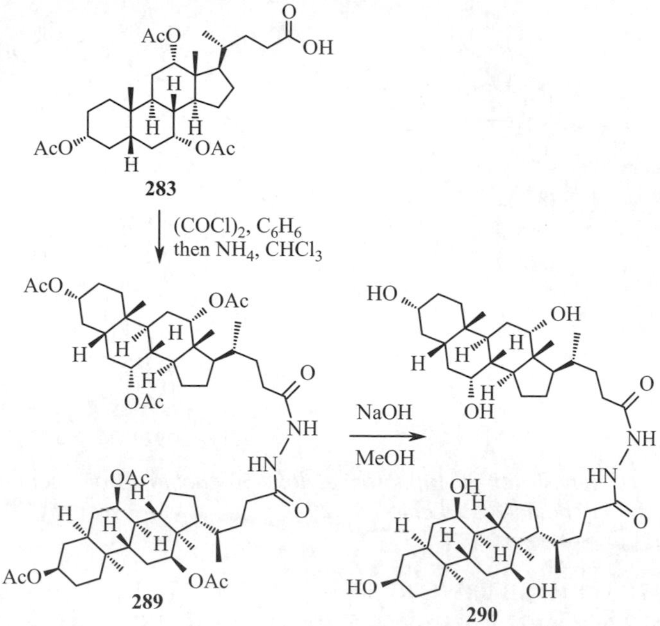

Scheme 2.4.30 *Synthesis of cholic acid-based dimers **289** and **290***

31.2, 34.3, 34.5, 34.6, 34.7, 37.7, 40.9, 45.0, 43.3, 47.4, 70.6, 74.0, 75.3, 170.2, 170.3, 170.4, 170.5. FABMS: *m*/*z* 1087 $[M + Na]^+$ (Łotowski and Guzmanski, 2006).

Similarly, *N*,*N*′-di(5β-cholan-3α,7α,12α-yl triacetate)hydrazine (**289**) was hydrolyzed to *N*,*N*′-di(5β-cholan-3α,7α,12α-triol)hydrazine (**290**) (*Scheme 2.4.30*), following the method outlined for triol **288**. After the usual cleanup, and crystallization from $CHCl_3$-MeOH afforded triol **290** (180 mg, 84%, mp: >350 °C). IR ($CHCl_3$): ν_{max} cm^{-1} 3398br (O–H and N–H), 1630s (C=O), 1075m (C–O), 607m. ^1H NMR (200 MHz, CD_3OD): δ 3.94 (m, 2H, 2 × 12β-CH), 3.80 (m, 2H, 2 × 7β-CH), 3.62 (m, 2H, 2 × 3β-CH), 1.01 (d, 6H, $J = 5.91$ Hz, 2 × 21-Me), 0.89 (s, 6H, 2 × 19-Me), 0.69 (s, 6H, 2 × 18-Me), a N–H singlet is observed at 9.63 ppm in DMSO-d_6. ^{13}C NMR (50 MHz, CD_3OD): δ 12.5, 17.2, 22.7, 23.0, 26.3, 27.4, 28.6, 30.3, 30.4, 34.5, 35.0, 35.2, 35.4, 39.3, 39.4, 40.2, 41.5, 41.6, 45.8, 46.3, 66.3, 70.5, 71.2, 171.7. FABMS: *m*/*z* 835 $[M + Na]^+$ (Łotowski and Guzmanski, 2006).

Lithocholic acid-based dimers, 3α-trifluoroacetyloxy-5β-cholan-24-oic acid ethane-1,2-diol diester (**293**) and 3α-hydroxy-5β-cholan-24-oic acid ethane-1,2-diol diester (**294**) were synthesized from lithocholic acid (**48**) in several steps (Kolehmainen *et al*., 1996). 3α-Trifluoroacetyloxy-5β-cholan-24-oic acid (**291**) was first prepared from **48** in the following manner (*Scheme 2.4.31*). $(CF_3CO)_2O$ (19 mL, 130 mmol) was added carefully to a cooled (−10 °C in an ice-salt bath) solution of lithocholic acid (**48**, 3.0 g, 8 mmol) in Na-dried THF (75 mL) over 15 min. The mixture was kept at −10 °C for 60 min, stirred for 2 h at room temperature. A mixture of Et_2O (120 mL) and ice (30 g) was added to the reaction mixture, the organic layer was sequentially washed with H_2O (60 mL), saturated aqueous $NaHCO_3$ to make the solution alkaline and neutralized with brine (60 mL), dried ($MgSO_4$) and evaporated under vacuum to obtain **291** (76 g, 100%, mp: 163–166 °C). ^{13}C NMR (67.7 MHz, $CDCl_3$): δ 34.8 (C-1), 26.3 (C-2), 79.4 (C-3), 31.7 (C-4), 41.9 (C-5), 27.0

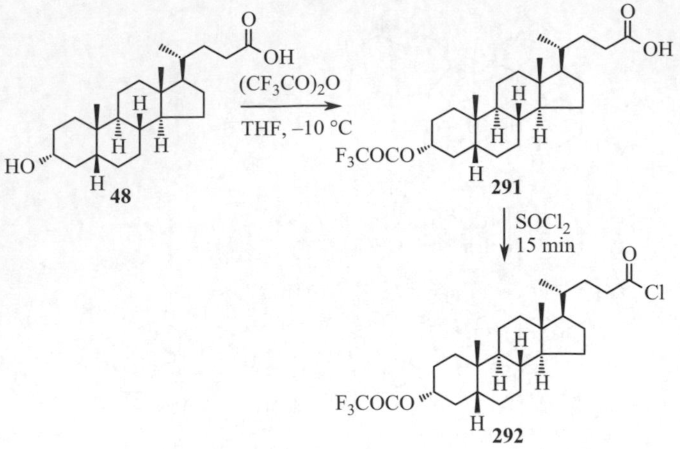

Scheme 2.4.31 *Synthesis of 3α-trifluoroacetyloxy-5β-cholan-24-oic acid (**291**) and 3α-trifluoroacetyloxy-5β-cholan-24-oyl chloride (**292**)*

(C-6), 26.2 (C-7), 35.8 (C-8), 40.5 (C-9), 34.6 (C-10), 20.9 (C-11), 40.1 (C-12), 42.8 (C-13), 56.5 (C-14), 24.2 (C-15), 28.2 (C-16), 56.5 (C-17), 11.8 (C-18), 23.2 (C-19), 35.6 (C-20), 18.1 (C-21), 31.2 (C-22), 30.8 (C-23), 180.7 (C-24), TFA: 157.1 (CO), 114.7 (CF_3). CIMS: *m*/*z* 472 $[M]^+$, 358 $[M - CF_3CO_2H]^+$ (Kolehmainen *et al*., 1996).

Compound **291** was converted to 3α-trifluoroacetyloxy-5β-cholan-24-oyl chloride (**292**) (*Scheme 2.4.31*). 3α-Trifluoroacetyloxy-5β-cholan-24-oic acid (**291**, 2.16 g, 4.60 mmol) and distilled $SOC1_2$ (15 mL) were refluxed for 15 min, excess of $SOC1_2$ was evaporated under pressure and the crude product was dissolved in $CC1_4$ (50 mL) and rotary evaporated to produce acetyl chloride **292** (2.25 g, 100%, mp: 98–101 °C). ^{13}C NMR (67.7 MHz, $CDCl_3$): δ 34.7 (C-1), 26.2 (C-2), 79.3 (C-3), 31.6 (C-4), 41.9 (C-5), 26.9 (C-6), 26.1 (C-7), 35.7 (C-8), 40.4 (C-9), 34.5 (C-10), 20.8 (C-11), 40.0 (C-12), 42.8 (C-13), 56.4 (C-14), 24.1 (C-15), 28.1 (C-16), 55.8 (C-17), 12.0 (C-18), 23.1 (C-19), 34.9 (C-20), 18.2 (C-21), 31.0 (C-22), 44.3 (C-23), 173.9 (C-24), TFA: 156.9 (CO), 114.6 (CF_3). CIMS: *m*/*z* 490 $[M]^+$, 386 $[M - CF_3Cl]^+$ (Kolehmainen *et al*., 1996).

3α-Trifluoroacetyloxy-5β-cholan-24-oyl chloride (**292**) was employed for the synthesis of 3α-trifluoroacetyloxy-5β-cholan-24-oic acid ethane-1,2-diol diester (**293**) (*Scheme 2.4.32*). A mixture of acetyl chloride **292** (5.19 g, 11.0 mmol) in $CC1_4$ (25 mL), $C_2H_6O_2$ (4.0 mL, 72.0 mmol) and C_5H_5N (1 mL, 12.0 mmol) was refluxed at 85 °C for 48 h. The lower layer of the reaction mixture was separated, $CHC1_3$ (10 mL) was added, the $CHCl_3$ layer was washed with saturated aqueous $NaHCO_3$ (3 × 10 mL) followed by H_2O (2 × 10 mL), the organic solvent was dried ($MgSO_4$) and evaporated to dryness. The crude dimer was purified by CC, eluted with 10% Me_2CO in $CHCl_3$, to yield dimer **293** (2.22 g, 40%, mp: 70–80 °C). ^{13}C NMR (67.7 MHz, $CDCl_3$): δ 34.6 (2 × C-1), 26.1 (2 × C-2), 79.1 (2 × C-3), 31.5 (2 × C-4), 41.8 (2 × C-5), 26.8 (2 × C-6), 26.0 (2 × C-7), 35.6 (2 × C-8), 40.3 (2 × C-9), 34.4 (2 × C-10), 20.7 (2 × C-11), 40.0 (2 × C-12), 42.6 (2 × C-13), 56.3 (2 × C-14), 24.0 (2 × C-15), 28.0 (2 × C-16), 55.9 (2 × C-17), 11.9 (2 × C-18), 23.0 (2 × C-19), 35.2 (2 × C-20), 18.1 (2 × C-21), 30.9 (2 × C-22), 30.8 (2 × C-23), 173.5 (2 × C-24), ethane-1,2-diol: 62.0 (C-1 and C-2), TFA: 156.9

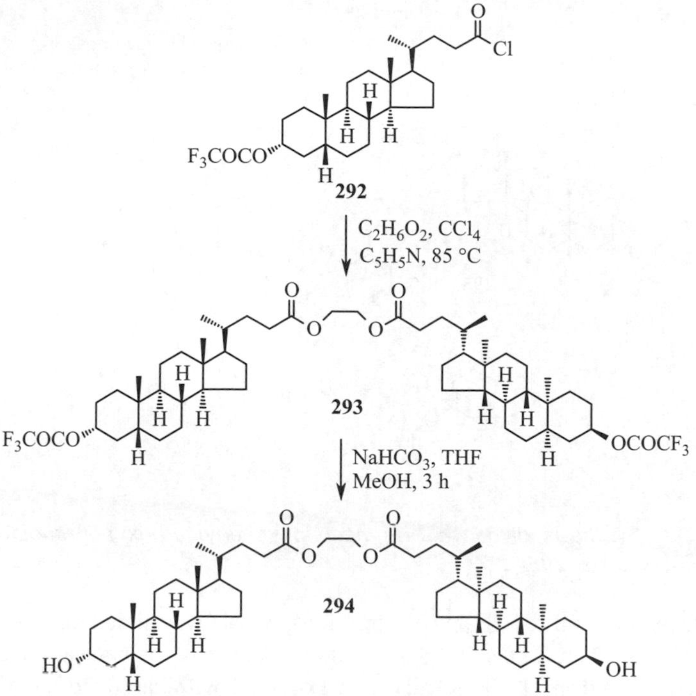

Scheme 2.4.32 *Synthesis of lithocholic acid-based dimers **293** and **294***

(2 × CO), 114.4 (2 × CF_3). CIMS: *m/z* 970 $[M]^+$, 856 $[M - CF_3CO_2H]^+$ (Kolehmainen *et al.*, 1996).

3α-Trifluoroacetyloxy-5β-cholan-24-oic acid ethane-1,2-diol diester (**293**, 2.22 g, 2.29 mmol) in a mixture of 50% THF in MeOH (90 mL) and saturated aqueous $NaHCO_3$ (22 mL) were stirred at room temperature for 60 min, an additional amount of saturated aqueous $NaHCO_3$ (4 mL) was added and the mixture was stirred for another 2 h (*Scheme 2.4.32*). A mixture of Et_2O (160 mL) and H_2SO_4 (80 mL, 0.6M) was added to the reaction mixture, the Et_2O layer was washed twice with H_2O (40 mL), dried ($MgSO_4$) and evaporated under pressure. The crude dimer was purified by CC as outlined above to afford lithocholic acid ethane-1,2-diol diester (**294**,1.20 g, 70%, mp: 90–100 °C). ^{13}C NMR (67.7 MHz, $CDCl_3$): δ 35.6 (2 × C-1), 30.5 (2 × C-2), 71.7 (2 × C-3), 36.4 (2 × C-4), 42.1 (2 × C-5), 27.2 (2 × C-6), 26.4 (2 × C-7), 35.8 (2 × C-8), 40.4 (2 × C-9), 34.5 (2 × C-10), 20.8 (2 × C-11), 40.2 (2 × C-12), 42.7 (2 × C-13), 56.5 (2 × C-14), 24.2 (2 × C-15), 28.1 (2 × C-16), 55.9 (2 × C-17), 12.0 (2 × C-18), 23.3 (2 × C-19), 35.3 (2 × C-20), 18.2 (2 × C-21), 31.1 (2 × C-22), 30.9 (2 × C-23), 173.9 (2 × C-24), ethane-1,2-diol: 62.0 (C-1 and C-2). CIMS: *m/z* 779 $[M]^+$, 761 $[M - HO_2]^+$ (Kolehmainen *et al.*, 1996).

Similarly, 3α,3′α-bis(n-acetyloxyphenylcarboxy)-5β-cholan-24-oic acid ethane-1,2-diol diesters (**295–297**) could be obtained by the reaction of an aroyl chloride with lithocholic acid ethane-1,2-diol diester (**294**) (Tamminen *et al.*, 2000b). Dimer **294** (1.0 g, 1.28 mmol) and freshly prepared 2-acetyloxybenzoyl chloride (770 mg, 3.88 mmol) were dissolved in

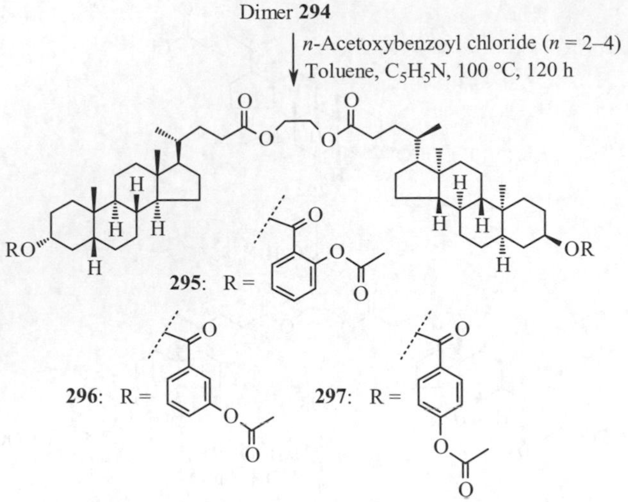

Dimer **294**

n-Acetoxybenzoyl chloride (*n* = 2–4)
Toluene, C_5H_5N, 100 °C, 120 h

295: R =

296: R =

297: R =

Scheme 2.4.33 *Synthesis of 3α,3′α-bis(n-acetyloxyphenylcarboxy)-5β-cholan-24-oic acid ethane-1,2-diol diesters (**295–297**)*

Na-dried toluene (130 mL), C_5H_5N (210 mg, 2.65 mmol) was added and the mixture was kept at 100 °C for 120 h (*Scheme 2.4.33*). The solvent was rotary evaporated, the crude product was dissolved in DCM (50 mL) and extracted with saturated aqueous $NaHCO_3$ solution (4 × 50 mL), washed with H_2O (50 mL), dried ($MgSO_4$) and evaporated to dryness. The crude dimer was purified by step gradient CC (mobile phase: 2% Me_2CO in DCM, then 10% EtOAc in *n*-hexane) to give 3α,3′α-bis(2-acetyloxyphenylcarboxy)-5β-cholan-24-oic acid ethane-1,2-diol diester (**295**, 36 mg, 2.5%). ^{13}C NMR (125 MHz, $CDCl_3$): δ 35.0 (2 × C-1), 26.7 (2 × C-2), 75.2 (2 × C-3), 32.3 (2 × C-4), 41.9 (2 × C-5), 27.0 (2 × C-6), 26.4 (2 × C-7), 35.8 (2 × C-8), 40.6 (2 × C-9), 34.6 (2 × C-10), 20.9 (2 × C-11), 40.2 (2 × C-12), 42.8 (2 × C-13), 56. 6 (2 × C-14), 24.2 (2 × C-15), 28.2 (2 × C-16), 56.0 (2 × C-17), 12.1 (2 × C-18), 23.3 (2 × C-19), 35.3 (2 × C-20), 18.3 (2 × C-21), 31.1 (2 × C-22), 30.9 (2 × C-23), 174.0 (2 × C-24), 62.0 (2 × C-25), 164.6 (2 × aroyl CO), 2-acetyloxybenzoic acid: 124.1 (2 × C-1), 150.5 (2 × C-2), 123.7 (2 × C-3), 133.5 (2 × C-4), 125.9 (2 × C-5), 131.6 (2 × C-6), 169.5 (2 × C-7), 21.1 (2 × C-8). MS (MALDI–TOF): *m/z* 1125.6 [M + Na]$^+$ (Tamminen *et al.*, 2000b).

The same procedure was followed for the synthesis of dimer **296** using 3-acetyloxybenzoyl chloride (*Scheme 2.4.33*). The crude dimer was purified by step gradient CC (mobile phase: 40% EtOAc in *n*-hexane, then 10% EtOAc in DCM) to obtain 3α,3′α-bis(3-acetyloxyphenylcarboxy)-5β-cholan-24-oic acid ethane-1,2-diol diester (**296**, 40 mg, 2.8%). ^{13}C NMR (125 MHz, $CDCl_3$): δ 35.1 (2 × C-1), 26.7 (2 × C-2), 75.4 (2 × C-3), 32.3 (2 × C-4), 42.0 (2 × C-5), 27.0 (2 × C-6), 26.3 (2 × C-7), 35.8 (2 × C-8), 40.5 (2 × C-9), 34.6 (2 × C-10), 20.9 (2 × C-11), 40.2 (2 × C-12), 42.8 (2 × C-13), 56.5 (2 × C-14), 24.2 (2 × C-15), 28.2 (2 × C-16), 56.1 (2 × C-17), 12.1 (2 × C-18), 23.4 (2 × C-19), 35.4 (2 × C-20), 18.3 (2 × C-21), 31.2 (2 × C-22), 30.9 (2 × C-23), 174.0 (2 × C-24), 62.0 (2 × C-25), 165.1 (2 × aroyl CO), 3-acetyloxybenzoic acid: 132.5 (2 × C-1), 122.7 (2 × C-2), 150.6 (2 × C-3), 126.0 (2 × C-4), 129.3 (2 × C-5), 127.0 (2 × C-6), 169.2 (2 × C-7), 21.0 (2 × C-8). MS (MALDI–TOF): *m/z* 1125.5 [M + Na]$^+$ (Tamminen *et al.*, 2000b).

Again, following the above method, 3α,3′α-bis(4-acetyloxyphenylcarboxy)-5β-cholan-24-oic acid ethane-1,2-diol diester (**297**) was synthesized from **294** by using a different aroyl chloride, 4-acetyloxybenzoyl chloride (*Scheme 2.4.33*). The crude dimer was purified by step gradient CC (mobile phase: 5% Me_2CO in DCM, then 20% EtOAc in DCM) to provide dimer **297** (30 mg, 2.1%). ^{13}C NMR (125 MHz, $CDCl_3$): δ 35.1 (2 × C-1), 26.8 (2 × C-2), 75.1 (2 × C-3), 32.4 (2 × C-4), 42.0 (2 × C-5), 27.1 (2 × C-6), 26.4 (2 × C-7), 35.8 (2 × C-8), 40.5 (2 × C-9), 34.6 (2 × C-10), 20.9 (2 × C-11), 40.2 (2 × C-12), 42.8 (2 × C-13), 56.6 (2 × C-14), 24.2 (2 × C-15), 28.2 (2 × C-16), 56.0 (2 × C-17), 12.1 (2 × C-18), 23.3 (2 × C-19), 35.3 (2 × C-20), 18.3 (2 × C-21), 31.1 (2 × C-22), 30.9 (2 × C-23), 174.0 (2 × C-24), 62.0 (2 × C-25), 165.3 (2 × aroyl CO), 4-acetyloxybenzoic acid: 128.5 (2 × C-1), 131.1 (2 × C-2), 121.5 (2 × C-3), 154.1 (2 × C-4), 121.5 (2 × C-5), 131.1 (2 × C-6), 168.8 (2 × C-7), 21.1 (2 × C-8). MS (MALDI–TOF): m/z 1125.4 $[M + Na]^+$ (Tamminen *et al.*, 2000b).

Several lithocholic acid-based dimers, 3β,3′β-bis(pyridine-n-carboxy)-5β-cholan-24-oic acid ethane-1,2-diol diesters (**298–300**), were prepared from lithocholic acid ethane-1,2-diol diester (**294**) in a single step (Kolehmainen *et al.*, 1999). Pyridine-n-carboxylic acid chloride was prepared freshly from pyridine-n-carboxylic acid and $SOCl_2$. A stirred mixture of **294** (500 mg, 0.64 mmol) in Na-dried toluene (65 mL) was treated with freshly prepared pyridine-n-carboxylic acid chloride (440 mg) in C_5H_5N (400 μL, 5 mmol) at 100 °C on an oil bath (*Scheme 2.4.34*). After 90 h, the solvent was rotary evaporated and the solid was dissolved in $CHCl_3$, the solution was washed with saturated aqueous $NaHCO_3$ (4 × 40 mL), followed by H_2O (40 mL), dried ($MgSO_4$) and evaporated under vacuum. The crude residue was purified by CC, eluted with 12% EtOAc in DCM, to yield dimers **298–300**.

3β,3′β-Bis(pyridine-2-carboxy)-5β-cholan-24-oic acid ethane-1,2-diol diesters (**298**, 110 mg, 17%). ^{13}C NMR (125 MHz, $CDCl_3$): δ 35.1 (2 × C-1), 26.5 (2 × C-2), 76.1 (2 × C-3), 32.2 (2 × C-4), 42.0 (2 × C-5), 27.1 (2 × C-6), 26.3 (2 × C-7), 35.8 (2 × C-8), 40.3 (2 × C-9), 34.7 (2 × C-10), 20.8 (2 × C-11), 40.2 (2 × C-12), 42.7 (2 × C-13), 56.5 (2 × C-14), 24.2 (2 × C-15), 28.2 (2 × C-16), 56.0 (2 × C-17), 12.0 (2 × C-18), 23.2 (2 × C-19), 35.3 (2 × C-20), 18.2 (2 × C-21), 31.1 (2 × C-22), 30.9 (2 × C-23), 173.9 (2 × C-24), ethane-1,2-diol: 62.0 (C-1 and C-2), pyridine-2-carboxylic acid: 164.7 (2 × CO), 148.7

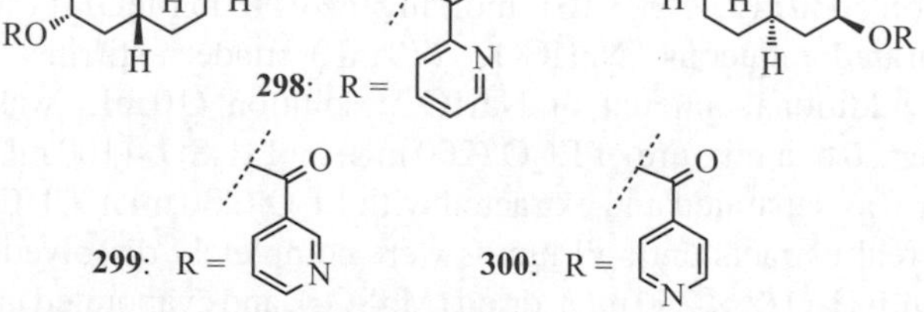

Dimer **294**

Pyridine-*n*-carboxylic acid chloride (*n* = 2–4)
Toluene, C_5H_5N, 100 °C, 120 h

298: R =
299: R =
300: R =

Scheme 2.4.34 *Synthesis of 3β,3′β-bis(pyridine-n-carboxy)-5β-cholan-24-oic acid ethane-1,2-diol diesters (**298–300**)*

(2 × C-1), 149.8 (2 × C-3), 126.6 (2 × C-4), 136.9 (2 × C-5), 125.1 (2 × C-6) (Kolehmainen *et al.*, 1999).

3β,3′β-Bis(pyridine-3-carboxy)-5β-cholan-24-oic acid ethane-1,2-diol diesters (**299**, 180 mg, 28%). ^{13}C NMR (125 MHz, $CDCl_3$): δ 34.9 (2 × C-1), 26.6 (2 × C-2), 75.6 (2 × C-3), 32.2 (2 × C-4), 41.9 (2 × C-5), 27.0 (2 × C-6), 26.3 (2 × C-7), 35.7 (2 × C-8), 40.4 (2 × C-9), 34.5 (2 × C-10), 20.8 (2 × C-11), 40.0 (2 × C-12), 42.7 (2 × C-13), 56.4 (2 × C-14), 24.1 (2 × C-15), 28.1 (2 × C-16), 55.9 (2 × C-17), 12.0 (2 × C-18), 23.3 (2 × C-19), 35.2 (2 × C-20), 18.2 (2 × C-21), 31.0 (2 × C-22), 30.8 (2 × C-23), 173.8 (2 × C-24), ethane-1,2-diol: 61.9 (C-1 and C-2), pyridine-3-carboxylic acid: 164.6 (2 × CO), 126.7 (2 × C-1), 150.6 (2 × C-2), 152.8 (2 × C-4), 123.2 (2 × C-5), 137.1 (2 × C-6) (Kolehmainen *et al.*, 1999).

3β,3′β-Bis(pyridine-4-carboxy)-5β-cholan-24-oic acid ethane-1,2-diol diesters (**300**, 120 mg, 19%). ^{13}C NMR (125 MHz, $CDCl_3$): δ 35.0 (2 × C-1), 26.7 (2 × C-2), 76.2 (2 × C-3), 32.3 (2 × C-4), 42.0 (2 × C-5), 27.0 (2 × C-6), 26.4 (2 × C-7), 35.8 (2 × C-8), 40.6 (2 × C-9), 34.7 (2 × C-10), 20.9 (2 × C-11), 40.2 (2 × C-12), 42.8 (2 × C-13), 56.5 (2 × C-14), 24.2 (2 × C-15), 28.2 (2 × C-16), 56.1 (2 × C-17), 12.1 (2 × C-18), 23.3 (2 × C-19), 35.4 (2 × C-20), 18.0 (2 × C-21), 31.1 (2 × C-22), 31.0 (2 × C-23), 174.0 (2 × C-24), ethane-1,2-diol: 62.1 (C-1 and C-2), pyridine-4-carboxylic acid: 164.5 (2 × CO), 138.3 (2 × C-1), 123.0 (2 × C-2), 150.3 (2 × C-3), 150.3 (2 × C-5), 123.0 (2 × C-6) (Kolehmainen *et al.*, 1999).

Five more lithocholic acid-based dimers **301–305** were synthesized from 3α-trifluoroacetyloxy-5β-cholan-24-oyl chloride (**292**) (Tamminen *et al.*, 2000c). For the synthesis of 3α,3′α-bis(trifluoroacetyloxy)-5β-cholan-24-oic acid piperazine diamide (**301**), piperazine (1.38 g, 16.02 mmol) and C_5H_5N (2.55 g, 32.24 mmol) were added to a stirred solution of **292** (15.65 g, 31.87 mmol) in CCl_4 (90 mL) and refluxed for 96 h (*Scheme 2.4.35*). The reaction mixture was cooled, $CHCl_3$ (75 mL) was added, and the $CHCl_3$ layer was extracted with saturated aqueous $NaHCO_3$ (3 × 150 mL), washed with H_2O (1 × 150 mL), dried ($MgSO_4$), and rotary evaporated. The crude diamide was purified by CC, eluted with 5% Me_2CO in $CHCl_3$, to provide dimer **301** (8.87 g, 56%). ^{13}C NMR (125 MHz, $CDCl_3$): δ 34.7 (2 × C-1), 26.2 (2 × C-2), 79.4 (2 × C-3), 31.6 (2 × C-4), 41.9 (2 × C-5), 26.9 (2 × C-6), 26.1 (2 × C-7), 35.8 (2 × C-8), 40.5 (2 × C-9), 34.5 (2 × C-10), 20.8 (2 × C-11), 40.1 (2 × C-12), 42.8 (2 × C-13), 56.4 (2 × C-14), 24.2 (2 × C-15), 28.3 (2 × C-16), 56.0 (2 × C-17), 12.0 (2 × C-18), 23.2 (2 × C-19), 35.6 (2 × C-20), 18.5 (2 × C-21), 31.3 (2 × C-22), 30.3 (2 × C-23), 172.5 (2 × C-24), piperazine: 45.4 (2 × C-1), 41.6 (2 × C-2), TGA: 157.0 (2 × CO), 114.6 (2 × CF_3). MS (MALDI–TOF): *m/z* 995.40 $[M + H]^+$ (Tamminen *et al.*, 2000c).

The synthesis of 3α,3′α-dihydroxy-5β-cholan-24-oic acid piperazine diamide (**302**) was prepared from a solution of **301** (3.15 g, 3.16 mmol) in 50% THF in MeOH (120 mL) by the treatment with saturated aqueous $NaHCO_3$ (70 mL) under stirring for 60 min (*Scheme 2.4.35*). An additional amount of $NaHCO_3$ solution (10 mL) was added and stirred continued. After 20 h, a mixture of Et_2O (200 mL) and H_2SO_4 (100 mL, 0.6M) was added, the lower layer was separated and extracted with Et_2O (230 mL). $CHCl_3$ (250 mL) was added to the ethereal extracts until all lumps were completely dissolved, the $CHCl_3$ solution was washed with H_2O (2 × 250 mL), dried ($MgSO_4$), and evaporated under vacuum to obtain dimer **302** (2.11 g, 83%). ^{13}C NMR (125 MHz, $CDCl_3$): δ 35.4 (2 × C-1), 30.6 (2 × C-2), 71.9 (2 × C-3), 36.5 (2 × C-4), 42.1 (2 × C-5), 27.2 (2 × C-6), 26.4 (2 × C-7),

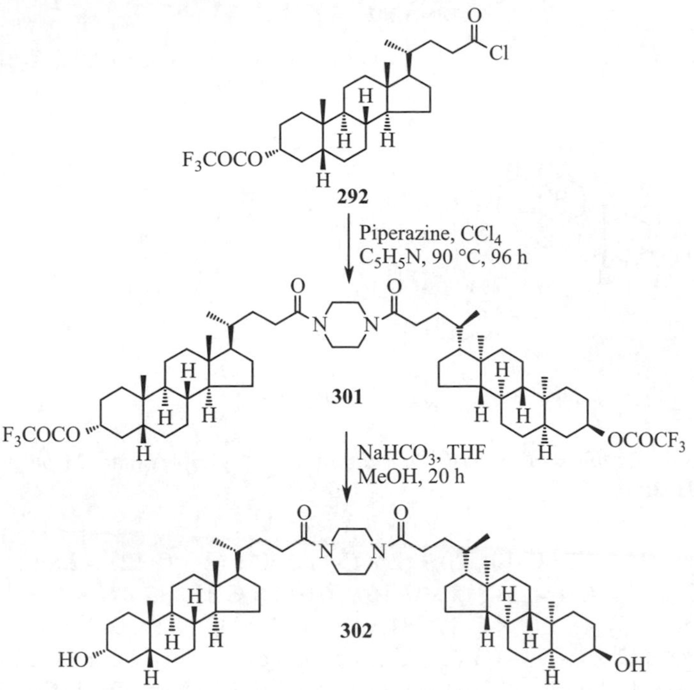

Scheme 2.4.35 *Synthesis of lithocolic acid-based dimers **301** and **302***

35.9 (2 × C-8), 40.5 (2 × C-9), 34.6 (2 × C-10), 20.8 (2 × C-11), 40.2 (2 × C-12), 42.8 (2 × C-13), 56.5 (2 × C-14), 24.2 (2 × C-15), 28.3 (2 × C-16), 56.0 (2 × C-17), 12.1 (2 × C-18), 23.4 (2 × C-19), 35.6 (2 × C-20), 18.5 (2 × C-21), 31.4 (2 × C-22), 30.3 (2 × C-23), 172.7 (2 × C-24), piperazine: 45.4 (2 × C-1), 41.6 (2 × C-2). MS (MALDI–TOF): *m/z* 803.89 $[M + H]^+$ (Tamminen *et al.*, 2000c).

The synthesis of lithocholic acid-based dimers, 3α,3′α-bis(pyridine-n-carboxy)-5β-cholan-24-oic acid piperazine diamides (**303–305**) (*Scheme 2.4.36*), following the same synthetic protocol as described earlier for lithocholic acid-based dimers **298–300**, was accomplished in a single step from hydroxy piperazine diamide **302** (500 mg, 0.64 mmol) (Tamminen *et al.*, 2000c).

3α,3′α-Bis(pyridine-2-carboxy)-5β-cholan-24-oic acid piperazine diamide (**303**, 170 mg, 13%). ^{13}C NMR (125 MHz, $CDCl_3$): δ 35.1 (2 × C-1), 26.5 (2 × C-2), 76.0 (2 × C-3), 32.1 (2 × C-4), 42.1 (2 × C-5), 27.0 (2 × C-6), 26.3 (2 × C-7), 35.8 (2 × C-8), 40.2 (2 × C-9), 34.6 (2 × C-10), 20.7 (2 × C-11), 40.1 (2 × C-12), 42.7 (2 × C-13), 56.4 (2 × C-14), 24.1 (2 × C-15), 28.2 (2 × C-16), 56.0 (2 × C-17), 12.0 (2 × C-18), 23.1 (2 × C-19), 35.5 (2 × C-20), 18.4 (2 × C-21), 31.3 (2 × C-22), 30.2 (2 × C-23), 172.4 (2 × C-24), piperazine: 45.3 (2 × C-1), 41.5 (2 × C-2), pyridine-2-carboxylic acid: 164.7 (2 × CO), 148.7 (2 × C-1), 149.8 (2 × C-3), 126.6 (2 × C-4), 136.8 (2 × C-5), 125.0 (2 × C-6). MS (MALDI–TOF): *m/z* 1014.34 $[M + H]^+$ (Tamminen *et al.*, 2000c).

3α,3′α-Bis(pyridine-3-carboxy)-5β-cholan-24-oic acid piperazine diamide (**304**, 610 mg, 48%). ^{13}C NMR (125 MHz, $CDCl_3$): δ 35.0 (2 × C-1), 26.7 (2 × C-2), 75.6 (2 × C-3), 32.3 (2 × C-4), 41.9 (2 × C-5), 27.0 (2 × C-6), 26.3 (2 × C-7), 35.8 (2 × C-8),

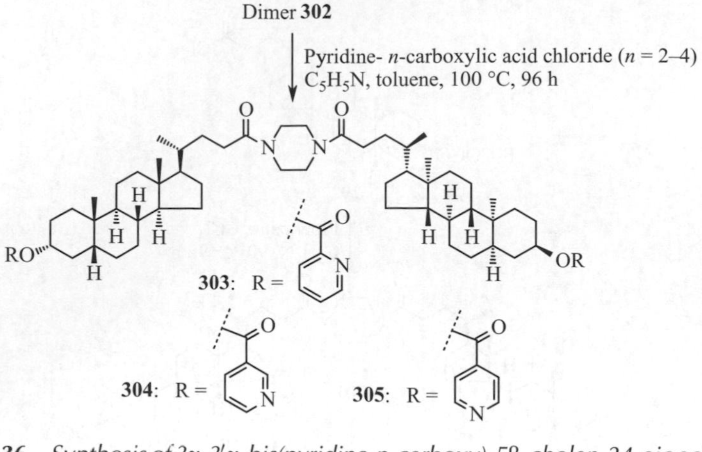

Dimer **302**

Pyridine- *n*-carboxylic acid chloride (n = 2–4)
C_5H_5N, toluene, 100 °C, 96 h

303: R =

304: R =

305: R =

Scheme 2.4.36 *Synthesis of 3α,3′α-bis(pyridine-n-carboxy)-5β-cholan-24-oic acid piperazine diamides (**303–305**)*

40.5 (2 × C-9), 34.6 (2 × C-10), 20.9 (2 × C-11), 40.1 (2 × C-12), 42.8 (2 × C-13), 56.4 (2 × C-14), 24.2 (2 × C-15), 28.3 (2 × C-16), 56.0 (2 × C-17), 12.0 (2 × C-18), 23.3 (2 × C-19), 35.6 (2 × C-20), 18.5 (2 × C-21), 31.3 (2 × C-22), 30.2 (2 × C-23), 172.4 (2 × C-24), piperazine: 45.3 (2 × C-1), 41.5 (2 × C-2), pyridine-3-carboxylic acid: 164.7 (2 × CO), 126.7 (2 × C-1), 150.9 (2 × C-2), 153.1 (2 × C-4), 123.1 (2 × C-5), 136.9 (2 × C-6). MS (MALDI–TOF): m/z 1014.42 $[M + H]^+$ (Tamminen *et al.*, 2000c).

3α,3′α-Bis(pyridine-4-carboxy)-5β-cholan-24-oic acid piperazine diamide (**305**, 640 mg, 51%). ^{13}C NMR (125 MHz, $CDCl_3$): δ 35.0 (2 × C-1), 26.6 (2 × C-2), 76.1 (2 × C-3), 31.2 (2 × C-4), 41.9 (2 × C-5), 27.0 (2 × C-6), 26.3 (2 × C-7), 35.8 (2 × C-8), 40.5 (2 × C-9), 34.6 (2 × C-10), 20.9 (2 × C-11), 40.1 (2 × C-12), 42.8 (2 × C-13), 56.5 (2 × C-14), 24.2 (2 × C-15), 28.3 (2 × C-16), 56.1 (2 × C-17), 12.0 (2 × C-18), 23.3 (2 × C-19), 35.6 (2 × C-20), 18.5 (2 × C-21), 31.3 (2 × C-22), 30.3 (2 × C-23), 172.4 (2 × C-24), piperazine: 45.3 (2 × C-1), 41.5 (2 × C-2), pyridine-4-carboxylic acid: 164.5 (2 × CO), 138.1 (2 × C-1), 122.8 (2 × C-2), 150.5 (2 × C-3), 150.5 (2 × C-5), 122.8 (2 × C-6). MS (MALDI–TOF): m/z 1014.15 $[M + H]^+$ (Tamminen *et al.*, 2000c).

The synthesis of new cholic acid-based *gemini* surfactants **308–310** were synthesized from cholic acid (**49**) (Alcalde *et al.*, 2008). Initially, 3α,7α,12α-trihydroxy-5β-cholan-24-amide (**306**) was prepared from **49** by reacting with a mixture of $EtCO_2Cl$, Et_3N, NH_3 in MeOH and dioxane for 2 h at room temperature (*Scheme 2.4.37*). ^{1}H NMR (300 MHz, CD_3OD): δ 3.95 (br s, 1H, 12α-CH), 3.79 (br s, 1H, 7β-CH), 3.34 (br s, 1H, 3β-CH), 0.91 (s, 3H, 19-Me), 0.71 (s, 3H, 18-Me). ^{13}C NMR (75 MHz, CD_3OD): δ 36.5 (C-1), 31.2 (C-2), 72.9 (C-3), 40.5 (C-4), 43.2 (C-5), 35.9 (C-6), 69.1 (C-7), 41.1 (C-8), 27.9 (C-9), 35.9 (C-10), 29.6 (C-11), 74.1 (C-12), 47.5 (C-13), 43.0 (C-14), 24.3 (C-15), 28.7 (C-16), 48.1 (C-17), 13.0 (C-18), 23.2 (C-19), 37.0 (C-20), 17.7 (C-21), 33.4 (C-22), 33.3 (C-23), 180.3 (C-24) (Alcalde *et al.*, 2008).

24-Cholanamide (**306**) was converted to 3α,7α,12α-trihydroxy-5β-cholan-24-amine (**307**) (*Scheme 2.4.37*). A mixture of **306** and $LiAlH_4$ in THF was refluxed for 24 h to provide 24-cholanamine (**307**). ^{1}H NMR (300 MHz, CD_3OD): δ 3.86 (br s, 1H, 12α-CH), 3.70 (br s, 1H, 7β-CH), 3.53 (br s, 1H, 3β-CH), 2.65 (m, 2H, 24-CH_2), 0.82 (s, 3H, 19-Me),

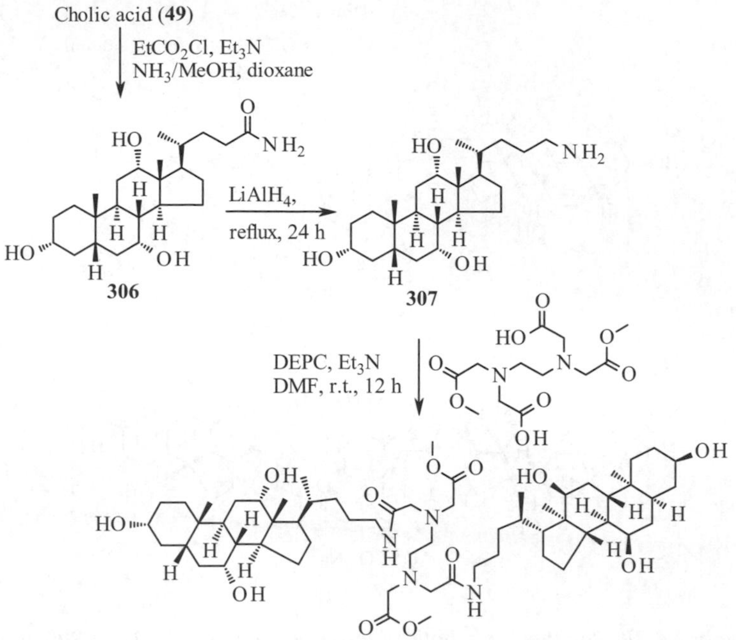

Cholic acid (**49**) → $EtCO_2Cl$, Et_3N, NH_3/MeOH, dioxane → **306** → $LiAlH_4$, reflux, 24 h → **307** → DEPC, Et_3N, DMF, r.t., 12 h → **308**

Scheme 2.4.37 *Synthesis of cholic acid* gemini *surfactant **308***

0.62 (s, 3H, 18-Me). ^{13}C NMR (75 MHz, CD_3OD): δ 40.5 (C-1), 31.2 (C-2), 72.9 (C-3), 40.5 (C-4), 43.2 (C-5), 35.9 (C-6), 69.1 (C-7), 41.0 (C-8), 27.9 (C-9), 35.9 (C-10), 29.6 (C-11), 74.1 (C-12), 47.5 (C-13), 43.0 (C-14), 24.3 (C-15), 28.7 (C-16), 48.1 (C-17), 13.0 (C-18), 23.2 (C-19), 37.1 (C-20), 18.0 (C-21), 34.1 (C-22), 27.8 (C-23), 42.2 (C-24) (Alcalde *et al.*, 2008).

24-Cholanamine (**307**) was transformed to cholic acid *gemini* surfactant **308** using dimethyl ester of EDTA (*Scheme 2.4.37*). A mixture of dimethyl ester of EDTA (600 mg, 1.87 mmol) in a mixture of dry DMF (5 mL) and dry THF (10 mL) and DEPC (650 μL, 4.28 mmol) solution was stirred at 0 °C for 30 min, a solution of amine **307** (1.55 g, 3.94 mmol) in a mixture of dry Et_3N (600 μL, 4.30 mmol) and dry THF (20 mL) was added dropwise and stirred at the same temperature for 1.5 h, the reaction temperature was allowed to rise to room temperature while stirring for another 6 h. The solvent was evaporated under pressure and the residue was dissolved in $CHCl_3$ (200 mL), washed with H_2O (2 × 50 mL) to remove DMF. The organic solvent was dried (Na_2SO_4), partially rotary evaporated and purified by CC, eluted with 30% MeOH in EtOAc, to give triol **308** (56%) (Alcalde *et al.*, 2008).

The surfactant **308** was refluxed with 1M KOH in MeOH at 80 °C to remove the methyl groups of the ester in the EDTA spacer (*Scheme 2.4.35*). After 60 min, the reaction mixture was rotary evaporated and the solid was redissolved in H_2O (200 mL) and acidified with HCl (pH 1), the solution was cooled in an ice bath to form precipitates, the resulting solids were filtered off and dried in a vaccum oven to obtain diacid **309**. ^{1}H NMR (300 MHz, CD_3OD): δ 7.95 (m, 4H, 2 × amide CH_2), 3.79 (br s, 2H, 2 × 12β-CH), 3.62 (br s, 2H, 7β-CH), 3.35(s,

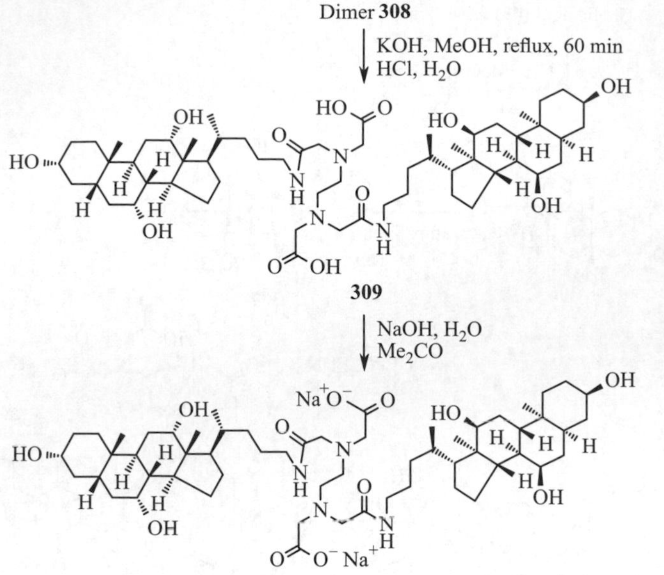

Dimer **308**

KOH, MeOH, reflux, 60 min
HCl, H_2O

309

NaOH, H_2O
Me_2CO

310

Scheme 2.4.38 *Synthesis of cholic acid-based* gemini *surfactants* ***309*** *and* ***310***

4H, 2 × CH_2-NH), 3.19 (s, 6H, 2 × CH_2-CO and 3β-CH), 3.04 (m, 4H, 2 × 24-CH_2), 2.70 (s, 4H, N-CH_2-CH_2-N), 0.85 (s, 6H, 2 × 19-Me), 0.59 (s, 6H, 2 × 18-Me). ^{13}C NMR (75 MHz, CD_3OD): δ 36.0 (2 × C-1), 31.1 (2 × C-2), 71.2 (2 × C-3), 40.3 (2 × C-4), 42.3 (2 × C-5), 35.6 (2 × C-6), 67.0 (2 × C-7), 40.2 (2 × C-8), 26.9 (2 × C-9), 35.1 (2 × C-10), 29.3 (2 × C-11), 71.8 (2 × C-12), 46.5 (2 × C-13), 42.0 (2 × C-14), 23.5 (2 × C-15), 28.0 (2 × C-16), 47.0 (2 × C-17), 13.0 (2 × C-18), 23.3 (2 × C-19), 37.8 (2 × C-20), 18.1 (2 × C-21), 33.6 (2 × C-22), 26.6 (2 × C-23), 39.6 (2 × C-24), 53.1 (N-CH_2-CH_2-N), 56.2 (CH_2-CO), 58.4 (CH_2-NH), 170.8 (C=O), 173.2 (C–NH) (Alcalde *et al.*, 2008).

The disodium salt **310** was prepared from diacid **309** by adding the stoichiometric amount of conc. NaOH (*Scheme 2.4.38*). Both the diacid and the disodium salts were repeatedly crystallized to yield the *gemini* surfactants **309** and **310**. The ^{1}H and ^{13}C NMR of these compounds were similar to that of *gemini* surfactant **308** (Alcalde *et al.*, 2008).

A C_2-symmetric polyhydroxylated dimer, bis-(20*S*)-5α-23,24-bisnorchol-16-en-3β,6α,7β-triol-22-terephthaloate (**325**) was prepared with a 22% yield from DHEA (**66**, androst-5-en-3β-ol-17-one) in a multistep sequence (Filippo *et al.*, 2003).

3β-[(*tert*-Butyldiphenylsilyl)oxy]-androst-5-en-17-one (**311**) was prepared from a mixture of DHEA (**66**, 500 mg, 1.97 mmol), DBU (460 μL, 3.0 mmol) and TPSCl (762 mg, 2.70 mmol) in DCM (5 mL), after stirring at room temperature for 2 h. Saturated aqueous NH_4Cl (2.5 mL) was added to the mixture and extracted with DCM. The extract was dried (Na_2SO_4), evaporated under pressure and purified by gradient FCC (mobile phase: 10–15% EtOAc in petroleum ether) to obtain DHEA derivative **311** (857 mg, 95%, mp: 119–121 °C). ^{1}H NMR (400 MHz, $CDCl_3$): δ 7.68 (m, 4H, Ph-CH), 7.40 (m, 6H, Ph-CH), 5.16 (m, 1H, 6-CH), 3.55 (m, 1H, 3α-CH), 1.08 [s, 9H, $C(CH_3)_3$], 1.02 (s, 3H, 19-Me), 0.86 (s, 3H, 18-Me). ^{13}C NMR (100 MHz, $CDCl_3$): δ 13.9, 19.0, 19.3, 20.1, 21.7, 26.5, 26.9, 30.6, 31.3, 31.7,

35.7, 36.5, 37.0, 42.4, 47.4, 50.0, 51.6, 72.9, 120.3, 127.4, 127.5, 129.4, 134.6, 134.8, 135.6, 142.3, 221.2. EIMS: *m/z* 526 $[M]^+$ (Filippo *et al.*, 2003).

A mixture of $C_2H_6O_2$ (16.6 mL, 190 mmol), $(EtO)_3CH$ [15.8 mL, 95.0 mmol] and *p*-TsOH·H_2O (543 mg, 28.0 mmol) were added to a stirred solution of **311** (10 g, 190 mmol) in DCM (3 mL), and allowed to stand overnight (*Scheme 2.4.39*). Et_3N (3 mL) and H_2O were added and the mixture was extracted with DCM. The organic layer was dried (Na_2SO_4), rotary evaporated and purified by FCC, eluted with 10% Et_2O in petroleum ether, to afford 3β-(*tert*-butyldiphenylsilyloxy)-17,17-(ethylenedioxy)-androst-5-ene (**312**, 11 g, 100%, mp: 120-122 °C). ^1H NMR (400 MHz, $CDCl_3$): δ 7.70 (m, 4H, Ph-CH), 7.40 (m, 6H, Ph-CH), 5.19 (m, 1H, 6-CH), 3.80–3.90 (m, 4H, OCH_2CH_2O), 3.62 (m, 1H, 3α-CH), 1.12 [s, 9H, $C(CH_3)_3$], 1.05 (s, 3H, 19-Me), 0.89 (s, 3H, 18-Me). ^{13}C NMR (100 MHz, $CDCl_3$): δ 13.2, 18.4, 19.5, 21.8, 26.5, 29.6, 30.2, 30.9, 31.2, 33.2, 35.5, 36.2, 41.5, 44.7, 48.9, 49.6, 63.5, 64.1, 72.2, 118.5, 120.0, 126.5, 128.5, 133.8, 134.8, 140.2. EIMS: *m/z* 570 $[M]^+$ (Filippo *et al.*, 2003).

Compound **312** was converted to 3β-(*tert*-butyldiphenylsilyloxy)-17,17-(ethylenedioxy)-androst-5-en-7-one (**313**) (*Scheme 2.4.39*). First, finely ground CrO_3 (175.2 g, 1.75 mol) was collected in a flask and dried under vacuum over P_2O_5 for 2 h. DCM (1000 mL) was added to the flask, the solution was cooled (−20 °C) and Ar was purged through it. Dimethylpyrazole (168.42 g, 1.75 mol) and **312** (50 g, 88 mmol) were added, sequentially, to the cooled solution and stirred at the same temperature (−20 °C) for 4 h. 5M NaOH (500 μL) was added and the reaction mixture was stirred at 0 °C for another 30 min. Et_2O (300 mL) was added and the solution was filtered through SiO_2 and $CaSO_4$ (10% in w/w), dried (Na_2SO_4) and evaporated under vacuum. The crude product was subjected to

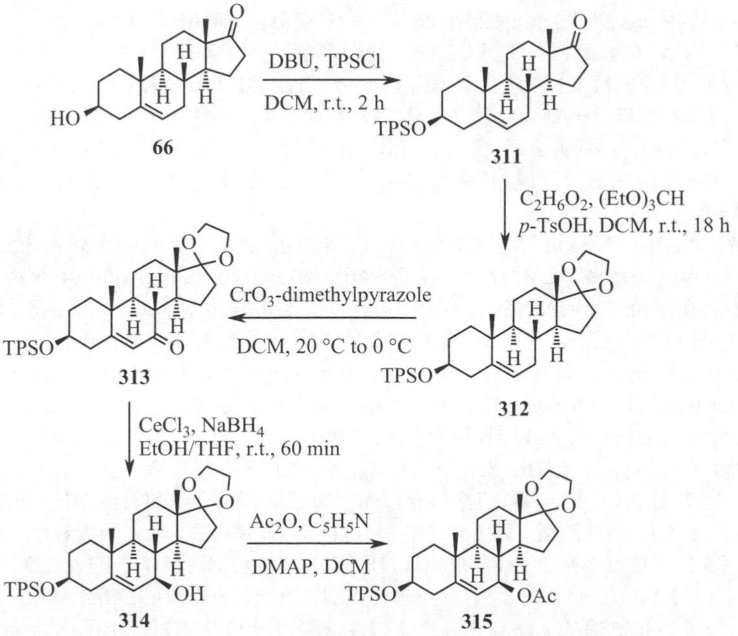

Scheme 2.4.39 *Synthesis of DHEA derivatives* ***311–315***

gradient FCC (mobile phase: 20–40% Et_2O in petroleum ether) to provide **313** (41.5 g, 81%, mp: 159–160 °C). ^1H NMR (400 MHz, $CDCl_3$): δ 7.65 (m, 4H, Ph-CH), 7.42 (m, 6H, Ph-CH), 5.46 (m, 1H, 6-CH), 3.80–3.90 (m, 4H, OCH_2CH_2O), 3.63 (m, 1H, 3α-CH), 1.07 [s, 9H, $C(CH_3)_3$], 1.16 (s, 3H, 19-Me), 0.84 (s, 3H, 18-Me). ^{13}C NMR (100 MHz, $CDCl_3$): δ 14.3, 17.3, 19.0, 20.6, 25.0, 26.9, 29.6, 31.4, 34.1, 36.2, 38.2, 42.1, 44.3, 45.2, 46.1, 49.7, 64.4, 65.1, 71.8, 118.6, 125.6, 127.5, 129.6, 129.7, 134.0, 134.2, 135.6, 166.0, 201.6. EIMS: *m/z* 584 $[M]^+$ (Filippo *et al.*, 2003).

3β-[(*tert*-Butyldiphenylsilyl)oxy]-17,17-(ethylenedioxy)-androst-5-en-7β-ol (**314**) was then prepared from a solution of **313** (40 g, 68 mmol) in a mixture of THF (400 mL) and EtOH (800 mL) at 0 °C after treating with $CeCl_3$ (12.7 g, 37 mmol) and $NaBH_4$ (5.17 g, 137 mmol) (*Scheme 2.4.39*). The reaction mixture was stirred at room temperature for 60 min, H_2O was added and the solvents were removed under pressure. The reduced residue was extracted with EtOAc, the solvent was dried (Na_2SO_4), rotary evaporated and purified by FCC, eluted with 20% Et_2O in petroleum ether, to give **314** (32.1 g, 81%, mp: 188-189 °C). ^1H NMR (400 MHz, $CDCl_3$): δ 7.66 (m, 4H, Ph-CH), 7.41 (m, 6H, Ph-CH), 5.03 (m, 1H, 6-CH), 3.80-3.90 (m, 4H, OCH_2CH_2O), 3.78 (bd, 1H, $J = 8.5$ Hz, 7-CH), 3.55 (m, 1H, 3α-CH), 1.05 [s, 9H, $C(CH_3)_3$], 1.02 (s, 3H, 19-Me),0.84 (s, 3H, 18-Me). ^{13}C NMR (100 MHz, $CDCl_3$): δ 14.2, 19.1, 20.3, 24.9, 26.9, 30.1, 31.7, 34.2, 36.5, 36.8, 41.0, 41.8, 46.0, 47.9, 49.8, 64.5, 65.1, 72.7, 73.4, 118.9, 125.1, 127.5, 129.5, 134.6, 135.7, 143.8. EIMS: *m/z* 586 $[M]^+$ (Filippo *et al.*, 2003).

Simple acetylation of **314** produced 3β-[(*tert*-butyldiphenylsilyl)oxy]-17,17-(ethylene dioxy)-7β-acetyloxy-androst-5-ene (**315**) (*Scheme 2.4.39*). Ac_2O (91 μL, 0.96 mmol) and a catalytic amount of DMAP were added to a stirred solution of **314** (141 mg, 0.240 mmol) in 10% DCM in C_5H_5N (1 mL) at room temperature. After 4 h, the solvent was evaporated under pressure and purified by FCC (mobile phase: 5% EtOAc in petroleum ether) to produce **315** (149 mg, 98%, mp: 124–126 °C). ^1H NMR (400 MHz, $CDCl_3$): δ 7.66 (m, 4H, Ph-CH), 7.39 (m, 6H, Ph-CH), 5.02 (m, 1H, 6-CH), 4.97 (bd, 1H, $J = 8.5$ Hz, 7-CH), 3.80–3.90 (m, 4H, OCH_2CH_2O), 3.54 (m, 1H, 3α-CH), 2.00 (s, 3H, s, $COCH_3$), 1.07 [s, 9H, $C(CH_3)_3$], 1.07 (s, 3H, 19-Me), 0.85 (s, 3H, 18-Me). ^{13}C NMR (100 MHz, $CDCl_3$): δ 14.1, 19.0, 20.3, 21.5, 23.7, 26.9, 29.9, 31.6, 34.0, 36.4, 36.6, 41.8, 45.9, 47.7, 49.3, 64.4, 65.1, 72.5, 75.7, 118.6, 120.8, 127.4, 129.5, 134.6, 135.6, 145.6, 170.9. EIMS: *m/z* 628 $[M]^+$ (Filippo *et al.*, 2003).

A mixture of $BH_3{\cdot}SMe_2$ in 2M THF (21.4 mL, 42.8 mmol) and **315** (5.38 g, 8.55 mmol) in THF (32 mL) was stirred at 0 °C (*Scheme 2.4.40*). After 10 min, the solution was warmed to room temperature and stirred for another 30 h. The solution was then cooled at 0 °C, and absolute EtOH (170 mL), a solution of 3M NaOH (59 mL) and H_2O_2 (60 mL, 30% in H_2O) were added and stirred for 2 h, concentrated under pressure and extracted with EtOAc. The organic layer was dried (Na_2SO_4), rotary evaporated and purified by FCC, eluted with 20% Et_2O in petroleum ether, to yield 3β-[(*tert*-butyldiphenylsilyl)oxy]-17,17-(ethylene-dioxy)-5α-androstan-6α,7β-diol (**316**, 3.71 g, 72%, mp: 177–178 °C). ^1H NMR (400 MHz, $CDCl_3$): δ 7.67 (m, 4H, Ph-CH), 7.38 (m, 6H, Ph-CH), 3.80–3.90 (m, 4H, OCH_2CH_2O), 3.55 (m, 1H, 3α-CH), 3.17 (m, 1H, 6β-CH), 3.05 (bd, 1H, $J = 8.5$ Hz, 7α-CH), 1.05 [s, 9H, $C(CH_3)_3$], 1.05 (s, 3H, 19-Me), 0.84 (s, 3H, 18-Me). ^{13}C NMR (100 MHz, $CDCl_3$): δ 13.5, 14.4, 19.0, 20.5, 25.3, 26.9, 30.1, 31.2, 32.1, 34.1, 35.6, 37.2, 40.9, 46.4, 47.5, 49.6, 51.6, 64.4, 65.1, 72.4, 74.7, 80.7, 118.6, 127.4, 129.4, 134.5, 134.7, 135.7. EIMS: *m/z* 604 $[M]^+$ (Filippo *et al.*, 2003).

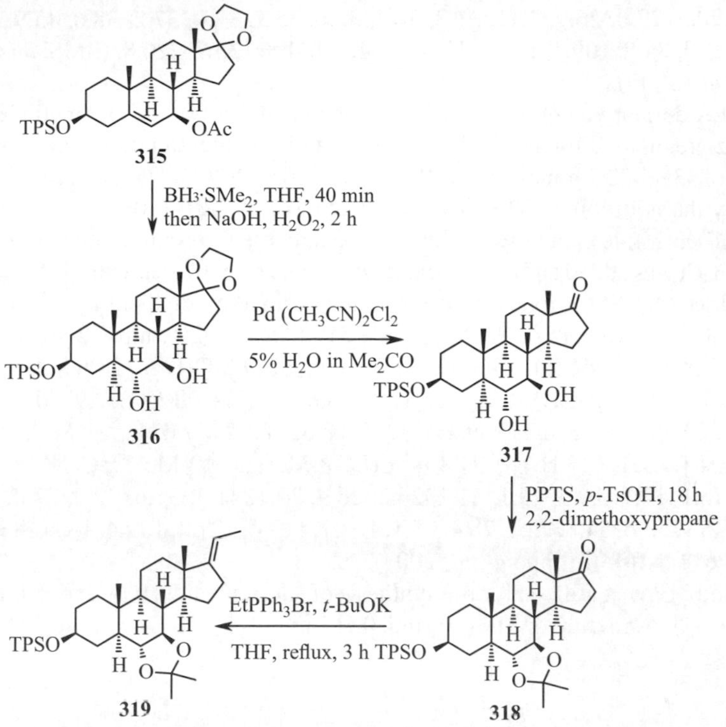

Scheme 2.4.40 Synthesis of DHEA derivatives **316–319**

3β-[(*tert*-Butyldiphenylsilyl)oxy]-5α-androstan-6α,7β-diol-17-one (**317**) was prepared from a stirred mixture of $Pd(MeCN)_2Cl_2$ (9.0 mg, 0.03 mmol) and **316** (100 mg, 0.165 mmol) in 5% H_2O in Me_2CO (3 mL) (*Scheme 2.4.40*). The reaction mixture was heated at 40 °C, then stirred for 4 h at room temperature and evaporated to dryness followed by purification of the residue on FCC (mobile phase: 40% Et_2O in petroleum ether) to afford **317** (78 mg, 84%, mp: 160–161 °C). ^1H NMR (400 MHz, $CDCl_3$): δ 7.66 (m, 4H, Ph-CH), 7.40 (m, 6H, Ph-CH), 3.80–3.90 (m, 4H, OCH_2CH_2O), 3.54 (m, 1H, 3α-CH), 3.19 (dd, 1H, $J = 8.6$, 11.1 Hz, 6β-CH), 3.12 (dd, 1H, $J = 8.6$ and 10.0 Hz, 7α-CH), 1.05 [s, 9H, $C(CH_3)_3$], 1.05 (s, 3H, 19-Me), 0.86 (s, 3H, 18-Me). ^{13}C NMR (100 MHz, $CDCl_3$): δ 13.6, 14.0, 19.1, 20.5, 24.7, 26.9, 31.2, 31.3, 32.1, 35.8, 35.9, 37.1, 40.2, 47.7, 48.2, 51.0, 51.9, 72.3, 74.9, 80.0, 127.4, 127.5, 129.5, 134.5, 134.6, 135.7, 221.3. FABMS: *m/z* 583 $[M + Na]^+$ (Filippo *et al.*, 2003).

In the next step, PPTS (89 mg 0.35 mmol) was added to a solution of **317** (1.98 g, 3.54 mmol) in 2,2-dimethoxypropane (10 mL) at room temperature and stirred overnight. Et_3N was added and the mixture was evaporated under pressure (*Scheme 2.4.40*). The residue was subjected to FCC, eluted with 15% Et_2O in petroleum ether, to obtain 3β-[(*tert*-butyl diphenylsilyl)oxy]-6α,7β-[(methylethyldene)-bisoxy]-5α-androstan-17-one (**318**, 1.98 g, 93%, mp: 104–105 °C). ^1H NMR (400 MHz, $CDCl_3$): δ 7.67 (m, 4H, Ph-CH), 7.38 (m, 6H, Ph-CH), 3.59 (m, 1H, 3α-CH), 3.22 (dd, 1H, $J = 8.6$ and 11.1 Hz, 6β-CH), 3.01 (dd, 1H, $J = 8.6$ and 10.0 Hz, 7α-CH), 1.96 (s, 3H, MeCO), 1.35 (s, 3H, MeCO), 1.03 [s, 9H, $C(CH_3)_3$], 0.90 (s, 3H, 19-Me), 0.87 (s, 3H, 18-Me). ^{13}C NMR (100 MHz, $CDCl_3$): δ 13.8,

14.7, 19.1, 20.3, 23.6, 26.9, 27.1, 31.2, 31.4, 32.6, 35.7, 37.0, 37.2, 38.6, 45.9, 47.5, 50.1, 52.6, 71.9, 78.3, 84.8, 109.1, 127.4, 129.4, 134.3, 134.8, 135.7, 220.8. EIMS: *m/z* 600 $[M]^+$ (Filippo *et al.*, 2003).

The DHEA derivative (*Z*)-3β-[(*tert*-butyldiphenylsilyl)oxy]-6α,7β-[(methylethyldene)-bis-oxy]-5α-pregn-17(20)-ene (**319**) was prepared by the treatment of a solution of $EtPPh_3Br$ (8.43 g, 22.7 mmol) in THF (36.5 mL) with *t*-BuOK, (2.33 g, 20.8 mmol), followed by the addition of **318** (3.98 g, 6.63 mmol) in THF (10 mL) under stirring for 10 min (*Scheme 2.4.40*), and then refluxed for 3 h, cooled to room temperature. Saturated aqueous NH_4Cl was added and concentrated *in vacuo*, and extracted with Et_2O. The ethereal layer was dried (Na_2SO_4), rotary evaporated and purified by gradient FCC (mobile phase: 10–30% Et_2O in petroleum ether) to provide **319** (3.76 g, 92%, mp: 72-74 °C). ^1H NMR (400 MHz, $CDCl_3$): δ 7.66 (m, 4H, Ph-CH), 7.38 (m, 6H, Ph-CH), 5.12 (br q, 1H, $J = 7.0$ Hz, 20-CH), 3.60 (m, 1H, 3α-CH), 3.21 (dd, 1H, $J = 8.6$, 11.1 Hz, 6β-CH), 2.97 (dd, 1H, $J = 8.6$, 10.0 Hz, 7α-CH), 1.96 (3H, s, MeCO), 1.62 (br d, 3H, $J = 7.0$ Hz, 21-Me), 1.39 (s, 3H, MeCO), 1.04 [s, 9H, $C(CH_3)_3$], 0.88 (s, 6H, 18-Me and 19-Me). ^{13}C NMR (100 MHz, $CDCl_3$): δ 13.1, 14.7, 16.9, 19.1, 21.2, 26.2, 26.9, 27.1, 31.3, 31.6, 32.7, 37.0, 37.2, 38.5, 40.5, 44.2, 45.9, 52.6, 54.8, 72.1, 78.4, 85.1, 108.7, 113.6, 127.4, 129.4, 133.6, 135.8, 149.5. EIMS: *m/z* 612 $[M]^+$ (Filippo *et al.*, 2003).

Compound **319** was utilized for the synthesis of subsequent DHEA derivatives **320–323** (*Scheme 2.4.41*). A mixture of **319** (100 mg, 0.163 mmol) in 5% H_2O in Me_2CO (3 mL) and

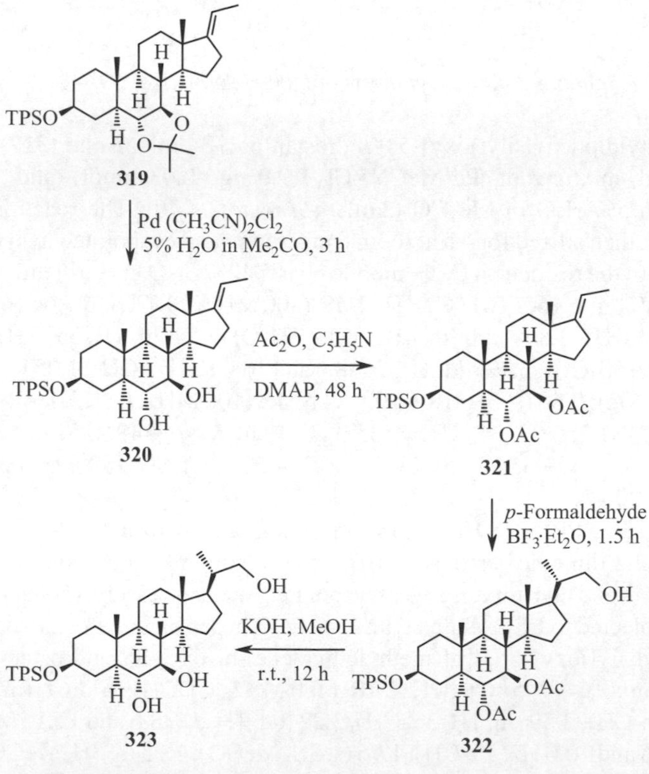

Scheme 2.4.41 *Synthesis of DHEA derivatives* ***320–323***

$Pd(MeCN)_2Cl_2$ (2.0 mg, 8 μmol) was heated at 40 °C and stirred for 3 h, the solvent was removed under pressure, followed by purification on FCC, eluted with 10% Et_2O in petroleum ether, to give (*Z*)-3β-[(*tert*-butyldiphenylsilyl)oxy]-5α-pregn-17(20)-en-6α,7β-diol (**320**, 86 mg, 92%, mp: 98–99 °C). ^1H NMR (400 MHz, $CDCl_3$): δ 7.66 (m, 4H, Ph-CH), 7.40 (m, 6H, Ph-CH), 5.13 (br q, 1H, *J* = 7.0 Hz, 20-CH), 3.56 (m, 1H, 3α-CH), 3.04 (dd, 1H, *J* = 8.6. 10.0 Hz, 7α-CH), 3.19 (dd, 1H, *J* = 11.1, 8.6 Hz, 6β-CH), 1.96 (3H, s, MeCO), 1.62 (br d, 3H, *J* = 7.0 Hz, 21-Me), 1.39 (s, 3H, MeCO), 1.05 [s, 9H, $C(CH_3)_3$], 0.87 (s, 3H, 19-Me), 0.84 (s, 3H, 18-Me). ^{13}C NMR (100 MHz, $CDCl_3$): δ 13.1, 13.5, 16.9, 19.0, 21.4, 26.9, 27.2, 31.2, 32.0, 32.2, 35.6, 36.8, 37.1, 40.5, 44.9, 47.6, 52.0, 55.6, 72.5, 74.8, 80.2, 113.7, 127.4, 129.4, 134.5, 134.8, 135.7, 149.1. EIMS: *m/z* 572 $[M]^+$ (Filippo *et al.*, 2003).

The diol **320** (4.00 g, 6.99 mmol) in DCM (20 mL) was acetylated using C_5H_5N (5.5 mL, 68 mmol), Ac_2O (5.16 mL, 54.7 mmol) and a catalytic amount of DMAP at room temperature to produce the diacetoxy product (*Scheme 2.4.41*), which was purified as above, to afford (*Z*)-6α,7β-(diacetoxy)-3β-[(*tert*-butyldiphenylsilyl)-oxy]-5α-pregn-17(20)-ene (**321**, 4.39 g, 96%, mp: 88–89 °C). ^1H NMR (400 MHz, $CDCl_3$): δ 7.66 (m, 4H, Ph-CH), 7.40 (m, 6H, Ph-CH), 5.13 (br q, 1H, *J* = 7.0 Hz, 20-CH), 3.56 (m, 1H, 3α-CH), 3.19 (dd, 1H, *J* = 8.6, 11.1 Hz, 6β-CH), 3.04 (dd, 1H, *J* = 8.6, 10.0 Hz, 7α-CH), 1.96 (s, 3H, MeCO), 1.87 (s, 3H, MeCO), 1.62 (br d, 3H, *J* = 7.1 Hz, 21-Me), 1.04 [s, 9H, $C(CH_3)_3$], 0.94 (s, 3H, 19-Me), 0.86 (s, 3H, 18-Me). ^{13}C NMR (100 MHz, $CDCl_3$): δ 13.2, 13.4, 16.9, 19.1, 20.7, 21.5, 25.4, 27.0, 31.1, 31.8, 32.2, 35.8, 36.6, 37.0, 38.7, 45.0, 46.0, 51.9, 54.7, 72.1, 74.6, 77.8, 113.9, 127.5, 129.5, 134.5, 134.6, 135.7, 148.6, 170.7, 171.0. EIMS: *m/z* 656 $[M]^+$ (Filippo *et al.*, 2003).

Compound **321** was transformed to 6α,7β-(diacetoxy)-3β-[(*tert*-butyldiphenylsilyl)-oxy]-5α-23,24-bisnorchol-16-en-22-ol (**322**) (*Scheme 2.4.41*). *p*-Formaldehyde (58 mg, 1.95 mmol) and $BF_3{\cdot}Et_2O$ (5 mL, 0.039 mmol) were added to a solution of **321** (263 mg, 0.40 mmol) in DCM (24 mL), and stirred at room temperature. After 1.5 h, H_2O was added and extracted with DCM, the organic solvent was dried (Na_2SO_4), evaporated to dryness and purified by FCC (mobile phase: 20% EtOAc in petroleum ether) to obtain **322** (217 mg, 79%, mp: 102–104 °C). ^1H NMR (400 MHz, $CDCl_3$): δ 7.63 (m, 4H, Ph-CH), 7.37 (m, 6H, Ph-CH), 5.34 (br s,1H, 16-CH), 4.79 (dd, 1H, *J* = 11.1, 8.6 Hz, 6α-CH), 4.67 (dd, 1H, *J* = 8.6, 10.0 Hz, 7β-CH), 3.56 (m, 1H, 3α-CH), 3.54 (m, 3H, 3-CH, 22-CH and 22-OH), 1.97 (s, 3H, $COCH_3$), 1.87 (s, 3H, $COCH_3$), 1.03 [s, 9H, $C(CH_3)_3$], 0.98 (d, 3H, *J* = 6.9 Hz, 21-Me), 0.95 (s, 3H, 19-Me), 0.76 (s, 3H, 18-Me). ^{13}C NMR (100 MHz, $CDCl_3$): δ 13.3, 16.0, 18.2, 19.0, 20.6, 20.9, 21.4, 26.9, 31.1, 32.0, 32.1, 34.2, 34.9, 36.0, 36.8, 37.9, 46.4, 47.8, 52.0, 54.8, 66.5, 72.0, 74.2, 77.6, 122.8, 127.4, 129.5, 134.5, 135.7, 156.2, 170.6, 170.7. EIMS: *m/z* 686 $[M]^+$ (Filippo *et al.*, 2003).

Alkaline hydrolysis of compound **322** produced 3β-[(*tert*-butyldiphenylsilyl)oxy]-5α-23,24-bisnorchol-16-en-6α,7β,22-triol (**323**) (*Scheme 2.4.41*). 6α,7β-(Diacetoxy)-3β-[(*tert*-butyldiphenylsilyl)-oxy]-5α-23,24-bisnorchol-16-en-22-ol (**322**, 226 mg, 0.330 mmol) was treated with 5% KOH in MeOH solution (3 mL) at room temperature. $CHCl_3$ was added and filtered through a pad of Celite and evaporated under pressure. The crude solid was purified by gradient FCC, eluted with 0–3% MeOH in $CHCl_3$, to yield **323** (171 mg, 86%, mp: 184–186 °C). ^1H NMR (400 MHz, $CDCl_3$): δ 7.63 (m, 4H, Ph-CH), 7.37 (m, 6H, Ph-CH), 5.41 (br s,1H, 16-CH), 3.57 (m, 3H, 3-CH, 22-CH and 22-OH), 3.56 (m, 1H, 3α-CH), 3.20 (dd, 1H, *J* = 8.6, 11.1 Hz, 6α-CH), 3.03 (dd, 1H, *J* = 8.6, 10 Hz, 7β-CH), 1.08 [s, 9H, $C(CH_3)_3$], 1.02 (d, 3H, *J* = 6.8 Hz, 21-Me), 0.88 (s, 3H, 19-Me), 0.78 (s, 3H, 18-Me). ^{13}C NMR

(100 MHz, $CDCl_3$): δ 13.5, 16.1, 18.2, 19.0, 20.9, 26.9, 31.2, 32.2, 33.8, 34.5, 34.8, 35.7, 36.9, 39.6, 47.5, 47.9, 52.1, 56.0, 66.4, 72.5, 74.5, 80.0, 123.3, 127.3, 129.3, 134.4, 134.7, 135.6, 156.2. FABMS: *m/z* 625 $[M + Na]^+$ (Filippo *et al.*, 2003).

Compound **323** was employed to synthesize bis-(20*S*)-3β-[(*tert*-butyldiphenylsilyl)oxy]-5α-23,24-bisnorchol-16-en-6α,7β-diol-22-terephthaloate (**324**) (*Scheme 2.4.42*). Terephthaloyl chloride (87 mg, 0.427 mmol) in DCM (1 mL) was added to a mixture of **323** (515 mg, 0.855 mmol) in DCM (3 mL), DMAP (313 mg, 0.257 mmol) and stirred for 48 h at room temperature. H_2O was added and extracted with $CHCl_3$, the solvent was dried (Na_2SO_4), rotary evaporated and the crude dimer was purified by gradient FCC (mobile phase: 1-2% MeOH in $CHCl_3$) to give dimer **324** (401 mg, 70%, mp: 177–178 °C). ^{1}H NMR (400 MHz, $CDCl_3$): δ 8.05 (m, 4H, terephthaloate Ph-CH), 7.69 (m, 8H, Ph-CH), 7.38 (m, 12H, Ph-CH), 5.49 (br s, 2H, 2 × 16-CH), 4.20 (dd, 2H, *J* = 8.3, 10.2 Hz, 22-CH_2) and 4.39 (dd, 2H, *J* = 8.3, 10.2 Hz, 22-CH_2), 3.55 (m, 2H, 2 × 3α-CH), 3.21 (dd, 2H, *J* = 8.6, 11.1 Hz, 2 × 6β-CH), 3.04 (dd, 2H, *J* = 8.6, 10.0 Hz, 2 × 7α-CH), 1.05 [s, 18H, 2 × $C(CH_3)_3$], 1.12 (d, 6H, *J* = 6.7 Hz, 2 × 21-Me), 0.87 (s, 6H, 2 × 19-Me), 0.77 (s, 6H, 2 × 18-Me). ^{13}C NMR (100 MHz, $CDCl_3$): δ 13.6, 16.2, 18.8, 19.1, 20.9, 26.9, 31.2, 31.4, 32.1, 33.9, 34.5, 35.9, 37.0, 39.7, 47.8, 47.9, 52.1, 55.7, 69.2, 72.5, 74.7, 80.2, 123.5, 127.4, 129.3, 134.4, 134.7, 135.7, 155.4, 165.7. FABMS: *m/z* 1357 $[M + Na]^+$ (Filippo *et al.*, 2003).

A solution of HF in C_5H_5N (275 μL) was added to a solution of **324** (49 mg, 0.037 mmol) in C_5H_5N (2 mL), and stirred overnight at room temperature and concentrated under N_2 (*Scheme 2.4.42*) The crude dimer was purified by FCC, eluted with 5% MeOH in $CHCl_3$, to

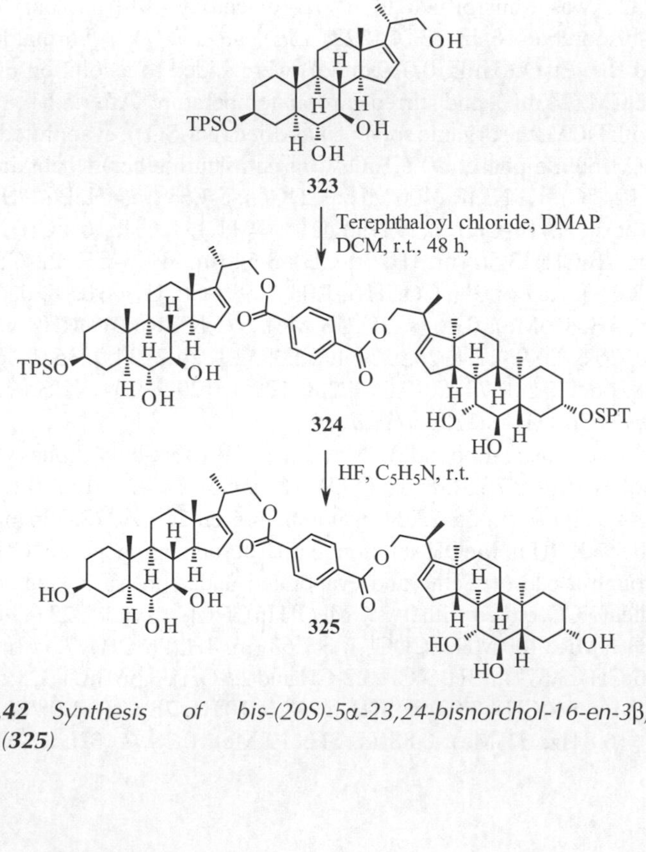

Scheme 2.4.42 *Synthesis of bis-(20S)-5α-23,24-bisnorchol-16-en-3β,6α,7β-triol-22-terephthaloate (**325**)*

provide bis-(20*S*)-5α-23,24-bisnorchol-16-en-3β,6α,7β-triol-22-terephthaloate (**325**, 25 mg, 80%, mp: 106–107 °C). ^{1}H NMR (400 MHz, $CDCl_3$): δ 8.10 (m, 2H, Ph-CH), 7.69 (m, 2H, Ph-CH), 5.57 (bs, 2H, 2 × 16-CH), 4.45 (dd, 2H, J = 10.6, 6.5 Hz, 22-CH_2) and 4.24 (dd, 2H, J = 7.6, 10.6 Hz, 22-CH_2), 3.51 (m, 2H, 2 × 3α-CH), 3.16 (dd, 1H, J = 8.6, 11.1 Hz, 2 × 6β-CH), 3.04 (dd, 1H, J = 8.6, 10.0 Hz, 2 × 7α-CH), 2.62 (m, 2H, 2 × 20-CH), 2.22 (m, 4H, 2 × 15-CH_2), 2.15 (m, 4H, 4-CH_2), 1.18 (d, 6H, J = 6.9 Hz, 2 × 21-Me), 0.92 (s, 6H, 2 × 19-Me), 0.84 (s, 6H, 2 × 18-Me). ^{13}C NMR (100 MHz, $CDCl_3$): δ 13.9, 16.7, 19.3, 22.4, 31.8, 32.8, 33.2, 35.0, 36.0, 36.9, 38.4, 41.2, 49.0 (overlapping with $^{13}CD_3OD$), 53.9, 57.8, 70.4, 71.8, 75.9, 81.0, 125.0, 130.6, 135.5, 156.9, 167.0. FABMS: *m/z* 881 [M + Na]$^+$ (Filippo *et al.*, 2003).

A series of C_2-symmetric 17β-estradiol homodimers **329–340**, which are linked at 17α position of the steroid nucleus with either an alkyl chain or a PEG chain, were prepared from estrone (**64**) through a multistep sequence (Berube *et al.*, 2006). Estrone (**64**) was converted to 3-benzyloxy-1,3,5(10)-estratrien-17-one (**326**) by treating a stirred solution of **64** (1.0 g, 3.70 mmol) in DCM (10 mL), with BnBr (530 μL, 4.44 mmol), TBAHS (100 mg), and a solution of NaOH (10% w/v, 5 mL) (*Scheme 2.4.43*). The mixture was refluxed for 24 h, diluted with Et_2O (30 mL) and H_2O (30 mL), and washed with H_2O (4 × 75 mL). The solvent was dried ($MgSO_4$), evaporated under pressure and triturated with *n*-hexane to produce benzylated estrone **326** (99%, mp: 128–129 °C) which was used without further purification at the next step. IR (KBr): ν_{max} cm^{-1} 1731s (C=O), 1614s (C=C), 1230s (C–O), 1008s (C–O). ^{1}H NMR (500 MHz, $CDCl_3$): δ 7.41 (d, 2H, J = 7.6 Hz, CHa), 7.36 (t, 2H, J = 7.5 Hz, CHb), 7.29 (t, 1H, J = 7.2 Hz, CHc), 7.18 (d, 1H, d, J = 8.6 Hz, 1-CH), 6.78 (dd, 1H, J = 2.5, 8.5 Hz, 2-CH), 6.71 (d, 1H, J = 2.1 Hz, 4-CH), 5.01 (s, 2H, Ph-CH_2), 2.88 (m 2H, 6-CH_2), 2.50-1.39 (m, 13H, 3 × CH and 5 × CH_2), 0.89 (s, 3H, 18-Me). ^{13}C NMR (125 MHz, $CDCl_3$): δ 13.8, 21.5, 25.8, 26.5, 29.6, 31.5, 35.8, 38.3, 43.9, 47.9, 50.3, 69.8, 112.3, 114.8, 126.2, 127.3, 127.7, 128.4, 132.2, 137.2, 137.7, 156.8, 220.5. EIMS: *m/z* 360 [M]$^+$ (Berube *et al.*, 2006).

3-Benzyloxy-1,3,5(10)-estratrien-17-one (**326**) was utilized for the synthesis of 3-*O*-benzyl-17α-(prop-2′-enyl)-1,3,5(10)-estratrien-17β-ol (**327**) (*Scheme 2.4.43*). A solution of allylmagnesium bromide in Et_2O (53 mL, 41.70 mmol), freshly prepared from allyl bromide and Mg, was added to a solution of **326** (1.50 g, 4.17 mmol) in dry Et_2O (50 mL) and stirred for 30 min. Et_2O (30 mL) was added and washed with 4% HCl (75 mL) and finally with H_2O (4 × 75 mL). The organic phase was dried ($MgSO_4$), rotary evaporated and purified by FCC (mobile phase: 20% Me_2CO in *n*-hexane) to produce **327** (1.60 g, 95%). IR (KBr): ν_{max} cm^{-1} 3558br (O–H), 3470br (O–H), 1601s (C=C), 1222s (C–O), 1025s (C–O). ^{1}H NMR (500 MHz, $CDCl_3$): δ 7.39 (d, 2H, J = 7.5 Hz, CHa), 7.34 (t, 2H, J = 7.5 Hz, CHb), 7.28 (t, 1H, J = 7.4 Hz, CHc), 7,17 (d, 1H, J = 8.5 Hz, 1-CH), 6.75 (dd, 1H, J = 8.5, 2.5 Hz, 2-CH), 6.69 (d, 1H, J = 2.1 Hz, 4-CH), 6.00 (m, 1H, $CH_2CH{=}CH_2$), 5.18 (dd, 2H, J = 11.0, 18.1 Hz, $CH_2CH{=}CH_2$), 4.98 (2H, s, Ph-CH_2), 2.84 (m, 2H, 6-CH_2), 2.50–1.22 (m, 16 H, OH, 3 × CH and 6 × CH_2), 0.90 (s, 3H, 18-Me). ^{13}C NMR (125 MHz, $CDCl_3$): δ 14.3, 23.4, 26.3, 27.5, 29.8, 31.8, 34.9, 39.5, 41.8, 43.8, 46.5, 49.6, 69.9, 82.4, 112.2, 114.8, 119.0, 126.2, 127.4, 127.7, 128.5, 132.9, 134.9, 137.3, 137.9, 156.7. EIMS: *m/z* 402 [M]$^+$ (Berube *et al.*, 2006).

A solution of alkene **327** (2.20 g, 5.47 mmol) in dry THF (22 mL) was treated with a 1M solution of BH_3.THF (21.89 mL, 21.89 mmol) at 0 °C under N_2 with constant stirring for 3 h after which H_2O (1 mL) was added (*Scheme 2.4.43*). The mixture was treated with a 30%

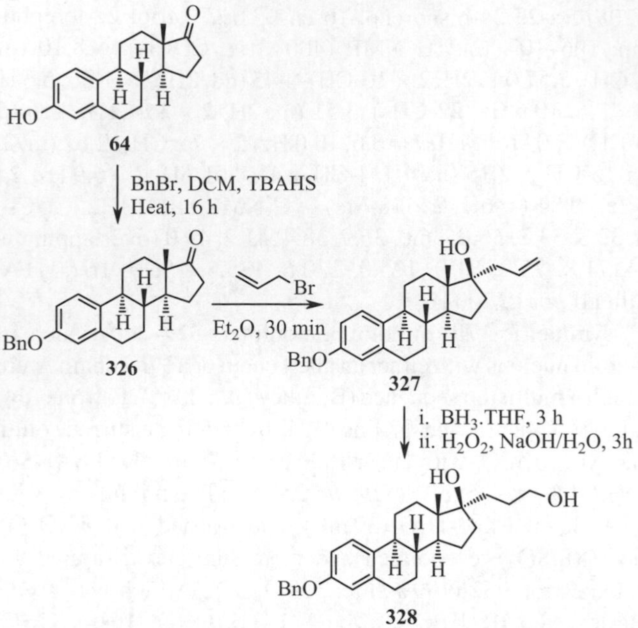

HO
64
BnBr, DCM, TBAHS
Heat, 16 h
BnO
326
Br
Et_2O, 30 min
HO
BnO
327
i. BH_3.THF, 3 h
ii. H_2O_2, NaOH/H_2O, 3h
HO
OH
BnO
328

Scheme 2.4.43 *Synthesis of 17β-estradiol derivatives **326–328***

H_2O_2 solution (2.47 mL, 21.89 mmol), then with 3% NaOH (29 mL, 21.89 mmol) at 22 °C and stirred for 60 min. EtOAc (40 mL) was added and the mixture was washed with saturated aqueous NH_4Cl (50 mL), followed by H_2O (4 × 50 mL). The solvent was dried ($MgSO_4$), evaporated under vacuum and purified by FCC, eluted with 33% Me_2CO in hexane, to obtain 3-*O*-benzyl-17α-(3-hydroxypropyl)-1,3,5(10)-estratrien-17β-ol (**328**, 75%). IR (KBr): ν_{max} cm^{-1} 3360br (O–H), 1600s (C=C), 1254s (C–O), 1017s (C–O). ^{1}H NMR (500 MHz, $CDCl_3$): δ 7.41 (d, 2H, J = 7.0 Hz, CHa), 7.36 (t, 2H, J = 7.2 Hz, CHb), 7.30 (d, 1H, J = 6.7 Hz, CHc), 7.18 (d, 1H, J = 8.5 Hz, 1-CH), 6.76 (dd, 1H, J = 2.0, 7.9 Hz, 2-CH), 6.71 (s, 1H, 4-CH), 5,00 (s, 2H, Ph-CH_2), 3.67 (m, 2H, 3′-CH_2OH), 2.86 (m, 2H, 6-CH_2), 2.59 (s, 2H, 2 × OH), 2.31-1.20 (m, 17 H, 3 × CH and 7 × CH_2), 0.90 (3H, s, 18-Me). ^{13}C NMR (125 MHz, $CDCl_3$): δ 14.3, 23.4, 26.3, 26.9, 27.5, 29.8, 31.5, 33.4, 34.4, 39.6, 43.8, 46.7, 49.5, 63.3, 69.9, 83.2, 112.2, 114.8, 126.2, 127.4, 127.8, 128.5, 132.9, 137.3, 138.0, 156.7. EIMS: *m/z* 420 $[M]^+$ (Berube *et al.*, 2006).

For the synthesis of C_2-symmetric 17β-estradiol homodimers **329–333**, the following general procedure was employed (*Scheme 2.4.44*). 17β-Estradiol derivative **328** (200 mg, 0.48 mmol) in a mixture of 30% DMF in THF (3 mL) was treated with NaH (48 mg, 1.19 mmol) at 22 °C for 30 min. The appropriate dibromo-PEG (0.24 mmol) was added and the resulting solution was stirred for 24 h at the same temperature. EtOAc (25 mL) was added and extracted as above, the solvent was dried ($MgSO_4$), evaporated to dryness and purified by FCC (mobile phase: 40% Me_2CO in *n*-hexane) to yield C_2-symmetric 17β-estradiol homodimers **329–333** (40–52%) (Berube *et al.*, 2006).

17β-Estradiol homodimer **329** (40%). IR (KBr): ν_{max} cm^{-1} 3443br (O–H), 1598s (C=C), 1248s (C–O), 1103s (C–O). ^{1}H NMR (500 MHz, $CDCl_3$): δ 7.43 (d, 4H, J = 7.5 Hz, 2 × CHa), 7.38 (t, 4H, J = 7.4 Hz, 2 × CHb), 7.32 (t, 2H, J = 7.1 Hz, 2 × CHc), 7.20 (d, 2H,

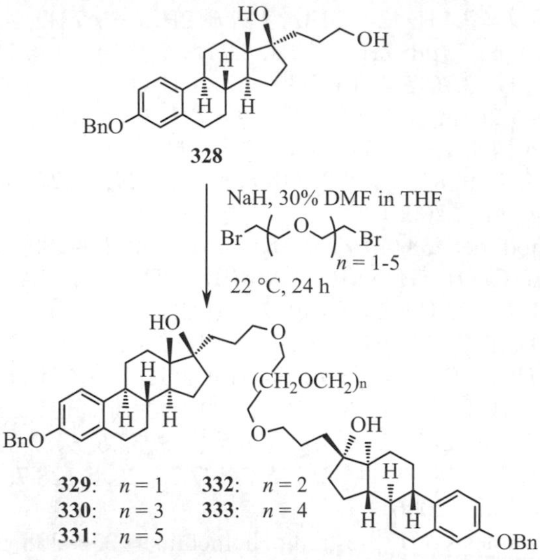

328

NaH, 30% DMF in THF

$Br\text{-}(\text{-}O\text{-})_n\text{-}Br$, $n = 1\text{-}5$

22 °C, 24 h

$(CH_2OCH_2)_n$

329: $n = 1$ **332**: $n = 2$
330: $n = 3$ **333**: $n = 4$
331: $n = 5$

Scheme 2.4.44 *Synthesis of protected 17β-estradiol homodimers **329–333***

$J = 8.7$ Hz, 2 × 1-CH), 6.77 (dd, 2H, $J = 2.4$, 8.7 Hz, 2 × 2-CH), 6.73 (s, 2H, 2 × 4-CH), 5.03 (s, 4H, 2 × Ph-CH_2), 3.68-3.52 (m, 12H, 6 × CH_2O), 2.84 (m, 4H, 2 × 6-CH_2), 2.40-1,20 (m, 34H, 14 × CH_2, 6 × CH), 0.92 (s, 6H, 2 × 18-Me). ^{13}C NMR (125 MHz, $CDCl_3$): δ 14.4, 23.4, 24.0, 26.3, 27.5, 29.8, 31.6, 33.4, 34.3, 39.6, 43.8, 46.7, 49.5, 69.9, 70.0, 70.5, 71.9, 82.9, 112.2, 114.8, 126.2, 127.4, 127.8, 128.5, 133.0, 137.3, 138.0, 156.7. EIMS: *m/z* 892 $[M - H_2O]^+$ (Berube *et al.*, 2006).

17β-Estradiol homodimer **330** (42%). IR (KBr): ν_{max} cm^{-1} 3418br (O–H), 1614s (C=C), 1230s (C–O), 1092s (C–O). ^1H NMR (500 MHz, $CDCl_3$): δ 7.43 (d, 4H, $J = 7.5$ Hz, 2 × CHa), 7.38 (t, 4H, $J = 7.4$ Hz, 2 × CHb), 7.32 (t, 2H, $J = 7.1$ Hz, 2 × CHc), 7.20 (d, 2H, $J = 8.7$ Hz, 2 × 1-CH), 6.78 (dd, 2H, $J = 8.7$ and 2.4 Hz, 2 × 2-CH), 6.72 (s, 2H, 2 × 4-CH), 5.03 (s, 4H, 2 × Ph-CH_2), 3.80-3.50 (s and m, 16H, 8 × CH_2O), 2.84 (m, 4H, 2 × 6-CH_2), 2.66 (s, 2H, 2 × OH), 2.40-1.20 (m, 34H, 14 × CH_2, 6 × CH), 0.92 (s, 6H, 2 × 18-Me). ^{13}C NMR (125 MHz, $CDCl_3$): δ 14.4, 23.4, 24.0, 26.3, 27.5, 29.8, 31.7, 33.4, 34.3, 39.6, 43.8, 46.8, 49.5, 69.9, 70.0, 70.5, 70.6, 72.0, 83.0, 112.2, 114.8, 126.3, 127.4, 127.8, 128.5, 133.0, 137.3, 138.0, 156.7. EIMS: *m/z* 937 $[M - H_2O]^+$ (Berube *et al.*, 2006).

17β-Estradiol homodimer **331** (43%). IR (KBr): ν_{max} cm^{-1} 3426br (O–H), 1607s (C=C), 1234s, (C–O), 1025s (C–O). ^1H NMR (500 MHz, $CDCl_3$): δ 7.42 (d, 4H, $J = 7.5$ Hz, 2 × CHa), 7.38 (t, 4H, $J = 7.4$ Hz, 2 × CHb), 7.32 (t, 2H, $J = 7.1$ Hz, 2 × CHc), 7.20 (d, 2H, $J = 8.7$ Hz, 2 × 1-CH), 6.77 (dd, 2H, $J = 2.4$, 8.7 Hz, 2 × 2-CH), 6.72 (s, 2H, 2 × 4-CH), 5.03 (s, 4H, 2 × Ph-CH_2), 3.75-3.40 (m, 20H, 10 × CH_2O), 2.84 (m, 4H, 2 × 6-CH_2), 2.48 (s, 2H, 2 × OH), 2.35-1.20 (m, 34H, 14 × CH_2, 6 × CH), 0.91 (s, 6H, 2 × 18-Me). ^{13}C NMR (125 MHz, $CDCl_3$): δ 14.4, 23.4, 24.0, 26.3, 27.5, 29.8, 31.6, 33.5, 34.4, 39.6, 43.8, 46.7, 49.5, 69.9, 70.0, 70.5, 70.6, 72.0, 83.0, 112.2, 114.8, 126.2, 127.4, 127.8, 128.5, 133.0, 137.3, 138.0, 156.7. EIMS: *m/z* 980 $[M - H_2O]^+$ (Berube *et al.*, 2006).

17β-Estradiol homodimer **332** (40%). IR (KBr): ν_{max} cm^{-1} 3448br (O–H), 1608s (C=C), 1235s (C–O), 1104s (C–O). ^1H NMR (500 MHz, $CDCl_3$): δ 7.42 (d, 4H, $J = 7.5$ Hz,

2 × CHa), 7.37 (t, 4H, $J = 7.4$ Hz, 2 × CHb), 7.31 (t, 2H, $J = 7.2$ Hz, 2 × CHc), 7.20 (d, 2H, $J = 8.6$ Hz, 2 × 1-CH), 6.77 (dd, 2H, $J = 2.4$, 8.7 Hz, 2 × 2-CH), 6.71 (s, 2H, 2 × 4-CH), 5.03 (s, 4H, 2 × Ph-CH_2), 3.70–3.45 (m, 24H, 12 × CH_2O), 2.83 (m, 4H, 2 × 6-CH_2), 2.42 (s, 2H, 2 × OH), 2.35–1.20 (m, 34H, 14 × CH_2, 6 × CH), 0.91 (s, 6H, 2 × 18-Me). ^{13}C NMR (125 MHz, $CDCl_3$): δ 14.4, 23.4, 24.0, 26.3, 27.5, 29.8, 31.6, 33.5, 34.4, 39.6, 43.8, 46.7, 49.5, 69.9, 70.0, 70.5, 72.0, 82.9, 112.2, 114.8, 126.2, 127.4, 127.8, 128.5, 133.0, 137.3, 138.0, 156.7 (Berube *et al*., 2006).

17β-Estradiol homodimer **333** (40%). IR (KBr): ν_{max} cm^{-1} 3458br (O–H), 1603s (C=C), 1252s (C–O), 1025s (C–O). ^{1}H NMR (500 MHz, $CDCl_3$): δ 7.42 (d, 4H, $J = 7.3$ Hz, 2 × CHa), 7.38 (t, 4H, $J = 7.4$ Hz, 2 × CHb), 7.31 (t, 2H, $J = 7.3$ Hz, 2 × CHc), 7.20 (d, 2H, $J = 8.7$ Hz, 2 × 1-CH), 6.77 (dd, 2H, $J = 2.4$, 8.7 Hz, 2 × 2-CH), 6.72 (d, 2H, $J = 2.3$ Hz, 2 × 4-CH), 5.03 (s, 4H, 2 × Ph-CH_2), 3.70–3.45 (m, 28H, 12 × CH_2O), 2.84 (m, 4H, 2 × 6-CH_2), 2.40-1.20 (m, 36H, 14 × CH_2, 6 × CH, 2 × OH), 0.91 (s, 6H, 2 × 18-Me). ^{13}C NMR (125 MHz, $CDCl_3$): δ 14.4, 24.4, 25.0, 27.3, 28.5, 30.8, 32.6, 34.5, 35.4, 40.6, 44.8, 47.7, 50.5, 70.9, 71.0, 71.5, 73.0, 83.9, 113.2, 115.7, 127.2, 128.4, 128.7, 129.5, 134.0, 138.3, 139.0, 157.7 (Berube *et al*., 2006).

Similarly, for the synthesis of 17β-estradiol homodimers **334–338**, the following general method was followed (Berube *et al*., 2006). To a stirred solution of dimers **329–333** (0.07 mmol) in dry THF (2.5 mL), Pd-C (10 mg, 10% on carbon) was added and H_2 bubble was passed into the reaction mixture for 3 min (*Scheme 2.4.45*). The hydrogenolysis was performed under 1 atmosphere of H_2 for 24 h at 22 °C. The catalyst was filtered off and the filtrate was rotary evaporated to afford 17β-estradiol homodimers **334–338** in a quantitative yield.

17β-Estradiol homodimer **334**: IR (NaCl): ν_{max} cm^{-1} 3372br (O–H), 1609s (C=C), 1240s (C–O), 1093s (C–O). ^{1}H NMR (500 MHz, CD_3OD): δ 7.06 (d, 2H, $J = 8.4$ Hz, 2 × 1-CH), 6.58 (dd, 2H, $J = 1.6$, 8.6 Hz, 2 × 2-CH), 6.52 (s, 2H, 2 × 4-CH), 3.65–3.40 (m, 12H, 6 × CH_2O), 3.03 (s, 4H, 3-OH and 17β-OH), 2.76 (m, 4H, 2 × 6-CH_2), 2.30-1.15 (m, 34H,

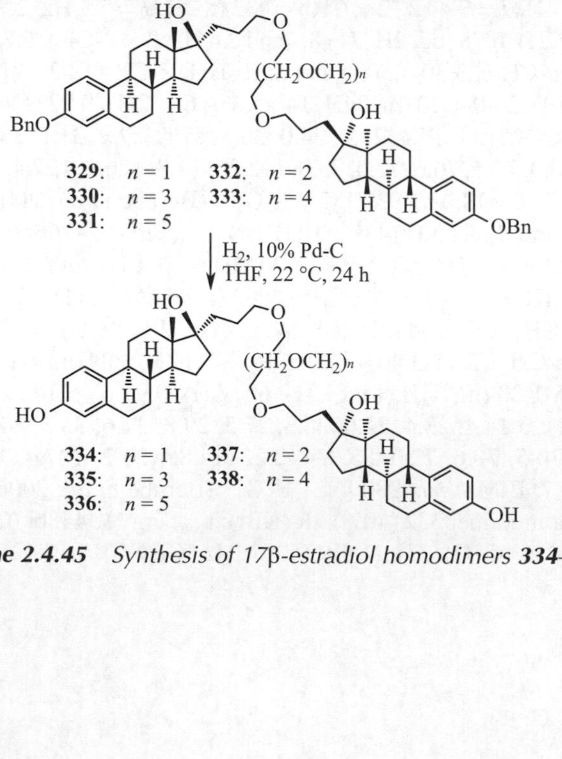
HO
H
H H
BnO
(CH2OCH2)n
O
O
OH
H
H H
OBn
329: n = 1
330: n = 3
331: n = 5
332: n = 2
333: n = 4
H2, 10% Pd-C
THF, 22 °C, 24 h
HO
HO
OH
OH
334: n = 1
335: n = 3
336: n = 5
337: n = 2
338: n = 4

Scheme 2.4.45 *Synthesis of 17β-estradiol homodimers **334–338***

$14 \times CH_2$, $6 \times CH$), 0.85 (s, 6H, $2 \times$ 18-Me). ^{13}C NMR (125 MHz, CD_3OD): δ 14.4, 23.3, 23.8, 26.3, 27.4, 29.6, 31.5, 33.2, 33.8, 39.6, 43.7, 46.7, 49.5, 69.9, 70.3, 72.0, 82.9, 112.6, 115.1, 126.1, 131.8, 137.9, 154.1. EIMS: *m/z* 712 $[M - H_2O]^+$ (Berube *et al.*, 2006).

17β-Estradiol homodimer **335**: IR (NaCl): ν_{max} cm^{-1} 3346br (O–H), 1608s (C=C), 1240s (C–O), 1092s (C–O). ^1H NMR (500 MHz, acetone-d_6, CD_3OD): δ 7.07 (d, 2H, $J = 8.4$ Hz, $2 \times$ 1-CH), 6.56 (dd, 2H, $J = 2.5$, 8.6 Hz, $2 \times$ 2-CH), 6.50 (d, 2H, $J = 1.5$ Hz, $2 \times$ 4-CH), 3.65-3.40 (m, 16H, $8 \times CH_2O$), 3.28 (s, 4H, 3-OH and 17β-OH), 2.78 (m, 4H, $2 \times$ 6-CH_2), 2.35-1.20 (m, 34H, $14 \times CH_2$ and $6 \times CH$), 0.91 (s, 6H, $2 \times$ 18-Me). ^{13}C NMR (125 MHz, acetone-d_6, CD_3OD): δ 15.1, 24.2, 25.1, 27.4, 28.5, 29.6, 32.6, 34.2, 34.7, 40.9, 44.8, 47.7, 50.5, 70.9, 71.3, 71.4, 72.7, 83.1, 113.5, 115.9, 127.0, 132.1, 138.5, 155.9. EIMS: *m/z* 756 $[M - H_2O]^+$ (Berube *et al.*, 2006).

17β-Estradiol homodimer **336**: IR (NaCl): ν_{max} cm^{-1} 3385br (O–H), 1610s (C=C), 1287s and 1036s (C–O). ^1H NMR (500 MHz, $CHCl_3$): δ 7.05 (d, 2H, $J = 8.5$ Hz, $2 \times$ 1-CH), 6.59 (d, 2H, $J = 2.5$, 8.2 Hz, $2 \times$ 2-CH), 6.51 (d, 2H, $J = 1.4$ Hz, $2 \times$ 4-CH), 3.65–3.42 (m, 20H, $10 \times CH_2O$), 3.06 (br s, 4H, 3-OH and 17β-OH), 2.76 (m, 4H, $2 \times$ 6-CH_2), 2.30–1.15 (m, 34H, $14 \times CH_2$, $6 \times CH$), 0.84 (s, 6H, $2 \times$ 18-Me). ^{13}C NMR (125 MHz, $CHCl_3$): δ 15.4, 24.3, 24.8, 27.1, 28.4, 30.6, 32.5, 34.2, 34.8, 40.6, 44.6, 47.7, 50.4, 70.8, 71.3, 71.4, 73.0, 84.0, 113.6, 116.2, 127.1, 132.7, 138.9, 155.1. EIMS: *m/z* 800 $[M - H_2O]^+$ (Berube *et al.*, 2006).

17β-Estradiol homodimer **337**: IR (NaCl): ν_{max} cm^{-1} 3383br (O–H), 1611s (C=C), 1248s (C–O), 1100s (C–O). ^1H NMR (500 MHz, $CHCl_3$): δ 7.08 (d, 2H, $J = 8.6$ Hz, $2 \times$ 1-CH), 6.61 (dd, 2H, $J = 2.5$, 8.6 Hz, $2 \times$ 2-CH), 6.54 (d, 2H, $J = 1.4$ Hz, $2 \times$ 4-CH), 3.68–3.48 (m, 24H, $12 \times CH_2O$), 2.76 (m, 4H, $2 \times$ 6-CH_2), 2.37–1.10 (m, 38H, 3-OH, $6 \times CH$, 17β-OH and $14 \times CH_2$), 0.87 (s, 6H, $2 \times$ 18-Me). ^{13}C NMR (125 MHz, $CHCl_3$): δ 14.4, 23.3, 23.9, 26.3, 27.4, 29.7, 31.5, 33.5, 34.2, 39.6, 43.6, 46.7, 49.4, 70.0, 70.5, 72.1, 83.2, 112.8, 115.3, 126.2, 132.0, 137.9, 154.0. EIMS: *m/z* 844 $[M - H_2O]^+$ (Berube *et al.*, 2006).

17β-Estradiol homodimer **338**: IR (NaCl): ν_{max} cm^{-1} 3428br (O–H), 1613s (C=C), 1286s (C–O), 1103s (C–O). ^1H NMR (500 MHz, $CHCl_3$): δ 7.08 (d, 2H, $J = 8.6$ Hz, $2 \times$ 1-CH), 6.73 (s, 2H, 3-OH), 6.62 (s, 2H, $J = 8.1$ Hz, $2 \times$ 2-CH), 6.54 (s, 2H, $2 \times$ 4-CH), 3.68-3.48 (m, 28H, $14 \times CH_2O$), 2.76 (m, 4H, $2 \times$ 6-CH_2), 2.37–1.10 (m, 36H, 17β-OH, $6 \times CH$ and $14 \times CH_2$), 0.87 (s, 6H, $2 \times$ 18-Me). ^{13}C NMR (125 MHz, $CHCl_3$): δ 14.4, 23.3, 23.9, 26.3, 27.4, 29.7, 31.5, 33.6, 34.2, 39.6, 43.6, 46.7, 49.4, 70.0, 70.4, 72.1, 83.2, 112.8, 115.3, 126.2, 132.0, 137.9, 154.0. EIMS: *m/z* 888 $[M - H_2O]^+$ (Berube *et al.*, 2006).

NaH (56 mg, 1.41 mmol) was added to a stirred solution of diol **338** (237 mg, 0.56 mmol) in 30% DMF in THF (3 mL) at 22 °C (*Scheme 2.4.46*). After 30 min, 1,6-dibromohexane (69 mg, 0.28 mmol) was added and stirred for 24 h at the same temperature. After the usual workup, as described above for dimers **329–333**, and purification by FCC (mobile phase: 20% Me_2CO in *n*-hexane) provided 17β-estradiol homodimer **339** (47%). IR (NaCl): ν_{max} cm^{-1} 3417br (O–H), 1604s (C=C), 1230s (C–O), 1025s (C–O). ^1H NMR (500 MHz, $CDCl_3$): δ 7.42 (d, 4H, $J = 7.5$ Hz, $2 \times$ CHa), 7.37 (t, 4H, $J = 7.4$ Hz, $2 \times$ CHb), 7.31 (t, 2H, $J = 7.1$ Hz, $2 \times$ CHc), 7.20 (d, 2H, $J = 8.7$ Hz, $2 \times$ 1-CH), 6.77 (dd, 2H, $J = 2.4$, 8.7 Hz, $2 \times$ 2-CH), 6.71 (s, 2H, $2 \times$ 4-CH), 5.03 (s, 4H, $2 \times$ Ph-CH_2), 3.45 (8H, m, $2 \times CH_2OCH_2$), 2.40–1.10 (m, 44H, $18 \times CH_2$, $6 \times CH$, $2 \times OH$), 0.91 (s, 6H, $2 \times$ 18-Me). ^{13}C NMR (125 MHz, $CDCl_3$): δ 14.4, 23.4, 24.1, 26.3, 27.5, 29.8, 31.7, 33.8, 34.6, 39.6, 43.8, 46.7, 49.5, 69.9, 70.9, 70.5, 71.5, 82.9, 112.2, 114.8, 126.2, 127.4, 127.8, 128.5, 133.0, 137.3, 138.0, 156.7. EIMS: *m/z* 905 $[M + H_2O]^+$ (Berube *et al.*, 2006).

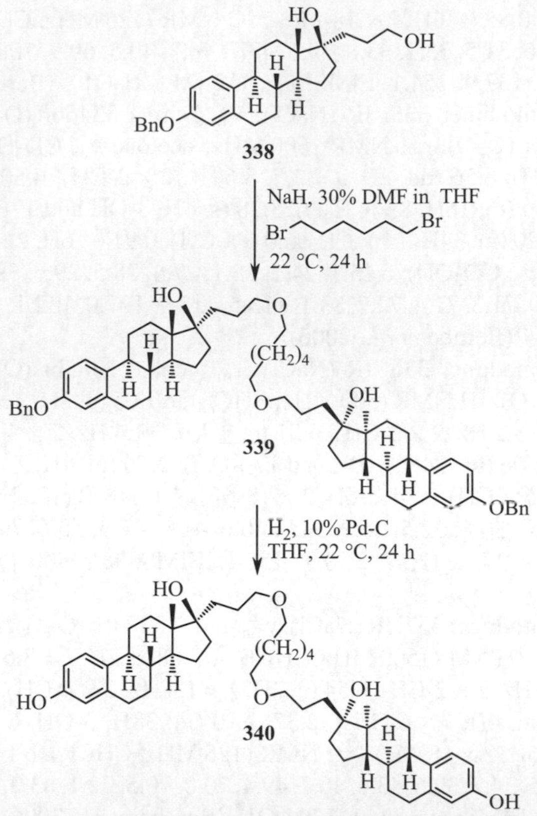

338

NaH, 30% DMF in THF
Br~~~~~~Br
22 °C, 24 h

339

H_2, 10% Pd-C
THF, 22 °C, 24 h

340

Scheme 2.4.46 *Synthesis of deprotected 17β-estradiol homodimers **339** and **340***

17β-Estradiol homodimer **339** was reduced to dimer **340** (*Scheme 2.4.46*), Pd-C (10 mg, 10% on carbon) was added to a stirred solution of dimer **339** (50 mg, 0.05 mmol) in dry THF (1.5 mL), and H_2 was bubbled through. The hydrogenolysis was performed under 1 atmosphere of H_2 for 24 h at 22 °C. The catalyst was filtered off and the filtrate was rotary evaporated to furnish dimer **340** in a quantitative yield. IR (NaCl): ν_{max} cm^{-1} 3333br (O–H), 1610s (C=C), 1248s (C–O), 1100s (C–O). ^{1}H NMR (500 MHz, acetone-d_6, CD_3OD): δ 7.08 (d, 2H, J = 8.3 Hz, 2 × 1-CH), 6.56 (dd, 2H, J = 1.7, 8.3 Hz, 2 × 2-CH), 6.50 (s, 2H, 2 × 4-CH), 3.43-3.38 (m, 8H, 2 × CH_2OCH_2), 3.30–3.10 (s, 2H, 3-OH and 17-OH), 2.40–1.10 (m, 42H, 18 × CH_2, 6 × CH), 0.91 (s, 6H, 2 × 18-Me). ^{13}C NMR (125 MHz, $CDCl_3$): δ 15.1, 24.2, 25.1, 26.9, 27.4, 28.5, 30.6, 32.6, 34.3, 34.8, 40.9, 44.8, 47.7, 50.5, 71.2, 72.3, 83.0, 113.5, 115.9, 127.0, 132.2, 138.5, 155.9 (Berube *et al.*, 2006).

Ra *et al.* (2000) reported the synthesis of several lithocholic acid-based dimers **345–348** with ethane-1,2-diol bridges. Initially, 3α-hydroxy-5β-cholan-24-yl ethane-1,2-diol mono THP ether (**341**) was prepared from 5β-cholane-3α,24-diol (**111**) (*Scheme 2.4.47*). Ethane-1,2-diol mono *p*-tosylate THP ether (82.8 mg, 0.276 mmol) was added to a stirred mixture of diol **111** (49.6 mg, 0.137 mmol) and 60% NaH (23 mg, 0.477 mmol). The mixture was heated at 78 °C (water bath) for 24 h, after which, H_2O was added and mixture was extracted into Et_2O. The ethereal solution was dried ($MgSO_4$), concentrated under vacuum and the residue was purified by FCC (mobile phase: 20% EtOAc in *n*-hexane) to produce **341** (38 mg, 56%). ^{1}H NMR (300 MHz, $CDCl_3$): δ 4.62 (t, 1H, $J = 3.5$ Hz), 3.88–3.55 (m, 9H), 2.15–1.01 (m, 32H), 0.10–0.86 (m, 9H), 0.61 (s, 3H) (Ra *et al.* 2000).

3α-Hydroxy-5β-cholan-24-yl ethane-1,2-diol (**342**) was prepared from **341** (*Scheme 2.4.47*). A mixture of **341** (175 mg, 0.356 mmol) and *p*-TsOH (6.8 mg, 36 µmol) in MeOH (20 mL) was stirred at room temperature for 5 h. Usual cleanup and purification as described above gave **342** (143 mg, 98.6%). ^{1}H NMR (300 MHz, $CDCl_3$): δ 3.71 (br, 2H), 3.64 (m, IH), 3.50 (t, 2H, $J = 2.1$ Hz), 3.42 (m, 2H), 2.15–1.03 (m, 27H), 0.89 (m, 9H), 0.62 (s, 3H) (Ra *et al.*, 2000).

3α-hydroxy-5β-cholan-24-yl ethane-1,2-diol tosylate (**343**) was prepared from 3α-hydroxy-5β-cholan-24-yl ethane-1,2-diol (**342**) (*Scheme 2.4.47*). *p*-TsCl (12.4 mg, 0.065 mmol) was added to a cold (0 °C), magnetically stirred solution of **342** (20 mg, 5 µmol) and Et_3N (15.2 mg, 0.15 mmol) in dry DCM (1 mL), the mixture was warmed to

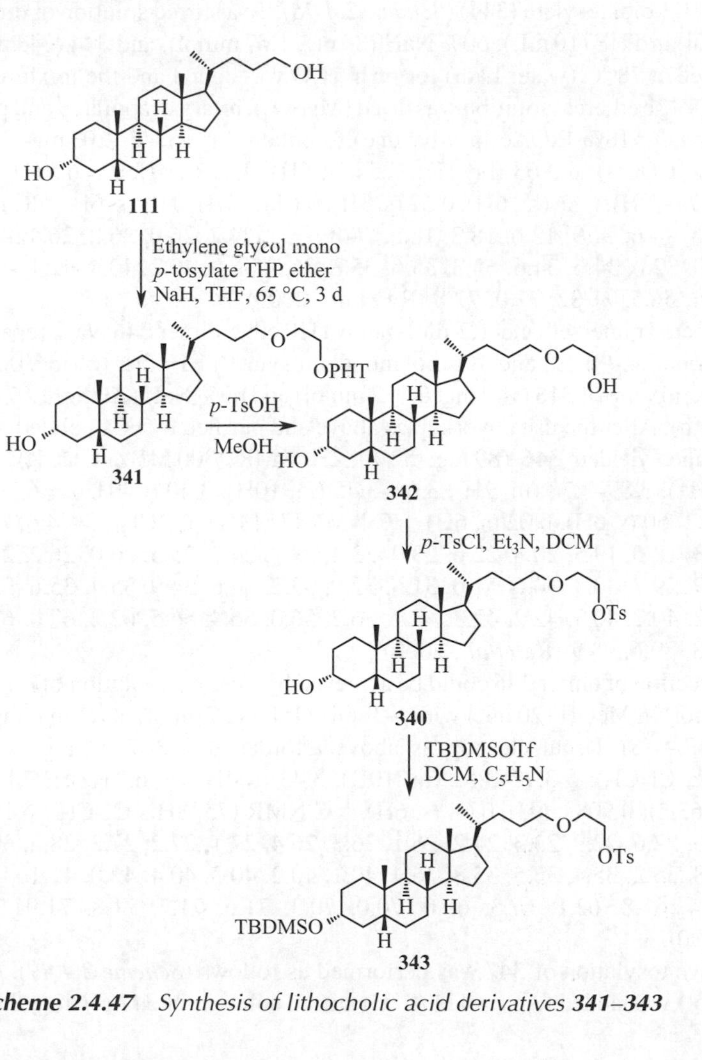

Scheme 2.4.47** Synthesis of lithocholic acid derivatives **341–343

room temperature and stirred for 5 h. H_2O was added and the mixture was extracted with DCM (50 mL), the solvent was washed with saturated aqueous $NaHCO_3$ (5 mL) followed by brine, dried ($MgSO_4$), rotary evaporated and purified by FCC, eluted with 20% EtOAc in *n*-hexane, to provide **343** (23 mg, 81%). ^{1}H NMR (300 MHz, $CDCl_3$): δ 7.78 (d, 2H, $J=8.4$ Hz), 7.32 (d, 2H, $J=8.0$ Hz), 4.13 (t, 2H, $J=4.8$ Hz), 3.59 (br, 1H), 3.58 (t, 2H, $J=4.8$ Hz), 3.32 (t, 2H, $J=6.6$ Hz), 2.43 (s, 3H), 1.96–0.90 (m, 34H), 0.86 (s, 3H), 0.61(s, 3H) (Ra *et al.*, 2000).

3β-*tert*-Butyldimethylsiloxy-5β-cholan-24-yl ethane-1,2-diol tosylate (**344**) was prepared from **343** (*Scheme 2.4.48*). TBDMSOTf (39 mg, 0.147 mmol) in DCM (5 mL) was added to a cold (0 °C) solution of **343** (55 mg, 98 μmol) in C_5H_5N (15.5 mg, 0.20 mmol), and stirred for 30 min. H_2O was added and the mixture was extracted with Et_2O (20 mL), washed with brine, dried ($MgSO_4$) and evaporated to dryness to obtain compound **344** (65 mg, 99%), which was used in the next step without further purification (Ra *et al.*, 2000).

Dimer **345** was prepared from the diol **111** and 3-*tert*-butyldimethylsiloxy-5-cholan-24-yl ethane-1,2-diol tosylate (**344**) (*Scheme 2.4.48*). To a stirred solution of diol **111** (130 mg, 0.36 mmol) in THF (10 mL), 60% NaH (80 mg, 1.67 mmol), and **344** (243 mg, 0.36 mmol) were added at 78 °C (water bath) for 66 h. H_2O was added and the mixture was extracted with Et_2O. The ethereal solution was dried ($MgSO_4$), rotary evaporated and purified by FCC (mobile phase: 10% EtOAc in *n*-hexane) to obtain dimer **345** (103 mg, 33%). ^{1}H NMR (300 MHz, $CDCl_3$): δ 3.63 (br, 1H), 3.55 (s, 6H), 3.41 (t, 4H, $J=6.5$ Hz), 2.15-0.95 (m, 62H), 0.87 (s, 9H), 0.86 (m, 6H), 0.62 (s, 3H), 0.61 (s, 3H), 0.03 (s, 6H). ^{13}C NMR (75 MHz, $CDCl_3$): δ −4.6, 11.9, 12.0, 18.3, 18.6, 20.8, 23.4, 24.2, 26.0, 26.2, 26.4, 27.2, 27.3, 28.3, 30.5, 31.0, 32.0, 34.6, 34.6, 35.3, 35.6, 35.8, 36.5, 36.9, 40.2, 40.4, 42.1, 42.3, 42.7, 56.1, 56.2, 56.4, 56.5, 71.82, 72.0, 72.8 (Ra *et al.*, 2000).

A protected dimer, ethane-1,2-diol mono THP ether dimer **346** was prepared from dimer **345** (*Scheme 2.4.48*). Ethane-1,2-diol mono *p*-tosylate THP ether (65 mg, 0.215 mmol) was added to a mixture of **345** (149 mg, 0.172 mmol) and 60% NaH (43 mg, 1.07 mmol) at 78 °C (water bath). After the usual workup as above and purified by FCC, eluted with 5% EtOAc in *n*-hexane, yielded **346** (89 mg, 52%). ^{1}H NMR (300 MHz, $CDCl_3$): δ 4.62 (t, 1H, $J=3.48$ Hz), 3.89–3.78 (m, 2H), 3.64–3.45 (m, 10H), 3.40 (t, 4H, $J=6.7$ Hz), 1.93–0.92 (m, 63H), 0.60 (s, 6H), 0.02 (s, 6H). ^{13}C NMR (75 MHz, $CDCl_3$): δ −4.6, 10.9, 12.0, 14.0, 14.1, 18.3, 18.6, 19.5, 20.8, 22.6, 23.0, 23.4, 23.7, 24.2, 25.5, 26.0, 26.2, 26.4, 27.2, 27.3, 28.3, 28.9, 29.7, 30.3, 30.6, 31.0, 31.6, 32.0, 33.2, 34.6, 34.9, 35.4, 35.6, 35.8, 35.9, 36.9, 38.7, 40.2, 40.2, 40.3, 42.2, 42.3, 42.7, 56.2, 56.0, 56.4, 56.5, 62.2, 67.0, 67.3, 68.1, 70.1, 72.1, 72.8, 79.6, 98.9 (Ra *et al.*, 2000).

Deprotection of dimer **346** could be achieved by treating a solution of dimer **346** (72 mg, 0.072 mmol) in MeOH (20 mL) with *p*-TsOH (1.4 mg, 7 μmol) at room temperature for 5 h (*Scheme 2.4.48*). Usual cleanup as above afforded diol **347** (57 mg, 99%). ^{1}H NMR (300 MHz, $CDCl_3$): δ 3.72–3.52 (m, 10H), 3.43 (t, 4H, $J=6.7$ Hz), 3.30 (m, 1H), 2.08-0.96 (m, 63H), 0.90 (s, 6H), 0.64 (s, 6H). ^{13}C NMR (75 MHz, $CDCl_3$): δ 12.0, 18.6, 19.1, 20.8, 21.0, 22.6, 22.9, 23.3, 24.2, 26.1, 26.3, 26.4, 27.1, 27.2, 27.3, 28.2, 30.5, 32.0, 33.1, 34.5, 34.8, 35.2, 35.4, 35.5, 35.8, 36.4, 40.1, 40.2, 40.3, 40.4, 42.0, 42.1, 42.6, 56.1, 56.4, 56.4, 60.4, 61.8, 62.1, 67.3, 69.0, 70.0, 70.3, 71.6, 71.7, 71.8, 71.9, 72.0, 79.7 (Ra *et al.*, 2000).

Selective tosylation of **347** was performed as follows (*Scheme 2.4.49*). *p*-TsCl (20 mg, 0.11 mmol) was added to a cooled (0 °C) stirred solution of **347** (56 mg, 0.070 mmol) and

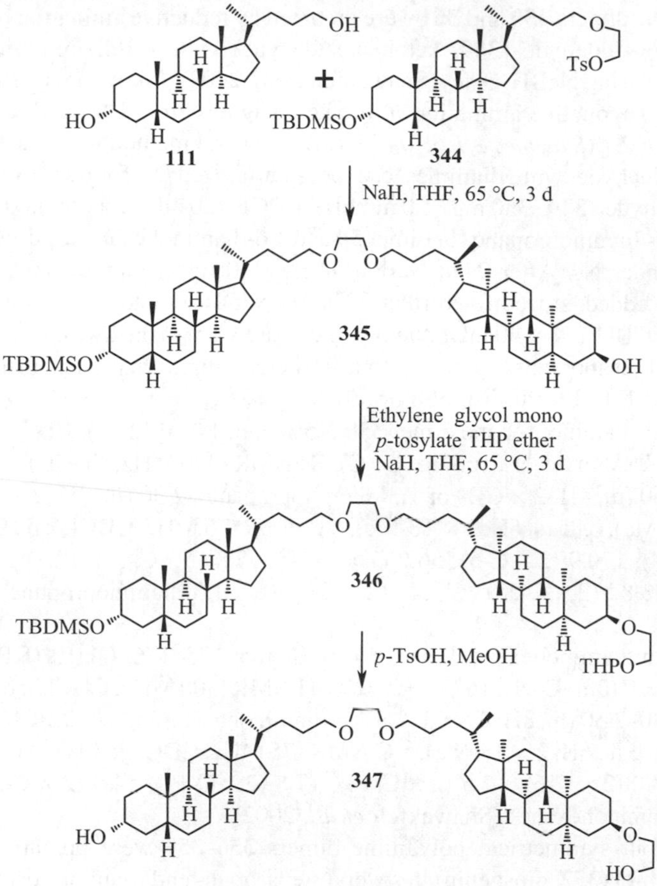

OH H H H HO H **111** + TsO O H H H TBDMSO H **344**

NaH, THF, 65 °C, 3 d

O O H H H TBDMSO H **345** H H H OH H

Ethylene glycol mono
p-tosylate THP ether
NaH, THF, 65 °C, 3 d

O O H H H TBDMSO H **346** H H H O H THPO

p-TsOH, MeOH

O O H H H HO H **347** H H H O H HO

Scheme 2.4.48 *Synthesis of lithocholic acid-based dimers **345–347***

Et_3N (21 mg, 0.21 mmol) in dry DCM (10 mL). The reaction mixture was warmed to room temperature and stirred for 5 h. Usual workup and purification on FCC, eluted with 20% EtOAc in *n*-hexane, produced dimer **348** (30 mg, 45%). ^1H NMR (300 MHz, $CDCl_3$): δ 7.82 (d, 2H, J = 8.2 Hz), 7.36 (d, 2H, J = 7.9 Hz, 2H), 3.82 (br, 2H), 3.64-3.41 (m, 8H), 2.46 (s, 3H), 1.96–0.88 (m, 67H), 0.86 (s, 6H), 0.64 (s, 6H) (Ra *et al.*, 2000).

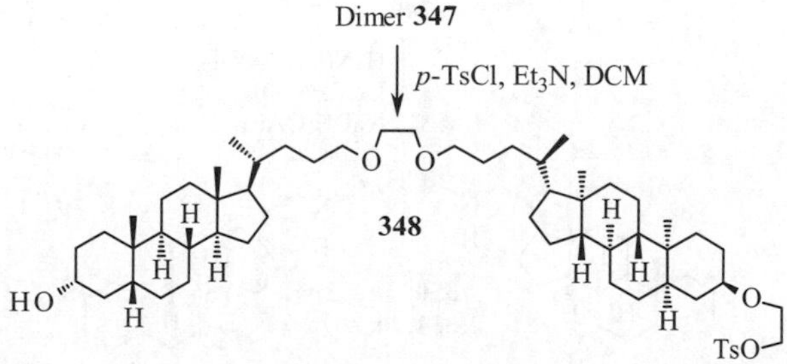

Dimer **347**

p-TsCl, Et_3N, DCM

O O H H H HO H **348** H H H O H TsO

Scheme 2.4.49 *Synthesis of lithocholic acid-based dimer **348***

Two pregnene dimers **350** and **351** were prepared by reductive amination of 3-oxopregn-4-ene-20β-carboxaldehyde (**349**, ketobisnoaldehyde) with 1,3-diaminopropane or 1,6-diaminohexane using NaBH(OAc) (Shawakfeh *et al.*, 2002), following the same procedure as described for reductive amination of the keto stigmasterol (**127**) and keto cholesterol (**128**) in *Section 2.1* (*Schemes 2.1.68 and 2.1.69*). General method for reductive amination of ketobisnoaldehyde with diamines can be summarized as follows (*Scheme 2.4.50*). Ketobisnoaldehyde (**349**, 350 mg, 1.0 mmol) in DCE (20 mL) was treated with diamine (1.0 mmol, 1,3-diaminopropane for dimer **50** and 1,6-diaminohexane for dimer **51**) at room temperature under N_2. After 24 h, $NaBH(OAc)_3$ (430 mg, 2.0 mmol) and glacial AcOH (500 μL) were added, stirred for further 72 h. 1N NaOH was added and the mixture was extracted with $CHCl_3$ (4 × 40 mL), the organic phase was washed with brine (2 × 25 mL), dried ($MgSO_4$), evaporated to dryness to afford crude dimer that was purified by recrystallization from EtOH to furnish **350** or **351**.

Pregnene 1,3-diaminopropane dimer **350** (65%, mp: 140–142 °C). IR (KBr): ν_{max} cm^{-1} 3455m (N–H), 2945m (C–H), 1667m (C=C). ^{1}H NMR (300 MHz, $CDCl_3$): δ 5.70 (br s, 2H, 2 × 4-CH), 2.60 (m, 4H, 2 × CH_2 of 1,3-diaminopropane), 2.30 (m, 4H, 2 × 22-CH_2), 1.21 (s, 6H, 2 × 19-Me), 0.85 (s, 6H, 2 × 18-Me). ^{13}C NMR (75 MHz, $CDCl_3$): δ 199.4 (2 × C-3), 123.8 (2 × C-4), 170.9 (2 × C-5), 56.0 (2 × C-14), 55.8 (2 × C-17), 12.1 (2 × C-18), 17.1 (2 × C-19), 53.8 (2 × C-22), 39.4 (2 × CH_2 of 1,3-diaminopropane) (Shawakfeh *et al.*, 2002).

Pregnene 1,6-diaminohexane dimer **351** (64%, mp: 175–176 °C). IR (KBr): ν_{max} cm^{-1} 3414m (N–H), 2940m (C–H), 1673m (C=C). ^{1}H NMR (300 MHz, $CDCl_3$): δ 5.70 (br s, 2H, 2 × 4-CH), 2.80–2.30 (m, 8H, 2 × CH_2 of 1,6-diaminohexane and 2 × 22-CH_2), 1.26 (s, 6H, 2 × 19-Me), 0.85 (s, 6H, 2 × 18-Me). ^{13}C NMR (75 MHz, $CDCl_3$): δ 199.0 (2 × C-3), 170.0 (2 × C-5), 123.0 (2 × C-6), 12.5 (2 × C-18), 17.8 (2 × C-19), 54.5 (2 × C-22), 39.0 (2 × CH_2 of 1,6-diaminohexane) (Shawakfeh *et al.*, 2002).

Similarly, four symmetrical polyamine dimers **356–359** were prepared from (25*R*)-spirost-5-en-3β-ol (**352**, diosgenin) in several steps *via* di- and triamine spacers under mild conditions in relatively high yields (Shawakfeh *et al.*, 2010).

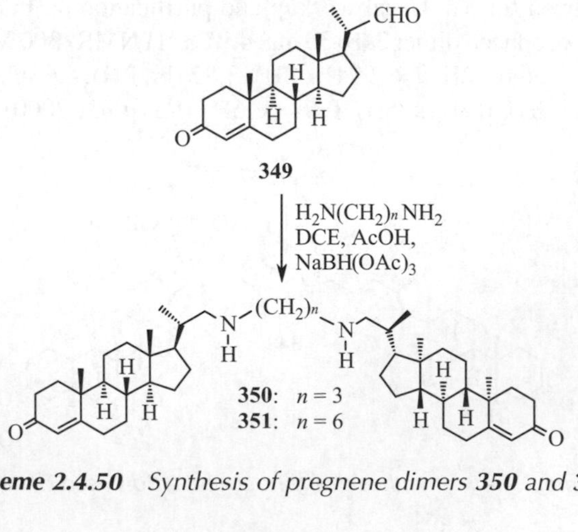

Scheme 2.4.50** Synthesis of pregnene dimers **350** and **351

First, diosgenin (**352**) was converted to its acetate, (25*R*)-spirost-5-en-3β-yl acetate (**353**) (Elgendya and Al-Ghamdy, 2007; Leng *et al.*, 2010), by treating a solution of **353** (20 g, 0.48 mol) in C_5H_5N (1 mL) with Ac_2O (150 mL), and heating the mixture and stirring at 90 °C for 60 min (*Scheme 2.4.51*). A mixture of ice-H_2O was added to form precipitate, the solid was filtered off, the filtrate was washed with H_2O, dried ($MgSO_4$) and rotary evaporated to obtain the crude residue which was recrystallized with MeOH to give diosgenin acetate (**353**, 20.9 g, 95%, mp: 193.6–194.5 °C). IR (KBr): ν_{max} cm^{-1} 2906s (C–H), 2892s (C–H), 1721s (C=O), 1600s (C=C), 1132m (C–O), 1097s (C–O). ^{1}H NMR (500 MHz, $CDCl_3$): δ 5.37 (d, 1H, 6-CH), 4.60 (m, 1H, 3α-CH), 4.41 (q, 1H, 16-CH), 3.46 (m, 1H, 26-CHb), 3.37 (t, 1H, J = 10.8 Hz, 26-CHa), 2.03 (s, 3H, 3β-MeCO), 1.04 (s, 3H, 19-Me), 0.98 (d, 3H, J = 7.6, 21-Me), 0.79 (s, 3H, 18-Me), 0.78 (d, 3H, J = 5.2 Hz, 27-Me). ^{13}C NMR (125 MHz, $CDCl_3$): δ 32.1 (C-1), 38.1 (C-2), 73.9 (C-3), 170.5 (3β-CO), 16.3 (3β-MeCO), 39.8 (C-4), 139.7 (C-5), 122.4 (C-6), 31.5 (C-7), 30.3 (C-8), 56.5 (C-9), 36.8 (C-10), 20.8 (C-11), 31.9 (C-12), 40.3 (C-13), 49.9 (C-14), 27.8 (C-15), 80.8 (C-16), 62.2 (C-17), 21.4 (C-18), 19.9 (C-19), 41.5 (C-20), 14.5 (C-21), 109.3 (C-22), 37.0 (C-23), 28.8 (C-24), 31.4 (C-25), 66.8 (C-26), 17.1 (C-27). EIMS: *m/z* 456 [M]$^+$ (Elgendya and Al-Ghamdy, 2007; Leng *et al.*, 2010).

(22β,25*R*)-3β-Acetyloxyfurost-5-en-26-ol (**354**) was prepared (*Scheme 2.4.48*) from disogenin acetate (**353**) by treating it with $NaBH_3CN$ and AcOH for 8 h, following the same procedure as described for the synthesis of (3β,22β,25*R*)-furost-5-ene-3,26-diol (Basler *et al.*, 2000). Alcohol **354** was converted to its aldehyde, (22β,25*R*)-3β-acetyloxyfurost-5-en-26-al (**355**) by oxidation (*Scheme 2.4.51*). To a mixture of powdered $CaCO_3$ (1.3 g,

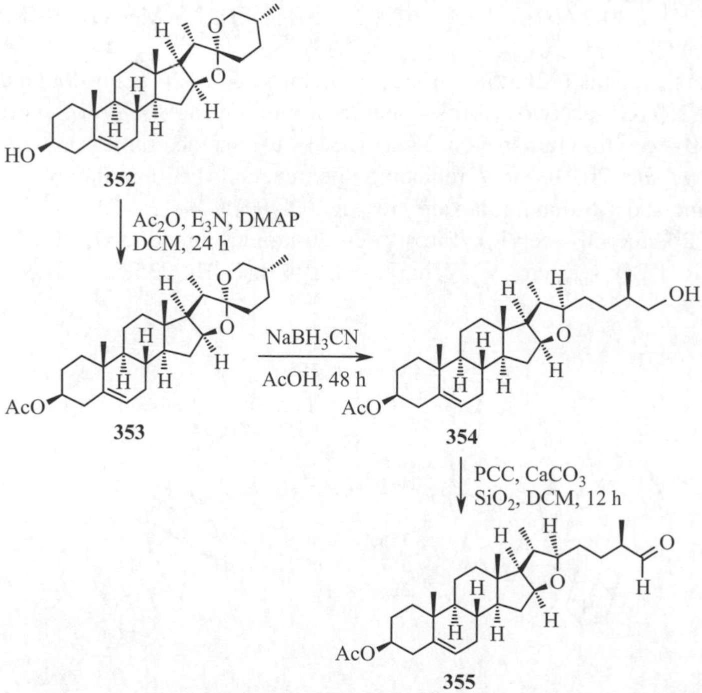

352

Ac_2O, E_3N, DMAP
DCM, 24 h

353

$NaBH_3CN$
AcOH, 48 h

354

PCC, $CaCO_3$
SiO_2, DCM, 12 h

355

Scheme 2.4.51 *Synthesis of (25R)-3β-acetyloxyfurost-5-en-26-al (**355**)*

3.0 mmol), SiO_2 (1.0 g) and **354** (2.5 g, 5.5 mmol) in DCM (50 mL), PCC (700 mg, 3.0 mmol) was added at room temperature and then stirred for 12 h. Et_2O (50 mL) was added and the mixture was passed through a short column of Florisil. The solvent was evaporated to dryness and the crude product was subjected to CC, eluted with 10% EtOAc in *n*-hexane, to yield **355** (1.5 g, 60%, mp: 122–125 °C). IR (KBr): ν_{max} cm^{-1} 2947s (C–H), 2893s (C–H), 2832s (C–H), 2717s (C–H), 1736s (C=O), 1463m (C–H), 1380m (C–H), 1243s (C–O), 1035s (C–O). ^{1}H NMR (400 MHz, $CDCl_3$): δ 9.70 (d, 1H, *J* = 2.0 Hz, 26-CH), 5.30 (d, 1H, *J* = 5.3 Hz, 6-CH), 4.60 (m, 1H, 3-CH), 4.30 (m, 1H, 16-CH), 2.00 (s, 3H, MeCO), 3.30 (m, 1H, 22-CH) (Shawakfeh *et al.*, 2010).

(22β,25*R*)-3β-Acetyloxyfurost-5-en-26-al (**355**) was utilized for the synthesis of dimer, 1,3-bis[(22β,25*R*)-3β-acetyloxyfurost-5-en-26-amino]propane (**356**) (*Scheme 2.4.52*). Glacial AcOH (2 mL) was added to a stirred solution of **355** (300 mg, 0.7 mmol) in DCE (10 mL) under N_2 at room temperature. After 60 min, 1,3-Diaminopropane (78 mg, 1.05 mmol) was added, stirred under the same conditions for further 48 h. $NaBH(OAc)_3$ (280 mg, 1.3 mmol), followed by glacial AcOH (0.5 mL) were added, stirred under N_2 for another 48 h. 1N NaOH was added and the mixture was extracted with $CHCl_3$ (2 × 20 mL). The solvent was washed with brine (2 × 20 mL), dried (Na_2SO_4), rotary evaporated to afford crude oil, which was washed again with Et_2O, and solvent was removed under vacuum to produce the crude product, which was further purified by prep-TLC (mobile phase: 5% NH_3 solution in EtOH) to afford dimer **356** (480 mg, 71%, mp: 71–74 °C). IR (KBr): ν_{max} cm^{-1} 3459m (N–H), 2956m (C–H), 2334s (C–H), 1734s (C=O), 1561m (N–H), 1450m (C–H), 1377m (C–H), 1252s (C–O), 1040s (C–O). ^{1}H NMR (400 MHz, $CDCl_3$): δ 5.40 (d, *J* = 4.5 Hz, 2H, 2 × 6-CH), 4.60 (m, 2H, 2 × 3-CH), 4.30 (m, 2H, 2 × 16-CH), 3.30 (m, 2H, 2 × 22-CH), 2.30-2.60 (m, 4H, 2 × CH_2N), 2.0 (s, 6H, 2 × MeCO). ESIMS: *m/z* 955.9 $[M]^+$ (Shawakfeh *et al.*, 2010).

The dimers, 1,4-bis[(22β,25*R*)-3β-acetyloxyfurost-5-en-26-amino]butane (**357**) and 1,6-bis[(22β,25*R*)-3β-acetyloxyfurost-5-en-26-amino]hexane (**358**) were prepared from (22β,25*R*)-3β-Acetyloxyfurost-5-en-26-al (**355**), using the same protocol as above (Shawakfeh *et al.*, 2010), only replacing the reactant 1,3-diaminopropane with 1,4-diaminobutane and 1,6-diaminohexane, respectively (*Scheme 2.4.52*).

1,4-Bis[(22β,25*R*)-3β-acetyloxyfurost-5-en-26-amino]butane (**357**, 400 mg, 59%, mp: 95–98 °C). IR (KBr): ν_{max} cm^{-1} 3377m (N–H), 2943s (C–H), 2358s (C–H), 1662s (C=C),

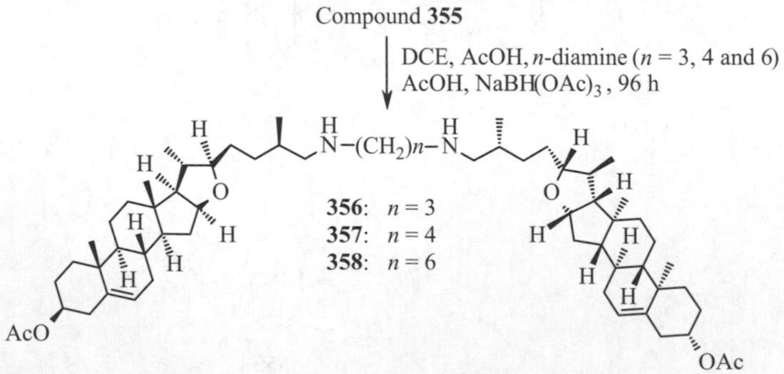

Compound **355**

DCE, AcOH, *n*-diamine (*n* = 3, 4 and 6)
AcOH, $NaBH(OAc)_3$, 96 h

356: *n* = 3
357: *n* = 4
358: *n* = 6

Scheme 2.4.52 *Synthesis of diosgenin polyamine dimer 356–358*

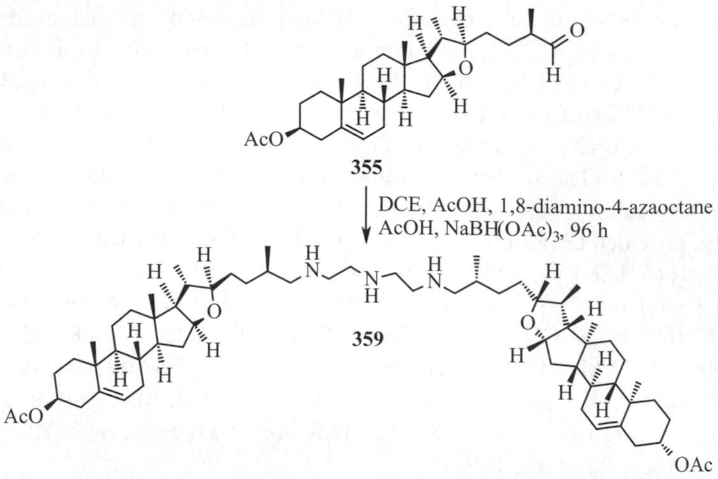

Scheme 2.4.53 *Synthesis of diosgenin polyamine dimer* ***359***

1598m (N–H), 1449m (C–H), 1235s (C–O), 1053s. ^{1}H NMR (400 MHz, $CDCl_3$): δ 5.40 (d, $J = 4.5$ Hz, 2H, 2 × 6-CH), 4.60 (m, 2H, 2 × 3-CH), 4.30 (m, 2H, 2 × 16-CH), 3.30 (m, 2H, 2 × 22-CH), 2.30-2.90 (m, 4H, 2 × CH_2N), 2.00 (s, 6H, 2 × MeCO). ESIMS: *m/z* 972.1 $[M]^+$ (Shawakfeh *et al.*, 2010).

1,6-Bis[(22β,25*R*)-3β-acetyloxyfurost-5-en-26-amino]hexane (**358**, 600 mg, 86%, mp: 67 °C). IR (KBr): ν_{max} cm^{-1} 3467m (N–H), 2942s (C–H), 2367s (C–H), 1747s (C=O), 1572m (N–H), 1424m (C–H), 1243s (C–O), 1024s. ^{1}H NMR (400 MHz, $CDCl_3$): δ 5.30 (d, $J = 4.5$ Hz, 2H, 2 × 6-CH), 4.60 (m, 2H, 2 × 3-CH), 4.30 (dt, 2H, $J = 7.5$, 13.0 Hz, 2 × 16-CH), 3.30 (m, 2H, 2 × 22-CH), 2.30-2.70 (m, 4H, 2 × CH_2N), 2.00 (s, 6H, 2 × MeCO). ESIMS: *m/z* 999.3 $[M]^+$ (Shawakfeh *et al.*, 2010).

1,8-Bis[(22β,25*R*)-3β-acetyloxyfurost-5-en-26-amino]-4-azaoctane (**359**) was prepared from **355** (300 mg, 0.7 mmol) by the reaction with 1,8-diamino-4-azaoctane (150 mg, 1.05 mmol) and $Na(OAc)_3BH$ (280 mg, 1.3 mmol) (*Scheme 2.4.53*). Usual cleanup and purification as described above furnished dimer **359** (620 mg, 86%, mp: 227–230 °C). IR (KBr): ν_{max} cm^{-1} 3480m (N–H), 2954m (C–H), 2356s (C–H), 1737s (C=O), 1574m (N–H), 1444m (C–H), 1248s (C–O), 1139s. ^{1}H NMR (400 MHz, $CDCl_3$): δ 5.40 (d, $J = 4.5$ Hz, 2H, 2 × 6-CH), 4.60 (m, 2H, 2 × 3-CH), 4.30 (dt, 2H, $J = 13.0$ and 7.5 Hz, 2 × 16-CH), 3.30 (m, 2H, 2 × 22-CH), 2.30-3.00 (m, 4H, 2 × CH_2N), 2.00 (s, 6H, 2 × MeCO). ESIMS: *m/z* 1026.7 $[M]^+$ (Shawakfeh *et al.*, 2010).

The synthesis of dimeric steroids **363** and **364**, as components of artificial lipid bilayers, was accomplished from diosgenin (**352**) in several steps (Morzycki *et al.*, 1997). In order to synthesize these dimers, diosgenin (**352**) was first transformed to (25*R*)-spirost-5-en-3β-*tert*-butyldimethylsilyloxy (**360**) (*Scheme 2.4.54*). Imidazole (800 mg, 12.0 mmol) was added to a solution of TBDMSCl (900 mg, 6.0 mmol) in dry DMF (10 mL), stirred at room temperature for 15 min under N_2, **352** (500 mg, 1.2 mmol) was added to the mixture and stirred at room temperature for a further 16 h. The reaction mixture was filtered off and the solid was subjected to FCC, eluted with 5% EtOAc in *n*-hexane, to provide diosgenin TBDMS ether **360**.

Reductive ring F fission of diosgenin TBDMS ether **360** was achieved as follows (*Scheme 2.4.54*). $LiAlH_4$ (3.15 g, 0.08 mmol) was added carefully to a stirred ice-cold solution of $AlCl_3$ (44.8 g, 0.33 mol) in Et_2O. After 30 min, a solution of **360** (4.41 g, 8.4 mmol) in Et_2O (150 mL) was added and stirred for another 18 h. The excess hydride was decomposed with Me_2CO, an aqueous solution of $AlCl_3$ was added to the mixture and extracted with $CHCl_3$. The solvent was removed under pressure and the crude solid was purified by CC (mobile phase: 10% EtOAc in C_6H_6), followed by crystallization from Me_2CO-C_6H_6 provided (25*R*)-furost-5-en-26-ol-3β-TBDMS ether (**361**, 2.65 g, 60%, mp: 134–136 °C). IR ($CHCl_3$): ν_{max} cm^{-1} 3660br (O–H), 3440m (O–H), 1090s (C–O). ^{1}H NMR (200 MHz, $CDCl_3$): δ 5.35 (m, 1H, 6-CH), 4.30 (m, 1H, 16α-CH), 3.47 (m, 3H, 3α-CH and 26-CH_2), 3.32 (m, 1H, 22-CH), 1.01 (s, 3H, 18-Me), 1.00 (d, 3H, $J = 8.8$ Hz, 21-Me), 0.92 (d, $J = 6.7$ Hz, 3H, 27-Me), 0.89 (s, 9H, *t*-Bu-Si), 0.81 (s, 3H, 19-Me), 0.06 (s, 6H, Me-Si). ^{13}C NMR (50 MHz, $CDCl_3$): δ −4.6, 16.4, 16.6, 18.3, 18.9, 19.4, 20.7, 25.9, 30.2, 30.5, 31.6, 32.1, 32.2, 35.8, 36.7, 37.4, 37.9, 39.5, 40.7, 42.8, 50.2, 57.0, 65.1, 68.1, 72.6, 83.2, 90.4, 120.9, 141.6 (Morzycki *et al*., 1997).

Iodination of (25*R*)-furost-5-en-26-ol-3β-TBDMS ether (**361**) could be carried out as follows (*Scheme 2.4.54*). To a stirred solution of I_2 (333 mg, 1.3 mmol) and Ph_3P (300 mg, 1.14 mmol) in C_6H_6 (30 mL) and C_5H_5N (1.1 mL), **361** (200 mg, 0.38 mmol) was added and stirred for 2 h at room temperature. The reaction mixture was stirred at room temperature for 2 h, the solvent was evaporated to dryness and the crude product was purified by CC, eluted with C_6H_6, to obtain diosgenin iodide **362** (230 mg, 96%, mp: 108–110 °C). IR ($CHCl_3$): ν_{max} cm^{-1} 2920m (C–H), 1090s (C–O). ^{1}H NMR (200 MHz, $CDCl_3$): δ 5.31 (m, 1H, 6-CH), 4.31 (m, 1H, 16α-CH), 3.49 (m, 3H, 3α-CH), 3.29 (m, 1H, 22-CH), 3.22 (m, 2H, 26-CH_2),

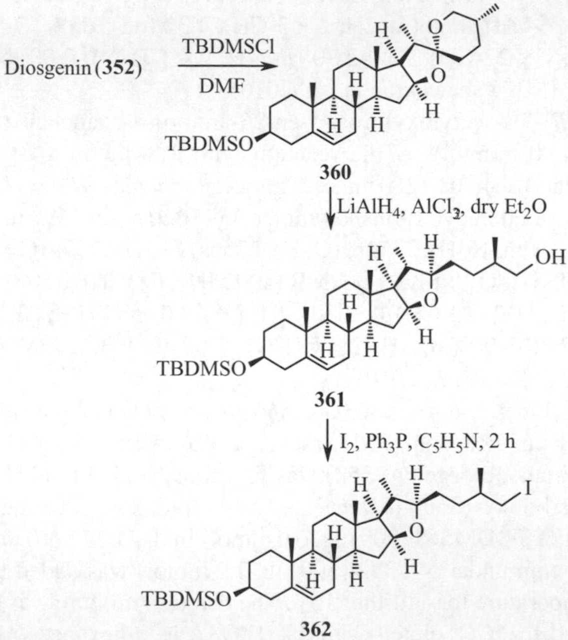

Scheme 2.4.54** Synthesis of diosgenin derivatives **360–362

1.01 (s, 3H, 18-Me), 0.96-1.01 (m, 6H, 21-Me and 27-Me), 0.89 (s, 9H, *t*-Bu-Si), 0.80 (s, 3H, 19-Me), 0.05 (s, 6H, Me-Si). ^{13}C NMR (50 MHz, $CDCl_3$): δ −4.6, 16.4, 17.8, 18.2, 19.0, 19.4, 20.5, 20.7, 25.9, 30.9, 31.6, 32.0, 32.2, 33.6, 34.8, 36.7, 37.4, 37.9, 39.4, 40.7, 42.8, 50.2, 57.0, 65.1, 72.5, 83.2, 90.0, 120.9, 141.6 (Morzycki *et al.*, 1997).

The Wurtz reaction of diosgenin iodide (**362**) produced the dimer **363** (*Scheme 2.4.55*). A solution of **362** (160 mg, 0.25 mmol) in dry toluene (3 mL) was treated with Na (30 mg, 1.3 mmol) and the mixture was refluxed for 20 h. C_6H_6 was added and the solution was washed with H_2O, dried ($MgSO_4$), the solvent was evaporated to dryness and the crude dimer was purified by CC (mobile phase: 25% *n*-hexane in C_6H_6), followed by crystallization from Me_2CO–hexane to produce 3β-TBDMS ether dimer **363** (20 mg, 16%, mp: 192–194 °C). IR ($CHCl_3$): ν_{max} cm^{-1} 2920m (C–H), 1090s (C–O). ^1H NMR (200 MHz, $CDCl_3$): δ 5.31 (m, 2H, 2 × 6-CH), 4.30 (m, 2H, 2 × 16α-CH), 3.46 (m, 6H, 2 × 3α-CH), 3.30 (m, 2H, 2 × 22-CH), 1.01 (s, 6H, 2 × 18-Me), 1.00 (d, 6H, J = 7.4 Hz, 2 × 21-Me), 0.89 (s, 9H, *t*-Bu-Si), 0.85 (d, 6H, J = 5.5 Hz, 2 × 27-Me), 0.80 (s, 6H, 2 × 19-Me), 0.06 (s, 6H, Me-Si). EIMS: *m/z* 1026 $[M]^+$ (Morzycki *et al.*, 1997).

Di(3β-hydroxyfurost-5-en-26-yl) (**364**) was then prepared from dimer **363** (*Scheme 2.4.55*). A solution of dimer **363** (210 mg, 0.205 mmol) in $CHCl_3$ (2 mL) and MeOH (2 mL), was added to a mixture of MeOH:H_2O:HCl (0.50:0.25:0.25, 1 mL), the reaction mixture was allowed to stand at room temperature. After 15 min, saturated aqueous $NaHCO_3$ was added and the mixture was extracted with $CHCl_3$. The solvent was removed by evaporation under pressure and the crude dimer was purified by CC, eluted with 25% EtOAc in C_6H_6, to yield diol **364** (142 mg, 87%, mp: 123–126 °C). IR ($CHCl_3$): ν_{max} cm^{-1} 3599br (O–H), 3389br (O–H), 1096s (C–O), 1046s (C–O). ^1H NMR (200 MHz, $CDCl_3$): δ 5.34 (m, 2H, 2 × 6-CH), 4.30 (m, 2H, 2 × 16α-CH), 3.52 (m, 6II, 2 × 3α-CH),

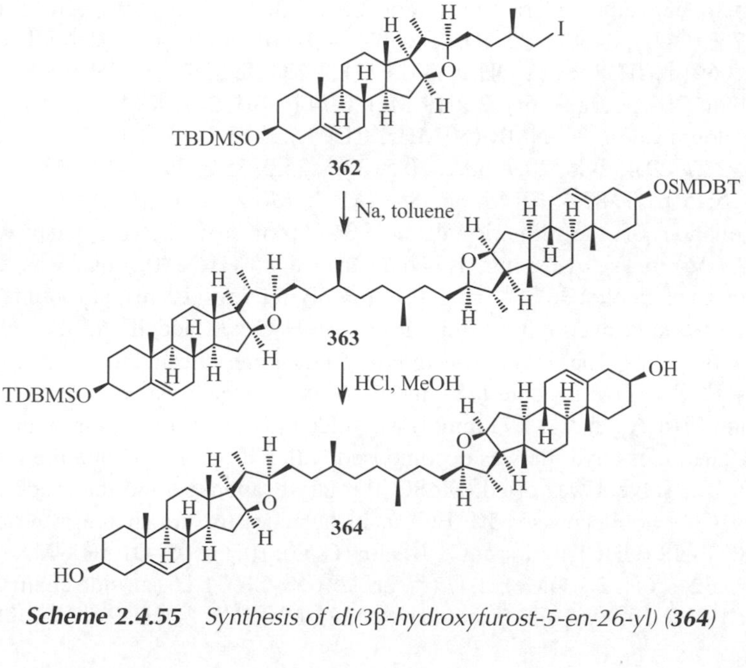

Scheme 2.4.55 *Synthesis of di(3β-hydroxyfurost-5-en-26-yl) (**364**)*

3.30 (m, 2H, 2 × 22-CH), 1.02 (s, 6H, 2 × 18-Me), 1.00 (d, 6H, $J = 7.4$ Hz, 2 × 21-Me), 0.85 (d, 6H, $J = 5.5$ Hz, 2 × 27-Me), 0.81 (s, 6H, 2 × 19-Me). ^{13}C NMR (50 MHz, $CDCl_3$): δ 16.4, 19.1, 19.4, 19.5, 20.7, 31.1, 31.6, 31.9, 32.0, 32.3, 33.3, 34.0, 34.1, 36.6, 37.2, 37.9, 39.4, 40.7, 42.2, 50.1, 57.0, 65.2, 71.7, 83.1, 90.5, 121.5, 140.7. EIMS: *m/z* 798 $[M]^+$ (Morzycki *et al.*, 1997).

Several pregnane-based dimers **366–369** were synthesized from pregnanoic ester (**365**) by an *'alkylation-reduction'* method using 1,14-diiodotetradecane (Morzycki *et al.*, 1997). A cooled (–20 °C) solution of **365** (552 mg, 1.48 mmol) in dry THF (1.5 mL) was treated with 2M solution of LDA in THF:ethylbenzene:heptane (800 μL) under Ar (*Scheme 2.4.56*). The reaction mixture was stirred for 15 min, then a solution of 1,14-diiodotetradecane (332 mg, 0.74 mmol) in THF (1 mL) was added dropwise at −20 °C under Ar. After 2.5 h, aqueous EtOH was added and the mixture was extracted with $CHCl_3$, the extract was dried ($MgSO_4$), concentrated *in vacuo*, and the crude dimer was purified by CC (mobile phase: 5% EtOAc in heptane) to afford dimeric ester **366** (180 mg, 26%). IR ($CHCl_3$): ν_{max} cm^{-1} 1720s (C=O), 1093m (C–O). ^{1}H NMR (200 MHz, $CDCl_3$): δ 4.10 (q, 4H, $J = 7.1$ Hz, 2 × OCH_2CH_3), 3.32 (s, 6H, 2 × OMe), 2.76 (m, 2H, 2 × 6α-CH), 2.25 (m, 2H, 2 × 20-CH), 1.26 (t, 6H, $J = 7.1$ Hz, 2 × OCH_2CH_3), 1.24 (m, side chain CH_2), 1.01 (s, 6H, 2 × 19-Me), 0.75 (s, 6H, 2 × 18-Me), 0.64 and 0.43 (m, 2 × 2H, cyclopropane). ^{13}C NMR (50 MHz, $CDCl_3$): δ 12.4, 13.1, 14.2, 19.3, 21.5, 22.6, 23.7, 24.9, 27.2, 27.3, 29.5, 29.6, 30.5, 32.1, 33.3, 35.0, 35.2, 38.0, 42.4, 43.4, 47.4, 48.0, 52.8, 55.8, 56.5, 59.6, 82.3, 176.3. EIMS: *m/z* 942 $[M]^+$ (Morzycki *et al.*, 1997).

The dimeric ester **366** was reduced to the dimeric alcohol **367** (*Scheme 2.4.56*). $LiAlH_4$ (60 mg, 1.58 mmol) was added to a solution of **366** (175 mg, 0.19 mmol) in THF (7 mL) at room temperature and stirred at room temperature overnight. The excess hydride was decomposed with a drop of H_2O and the mixture was filtered off. The solvent was dried ($MgSO_4$), rotary evaporated and the crude dimer was subjected to CC, eluted with 20% EtOAc in petroleum ether, to provide dimeric alcohol **367** (142 mg, 89%). IR ($CHCl_3$): ν_{max} cm^{-1} 3627br (O–H), 3430br (O–H), 1094m (C–O), 1076m (C–O). ^{1}H NMR (200 MHz, $CDCl_3$): δ 3.69 (m, 4H, 2 × CH_2OH), 3.33 (s, 6H, 2 × OMe), 2.77 (m, 2H, 2 × 6α-CH), 1.25 (m, side chain CH_2), 1.02 (s, 6H, 2 × 19-Me), 0.74 (s, 6H, 2 × 18-Me), 0.65 and 0.43 (m, 2 × 2H, cyclopropane). ^{13}C NMR (50 MHz, $CDCl_3$): δ 12.5, 13.1, 19.3, 21.5, 22.7, 24.0, 24.9, 26.4, 27.7, 29.4, 29.6, 29.7, 30.2, 30.5, 33.3, 35.0, 35.2, 39.7, 42.4, 42.5, 43.4, 48.0, 50.5, 56.4, 56.5, 62.7, 82.4. EIMS: *m/z* 858 $[M]^+$ (Morzycki *et al.*, 1997).

Deoxygenation of the dimeric alcohol **367** furnished the pregnane dimer **368** (*Scheme 2.4.56*). *p*-TsCl (255 mg, 1.34 mmol) in dry C_5H_5N (0.7 mL) was added to a stirred solution of alcohol **367** (140 mg, 0.16 mmol) in DCM (10 mL) at room temperature and allowed to stand overnight. A mixture of ice-H_2O was added and the solution was extracted with $CHCl_3$. The solvent was removed under pressure and the crude tosylate was purified by FCC to give tosylate (182 mg, 0.16 mmol) which was dissolved in dry THF (10 mL), and $LiAlH_4$ (65 mg, 1.71 mmol) was added to it, and stirred at room temperature overnight. The excess hydride was decomposed with a drop of H_2O and the mixture was filtered off. The solvent was dried ($MgSO_4$), rotary evaporated and the crude dimer was purified by CC (mobile phase: 5% EtOAc in heptane) to furnish pregnane dimer **368** (103 mg, 80%). IR ($CHCl_3$): ν_{max} cm^{-1} 1093m (C–O), 1076m (C–O). ^{1}H NMR (200 MHz, $CDCl_3$): δ 3.32 (s, 6H, 2 × OMe), 2.77 (m, 2H, 2 × 6α-CH), 1.26 (m. side chain CH_2), 1.02 (s, 6H, 2 × 19-Me), 0.90 (d, 6H, $J = 6.4$ Hz, 2 × 21-Me), 0.71 (s, 6H, 2 × 18-Me), 0.64 and

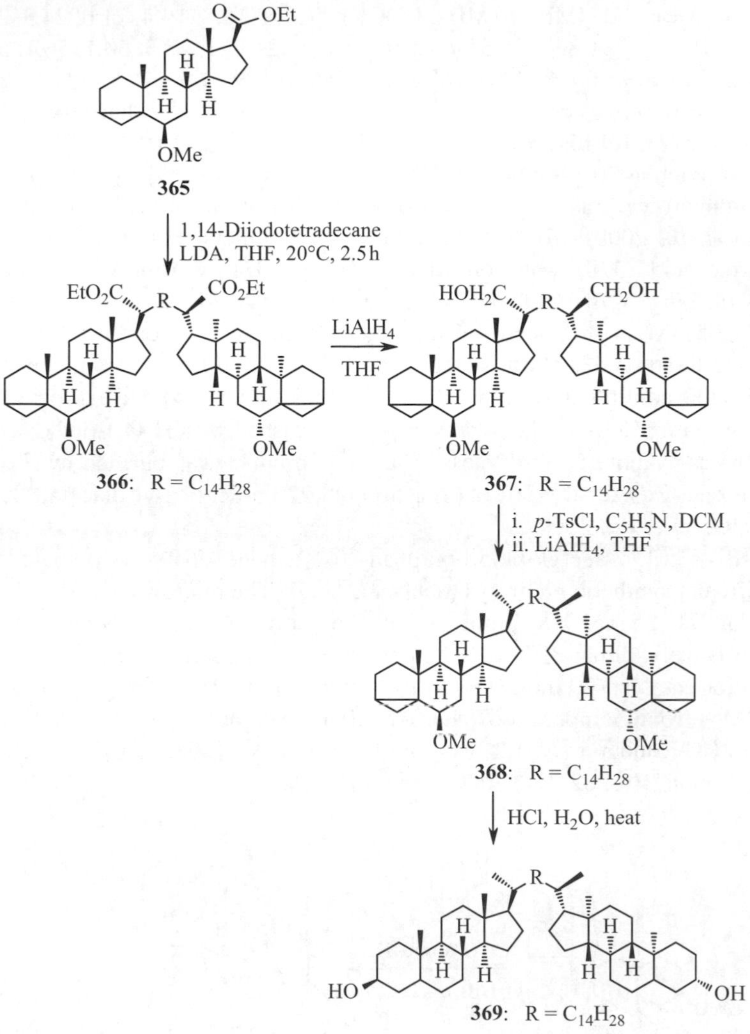

365

1,14-Diiodotetradecane
LDA, THF, 20°C, 2.5 h

366: $R = C_{14}H_{28}$

$LiAlH_4$
THF

367: $R = C_{14}H_{28}$

i. *p*-TsCl, C_5H_5N, DCM
ii. $LiAlH_4$, THF

368: $R = C_{14}H_{28}$

HCl, H_2O, heat

369: $R = C_{14}H_{28}$

Scheme 2.4.56 *Synthesis of pregnane-based dimers **366–369***

0.43 (m, 2 × 2H, cyclopropane). ^{13}C NMR (50 MHz, $CDCl_3$): δ 12.2, 13.0, 18.7, 19.3, 21.5, 22.8, 24.2, 25.0, 26.1, 28.3, 29.7, 29.8, 30.2, 30.5, 33.3, 35.0, 35.3, 35.8, 35.9, 40.3, 42.8, 43.4, 48.0, 56.3, 56.5, 82.4 (Morzycki *et al*., 1997).

p-TsOH (15 mg, 0.08 mmol) in H_2O (3 mL) was added to a solution of pregnane dimer **368** (103 mg, 0.125 mmol) in dioxane (10 mL) at 80 °C. After 2 h, H_2O was added and the mixture was extracted with $CHCl_3$ (*Scheme 2.4.56*). The solvent was dried ($MgSO_4$), concentrated *in vacuo* and the crude dimer was subjected to CC, eluted with 35% EtOAc in petroleum ether, followed by crystallization from DCM-*n*-hexane to give pregnane dimer **369** (82 mg, 82%, mp: 182–185 °C). IR ($CHCl_3$): ν_{max} cm^{-1} 3600br (O–H), 2931m (C–H), 958m. ^1H NMR (200 MHz, $CDCl_3$): δ 5.35 (m, 2H, 2 × 6α-CH), 3.52 (m, 2H, 2 × 3α-CH), 1.25 (m, side chain CH_2), 1.01 (s, 6H, 2 × 19-Me), 0.91 (d, 6H, J = 6.4 Hz, 2 × 21-Me), 0.67

(s, 6H, 2 × 18-Me). ^{13}C NMR (50 MHz, $CDCl_3$): δ 11.8, 18.7, 19.4, 21.0, 24.3, 26.1, 28.2, 29.6, 29.7, 30.1, 31.6, 31.8, 31.9, 35.7, 35.9, 36.4, 37.2, 39.7, 42.3, 50.1, 56.1, 56.7, 71.7, 121.6, 140.7. EIMS: *m/z* 798 $[M]^+$ (Morzycki *et al.*, 1997).

The synthesis of progesterone homodimers, (*E*)-1,2-bis-(3α,12α-diacetoxy-5β-pregnan-20-yl)-ethene (**372**), 1,4-bis-(3,12-dioxo-5β-pregnan-20-yl)-2*E*-butene (**375**) and (*E*)-1,2-bis-(3-oxo-4-pregnen-20-yl)-ethene (**378**) were accomplished by olefin metathesis either using microwave assisted metathesis (MWAM) or Grubbs 2nd-generation catalyst (Edelsztein *et al.*, 2009). To synthesize these homodimers, first 3α,12α-diacetoxy-5β-cholan-24-oic acid (**370**) was converted to alkene **371** as follows (*Scheme 2.4.57*). $Cu(AcO)_2$ (422 mg, 1.75 mmol) in C_5H_5N (660 μL) was added to a stirred solution of **370** (5.4 g, 8.75 mM) in dry toluene (53 mL) under Ar at room temperature. After 15 min, the mixture was refluxed and DIB was added portionwise (5 × 430 mg, 5.0 mmol) every 90 min, while the heating continued. After 8 h, the reaction mixture was cooled to room temperature and extracted with 5% HCl. The organic layer was washed with H_2O, dried (Na_2SO_4), and the solvent was evaporated to dryness. The crude product was purified by FCC (mobile phase: *n*-hexane-EtOAc) to yield 24-nor-5β-chol-22-en-3α,12α-yl diacetate (**371**, 2.8 g, 57%) (Edelsztein *et al.*, 2009).

(*E*)-1,2-Bis-(3α,12α-acetyloxy-5β-pregnan-20-yl)ethene (**372**) was prepared by applying two different synthetic pathways (*Scheme 2.4.57*). The microwave assisted metathesis (MWAM) of **371** (25 mg, 0.06 mmol) afforded dimer **372** (24 mg, 97%) with an excellent yield. Alternatively, dimer **372** could be prepared with moderate yield by adding Grubbs 2nd-generation catalyst (20 mol%) to a stirred solution of **371** (128 mg, 0.30 mmol) in dry DCM (0.2M) at room temperature under Ar. After 6.5 h, the reaction mixture was purified directly to FCC to obtain (*E*)-1,2-bis-(3α,12α-acetyloxy-5β-pregnan-20-yl)-ethene (**372**, 65 mg, 52%, mp: 201–202 °C) (Edelsztein *et al.*, 2009).

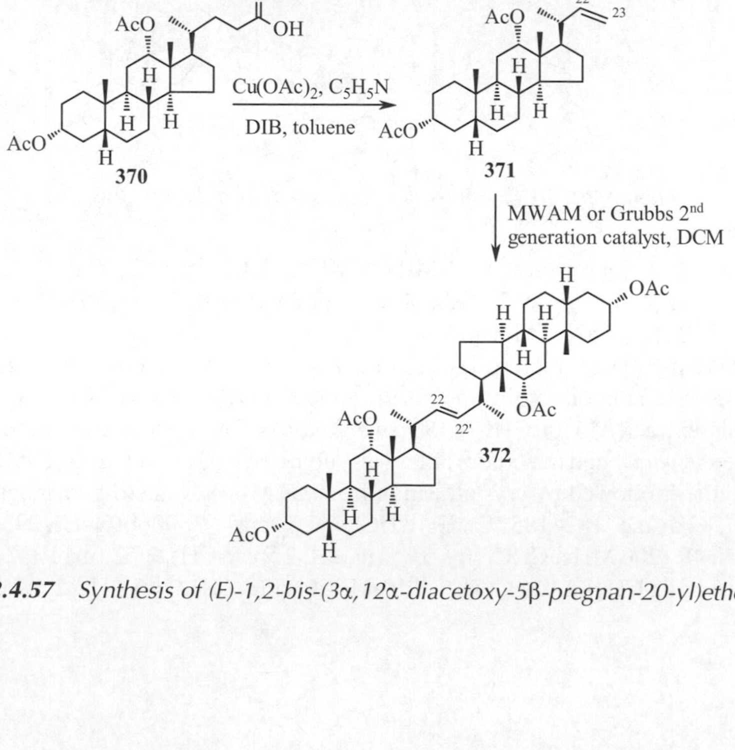

Scheme 2.4.57 *Synthesis of (E)-1,2-bis-(3α,12α-diacetoxy-5β-pregnan-20-yl)ethene (**372**)*

IR (KBr): ν_{max} cm^{-1} 2940m (C–H), 2857m (C–H), 1729s (C=O), 1234s (C–O), 1125m (C–C). ^{1}H NMR (500 MHz, $CDCl_3$): δ 5.08 (m, 1H, 22-CH), 5.06 (m, 1H, 12-CH), 4.71 (m, 1H, 3-CH), 2.11 (s, 3H, 3α-MeCO), 2.04 (s, 3H, 12α-MeCO), 0.91 (s, 3H, 19-Me), 0.86 (d, *J* = 6.7 Hz, 3H, 21-Me), 0.73 (s, 3H, 18-Me). ^{13}C NMR (125 MHz, $CDCl_3$): δ 34.5 (C-1), 27.9 (C-2), 74.2 (C-3), 170.6 (3α-CO), 21.5 (3α-MeCO), 32.3 (C-4), 41.9 (C-5), 26.9 (C-6), 26.7 (C-7), 34.1 (C-8), 35.7 (C-9), 34.8 (C-10), 25.7 (C-11), 75.8 (C-12), 170.5 (12α-CO), 21.4 (12α-MeCO), 44.9 (C-13), 49.5 (C-14), 23.6 (C-15), 25.9 (C-16), 47.8 (C-17), 12.7 (C-18), 23.1 (C-19), 39.4 (C-20), 20.4 (C-21), 134.2 (C-22). HRESIMS: *m/z* 855.5745 calculated for $C_{52}H_{80}O_8Na$, found: 855.5729 (Edelsztein *et al.*, 2009).

3,12-Dioxo-5β-cholan-24-al (**373**) was synthesized from methyl 3α,12α-dihydroxy-5β-cholan-24-oate (**100**) by reduction (*Scheme 2.4.58*). $LiAlH_4$ (93 mg, 2.46 mmol) was added to a stirred solution of **100** (500 mg, 1.23 mmol) in dry THF (25 mL) at 0 °C under Ar. After 15 min, sequentially, Me_2CO, HCl 10%, and sodium and potassium tartrate saturated solution were added to the mixture at room temperature and the mixture was evaporated under pressure. The residue was dissolved in DCM, successively washed with $NaHCO_3$, brine and H_2O. The solvent was dried (Na_2SO_4) and rotary evaporated to obtain 5β-cholane-3α,12α,24-triol, which was oxidized without further purification with PCC (1.2 g, 5.54 mmol), barium carbonate ($BaCO_3$, 631 mg, 3.20 mmol), and 4 Å molecular sieves in dry DCM (20 mL) at room temperature under N_2 for 30 min. Et_2O was added to the reaction mixture, the solvent was evaporated under vacuum, the crude solid was redissolved in minimum amount of Et_2O and the resulting solution was purified by FCC, eluted with Et_2O, to produce dione **373** (368 mg, 83%, mp: 108–110 °C). IR (KBr): ν_{max} cm^{-1} 2967m (C–H), 1707s (C=O), 1657m (C=C), 1460m (C–H), 1219m (C–O), 1083m (C–C), 891m. ^{1}H NMR (500 MHz, $CDCl_3$): δ 9.78 (t, 1H, *J* = 1.7 Hz, 24-CH), 2.59 (dd, 1H, *J* = 13.2, 15.0 Hz, 4-CHa), 2.50 (ddd, 1H, *J* = 0.8, 5.4, 9.5 Hz, 23-CHa), 2.43 (dt, 1H, *J* = 1.0, 7.7 Hz, 23-CHb), 2.33 (dd, 1H, *J* = 5.4, 14.7 Hz, 2-CHa), 2.19 (ddd, 1H, *J* = 3.1, 6.4, 14.7 Hz, 2-CHb), 1.12 (s, 3H, 19-Me), 1.06 (s, 3H, 18-Me), 0.86 (d, 3H, *J* = 6.6 Hz, 21-Me). ^{13}C NMR (125 MHz, $CDCl_3$): δ 36.8 (C-1), 36.9 (C-2), 212.1 (C-3), 42.1 (C-4), 43.7 (C-5), 25.5 (C-6), 27.4 (C-7), 35.4 (C-8), 58.5 (C-9), 35.6 (C-10), 38.4 (C-11), 214.1 (C-12), 57.6 (C-13), 44.3 (C-14), 26.6 (C-15), 24.3 (C-16), 46.5 (C-17), 11.7 (C-18), 22.1 (C-19), 35.6 (C-20), 18.7 (C-21), 27.6 (C-22), 41.2 (C-23), 203.0 (C-24). EIMS: *m/z* 372 $[M]^+$ (Edelsztein *et al.*, 2009).

3,12-Dioxo-5β-chol-24-ene (**374**) was prepared from 3,12-dioxo-5β-cholan-24-al (**373**, 500 mg, 1.35 mmol) following the general Wittig reaction, and compound **374** (424 mg, 85%, mp: 121–122 °C) was prepared as a white crystalline solid (*Scheme 2.4.58*). IR (KBr): ν_{max} cm^{-1} 2970m (C–H), 2948m (C–H), 2906m (C–H), 2869m (C–H), 1709s (C=O), 1507m (C–H), 1452m (C–H), 1028m (C–C). ^{1}H NMR (500 MHz, $CDCl_3$): δ 5.79 (ddt, 1H, *J* = 6.7, 10.2, 17.1 Hz, 24-CH), 4.98 (ddt, 1H, *J* = 1.7, 2.0, 17.2 Hz, 25-CHa), 4.91 (ddt, 1H, *J* = 1.2, 2.1, 10.1 Hz, 25-CHb), 2.61 (t, 1H, *J* = 12.5 Hz, 11-CHb), 2.59 (dd, 1H, *J* = 13.6, 14.8 Hz, 4-CHa), 1.11 (s, 3H, 19-Me), 1.05 (s, 3H, 18-Me), 0.86 (d, 3H, *J* = 6.6 Hz, 21-Me). ^{13}C NMR (125 MHz, $CDCl_3$): δ 36.8 (C-1), 36.9 (C-2), 212.1 (C-3), 42.1 (C-4), 43.7 (C-5), 25.5 (C-6), 27.6 (C-7), 35.5 (C-8), 58.6 (C-9), 35.6 (C-10), 38.4 (C-11), 214.2 (C-12), 57.6 (C-13), 44.3 (C-14), 26.6 (C-15), 24.4 (C-16), 46.8 (C-17), 11.7 (C-18), 22.2 (C-19), 35.7 (C-20), 18.8 (C-21), 34.8 (C-22), 30.9 (C-23), 139.5 (C-24), 114.1 (C-25). EIMS: *m/z* 370 $[M]^+$ (Edelsztein *et al.*, 2009).

The synthesis of 1,4-bis-(3,12-dioxo-5β-pregnan-20-yl)-2*E*-butene (**375**) was accomplished from 3,12-dioxo-5β-chol-24-ene (**374**, 20 mg, 0.05 mmol) following the general

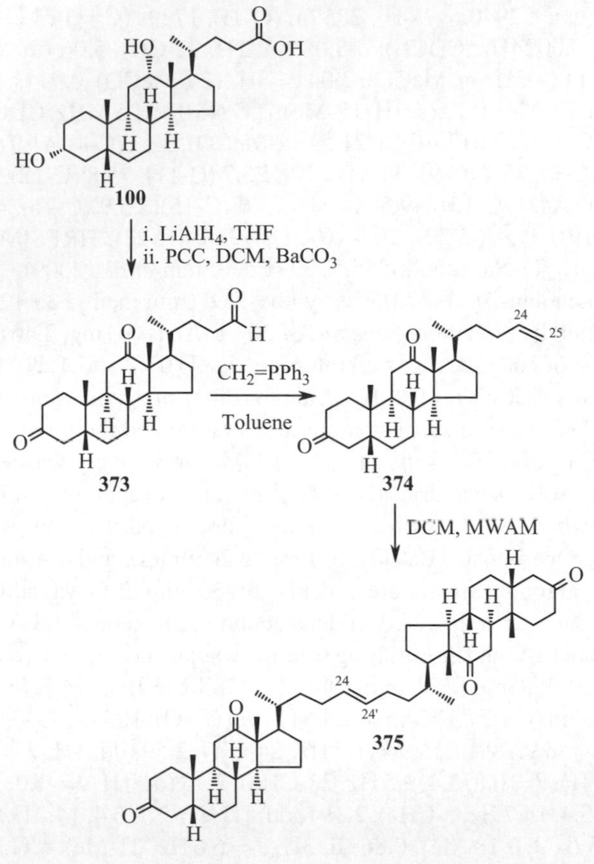

100

i. $LiAlH_4$, THF
ii. PCC, DCM, $BaCO_3$

373

CH_2=PPh_3
Toluene

374

DCM, MWAM

375

Scheme 2.4.58 *Synthesis of 1,4-bis-(3,12-dioxo-5β-pregnan-20-yl)-2E-butene (**375**)*

MWAM reaction (*Scheme 2.4.58*). Usual workup and purification as described above provided dimer **375** (16 mg, 89%, mp: 246–249 °C). IR (KBr): ν_{max} cm^{-1} 2926m (C–H), 2862m (C–H), 2249m (C–H), 1705s (C=O), 1646s (C=C), 1460m (C–H), 1383m (C–H), 1272m (C–O), 919m. ^1H NMR (500 MHz, $CDCl_3$): δ 5.38 (m, 1H, 24-CH), 2.62 (2H, m, 4-CHa and 11-CHb), 1.12 (s, 3H, 19-Me), 1.06 (s, 3H, 18-Me), 0.86 (d, 3H, J = 6.6 Hz, 21-Me). ^{13}C NMR (125 MHz, $CDCl_3$): δ 36.8 (C-1), 36.9 (C-2), 212.2 (C-3), 42.2 (C-4), 43.7 (C-5), 25.5 (C-6), 27.6 (C-7), 35.5 (C-8), 58.6 (C-9), 35.6 (C-10), 38.4 (C-11), 214.3 (C-12), 57.6 (C-13), 44.3 (C-14), 26.6 (C15), 24.4 (C-16), 46.8 (C-17), 11.8 (C-18), 22.2 (C-19), 35.8 (C-20), 18.9 (C-21), 35.5 (C-22), 29.7 (C-23), 130.5 (C-24). EIMS: *m/z* 671 [M + ketene]$^+$ (Edelsztein *et al.*, 2009).

20-Ethenyl-pregn-4-en-3-one (**377**) was prepared from pregn-4-en-20β-al-3-one (**376**, 500 mg, 1.5 mmol) following the general Wittig reaction (*Scheme 2.4.59*), and compound **377** (426 mg, 87%, mp: 116–117 °C) was prepared as a white crystalline solid. IR (KBr): ν_{max} cm^{-1} 2951m (C–H), 2865m (C–H), 2843m (C–H), 1671s (C=C), 1452m (C–H), 1230m, 911m, 864m. ^1H NMR (500 MHz, $CDCl_3$): δ 5.72 (s, 1H, 4-H), 5.66 (ddd, 1H, J = 8.4, 10.2, 17.1 Hz, 22-CH), 4.91 (dd, 1H, J = 2.0, 17.1 Hz, 23-CHa), 4.82 (dd, 1H, J = 2.0, 10.2 Hz, 23-CHb), 1.19 (s, 3H, 19-Me), 1.04 (d, 3H, J = 6.6 Hz, 21-Me),

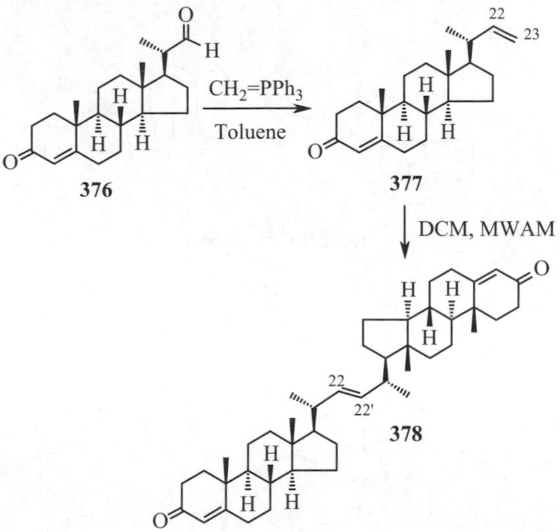

Scheme 2.4.59 *Synthesis of (E)-1,2-bis-(3-oxo-pregn-4-en-20-yl)ethene (**378**)*

0.74 (s, 3H, 18-Me). ^{13}C NMR (125 MHz, $CDCl_3$): δ 35.7 (C-1), 34.0 (C-2), 199.6 (C-3), 123.8 (C-4), 171.6 (C-5), 32.9 (C-6), 32.0 (C-7), 35.6 (C-8), 53.8 (C-9), 38.6 (C-10), 21.0 (C-11), 39.5 (C-12), 42.4 (C-13), 55.9 (C-14), 24.2 (C-15), 28.3 (C-16), 55.4 (C-17), 12.2 (C-18), 17.4 (C-19), 41.2 (C-20), 20.1 (C-21),145.1 (C-22), 111.7 (C-23). EIMS: *m/z* 326 $[M]^+$ (Edelsztein *et al.*, 2009).

(*E*)-1,2-Bis-(3-oxo-4-pregnen-20-yl)ethene (**378**) was prepared from 20-ethenyl-pregn-4-en-3-one (**377**, 82 mg, 0.25 mmol) following the general MWAM reaction (*Scheme 2.4.59*). After purification by CC (mobile phase: *n*-hexane-EtOAc) furnished dimeric ketone **378** (38 mg, 49%, mp: 250–254 °C). IR (KBr): ν_{max} cm^{-1} 2968m (C–H), 2937m (C–H), 2865m (C–H), 1635m (C=C), 1458m (C–H), 1269m (C–C), 1227m (C–C). ^{1}H NMR (500 MHz, $CDCl_3$): δ 5.73 (s, 1H, 4-CH), 5.11 (m, 1H, 22-CH), 1.19 (s, 3H, 19-Me), 0.98 (d, 3H, $J = 6.6$ Hz, 21-Me), 0.72 (s, 3H, 18-Me). ^{13}C NMR (125 MHz, $CDCl_3$): δ 35.7 (C-1), 34.0 (C-2), 199.7 (C-3), 123.8 (C-4), 171.7 (C-5), 33.0 (C-6), 32.0 (C-7), 35.6 (C-8), 53.8 (C-9), 38.6 (C-10), 21.0 (C-11), 39.5 (C-12), 42.3 (C-13), 56.0 (C-14), 24.3 (C-15), 28.8 (C-16), 56.0 (C-17), 12.1 (C-18), 17.4 (C-19), 40.1 (C-20), 20.9 (C-21), 134.3 (C-22). EIMS: *m/z* 624 $[M]^+$ (Edelsztein *et al.*, 2009).

Several other estradiol-based ring D-ring D dimers through side chain and spacer groups (**379–385**) have been reported to date (Fournier and Poirier, 2009).

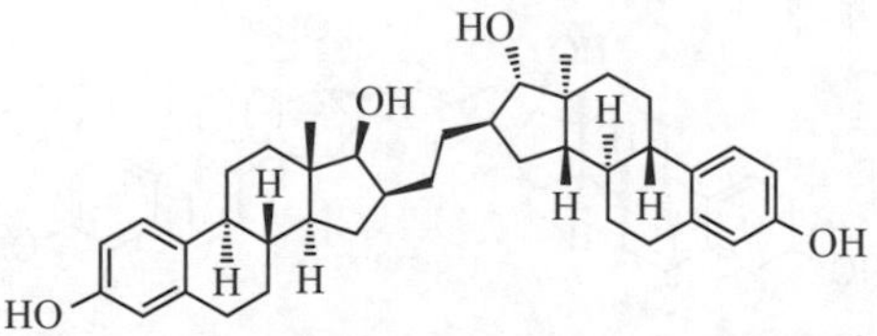

Estradiol dimer **379**

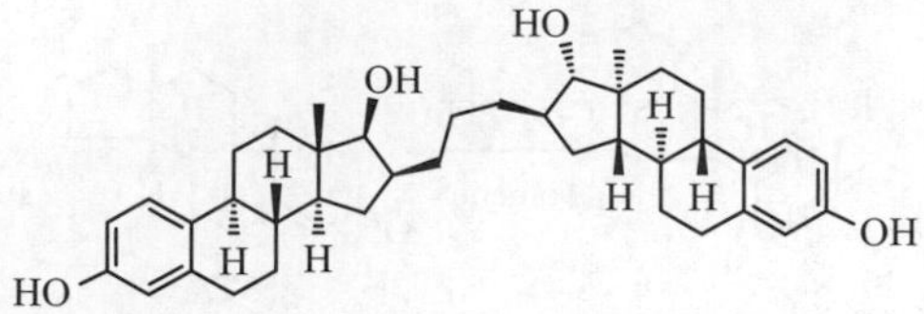

Estradiol dimer **380**

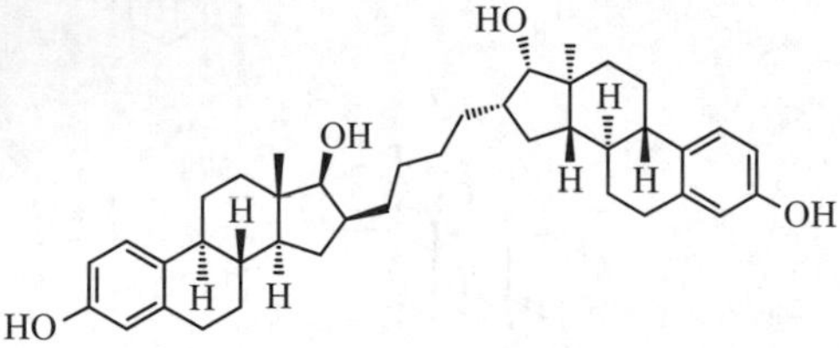

Estradiol dimer **381**

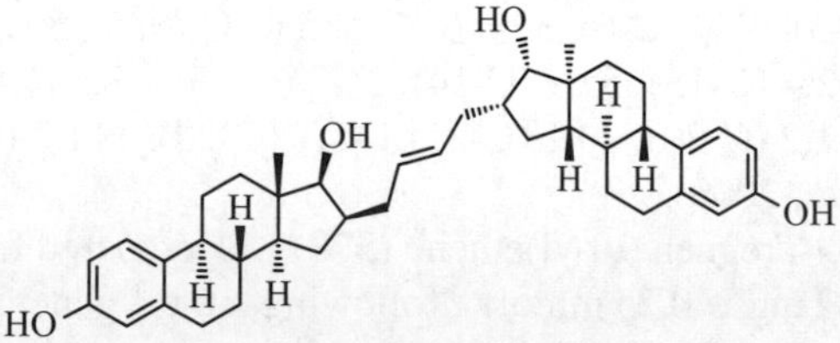

Estradiol dimer **382**

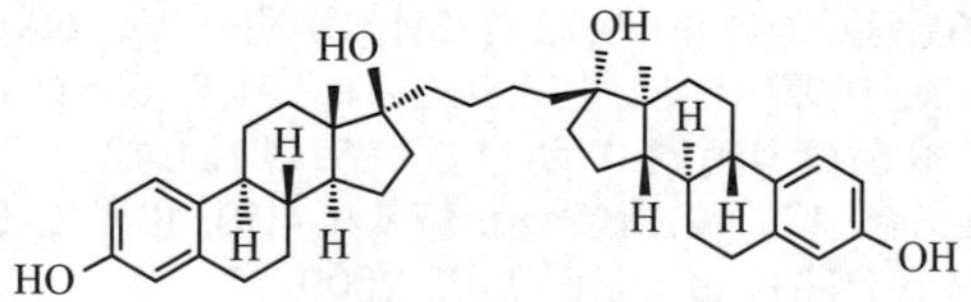

Estradiol dimer **383**

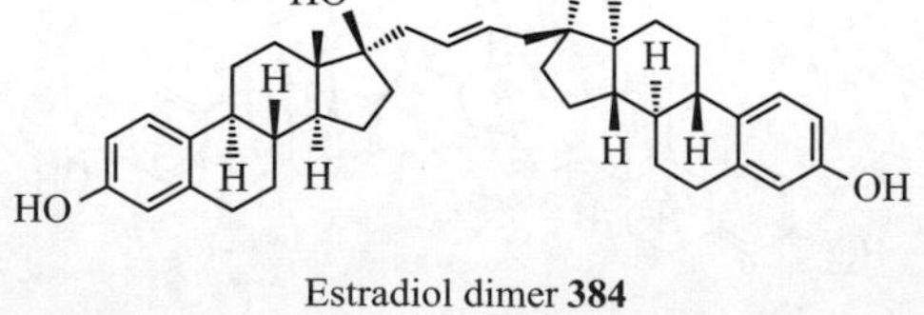

Estradiol dimer **384**

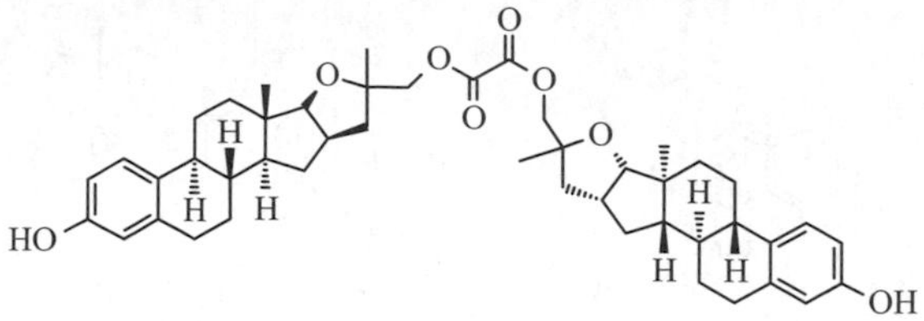

Estradiol dimer **385**

2.5 Dimers *via* Ring A–Ring D Connection

2.5.1 Direct Connection

Irradiation of androstenedione (**386**, androst-4-ene-3,17-dione) yielded dimer **387**, where the connection is in between C-16 position of one molecule and C-3 position of the other (DellaGreca *et al.*, 2002). Androstenedione (**386**, 200 mg), solidified onto the surface of a Pyrex flask from a $CHCl_3$ solution, was irradiated for 4 h to yield crude dimer (58 mg), which was purified by FCC, eluted with 20% Me_2CO in $CHCl_3$, to afford dimer **387** (29%) (*Scheme 2.5.1*). IR ($CHCl_3$): ν_{max} cm^{-1} 3550br (O–H), 1746s (C=O), 1690s (C=C). ^1H NMR (500 MHz, $CDCl_3$): δ 5.76 (s, 1H, 4-CH), 5.34 (s, 1H, 4′-CH), 2.75 (dd, 1H, 16β-CH), 1.19 (s, 3H, 19-Me), 1.06 (s, 3H, 19′-Me), 0.97 (s, 3H, 18-Me), 0.88 (s, 3H, 18′-Me). ^{13}C NMR (125 MHz, $CDCl_3$): δ 35.6 (C-1), 32.4 (C-1′), 33.9 (C-2), 32.8 (C-2′), 199.1 (C-3), 73.0 (C-3′), 124.3 (C-4), 124.0 (C-4′), 169.8 (C-5), 148.6 (C-5′), 32.0 (C-6), 32.2 (C-6′), 30.9 (C-7), 31.5 (C-7′), 35.2 (C-8), 35.6 (C-8′), 53.5 (C-9), 52.6 (C-9′), 38.6 (C-10), 37.4 (C-10′), 20.8 (C-11), 20.1 (C-11′), 30.3 (C-12), 30.1 (C-12′), 48.9 (C-13), 47.8 (C-13′), 49.4 (C-14), 51.3 (C-14′), 25.4 (C-15), 21.7 (C-15′), 53.0 (C-16), 35.7 (C-16′), 222.7 (C-17), 220.8 (C-17′), 14.8 (C-18), 13.8 (C-18′), 17.3 (C-19), 20.2 (C-19′). EIMS: *m/z* 554 $[M - H_2O]^+$ (DellaGreca *et al.*, 2002).

The synthesis of two estradiol dimers **388** and **389**, which formed a new class of steroid sulphatase reversible inhibitors, was achieved *via* ring A–ring D connections of two molecules of estra-1,3,5(10)-triene-3,17β-diol (estradiol) (Fournier and Poirier, 2009).

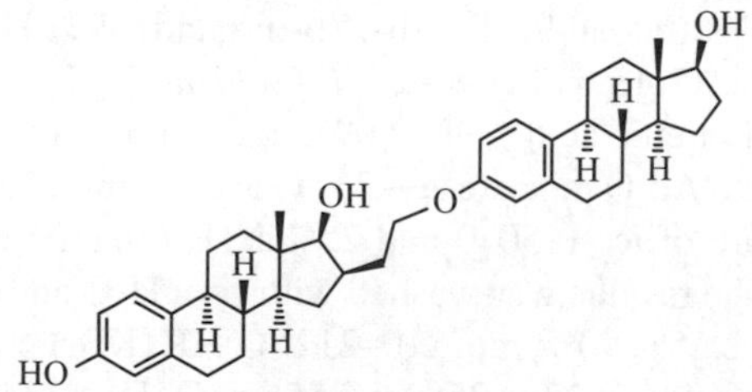

Estradiol dimer **388**

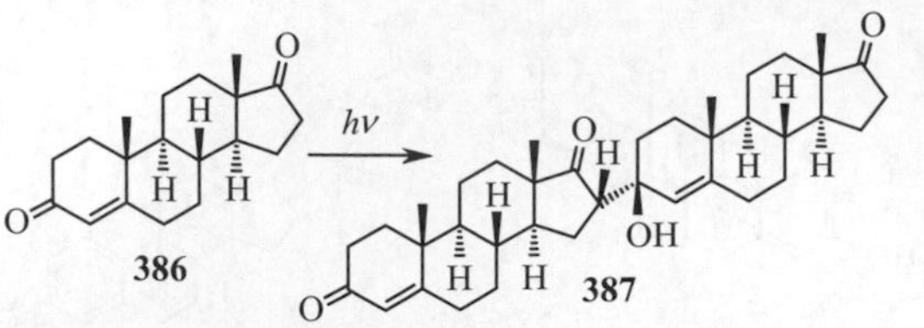

hν

386

387

Scheme 2.5.1 *Synthesis of androstenedione dimer **387***

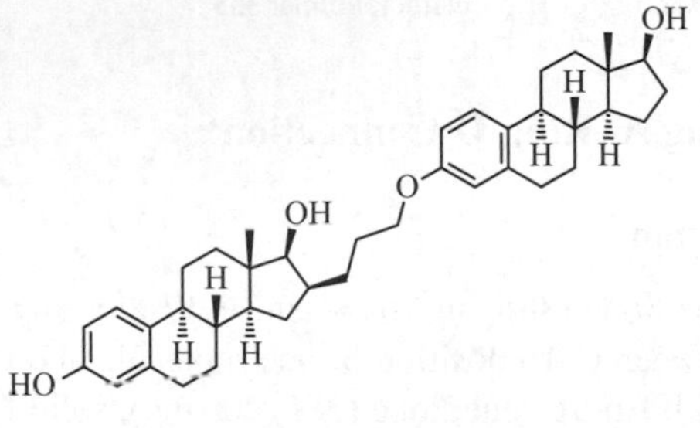

Estradiol dimer **389**

Two androstalondione dimers, methyl 3′β-[(3β-nitratoandrost-5-ene-17β-carbonyl) oxy]-androst-5′-en-17′β-oate (**393**) and 3′β-[(3β-hydroxyandrost-5-ene-17β-carbonyl) oxy]-androst-5′-en-17′β-oate (**394**) were synthesized from etienic acid (**390**, 3β-hydroxyandrost-5-en-17β-oic acid) (Reschel *et al.*, 2002). Direct self-esterification of **390** using acid catalysis and different dehydration agents did not lead to reasonable results. The same condensation with DCC showed some conversion with relatively poor yield as the main product was difficult to separate from nitrogen containing byproducts.

Etienic acid (**390**) was converted to methyl 3β-hydroxyandrost-5-en-17β-oate (**391**) as well as 3β-nitratoandrost-5-en-17β-oic acid (**392**) (*Scheme 2.5.2*). Methyl etienic carboxylate (**391**) was prepared from **390** by esterification using MeOH and AcCl at room temperature for 24 h. ^{1}H NMR (250 MHz, $CDCl_3$): δ 5.35 (dt, 1H, 6-CH), 3.52 (m, 1H, 3-CH), 3.67 (s, 3H, 20-OMe), 1.01 (s, 3H, 19-Me), 0.67 (s, 3H, 18-Me). ^{13}C NMR (62.5 MHz, $CDCl_3$): δ 37.3 (C-1), 31.6 (C-2), 71.7 (C-3), 42.3 (C-4), 140.8 (C-5), 121.4 (C-6), 31.8 (C-7), 32.0 (C-8), 50.1 (C-9), 36.6 (C-10), 20.9 (C-11), 38.2 (C-12), 44.0 (C-13), 56.1 (C-14), 24.6 (C-15), 23.6 (C-16), 55.2 (C-17), 13.3 (C-18, 19.4 (C-19), 174.6 (C-20), 51.2 (20-OMe) (Reschel *et al.*, 2002).

For the synthesis of 3β-nitratoandrost-5-en-17β-oic acid (**392**), HNO_3 (2.7 mL, 39 mmol) was added dropwise to Ac_2O (13.6 mL) at −25 °C (*Scheme 2.5.2*). After stirring and cooling for 10 min, a solution of etienic acid (**390**, 2.09 g, 6.6 mmol) in DCM (30 mL) was added dropwise to the HNO_3 and Ac_2O mixture at −25 °C over 30 min. The mixture was stirred at −25 °C for 4 h, a mixture of ice (150 g) and 25% NH_3 (30 mL) was added, kept at 5 °C overnight and filtered. The residue was washed with cool H_2O and dried in an oven to give etienic acid nitrate (**392**, 2.15 g, 90%, mp: 210–212 °C). IR (KBr): ν_{max} cm^{-1} 3510br (O–H), 2749m (C–H), 2719m, 2664m, 2627m, 2583m, 2556m (O–H), 1734s (C=O), 1700s (C=O), 1670s (C=C), 1626s (NO_2), 1419m (C–O), 1276s (NO_2), 973m (NOC), 868m (N-O), 598m (NO_2). ^{1}H NMR (250 MHz, $CDCl_3$): δ 5.46 (bd, 1H, J = 4.9 Hz, 6-CH), 4.81 (m, 1H, 3-CH), 1.03 (s, 3H, 19-Me), 0.76 (s, 3H, 18-Me). EIMS: m/z 363 $[M]^+$ (Reschel *et al.*, 2002).

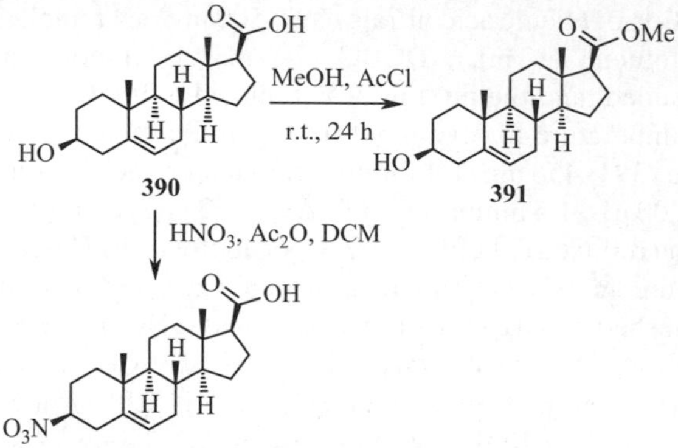

MeOH, AcCl
r.t., 24 h
HNO$_3$, Ac$_2$O, DCM

Scheme 2.5.2 *Synthesis of etienic acid derivatives **391** and **392***

3β-Nitratoandrost-5-en-17β-oic acid (**392**) was transformed *in situ* to the mixed anhydride, which was used for condensation with methyl etienic carboxylate (**391**) to furnish methyl 3′β-[(3β-nitratoandrost-5-ene-17β-carbonyl)oxy]-androst-5′-en-17′β-oate (**393**). Then *O*-nitro group was removed by zinc in acidic medium to afford 3′β-[(3β-hydroxyandrost-5-ene-17β-carbonyl)oxy]-androst-5′-en-17′β-oate (**394**) (*Scheme 2.5.3*).

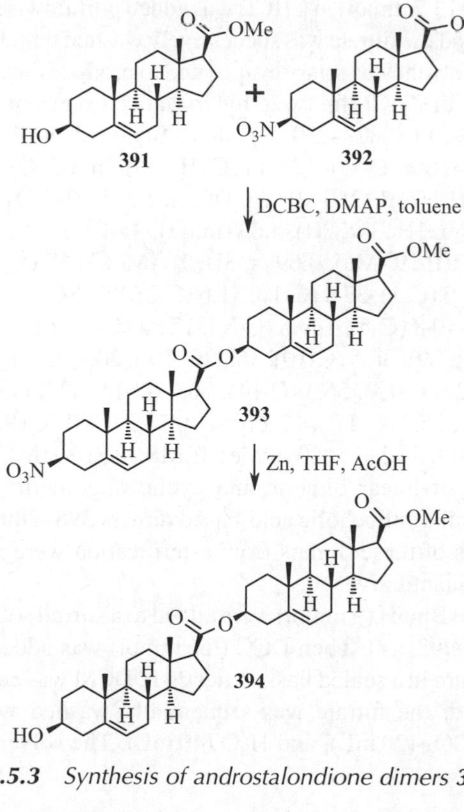

DCBC, DMAP, toluene
Zn, THF, AcOH

Scheme 2.5.3 *Synthesis of androstalondione dimers **393** and **394***

To a stirred solution of etienic acid nitrate (**392**, 500 mg, 1.4 mmol, dried under vacuum for 2 h), in dry toluene (12 mL), DCBC (200 μL, 1.4 mmol) and DMAP (40 mg, 0.33 mmol) were added, and the mixture was refluxed at 90 °C (oil bath) for 2 h. After cooling to room temperature for overnight to obtain the mixed anhydride, then methyl etienic carboxylate (**391**, 458 mg, 1.4 mmol, dried under vacuum for 2 h) in dry toluene (10 mL), DCBC (200 μL, 1.5 mmol) and DMAP (2.62 g, 21.5 mmol) were added to the reaction mixture and refluxed as before at 90 °C for another 2 h. The solution was cooled to the room temperature and H_2O (50 mL) was added, extracted with Et_2O and the extract was sequentially washed with H_2O, saturated aqueous $NaHCO_3$ and H_2O, the solvent was evaporated to dryness. The crude dimer was purified by CC, eluted with toluene and recrystallized from toluene-Et_2O to yield pure methyl 3′β-[(3β-nitratoandrost-5-ene-17β-carbonyl)oxy]-androst-5′-en-17′β-oate (**393**, 780 mg, 84%, mp: 228–230 °C). IR (KBr): ν_{max} cm^{-1} 1720s (C=O), 1670s (C=C), 1626m (NO_2), 1438m (C–O), 1382m (C–H), 1372m (C–H), 1292m (C–H), 1288m (C–O), 1276m (C–O), 1198m (C–O), 1171m (C–O), 975m (NOC), 888m, 868m, 852m (NO), 598m (NO_2). $^1H^1H$ NMR (250 MHz, $CDCl_3$): δ 5.45 (bd, 1H, J = 4.9 Hz, 6′-CH), 5.38 (bd, 1H, J = 4.3 Hz, 6-CH), 4.80 (m, 1H, 3′α-CH), 4.63 (m, 1H, 3α-CH), 3.68 (s, 3H, 20′-OMe), 1.03 (s, 6H, 2 × 19-Me), 0.69 (s, 3H, 18′-Me), 0.67 (s, 3H, 18-Me). EIMS: *m/z* 678 $[M]^+$ (Reschel *et al.*, 2002).

Androstalondione dimer **393** was employed for the synthesis of 3′β-[(3β-hydroxyandrost-5-ene-17β-carbonyl)oxy]-androst-5′-en-17′β-oate (**394**) (*Scheme 2.5.3*). To a stirred solution of **393** (569 mg, 0.8 mmol) in THF (24 mL), AcOH (6 mL) and H_2O (1.2 mL) were added, then Zn powder (900 mg, 13.7 mmol) in $CHCl_3$ was added portionwise. After 3 h, the reaction mixture was filtered off and the filtrate was successively washed with H_2O, saturated aqueous $NaHCO_3$ and H_2O. The solvent was rotary evaporated to produce the crude dimer, which was purified by CC, eluted with $CHCl_3$ followed by crystallization from toluene-Et_2O to obtain dimer **394** (780 mg, 84%, mp: 228–230 °C). IR (KBr): ν_{max} cm^{-1} 3608br (O–H), 1720s (C=O), 1669s (C=C), 1436m (C–O), 1383m (C–H), 1198m (C–O), 1173m (C–O), 1054m (C–O), 1044m (C–O). ^1H NMR (250 MHz, $CDCl_3$): δ 5.35 (bt, 1H, 6′-CH), 5.37 (bd, 1H, J = 4.3 Hz, 6-CH), 3.52 (m, 1H, 3′α-CH), 4.63 (m, 1H, 3α-CH), 3.67 (s, 3H, 20′-OMe), 1.01 (s, 3H, 19-Me), 1.03 (s, 3H, 19′-Me), 0.69 (s, 3H, 18′-Me), 0.67 (s, 3H, 18-Me). ^{13}C NMR (62.5 MHz, $CDCl_3$): δ 37.3 (C-1), 37.0 (C-1′), 31.6 (C-2), 27.9 (C-2′), 71.7 (C-3), 73.4 (C-3′), 42.2 (C-4), 38.5 (C-4′), 140.8 (C-5), 139.8 (C-5′), 121.4 (C-6), 122.2 (C-6′), 31.8 (2 × C-7), 32.0 (2 × C-8), 50.1 (2 × C-9), 36.5 (C-10), 36.7 (C-10′), 20.9 (C-11), 21.0 (C-11′), 38.2 (C-12), 38.4 (C-12′), 43.9 (2 × C-13), 56.3 (C-14), 56.1 (C-14′), 24.5 (2 × C-15), 23.5 (C-16), 23.6 (C-16′), 55.3 (C-17), 55.2 (C-17′), 13.2 (C-18), 13.3 (C-18′), 19.3 (C-19), 19.4 (C-19′), 173.4 (C-20), 174.6 (C-20′), 51.2 (2 × 20′-OMe). EIMS: *m/z* 633 $[M]^+$ (Reschel *et al.*, 2002).

During the synthesis of linear dimeric and cyclic oligomeric cholate derivatives of lithocholic acid (**48**), similar lithocholic acid-based dimers **398**–**400** were prepared (Li and Dias, 1997b). The yields of these dimers from esterification were generally good, and in some cases, they were quantitative.

DMAP (1.2 mmol) and BnOH (9 mmol) were added to a stirred solution of **48** (6 mmol) in dry DCM (8 mL) (*Scheme 2.5.4*). Then DCC (6.6 mmol) was added and the mixture was stirred at room temperature in a sealed flask. After 24 h, DCM was added to the mixture and the solution was filtered, the filtrate was sequentially washed with 5% HCl (20 mL), saturated aqueous $NaHCO_3$ (20 mL), and H_2O (40 mL). The solvent was dried ($MgSO_4$),

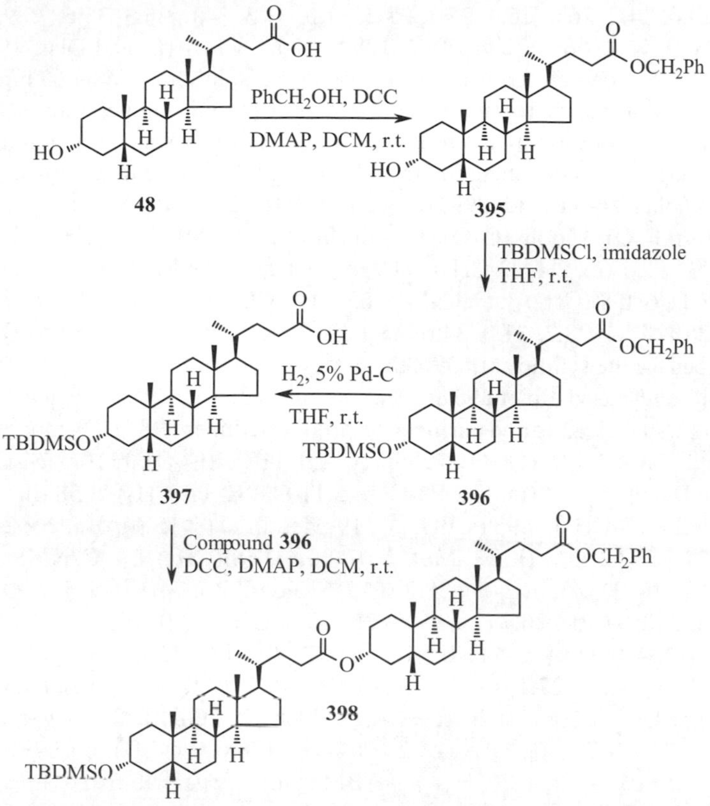

Scheme 2.5.4 *Synthesis of lithocholic acid-based dimer **398***

evaporated to dryness and the crude residue was purified by CC and crystallization from *n*-hexane-EtOAc to provide benzyl lithocholate (**395**, 84%, mp: 116–118 °C). IR (KBr): ν_{max} cm^{-1} 3530m (O–H), 1717s (C=O), 1315m (C–O), 1167s (C–O), 770m (=C–H), 715m (=C–H). ^{1}H NMR (250 MHz, $CDCl_3$): δ 7.35 (m, 5H, Ph-CH), 5.11 (s, 2H, Ph-CH_2), 3.62 (m, 1H, 3-CH), 2.37 (m, 2H, 23-CH_2), 0.91 (s, 3H, 19-Me), 0.90 (d, J = 6.1 Hz, 3H, 21-Me), 0.62 (s, 3H, 18-Me). ^{13}C NMR (62.5 MHz, $CDCl_3$): δ 12.1, 18.3, 20.9, 23.4, 24.2, 26.5, 27.2, 28.2, 30.6, 31.1, 31.4, 34.6, 35.4, 35.9, 36.5, 40.2, 40.5, 42.2, 42.8, 56.0, 56.5, 66.1, 71.9, 128.2, 128.5, 136.2, 174.1 (Li and Dias, 1997b).

Benzyl 3α-*tert*-butyldimethylsilyloxy-5β-cholan-24-oate (**396**) was prepared from a stirred solution of **395** (3 mmol) and imidazole (7.5 mmol) in dry THF (6 mL), by the addition of TBDMSCl (3.6 mmol) and stirring at room temperature (*Scheme 2.5.4*). After 24 h, H_2O (100 mL) was added and the mixture was extracted with $CHCl_3$ (3 × 30 mL). The pooled extracts were dried ($MgSO_4$), rotary evaporated and the crude product was purified by CC to give **393** (92%, mp: 68–70 °C). IR (KBr): ν_{max} cm^{-1} 1735s (C=O), 1255m (C–O), 1175s (C–O), 1110s (Si-O). ^{1}H NMR (250 MHz, $CDCl_3$): δ 7.37 (m, 5H, Ph), 5.11 (s, 2H, Ph-CH_2), 3.58 (m, 1H, 3-CH), 2.33 (m, 2H, 23-CH_2), 0.89 (s, 15H, *t*-Bu, 19-Me and 21-Me), 0.61 (s, 3H, 18-Me), 0.06 [s, 6H, Si(CH_3)$_2$]. ^{13}C NMR (62.5 MHz, $CDCl_3$): δ -4.6, 12.0,

18.3, 20.8, 23.4, 24.2, 26.0, 26.4, 27.3, 28.2, 31.0, 31.3, 34.6, 35.3, 35.6, 35.9, 37.0, 40.2, 42.4, 42.7, 56.0, 56.4, 66.1, 72.9, 128.2, 128.5, 136.2, 174.1 (Li and Dias, 1997b).

Pd-C (1.27 g, 5% on carbon) was added to a stirred solution of **396** (2.7 mmol) in THF (27 mL), under H_2 atmosphere, and stirred the mixture at room temperature (*Scheme 2.5.4*). After 60 min, the catalyst was filtered and the filtrate was rotary evaporated. The crude product was purified by crystallization from EtOAc-$CHCl_3$ to yield 3α-*tert*-butyldimethylsilyloxy-5β-cholan-24-oic acid (**397**, 100%, mp: 200-202 °C). IR (KBr): ν_{max} cm^{-1} 3700m (O–H), 1690s (C=O), 1260m (C–O), 1105m (Si-O). ^1H NMR (250 MHz, $CDCl_3$): δ 3.58 (m, 1H, 3-CH), 2.31 (m, 2H, 23-CH_2), 0.89 (s, 15H, *t*-Bu, 19-Me and 21-Me), 0.64 (s, 3H, 18-Me), 0.06 [s, 6H, $Si(CH_3)_2$]. ^{13}C NMR (63 MHz, $CDCl_3$): δ -4.6, 12.1, 18.3, 20.9, 23.4, 24.3, 26.0, 26.4, 27.4, 28.2, 31.0, 34.6, 35.4, 40.3, 56.0, 56.5, 72.9 (Li and Dias, 1997b).

As described for the synthesis of **396**, the same procedure was followed for the synthesis of lithocholic acid-based dimer **398** using compounds **396** and **397** (*Scheme 2.5.4*) (Li and Dias, 1997b). Usual cleanup and purification produced dimer **398** (71%, mp: 140–142 °C). IR (KBr): ν_{max} cm^{-1} 1730s (C=O), 1255m (C–O), 1175s (C–O), 1110m (Si-O). ^1H NMR (250 MHz, $CDCl_3$): δ 7.37 (m, 5H, Ph-CH), 5.11 (s, 2H, Ph-CH_2), 3.58 (m, 1H, 3′-CH), 2.31 (m, 4H, 2 × 23-CH_2), 0.93 (s, 6H, 2 × 19-Me), 0.90 (s, 15H, *t*-Bu, 2 × 21-Me), 0.64 (s, 3H, 18-Me), 0.62 (s, 3H, 18′-Me), 0.07 [s, 6H, $Si(CH_3)_2$]. ^{13}C NMR (62.5 MHz, $CDCl_3$): δ 35.6 (C-1), 35.4 (C-1′), 28.2 (C-2), 30.6 (C-2′), 74.1 (C-3′), 72.9 (C-3), 40.2 (2 × C-4), 42.3 (2 × C-5), 26.7 (2 × C-6), 26.4 (2 × C-7), 37.0 (2 × C-8), 41.9 (2 × C-9), 32.3 (2 × C-10), 420.9 (2 × C-11), 0.5 (2 × C-12), 42.8 (2 × C-13), 56.5 (2 × C-14), 24.2 (2 × C-15), 27.4 (C-16), 27.1 (C-16′), 56.1 (2 × C-17), 12.0 (2 × C-18), 23.4 (2 × C-19), 35.1 (C-20), 34.3 (C-20′), 18.3 (2 × C-21), 31.8 (2 × C-22), 31.4 (C-23), 31.1 (C-23′),173.8 (C-24′), 174.0 (C-24), 128.2 (Ph-CH), 128.6 (Ph-CH), 66.1 (Ph-CH_2), 26.3 (*t*-Bu), 26.0 (*t*-Bu), −4.6 [$Si(CH_3)_2$]. FABMS (LiI): *m/z* 945.6 $[M + Li]^+$ (Li and Dias, 1997b).

To a solution of **398** (2.0 mmol) in THF (50 mL), 50% HF (7.5 mL) was added and stirred at room temperature (*Scheme 3.5.5*). After 60 min, $CHCl_3$ and H_2O were added and the organic layer was separated, dried ($MgSO_4$) and evaporated to dryness. The crude dimer was purified by crystallization from *n*-hexane-EtOAc to afford dimer **399** (100%, mp: 166–168 °C). IR (KBr): ν_{max} cm^{-1} 3540 (O–H), 1725s (C=O), 1223m (C–O), 1180s (C–O), 1110m (C–O). ^1H NMR (250 MHz, $CDCl_3$): δ 7.35 (m, 5H, Ph-CH), 5.11 (s, 2H, Ph-CH_2), 4.72 (m, 1H, 3′-CH), 3.62 (m, 1H, 3-CH), 2.30 (m, 4H, 2 × 23-CH_2), 0.91 (s, 6H, 2 × 19-Me), 0.90 (d, 6H, 2 × 21-Me), 0.64 (s, 3H, 18-Me), 0.62 (s, 3H, 18′-Me). ^{13}C NMR (62.5 MHz, $CDCl_3$): δ 35.9 (C-1), 35.4 (C-1′), 28.2 (2 × C-2), 71.9 (C-3), 74.1 (C-3′), 40.2 (2 × C-4), 42.2 (2 × C-5), 26.7 (2 × C-6), 26.3 (2 × C-7), 36.5 (2 × C-8), 41.9 (2 × C-9), 32.3 (2 × C-10), 20.9 (2 × C-11), 42.8 (2 × C-13), 40.5 (2 × C-12), 56.5 (2 × C-14), 24.2 (2 × C-15), 27.2 (C-16), 27.1 (C-16′), 56.5 (2 × C-17), 12.1 (2 × C-18), 23.4 (2 × C-19), 35.1 (C-20), 34.6 (C-20′), 18.3 (2 × C-21), 31.8 (2 × C-22), 31.4 (C-23), 31.1 (C-23′), 174.1 (C-24), 173.8 (C-24′), 66.1 (Ph-CH_2), 128.6 (Ph-CH), 128.2 (Ph-CH). FABMS (LiI): *m/z* 831.7 $[M + Li]^+$ (Li and Dias, 1997b).

Dimer **399** was utilized for the synthesis of dimer **400** (*Scheme 2.5.5*), following the same method as described for the synthesis of **399** (*Scheme 3.5.4*). Usual workup and purification as described above provided dimer **400** (100%, mp: 253–255 °C). IR (KBr): ν_{max} cm^{-1} 3425br (O–H), 1730s (C=O), 1190m (C–O), 1050m (C–O). ^1H NMR (250 MHz, $CDCl_3$): δ 4.68 (m, 1H, 3′β-CH), 3.49 (m, 1H, 3β-CH), 2.24 (m, 4H, 2 × 23-CH_2), 0.92 (s, 6H,

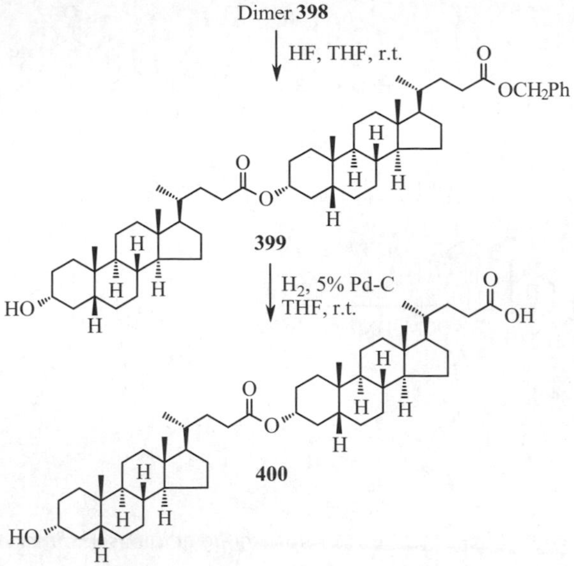

Scheme 2.5.5 *Synthesis of lithocholic acid-based dimers **399** and **400***

2 × 19-Me), 0.90 (d, 6H, 2 × 21-Me), 0.64 (s, 3H, 2 × 18-Me). ^{13}C NMR (62.5 MHz, $CDCl_3$): δ 35.3 (2 × C-1), 30.3 (2 × C-2), 73.7 (C-3′), 70.5 (C-3), 36.3 (2 × C-4), 41.2 (2 × C-5), 27.1 (2 × C-6), 26.3 (2 × C-7), 35.7 (2 × C-8), 41.6 (2 × C-9), 34.4 (2 × C-10), 20.6 (2 × C-11), 39.2 (2 × C-12), 42.5 (2 × C-13), 56.3 (2 × C-14), 24.0 (2 × C-15), 28.0 (2 × C-16), 55.8 (2 × C-17), 12.0 (2 × C-18), 23.3 (2 × C-19), 35.0 (2 × C-20), 18.2 (2 × C-21), 31.5 (2 × C-22), 30.8 (2 × C-23). FABMS (LiI): *m/z* 741.6 $[MLi]^+$ (Li and Dias, 1997b).

Three more lithocholic acid-based dimers **402–404** were prepared by simple esterification between the side-chain acid group (ring D) of one lithocholic acid monomer and the 3α-hydroxyl group (ring A) of another lithocholic acid monomer (Nahar and Turner, 2003). Esterification of lithocholic acid (**48**, 100 mg, 0.27 mmol) was carried out using MeOH and AcCl at room temperature for 24 h to obtain methyl lithocholate (**50**, 104 mg, 99%, mp: 116–117 °C) (*Scheme 2.5.6*).

Lithocholic acid (**48**) was oxidized to 3-oxo-5β-cholan-24-oic acid (**401**), using the following protocol as outlined above. Jones' reagent (1 mL) was added dropwise to a solution of **48** (100 mg, 0.27 mmol) in Me_2CO (10 mL) and stirred at 4 °C for 30 min (*Scheme 2.5.6*). The reaction mixture was concentrated *in vacuo*, diluted with Et_2O (15 mL), washed with saturated aqueous $NaHCO_3$ (2 × 20 mL) followed by H_2O (2 × 20 mL), and dried ($MgSO_4$). Evaporation of the solvent yielded ketone **401** (94 mg, 93%, mp: 122–123 °C). ^{1}H NMR (400 MHz, $CDCl_3$): δ 0.68 (s, 3H, 18-Me), 0.92 (d, 3H, $J = 6.1$ Hz, 21-Me), 1.01 (s, 3H, 19-Me). ^{13}C NMR (100 MHz, $CDCl_3$): δ 37.5 (C-1), 37.3 (C-2), 203.8 (C-3), 42.6 (C-4), 44.6 (C-5), 26.1 (C-6), 26.9 (C-7), 35.6 (C-8), 41.1 (C-9), 35.2 (C-10), 21.5 (C-11), 40.3 (C-12), 43.1 (C-13), 56.7 (C-14), 24.4 (C-15), 28.4 (C-16), 56.3 (C-17), 12.4 (C-18), 22.9 (C-19), 35.8 (C-20), 18.5 (C-21), 31.2 (C-22), 31.0 (C-23), 180.1 (C-24). ESIMS: *m/z* 397 $[M + Na]^+$ (Nahar and Turner, 2003).

To a stirred solution of **401** (97 mg, 0.26 mmol) in dry DCM (5 mL), DMAP (1.6 mg, 0.027 mmol) and followed by methyl lithocholate (**50**, 102 mg, 0.26 mmol) were added at

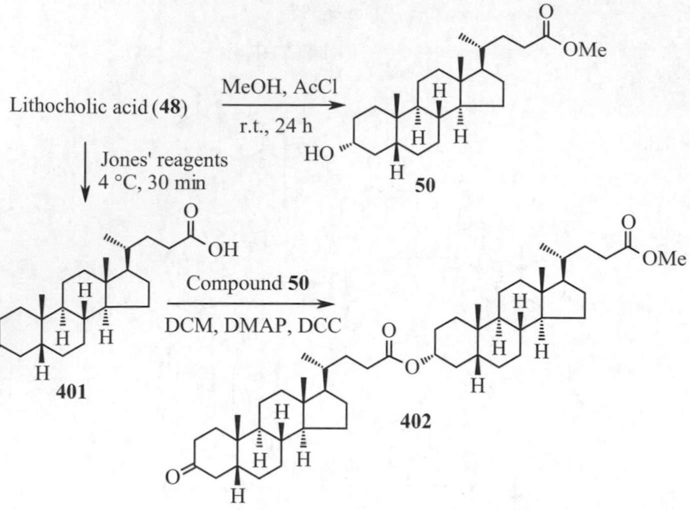

Scheme 2.5.6 *Synthesis of lithocholic acid-based dimer* ***402***

room temperature (*Scheme 2.5.6*). After cooling to 4 °C, a solution of DCC (54 mg, 0.27 mmol) in DCM (1 mL) was added slowly to the reaction mixture and stirred for another 20 min at 4 °C, then 72 h at room temperature. The resulting precipitate (dicyclohexylurea) was filtered off and the filtrate was rotary evaporated. The residue was dissolved in DCM and washed twice with 0.5N HCl followed by saturated aqueous $NaHCO_3$, and dried ($MgSO_4$). The solvent was removed by evaporation to produce crude dimer, which was purified by prep-TLC (mobile phase: 15% EtOAc in petroleum ether) to yield 3-oxo-5β-cholan-24-oic acid (cholan-24-oic acid methyl ester)-3α-yl ester (**402**, 99 mg, 51%, mp: 201–202 °C). IR ($CHCl_3$): ν_{max} cm^{-1} 2933s (O–H), 2860s (O–H), 1736s (C=O), 1715m (C=O), 1453m (C–H), 1384m (C–H), 1257s (C–O), 1222m (C–O), 1175s (C–O), 1118m (C–O), 840m, 779m. ^1H NMR (400 MHz, $CDCl_3$): δ 1.75 and 0.95 (s, 4H, 2 × 1-CH_2), 1.29 and 1.60 (s, 4H, 2 × 2-CH_2), 4.72 (m, 1H, 3′β-CH), 1.71 and 1.45 (s, 4H, 2 × 4-CH_2), 1.35 (s, 2H, 2 × 5β-CH), 1.23 and 1.83 (s, 4H, 2 × 6-CH_2), 1.09 and 1.39 (s, 4H, 2 × 7-CH_2), 1.38 (s, 2H, 2 × 8β-CH), 1.41 (s, 2H, 2 × 9α-CH), 1.38 and 1.23 (s, 4H, 2 × 11-CH_2), 1.14 and 1.96 (s, 4H, 2 × 12-CH_2), 1.05 (s, 2H, 2 × 14α-CH), 1.04 and 1.56 (s, 4H, 2 × 15-CH_2), 1.85 and 1.27 (s, 4H, 2 × 16-CH_2), 1.10 (s, 4H, 2 × 17α-CH_2), 0.69 (s, 3H, 18-Me), 0.65 (s, 3H, 18′-Me), 1.02 (s, 3H, 19-Me), 0.93 (s, 3H, 19′-Me), 1.41 (s, 2H, 2 × 20β-CH), 0.93 (d, 2H, J = 6.2 Hz, 2 × 21α-CH), 0.91 (d, 2H, J = 6.2 Hz, 2 × 21α-CH), 1.75 and 1.29 (s, 4H, 2 × 22-CH_2), 2.32 and 2.12 (s, 2H, 23-CH_2), 2.25 and 2.25 (s, 2H, 23′-CH_2). ^{13}C NMR (100 MHz, $CDCl_3$): δ 37.2 (C-1), 35.1 (C-1′), 37.1 (C-2), 26.4 (C-2′), 213.4 (C-3), 74.1 (C-3′), 42.4 (C-4), 32.3 (C-4′), 44.4 (C-5), 40.5 (C-5′), 25.8 (C-6), 27.1 (C-6′), 26.7 (2 × C-7), 35.8 (C-8), 35.6 (C-8′), 41.9 (C-9), 40.8 (C-9′), 34.9 (C-10), 34.5 (C-10′), 21.2 (C-11), 20.9 (C-11′), 40.2 (C-12), 40.1 (C-12′), 42.8 (2 × C-13), 56.6 (C-14), 56.5 (C-14′), 24.2 (2 × C-15), 28.2 (2 × C-16), 56.1 (2 × C-17), 12.1 (2 × C-18), 22.7 (C-19), 23.4 (C-19′), 35.4 (C-20), 35.3 (C-20′), 18.3 (2 × C-21), 31.8 (C-22), 31.1 (C-22′), 31.8 (2 × C-23), 173.8 (C-24), 174.8

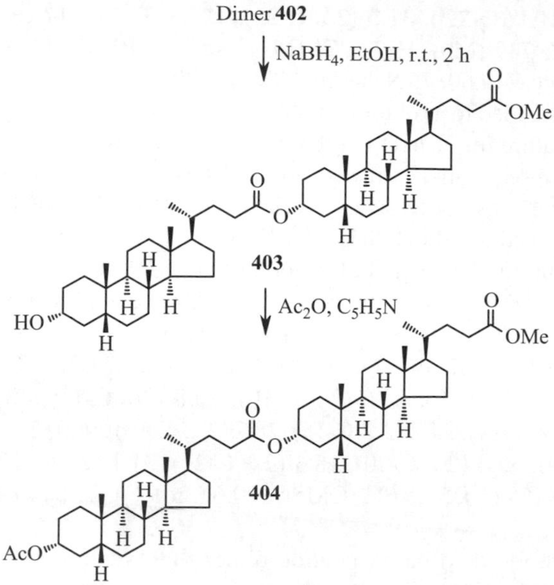

Dimer **402**

$NaBH_4$, EtOH, r.t., 2 h

403

Ac_2O, C_5H_5N

404

Scheme 2.5.7 *Synthesis of lithocholic acid-based dimers* ***403*** *and* ***404***

(C-24′), 51.5 (24′-OMe). FABMS: *m/z* 747 $[M + H]^+$, 769 $[M + Na]^+$. HRFABMS: 747.5927 calculated for $C_{49}H_{79}O_5$, found: 747.5929 (Nahar and Turner, 2003).

3α-Hydroxy-5β-cholan-24-oic acid (cholan-24-oic acid methyl ester)-3α-yl ester (**403**) was prepared from a stirred solution of **402** (89 mg, 0.12 mmol) in EtOH (10 mL) at room temperature, treated with $NaBH_4$ (9 mg, 2 equiv.) and stirred for 2 h (*Scheme 2.5.7*). H_2O (15 mL) was added to the solution, and extracted with EtOAc (3 × 15 mL). The pooled organic extracts were dried ($MgSO_4$). Evaporation of the solvent gave oily product, which was purified by prep-TLC, eluted with 18% EtOAc in petroleum ether, to provide dimer **403** (69 mg, 77%, mp: 203–204 °C). IR ($CHCl_3$): ν_{max} cm^{-1} 3342br (O–H), 2931vs (O–H), 2864s (O–H), 1735s (C=O), 1450m (C–H), 1379m (C–H), 1271s (C–O), 1168s (C–O), 1125m (C–O), 1070m (C–O), 892m, 778m. ^1H NHR (400 MHz, $CDCl_3$): δ 0.76 and 0.96 (s, 4H, 2 × 1-CH_2), 1.30 and 1.62 (s, 4H, 2 × 2-CH_2), 4.67 (m, 1H, 3β-CH), 4.72 (m, 1H, 3′β-CH), 1.72 and 1.48 (s, 4H, 2 × 4-CH_2), 1.36 (s, 2H, 2 × 5β-CH), 1.25 and 1.84 (s, 4H, 2 × 6-CH_2), 1.10 and 1.42 (s, 4H, 2 × 7-CH_2), 1.40 (s, 2H, 2 × 8β-CH), 1.42 (s, 2H, 2 × 9α-CH), 1.39 and 1.25 (s, 4H, 2 × 11-CH_2), 1.15 and 1.98 (s, 4H, 2 × 12-CH_2), 1.07 (s, 2H, 2 × 14α-CH), 1.05 and 1.56 (s, 4H, 2 × 15-CH_2), 1.87 and 1.29 (s, 4H, 2 × 16-CH_2), 1.09 (s, 4H, 2 × 17α-CH_2), 0.60 (s, 6H, 2 × 18-Me), 0.86 (s, 6H, 2 × 19-Me), 1.43 (s, 2H, 2 × 20β-CH), 0.87 (d, 2H, J = 6.2 Hz, 2 × 21α-CH), 0.91 (d, 2H, J = 6.2 Hz, 2 × 21α-CH), 1.76 and 1.32 (s, 4H, 2 × 22-CH_2), 2.25 and 2.25 (s, 2H, 2 × 23-CH_2). ^{13}C NMR (100 MHz, $CDCl_3$): δ 35.3 (C-1), 35.0 (C-1′), 30.5 (C-2), 26.3 (C-2′), 71.8 (C-3), 74.1 (C-3′), 36.5 (C-4), 32.3 (C-4′), 42.1 (C-5), 41.9 (C-5′), 27.2 (C-6), 27.0 (C-6′), 26.4 (C-7), 26.7 (C-7′), 35.8 (C-8), 35.4 (C-8′), 40.4 (2 × C-9), 34.6 (2 × C-10), 20.8 (C-11), 20.9 (C-11′), 40.2 (2 × C-12), 42.7 (2 × C-13), 56.6 (2 × C-14), 24.2 (2 × C-15), 28.2 (2 × C-16), 56.0 (C-17), 56.1 (C-17′), 12.0 (C-18), 12.1 (C-18′), 23.3 (C-19), 23.4 (C-19′), 35.8 (C-20), 35.4 (C-20′), 18.3 (2 × C-

21), 31.7 (C-22), 31.1 (C-22′), 31.0 (2 × C-23), 173.8 (C-24), 174.8 (C-24′), 51.5 (24′-OMe). FABMS: *m/z* 749 $[M + H]^+$, 771 $[M + Na]^+$. HRFABMS: 749.6084 calculated for $C_{49}H_{79}O_5$, found: 749.6087 (Nahar and Turner, 2003).

Ac_2O (1 mL) was added to a solution of **403** (90 mg, 0.12 mmol) in C_5H_5N (500 μL), and left at room temperature for 36 h, H_2O (10 mL) was added to quench the excess Ac_2O, and the mixture was subsequently extracted with $CHCl_3$ (3 × 15 mL) (*Scheme 2.5.7*). The pooled organic layer was successively washed with H_2O, 1M HCl, saturated aqueous $NaHCO_3$ and brine, and dried ($MgSO_4$). The mixture was evaporated to dryness, and the resulting solid was purified by prep-TLC (mobile phase: 20% EtOAc in petroleum ether) to afford 3α-acetyloxy-5β-cholan-24-oic acid (cholan-24-oic acid methyl ester)-3β-yl ester (**404**, 68 mg, 72%, mp: 210–212 °C). ^{13}C NMR (100 MHz, $CDCl_3$): δ 35.0 (2 × C-1), 26.3 (2 × C-2), 74.4 (2 × C-3), 32.3 (2 × C-4), 41.9 (2 × C-5), 27.0 (2 × C-6), 26.6 (2 × C-7), 35.8 (2 × C-8), 40.4 (2 × C-9), 34.6 (2 × C-10), 20.8 (2 × C-11), 40.2 (2 × C-12), 42.7 (2 × C-13), 56.5 (2 × C-14), 24.2 (2 × C-15), 28.2 (2 × C-16), 56.0 (2 × C-17), 12.0 (2 × C-18), 23.3 (2 × C-19), 35.4 (2 × C-20), 18.3 (2 × C-21), 31.1 (2 × C-22), 31.0 (2 × C-23), 173.8 (C-24), 174.8 (C-24′), 51.5 (24′-OMe), 170.6 (3α-CO), 21.5 (3α-MeCO) (Nahar and Turner, 2003).

A tailor-made cholic acid-based peptide dimer **409** was prepared from cholic acid (**49**) and amino acid glycine in six steps (Tian *et al.*, 2002). Initially, cholic acid glycine conjugate, ethyl 3α,7α,12α-trihydroxy-5β-glycocholate (**405**) was prepared by refluxing ethyl glycinate hydrochloride (1.4 equiv.), EEDQ (1.4 equiv.), cholic acid (**49**, 1 equiv.), and Et_3N in EtOAc solution overnight (*Scheme 2.5.8*). Once **405** had been prepared, Ac_2O (21 mL, 222 mmol, 8.43 equiv.) and DMAP (1.9 g, 15.5 mmol, 0.6 equiv.) were added, sequentially, to a cooled (0 °C) solution of **405** (13 g, 26.4 mmol) in C_5H_5N (31 mL). The reaction mixture was stirred at 23 °C for 2 h, and the solvent was reduced *in vacuo*. The residue was poured into 1M HCl (1000 mL) and extracted with EtOAc (3 × 200 mL). The pooled organic layers were washed with aqueous $NaHCO_3$ followed by brine, dried (Na_2SO_4) and concentrated under vacuum to give ethyl 3α,7α,12α-triacetoxy-5β-glycocholate (**406**, 14 g, 86%). ^1H NMR (250 MHz, $CDCl_3$): δ 6.44 (br s, 1H, $NHCH_2CO_2Et$), 5.09 (s, 1H, 12β-CH), 4.90 (s, 1H, 7β-CH), 4.57 (br s, 1H, 3β-CH), 4.20 (q, 2H, CH_2CH_3), 4.00 (br s, 2H, $NHCH_2CO_2Et$), 2.14(s, 3H, 12α-MeCO), 2.09 (s, 3H, 7α-MeCO), 2.06 (s, 3H, 3α-MeCO), 1.40 (t, 3H, CH_2CH_3), 0.93 (s, 3H, 19-Me), 0.82 (d, $J = 6.1$ Hz, 3H, 21-Me), 0.73 (s, 3H, 18-Me). FABMS: *m/z* 620 $[M + H]^+$ (Tian *et al.*, 2002).

In the next step, 3α,7α,12α-triacetoxy-5β-glycocholic acid (**407**) was prepared from **406** (*Scheme 2.5.8*). K_2CO_3 (100 mL, 10% aqueous solution) was added carefully to a stirred solution of **406** (14g, 22.0 mmol) in EtOH (50 mL) at 23 °C. After 17 h, 1M HCl was added to the mixture and extracted with EtOAc, the organic layer was washed with saturated aqueous $NaHCO_3$ followed by brine, dried (Na_2SO_4) and evaporated under vacuum to obtain glycocholic acid **407** (12.5 g, 93%). ^1H NMR (250 MHz, $CDCl_3$): δ 10.30 (s, 1H, COOH), 6.73 (br s, 1H, $NHCH_2COOH$), 5.09 (s, 1H, 12β-CH), 4.91 (s, 1H, 7β-CH), 4.57 (br s, 1H, 3β-CH), 4.03 (br s, 2H, $NHCH_2CO_2H$), 2.15 (s, 3H, 12α-MeCO), 2.10 (s, 3H, 7α-MeCO), 2.06 (s, 3H, 3α-MeCO), 0.93 (s, 3H, 19-Me), 0.81 (d, $J = 6.1$ Hz, 3H, 21-Me), 0.73 (s, 3H, 18-Me). FABMS: *m/z* 592 $[M + 1]^+$ (Tian *et al.*, 2002).

3α,7α,12α-Triacetoxy-5β-glycocholic acid (**406**) was also employed to synthesize ethyl 7α,12α-diacetoxy-3α-hydroxy-5β-glycocholate (**408**) (*Scheme 2.5.8*). EtONa (1M in

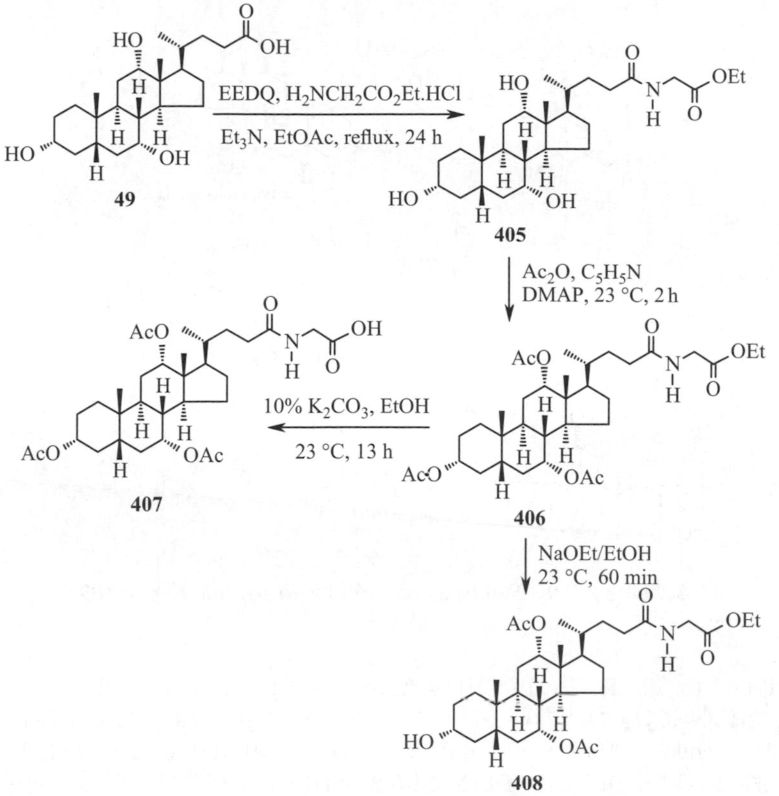

*Scheme 2.5.8 Synthesis of cholic acid derivatives **405–408***

EtOH, 50 mL) was added to a stirred solution of **406** (23 g, 37.0 mmol) in THF (400 mL) at 23 °C. After 2 h, and the solvent was rotary evaporated and the residue was poured into 1M HCl (1000 mL) and extracted with EtOAc (3 × 300 mL). The organic layer was washed with saturated aqueous $NaHCO_3$ followed by brine, dried (Na_2SO_4), evaporated under pressure and the crude product was purified by FCC to produce alcohol **408** (16 g, 75%). ^1H NMR (250 MHz, $CDCl_3$): δ 6.34 (br s, 1H, $NHCH_2CO_2Et$), 5.10 (s, 1H, 12β-CH), 4.90 (s, 1H, 7β-CH), 4.22 (q, 2H, CH_2CH_3), 4.03 (br s, 2H, $NHCH_2CO_2Et$), 3.50 (br s, 1H, 3β-CH), 2.13 (s, 3H, 12α-MeCO), 2.09 (s, 3H, 7α-MeCO), 1.29 (t, 3H, CH_2CH_3), 0.91 (s, 3H, 19-Me), 0.82 (d, J = 6.1 Hz, 3H, 21-Me), 0.73 (s, 3H, 18-Me). EIMS: m/z 577.4 $[M]^+$ (Tian *et al.*, 2002).

Finally, the cholic acid peptide dimer **409** was prepared from the cholic acid derivatives glycocholic acid **407** and alcohol **408** (*Scheme 2.5.9*). To a stirred mixture of 3α,7α,12α-triacetoxy-5β-glycocholic acid (**407**, 6.7 g, 11.3 mmol), ethyl 7α,12α-diacetoxy-3α-hydroxy-5β-glycocholate (**408**, 6.5 g, 11.3 mmol) and DMAP (5.5 g, 45.0 mmol, 3.98 equiv.) in dry toluene (366 mL), 2,6-dichlorobenzoyl chloride (2 mL, 14.0 mmol, 1.2 equiv.) was added and refluxed for 48 h. The solvent was rotary evaporated, and the crude solid was purified on FCC to yield peptide dimer **409** (57%, mp: 128–130 °C). ^1H NMR (250 MHz, $CDCl_3$): δ 6.35 (br s, 2H, $NHCH_2CO$), 5.10 (br s, 2H, 2 × 12β-CH), 4.91 (br s, 2H, 2 × 7β-CH),

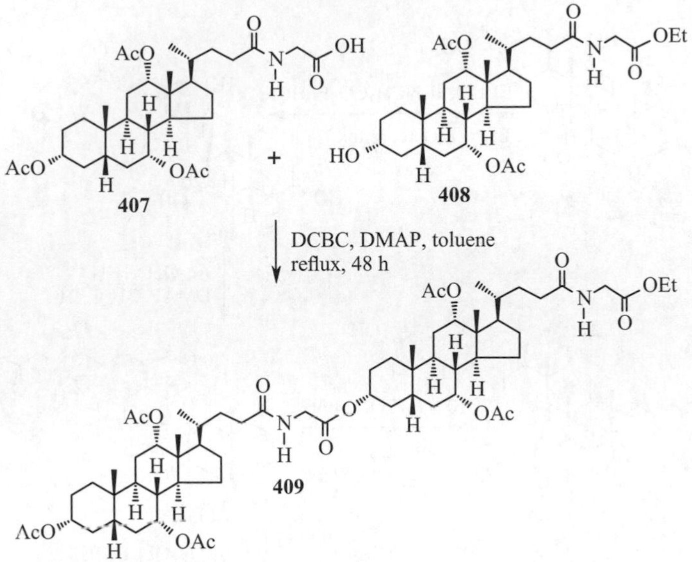

407 + 408

DCBC, DMAP, toluene reflux, 48 h

409

Scheme 2.5.9 *Synthesis of cholic acid peptide dimer* ***409***

4.57 and 4.62 (br d, 2H, 2 × 3β-CH), 4.20 (q, 2H, CH_2CH_3), 4.02 (d, 2H, $NHCH_2CO$, 4.00 (d, 2H, $NHCH_2CO$), 2.05-2.14 (m, 15H, 3α-CH_3CH_2CO, 7α-CH_3CH_2CO, 12α-CH_3CH_2CO), 1.28 (t, 3H, CH_2CH_3), 0.93 (s, 6H, 2 × 19-Me), 0.82 (d, 6H, 2 × 21-Me), 0.73 (s, 6H, 2 × 18-Me). ^{13}C NMR (62.5 MHz, $CDCl_3$): δ 34.7 (2 × C-1), 26.9 (2 × C-2), 74.0 (2 × C-3), 34.7 (2 × C-4), 40.8 (2 × C-5), 31.1 (2 × C-6), 70.6 (C-7), 70.7 (C-7′), 37.6 (2 × C-8), 28.8 (2 × C-9), 34.2 (2 × C-10), 25.5 (2 × C-11), 75.3 (2 × C-12), 45.0 (2 × C-13), 43.3 (2 × C-14), 22.8 (2 × C-15), 27.2 (2 × C-16), 47.4 (2 × C-17), 12.2 (2 × C-18), 22.5 (2 × C-19), 34.5 (2 × C-20), 17.5 (2 × C-21), 31.3 (2 × C-22), 33.0 (2 × C-23), 169.7 (2 × C-24), 41.3 (24-NCH_2CO), 41.5 (24-NCH_2CO_2), 173.6 (24′-CO), 14.1 (24′-OCH_2CH_3), 61.4 (24′-O**C**H_2CH_3), 170.1 (7α-CO), 170.4 (12α-CO), 21.4 (7α-MeCO and 12α-MeCO), 170.5 (3α-CO), 21.6 (3α-MeCO). FABMS: *m/z* 1151 $[M + 1]^+$ (Tian *et al.*, 2002).

Applying a similar protocol as above, a chenodeoxycholic acid-based dimer **413** was achieved from chenodeoxycholic acid (**410**, 3α,7α-dihydroxy-5β-cholanic acid or 3α,7α-dihydroxy-5β-cholan-24-oic acid) in three steps (Karabulut *et al.*, 2007). First, chenodeoxycholic acid (**410**) was converted to its diacetate, 3α,7α-diacetoxy-5β-cholan-24-oic acid (**411**) (*Scheme 2.5.10*). Ac_2O (4.0 mL, 42.0 mmol) was added to a stirred solution of **410** (2.0 g, 5.11 mmol) in C_5H_5N (6 mL) at room temperature. After 24 h, a mixture of ice (20 g) containing conc. HCl (5 mL) was added to the reaction mixture to form the precipitate, which was filtered off to obtain diacetate **411** (2.40 g, 99%, mp: 215–218 °C) was obtained. ^1H NMR (250 MHz, $CDCl_3$): δ 4.89 (s, 1H, 7β-CH), 4.59 (m, 1H, 3β-CH), 2.40 (m, 2H, 23-CH_2), 2.05 (s, 3H, 7α-MeCO), 2.03 (s, 3H, 3α-MeCO), 0.95 (s, 3H, 19-Me), 0.93 (d, $J = 6.1$ Hz, 3H, 21-Me), 0.64 (s, 3H, 18-Me). ^{13}C NMR (62.5 MHz, $CDCl_3$): δ 35.2 (C-1), 26.8 (C-2), 74.2 (C-3), 170.6 (3α-CO), 21.6 (3α-MeCO), 34.6 (C-4), 40.9 (C-5), 31.3 (C-6), 71.3 (C-7), 170.8 (7α-CO), 21.5 (7α-MeCO), 37.9 (C-8), 34.0 (C-9), 34.9

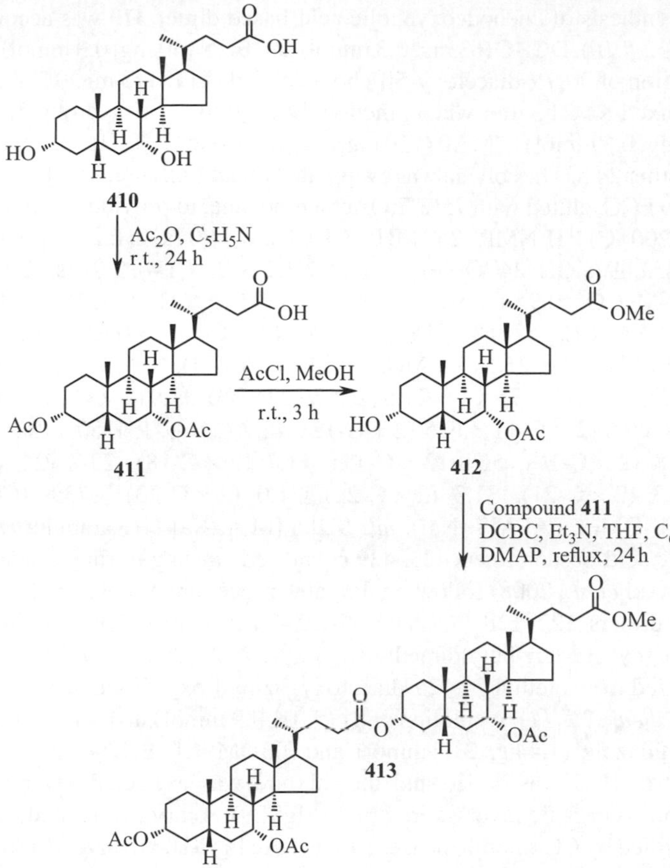

410

Ac_2O, C_5H_5N
r.t., 24 h

411

AcCl, MeOH
r.t., 3 h

412

Compound **411**
DCBC, Et_3N, THF, C_6H_6,
DMAP, reflux, 24 h

413

Scheme 2.5.10 *Synthesis of chenodeoxycholic acid-based dimer **413***

(C-10), 20.6 (C-11), 39.5 (C-12), 42.7 (C-13), 50.4 (C-14), 23.5 (C-15), 28.0 (C-16), 55.7 (C-17), 11.7 (C-18), 22.7 (C-19), 35.1 (C-20), 18.2 (C-21), 30.9 (C-22), 30.7 (C-23), 180.0 (C-24). EIMS: *m/z* 476.5 $[M]^+$ (Karabulut *et al.*, 2007).

Methyl 7α-acetyloxy-3α-hydroxy-5β-cholan-24-oate (**412**) was prepared from 3α,7α-diacetoxy-5β-cholan-24-oic acid (**411**) (*Scheme 2.5.10*). To a ice-cooled solution of AcCl (1 mL) in MeOH (10 mL), diacetate **411** was added and stirring was continued at room temperature. After 3 h, *n*-hexane (1 mL) added to the reaction mixture and the solvent was evaporated to dryness to produce **412** (mp: 105–110 °C) (Karabulut *et al.*, 2007). ^{1}H NMR (250 MHz, $CDCl_3$): δ 4.88 (s, 1H, 7β-CH), 3.66 (s, 3H, 24-OMe), 3.48 (m, 1H, 3β-CH), 2.40 (m, 2H, 23-CH_2), 2.05 (s, 3H, 7α-MeCO), 0.92 (s, 3H, 19-Me), 0.91 (d, J = 6.1 Hz, 3H, 21-Me), 0.64 (s, 3H, 18-Me). ^{13}C NMR (62.5 MHz, $CDCl_3$): δ 35.4 (C-1), 30.7 (C-2), 71.9 (C-3), 39.0 (C-4), 41.3 (C-5), 31.3 (C-6), 71.5 (C-7), 170.9 (7α-CO), 21.8 (7α-MeCO), 38.1 (C-8), 34.3 (C-9), 34.9 (C-10), 20.8 (C-11), 39.7 (C-12), 42.8 (C-13), 50.5 (C-14), 23.7 (C-15), 28.2 (C-16), 55.8 (C-17), 11.9 (C-18), 22.9 (C-19), 35.3 (C-20), 18.4 (C-21), 31.1 (C-22), 31.0 (C-23), 174.9 (C-24), 51.7 (24-OMe). EIMS: *m/z* 448.4 $[M]^+$ (Karabulut *et al.*, 2007).

Finally, the synthesis of chenodeoxycholic acid-based dimer **413** was accomplished as follows (*Scheme 2.5.10*). DCBC (63 mg, 0.3 mmol) and Et_3N (30 mg, 0.3 mmol) were added to a stirred solution of 3α,7α-diacetoxy-5β-cholanic acid (**411**, 143 mg, 0.3 mmol) in THF (5 mL) and refluxed for 2 h, after which, methyl 7α-acetyloxy-3α-hydroxy-5β-cholan-24-oate (**412**, 134 mg, 0.3 mmol), DMAP (150 mg, 1.23 mmol) in C_6H_6 (10 mL) was added and refluxed for another 24 h. The solvent was evaporated under pressure, and the crude product was subjected to FCC, eluted with 75% EtOAc in *n*-hexane, to provide dimer **413** (249 mg, 92%, mp: 195–200 °C). ^{1}H NMR (250 MHz, $CDCl_3$): δ 4.90 (s, 2H, 2 × 7β-CH), 4.63 (m, 2H, 2 × 3β-CH), 3.69 (s, 3H, 24′-OMe), 2.30 (m, 2H, 2 × 23-CH_2), 2.07 (s, 3H, 7α-MeCO), 2.05 (s, 3H, 7α-MeCO), 0.95 (s, 6H, 2 × 19-Me), 0.93 (d, 6H, 2 × 21-Me), 0.67 (s, 6H, 2 × 18-Me). ^{13}C NMR (62.5 MHz, $CDCl_3$): δ 35.3 (2 × C-1), 28.0 (2 × C-2), 73.9 (C-3), 74.2 (C′-3), 170.7 (3′α-CO), 21.6 (3′α-MeCO), 34.6 (2 × C-4), 40.9 (2 × C-5), 31.6 (2 × C-6), 71.2 (2 × C-7), 170.4 (2 × 7α-CO), 21.5 (2 × 7α-MeCO), 37.9 (2 × C-8), 34.0 (2 × C-9), 34.8 (2 × C-10), 20.6 (2 × C-11), 39.5 (2 × C-12), 42.7 (2 × C-13), 50.4 (2 × C-14), 23.5 (2 × C-15), 26.8 (2 × C-16), 55.8 (2 × C-17), 11.7 (2 × C-18), 22.7 (2 × C-19), 34.9 (2 × C-20), 18.3 (2 × C-21), 31.3 (2 × C-22), 31.0 (2 × C-23), 173.8 (C-24), 174.7 (C-24), 51.5 (24′-OMe). FABMS (NaI): *m/z* 929.6 $[M + Na]^+$ (Karabulut *et al.*, 2007).

Several cholic acid-based dimers **425–429** connected *via* ring A-ring D side chain were prepared (Łotowski *et al.*, 2008) following the same procedure as described previously for ring A–ring A dimers **122–128** (*Schemes 2.1.55–2.1.57*). In order to synthesize these dimers, first, methyl 3α-*tert*-butyldimethylsilyloxy-7α,12α-diacetoxy-5β-cholan-24-oate (**414**) was prepared from methyl 7α,12α-diacetoxy-3α-hydroxy-5β-cholan-24-oate (**81**), as the precursor (*Scheme 2.5.11*). A solution of **81** (5.0 g, 9.9 mmol) in dry DMF (120 mL) was treated with imidazole (1.44 g, 21.2 mmol) and TBDMSCl (3.12 g, 20.8 mmol) under stirring. After 5 h, H_2O was added and the mixture was extracted several times with C_6H_6, the combined organic layers were dried ($MgSO_4$), rotary evaporated and the crude product was purified by CC (mobile phase: 10% EtOAc in C_6H_6) to give **414** (6.01 g, 98%). ^{1}H NMR (200 MHz, $CDCl_3$): δ 5.07 (m, 1H, 12β-CH), 4.88 (m, 1H, 7β-CH), 3.67 (s, 3H, 24-OMe), 3.44 (m, 1H, 3β-CH) 2.15 (s, 3H, 12α-MeCO), 2.08 (s, 3H, 7α-MeCO), 0.89 (m, 12H, 19-Me and *t*-Bu), 0.81 (d, 3H, J = 6.0 Hz, 21-Me), 0.73 (s, 3H, 18-Me), 0.07 (s, 6H, $(Me)_2$-Si) (Łotowski *et al.*, 2008).

In the next step, methyl 3α-*tert*-butyldimethylsilyloxy-7α,12α-diacetoxy-5β-cholan-24-oate (**414**) was reduced to alcohol **415** (*Scheme 2.5.11*). $LiAlH_4$ (330 mg, 8.68 mmol) was added to a stirred solution of **414** (4.5 g, 7.26 mmol) in dry THF (30 mL) at 0 °C. After 2 h, the excess hydride was decomposed with a few drops of H_2O and the mixture was filtered off. The organic layer was dried ($MgSO_4$), solvent was evaporated to dryness and the crude solid was purified by CC, eluted with 80% EtOAc in *n*-hexane, to obtain 3α-*tert*-butyldimethylsilyloxy-24-hydroxy-5β-cholan-7α,12α-yl diacetate (**415**, 3.38 g, 79%). IR ($CHCl_3$): ν_{max} cm^{-1} 3627br (O–H), 3480m (O–H), 1720s (C=O), 1255m (C–O), 1076m (C–O), 837m. ^{1}H NMR (200 MHz, $CDCl_3$): δ 5.07 (m, 1H, 12β-CH), 4.87 (m, 1H, 7β-CH), 3.60 (t, 2H, J = 6.1 Hz, CH_2OH), 3.43 (m, 1H, 3β-CH), 2.14 (s, 3H, 12α-MeCO), 2.07 (s, 3H, 7α-MeCO), 0.88 (m, 12H, 19-Me, *t*-Bu), 0.81 (d, 3H, J = 6.4 Hz, 21-Me), 0.72 (s, 3H, 18-Me), 0.06 (s, 6H, $(Me)_2$-Si) (Łotowski *et al.*, 2008).

Iodination of alcohol **415**, following the same method as described for iodide **119** but using half the amount of iodinating agent (*Scheme 2.1.54*), produced 3α-hydroxy-24-iodo-5β-cholan-7α,12α-yl diacetate (**416**) (*Scheme 2.5.11*). After the usual cleanup as above and

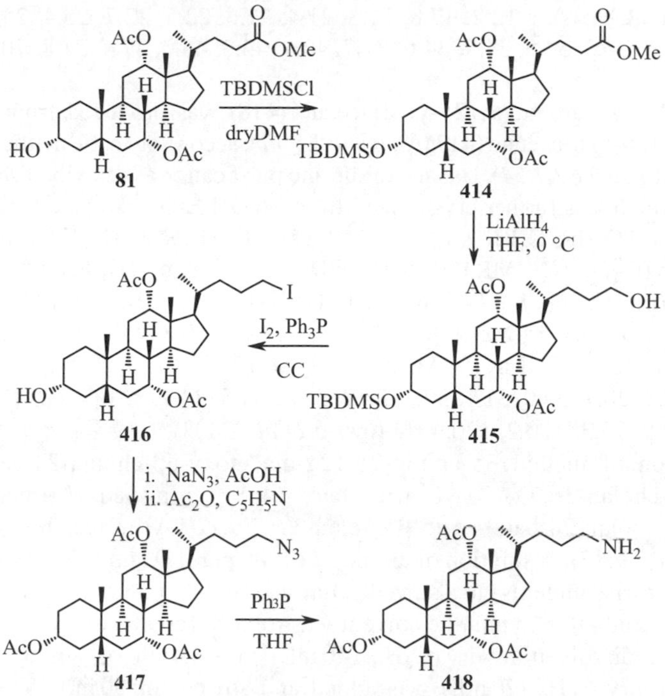

Scheme 2.5.11 *Synthesis of cholic acid derivatives* ***414–418***

CC purification (mobile phase: 70% EtOAc in *n*-hexane) provided iodide **416** (2.80 g, 94%). IR ($CHCl_3$): ν_{max} cm^{-1} 3606br (O–H), 3435br (O–H), 1721s (C=O), 1255m (C–O), 1025m (C–O), 850m. ^{1}H NMR (200 MHz, $CDCl_3$): δ 5.09 (m, 1H, 12β-CH), 4.90 (m, 1H, 7β-CH), 3.51 (m, 1H, 3β-CH), 3.14 (m, 2H, CH_2I), 2.13 (s, 3H, 12α-MeCO), 2.09 (s, 3H, 7α-MeCO), 0.91 (m, 3H, 19-Me), 0.83 (d, 3H, $J = 6.3$ Hz, 21-Me), 0.73 (s, 3H, 18-Me). ^{13}C NMR (50 MHz, $CDCl_3$): δ 7.5, 12.0, 17.7, 20.8, 21.3, 21.4, 22.4, 22.6, 25.3, 27.0, 28.7, 29.9, 30.2, 31.1, 34.1, 34.7, 36.4, 37.5, 38.4, 40.8, 43.1, 44.8, 47.2, 70.7, 71.3, 75.2, 170.4 (Łotowski *et al.*, 2008).

24-Azido-5β-cholan-3α,7α,12α-yl triacetate (**417**) was prepared from 3α-hydroxy-24-iodo-5β-cholan-7α,12α-yl diacetate (**416**) (*Scheme 2.5.11*) (Łotowski *et al.*, 2008), utilizing the same protocol as described for azide **120** (*Scheme 2.1.54*), and then acetylated the crude azide *in situ*. The solid was dissolved in dry C_5H_5N (15 mL) and Ac_2O (5 mL) was added and the mixture was stirred at room temperature overnight, after which, acidified H_2O was added and the mixture was extracted twice with DCM. The pooled organic layers were extracted with saturated aqueous $NaHCO_3$, dried ($MgSO_4$) and the solvent was evaporated under vacuum. The crude product was purified by CC as above and crystallized from heptane-EtOAc to yield azide **417** (2.80 g, 94%, mp: 142–145 °C). IR ($CHCl_3$): ν_{max} cm^{-1} 3395br (O–H), 2099m (N_3), 1724s (C=O), 1254m (C–O), 1121m (C–O). ^{1}H NMR (200 MHz, $CDCl_3$): δ 5.07 (m, 1H, 12β-CH), 4.89 (m, 1H, 7β-CH), 4.56 (m, 1H, 3β-CH), 3.23 (m, 2H, CH_2N_3), 2.13 (s, 3H, 12α-MeCO), 2.07 (s, 3H, 7α-MeCO), 2.03 (s, 3H, 3α-MeCO), 0.90 (s, 3H, 19-Me), 0.81 (d, 3H, $J = 6.4$ Hz, 21-Me), 0.72 (s, 3H, 18-Me). ^{13}C

NMR (50 MHz, $CDCl_3$): δ 12.2, 17.8, 21.3, 21.4, 21.6, 22.5, 22.7, 25.4, 25.5, 26.8, 27.2, 28.8, 31.2, 32.6, 34.3, 34.5, 34.7, 34.6, 37.7, 40.9, 43.3, 45.0, 47.4, 51.8, 70.6, 74.0, 75.3, 170.3, 170.5.

24-Amino-5β-cholan-3α,7α,12α-yl triacetate (**418**) was produced from 24-azido-5β-cholan-3α,7α,12α-yl triacetate (**417**) (*Scheme 2.5.11*), according to the method as described for amine **121** (*Scheme 2.1.54*). Usual workup and purification as described above furnished crude solid, which was further crystallized from MeOH-EtOAc to give amine **418** (2.1 g, 95%, mp: 83–84 °C). IR ($CHCl_3$): ν_{max} cm^{-1} 3395br (O–H and N–H), 1724s (C=O), 1254m (C–O), 1025m (C–O). ^1H NMR (200 MHz, $CDCl_3$): δ 5.08 (m, 1H, 12β-CH), 4.90 (m, 1H, 7β-CH), 4.56 (m, 1H, 3β-CH), 2.65 (m, 2H, CH_2NH_2), 2.12 (s, 3H, 12α-MeCO), 2.07 (s, 3H, 7α-MeCO), 2.03 (s, 3H, 3α-MeCO), 0.91 (s, 3H, 19-Me), 0.82 (d, 3H, $J = 6.3$ Hz, 21-Me), 0.72 (s, 3H, 18-Me). ^{13}C NMR (50 MHz, $CDCl_3$): δ 12.1, 17.7, 21.2, 21.3, 21.5, 22.4, 22.7, 25.4, 26.7, 27.1, 28.7, 29.0, 31.1, 32.7, 34.2, 34.4, 34.5, 34.7, 37.6, 40.8, 41.9, 43.2, 44.9, 47.4, 70.6, 73.9, 75.3, 170.2, 170.4 (Łotowski *et al.*, 2008).

Esterification of methyl 3α-amino-7α,12α-diacetoxy-5β-cholan-24-oate (**121**) and 24-amino-5β-cholan-3α,7α,12α-yl triacetate (**418**) furnished *N*-(methyl 7α,12α-diacetoxy-5β-cholan-24-oate)-*N′*-(5′β-cholan-3′α,7′α,12′α-yl triacetate)3α,24′-oxamide (**419**) (*Scheme 2.5.12*). A solution of amine **121** (500 mg, 1.0 mmol) in dry C_6H_6 (40 mL) was added dropwise under N_2 to a stirred solution of $(COCl)_2$ (1.5 mL, 17.5 mmol) in dry C_6H_6 (60 mL), and after 15 min the solvent was rotary evaporated to dryness. The crude product was redissolved in dry C_6H_6 (40 mL), a solution of amine **418** (470 mg, 0.91 mmol) in dry C_6H_6 (20 mL) was added and stirred for 30 min. The solvent was removed under pressure and the crude residue was purified to CC, eluted with 20% EtOAc in *n*-hexane, and recrystallized from heptane-EtOAc to yield oxamide dimer **419**

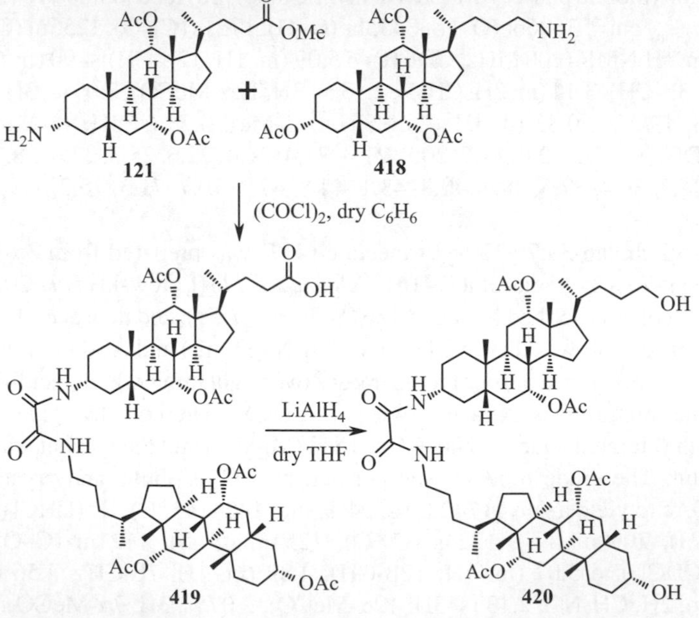

Scheme 2.5.12 *Synthesis of cholic acid-based dimers* ***419*** *and* ***420***

(450 mg, 46%, mp: 134–136 °C). IR ($CHCl_3$): ν_{max} cm^{-1} 3608br (O–H), 3394m (N–H), 1725s (C=O), 1673s (C=O), 1254m (C–O), 1024m (C–O). ^{1}H NMR (200 MHz, $CDCl_3$): δ 7.53 (t, 1H, J = 6.0 Hz, CH_2NH), 7.36 (d, 1H, J = 8.5 Hz, 3α-NH), 5.08 (m, 2H, 2 × 12β-CH), 4.91 (m, 2H, 7β-CH), 4.57 (m, 1H, 3′β-CH–O), 3.66 (m, 4H, 24-OMe, 3β-CH–N), 3.28 (m, 2H, CH_2NH), 2.17 (s, 3H, 12α-MeCO), 2.13 (s, 3H, 12′α-MeCO), 2.08 (s, 3H, 7α-MeCO), 2.06 (s, 3H, 7′α-MeCO), 2.04 (s, 3H, 3′α-MeCO), 0.93 (s, 3H, 19-Me), 0.91 (s, 3H, 19′-Me), 0.81 (d, 6H, J = 6.1 Hz, 2 × 21-Me), 0.72 (s, 6H, 2 × 18-Me). ^{13}C NMR (50 MHz, $CDCl_3$): δ 12.0, 12.1, 14.0, 17.4, 17.7, 21.3, 21.4, 21.5, 21.6, 22.4, 22.5, 22.6, 25.4, 25.6, 26.8, 27.0, 27.1, 27.4, 28.7, 30.6, 30.7, 31.1, 31.7, 32.7, 34.1, 34.2, 34.4, 34.5, 34.6, 35.2, 35.3, 37.6, 40.1, 40.8, 41.3, 43.2, 44.8, 44.9, 47.2, 47.3, 49.8, 51.4, 70.6, 73.9, 75.2, 75.3, 158.7, 160.0, 170.2, 170.3, 170.4, 170.5, 174.4. EIMS: m/z 1078 $[M]^+$ (Łotowski *et al.*, 2008).

$LiAlH_4$ reduction of oxamide dimer **419**, following the same procedure as described earlier for diol **125** (*Scheme 2.1.56*), produced *N*-(24-hydroxy-5β-cholan-7α,12α-yl diacetate)-*N*′-(3′-hydroxy-5′β-cholan-7′α,12′α-yl diacetate)-3α,24′-oxamide (**420**) (*Scheme 2.5.12*). After crystallization from heptane-EtOAc, diol **420** (340 mg, 87%, mp: 150–153 °C) was prepared. IR ($CHCl_3$): ν_{max} cm^{-1} 3616br (O–H), 3394m (N–H), 1723s (C=O), 1673s (C=O), 1511m (N–H), 1255m (C=O), 1024m (C–O). ^{1}H NMR (200 MHz, $CDCl_3$): δ 7.51 (t, 1H, J = 6.0 Hz, CH_2NH), 7.33 (d, 1H, J = 8.6 Hz, 3α-NH), 5.07 (m, 2H, 2 × 12β-CH), 4.89 (m, 2H, 7β-CH), 3.65 (m, 4H, CH_2OH, 3′β-CH–O and 3β-CH–N), 3.24 (m, 2H, CH_2NH), 2.16 (s, 3H, 12′α-MeCO), 2.11 (s, 3H, 12α-MeCO), 2.07 (s, 3H, 7α-MeCO), 2.06 (s, 3H, 7′α-MeCO), 0.92 (s, 3H, 19-Me), 0.89 (s, 3H, 19′-Me), 0.82 (m, 6H, 2 × 21-Me), 0.72 (s, 3H, 18-Me), 0.71 (s, 3H, 18′-Me). ^{13}C NMR (50 MHz, $CDCl_3$): δ 12.2, 17.8, 17.9, 21.4, 21.5, 21.6, 21.7, 22.6, 22.7, 22.8, 22.9, 25.5, 25.7, 27.3, 27.6, 28.9, 29.2, 29.7, 30.5, 31.3, 31.4, 31.6, 32.9, 34.3, 34.7, 34.8, 35.3, 35.5, 37.7, 37.8, 38.7, 40.2, 41.0, 41.4, 43.3, 43.4, 45.0, 47.4, 47.5, 49.9, 63.4, 70.7, 70.8, 71.7, 75.3, 75.4, 158.8, 160.1, 170.4, 170.6 (Łotowski *et al.*, 2008).

Iodination of diol **420** produced *N*-(24-iodo-5β-cholan-7α,12α-yl diacetate)-*N*′-(3′α-iodo-5′β-cholan-7′α,12′α-yl diacetate)-3α,24′-oxamide (**421**) (*Scheme 2.5.13*), and the method was similar to that described earlier for diiodide **126** (*Scheme 2.1.56*). Crystallization from the same solvent system as above provided diiodide **421** (325 mg, 79%, mp: 145–147 °C) as crystals. IR ($CHCl_3$): ν_{max} cm^{-1} 3616br (O–H), 3394m (N–H), 1724s (C=O), 1673s (C=O), 1511m (N–H), 1254m (C–O), 1023m (C–O). ^{1}H NMR (200 MHz, $CDCl_3$): δ 7.49 (t, 1H, J = 6.0 Hz, CH_2NH), 7.34 (d, 1H, J = 8.7 Hz, 3α-NH), 5.09 (m, 2H, 2 × 12β-CH), 4.91 (m, 3H, 2 × 7β-CH, 3β-CH), 3.62 (m, 1H, 3β-CH–N), 3.14 (m, 4H, CH_2NH, 2 × CH_2I), 2.18 (s, 3H, 12α-MeCO), 2.10 (s, 3H, 12′α-MeCO), 2.07 (s, 3H, 7α-MeCO), 2.06 (s, 3H, 7′α-MeCO), 1.01 (s, 3H, 19-Me), 0.93 (s, 3H, 19′-Me), 0.81 (m, 6H, 2 × 21-Me), 0.73 (s, 6H, 2 × 18-Me). ^{13}C NMR (50 MHz, $CDCl_3$): δ 7.6, 12.2, 17.8, 18.0, 21.0, 21.4, 21.5, 21.7, 22.7, 22.8, 25.5, 25.7, 27.2, 27.5, 28.9, 29.9, 30.1, 30.9, 31.3, 32.9, 32.9, 34.2, 34.3, 34.7, 35.2, 35.3, 35.5, 36.6, 37.6, 37.7, 38.2, 38.3, 38.9, 40.2, 41.4, 43.3, 45.0, 45.1, 47.4, 49.9, 70.7, 75.3, 75.4, 158.8, 160.1, 170.1, 170.3, 170.4, 170.6 (Łotowski *et al.*, 2008).

The synthesis of *N*-(24-azido-5β-cholan-7α,12α-yl diacetate)-*N*′-(3′α-azido-5′β-cholan-7′α,12′α-yl diacetate)-3α,24′-oxamide (**422**) was achieved from diiodide **421** (*Scheme 2.5.13*), following the procedure as described for diazide **127** (*Scheme 2.1.57*). After crystallization from heptane-EtOAc diazide **422** (261 mg, 93%, mp: 152–155 °C)

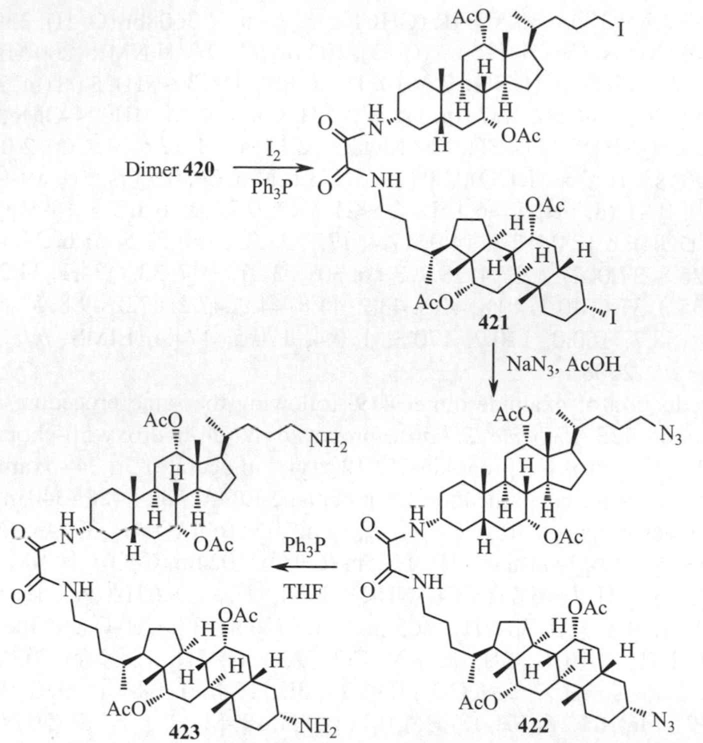

Scheme 2.5.13 *Synthesis of cholic acid-based dimers* ***421–423***

could be prepared. IR ($CHCl_3$): ν_{max} cm^{-1} 3616br (O–H) 3393m (N–H), 2098m (N_3), 1724s (C=O), 1674s (C=O), 1510m (N–H), 1254m (C–O). ^{1}H NHR (200 MHz, $CDCl_3$): δ 7.50 (t, 1H, J = 6.1 Hz, CH_2NH), 7.34 (d, 1H, J = 8.63 Hz, 3α-NH), 5.09 (m, 2H, 2 × 12β-CH), 4.91 (m, 2H, 2 × 7β-CH), 3.62 (m, 1H, 3β-CHN), 3.25 (m, 4H, CH_2NH, CH_2N_3), 2.19 (s, 3H, 12α-MeCO), 2.13 (s, 3H, 12′α-MeCO), 2.10 (s, 3H, 7α-MeCO), 2.07 (s, 3H, 7′α-MeCO), 0.94 (s, 3H, 19-Me), 0.93 (s, 3H, 19′-Me), 0.84 (m, 6H, 2 × 21-Me), 0.74 (s, 3H, 18-Me), 0.73 (s, 3H, 18′-Me). ^{13}C NMR (50 MHz, $CDCl_3$): δ 12.1, 17.7, 17.8, 21.3, 21.4, 21.5, 21.6, 22.5, 22.6, 22.7, 25.3, 25.4, 25.5, 25.8, 26.6, 27.2, 27.5, 28.1, 28.5, 28.7, 28.8, 31.1, 31.2, 32.6, 32.8, 34.2, 34.3, 34.6, 35.0, 35.4, 37.6, 40.1, 41.2, 41.3, 43.2, 43.3, 44.9, 45.0, 47.4, 49.8, 51.8, 60.7, 70.5, 70.6, 75.3, 158.7, 160.0, 170.2, 170.3, 170.4, 170.5 (Łotowski *et al.*, 2008).

Another oxamide dimer, *N*-(24-amino-5β-cholan-7α,12α-yl diacetate)-*N*′-(3′α-amino-5′β-cholan-7′α,12′α-yl diacetate)-3α,24′-oxamide (**423**), was prepared from diazide **422** (*Scheme 2.5.13*), following the method as described for diamine **128** (*Scheme 2.1.57*). Crystallization from heptane-EtOAc-MeOH gave diamine **423** (202 mg, 81%, mp: 230 °C). IR ($CHCl_3$): ν_{max} cm^{-1} 3615m (N–H), 339br (N–H), 1724s (C=O), 1673s (C=O), 1511m (N–H), 1254m (C–O), 910m. ^{1}H NMR (200 MHz, $CDCl_3$): δ 7.52 (t, 1H, J = 4.50 Hz, CH_2NH), 7.34 (d, 1H, J = 7.44 Hz, 3′α-NH), 5.08 (m, 2H, 2 × 12β-CH), 4.90 (m, 2H, 2 × 7β-CH), 3.68 (m, 1H, 3β-CHN), 3.27 (m, 2H, CH_2NH), 2.82 (m, 3H, 3β-CHNH,

CH_2NH), 2.18 (s, 3H, 12α-MeCO), 2.15 (s, 3H, 12′α-MeCO), 2.12 (s, 3H, 7α-MeCO), 2.07 (s, 3H, 7′α-MeCO), 0.92 (s, 6H, 2 × 19-Me), 0.82 (m, 6H, 2 × 21-Me), 0.72 (s, 6H, 2 × 18-Me). ^{13}C NMR (50 MHz, $CDCl_3$): δ 12.2, 17.7, 21.5, 21.6, 21.7, 22.5, 22.6, 22.8, 25.5, 25.7, 25.8, 27.4, 27.8, 28.7, 28.9, 29.6, 30.2, 30.9, 31.2, 32.6, 32.8, 34.2, 34.7, 35.2, 35.8, 37.6, 37.8, 40.2, 40.8, 41.2, 41.3, 43.1, 43.6, 44.9, 45.0, 47.3, 49.9, 51.5, 53.4, 70.6, 75.3, 158.8, 160.1, 170.3, 170.4, 170.6, 170.7 (Łotowski *et al.*, 2008).

2.6 Dimers *via* Connection of C-19

Steroidal dimer resulting from the connection between C-19 of two monomers, with or without spacers, is extremely uncommon. The cholestane homodimer **425** is one of such dimer (Fajkos and Joska, 1979), and was purified from the reaction mixture of the Simmons–Smith methylenation of 19-hydroxy-cholest-5-en-3β-yl acetate (**424**) (*Scheme 2.6.1*). It can be noted that the stereospecific transformation of olefins into cyclopropanes by means of the treatment with methylene diiodide and zinc-copper couple is generally known as the Simmons–Smith reaction. The crude dimer was purified by CC, eluted with ligroin: Et_2O = 33:1, followed by recrystallization from $CHCl_3$-MeOH to produce diacetate **425** (2.2 g, mp: 153–155 °C). IR (KBr): ν_{max} cm^{-1} 3070s (C–H), 1735s (C=O), 1730s (C=O), 1247m (C–O), 1115m (C–O), 1100m (C–O), 1035m (C–O), 1025m (C–O). ^1H NMR (100 MHz, $CDCl_3$): δ 4.96 (m, 2H, 2 × 3α-CH), 4.65 (s, 2H, O-CH_2-O), 3.54 and 3.35 (two d, 4H, J = 10 Hz, 2 × 19-CH_2O), 2.00 (s, 6H, 2 × MeCO), 0.89 (d, 12H, J = 6.0 Hz, 2 × 21-Me) 0.86 (d, 12H, J = 6.5 Hz, 2 × 26-Me and 2 × 27-Me), 0.72 (m, 4H, cyclopropane 2 × CH_2), 0.64 (s, 6H, 2 × 18-Me), 0.03 (m, 2H, cyclopropane 2 × CH). MS: m/z 868 $[M-60]^+$, 808 $[M-120]^+$ (Fajkos and Joska, 1979).

Alkaline hydrolysis of the diacetate **425** gave the diol **426** (*Scheme 2.6.1*). After the usual cleanup, the crude dimer was crystallized from MeOH to provide diol **426** (205 mg, mp: 134–137 °C) (Fajkos and Joska, 1979).

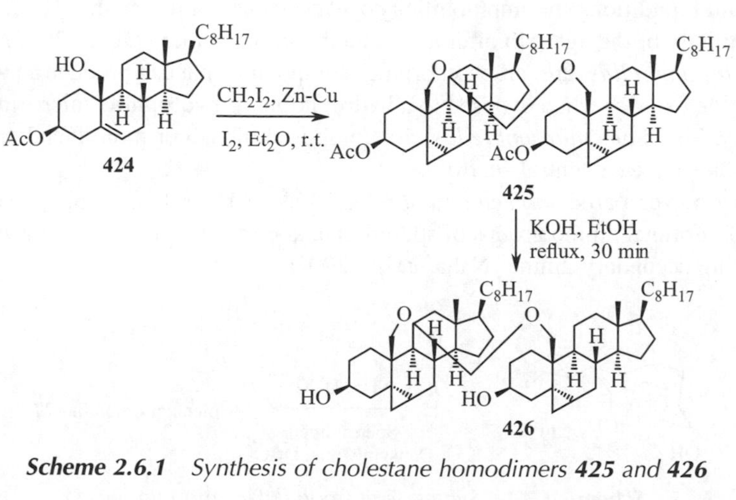

Scheme 2.6.1 *Synthesis of cholestane homodimers **425** and **426***

2.7 Molecular Umbrellas

Regen's group first introduced an interesting class of surfactant molecules, *'molecular umbrellas'* *e.g.*, **427–429** (Janout *et al.*, 1996). *'Molecular umbrellas'*, a unique class of amphiphiles, have opened up a promising avenue for the design of the vectors for intracellular delivery of hydrophilic therapeutic agents.

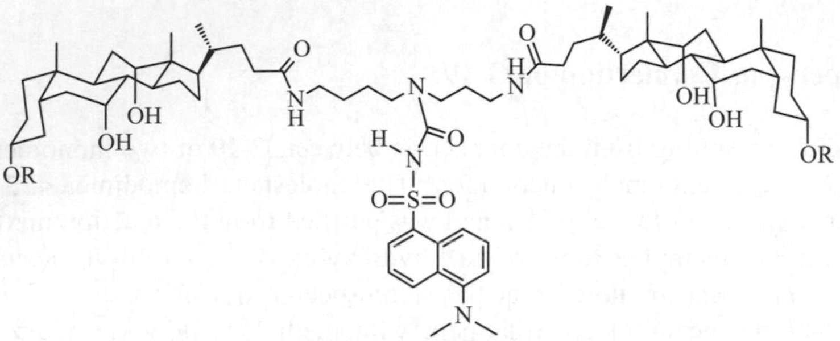

427 R = H **428** R = $CONHCH_2CH_2OH$ **429** R = $CONHCH_2CH_2OCH_3$

'Molecular umbrella' **427** was prepared from cholic acid (**49**) as 'wall material', spermidine as the scaffold, and an environmentally friendly fluorescent probe, 5-dimethyl-amino-1-napthalenesulphonyl (dansyl) as the reagent (*Scheme 2.7.1*) (Janout *et al.*, 1996). Cholic acid (**49**) was converted to its *N*–hydroxysuccinimide ester, followed by condensation with the primary amino groups of spermidine, and finally the coupling with dansyl-glycine furnished **427**.

Since then, several *'molecular umbrellas'* have been prepared involving two steroidal molecules (*e.g.*, cholic acid, **49**) to date (Nahar *et al.*, 2007a). The synthesis of *'molecular umbrellas'* involves the use of amphiphilic molecules that maintain a hydrophobic as well as a hydrophilic face. Two or more such amphiphiles are covalently coupled to a suitable scaffold either before or after a selected agent is attached to a central location. Under suitable environmental conditions, the amphiphilicity of each wall combines with the hydrophobicity or hydrophilicity of the agent to offer a 'shielded' conformation (*Figure 2.7.1*).

'Molecular umbrellas' are 'amphomorphic' compounds that can produce a hydrophobic or hydrophilic exterior when exposed to a hydrophobic or hydrophilic microenvironment, respectively. So, *'molecular umbrellas'* are composed of two or more facial amphiphiles that are connected to a central scaffold.

Regen's group prepared *'molecular umbrellas'* **430–433** by direct coupling of cholic acid (**49**) to both terminal amino groups of spermidine, and attachment of the dansyl moiety to the remaining secondary amine (Nahar *et al.*, 2007a).

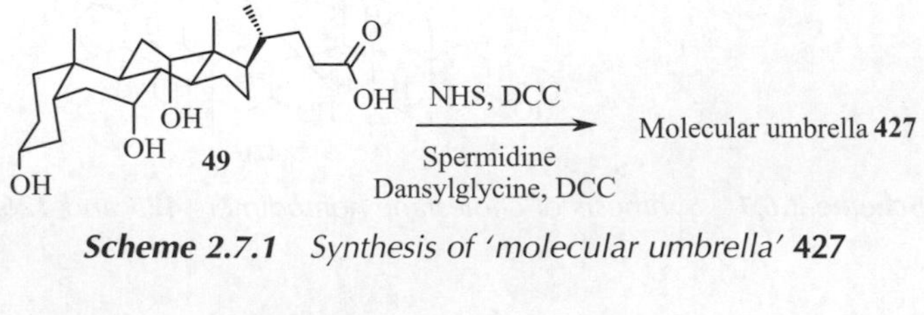

49 — NHS, DCC; Spermidine; Dansylglycine, DCC → Molecular umbrella **427**

Scheme 2.7.1 *Synthesis of 'molecular umbrella'* **427**

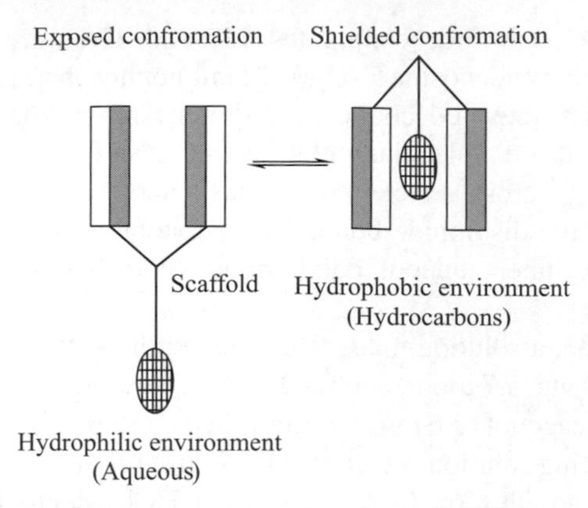

Figure 2.7.1 *'Molecular umbrella'*

430 R = H **431** R = $CONHCH_2CH_2OH$ **432** R = $CONHCH_2CH_2OCH_3$

The *di*-walled *'molecular umbrella'* **430** was prepared by acylation of the terminal amino groups of spermidine with cholic acid (**49**), followed by condensation with bis(3-*O*-[N=1,2,3-benzotriazin-4(*3H*)-one]-yl)-5,5′-dithiobis-2-nitrobenzoate and displacement with glutathione (Janout *et al.*, 2001; Nahar *et al.*, 2007a).

Glutathione is covalently attached to the chlolate-spermidine-Ellman's reagent derivative **433**, and could be transported across bilayers *via* an umbrella mechanism. The facial amphiphilicity of the cholate derivatives is capable of shielding the glutathione moiety from the hydrophobic core as it enters the membrane. The hexa-sulphate analogue **434** could be prepared by replacing the hydroxyls of the steroid units of **433** with sulphate groups, before the displacement with glutathione.

Chlolate-spermidine-Ellmans reagent

433 R = H **434** R = SO_3^-

Molecular umbrella–nucleoside conjugates **435** and **436** were prepared *via* thiolate disulphide displacement by adenosine 5′-*O*-(3-thiomonophosphate) and adenosine 5′-*O*-(3-thiotriphosphate) on an activated cholic acid dimer, spermidine and 5,5′-dithiobis-(2-nitrobenzoic acid), and those molecular umbrella–nucleoside conjugates **435** and **436** were transported successfully across vesicle membranes (Janout *et al.*, 2002). It was observed that after reduction of the disulphide bond, 3-thiomonophosphate and 3-thiotriphosphate were released into the inner aqueous compartment with half-lives of 20 and 120 min, respectively.

To prepare dimer **435**, a solution made from the tetralithium salt of adenosine 5′-*O*-(3-thiotriphosphate) (3.12 mg, 5.7 μmol) and H_2O (150 μL) was added to a solution of chlolate-spermidine-Ellman's reagent (12.6 mg, 5.7 μmol) in a mixture of MeOH (400 μL) and H_2O (150 μL), and the resulting solution was stirred for 72 h at room temperature. Evaporation of solvents, followed by prep-TLC on SiO_2, eluted with $CHCl_3$:MeOH:H_2O = 6:4:1, produced molecular umbrella-ATP conjugate **435** (2.51 mg, 24.9%). ^{1}H NMR (360 MHz, CD_3OD): δ 8.42 (d, 1H), 8.19 (s, 1H), 8.06 (m, 1H), 7.70 (m, 2H), 6.05 (m, 1H), 4.67 (m, 1H), 4.45 (m, 1H), 4.25 (m, 3H), 3.92 (m, 2H), 3.78 (s, 2H), 3.35 (m, 6H), 3.00 (m, 4H), 0.89–2.20 (m, 66H), 0.68 (d, 6H). HRMS: *m/z* 1710.6006 calculated for $C_{72}H_{110}N_9O_{24}P_3S_2Na_3$, found: 1710.6094 (Janout *et al.*, 2002).

Similar procedure, as outlined above for the synthesis of the molecular umbrella-ATP conjugate **435**, was applied for the preparation of the molecular umbrella-AMP conjugate **436**, which was purified by prep-TLC on SiO_2 (mobile phase: $CHCl_3$:MeOH:H_2O = 65:25:4) to give **436** (53%). ^{1}H NMR (360 MHz, CD_3OD): δ 8.41 (d, 1H), 8.19 (s, 1H), 8.03 (m, 1H), 7.68 (m, 2H), 6.06 (m, 1H), 4.66 (m, 1H), 4.30 (m, 1H), 4.20 (m, 3H), 3.92 (m, 2H), 3.78 (s, 2H), 3.35 (m, 6H), 3.00 (m, 4H), 0.90-2.25 (m, 66H), 0.69 (d, 6H). HRMS: *m/z* 1512.6910 calculated for $C_{72}H_{109}N_9O_{17}PS_2Na_2$, found: 1512.6959 (Janout *et al.*, 2002).

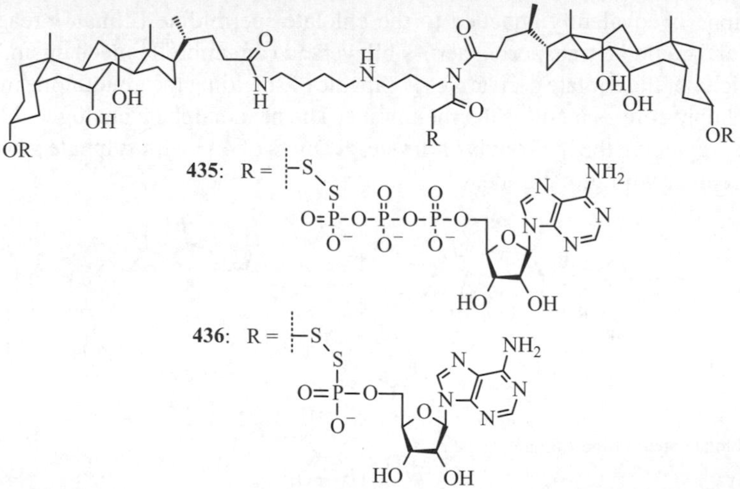

'Molecular umbrellas', comprising bile acids as *'umbrella walls'*, polyamines, *e.g.*, spermidine and spermine, as *'scaffold material'*, and L-lysine as *'branches'*, are capable of

transporting certain hydrophilic peptides, nucleotides, and oligonucleotides across liposomal membranes by passive diffusion. These molecules also increase H_2O solubility and hydrolytic stability of any hydrophobic drug, and show significant antiviral activity (Janout *et al.*, 1996, 2001, 2002; Nahar *et al.*, 2007a).

In order to probe transbilayer movement, the diwalled *'molecular umbrella'* **437** was prepared from **49** by reaction of N^1,N^3-spermidine-bis(cholic acid amide) (**430**) with [*N*-1,2,3-benzotriazin-4(3H)one-yl]-3-(2-pyridyldithio) propionate (BPDP) (*Scheme 2.7.2*). BPDP (32 mg, 89 µmol) was added to a mixture of **430** (8.0 mg, 86 µmol), DCM (2 mL) and DIPEA (44 mg, 0.34 mmol), stirred for 60 min at room temperature, the solvent was removed *in vacuo* and the solid mass was dissolved in MeOH (2 mL). It was then poured into a mixture of saturated aqueous $NaHCO_3$ (20 mL) and 10% Na_2CO_3 (1 mL) to induce precipitation, which was purified by prep-TLC on SiO_2 (mobile phase: $CHCl_3$:MeOH:H_2O = 103:27:3) to yield **437** (71 mg, 73%). ^{1}H NMR (360 MHz, $CDCl_3$-CD_3OD = 2:1): δ 8.38 (d, 1H), 7.75 (m, 2H), 7.18 (m, 1H), 3.92 (br s, 2H), 3.78 (br s, 2H), 3.35 (s, 2H), 2.96-3.30 (m, 8H), 2.80 (t, 2H), 2.36 (t, 2H), 0.80-2.30 (m, 60H), 0.66 (s, 6H). HRMS: *m/z* 1123.7167 calculated for $C_{63}H_{103}N_4O_{97}S_2$, found: 1123.7165 (Shawaphun *et al.*, 1999).

To distinguish between two possibilities, *i.e.*, *'molecular umbrellas'* that have more walls could exhibit higher transport rates due to a greater shielding capacity, and alternatively, they could display lower transport rates due to a decreased diffusion coefficient, and to study the consequences of lipophilicity of the *'molecular umbrella'* on membrane transport, the permeation behaviour of a series of *di-*, *tetra-* and *octa*-walled conjugates and a nonumbrella derivative of Cascade Blue was investigated (Mehiri *et al.*, 2009). Two *bi*-walled 'molecular umbrellas' **438** and **439**, synthesis of which were reported earlier, were used in that study (*Scheme 2.7.3*) (Kondo *et al.*, 2008).

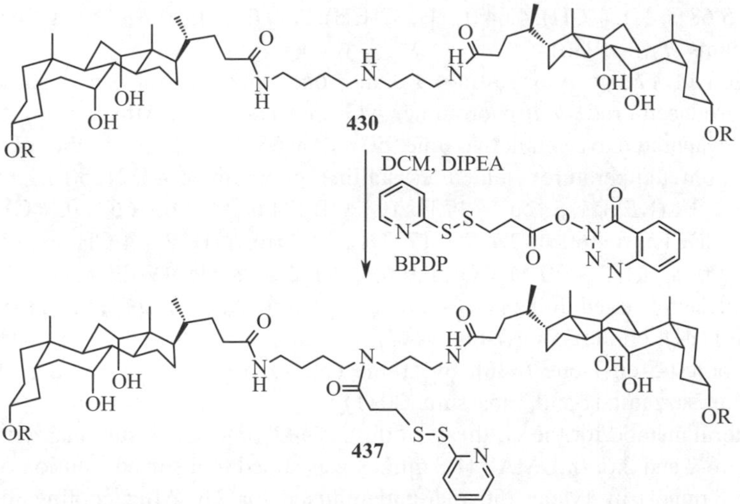

Scheme 2.7.2 *Synthesis of 'molecular umbrella'* **437**

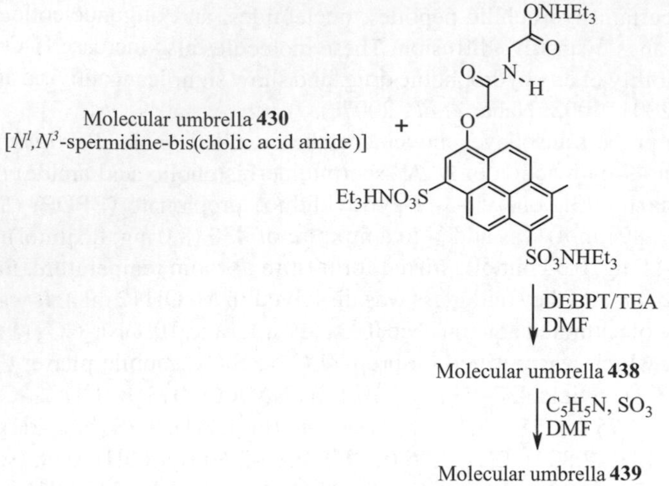

Molecular umbrella **430**
[N^1,N^3-spermidine-bis(cholic acid amide)] + ONHEt$_3$ ... Et$_3$HNO$_3$S ... SO$_3$NHEt$_3$
DEBPT/TEA, DMF
Molecular umbrella **438**
C_5H_5N, SO_3, DMF
Molecular umbrella **439**

Scheme 2.7.3 *Synthesis of 'molecular umbrellas'* ***438*** *and* ***439***

2.8 Miscellaneous

Williams *et al*. (1986) reported *anti*-photodimer **443** from 17β-hydroxy-5(10)-estren-3-one (**440**) in three steps. First, the estrone derivative **440** was brominated and dehydrobrominated with pyridinium bromide perbromide to afford 17β-hydroxyestra-4,9-dien-3-one (**441**, 74%, mp: 175–176 °C) (*Scheme 2.8.1*).

Acetylation of 17β-hydroxyestra-4,9-dien-3-one (**441**) with Ac_2O in C_5H_5N yielded crude acetylated product that after crystallization from *n*-hexane-Et_2O gave 17β-acetyloxyestra-4,9-dien-3-one (**442**, mp: 106–107 °C) (*Scheme 2.8.1*). IR (KBr): ν_{max} cm^{-1} 1724s (C=O), 1660s (C=O), 1607m (C=C). UV (MeOH): λ_{max} nm 303. ^{1}H NMR (100 MHz, $CDCl_3$): δ 5.68 (s, 1H, 4-CH), 4.64 (t, 1H, 17-CH), 2.30 (s, 3H, 20-MeCO), 0.96 (9, 3H, 18-Me) (Williams *et al*., 1986).

Irradiation of 17β-acetyloxyestra-4,9-dien-3-one (**442**, 450 mg) in degassed MeOH (450 mL) produced crude *anti*-photodimer **443** (*Scheme 2.8.1*). After 40 min, the solvent was rotary evaporated to a smaller volume (50 mL) at 35 °C, and cooling the solvent slowly down to room temperature yielded crystalline photodimer **443** (250 mg, 55%, mp: 108–109 °C). IR (KBr): ν_{max} cm^{-1} 1737s (C=O), 1714s (C=O), 1667s (C=C). ^{1}H NMR (100 MHz, $CDCl_3$): δ 4.60 (t, 2H, 2 × 17-CH), 2.92 (m, 11H, 2 × 4-CH and cyclobutane protons), 2.00 (s, 6H, 2 × 20-MeCO), 0.98 (s, 6H, 2 × 18-Me) (Williams *et al*., 1986).

Four cholesterol-based dimers **447–450** have been prepared, respectively, from cholest-5-en-3-one (**132**), cholest-5-en-7-one (**444**), 3β-chlorocholest-5-en-7-one (**445**) and 3β-acetyloxycholest-5-en-7-one (**446**), by amine catalyzed dimerization using DMAP and xylene (Shamsuzzamana and Tabassum, 2011).

The general method for the synthesis of dimers **447–450** can be summarized as follows (*Schemes 2.8.2* and *2.8.3*). DMAP (1.3 mmol) was added to a stirred solution of steroidal ketones (1.3 mmol) in xylene (50 mL) and refluxed for 2 h. After cooling the reaction mixture to room temperature, the solvent was distilled off *in vacuo* to obtain the oily residue, which was dissolved in Et_2O and sequentially washed with H_2O, 5% $NaHCO_3$ and again

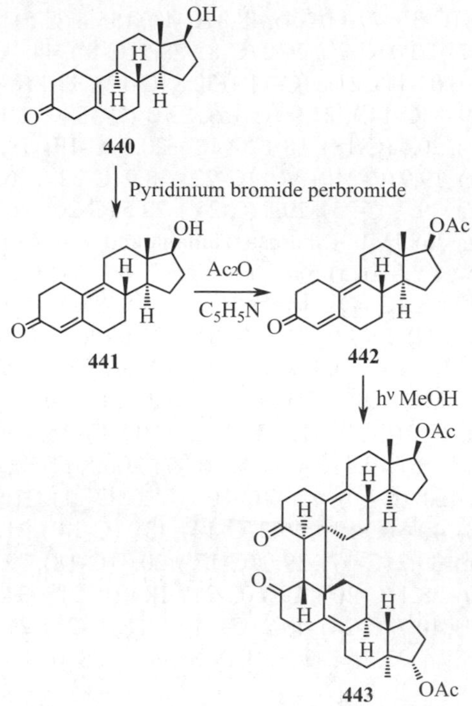

440

Pyridinium bromide perbromide

441

Ac_2O

C_5H_5N

442

hν MeOH

443

Scheme 2.8.1 *Synthesis of anti-photodimer* ***443***

with H_2O. The etheral layer was dried (Na_2SO_4) and evaporated to dryness. The crude dimer was purified by CC (mobile phase: *n*-hexane-Et_2O = 5:1) and recrystallized from MeOH to produce corresponding steroidal dimers **447–450** (Shamsuzzamana and Tabassum, 2011).

Cholest-5-en-3-spiro-(6′α,5′-oxa)-5′α-cholest-3′-one (**447**, 52%, mp: 230–233 °C). IR (KBr): ν_{max} cm^{-1} 1730s (C=O), 1628s (C=C), 1048s (C–O). ^1H NMR (400 MHz, $CDCl_3$): δ 5.59 (dd, 1H, *J* = 5.2, 7.6 Hz, 6-CH), 2.31 (dd, 1H, *J* = 4.8, 6.3 Hz, 6′-CH), 2.47 (s, 2H, 4′-CH_2), 2.16 (s, 2H, 4-CH_2), 1.18 (s, 3H, 10-Me), 0.77 (s, 3H, 13-Me), 1.14 (s, 3H, 10′-Me), 0.73 (s, 3H, 13′-Me), 0.92 and 0.90 (s, 6H, 21-Me and 21′-Me), 0.84 and 0.80 (s, 6H, 18-Me and 18′-Me). ^{13}C NMR (100 MHz, $CDCl_3$): δ 31.5 (C-1), 27.0 (C-1′), 38.6 (C-2), 48.5

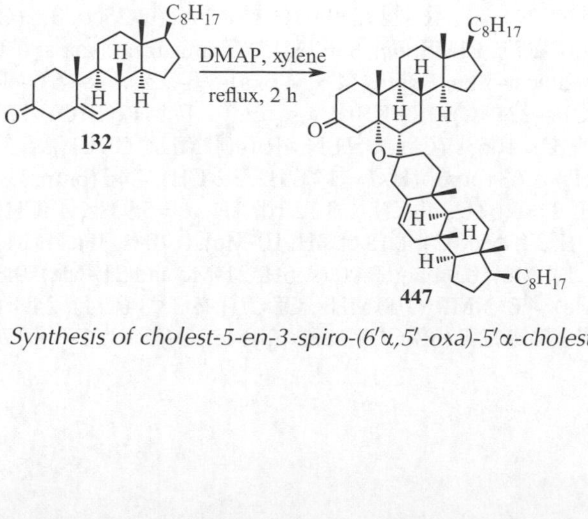

132

DMAP, xylene

reflux, 2 h

447

Scheme 2.8.2 *Synthesis of cholest-5-en-3-spiro-(6′α,5′-oxa)-5′α-cholest-3′-one (**447**)*

(C-2′), 95.6 (C-3), 204.8 (C-3′), 49.1 (C-4), 48.8 (C-4′), 155.6 (C-5), 133.5 (C-5′), 131.1 (C-6), 53.0 (C-6′), 31.3 (C-7), 22.0 (C-7′), 29.2 (C-8), 28.5 (C-8′), 44.4 (C-9), 37.1 (C-9′), 40.7 (C-10), 43.3 (C-10′), 21.0 (C-11), 21.2 (C-11′), 32.8 (C-12), 32.1 (C-12′), 38.5 (C-13), 40.6 (C-13′), 47.9 (C-14), 49.4 (C-14′), 21.9 (C-15), 22.0 (C-15′), 21.1 (C-16), 21.1 (C-16′), 46.9 (C-17), 48.6 (C-17′), 20.4 (C-18), 18.1 (C-18′), 20.5 (C-19), 19.1 (C-19′), 31.0 (C-20), 29.1 (C-20′), 19.9 (C-21), 19.9 (C-21′), 37.1 (C-22), 38.9 (C-22′), 26.3 (C-23), 26.0 (C-23′), 38.8 (C-24), 41.1 (C-24′), 29.9 (C-25), 29.5 (C-25′), 22.5 (C-26), 22.4 (C-26′), 22.7 (C-27), 22.9 (C-27′). EIMS: *m/z* 768 $[M]^+$ (Shamsuzzamana and Tabassum, 2011).

Cholest-5-en-7-spiro-(4′α,5′-oxa)-5′α-cholestan-7′-one (**448**, 57%, mp: 224–226 °C). IR (KBr): ν_{max} cm^{-1} 1719s (C=O), 1625s (C=C), 1042s (C–O). ^{1}H NMR (400 MHz, $CDCl_3$): δ 5.31 (s, 1H, 6-CH), 2.74 (s, 2H, 6′-CH_2), 2.19 (dd, 2H, J = 4.8, 6.1 Hz, 4-CH_2), 2.41 (dd, 1H, J = 4.2, 5.7 Hz, 4′-CH), 1.19 (s, 3H, 10-Me), 0.76 (s, 3H, 13-Me), 1.15 (s, 3H, 10′-Me), 0.74 (s, 3H, 13′-Me), 0.96 and 0.92 (s, 6H, 21-Me and 21′-Me), 0.85 and 0.80 (s, 6H, 18-Me and 18′-Me). 25.7 (C-2′), 31.2 (C-3), 23.2 (C-3′), 35.8 (C-4), 58.1 (C-4′), 152.9 (C-5), 97.4 (C-5′), 129.1 (C-6), 49.4 (C-6′), 95.3 (C-7), 206.5 (C-7′), 40.0 (C-8), 45.2 (C-8′), 35.0 (C-9), 30.7 (C-9′), 42.6 (C-10), 42.4 (C-10′), 21.6 (C-11), 21.9 (C-11′), 32.0 (C-12), 33.2 (C-12′), 44.1 (C-13), 40.3 (C-13′), 49.2 (C-14), 48.1 (C-14′), 21.4 (C-15), 21.0 (C-15′), 23.0 (C-16), 21.0 (C-16′), 39.2 (C-17), 39.9 (C-17′), 20.3 (C-18), 19.1 (C-18′), 20.7 (C-19), 19.9 (C-19′), 30.4 (C-20), 28.1 (C-20′), 19.7 (C-21), 18.8 (C-21′), 34.1 (C-22), 35.4 (C-22′), 27.4 (C-23), 26.1 (C-23′), 40.7 (C-24), 40.2 (C-24′), 28.8 (C-25), 29.9 (C-25′), 22.4 (C-26), 22.3 (C-26′), 22.9 (C-27), 23.0 (C-27′). EIMS: *m/z* 768 $[M]^+$ (Shamsuzzamana and Tabassum, 2011).

3β-Chlorocholest-5-en-7-spiro-(4′α,5′-oxa)-3′β-chloro-5′α-cholestan-7′-one (**449**, 62%, mp: 240–242 °C). IR (KBr): ν_{max} cm^{-1} 1717s (C=O), 1622s (C=C), 1052s (C–O), 728s (C–Cl), 736s (C–Cl). ^{1}H NMR (400 MHz, $CDCl_3$): δ 5.49 (s, 1H, 6-CH), 2.42 (s, 2H, 6′-CH_2), 3.86 (br m, 1H, J = 17.3 Hz, 3α-CH), 3.59 (br m, 1H, J = 16.6 Hz, 3′α-CH), 2.13 (d, 2H, J = 6.1 Hz, 4-CH_2), 2.86 (d, 1H, J = 5.8 Hz, 4′-CH), 1.16 (s, 3H, 10-Me), 0.75 (s, 3H, 13-Me), 1.12 (s, 3H, 10′-Me), 0.70 (s, 3H, 13′-Me), 0.96 and 0.94 (s, 6H, 21-Me and 21′-Me), 0.87 and 0.81 (s, 6H, 18-Me and 18′-Me). ^{13}C NMR (100 MHz, $CDCl_3$): δ 34.2 (C-1), 25.3 (C-1′), 37.7 (C-2), 29.8 (C-2′), 56.9 (C-3), 48.2 (C-3′), 45.0 (C-4), 57.6 (C-4′), 161.0 (C-5), 93.1 (C-5′), 125.4 (C-6), 46.1 (C-6′), 90.6 (C-7), 205.3 (C-7′), 38.4 (C-8), 43.9 (C-8′), 35.3 (C-9), 31.6 (C-9′), 42.5 (C-10), 43.7 (C-10′), 21.2 (C-11), 21.2 (C-11′), 32.2 (C-12), 32.4 (C-12′), 43.1 (C-13), 40.9 (C-13′), 48.1 (C-14), 48.1 (C-14′), 21.3 (C-15), 21.1 (C-15′), 22.0 (C-16), 21.5 (C-16′), 39.2 (C-17), 39.0 (C-17′), 20.1 (C-18), 20.0 (C-18′), 20.4 (C-19), 20.7 (C-19′), 30.2 (C-20), 29.8 (C-20′), 19.0 (C-21), 19.1 (C-21′), 36.4 (C-22), 36.2 (C-22′), 26.6 (C-23), 27.2 (C-23′), 39.7 (C-24), 40.3 (C-24′), 29.9 (C-25), 29.4 (C-25′), 22.5 (C-26), 22.2 (C-26′), 23.0 (C-27), 22.7 (C-27′). EIMS: *m/z* 836 $[M]^+$ (Shamsuzzamana and Tabassum, 2011).

3β-Acetyloxy-cholest-5-en-7-spiro-(4′α,5′-oxa)-3′β-acetyloxy-5′α-cholestan-7′-one (**450**, 58%, mp: 219–221 °C). IR (KBr): ν_{max} cm^{-1} 1714s (C=O), 1739s (C=O), 1732s (C=O), 1621s (C=C), 1063s (C–O). ^{1}H NMR (400 MHz, $CDCl_3$): δ 5.63 (s, 1H, 6-CH), 2.38 (s, 2H, 6′-CH_2), 4.63 (br m, 1H, J = 17.6 Hz, 3α-CH), 4.46 (br m, 1H, J = 16.8 Hz, 3′α-CH), 2.51 (d, 2H, J = 6.6 Hz, 4′-CH_2), 3.12 (d, 1H, J = 5.9 Hz, 4′-CH), 2.04 (s, 3H, 3β-MeCO), 2.01 (s, 3H, 3′β-MeCO), 1.18 (s, 3H, 10-Me), 0.70 (s, 3H, 13-Me), 1.16 (s, 3H, 10′-Me), 0.74 (s, 3H, 13′-Me), 0.94 and 0.90 (s, 6H, 21-Me and 21′-Me), 0.83 and 0.81 (s, 3H, 18-Me and 18′-Me). ^{13}C NMR (100 MHz, $CDCl_3$): δ 33.3 (C-1), 24.1 (C-1′), 33.9 (C-2), 26.8 (C-2′), 82.4 (C-3), 173.0 (3β-CO), 18.6 (3β-MeCO), 70.8 (C-3′), 170.5 (3′β-CO), 19.3

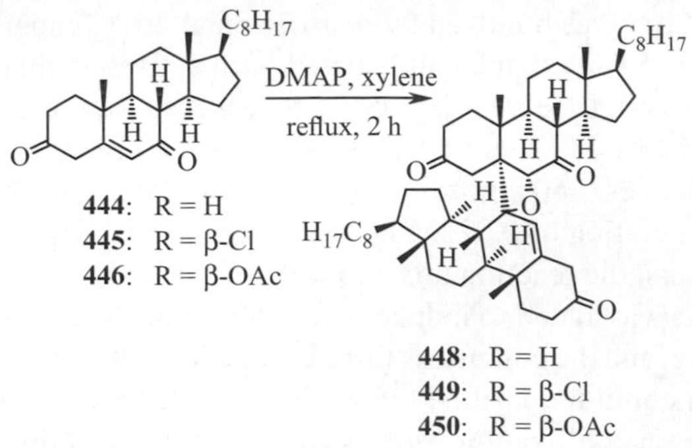

C_8H_{17} DMAP, xylene reflux, 2 h C_8H_{17} $H_{17}C_8$

444: R = H
445: R = β-Cl
446: R = β-OAc

448: R = H
449: R = β-Cl
450: R = β-OAc

Scheme 2.8.3 *Synthesis of steroidal dimers **448–450***

(3′β-MeCO), 41.9 (C-4), 61.4 (C-4′), 158.3 (C-5), 96.1 (C-5′), 129.8 (C-6), 46.0 (C-6′), 88.0 (C-7), 208.1 (C-7′), 39.0 (C-8), 44.9 (C-8′), 35.3 (C-9), 30.6 (C-9′), 36.0 (C-10), 43.8 (C-10′), 21.5 (C-11), 20.7 (C-11′), 32.0 (C-12), 32.0 (C-12′), 42.1 (C-13), 41.5 (C-13′), 49.2 (C-14), 48.1 (C-14′), 21.3 (C-15), 21.2 (C-15′), 21.9 (C-16), 21.4 (C-16′), 38.8 (C-17), 40.7 (C-17′), 20.4 (C-18), 19.6 (C-18′), 20.8 (C-19), 20.1 (C-19′), 30.2 (C-20), 29.8 (C-20′), 19.6 (C-21), 19.2 (C-21′), 36.0 (C-22), 37.0 (C-22′), 26.2 (C-23), 24.1 (C-23′), 41.6 (C-24), 39.6 (C-24′), 28.9 (C-25), 23.5 (C-25′), 22.4 (C-26), 22.2 (C-26′), 22.8 (C-27), 22.6 (C-27′). EIMS: *m/z* 884 $[M]^+$ (Shamsuzzamana and Tabassum, 2011).

Several rather unusual dimers, trioxabicyclononanes **453–459** were synthesized from cholesteryl benzoate (**451**) (Vonwiller, 1997). First, cholesteryl benzoate aldol **452** was prepared from **451** (*Scheme 2.8.4*). A stirred solution of cholesteryl benzoate (**451**, 1.0 g, 2.04 mmol) in C_5H_5N (45 mL) containing hematoporphyrin sensitizer was irradiated

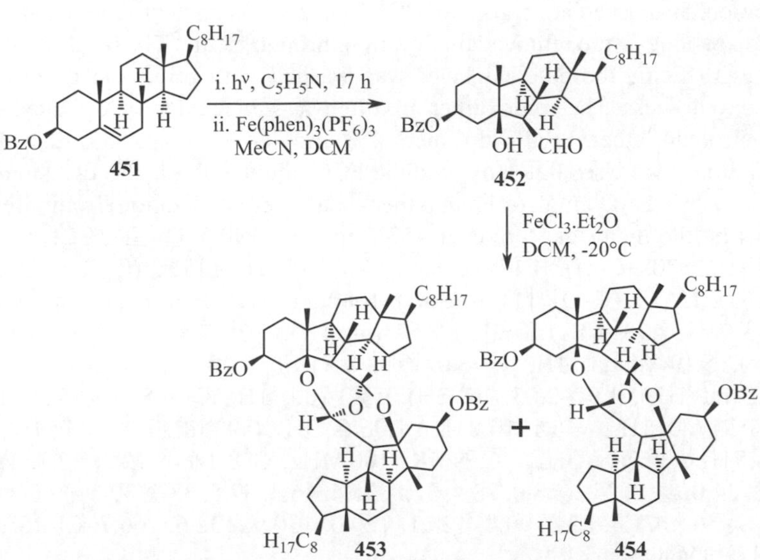

C_8H_{17} i. h^{ν}, C_5H_5N, 17 h; ii. $Fe(phen)_3(PF_6)_3$ MeCN, DCM

BzO **451** → BzO OH CHO **452**

$FeCl_3.Et_2O$ DCM, -20°C

453 + **454**

Scheme 2.8.4 *Synthesis of trioxabicyclononanes **453** and **454***

at 20 °C for 17 h. C_5H_5N was removed by distillation at 20 °C under vacuum. The crude solid was dissolved in Et_2O (50 mL) and stirred with activated charcoal for 30 min, the mixture was filtered through Celite followed by rotary evaporation gave a viscous oil. The crude oil was purified by FCC, eluted with 20% Et_2O in petroleum ether, to afford a mixture of hydroperoxides (860 mg), which was dissolved in DCM (45 mL), and cooled to −15 °C under N_2. A solution of 0.022M $Fe(phen)_3(PF_6)_3$ in MeCN (7.5 mL, 0.17 mmol) was added dropwise and the reaction mixture was stirred for 1.75 h with slow warming to 10 °C. Et_2O (150 mL) was added to induce precipitation of the catalyst, the red solid was filtered off with Celite, and the colourless filtrate was evaporated under vacuum to obtain a solid residue, which could be purified by FCC (mobile phase: 25% Et_2O in petroleum ether), followed by recrystallization from petroleum ether yielding aldol **452** (530 mg, 50%, mp: 143–145 °C). IR ($CHCl_3$): ν_{max} cm^{-1} 3600br (O–H), 3510br (O–H), 2954s (C–H), 2936s (C–H), 2869m (C–H), 2725w (C–H), 1714s (C=O)), 1280s (C–O), 1120m. ^{1}H NMR (400 MHz, $CDCl_3$): δ 9.71 (d, 1H, J = 2.8 Hz), 7.96–7.99 (m, 2H), 7.55–7.59 (m, 1H), 7.43–7.47 (m, 2H), 5.34 (dddd, 1H, J = 3.4 Hz), 2.51 (s, 1H, OH), 2.32 (dd, 1H, J = 3.2, 15.5 Hz), 2.22–2.27 (m, 2H), 2.09 (ddd, 1H, J = 3.2, 12.9 Hz), 2.00 (dd, 1H, J = 3.9, 15.5 Hz), 1.79–1.89 (m, 3H), 1.68 (ddd, 1H, J = 6.1, 9.1, 15.3 Hz), 1.07–1.55 (m, 18H), 1.01 (s, 3H, s), 0.93 (d, 3H, J = 6.5 Hz), 0.87 (d, 3H, J = 6.6 Hz), 0.86 (d, 3H, J = 6.6 Hz), 0.74 (s, 3H). ^{13}C NMR (100 MHz, $CDCl_3$): δ 12.5, 18.3, 18.7, 21.6, 22.5, 22.8, 23.8, 24.5, 24.8, 28.0, 28.2, 28.3, 35.6, 36.2, 39.5, 39.5, 39.6, 42.3, 44.7, 45.6, 51.3, 55.7, 56.0, 70.7, 83.6, 128.6, 129.4, 130.2, 133.2, 165.5, 203.9. MS: m/z 505 $[M - OH]^+$ (Vonwiller, 1997).

Formation of the dimers **453** and **454** was performed using **452** as follows (*Scheme 2.8.4*). A solution of 0.17M $FeCl_3.Et_2O$ in DCM (10.4 mL, 1.80 mmol, 2 equiv.) was added to a stirred solution of **452** (468.5 mg, 0.90 mmol) in DCM (21 mL) at −20 °C under N_2. A deep orange coloured solution indicated the formation of dimers. After 18 h, anhydrous Na_2SO_4 (1.0 g) was added and stirring was continued for another 7 h. An ice-cold saturated aqueous $NaHCO_3$ solution was added to the reaction mixture and the iron residues were removed by filtration through Celite. The resulting filtrate was extracted with Et_2O, the ethereal layer was washed with brine and dried (Na_2SO_4), evaporated to dryness. The crude dimer mixture was subjected to FCC, eluted with 15% Et_2O in petroleum ether, to afford dimers **453** and **454** (381 mg, 83%, 2:1 ratio). The individual dimers were separated by gradient FCC (mobile phase: 2% Et_2O in petroleum ether, then 7.5% Et_2O in petroleum ether) and recrystallization from MeCN-Et_2O providing first the major diastereomer **453** (mp: 189–190.5 °C). IR (CCl_4): ν_{max} cm^{-1} 2951s (C–H), 2870s (C–H), 1717s (C=O), 1468m (C–H), 1452m (C–H), 1383m (C–H), 1277s (C–O), 1161m (C–O), 1117s (C–O), 1082m (C–O), 1071m. ^{1}H NMR (400 MHz, $CDCl_3$): δ 7.52–7.98-8.01 (m, 2H), 7.57 (m, 1H), 7.39-7.44 (m, 2H), 5.29 (d, 1H, J = 5.3 Hz), 5.19 (dddd, 1H, J = 3.0, 6.5, 8.5 Hz), 2.40 (dd, 1H, J = 3.2, 15 Hz), 2.09–2.18 (m, 1H,), 1.97–2.05 (m, 2H), 1.80 (dd, 1H, J = 6.5, 15.0 Hz), 1.63–1.73 (m, 2H), 1.54 (dd, 1H, J = 5.1, 10.8 Hz), 1.06 (s, 3H), 0.92 (d, 3H, J = 6.5 Hz), 0.88 (d, 6H, J = 6.5 Hz), 0.67 (s, 3H). ^{13}C NMR (100 MHz, $CDCl_3$): δ 12.2, 17.7, 18.9, 21.2, 22.5, 22.8, 24.0, 24.8, 25.3, 28.0, 28.4, 32.8, 35.6, 36.3, 37.8, 38.7, 39.5, 39.6, 44.0, 46.8, 54.45, 55.9, 56.8, 71.4, 82.2, 90.4, 128.1, 129.6, 130.9, 132.6, 166.7. CIMS: m/z 1045 $[M + NH_4]^+$ (Vonwiller, 1997).

After crystallization from MeCN, the more polar minor diastereomer **454** could be obtained (mp: 102–104.5 °C). IR (KBr): ν_{max} cm^{-1} 2946s (C–H), 2867s (C–H), 1720s (C=O), 1467m (C–H), 1451m (C–H), 1383m (C–H), 1277s (C–O), 1188s (C–O), 1117s (C–O). ^{1}H NMR (400 MHz, $CDCl_3$): δ 8.04–8.06 (m, 2H), 7.49–7.53 (m, 1H), 7.35–7.39 (m, 2H), 5.04 (br m, 1H, *J* = 11.7 Hz), 5.03 (s, 1H), 2.38 (d, 1H, *J* = 15.4 Hz), 2.02 (d, 1H, *J* = 12.5 Hz), 1.01 (s, 3H), 0.92 (d, 3H, *J* = 6.4 Hz), 0.88 (d, 6H, *J* = 6.6 Hz), 0.65 (s, 3H). ^{13}C NMR (100 MHz, $CDCl_3$): δ 12.5, 18.8, 22.5, 22.8, 24.0, 24.5, 24.8, 28.0, 28.6, 30.1, 35.7, 36.3, 39.5, 39.7, 39.8, 43.8, 44.3, 46.8, 51.9, 54.7, 55.5, 56.6, 70.1, 82.7, 94.0, 128.0, 129.6, 131.5, 132.3, 166.1. CIMS: *m*/*z* 1045 [M + NH_4]$^+$ (Vonwiller, 1997).

Hydrolysis of a mixture of trioxabicyclononanes **453** and **454** furnished a mixture of diol dimers **455** and **456** (*Scheme 2.8.5*). Methanolic KOH (7 mL, 15% KOH in MeOH) was added to a stirred solution of dimers **453** and **454** (381 mg, 0.37 mmol) in DME (8 mL) room temperature. After 10 min, the reaction mixture was refluxed for 60 min, cooled and poured onto H_2O. The mixture was extracted with Et_2O, the ethereal layer was washed with brine, dried (Na_2SO_4), and evaporated under pressure. The crude dimers were purified by FCC, eluted with 20% petroleum ether in Et_2O, to give the less-polar major isomer **455** (178 mg, 59%, mp: 186–188 °C). IR (KBr): ν_{max} cm^{-1} 2947s (C–H), 1467m (C–H), 1158s (C–O), 1102s (C–O), 979m. ^{1}H NMR (400 MHz, $CDCl_3$): δ 0.67 (s, 3H), 1.03 (s, 3H), 0.74 (ddd, 1H, *J* = 5.1, 11.8 Hz), 0.86 (d, 3H, *J* = 6.6 Hz), 0.86 (d, 3H, *J* = 6.6 Hz), 0.91 (d, 3H, *J* = 6.5 Hz), 1.03 (s, 3H), 1.45 (dd, 1H, *J* = 5.2 and 15.4 Hz), 1.54 (dd, 1H, *J* = 5.5, 11.0 Hz), 1.86 (d, 1H, *J* = 10.4 Hz), 1.96 (ddd, 1H, *J* = 11.7 Hz), 2.20 (dddd, 1H, *J* = 3.6, 10.2, 13.8 Hz), 2.38 (dd, 1H, *J* = 2.2, 15.1 Hz), 3.97 (br m, 1H, *J* = 19.6 Hz), 5.45 (d, 1H, *J* = 5.3 Hz). ^{13}C NMR (100 MHz, $CDCl_3$): δ 12.2, 16.9, 18.9, 21.0, 22.5, 22.8, 23.9, 25.4, 28.0, 29.9, 33.7, 35.6, 36.2, 37.3, 39.4, 39.5, 40.5, 43.9, 46.8, 55.5, 55.59, 55.61, 57.2, 67.4, 83.5, 89.8. CIMS: *m*/*z* 820 [M + H]$^+$ (Vonwiller, 1997).

More polar minor isomer **456** (79 mg, 26%, mp: 95–97 °C) was afforded by changing the eluent to 100% Et_2O. IR (KBr): ν_{max} cm^{-1} 2942s (C–H), 2920s (C–H), 1468m (C–H), 1181s (C–O), 1100m (C–O), 997m, 969m. ^{1}H NMR (400 MHz, $CDCl_3$): δ 5.10 (s, 1H), 3.93 (br m, 1H, *J* = 17.8 Hz), 2.87 (d, 1H, *J* = 10.1 Hz), 2.25 (dd, 1H, *J* = 1.7, 14.8 Hz), 2.03 (ddd, 1H, *J* = 3.2, 12.8 Hz), 1.86 (br m, 1H, *J* = 22 Hz), 1.68 (ddd, 1H, *J* = 10.2, 11.6 Hz), 1.60 (dd, 1H, *J* = 4.8, 14.8 Hz), 1.47 (d, 1H, *J* = 10.0 Hz), 0.91 (d, 3H, *J* = 6.5 Hz), 1.02 (s, 3H), 0.87 (d, 3H, *J* = 6.6 Hz), 0.86 (d, 3H, *J* = 6.6 Hz), 0.66 (s, 3H). ^{13}C NMR (100 MHz, $CDCl_3$): δ 12.5, 18.0, 18.8, 21.2, 22.5, 22.8, 23.9, 24.6, 28.0, 28.5, 28.9, 30.7, 35.7, 36.2, 39.5, 39.6, 39.9, 44.4, 46.4, 46.9, 53.0, 54.9, 55.4, 56.1, 67.6, 85.7, 94.3. CIMS: *m*/*z* 820 [M + 1]$^+$ (Vonwiller, 1997).

Methylation of the diol **455** provided the dimethoxy **457** as follows (*Scheme 2.8.5*). At first, 60% NaH (43.2 mg, 1.08 mmol) was washed with dry pentane (2 × 2 mL), dried under N_2 stream, and dissolved in dry THF (6 mL). To this mixture, sequentially, diol **455** (122.8 mg, 0.15 mmol) in THF (5 mL) and MeI (56 μL, 0.90 mmol) were added, and refluxed for 24 h. After cooling in ice, H_2O was added to the reaction mixture and extracted with Et_2O. The ethereal extract was washed with brine, dried (Na_2SO_4) and rotary evaporated, the crude dimer was purified by FCC (mobile phase: 10% Et_2O in petroleum ether) and crystallized from MeOH to obtain dimethoxy **457** (118 mg, 92%, mp: 79–82 °C). IR (KBr): ν_{max} cm^{-1} 3469br (O–H), 2946s (C–H), 1469m (C–H), 1088m (C–O), 981m. ^{1}H NMR (400 MHz, $CDCl_3$): δ 5.28 (d, 1H, *J* = 5.1 Hz), 3.47 (m, 1H, *J* = 18.0 Hz), 3.25 (s, 3H), 2.15 (dd, 1H, *J* = 4.4, 14.4 Hz), 2.01 (ddd, 1H, *J* = 3.4, 12.7 Hz), 1.98 (ddd, 1H, *J* = 11.6 Hz), 1.85 (dddd,

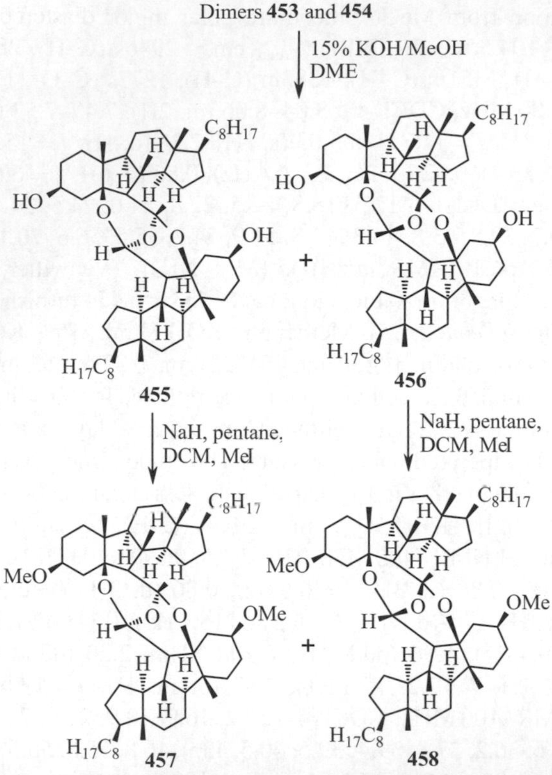

Dimers **453** and **454**

15% KOH/MeOH
DME

455 + **456**

NaH, pentane, DCM, MeI

457 + **458**

Scheme 2.8.5 *Synthesis of trioxabicyclononanes **455–458***

1H, J = 3.6, 5.3, 9.0, 13.0 Hz), 1.64 (dd, 1H, J = 5.5, 14.5 Hz), 1.51 (dd, 1H, J = 5.0, 10.5 Hz), 0.98 (s, 3H), 0.92 (d, 3H, J = 6.5 Hz), 0.87 (d, 3H, J = 6.6 Hz), 0.86 (d, 3H, J = 6.6 Hz), 0.75 (ddd, 1H, J = 4.4, 11.9 Hz), 0.65 (s, 3H). ^{13}C NMR (50 MHz, $CDCl_3$): δ 12.1, 17.6, 18.9, 21.1, 22.6, 22.8, 23.8, 24.6, 25.3, 28.0, 32.4, 35.6, 36.3, 37.4, 39.5, 39.6, 43.8, 46.6, 53.8, 55.6, 55.7, 56.1, 56.5, 75.7, 82.0, 90.3. CIMS: *m/z* 847 $[M + H]^+$ (Vonwiller, 1997).

Dimethoxy **458** was prepared by methylation of the diol **456**, following the procedure as described for dimethoxy **457** (*Scheme 2.8.5*). The minor diol **456** (93.0 mg, 114 μmol) in THF (5 mL) gave crude dimer, which was purified by FCC, eluted with 20% Et_2O in petroleum ether, followed by crystallization from MeOH to yield dimethoxy **458** (81.1 mg, 84%, mp: 227–231 °C). IR (KBr): ν_{max} cm^{-1} 3567br (O–H), 3370br (C–H), 2946s (C–H), 1468m (C–H), 1181s (C–O), 1079m, 970m. ^{1}H NMR (400 MHz, $CDCl_3$): δ 5.09 (s, 1H), 3.44 (m, 1H, J = 16.0 Hz), 3.33 (s, 3H), 2.16 (dd, 1H, J = 5.2, 14.4 Hz), 2.02 (ddd, 1H, J = 2.9, 12.7 Hz), 1.72 (dd, 1H, J = 4.9, 14.4 Hz), 1.61 (d, 1H, J = 10 Hz), 0.96 (s, 3H), 0.91 (d, 3H, J = 6.5 Hz), 0.87 (d, 3H, J = 6.6 Hz), 0.86 (d, 3H, J = 6.6 Hz), 0.66 (3H, s). ^{13}C NMR (50 MHz, $CDCl_3$): δ 12.6, 18.5, 18.8, 21.4, 22.5, 22.8, 23.9, 24.7, 25.5, 28.0, 28.6, 30.4, 35.7, 36.3, 39.5, 39.6, 39.8, 43.4, 44.4, 47.3, 52.1, 54.7, 55.4, 56.1, 56.3, 75.5, 84.0, 94.2. CIMS: *m/z* 847 $[M + 1]^+$ (Vonwiller, 1997).

Park *et al.* (1998) were also reported the synthesis of dimers **455**–**458** as well as a new dimer **459** and studied the chiral inducing power of ethers in a nematic mesophase.

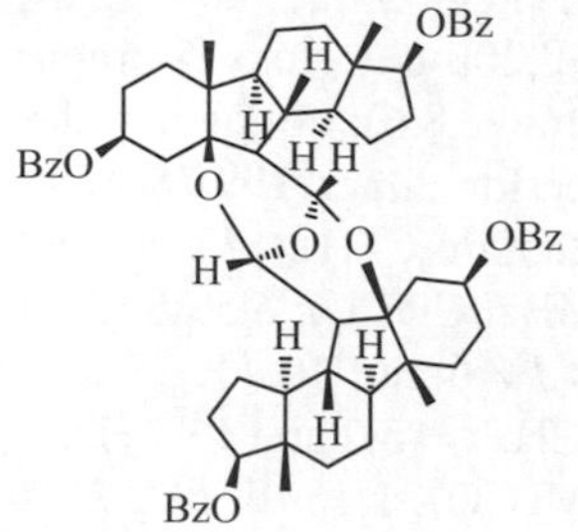

Trioxabicyclononan **459**

(*E*)-1,2-Bis-(3β,20β-diacetoxy-pregn-5-en-19-yl)ethene (**462**) was accomplished in three steps from 3β,20β-diacetoxy-pregn-5-en-19-al (**460**) (Edelsztein *et al.*, 2009). To achieve the synthesis of this dimer, 3β,20β-diacetoxy-pregn-5-en-19-al (**460**, 40 mg, 0.1 mmol) was converted to 3β,20β-diacetoxy-19-ethenyl-pregn-5-ene (**461**, 30 mg, 76%, mp: 125–127 °C) using the general Wittig reaction (CH_2=PPh_3, toluene) (*Scheme 2.8.6*).

IR (KBr): ν_{max} cm^{-1} 2956m (C–H), 2868m (C–H), 1740s (C=O), 1727s (C=O), 1370m (C–H), 1034m (C–O), 737m. ^1H NMR (500 MHz, $CDCl_3$): δ 5.76 (ddt, 1H, *J* = 5.0, 10.0, 22.0 Hz, 22-CH), 5.57 (td, 1H, *J* = 1.8, 5.2 Hz, 6-CH), 5.05 (ddd, 1H, *J* = 1.5, 2.1, 17.2 Hz, 23-CHb), 4.96 (dt, 1H, *J* = 2.0, 10.1 Hz, 23-CHa), 4.83 (dq,1H, *J* = 6.0, 10.9 Hz, 20α-CH), 4.63 (tt, 1H, *J* = 5.0, 11.5 Hz, 3α-CH), 2.57 (ddt, 1H, *J* = 1.9, 5.0, 14.5 Hz, 1H, 19-CHa), 2.09 (dd, 1H, *J* = 15.1, 9.8 Hz, 19-CHb), 2.04 (s, 3H, 3β-MeCO), 2.03 (s, 3H, 20β-MeCO), 1.15 (d, 3H, *J* = 6.1 Hz, 21-Me), 0.62 (s, 3H, 18-Me). ^{13}C NMR (125 MHz, $CDCl_3$): δ 37.2 (C-1), 27.7 (C-2), 73.9 (C-3), 170.6 (3β-CO), 21.6 (3β-MeCO), 38.2 (C-4), 136.8 (C-5), 124.7 (C-6), 31.4 (C-7), 32.5 (C-8), 51.4 (C-9), 39.7 (C-10), 21.2 (C-11), 39.5 (C-12), 42.4 (C-13), 57.2 (C-14), 24.2 (C-15), 25.4 (C-16), 54.8 (C-17), 12.7 (C-18), 36.5 (C-19), 72.9 (C-20), 170.5 (20β-CO), 21.4 (20β-MeCO), 19.9 (C-21), 137.3 (C-22), 115.3 (C-23). EIMS: *m/z* 387 [M + allyl]$^+$ (Edelsztein *et al.*, 2009).

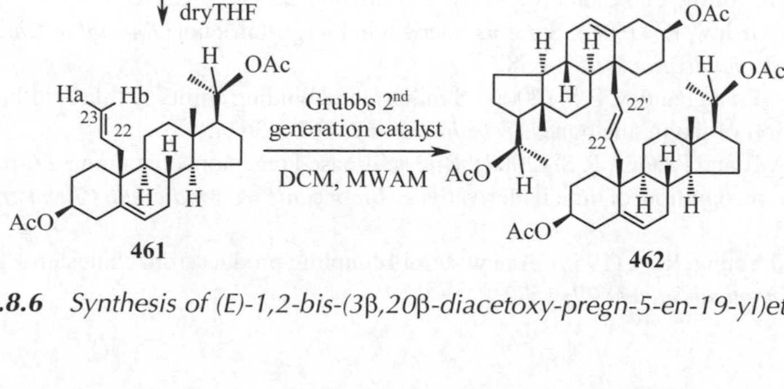

Scheme 2.8.6 Synthesis of (E)-1,2-bis-(3β,20β-diacetoxy-pregn-5-en-19-yl)ethene (**462**)

3β,20β-Diacetoxy-19-ethenyl-pregn-5-ene (**461**, 22 mg, 0.05 mmol) was employed for the synthesis of (*E*)-1,2-bis-(3β,20β-diacetoxy-5-pregnen-19-yl)-ethene (**462**, 12.4 mg, 64%, mp: 195–198 °C) (Scheme 2.8.6), following the general MWAM reaction and purification as described earlier for dimer **199** (*Scheme 2.2.15*). IR (KBr): ν_{max} cm^{-1} 2939m (C–H), 2873m (C–H), 2850m (C–H), 1733S (C=O), 1367m (C–H), 1247s (C–O), 1031m, 926m. ^{1}H NMR (500 MHz, $CDCl_3$): δ 5.48 (d, 2H, $J = 5$ Hz, 2 × 6-CH), 5.40 (m, 2H, 2 × 22-CH), 4.83 (dq, 2H, $J = 6$ and 10.4 Hz, 2 × 20α-CH), 4.62 (tt, 2H, $J = 5$ and 11.7 Hz, 2 × 3α-CH), 2.44 (dd, 2H, $J = 5.2$ and 15.7 Hz, 2 × 19-CHa), 2.05 (s, 6H, 2 × 3β-MeCO), 2.04 (s, 6H, 2 × 20β-MeCO), 1.16 (d, 6H, $J = 6.1$ Hz, 2 × 21-Me), 0.65 (s, 6H, 2 × 18-Me). ^{13}C NMR (125 MHz, $CDCl_3$): δ 37.1 (2 × C-1), 27.7 (2 × C-2), 73.9 (2 × C-3), 170.6 (2 × 3β-CO), 21.5 (2 × 3β-MeCO), 38.4 (2 × C-4), 137.1 (2 × C-5), 124.4 (2 × C-6), 31.6 (2 × C-7), 32.7 (2 × C-8), 51.3 (2 × C-9), 40.1 (2 × C-10), 21.5 (2 × C-11), 39.4 (2 × C-12), 42.3 (2 × C-13), 57.1 (2 × C-14), 24.2 (2 × C-15), 54.9 (2 × C-17), 25.6 (2 × C-16), 12.7 (2 × C-18), 35.5 (2 × C-19), 72.9 (2 × C-20), 170.5 (2 × 20β-CO), 21.5 (2 × 20β-MeCO), 19.9 (2 × C-21), 129.2 (2 × C-22). ESIMS: *m/z* 387 $[M + H]^+$ (Edelsztein *et al.*, 2009).

References

Alcalde, A. M., Antelo, A., Jover, A., Meijide, F. and Tato, J. V. (2008). Synthesis and characterization of a new *gemini* surfactant derived from 3α,7α,12α-trihydroxy-5β-cholan-24-amine (steroid residue) and ethylenediamintetraacetic acid (spacer). *12th International Electronic Conference on Synthetic Organic Chemistry* (ECSOC-12).

Basler, S., Bruncle, A., Jautelat, R. and Winterfeildt, E. (2000). Synthesis of cytostatic tetradecacyclic pyrazines and a novel reduction-oxidation sequence for spiro ketal opening in sapogenins. *Helvetica Chimica Acta* **83**, 1854–1880.

Bastien, D., Leblanc, V., Asselin, E. and Bérubé, G. (2010). First synthesis of separable isomeric testosterone dimers showing differential activities on prostate cancer cells. *Bioorganic and Medicinal Chemistry Letters* **20**, 2078–2081.

Berube, G., Rabouin, D., Perron, V., N'Zemba, B., Gaudreault, R.-C., Parenta, S. and Asselin, E. (2006). Synthesis of unique 17β-estradiol homo-dimers, estrogen receptors binding affinity evaluation and cytocidal activity on breast, intestinal and skin cancer cell lines. *Steroids* **71**, 911–921.

Bladon, P., Cornforth, J. W. and Jaeger, R. H. (1958). Electrolytic reduction of some α,β-unsaturated steroid ketones. *Journal of Chemical Society*, 863–871.

Bladon, P., McMeekin, W. and Williams, I. A. (1963). Steroids derived from hecogenin. Part III. The photochemistry of hecogenin acetate. *Journal of Chemical Society*, 5727–5737.

Blunt, J. W., Hartshorn, M. P. and Kirk, D. N. (1966). Reaction of epoxides-XI. A novel 'backbone rearrangement' of the cholestane skeleton. *Tetrahedron* **22**, 3195–3202.

Crabbe, P. and Mislow, K. (1968). Stereoisomerism in biergostatrienol. *Journal of Chemical Society Chemical Communications*, 657–658.

Chattopadhyay, P. and Pandey, P. S. (2006). Synthesis and binding ability of bile acid-based receptors for recognition of flavin analogues. *Tetrahedron* **62**, 8620–8624.

Chattopadhyay, P. and Pandey, P. S. (2007). Bile acid-based receptors containing 2,6-bis(acylamino) pyridine for recognition of uracil derivatives. *Bioorganic & Medicinal Chemistry Letters* **17**, 1553–1557.

Corey, E. J. and Young, R. L. (1955). A new steroid coupling product from cholesterol. *Journal of the American Chemical Society* **77**, 1672.

Daughenbaugh, P. J. and Allison, J. B. (1929). The action of thionyl chloride upon cholesterol and certain other alcohols. *Journal of the American Chemical Society* **51**, 3665–3667.

DellaGreca, M., Iesce, M. R., Previtera, L., Temussi, F. and Zarrelli, A. (2002). Solid-state photodimerization of steroid enones. *Journal of Organic Chemistry* **67**, 9011–9015.

Devaquet, A. and Salem, L. (1969). Intermolecular orbital theory. III. Thermal and photochemical dimerization of unsaturated ketone. *Journal of the American Chemical Society* **91**, 3793–3800.

Duddcck, H., Frclek, J., Snatzkc, G., Szczcpck, W. J. and Wagner, P. (1992). Synthesis and spectroscopic characterization of dimeric steroidal oximes. *Liebigs Annalen der Chemie* **7**, 715–718.

Dulou, R., Chopin, J. and Raoul, Y. (1951). Structures of the two bicholestadienes formed in the Salkowski test for cholesterol. *Bulletin de la Société Chimique de France* **18**, 616–621.

Edelsztein, V. C., Di Chenna, P. H. and Burton, G. (2009). Synthesis of C–C bonded dimeric steroids by olefin metathesis. *Tetrahedron* **65**, 3615–3623.

Elgendya, E. M. and Al-Ghamdy, H. (2007). Thermal and photooxidation reactions of the steroids: Sitosterol, stigmasterol and diosgenin. *Taiwan Pharmaceutical Journal* **59**, 113–132.

Fajkos, J. and Joska, J. (1979). Steroids. 215. Simmons-Smith methylenation of the 5,6-double bond in 19-hydroxylated steroids. *Collection of Czechoslovak Chemical Communications* **44**, 251–259.

Fieser, L. F. and Rajagopalan, S. (1950). Oxidation of steroids. 111. Selective oxidations and acylations in the bile acid series. *Journal of the American Chemical Society* **72**, 5530–5536.

Filippo, M. D., Izzo, I., Savignano, L., Tecilla, P. and Riccardis, F. D. (2003). Synthesis of a transmembrane ionophore based on a C_2-symmetric polyhydroxysteroid derivative. *Tetrahedron* **59**, 1711–1717.

Fournier, D. and Poirier, D. (2009). Estradiol dimers as a new class of steroid sulfatase reversible inhibitors. *Bioorganic and Medicinal Chemistry Letters* **19**, 693–696.

Gao, H. and Dias, J. R. (1998). Synthesis of cyclocholates and derivatives. III. Cyclocholates with 12-oxo and 7,12-oxo groups. *European Journal of Organic Chemistry* 719–724.

Gouin, S. and Zhu, X. X. (1996). Synthesis of 3α and 3β-dimers from selected bile acids. *Steroids* **61**, 664–669.

Guest, I. G., Jones, J. G. and Marples, B. A. (1971). Rearrangements of 10β-ethenyl steroids. *Tetrahedron Letters* **22**, 1979–1982.

Guthrie, J. P., Cossar, J. and Darson, B. A. (1986). A water soluble dimeric steroid with catalytic properties: Rate enhancements from hydrophobic binding. *Canadian Journal of Chemistry* **64**, 2456–2469.

Harmatha, J., Budesinsky, M. and Vokac, K. (2002). Photochemical transformation of 20-hydroxyecdysone: production of monomeric and dimeric ecdysteroid analogues. *Steroids* **67**, 127–135.

Hashimoto, T., Kato, Y., Shibahara, H., Toyooka, K., Ohta, T. and Satoh, D. (1979). Studies on digitalis glycosides XXXV. Diels–Alder type reaction of **16**,17-dehydrodigitoxgenin-3-acetate. (1) Dimerization. *Chemical and Pharmaceutical Bulletin* **27**, 1975–1979.

Hoffmann, S. and Kumpf, W. (1986). Estrogen-mesogens - dimeric estrogen derivatives. *Z. Chem.* **8**, 293–295.

Iriarte, J., Ponce, H., Saldana, E. and Crabbe, P. (1972). Synthesis of 17-iodoandrosta-5,16-dien-3β-ol and estrone 3-methyl ether dimers. *Journal of Mexican Chemical Society* 192–194.

Jautelat, R., Winterfelt, E. and Muller-Fahrnow, A. (1996). Photochemistry of hecogeninacetate revisited. *Journal fur praktische Chemie* **338**, 695–701.

Janout, V., Lanier, M. and Regen, S. L. (1996). Molecular umbrellas. *Journal of the American Chemical Society* **118**, 1573–1574.

Janout, V., Staina, I. V., Bandyopadhyay, P. and Regen, S. L. (2001) Evidence for an umbrella mechanism of bilayer transport. *Journal of the American Chemical Society* **123**, 9926–9927.

Janout, V., Jing, B. W. and Regen, S. L. (2002). Molecular umbrella-assisted transport of thiolated AMP and ATP across phospholipid bilayers. *Bioconjugate Chemistry* **13**, 351–356.

Joachimiak, R. and Paryzek, Z. (2004). Synthesis and alkaline metal ion binding ability of new steroid dimers derived from cholic and lithocholic acids. *Journal of Inclusion Phenomena and Macrocyclic Chemistry* **49**, 127–132.

Karabulut, H. R., Rashdan, S. A. and Dias, J. R. (2007). Notable chenodeoxycholic acid oligomers synthesis, characterization, and 7α-OR steric hindrance evaluation. *Tetrahedron* **63**, 5030–5035.

Kramer, W. and Kurz, G. (1983). Photolabile derivatives of bile salts. Synthesis and suitability for photoaffinity labeling. *Journal of Lipid Research* **24**, 910–923.

Karmas, G. (1968). A new type of steroid dimer. *Journal of Organic Chemistry* **33**, 2436–2440.

Knoll, S. M., Vu, B. D., Ba, D., Bourdon, S. and Bourdon, R. (1986). Structures of two cholesterol oxidation products by 2D-NMR spectroscopy. *Tetrahedron Letters* **27**, 2613–2616.

Kobuke, Y. and Nagatani, T. (2001). Transmembrane ion channels constructed of cholic acid derivatives. *Journal of Organic Chemistry* **66**, 5094–5101.

Kolehmainen, E., Tamminen, J., Lappalainen, K., Torkkel, T. and Seppala, R. (1996). Substituted methyl 5β-cholan-24-oates. Part III: Synthesis of a novel cholaphane ethylene glycol diester of lithocholic acid by cyclization with terephthalic acid. *Synthesis*, 1082–1084.

Kolehmainen, E., Tamminen, J., Kauppinen, R. and Linnanto, J. (1999). Silver (I) cation complexation with 3α,3α′alpha-bis(pyridine-*n*-carboxy) lithocholic acid 1,2-ethanediol diesters (n = 2-4): ^{1}H, ^{13}C and ^{15}N NMR spectral studies and molecular orbital calculations. *Journal of Inclusion Phenomena and Macrocyclic Chemistry* **35**, 75–84.

Kondo, M., Mehiri, M. and Regen, S. L. (2008). Viewing membrane-bound molecular umbrellas by parallax analyses. *Journal of the American Chemical Society* **130**, 13771–13777.

Leng, T. D. Zhang, J. X., Xie, J., Zhou, S. J., Huang, Y. J., Zhou, Y. H., Zhu, W. B. and Yan, G. M. (2010). Synthesis and anti-glioma activity of (25*R*)-spirostan-3β,5α,6β,19-tetrol. *Steroids* **75**, 224–229.

Li, Y. X. and Dias, J. R. (1997a). Dimeric and oligomeric steroids. *Chemical Review* **97**, 283–304.

Li, Y. X. and Dias, J. R. (1997b). Syntheses of linear dimeric and cyclic oligomeric cholate ester derivatives. *Synthesis*, 425–430.

Łotowski, Z., and Guzmanski, D. (2006). New acyclic dimers of cholic acid with oxamide and hydrazide spacers. *Monatshefte fur Chemie* **137**, 117–124.

Łotowski, Z., Piechowska, J. and Jarocki, Ł. (2008). Dimeric cholaphanes with oxamide spacers. *Monatshefte fur Chemie* **139**, 213–222.

Lund, H. (1957). Electroorganic preparations 1. Reduction of steroid ketones. *Acta Chemica Scandinavica* **11**, 283–290.

Martin, D. F. and Hartney, T. C. (1978). Clinically significant constituents. I. cholesterol. *Journal of Chemical Education* **53**, 239–241.

Masakazu, M. A., Takeuchi, T. and Mori, K. (1987). A simple synthesis of steroidal 3α,5-cyclo-6-ones and their efficient transformation to steroidal 2-en-6-ones. *Synthesis*, 181–183.

McMurry, J. E. (1983). Titanium-induced dicarbony1-coupling reactions. *Accounts of Chemical Research* **16**, 405–411.

Mehiri, M., Chen, W-H., Janout, V. and Regen, S. L. (2009). Molecular umbrella transport: exceptions to the classic size/lipophilicity rule. *Journal of the American Chemical Society* **131**, 1338–1339.

Mori, H. (1962). Studies on steroidal compounds. VIII. A new synthesis of 4-chloro-4-en-3-oxo-steroids. *Chemical and Pharmaceutical Bulletin* **10**, 429–432.

Morzycki, J. W., Kalinowski, S., Łotowski, Z. and Rabiczko, J. (1997). Synthesis of dimeric steroids as components of lipid membranes. *Tetrahedron* **53**, 10579–10590.

Mosettig, E. and Scheer, I. (1952). Steroids with an aromatic B-ring. *Journal of Organic Chemistry* **17**, 764–769.

Nahar, L. and Turner, A. B. (2003). Synthesis of ester-linked lithocholic acid dimers. *Steroids* **68**, 1157–1161.

Nahar, L., Sarker, S. D. and Turner, A. B. (2006). Facile synthesis of new oxalate dimers of naturally occurring 3-hydroxysteroids. *Chemistry of Natural Compounds* **42**, 549–552.

Nahar, L., Sarker, S. D. and Turner, A. B. (2007a). A review on synthetic and natural steroid dimers: 1997–2006. *Current Medicinal Chemistry* **14**, 1349–1370.

Nahar, L., Sarker, S. D. and Turner, A. B. (2007b) Synthesis of 17β-hydroxy steroidal oxalate dimers from naturally occurring steroids, *Acta Chimica Slovenica* **54**, 903–906.

Nahar, L., Sarker, S. D. and Turner, A. B. (2008). Convenient synthesis of new pregnenolone oxyminyl oxalate dimers. *Chemistry of Natural Compounds* **44**, 315–318.

Oppenauer, R. V. (1937). Oppenauer oxidation. *Recueil des Travaux Chimiques des Pays-Bas* **56**, 137–144.

Opsenica, D., Pocsfalvi, G., Juranic, Z., Tinant, B., Declercq, J-P., Kyle, D. E., Milhous, W. K. and Solaja, B. A. (2000). Cholic acid derivatives as 1,2,4,5-tetraoxane carriers: Structure and antimalarial and antiproliferative activity. *Journal of Medicinal Chemistry* **43**, 3274–3282.

Pandey, P. S., Rai, R. and Singh, R. B. (2002). S Synthesis of cholic acid-based molecular receptors: head-to-head cholaphanes. *Journal of Chemical Society, Perkin Trans 1*, 918–923.

Park, J. J., Sternhell, S. and Vonwiller, S. C. (1998). Highly efficient chirality inducers based on steroid-derived 2,6,9-trioxabicyclo[3.3.1]nonanes. *Journal of Organic Chemistry* **63**, 6749–6751.

Paryzek, Z., Piasecka, M., Joachimiak, R., Luks, E. and Radecka-Paryzek, W. (2010). New steroid dimer derived from cholic acid as receptor for lanthanum (III) and calcium (II) nitrates. *Journal of Rare Earths* **28**, 56–60.

Ra, C. S., Cho, S. W. and Choi, J. W. (2000). A steroidal cyclic dimer with ethylene glycol bridges. *Bulletin of Korean Chemical Society* **21**, 342–344.

Rega, M., Jimenez, C. and Rodriguez, J. (2007). 6*E* Hydroximinosteroid homodimerization by cross-metathesis processes. *Steroids* **72**, 729–735.

Reschel, M., Budesfnsky, M., Cerny, I., Pouzar, V. and Drasar, P. (2002). Synthesis of linear steroid oligoesters based on etienic acid. *Collection of Czechoslovak Chemical Communications* **67**, 1709–1718.

Schmidt, A., Beckert, R. and Weiß, D. (1992). Simple procedure for reductive coupling of steroids with a cross conjugated dienone system. *Tetrahedron Letters* **33**, 4299–4300.

Shamsuzzamana, M. G. A. and Tabassum, S. (2011). Synthesis of steroidal dimers: selective amine catalysed steroidal dimerization. *Journal of Chemical Sciences* **123**, 491–495.

Shawaphun, S., Janout, V. and Regen, S. L. (1999). Chemical evidence for transbilayer movement of molecular umbrellas. *Journal of the American Chemical Society* **121**, 5860–5864.

Shawakfeh, K. Q., Al-Ajlouni, A. M. and Ibdah, A. (2002). Synthesis and selective catalytic oxidation of new dimeric steroids. *Acta Chimica Slovenica* **49**, 805–813.

Shawakfeh, K. Q., Al-Said, N. H. and Abboushi, E. K. (2010). Synthesis of new di- and triamine diosgenin dimers. *Tetrahedron* **66**, 1420–1423.

Squire, E. N. (1951). Bisteroids. I. Bicholestanyl, The structure of bicholesteryl and 3,3′-bis(3,5-cholestadiene). *Journal of the American Chemical Society* **73**, 2586–2589.

Stohrer, G. (1971) Derivatives of 6,6′-bicholestane from cholesterol. *Steroids* **17**, 587.

Sucrow, W. and Chondromatidis, G. (1970). Enhydrazines. 1. Symmetrical *N*-methylpyrroles *via* enhydrazines. *Chemische Berichte* **103**, 1759–1766.

Suginome, H. and Uchida, T. (1979). Photo-induced transformations. Part 44. Formation of lactams in the photolysis of some Steroidal acetylhydrazones in the presence of oxygen. *Journal of Chemical Society Perkin Trans 1*, 1356–1364.

Suginome, H., Uchida, T., Kizuka, K. and Masamune, T. (1980). Synthesis and photoreaction of D-nor-5α-androstan-16-one acetylhydrazone in the presence of oxygen. *Bulletin of the Chemical Society of Japan*, **53**, 2285–2291.

Tamminen, J. T., Kolehmainen, E. T., Haapala, M. H., Salo, H. T. and Linnanto, J. M. (2000a). Synthesis and ^{13}C NMR chemical shift assignments of 2,2′-bipyridine-4,4′-dicarboxylates of bile acid methyl esters. *Arkivoc*, 80–86.

Tamminen, J., Kolehmainen, E., Linnanto, J., Salo, H. and Manttari, P. (2000b). 3α,3′α-bis(n-acetoxyphenylcarboxy)-5β-cholan-24-oic acid ethane-1,2-diol diesters (n = 2–4): ^{13}C NMR chemical shifts, variable-temperature and NOE ^{1}H NMR measurements and MO calculations of novel bile acid-based dimers. *Magnetic Resonance in Chemistry* **38**, 877–882.

Tamminen, J., Kolehmainen, E., Haapala, M. and Linnanto, J. (2000c). Bile acid-piperazine diamides: Novel steroidal templates in syntheses of supramolecular hosts: Isomeric pyridine-*n*-carboxy containing dimers and a cholaphane. *Synthesis-Stuttgart* **10**, 1464–1468.

Tanaka, K., Itagaki, Y., Satake, M., Naoki, H., Yasumoto, T., Nakanishi, K. and Berova, N. (2005). Three challenges toward the assignment of absolute configuration of Gymnocin-B. *Journal of the American Chemical Society* **127**, 9561–9570.

Templeton, J. F., Majgier-Baranowska, H. and Marat, K. (1990). Zinc-AcOH reduction of the steroid-4-en-3-one: novel conversion of the 4-en-3-one to the 2-en-4-one *via* a vinyl chloride. *Journal of Chemical Society Perkin Trans 1*, 2581–2584.

Templeton, J. F., Majgier-Baranowska, H. and Marat, K. (2000). Steroid dimer formation: metal reduction of methylandrost-4-ene-3,17-dion-19-oate. *Steroids* **65**, 219–223.

Tian, Z., Cui, H. and Wang, Y. (2002). Synthesis of linear and cyclic dimeric ester derivatives from bile acids and glycine. *Synthetic Communication* **32**, 3821–3829.

Todorović, N. M., Stefanović, M., Tinant, B., Declercq, J-P., Makler, M. T. and Šolaja, B. A. (1996). Steroidal geminal dihydroperoxides and 1,2,4,5-tetraoxanes: structure determination and their antimalarial activity. *Steroids* **61**, 688–696.

Valkonen, A., Sievänen, E., Ikonen, S., Lukashev, N. V., Donez, P. A., Averin, A. D., Lahtinen, M. and Kolehmainen, E. (2007). Novel lithocholaphanes: Syntheses, NMR, MS, and molecular modeling studies. *Journal of Molecular Structure* **846**, 65–73.

Valverde, L. F., Cedillo, D. F., Tolosa, L., Maldonado, G. and Reyes, G. C. (2006). Synthesis of pregnenolone-pregnenolone dimer *via* ring A-ring A connection. *Journal of the Mexican Chemical Society* **50**, 42–45.

Vonwiller, S. C. (1997). Efficient preparation of chiral C_2-Symmetrical 2,6,9-trioxabicyclo[3.3.1] nonanes from cholesteryl benzoate. *Journal of Organic Chemistry* **62**, 1155–1158.

Williams, J. R., Mattei, P. L., Abdel-Magid, A. and Blount, J. F. (1986). Photochemistry of estr-4-en-3-ones: 17β-hydroxy-4-estren-3-one, 17β-acetoxy-4,9-estradien-3-one, 17β-hydroxy-4,9,1l-estra-trien-3-onea, and Norgestrel. Photochemistry of 5α-estran-3-ones. *Journal of Organic Chemistry* **57**, 769–773.

Yus, M., Soler, T. and Foubelo, F. (2001). Synthesis of functionalised enantiopure steroids from estrone and cholestanone through organolithium intermediates. *Tetrahedron: Asymmetry* **12**, 801–810.

3

Synthesis of Cyclic Steroid Dimers

One of the most important areas in synthetic chemistry is the synthesis of molecules that can recognize and bind specific molecules, and then catalyze transformations of the bound molecules *i.e.*, 'artificial enzymes', and the preparation of systems that can reproduce themselves or otherwise store and process information at the molecular level (Tamminen and Kolehmainen, 2001). In this field of research, probably the most significant challenge of all is the design and synthesis of '*molecular actuators*', which undergo changes in shape in response to external stimuli and thus perform mechanical work. As steroids possess rigid structural units, are commercially available and offer two options for substitution, *e.g.*, axial and equatorial, at most positions, common naturally occurring steroid molecules, *e.g.*, bile acids, have been utilized as the starting material for the synthesis of '*molecular actuators*'. Among the common steroids, bile acids and their derivatives, because of their chemically different hydroxyl groups, enantiomeric purity, unique amphiphilicity, availability, and low cost, have been extensively used as architectural components in supramolecular chemistry.

Novel macrocyclic synthetic receptors bearing molecular cavities, *e.g.*, steroid-derived cholaphanes and cyclocholates, have been one of the most attractive areas in supramolecular chemistry (Bell and Anslyn, 1996). These host molecules are able to serve as model compounds for more complex biological systems and are vital for molecular recognition of substrates in enzymatic processes. Typically macrocycles bind substrates either in their defined cavity or above their plane. In this chapter, the synthesis of such macrocyclic steroid dimers, *e.g.*, cholaphanes and cyclocholates, are discussed.

3.1 With Spacer Groups: Cholaphanes

Steroid dimers with spacer groups can form macrocyclic compounds, commonly known as cholaphanes. In fact, cholaphanes are bile acid derived macrocycles composed of two to four bile acid subunits linked together through spacers (Tamminen and Kolehmainen, 2001; Nahar *et al.*, 2007). However, the frameworks of dimeric cholaphanes, *e.g.*, cholaphane

Steroid Dimers: Chemistry and Applications in Drug Design and Delivery, First Edition. Lutfun Nahar and Satyajit D. Sarker. Published 2012 by John Wiley & Sons, Ltd.

cyclodimer of cholic acids with aromatic spacer groups (*e.g.*, cholaphane **2**), are more common than the cholaphanes comprising three or four bile acid subunits, and they are more rigid and preorganized for strong and selective binding to other molecules (drugs) in comparison with acyclic steroid dimers. Cholaphane frameworks have potential variability and might be useful in biomimetic chemistry (Li and Dias, 1997a). So, cholaphanes are considered to be '*enzyme mimics*' or '*synthetic molecular receptors*' (Nahar *et al.*, 2007). The design of novel macrocyclic synthetic receptors with molecular cavities is one of the most important challenges in supramolecular chemistry, and cholaphanes have been continuously on the focus of supramolecular chemistry owing to their versatile properties. Cholaphanes, their synthesis, NMR spectroscopy, molecular mechanics calculations, and their binding of carbohydrate derivatives in organic solvents have been studied extensively (Bonar-Law and Davies, 1989, 1993a,b; Bonar-Law *et al.*, 1990, 1993; Bhattarai *et al.*, 1992; Davies and Wareham, 1999; Tamminen and Kolehmainen, 2001; Nahar *et al.*, 2007).

Most cholaphanes are synthesized involving 'head-to-tail' combination of cholic acids as in cholaphanes **2** and **3**. The first group of cholaphanes have moderate flexibility and limited solubility in organic solvents, which reduce their binding properties. In order to improve the binding properties of cholaphanes, Bhattarai *et al.* (1997) prepared the next generation of cholaphanes with externally directed alkyl chains enhancing their solubility in $CHCl_3$, and with truncated side chains, reducing their conformational freedom. One such compound is cholaphane **3**.

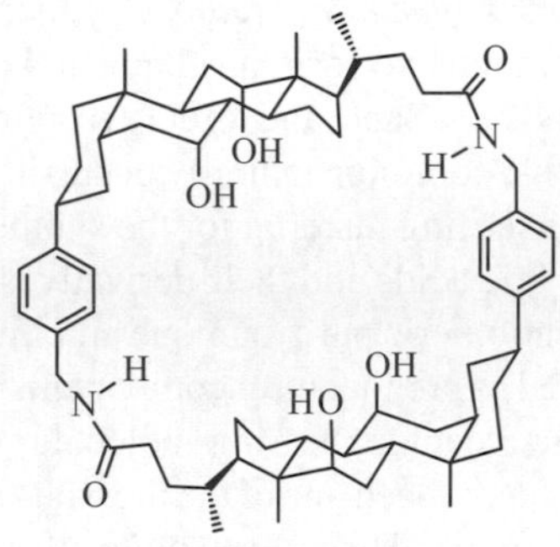

Cholaphane **2**

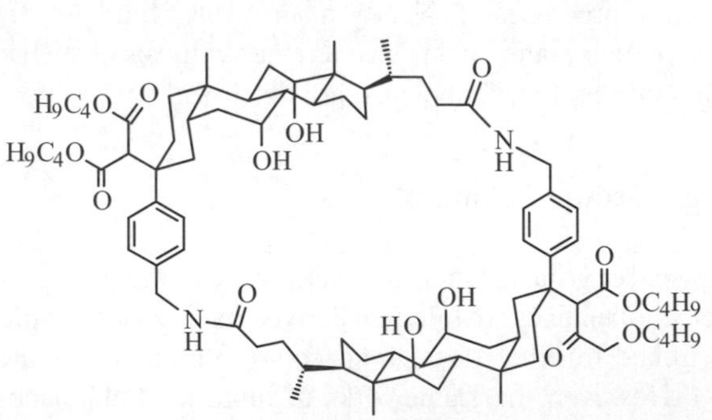

Cholaphane **3**

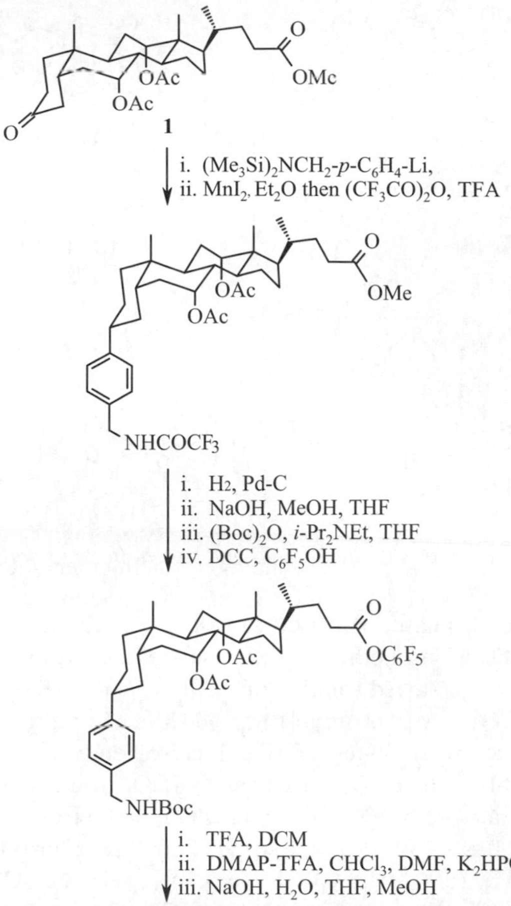

O
OAc
OMe
OAc
O
1

i. $(Me_3Si)_2NCH_2$-*p*-C_6H_4-Li,
ii. MnI_2, Et_2O then $(CF_3CO)_2O$, TFA

O
OAc
OMe
OAc
$NHCOCF_3$

i. H_2, Pd-C
ii. NaOH, MeOH, THF
iii. $(Boc)_2O$, *i*-Pr_2NEt, THF
iv. DCC, C_6F_5OH

O
OAc
OC_6F_5
OAc
NHBoc

i. TFA, DCM
ii. DMAP-TFA, $CHCl_3$, DMF, K_2HPO_4
iii. NaOH, H_2O, THF, MeOH

Cholaphane **2**

Scheme 3.1.1 *Synthesis of 'head-to-tail' cholaphane 2*

The 'head-to-tail' cholaphane **2** was obtained with an overall yield of 20% from methyl 3α,12α-diacetoxy-5β-cholan-24-oate (**1**) in a multistep sequence (*Scheme 3.1.1*) (Bonar-Law and Davies, 1989, 1993a).

The synthesis of macrocyclic cholaphanes **3–5** was accomplished from cholic acid (**6**) through a multistep sequence (Bhattarai *et al.*, 1997). To synthesize cholaphanes **3–5**, 3α,7α,12α-tris(formyloxy)-5β-cholan-24-oic acid (**8**) and 3α,7α,12α-tris(formyloxy)-24-nor-5β-cholan-23-oic acid (**9**) were obtained by formylation of cholic acid (**6**) and norcholic acid (**7**), respectively, using standard literature procedures (Schteingart and Hofmann, 1988; Tserng and Klein, 1977) (*Scheme 3.1.2*).

3α,7α,12α-Tris(formyloxy)-5β-cholan-24-oic (**8**) was converted to methyl 7α,12α-bis (formyloxy)-3α-hydroxy-5β-cholan-24-oate (**10**) (*Scheme 3.1.2*). To a stirred solution of **8** (20 g, 41 mmol) in DCM (400 mL) and MeOH (10 mL), CH_2N_2 was bubbled through while maintaining the temperature (−5 °C to 0 °C). After completion of the reaction,

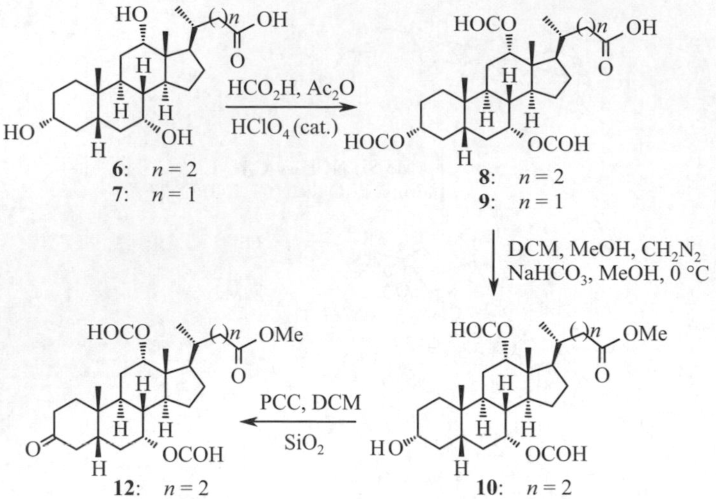

Scheme 3.1.2 *Synthesis of cholic acid and norcholic acid derivatives* ***8–13***

monitored by TLC (mobile phase: 30% EtOAc in *n*-hexane), the solvent was evaporated to dryness to afford corresponding methyl ester (not isolated). Deformylation was performed with this methyl ester. To a stirred solution of methyl ester (19.5 g, 39 mmol) in MeOH (300 mL), solid $NaHCO_3$ (5.0 g, 60 mmol) was added slowly at 0 °C. After 2 h, excess $NaHCO_3$ was neutralized with dilute AcOH and the solvent was evaporated to dryness. The residue was dissolved in Et_2O, washed with H_2O, dried (Na_2SO_4), and filtered. Evaporation of the filtrate produced crude gum, which was purified by FCC, eluted with Et_2O, to furnish **10** (8.0 g, 92%). IR ($CHCl_3$): ν_{max} cm^{-1} 3422br (O–H), 1718s (C=O), 1180m (C–O). ^1H NMR (300 MHz, $CDCl_3$): δ 8.14 (s, 1H, 12α-OCOH), 8.10 (s, 1H, 7α-OCOH), 5.26 (br s, 1H, 12β-CH), 5.05 (s, 1H, 7β-CH), 3.65 (s, 3H, 24-OMe), 3.49 (m, 1H, 3β-CH), 0.92 (s, 3H, 19-Me), 0.83 (d, 3H, $J = 6.2$ Hz, 21-Me), 0.75 (s, 3H, 18-Me). ^{13}C NMR (75 MHz, $CDCl_3$): δ 12.0, 17.3, 22.3, 22.7, 25.4, 27.0, 28.4, 30.2, 30.5, 30.8, 31.3, 34.1, 34.6, 34.8, 37.6, 38.4, 40.9, 42.8, 44.9, 47.1, 51.4, 70.8, 71.3, 75.2, 160.5, 160.7, 174.4 (Bhattarai *et al.*, 1997).

Methyl 7α,12α-bis(formyloxy)-3α-hydroxy-24-nor-5β-cholan-23-oate (**11**) was prepared from 3α,7α,12α-tris(formyloxy)-24-nor-5β-cholan-23-oic acid (**9**), according to the method described for **8** (*Scheme 3.1.2*). A stirred solution of **9** (15 g, 31 mmol) in DCM (200 mL) and MeOH (20 mL) was treated with CH_2N_2 to give the corresponding methyl ester, which was deformylated with $NaHCO_3$ (4.1 g, 49 mmol) in MeOH (300 mL), followed by FCC purification as above, yielded **11** (13.84 g, 95%). IR ($CHCl_3$): ν_{max} cm^{-1} 3422br (O–H), 2958m (C–H), 2878m (C–H), 1714s (C=O), 1186m (C–O), 908m, 734m. ^1H NMR (300 MHz, $CDCl_3$): δ 8.14 (s, 1H, 12α-OCOH), 8.09 (s, 1H, 7α-OCOH), 5.25 (br s, 1H, 12β-CH), 5.05 (br s, 1H, 7β-CH), 3.65 (s, 3H, 23-OMe), 3.49 (tt, 1H, 3β-CH), 0.92 (d, 3H, $J = 6.0$ Hz, 21-Me), 0.88 (s, 3H, 19-Me), 0.79 (s, 3H, 18-Me). ^{13}C NMR (75 MHz, $CDCl_3$): δ 12.0, 18.6, 22.2, 22.6, 25.4, 27.2, 28.4, 30.2, 31.4, 33.0, 34.1, 34.8, 37.6, 38.4, 40.8, 41.0, 42.9, 44.9, 47.1, 51.3, 70.7, 71.3, 75.1, 160.4, 160.9, 173.6 (Bhattarai *et al.*, 1997).

Methyl 7α,12α-bis(formyloxy)-3α-hydroxy-5β-cholan-24-oate (**10**) was oxidized to methyl 7α,12α-bis(formyloxy)- 3-oxo-5β-cholan-24-oate (**12**) (*Scheme 3.1.2*). SiO_2 (7.0 g) was dried overnight at 110 °C, added to the solution of PCC (5.6 g, 26 mmol) in DCM (50 mL) under N_2, and a solution of **10** (6.81 g, 14.2 mmol) in DCM (50 mL) was added dropwise. After 2 h, the mixture was filtered and the filtrate was rotary evaporated to obtain a greenish yellow solid, which was dissolved in Et_2O, washed with H_2O followed by brine, dried ($MgSO_4$), and filtered through a short pad of SiO_2. The solvent was evaporated to dryness and the crude solid was crystallized from *n*-hexane-DCM to give **12** (5.93 g, 87%). IR ($CHCl_3$): ν_{max} cm^{-1} 2965m (C–H), 2880m (C–H), 1716s (C=O), 1180m (C–O). ^1H NMR (300 MHz, $CDCl_3$): δ 8.16 (s, 1H, 12α-OCOH), 8.10 (s, 1H, 7α-OCOH), 5.31 (br s, 1H, 12β-CH), 5.16 (br s, 1H, 7β-CH), 3.66 (s, 3H, 24-OMe), 3.01 (t, 1H, $J = 15.0$ Hz), 1.04 (s, 3H, 19-Me), 0.85 (d, 3H, $J = 6.4$ Hz, 21-Me), 0.79 (s, 3H, 18-Me). ^{13}C NMR (75 MHz, $CDCl_3$): δ 12.1, 17.4, 21.5, 22.7, 25.8, 27.0, 29.4, 30.5, 30.8, 30.9, 33.3, 34.6, 36.1, 36.4, 37.6, 42.1, 42.8, 44.4, 44.9, 47.1, 51.4, 70.5, 75.0, 160.3, 160.3, 174.3, 211.5 (Bhattarai *et al.*, 1997).

Similarly, oxidation of methyl 7α,12α-bis(formyloxy)-3-oxo-24-nor-5β-cholan-23-oate (**13**) afforded methyl 7α,12α-bis(formyloxy)-3α-hydroxy-24-nor-5β-cholan-23-oate (**11**), following the procedure as described for **10** (*Scheme 3.1.2*). A stirred solution of **11** (4.95 g, 10.65 mmol) in DCM (80 mL) was treated with SiO_2 (5.0 g) and PCC (3.8 g, 18.6 mmol) to produce **13** (4.25 g, 86%). IR ($CHCl_3$): ν_{max} cm^{-1} 2963m (C–H), 2878m (C–H), 1715s (C=O), 1178m (C–O), 730m. ^1H NMR (300 MHz, $CDCl_3$): δ 8.16 (s, 1H, CH), 8.09 (s, 1H, CH), 5.30 (br s, 1H, 12β-CH), 5.15 (br d, 1H, 7β-CH), 3.65 (s, 3H, 23-OMe), 3.01 (t, 1H, $J = 12.0$ Hz), 1.04 (s, 3H, 19-Me), 0.91 (d, 3H, $J = 6.2$ Hz, 21-Me), 0.83 (s, 3H, 18-Me). ^{13}C NMR (75 MHz, $CDCl_3$): δ 12.1, 18.6, 21.4, 22.7, 25.7, 27.2, 29.4, 30.9, 33.0, 34.3, 36.0, 36.4, 37.6, 40.9, 42.0, 42.8, 44.4, 45.0, 47.1, 51.3, 70.4, 74.9, 160.2, 173.5, 211.5 (Bhattarai *et al.*, 1997).

Compound **14** was prepared from methyl 7α,12α-bis(formyloxy)-3-oxo-5β-cholan-24-oate (**12**) (*Scheme 3.1.3*). A stirred solution of **12** (4.0 g, 8.39 mmol) was reacted with dibutyl 2,2-dibromomalonate (4.4 g, 11.75 mmol) and Bu_3Sb (8.8 g, 7.7 mL, 29.4 mmol) in dry THF (30 mL) under N_2 at room temperature. The reaction temperature was increased to 60–65 °C (oil bath) and stirring was continued for 72 h, after which, the reaction mixture was cooled and stirred in the open air for another 2 h. The solvent was evaporated and the crude product was dissolved in a minimum amount of 20% Et_2O in *n*-hexane and purified by gradient CC (mobile phase: 100% *n*-hexane, 10% Et_2O in *n*-hexane, 20% Et_2O in *n*-hexane and 30% Et_2O in *n*-hexane) to accomplish methyl 3-[bis(butoxycarbonyl)methylidene]-7α,12α-bis(formyloxy)-5β-cholan-24-oate (**14**, 5.25 g, 93%). IR ($CHCl_3$): ν_{max} cm^{-1} 2962m (C–H), 2877m (C–H), 1720s (C=O), 1635s (C=C), 1273m (C–O), 1238m (C–O), 1201m (C–O), 1175m (C–O). ^1H NMR (300 MHz, $CDCl_3$): δ 8.13 (s, 1H, 12α-OCOH), 8.09 (s, 1H, 7α-OCOH), 5.28 (br s, 1H, 12β-CH), 5.11 (br s, 1H, 7β-CH), 4.15 (t, 4H, $J = 7.0$ Hz, $2 \times OCH_2$), 3.65 (s, 3H, 24-OMe), 0.95 (s, 3H, 19-Me), 0.84 (d, 3H, $J = 6.0$ Hz, 21-Me), 0.77 (s, 3H, 18-Me), ^{13}C NMR (75 MHz, $CDCl_3$): δ 12.0, 13.5, 17.3, 18.9, 21.6, 22.6, 25.6, 27.0, 29.1, 30.3, 30.5, 30.7, 31.2, 34.6, 34.6, 34.7, 37.2, 37.5, 42.7, 43.1, 44.8, 47.0, 51.3, 64.5, 64.6, 70.4, 75.0, 121.5, 160.3, 160.8, 165.4, 165.5, 174.2 (Davis and Bhattarai, 1995).

Methyl 3-[bis(butoxycarbonyl)methylidene]-7α,12α-bis(formyloxy)-24-nor-5β-cholan-23-oate) (**15**) was synthesized from **13**, using the method as outlined for **14** (*Scheme 3.1.3*).

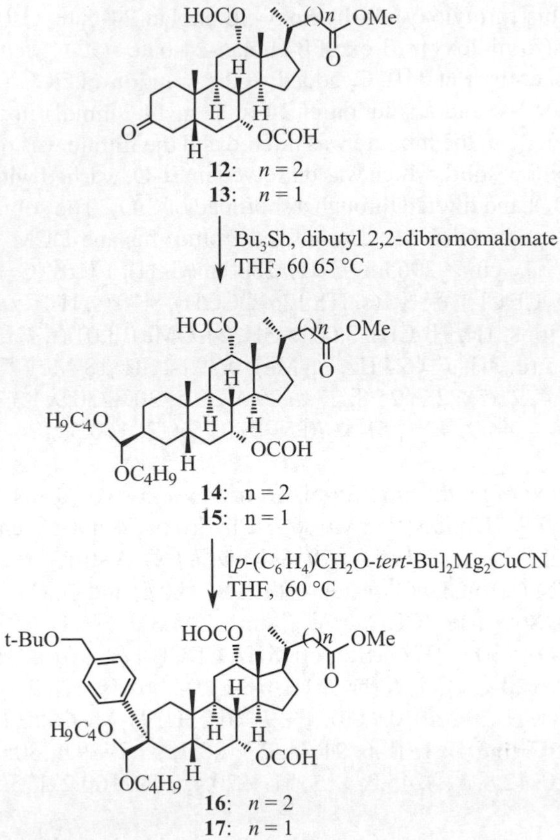

HOCO OMe O H H H O H OCOH

12: $n = 2$
13: $n = 1$

Bu_3Sb, dibutyl 2,2-dibromomalonate
THF, 60 65 °C

HOCO OMe O H H_9C_4O H H OCOH OC_4H_9 H

14: n = 2
15: n = 1

$[p\text{-}(C_6H_4)CH_2O\text{-}tert\text{-}Bu]_2Mg_2CuCN$
THF, 60 °C

t-BuO HOCO OMe O H H_9C_4O H H OCOH OC_4H_9 H

16: $n = 2$
17: $n = 1$

Scheme 3.1.3 *Synthesis of cholic acid and norcholic acid derivatives **14–17***

To a stirred solution of **13** (4.0 g, 8.65 mmol) and dibutyl 2,2-dibromomalonate (4.5 g, 12.04 mmol) in dry THF (30 mL), Bu_3Sb (8.82 g, 8 mL, 30.1 mmol) was injected slowly at room temperature under N_2. Usual workup and purification as described above provided **15** (5.50 g, 96%). IR ($CHCl_3$): ν_{max} cm^{-1} 2960m (C–H), 2878m (C–H), 1719s (C=O), 1634s (C=C), 1204m (C–O), 1174m (C–O), 1064m. ^1H NMR (300 MHz, $CDCl_3$): δ 8.14 (s, 1H, 12α-OCOH), 8.09 (s, 1H, 7α-OCOH), 5.28 (br s, 1H, 12β-CH), 5.11 (br s, 1H, 7β-CH), 4.15 (t, 4H, $J = 5.5$ Hz, $2 \times OCH_2$), 3.65 (s, 3H, 23-OMe), 1.04 (s, 3H, 19-Me), 0.85 (d, 3H, $J = 6.4$ Hz, 21-Me), 0.81 (s, 3H, 18-Me). ^{13}C NMR (75 MHz, $CDCl_3$): δ 12.0, 13.5, 18.6, 18.9, 21.7, 22.6, 25.6, 27.0, 27.1, 29.1, 30.3, 31.2, 33.0, 34.6, 34.7, 37.2, 37.5, 40.9, 42.8, 43.1, 44.9, 47.0, 51.3, 64.6, 64.7, 70.4, 74.9, 121.5, 160.3, 160.8, 165.4, 165.7, 173.4 (Bhattarai *et al.*, 1997).

Grignard reaction of methyl 3β-[bis(butoxycarbonyl)methylidene]-7α,12α-bis(formyloxy)-5β-cholan-24-oate (**14**) provided **16** (*Scheme 3.1.3*). In a flame-dried reaction condition, to a stirred solution of *tert*-butyl *p*-bromobenzyl ether (5.0 g, 20.56 mmol) in dry THF (30 mL), Mg turnings (700 mg, 28.5 mmol) was added under Ar, and the reaction was initiated by warming and sonication. After 60 min, the solution was transferred *via* cannula to another flame dried round bottom flask. The Mg-free Grignard reagent was cooled to −10 °C (ice-salt bath), and CuCN (895 mg, 10 mmol) was added under strong flow of Ar. Stirring was continued, and the resulting homogeneous solution was cooled to −60 °C. A solution of **14** (5.0 g, 7.4 mmol) in THF (20 mL) was added dropwise, and the reaction temperature was allowed to attain room temperature. Saturated aqueous NH_4Cl was added and the mixture was extracted with Et_2O. The ethereal extract was washed sequentially with H_2O and brine, dried ($MgSO_4$), and evaporated to produce a gummy residue, which was purified by FCC (mobile phase: 20% Et_2O in *n*-hexane followed by 30% Et_2O in *n*-hexane) to afford methyl 3β-[bis (butoxycarbonyl) methyl]-3α-[(*p*-*tert*-butoxymethyl)phenyl]-7α,12α-bis(formyl-oxy)-5β-cholan-24-oate (**16**, 5.39 g, 87%). IR ($CHCl_3$): ν_{max} cm^{-1} 2966m (C–H), 2876m (C–H), 1722s (C=O), 1618s (C=C), 1467m (C–H), 1439m (C–H), 1389m (C–H), 1364m (C–H), 1243m (C–O), 1178m (C–O), 1063m, 1020m. ^1H NMR (300 MHz, $CDCl_3$): δ 8.05 (s, 1H, 12α-OCOH), 8.01 (s, 1H, 7α-OCOH), 7.34 and 7.26 (q, 4H, $J = 8.6$ Hz, 4 × Ph-CH), 5.23 (br s, 1H, 12β-CH), 5.05 (br s, 1H, 7β-CH), 4.38 (s, 2H, OCH_2), 4.06 (s, 1H, OCH), 4.03-3.86 (m, 4H, 2 × OCH_2), 3.65 (s, 3H, 24-OMe), 1.28 (s, 9H), 1.03 (s, 3H, 19-Me), 0.89-0.80 (m, 9H, 21-Me and other CH_3), 0.73 (s, 3H, 18-Me). ^{13}C NMR (75 MHz, $CDCl_3$): δ 12.0, 13.5, 17.3, 18.9, 22.7, 22.8, 25.4, 27.1, 27.6, 28.5, 29.1, 30.2, 30.5, 30.8, 31.5, 32.1, 34.2, 34.6, 37.6, 37.9, 38.0, 42.5, 42.6, 44.8, 47.0, 51.4, 54.6, 63.8, 64.7, 70.6, 73.2, 74.9, 126.7, 127.1, 137.6, 144.5, 160.5, 160.6, 168.1, 174.4 (Bhattarai *et al.*, 1997).

Following the same procedure as described for **16**, the synthesis of methyl 3β-[bis(butoxycarbonyl)methyl]-3α-[(*p*-*tert*-butoxymethyl)phenyl]-7α,12α-bis(formyloxy)-24-nor-5β-cholan-23-oate (**17**, 5.7 g, 91%) was carried out using methyl 3β-[bis (butoxycarbonyl)methylidene]-7α,12α-bis(formyloxy)-24-nor-5β-cholan-23-oate) (**15**, 5.0 g, 7.6 mmol) as the starting material (*Scheme 3.1.3*). IR ($CHCl_3$): ν_{max} cm^{-1} 2966m (C–H), 2876m (C–H), 1722s (C=O), 1174m (C–O). ^1H NMR (300 MHz, $CDCl_3$): δ 8.05 (s, 1H, 12α-OCOH), 8.00 (s, 1H, 7α-OCOH), 7.35 and 7.26 (q, 4H, $J = 8.5$ Hz, 4 × Ph-CH), 5.23 (br s, 1H, 12β-CH), 5.05 (br d, 1H, 7β-CH), 4.38 (s, 2H, OCH_2), 4.05 (s, 1H, OCH), 4.03-3.86 (m, 4H, 2 × OCH_2), 3.64 (s, 3H, 24-OMe), 1.28 (s, 9H), 1.02 (s, 3H, 19-Me), 0.89-0.82 (m, 9H, 21-Me and other CH_3), 0.77 (s, 3H, 18-Me). ^{13}C NMR (75 MHz, $CDCl_3$): δ 12.0, 13.6, 18.6, 18.9, 19.0, 22.7, 22.8, 25.4, 27.2, 27.6, 30.3, 31.5, 32.1, 33.1, 34.3, 37.6, 37.9, 38.0, 41.0, 42.5, 42.7, 44.9, 47.2, 51.4, 54.6, 63.8, 66.8, 70.6, 73.2, 74.8, 126.7, 127.2, 137.6, 144.5, 160.6, 168.1, 168.2, 173.7 (Bhattarai *et al.*, 1997).

Methyl 3β-[bis(butoxycarbonyl)methyl]-3α-[(*p*-hydroxymethyl)phenyl]-7α,12α-bis (formyl oxy)-5β-cholan-24-oate (**18**) was prepared from methyl 3β-[bis(butoxycarbonyl)methyl]-3α-[(*p*-*tert*-butoxymethyl)phenyl]-7α,12α-bis(formyloxy)-5β-cholan-24-oate (**16**) (*Scheme 3.1.4*). TFA (5 mL, 65 mmol) was added to a stirred solution of **16** (5.08 g, 6.04 mmol) in dry DCM (50 mL) under N_2 and the mixture was heated at 50 °C for 5 h. The reaction mixture was poured into a biphasic mixture of Et_2O (400 mL) and aqueous NH_3 (100 mL, 4M) under vigorous stirring. After 5 h, the hydrolysis was complete, the phases were separated, and the aqueous layer was extracted with Et_2O (2 × 20 mL). The pooled extracts were washed with brine until neutralized, and dried ($MgSO_4$). The solvent was evaporated to dryness and the crude residue was purified by FCC (mobile phase: 20% *n*-hexane in Et_2O) to give **18** (3.87 g, 82%). IR ($CHCl_3$): ν_{max} cm^{-1} 3528br (O–H), 2962m (C–H), 2879m (C–H), 1721s (C–O), 1174m (C–O), 1062m (C–O), 1018m. ^1H NMR (300 MHz, $CDCl_3$): δ 8.04 (s, 1H, 12α-OCOH), 8.02 (s, 1H, 7α-OCOH), 7.40 and 7.31 (q, 4H, $J = 8.4$ Hz, 4 × Ph-CH), 5.24 (br s, 1H, 12β-CH), 5.06 (br s, 1H, 7β-CH), 4.67 (s, 2H,

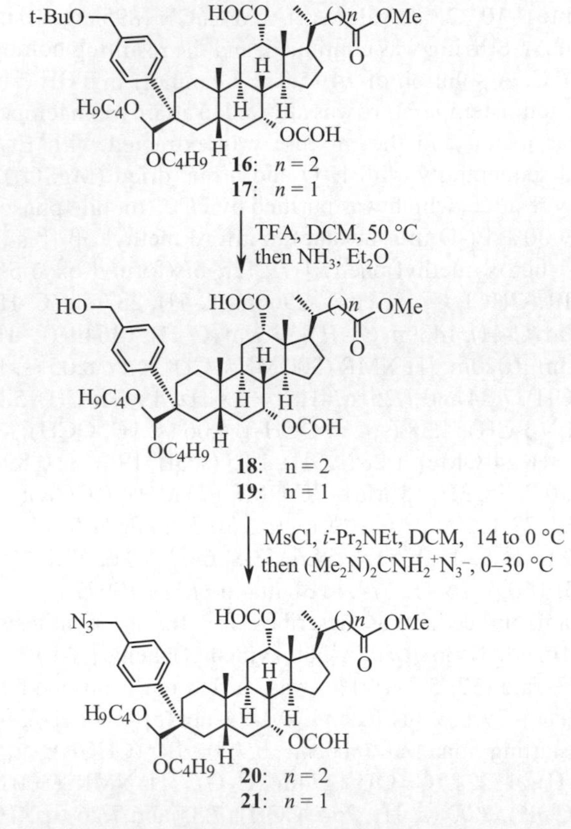

16: $n = 2$
17: $n = 1$

TFA, DCM, 50 °C
then NH_3, Et_2O

18: n = 2
19: n = 1

MsCl, *i*-Pr_2NEt, DCM, 14 to 0 °C
then $(Me_2N)_2CNH_2^+N_3^-$, 0–30 °C

20: $n = 2$
21: $n = 1$

Scheme 3.1.4 *Synthesis of cholic acid and norcholic acid derivatives* ***18–21***

OCH_2), 4.08 (s, 1H), 4.04-3.87 (m, 4H, 2 × OCH_2), 3.64 (s, 3H, 24-OMe), 1.04 (s, 3H, 19-Me), 0.89-0.80 (m, 9H, 21-Me and other CH_3), 0.73 (s, 3H, 18-Me). ^{13}C NMR (75 MHz, $CDCl_3$): δ 12.1, 13.6, 17.4, 19.0, 22.7, 22.9, 25.6, 27.1, 28.6, 29.3, 30.3, 30.6, 30.8, 31.6, 32.0, 34.3, 34.7, 37.7, 37.9, 38.1, 42.6, 42.8, 44.9, 47.1, 51.5, 53.8, 64.8, 65.0, 70.7, 75.1, 126.1, 127.4, 139.0, 145.1, 160.6, 168.0, 168.1, 174.4 (Bhattarai *et al.*, 1997).

The synthesis of methyl 3β-[bis(butoxycarbonyl)methyl]-3α-[(*p*-hydroxymethyl)phenyl]-7α,12α-bis(formyloxy)-24-nor-5β-cholan-23-oate (**19**, 4.8 g, 89%, mp: 120–123 °C) was accomplished from methyl 3β-[bis(butoxycarbonyl)methyl]-3α-[(*p-tert*-butoxymethyl) phenyl]-7α,12α-bis (formyloxy)-24-nor-5β-cholan-23-oate (**17**, 5.76 g, 6.98 mmol), according to the method described for **18** (*Scheme 3.1.4*). IR ($CHCl_3$): ν_{max} cm^{-1} 3548br (O–H), 1721s (C=O), 1177m (C–O). ^{1}H NMR (300 MHz, $CDCl_3$): δ 8.05 (s, 1H, 12α-OCOH), 8.03 (s, 1H, 7α-OCOH), 7.46 and 7.31 (q, 4H, $J = 8.6$ Hz, 4 × Ph-CH), 5.23 (br s, 1H, 12β-CH), 5.06 (br s, 1H, 7β-CH), 4.67 (s, 2H, OCH_2), 4.08 (s, 1H), 4.04–3.87 (m, 4H, 2 × OCH_2), 3.63 (s, 3H, 23-OMe), 1.04 (s, 3H, 19-Me), 0.89–0.82 (m, 9H, 21-Me and other CH_3), 0.78 (s, 3H, 18-Me). ^{13}C NMR (75 MHz, $CDCl_3$): δ 12.1, 13.6, 18.6, 18.9, 19.0, 22.7, 22.9, 25.5, 27.2, 28.6, 29.3, 30.3, 31.5, 32.0, 33.1, 34.3, 37.7, 37.9, 38.1, 41.0, 42.6, 42.9, 45.0, 47.2, 51.4, 54.4, 64.8, 64.9, 70.6, 74.9, 126.0, 127.4, 139.0, 145.1, 160.5, 168.0, 168.1, 173.6 (Bhattarai *et al.*, 1997).

In the next step, methyl 3β-[bis(butoxycarbonyl)methyl]-3α-[(*p*-hydroxymethyl)phenyl]-7α,12α-bis(formyloxy)-5β-cholan-24-oate (**18**) was transformed to methyl 3β-[bis (butoxycarbonyl)methyl]-3α-[(*p*-azidomethyl)phenyl]-7α,12α-bis(formyloxy)-5β-cholan-24-oate (**20**) (*Scheme 3.1.4*). DIEA (950 μL, 5.47 mmol) and MsCl (420 μL, 5.32 mmol) were added dropwise to a stirred cooled (ice-salt bath) solution of **18** (3.28 g, 4.19 mmol) in DCM (20 mL) under N_2. The mixture was allowed to warm to 0 °C. After 60 min, TMGA (1.29 g, 8.15 mmol) was added with stirring. The mixture was allowed to reach room temperature, and then warmed to 30 °C for 30 min. The mixture was diluted with Et_2O (40 mL) and filtered through a short pad of SiO_2, which was then washed with Et_2O. The combined filtrate and washings were washed with H_2O followed by brine, dried ($MgSO_4$), and evaporated to dryness to produce crude product (3.8 g). A portion of the solid (50 mg) was purified as above to obtain **20** (33 mg). IR ($CHCl_3$): ν_{max} cm^{-1} 2963m (C–H), 2879m (C–H), 2110m (N_3), 1722s (C=O), 1468m (N–H), 1243m (C–O), 1176m (C–O). ^1H NMR (300 MHz, $CDCl_3$): δ 8.06 (s, 1H, 12α-OCOH), 8.03 (s, 1H, 7α-OCOH), 7.42 and 7.25 (q, 4H, $J = 8.5$ Hz, 4 × Ph-CH), 5.24 (br s, 1H, 12β-CH), 5.06 (br s, 1H, 7β-CH), 4.33 (s, 2H, N_3CH_2), 4.07 (s, 1H), 4.03-3.84 (m, 4H, 2 × OCH_2), 3.64 (s, 3H, 24-OMe), 1.04 (s, 3H, 19-Me), 0.89-0.79 (m, 9H, 21-Me and other CH_3), 0.74 (s, 3H, 18-Me). ^{13}C NMR (75 MHz, $CDCl_3$): δ 12.0, 13.5, 17.3, 18.9, 22.7, 22.9, 25.6, 27.1, 28.6, 29.3, 30.2, 30.6, 30.8, 31.5, 32.0, 34.2, 34.7, 37.6, 37.8, 38.1, 42.7, 42.9, 44.9, 47.1, 51.4, 54.4, 64.8, 70.8, 75.0, 127.1, 127.7, 133.6, 145.8, 160.5, 168.0, 168.1, 174.4 (Bhattarai *et al.*, 1997).

The synthesis of methyl 3β-[bis(butoxycarbonyl)methyl]-3α-[*p*-(azidomethyl)phenyl]-7α,12α-bis(formyloxy)-24-nor-5β-cholan-23-oate (**21**) could be carried out using **19** as the starting material (*Scheme 3.1.4*), following the procedure as described for **20**. A stirred solution of **19** (3.57 g, 4.65 mmol) in DCM was reacted as above with MsCl (470 μL, 6.04 mmol) and DIEA (1.1 mL, 6.3 mmol) followed by TGMA (1.43 g, 9.06 mmol), to provide crude residue (3.96 g), which was purified by FCC (mobile phase: 40% EtOAc in *n*-hexane) to yield **21** (60 mg, mp 130–133 °C). IR ($CHCl_3$): ν_{max} cm^{-1} 2964m (C–H), 2880m (C–H), 2101m (N_3), 1723s (C=O), 1178m (C–O). ^1H NMR (300 MHz, $CDCl_3$): δ 8.06 (s, 1H, 12α-OCOH), 8.03 (s, 1H, 7α-OCOH), 7.43 and 7.25 (q, 4H, $J = 8.5$ Hz, 4 × Ph-CH), 5.24 (br t, 1H, 12β-CH), 5.07 (br d, 1H, 7β-CH), 4.33 (s, 2H, N_3CH_2), 4.08 (s, 1H), 4.07-3.84 (m, 4H, 2 × OCH_2), 3.63 (s, 3H, 23-OMe), 1.04 (s, 3H, 19-Me), 0.89–0.82 (m, 9H, 21-Me and other CH_3), 0.78 (s, 3H, 18-Me). ^{13}C NMR (75 MHz, $CDCl_3$): δ 12.0, 13.5, 18.6, 18.9, 22.7, 22.9, 25.5, 27.2, 28.6, 29.3, 30.2, 31.5, 32.0, 33.1, 34.2, 37.6, 37.8, 38.1, 41.0, 42.6, 42.9, 44.9, 47.2, 51.3, 54.4, 64.8, 70.5, 74.9, 127.1, 127.7, 133.6, 145.7, 160.5, 167.9, 168.1, 173.5 (Bhattarai *et al.*, 1997).

In the next reaction, methyl 3β-[bis(butoxycarbonyl)methyl]-3α-[(*p*-azidomethyl)phenyl]-7α,12α-bis(formyl oxy)-5β-cholan-24-oate (**20**) was employed to synthesize **22** (*Scheme 3.1.5*). Ph_3P (2.0 g, 7.6 mmol) and H_2O (3 mL) were added to a solution of **20** (3.75 g) in THF (20 mL), and the mixture was heated to 60 °C for 60 min, then $(Boc)_2O$ (1.47 g, 6.5 mmol) in THF (5 mL) was added, and the reaction mixture was cooled to room temperature, sonicated for 30 min, and stirring was continued overnight. The solvent was evaporated and the residue was dissolved in Et_2O (150 mL), washed with H_2O followed by brine, and dried ($MgSO_4$). The solvent was evaporated under pressure, the solid was dissolved in a minimum volume of Et_2O and cooled (ice-salt bath). Insoluble material was filtrated off and the filtrate was evaporated to dryness. The crude product was purified by

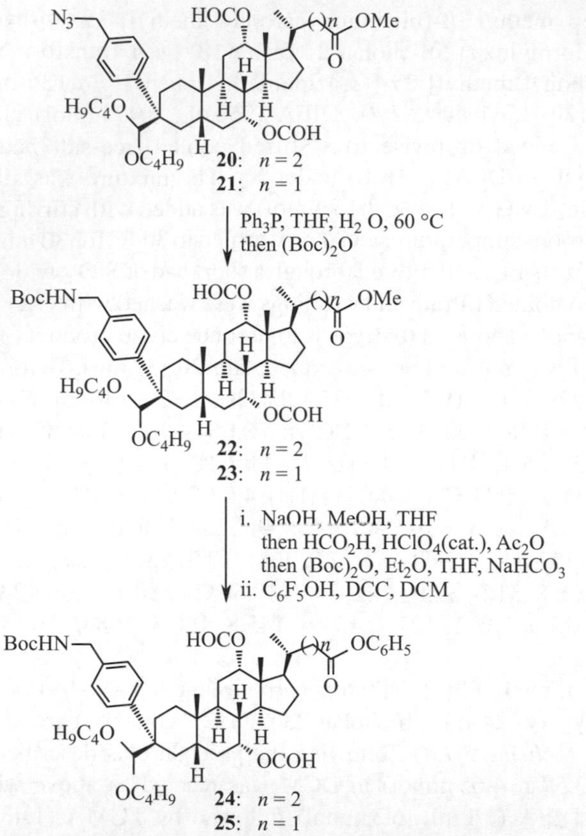

20: $n = 2$
21: $n = 1$

Ph_3P, THF, H_2O, 60 °C
then $(Boc)_2O$

22: $n = 2$
23: $n = 1$

i. NaOH, MeOH, THF
then HCO_2H, $HClO_4$(cat.), Ac_2O
then $(Boc)_2O$, Et_2O, THF, $NaHCO_3$
ii. C_6F_5OH, DCC, DCM

24: $n = 2$
25: $n = 1$

Scheme 3.1.5 *Synthesis of cholic acid and norcholic acid derivatives* ***22–25***

gradient FCC, eluted with *n*-hexane:EtOAc = 9:1 and 2:5, to produce methyl 3β-[bis (butoxycarbonyl)methyl]-3α-[([*p*-(tertbutoxycarbonyl)amino]methyl)phenyl]-7α,12α-bis (formyl oxy)-5β-cholan-24-oate (**22**, 3.21 g, 86%). IR ($CHCl_3$): ν_{max} cm^{-1} 3411m (N–H), 2963m (C–H), 2878m (C–H), 1722s (C=O), 1515m (N–H), 1367m (C–H), 1246m (C–H), 1174m (C–O). ^1H NMR (300 MHz, $CDCl_3$): δ 8.05 (s, 1H, 12α-OCOH), 8.02 (s, 1H, 7α-OCOH), 7.36 and 7.22 (q, 4H, $J = 8.5$ Hz, 4 × Ph-CH), 5.23 (br s. 1H, 12β-CH), 5.06 (br s, 1H, 7β-CH), 4.89 (br t, 1H), 4.29 (d, 2H, $J = 5.8$ Hz, $NHCH_2$), 4.07 (s, 1H, CH), 4.04–3.86 (m, 4H, 2 × OCH_2), 3.64 (s, 3H, 24-OMe), 1.46 (s, 9H), 1.04 (s, 3H, 19-Me), 0.89–0.79 (m, 9H, 21-Me and other CH_3), 0.74 (s, 3H, 18-Me). ^{13}C NMR (300 MHz, $CDCl_3$): δ 12.0, 13.5, 17.2, 18.9, 22.6, 22.9, 25.5, 27.0, 28.3, 28.5, 29.3, 30.2, 30.5, 30.7, 31.5, 32.0, 34.2, 34.6, 37.6, 37.8, 38.0, 42.5, 42.8, 44.1, 44.8, 47.0, 51.4, 54.3, 64.7, 70.6, 75.0, 79.2, 126.3, 127.4, 136.9, 144.6, 155.7, 160.5, 167.9, 168.0, 174.3 (Bhattarai *et al.*, 1997).

The synthesis of methyl 3β-[bis(butoxycarbonyl)methyl]-3α-[([*p*-(*tert*-butoxycarbonyl) amino]methyl)phenyl]-7α,12α-bis(formyloxy)-24-nor-5β-cholan-23-oate (**23**, 3.41 g, 84%) was performed from **21** (3.9 g), using the method as outlined for **22** (*Scheme 3.1.5*). IR ($CHCl_3$): ν_{max} cm^{-1} 3410m (N–H), 2965m (C–H), 2879m (C–H), 1723s (C=O), 1173m (C–O), 733m. ^1H NMR (300 MHz, $CDCl_3$): δ 8.05 (s, 1H, 12α-OCOH), 8.02 (s, 1H,

7α-OCOH), 7.36 and 7.22 (q, 4H, $J = 8.5$ Hz, 4 × Ph-CH), 5.23 (br s, 1H, 12β-CH), 5.06 (br s, 1H, 7β-CH), 4.94 (br t, 1H), 4.29 (d, 2H, $J = 5.8$ Hz, $NHCH_2$), 4.07 (s, 1H, CH), 4.04–3.86 (m, 4H, 2 × OCH_2), 1.04 (s, 3H, 19-Me), 0.89–0.82 (m, 9H, 21-Me and other CH_3), 0.78 (s, 3H, 18-Me). ^{13}C NMR (75 MHz, $CDCl_3$): δ 11.9, 13.4, 18.5, 18.8, 22.5, 22.8, 25.4, 27.1, 28.2, 28.5, 29.2, 30.1, 31.4, 31.9, 32.9, 34.1, 37.5, 37.7, 37.9, 40.9, 42.5, 42.7, 44.0, 44.8, 47.0, 51.2, 54.3, 64.6, 70.4, 74.8, 79.1, 126.2, 127.3, 136.9, 144.5, 155.7, 160.3, 167.8, 167.9, 173.4 (Bhattarai *et al.*, 1997).

Pentafluorophenyl 3β-[bis(butoxycarbonyl)methyl]-3α-[([*p*-(*tert*-butoxycarbonyl) amino]methyl) phenyl]-7α,12α-bis(formyloxy)-5β-cholan-24-oate (**24**) was synthesized from **22** (*Scheme 3.1.5*). To a stirred solution of **22** (500 mg, 0.57 mmol) in THF-MeOH (12 mL, 2:1), 1.5M NaOH (3 mL) was added and the mixture was stirred at room temperature. After 24 h, $NaHSO_4$ (400 mg, 3.3 mmol) was added, the organic layer was separated and evaporated to dryness. The residue was dissolved in Et_2O, washed with H_2O followed by brine, and dried (Na_2SO_4). The ethereal layer was rotary evaporated and the solid dihydroxy acid was dissolved in HCO_2H (5 mL), and a catalytic amount of 70% $HClO_4$ was added and heated at 50 °C (water bath) for 24 h. The mixture was cooled to 40 °C, Ac_2O (2 drops) was added, and the mixture was stirred for another 60 min. The mixture was evaporated under vacuum at room temperature. The solid was dissolved in THF-Et_2O (10 mL) and neutralized with a slight excess of 1% aqueous $NaHCO_3$. A solution of $(Boc)_2O$ (140 mg, 0.64 mmol) in Et_2O (2 mL) was added and stirring was continued overnight. The solution was diluted with Et_2O, dried ($MgSO_4$) and evaporated to dryness. The solid was purified by gradient FCC (mobile phase: 1% MeOH in $CHCl_3$ and 2% MeOH in $CHCl_3$) to furnish carboxylic acid (361 mg), which was dissolved in DCM (3 mL) with C_6F_5OH (84 mg, 0.456 mmol) and cooled to −10 °C under N_2. DCC (94 mg, 0.50 mmol) in DCM (2 mL) was added dropwise. The mixture was stirred overnight, after which, a drop of distilled H_2O was added and stirring was continued under an open air for 60 min. The mixture was cooled (−10 °C), precipitate was formed, which was filtered off, and the insoluble material was washed with cooled (−10 °C) *n*-hexane-Et_2O mixture, the filtrate was evaporated to dryness and the crude product was purified by FCC, eluted with *n*-hexane-Et_2O, to give **24** (327 mg, 56%). IR ($CHCl_3$): ν_{max} cm^{-1} 3414m (N–H), 2964m (C–H), 2879m (C–H), 1791s (C=O), 1755s (C=O), 1722s (C=O), 1519m (N–H), 1469m (C–H), 1174m (C–O). ^1H NMR (300 MHz, $CDCl_3$): δ 8.07 (s, 1H, 12α-OCOH), 8.03 (s, 1H, 7α-OCOH), 7.37 and 7.22 (q, 4H, $J = 8.4$ Hz, 4 × Ph-CH), 5.26 (br s, 1H, 12β-CH), 5.07 (br s, 1H, 7β-CH), 4.87 (br t, 1H), 4.30 (d, 2H, $J = 5.9$ Hz, $NHCH_2$), 4.04-3.86 (m, 4H, 2 × OCH_2), 1.46 (s, 9H), 1.05 (s, 3H, 19-Me), 0.89–0.83 (m, 9H), 0.77 (s, 3H, 18-Me) (Bhattarai *et al.*, 1997).

Similarly, the synthesis of pentafluorophenyl 3β-[bis(butoxycarbonyl)methyl]-3α-[([*p*-(*tert*-butoxy carbonyl)amino]methyl)phenyl]-7α,12α-bis(formyloxy)-24-nor-5β-cholan-23-oate (**25**) was accomplished from **23** (400 mg, 0.46 mmol), which was first converted to corresponding carboxylic acid (260 mg), according to the method described for **24** (*Scheme 3.1.5*). The carboxylic acid was treated as above with C_6F_5OH (62 mg, 0.33 mmol) and DCC (70 mg, 0.33 mmol) to yield **25** (262 mg, 57%). IR ($CHCl_3$): ν_{max} cm^{-1} 3414m (N–H), 2964m (C–H), 2880m (C–H), 2455m, 1789s (C=O), 1757s (C–O), 1722s (C=O), 1520m (N–H), 1176m (C–O). ^1H NMR (300 MHz, $CDCl_3$): δ 8.07 (s, 1H, 12α-OCOH), 8.03 (s, 1H, 7α-OCOH), 7.37 and 7.22 (q, 4H, $J = 8.3$ Hz, 4 × Ph-CH), 5.26 (br s, 1H, 12β-CH), 5.08 (br s, 1H, 7β-CH), 4.83 (br t, 1H), 4.30 (d, 2H, $NHCH_2$), 4.07 (s, 1H), 4.04–3.86

(m, 4H, 2 × OCH_2), 1.05 (s, 3H, 19-Me), 1.00 (d, 3H, J = 6.4 Hz, 21-Me), 0.90–0.83 (m, 6H), 0.81 (s, 3H, 18-Me). ^{13}C NMR (300 MHz, $CDCl_3$): δ 12.0, 13.5, 18.3, 18.9, 22.6, 22.9, 25.5, 27.3, 28.3, 28.8, 29.3, 30.2, 31.5, 32.0, 33.3, 34.2, 37.6, 37.8, 38.0, 40.1, 42.6, 42.9, 44.1, 45.0, 46.8, 54.3, 64.8, 70.5, 74.7, 79.4, 126.4, 127.4, 136.9, 144.7, 155.9, 160.5, 168.0, 168.1, 168.7 (Bhattarai *et al.*, 1997).

Pentafluorophenyl 3β-[bis(butoxycarbonyl)methyl]-3α-[([*p*-(*tert*-butoxycarbonyl) amino]methyl) phenyl]-7α,12α-bis(formyloxy)-5β-cholan-24-oate (**24**) was utilized for the synthesis of tetrakis(formyloxy)cholaphane **26** (*Scheme 3.1.6*). To a stirred cooled (0 °C) solution of **24** (327 mg, 0.31 mmol) in dry DCM (2 mL), TFA (1 mL) was added under Ar. The mixture was allowed to warm gradually to room temperature. After 4 h, the mixture was reduced under pressure, the residue was dissolved in CCl_4 (5 mL) and stirred at room temperature. The solvent was rotary evaporated and this method was repeated twice to provide the corresponding ammonium salt, which was dissolved in dry DCM (50 mL), DMAP (580 mg) in dry DCM (300 mL) was added at room temperature under Ar, and stirring was continued for 3 days. The solvent was removed under vacuum and the crude solid was purified by FCC, eluted with 30% EtOAc in $CHCl_3$, to afford macrocycle **26** (194 mg, 77%). IR ($CHCl_3$): ν_{max} cm^{-1} 3405m (N–H), 2961m (C–H), 2877m (C–H), 1724s (C=O), 1662s (C=C), 1519m (N–H), 1467m, 1178m (C–O). ^{1}H NMR (300 MHz, $CDCl_3$): δ 8.01 (s, 2H, 2 × 12α-OCOH), 7.99 (s, 2H, 2 × 7α-OCOH), 7.37 and 7.23 (q, 8H, J = 8.5 Hz, 8 × Ph-CH), 5.60 (t, 2H, J = 5.3 Hz, 2 × 12β-CH), 5.18 (br s, 2H, 2 × 7β-CH), 5.04 (br s, 2H), 4.40 and 4.26 (dq, 4H, J = 5.0, 5.5, 13.7 Hz, 2 × $NHCH_2$), 4.05 (s, 2H), 4.01-3.83 (m, 8H, 4 × OCH_2), 1.03 (s, 6H, 2 × 19-Me), 0.87–0.80 (m, 18H), 0.74 (s, 6H, 2 × 18-Me). ^{13}C NMR (75 MHz, $CDCl_3$): δ 12.0, 13.6, 17.5, 18.9, 22.7, 22.9, 25.7, 27.3, 28.5, 29.3, 30.3, 31.4, 31.6, 31.9, 32.0, 34.3, 34.3, 37.6, 37.7, 38.0, 42.6, 43.0, 43.6, 44.8, 45.9, 54.3, 64.9, 70.8, 75.5, 127.6, 127.8, 136.3, 145.2, 160.4, 167.9, 168.0, 172.8 (Bhattarai *et al.*, 1997).

In a similar fashion, another tetrakis(formyloxy)cholaphane **27** was synthesized from **25** *via* deprotection with TFA providing corresponding ammonium salt, which was reacted with DMAP (472 mg) in dry DCM (250 mL) to give macrocycle **27** (82 mg, 44%) (*Scheme 3.1.6*), following the same procedure as described for tetrakis(formyloxy) cholaphane **26**. IR ($CHCl_3$): ν_{max} cm^{-1} 3410m (N–H), 2961m (C–H), 2877m (C–H), 1754s (C=O), 1748s (C=O), 1727s (C=O), 1718s (C=O), 1680m (C=C), 1667m (C=C), 1514m (N–H), 1467m (N–H). ^{1}H NMR (300 MHz, $CDCl_3$): δ 8.12 (s, 2H, 2 × 12α-OCOH), 8.04 (s, 2H, 2 × 7α-OCOH), 7.33 and 7.13 (q, 8H, J = 8.3 Hz, 4 × Ph-CH), 5.36 (d, 2H, J = 7.4 Hz, 2 × 12β-CH), 5.11 (br s, 2H, 2 × 7β-CH), 5.01 (br s, 2H), 4.80 (q, 2H, J = 8.0, 12.9 Hz), 4.02–3.82 (br m, 12H, 4 × OCH_2 and other OCH_2), 1.00 (s, 6H, 2 × 19-Me), 0.88–0.78 (m, 24H, 2 × 21-Me, 2 × 18-Me and other CH_3). ^{13}C NMR (75 MHz, $CDCl_3$): δ 12.2, 12.3, 13.5, 17.6, 19.0, 22.3, 22.4, 26.2, 26.7, 29.7, 30.4, 31.5, 32.8, 33.8, 37.6, 38.0, 38.1, 42.6, 42.6, 43.7, 43.9, 44.2, 45.3, 49.2, 55.7, 64.8, 64.9, 71.0, 76.5, 128.0, 135.8, 146.0, 160.5, 161.4, 167.8, 168.0, 171.2. FABMS: *m/z* 1494 [M + Na]$^+$ (Bhattarai *et al.*, 1997).

Tetrakis(formyloxy) cholaphane **26** was converted to tetrahydroxy cholaphane **3** as follows (*Scheme 3.1.6*). K_2CO_3 in MeOH-H_2O (4:1, 3 mL, 3% solution) was added to a stirred solution of **26** (45 mg, 0.028 mmol) in THF (4 mL) at room temperature and the temperature was raised to 60 °C. After 24 h, the solvent was evaporated to dryness and the residue was dissolved in DCM with a small volume of 1% aqueous $NaHSO_4$. The organic

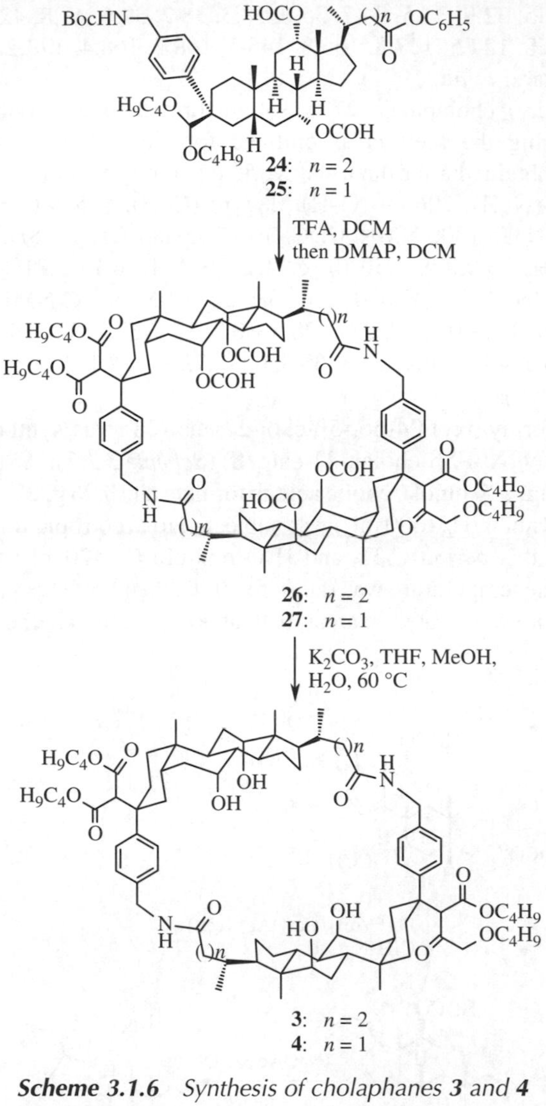
BocHN
HOCO
OC6H5
H9C4O
OC4H9
OCOH
24: n = 2
25: n = 1
TFA, DCM
then DMAP, DCM
26: n = 2
27: n = 1
K2CO3, THF, MeOH,
H2O, 60 °C
3: n = 2
4: n = 1

Scheme 3.1.6 *Synthesis of cholaphanes **3** and **4***

phase was washed with H_2O followed by brine, dried (Na_2SO_4), the solvent was rotary evaporated and the crude product was purified by gradient FCC (mobile phase: $CHCl_3$:EtOAc:MeOH = 44:5:1 and 40:5:5) to furnish cholaphane **3** (34 mg, 86%). IR ($CHCl_3$): ν_{max} cm^{-1} 3340m (N–H), 3332m (N–H), 3325m (N–H), 2963m (C–H), 2875m (C–H), 2336m (C–H), 1755s (C=O), 1722s (C=O), 1649s (C=C), 1535m (N–H), 1463m (C–H), 1415m (C–H), 754m. ^{1}H NMR (300 MHz, $CDCl_3$): δ 7.38 and 7.13 (q, 8H, J = 8.3 Hz, 8 × Ph-CH), 5.73 (br t, 2H, 2 × CH), 4.39 and 4.21 (dq, 4H, J = 4.6, 5.4, 13.9 Hz, 2 × 12β-CH and 2 × 7β-CH), 4.08 (s, 2H, 2 × CH), 4.02–3.80 (m, 8H, 4 × OCH_2), 1.04 (s, 6H, 2 × 19-Me), 0.93 (d, 6H, J = 6.0 Hz, 2 × 21-Me), 0.88-0.81 (m, 12H), 0.67 (s, 6H, 2 × 18-Me). ^{13}C NMR (75 MHz, $CDCl_3$): δ 12.5, 13.6, 17.4, 19.0, 22.9, 23.0, 26.9, 27.5, 28.4,

29.3, 30.3, 31.5, 32.5, 32.9, 34.5, 34.7, 35.1, 38.3, 38.7, 39.6, 41.8, 42.8, 43.7, 46.3, 46.4, 54.9, 64.8, 68.1, 72.8, 127.5, 127.8, 135.8, 145.5, 168.2, 168.4, 173.2. FABMS: *m/z* 1409 $[M + Na]^+$ (Bhattarai *et al.*, 1997).

Tetrakis(formyloxy) cholaphane **27** was transformed to tetrahydroxy cholaphane **4** (*Scheme 3.1.6*), using the method as outlined for cholaphane **3**. Pure cholaphane **4** (30 mg, 81%) was obtained after the usual workup and purification as above. IR ($CHCl_3$): ν_{max} cm^{-1} 3415m (N–H), 2966m (C–H), 2877m (C–H), 1755s (C=O), 1724s (C=O), 1662m (C=C). ^{1}H NMR (300 NMR, $CDCl_3$): δ 7.35 and 7.12 (q, 8H, $J = 8.4$ Hz, 8 × Ph-CH), 5.37 (br s, 2H, 2 × CH), 4.50 (br q, 2H, 2 × CH), 4.14-3.81 (m, 16H), 0.97-0.94 (d, 12H), 0.84 (q, 12H, $J = 7.1$ Hz), 0.72 (s, 6H, 2 × 18-Me). ^{13}C NMR (75 MHz, $CDCl_3$): δ 12.5, 13.6, 15.3, 18.1, 19.0, 22.7, 27.1, 28.5, 30.3, 32.7, 33.4, 34.3, 34.6, 38.4, 38.9, 39.3, 42.4, 43.9, 45.0, 46.5, 49.0, 54.9, 64.8, 65.8, 68.0, 72.8, 128.1, 135.6, 146.1, 168.1, 172.0 (Bhattarai *et al.*, 1997).

3α,7α,12α-Tris(formyloxy)-24-nor-5β-chol-22-ene (**28**) was synthesized from methyl 3α,7α,12α-tris(formyloxy)-5β-cholan-24-oate (**8**) (*Scheme 3.1.7*). A stirred mixture of Cu $(OAc)_2.H_2O$ (500 mg, 2.5 mmol), cholic acid triformate **8** (10.96 g, 22.2 mmol) and C_5H_5N (1.2 mL, 14.8 mmol) in C_6H_6 (600 mL) was subjected to azeotropic removal of H_2O under N_2 for 3 h. During this period C_6H_6 and H_2O mixture (3 × 20 mL) were collected in a Dean–Stark trap. The temperature was raised to 40 °C, then $Pb(OAc)_4$ (40 g, 86 mmol) was added and the mixture was heated in an oil bath at 90 °C. After 4 h, the mixture was cooled

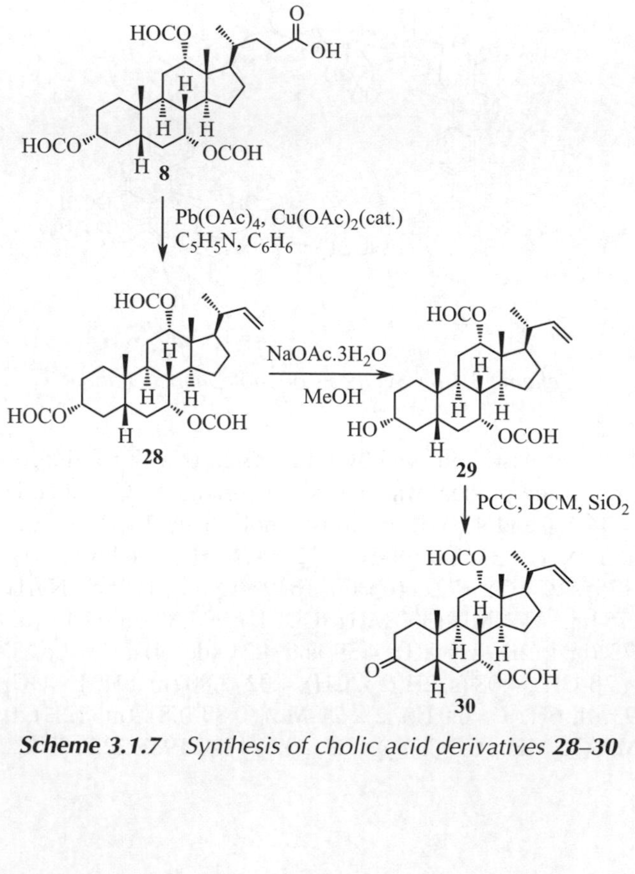

HOCO OH O H H H HOCO OCOH H **8**

$Pb(OAc)_4$, $Cu(OAc)_2$(cat.) C_5H_5N, C_6H_6

HOCO H H H HOCO OCOH H **28**

$NaOAc.3H_2O$ MeOH

HOCO H H H HO OCOH H **29**

PCC, DCM, SiO_2

HOCO H H H O OCOH H **30**

Scheme 3.1.7 *Synthesis of cholic acid derivatives **28–30***

and filtered through a pad of Celite. The insoluble Pb salts remaining on the filter were thoroughly washed with C_6H_6. The combined filtrates were washed with aqueous HCl followed by H_2O, dried ($MgSO_4$), evaporated to dryness and the solid was dissolved in Et_2O and filtered through a pad of silica, which was then washed with Et_2O. Filtrate and washings were combined, rotary evaporated and recrystallized from *n*-hexane to give **28** (7.32 g, 76%, mp: 185–188 °C). IR ($CHCl_3$): ν_{max} cm^{-1} 1720s (C=O), 1710s (C=O), 1605s (C=C). ^1H NMR (300 MHz, $CDCl_3$): δ 8.17 (s, 1H, 12α-OCOH), 8.09 (s, 1H, 7α-OCOH), 8.02 (s, 1H, 3α-OCOH), 5.61 (m, 1H, 3β-CH), 5.26 (br t, 1H, 12β-CH), 5.06 (br d, 1H, 7β-CH), 4.87 (m, 2H, 23-CH_2), 4.73 (m, 1H, 22-CH), 0.95 (s, 3H, 19-Me), 0.94 (d, 3H, $J = 6.5$ Hz, 21-Me), 0.78 (s, 3H, 18-Me). ^{13}C NMR (75 MHz, $CDCl_3$): δ 12.3, 19.5, 22.2, 22.7, 25.5, 26.5, 27.2, 28.6, 31.2, 34.2, 34.4, 34.4, 37.6, 40.4, 40.7, 42.9, 44.9, 46.8, 70.6, 73.6, 75.0, 112.2, 144.1, 160.4 (Bhattarai *et al.*, 1997).

3α,7α,12α-Tris(formyloxy)-24-nor-5β-chol-22-ene (**28**) was partially hydrolyzed to 3α-alcohol **29** (*Scheme 3.1.7*). To a stirred solution of **28** (100 mg, 0.22 mmol) in MeOH (24 mL), NaOAc.$3H_2O$ (50 mg, 0.37 mmol) was added. After 96 h, the solvent was evaporated under vacuum and the residue was dissolved in Et_2O, washed with H_2O followed by brine, dried ($MgSO_4$), and the solvent was rotary evaporated to provide 7α,12α-bis(formyloxy)-24-nor-5β-chol-22-en-3α-ol (**29**, 89 mg, 97.8%). ^1H NMR (300 MHz, $CDCl_3$): δ 8.13 (s, 1H, 12α-OCOH), 8.08 (s, 1H, 7α-OCOH), 5.80–5.40 (m, 1H, 12β-CH), 5.19 (br t, 1H, 7β-CH), 5.16-4.71 (m, 3H, 22-CH and 23-CH_2), 3.40 (m, 1H, 3β-CH), 0.88 (s, 3H, 19-Me), 0.74 (s, 3H, 18-Me) (Bhattarai *et al.*, 1997).

A solution of 7α,12α bis(formyloxy)-24-nor-5β-chol-22-en-3α-ol (**29**, 40 mg, 0.096 mmol) was oxidized with PCC (34 mg, 0.16 mmol), according to the same method as described earlier for **12**, to yield 7α,12α-bis(formyloxy)-24-nor-5β-chol-22-en-3-one (**30**, 35 mg, 85%) (*Scheme 3.1.7*). IR ($CHCl_3$): ν_{max} cm^{-1} 3025m (C–H), 2963m (C–H), 2878m (C–H), 1716s (C=O), 1642s (C=C), 1220m (C–O), 1179m (C–O). ^1H NMR (300 MHz, $CDCl_3$): δ 8.17 (s, 1H, 12α-OCOH), 8.09 (s, 1H, 7α-OCOH), 5.62 (m, 1H, 12β-CH), 5.30 (br t, 1H, 7β-CH), 5.15 (br d, 1H, 22-CH), 4.88 (m, 2H, 23-CH_2), 3.01 (t, 1H, $J = 14.0$ Hz, CH), 1.04 (s, 3H, 19-Me), 0.96 (d, 3H, $J = 6.6$ Hz, 21-Me), 0.81 (s, 3H, 18-Me). ^{13}C NMR (75 MHz, $CDCl_3$): δ 12.4, 19.6, 21.5, 22.8, 25.9, 27.2, 29.6, 31.0, 34.4, 36.2, 36.5, 37.7, 40.5, 42.2, 42.9, 44.5, 45.0, 46.9, 70.6, 75.0, 112.4, 144.1, 160.4, 160.3, 211.6 (Bhattarai *et al.*, 1997).

Next, 7α,12α-bis(formyloxy)-24-nor-5β-chol-22-en-3-one (**30**) was employed for the synthesis of 3β-[bis(butoxycarbonyl)methyl]-3α-[(*p*-(*tert*-butoxymethyl)phenyl]-7α,12α-bis(formyloxy)-24-nor-5β-chol-22-ene (**31**) (*Scheme 3.1.8*). A stirred solution of **30** (3.98 g, 6.47 mmol) was treated with *tert*-butyl *p*-bromobenzyl ether (4.23 g, 16.5 mmol), Mg (560 mg, 23 mmol), and CuCN (770 mg, 8.59 mmol), following the procedure as described for **16**. The crude residue was purified by gradient FCC (mobile phase: *n*-hexane-Et_2O = 9:1, 17:3, and 4:1) to yield **31** (3.51 g, 70%). IR ($CHCl_3$): ν_{max} cm^{-1} 2967m (C–H), 2877m (C–H), 1758s (C=O), 1724s (C=O), 1639s (C=C), 1467m (C–H), 1382m (C–H), 1364m (C–H), 1239m (C–O), 1188m (C–O), 1066m, 1021m, 959m, 733w. ^1H NMR (300 MHz, $CDCl_3$): δ 8.07 (s, 1H, 12α-OCOH), 8.05 (s, 1H, 7α-OCOH), 7.35 and 7.26 (q, 4H, $J = 8.5$ Hz, 4 × Ph-CH), 5.59 (m, 1H, 12β-CH), 5.23 (br s, 1H, 7β-CH), 5.05 (br s, 1H, CH), 4.85 (m, 2H, 23-CH_2), 4.39 (s, 2H), 4.05 (s, 1H, CH), 3.95 (m, 4H, 2 × OCH_2), 1.03 (s, 3H, 19-Me), 0.86 (d, 3H, $J = 6.9$ Hz, 21-Me), 0.75 (s, 3H, 18-Me). ^{13}C NMR (75 MHz, $CDCl_3$): δ 12.3, 13.6, 19.0, 19.5, 22.7, 22.8, 25.5, 27.2, 27.6, 28.7, 29.2, 30.3, 31.3,

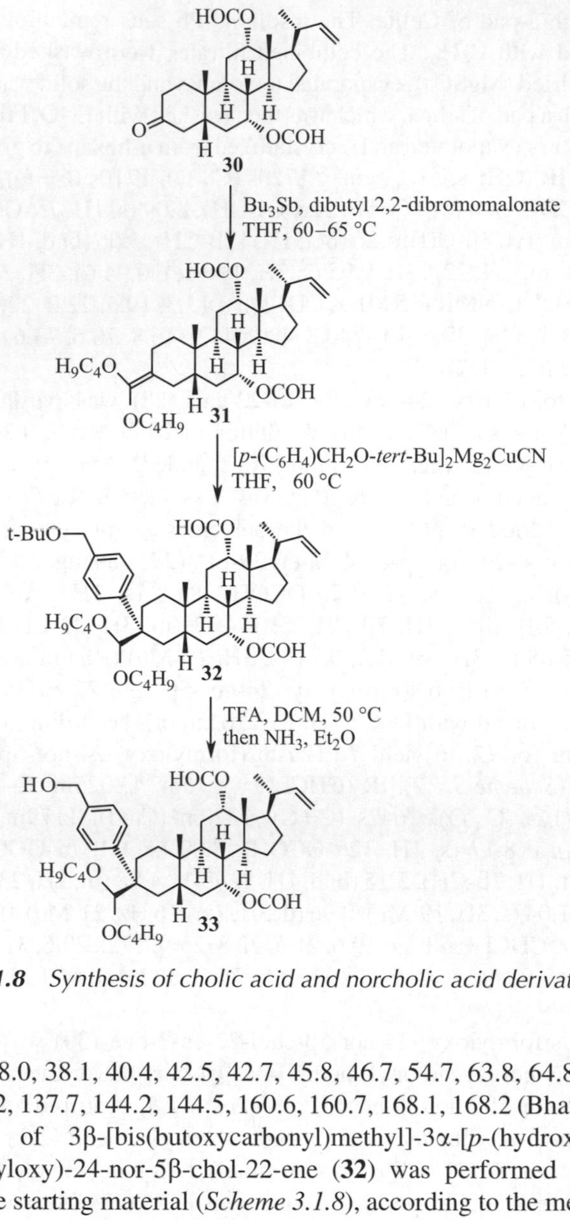

30

Bu_3Sb, dibutyl 2,2-dibromomalonate
THF, 60–65 °C

31

$[p\text{-}(C_6H_4)CH_2O\text{-}tert\text{-}Bu]_2Mg_2CuCN$
THF, 60 °C

32

TFA, DCM, 50 °C
then NH_3, Et_2O

33

Scheme 3.1.8 *Synthesis of cholic acid and norcholic acid derivatives* ***31–33***

32.2, 34.3, 37.7, 38.0, 38.1, 40.4, 42.5, 42.7, 45.8, 46.7, 54.7, 63.8, 64.8, 70.6, 73.2, 74.8, 112.2, 126.7, 127.2, 137.7, 144.2, 144.5, 160.6, 160.7, 168.1, 168.2 (Bhattarai *et al*., 1997).

The synthesis of 3β-[bis(butoxycarbonyl)methyl]-3α-[*p*-(hydroxymethyl)phenyl]-7α,12α-bis (formyloxy)-24-nor-5β-chol-22-ene (**32**) was performed using **31** (8.27 g, 10.60 mmol) as the starting material (*Scheme 3.1.8*), according to the method as described for **18**. The crude residue was purified by FCC, eluted with 20% EtOAc in *n*-hexane, to afford **32** (7.11 g, 92.7%). IR ($CHCl_3$): ν_{max} cm^{-1} 3546br (O–H), 2961m (C–H), 2876m (C–H), 1753s (C–H), 1720s (C–H), 1176m (C–O). ^1H NMR (300 MHz, $CDCl_3$): δ 8.06 (s, 1H, 12α-OCOH), 8.01 (s, 1H, 7α-OCOH), 7.40 and 7.31 (q, 4H, $J = 8.6$ Hz, 4 × Ph-CH), 5.59 (m, 1H, 12β-CH), 5.24 (br t, 1H, 7β-CH), 5.06 (br s, 1H, CH), 4.85 (m, 2H), 4.67 (s, 2H), 4.09 (s, 1H, CH), 3.96 (m, 4H, 2 × OCH_2), 1.05 (s, 3H, 19-Me), 0.86 (d, 3H, $J = 6.4$ Hz, 21-Me), 0.87 (t, 6H, $J = 6.7$ Hz), 0.76 (3H, s). ^{13}C NMR (75 MHz, $CDCl_3$): δ 12.3, 13.6, 19.0, 19.5, 22.7, 22.9, 25.6, 27.2, 28.7, 29.3, 30.3, 31.5, 32.0, 34.3, 37.6, 37.9,

38.1, 40.5, 42.6, 42.9, 44.8, 46.8, 54.4, 64.8, 64.9, 70.7, 74.9, 112.2, 126.1, 127.4, 139.0, 144.2, 145.1, 160.5, 160.6, 168.0, 168.1 (Bhattarai *et al.*, 1997).

3β-[Bis(butoxycarbonyl)methyl]-3α-[*p*-(azidomethyl)-phenyl]-7α,12α-bis(formyloxy)-24-nor-5β-chol-22-ene (**33**) was prepared from a stirred solution of **32** (510 mg, 0.705 mmol) in DCM, after treating with MsCl (90 mg, 61 μL, 0.77 mmol), and DIEA (140 μL, 0.80 mmol), followed by TGMA (220 mg, 1.39 mmol), following the procedure as described for **20** (*Scheme 3.1.8*). The crude azide (535 mg) was pure enough to use for the next steps. A portion (50 mg) was purified as above to give **33** (40 mg). IR ($CHCl_3$): ν_{max} cm^{-1} 2963m (C–H), 2879m (C–H), 2100m (N_3), 1756s (C=O), 1722s (C=O), 1639s (C=C), 1516m (N–H), 1468m (C–H), 1176m (C–O). ^1H NMR (300 MHz, $CDCl_3$): δ 8.07 (s, 1H, 12α-OCOH), 8.03 (s, 1H, 7α-OCOH), 7.43 and 7.23 (q, 4H, $J = 8.2$ Hz, 4 × Ph-CH), 5.64-5.52 (m, 1H, 12β-CH), 5.24 (br s, 1H, 7β-CH), 5.06 (br s, 1H), 4.91–4.79 (m, 2H, OCH_2), 4.33 (s, 2H, 23-CH_2), 4.08 (s, 1H, 22-CH), 3.99–3.86 (m, 4H, 2 × OCH_2), 1.05 (s, 3H, 19-Me), 0.92–0.82 (m, 9H), 0.76 (s, 3H, 18-Me). ^{13}C NMR (300 MHz, $CDCl_3$): δ 12.3, 13.6, 18.9, 19.4, 22.7, 22.9, 25.6, 27.2, 28.7, 29.2, 30.2, 31.5, 32.0, 34.2, 37.6, 37.7, 38.0, 40.4, 42.6, 44.8, 46.7, 54.3, 64.8, 70.6, 74.9, 112.2, 127.1, 127.6, 133.5, 144.1, 145.7, 160.5, 167.9, 168.0 (Bhattarai *et al.*, 1997).

3β-[Bis(butoxycarbonyl)methyl]-3α-[([*p*-(*tert*-butoxycarbonyl)amino]methyl)phenyl]-7α, 12α-bis(formyloxy)-24-nor-5β-chol-22-ene (**34**) was obtained from **33** (*Scheme 3.1.9*). To a stirred solution of **33** (6.48 g, 8.66 mmol) in THF-H_2O mixture (70 mL, 6:1), Ph_3P (2.75 g, 10.5 mmol) was added and the mixture was heated at 60 °C. After completion of the reduction (monitored by TLC), the mixture was cooled to room temperature, $(Boc)_2O$ (2.46 g, 13.06 mmol) in Et_3N (2.4 mL, 17.3 mmol) was added and stirring was continued overnight. The solvent was evaporated and the solid was dissolved in Et_2O (150 mL), washed with H_2O followed by brine, dried ($MgSO_4$) and evaporated under pressure. The solid was dissolved in minimum amount of Et_2O and cooled (ice-salt bath). Insoluble material (Ph_3PO) was removed by filtration, the filtrate was rotary evaporation and the crude product was purified by gradient FCC (mobile phase: 1% EtOAc in *n*-hexane and 20% EtOAc in *n*-hexane), to furnish **34** (6.10 g, 86%). IR ($CHCl_3$): ν_{max} cm^{-1} 3410m (N–H), 2965m (C–H), 2879m (C–H), 1721s (C=O), 1517m (N–H), 1468m (C–H), 1417m (C–H), 1245m (C–O), 1176m (C–O). ^1H NMR (300 MHz, $CDCl_3$): δ 8.07 (s, 1H, 12α-OCOH), 8.02 (s, 1H, 7α-OCOH), 7.37 and 7.22 (q, 4H, $J = 8.4$ Hz, 4 × Ph-CH), 5.65–5.28 (m, 1H, 12β-CH), 5.23 (br s, 1H, 7β-CH), 5.05 (br s, 1H), 4.91–4.79 (m, 2H), 4.30 (d, 2H, $J = 5.8$ Hz, N_3CH_2), 4.08 (s, 1H, CH), 4.04–3.86 (m, 4H, 2 × OCH_2), 1.46 (s, 9H), 1.04 (s, 3H, 19-Me), 0.97-0.83 (m, 9H), 0.76 (s, 3H, 18-Me). ^{13}C NMR (75 MHz, $CDCl_3$): δ 12.3, 13.6, 19.0, 19.5, 22.7, 23.0, 25.6, 26.8, 27.2, 28.4, 28.7, 29.3, 30.3, 31.5, 32.0, 34.3, 37.6, 37.8, 38.1, 40.5, 42.6, 42.8, 44.1, 44.8, 46.7, 54.3, 64.8, 70.7, 74.9, 79.4, 112.2, 126.4, 127.5, 137.0, 144.2, 144.7, 155.8, 160.6, 168.0, 168.1 (Bhattarai *et al.*, 1997).

3β-[Bis(butoxycarbonyl)methyl]-3α-[([*p*-(*tert*-butoxycarbonyl)amino]methyl)phenyl]-7α,12α-bis (formyloxy)-24-nor-5β-cholane-22(*R*,*S*)-23-diol (**35**) was synthesized from **34** (*Scheme 3.1.9*). A heterogeneous mixture of *N*-methylmorpholine *N*-oxide (1.51 g, 12.5 mmol) in C_5H_5N (100 μL, 1.2 mmol), aqueous Bu_4NOAc (1.4M, 2.5 mL, 3.5 mmol), KOAc (100 mg, 1 mmol) in Me_2CO-distilled H_2O (20 mL, 10:1, v/v), and 2% OsO_4 ($NaIO_4$) in dioxane-H_2O mixture (7 mL, 0.5 mmol) was stirred at room temperature. After 5 min, a solution of **34** (3.6 g, 4.37 mmol) in Me_2CO–H_2O (20 mL, 10:1) was added and stirring was maintained for 5 h at room temperature and then overnight at 40 °C. The reaction mixture

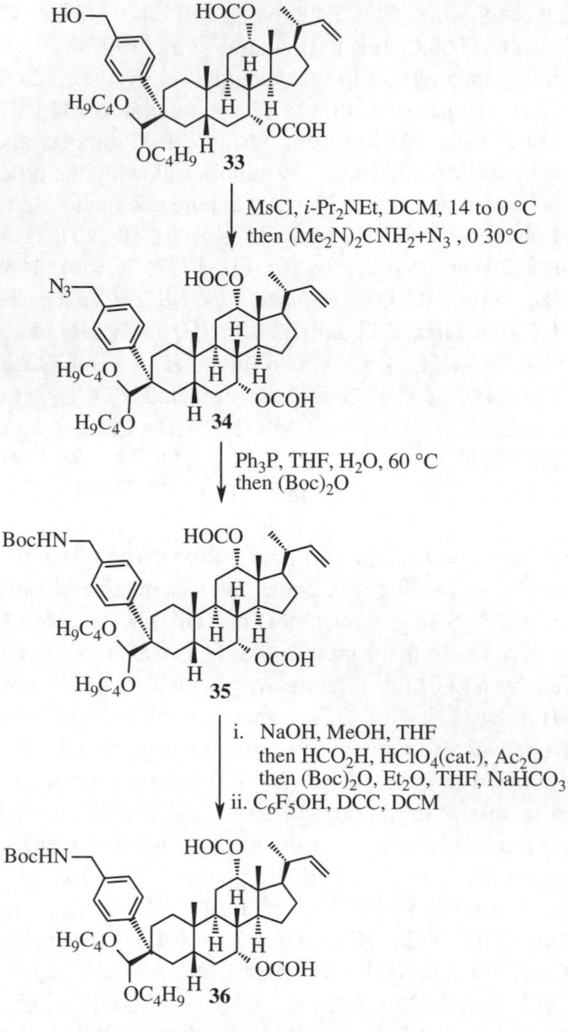

33

MsCl, *i*-Pr_2NEt, DCM, 14 to 0 °C
then $(Me_2N)_2CNH_2$+N_3 , 0 30°C

34

Ph_3P, THF, H_2O, 60 °C
then $(Boc)_2O$

35

i. NaOH, MeOH, THF
then HCO_2H, $HClO_4$(cat.), Ac_2O
then $(Boc)_2O$, Et_2O, THF, $NaHCO_3$
ii. C_6F_5OH, DCC, DCM

36

Scheme 3.1.9 *Synthesis of cholic acid and norcholic acid derivatives **34–36***

was cooled and $Na_2S_2O_5$ (2.7 g, 14.2 mmol) was added, and stirred for another 60 min. The mixture was diluted with Et_2O (50 mL), dried (Na_2SO_4), and evaporated to dryness to give **35** (4.1 g, 90%). A portion of which was purified by FCC, eluted with 40% *n*-hexane in EtOAc, to provide **35**. IR ($CHCl_3$): ν_{max} cm^{-1} 3410m (N–H), 2963m (C–H), 2877m (C–H), 1754s (C=O), 1721s (C=O), 1513m (N–H), 1466m (C–H), 1414m (C–H), 1384m (C–H), 1367m (C–H), 1245m (C–O), 1173m (C–O), 1063m, 755m. ^{1}H NMR (300 MHz, $CDCl_3$): δ 8.04 (s, 1H, 12α-OCOH), 8.02 (s, 1H, 7α-OCOH), 7.35 and 7.20 (q, 4H, $J = 8.3$ Hz, 4 × Ph-CH), 5.19 (br s, 1H, 12β-CH), 5.06 (br s, 1H, 7β-CH), 4.93 (br t, 1H, CH), 4.28 (d, 2H, $J = 5.6$ Hz, $NHCH_2$), 4.06 (s, 1H, CH), 4.03–3.86 (m, 4H, 2 × OCH_2), 3.69 (br d, 1H, 22-CH), 3.56–3.36 (br m, 2H, 23-CH_2), 1.04 (s, 3H, 19-Me), 0.89–0.80 (m, 9H), 0.75 (s, 3H, 18-Me). ^{13}C NMR (75 MHz, $CDCl_3$): δ 11.7, 12.0, 13.5, 18.9, 22.8, 25.5, 26.6, 28.3, 28.6,

29.3, 30.2, 31.5, 32.0, 34.2, 37.6, 37.8, 38.0, 39.3, 42.4, 42.6, 44.1, 44.2, 45.1, 54.3, 62.2, 64.7, 70.6, 73.5, 75.0, 79.3, 126.3, 127.3, 136.8, 144.6, 155.8, 160.4, 167.9, 168.0 (Bhattarai *et al.*, 1997).

The synthesis of pentafluorophenyl 3β-[bis(butoxycarbonyl)methyl]-3α-[([*p*-(*tert*-butoxy carbonyl)amino]methyl)phenyl]-7α,-12α-bis(formyloxy)-23,24-bisnor-5β-cholane-22-thiolate (**36**) was achieved from **35** as follows (*Scheme 3.1.10*). A mixture of **35** (1.01 g, 1.18 mmol) in MeCN–H_2O (20 mL, 3:1) and $NaIO_4$ (300 mg, 1.4 mmol) was stirred at room temperature. After complete oxidation of the diol (monitored on TLC), the reaction mixture was filtered, the filtrate was evaporated *in vacuo*, the residue was dissolved in Et_2O, washed, sequentially, with $NaHSO_4$, H_2O and brine, dried (Na_2SO_4), the solvent was removed under vacuum to obtain the 22-aldehyde. This crude aldehyde was dissolved in MeCN–H_2O (40 mL, 3:1) and cooled (−20 °C), NH_2SO_3H (250 mg, 2.57 mmol) was added with stirring over 5 min, a solution of $NaClO_2$

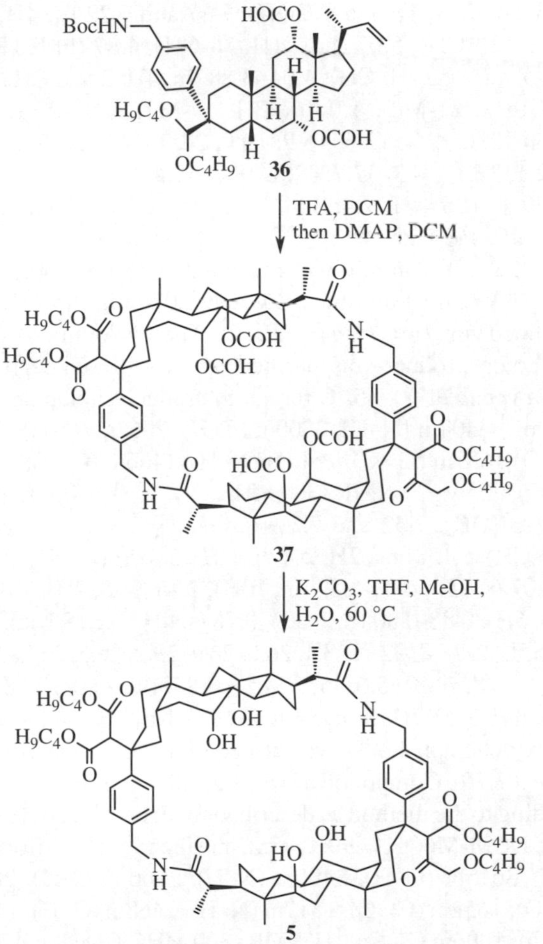

36

TFA, DCM
then DMAP, DCM

37

K_2CO_3, THF, MeOH,
H_2O, 60 °C

5

Scheme 3.1.10 *Synthesis of cholaphane 5*

(1.33 mmol) in H_2O (10 mL) was added dropwise and stirring was continued at 0 °C until TLC showed no trace of the aldehyde. $NaIO_4$ (2.0 g, 10.5 mmol) was added in one portion and the mixture was stirred for another 30 min. Et_2O was added to the mixture, filtered and the filtrate was rotary evaporated to dryness. The solid was dissolved in Et_2O (100 mL), washed with H_2O followed by brine, dried (Na_2SO_4), evaporated under vacuum to provide 22-carboxylic acid. A solution of this carboxylic acid and C_6F_5SH (360 mg, 0.24 mL, 1.74 mmol) in DCM (10 mL) was cooled (−10 °C) under N_2 and stirred for 10 min, then DCC (326 mg, 1.6 mmol) in DCM (3 mL) was added dropwise. The mixture was stirred overnight, a drop of water was added and the mixture was stirred again for 60 min and then cooled (−10 °C). The reaction mixture was filtered and the filtrate was washed with cooled (−10 °C) *n*-hexane-Et_2O mixture. The filtrate and evaporated under vacuum and purified by FCC, eluted with *n*-hexane-Et_2O, to afford **36** (1.12 g, 85%). IR ($CHCl_3$): ν_{max} cm^{-1} 3419m (N–H), 2968m (C–H), 2939m (C–H), 2880m (C–H), 1757s (C=O), 1724s (C=O), 1689m (C=C), 1516m (N–H), 1495m (C–H), 1472m (C–H), 1176m (C–O), 983m, 962m. ^1H NMR (300 MHz, $CDCl_3$): δ 8.09 (s, 1H, 12α-OCOH), 8.02 (s, 1H, 7α-OCOH), 7.37 and 7.22 (q, 4H, $J = 8.3$ Hz, 4 × Ph-CH), 5.20 (br s, 1H, 12β-CH), 5.07 (br s, 1H, 7β-CH), 4.87 (br t, 1H, CH), 4.29 (d, 2H, $J = 5.8$ Hz, $NHCH_2$), 4.07 (s, 1H, CH), 4.04–3.88 (m, 4H, 2 × OCH_2), 2.68 (m, 1H, CH), 1.19 (d, 3H, $J = 7.1$ Hz, 21-Me), 1.06 (s, 3H, 19-Me), 0.88-0.81 (m, 9H). ^{13}C NMR (75 MHz, $CDCl_3$): δ 12.3, 12.4, 13.6, 16.9, 19.0, 22.9, 24.7, 25.6, 26.1, 28.4, 28.7, 28.8, 29.3, 30.3, 31.5, 32.1, 32.1, 34.3, 37.7, 37.8, 38.1, 42.4, 42.6, 44.2, 44.6, 45.4, 51.3, 54.4, 64.8, 70.4, 74.3, 79.4, 126.4, 127.4, 134.7, 137.0, 144.7, 155.8, 160.4, 168.0, 168.1, 194.9 (Bhattarai *et al.*, 1997).

Compound **36** (1.123 g, 1.10 mmol) was converted to tetrakis(formyloxy)cholaphane **37**, by treating it with TFA (2 mL) and then DMAP (3.07 g) in dry DCM (1200 mL), using the method as outlined for **26** (*Scheme 3.1.10*). The cyclodimerization was performed over 6 days. The crude product was purified by FCC (mobile phase: 40% EtOAc in $CHCl_3$) and dried *in vacuo* at 50–55 °C for 4 h to produce cholaphane **37** (550 mg, 67%). IR ($CHCl_3$): ν_{max} cm^{-1} 3404m (N–H), 3300m (C–H), 2962m (C–H), 2876m (C–H), 1753s (C=O), 1722s (C=O) 1659m (C=C), 1517m (N–H), 1466m (C–H), 1414m (C–H), 1384m (C–H), 1179m (C–O), 755w. ^1H NMR (300 MHz, $CDCl_3$): δ 8.06 (s, 2H, 2 × 12α-OCOH), 7.88 (s, 2H, 2 × 7α-OCOH), 7.42 and 7.23 (q, 8H, $J = 8.3$ Hz, 8 × Ph-CH), 5.62 (d, 2H, $J = 5.6$ Hz, 2 × 12β-CH), 5.15 (br s, 2H, 2 × 7β-CH), 5.02 (br d, 2H, 2 × CH), 4.68 (dq, 2H, $J = 7.1$, 13.8 Hz), 4.07 (s, 2H), 4.03-3.82 (m, 10H), 2.45 (br d, 2H), 1.04 (d, 6H, $J = 6.1$ Hz, 2 × 19-Me), 1.03 (s, 6H), 0.88-0.80 (m, 12H), 0.78 (s, 6H, 2 × 18-Me). ^{13}C NMR (75 MHz, $CDCl_3$): δ 12.2, 13.5, 17.2, 18.8, 22.7, 23.1, 26.0, 26.4, 29.2, 29.5, 30.2, 31.3, 31.9, 34.1, 37.5, 37.7, 38.1, 42.4, 43.4, 43.8, 44.8, 45.0, 45.2, 54.0, 64.8, 71.0, 75.0, 127.8, 127.9, 136.1, 145.2, 160.0, 160.3, 167.8, 175.3. FABMS: *m/z* 1465 $[M + Na]^+$ (Bhattarai *et al.*, 1997).

Tetrakis(formyloxy)cholaphane **37** was utilized for the synthesis of tetrahydroxy cholaphane (**5**) (*Scheme 3.1.10*). Compound **37** (87 mg, 0.056 mmol) was treated with K_2CO_3 in MeOH-H_2O, according to the method as described for **3**, and the crude solid was purified by FCC (mobile phase: DCM-MeOH = 99:1, 98:2, and then 95:5) to afford **5** (72 mg, 90%). IR ($CHCl_3$): ν_{max} cm^{-1} 3615br (O–H), 3400m (N–H), 2966m (C–H), 2880m (C–H), 1756s (C=O), 1724s (C=O), 1652m (C=C), 1517m (N–H), 1468m (C–H), 1438m (C–H), 1383m (C–H), 1144m (C–O), 1096m (C–O). ^1H NMR (300 MHz, $CDCl_3$): δ 7.42 and 7.16 (q, 8H, $J = 8.3$ Hz, 8 × Ph-CH), 5.71 (br d, 2H, 2 × 12β-CH), 4.75 (dq, 2H, $J = 7.1$, 13.8 Hz), 4.11

(s, 2H, 2 × 7β-CH), 4.01-3.84 (m, 14H), 1.17 (d, 6H, $J = 6.9$ Hz, 2 × 21-Me), 1.00 (s, 6H, 2 × 19-Me), 0.84 (m, 12H), 0.69 (s, 6H, 2 × 18-Me). ^{13}C NMR (75 MHz, $CDCl_3$): δ 12.6, 13.6, 17.0, 18.9, 19.0, 23.0, 26.7, 26.8, 28.6, 29.6, 30.2, 32.4, 34.3, 34.7, 38.4, 38.5, 39.5, 41.7, 42.9, 43.6, 44.8, 45.3, 46.6, 54.6, 64.8, 68.0, 72.6, 127.5, 127.7, 135.7, 145.6, 168.3, 168.4, 176.1. FABMS: *m/z* 1354 $[M + Na]^+$ (Bhattarai *et al.*, 1997).

A cyclic cholapeptide **44** was synthesized from 7α,12α-diacetoxy-3α-hydroxy-24-nor-5β-cholan-23-oic acid (**38**) in a multistep sequence (Tian *et al.*, 2002). In order to synthesize cholaphane **44**, the precursor benzyl 7α,12α-diacetoxy-3α-hydroxy-24-nor-5β-cholan-23-oate (**39**) had to be prepared from **38** (*Scheme 3.1.11*). To a stirred solution of **38** (12.75 g, 26.7 mmol, 1 equiv.), BnOH (3.8 mL, 36.7 mmol, 1.37 equiv.), DCC (6.5 g, 31.5 mmol, 1.2 equiv.) and DMAP (900 mg, 7.37 mmol, 0.28 equiv.) in dry DCM (150mL) were added at 23 °C. After 24 h, the mixture was diluted with DCM and filtered. The filtrate was washed sequentially with 5% aqueous HCl, saturated aqueous $NaHCO_3$, brine,

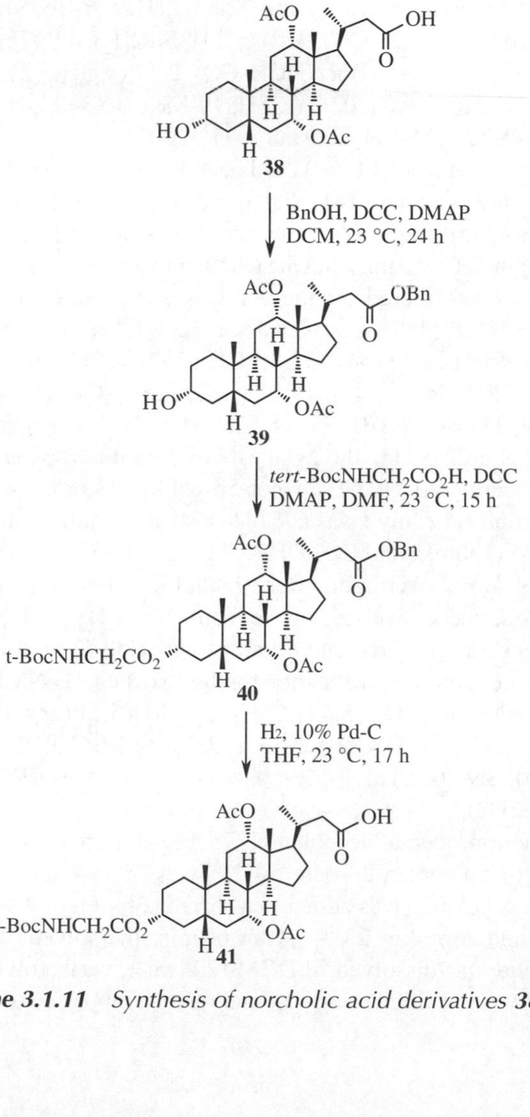
AcO
OH
O
H
H
H
HO
OAc
H
38
BnOH, DCC, DMAP
DCM, 23 °C, 24 h
AcO
OBn
O
HO
OAc
39
tert-BocNHCH2CO2H, DCC
DMAP, DMF, 23 °C, 15 h
AcO
OBn
t-BocNHCH2CO2
OAc
40
H2, 10% Pd-C
THF, 23 °C, 17 h
AcO
OH
t-BocNHCH2CO2
OAc
41

Scheme 3.1.11 *Synthesis of norcholic acid derivatives* ***38–41***

dried ($MgSO_4$) and evaporated to dryness. The residue was purified by CC to furnish **39** in a quantitative yield. ^{1}H NMR (250 MHz, $CDCl_3$): δ 7.33 and 7.35 (br s, 5H, 5 × Ph-CH), 5.10 (s, 2H, Ph-CH_2), 4.87 (br s, 1H, 12β-CH), 4.65 (br s, 1H, 7β-CH), 3.40 (br s, 1H, 3β-CH), 2.42 (d, 2H, 22-CH_2), 2.09 (s, 3H, 12α-MeCO), 2.07 (s, 3H, 7α-MeCO), 0.89 (s, 3H, 19-Me), 0.86 (d, 3H, 21-Me), 0.75 (s, 3H, 18-Me). FABMS: *m/z* 591 $[M + Na]^+$ (Tian *et al.*, 2002).

Benzyl 7α,12α-diacetoxy-3α-hydroxy-24-nor-5β-cholan-23-oate (**39**) was transformed to benzyl 7α,12α-diacetoxy-3α-(*N*-[(*tert*-butoxy)carbonyl]-glycyl)-24-nor-5β-cholan-23-oate (**40**) (*Scheme 3.1.11*). A mixture of *tert*-Boc glycine (2.7 g, 15.4 mmol, 1.2 equiv.), DMAP (640 mg, 5.2 mmol, 0.4 equiv.) and DCC (3.2 g, 15.5 mmol, 1.2 equiv.) was added to a stirred solution of **39** (7.4 g, 13 mmol, 1 equiv.) in dry DMF and stirred at 23 °C. After 24 h, the solvent was rotary evaporated and the solid was dissolved in EtOAc (400 mL), extracted successively with 1M HCl, saturated aqueous $NaHCO_3$ and brine, dried (Na_2SO_4) and evaporated to dryness. The crude product was purified by CC to give **40** (8.9 g, 94%, mp: 98–100 °C). ^{1}H NMR (250 MHz, $CDCl_3$): δ 7.58 (t, 1H, *tert*-BocNHCH_2CO), 7.34 (br s, 5H, 5 × Ph-CH), 5.16 (s, 1H, 12β-CH), 5.10 (s, 2H, PhCH_2), 4.90 (br s, 1H, 7β-CH), 4.63 (m, 1H, 3β-CH), 3.97 (d, 2H, *tert*-BocNHCH_2CO), 2.11 (s, 3H, 12α-MeCO), 2.08 (s, 3H, 7α-MeCO), 1.44 (s, 9H, $(CH_3)_3C$), 0.92 (s, 3H, 19-Me), 0.87 (d, 3H, 21-Me), 0.76 (s, 3H, 18-Me). FABMS: *m/z* 724 $[M - H]^+$ (Tian *et al.*, 2002).

In the next step, the synthesis of 7α,12α-diacetoxy-3α-(*N*-[(*tert*-butoxy)carbonyl]-glycyl)-24-nor-5β-cholan-23-oic acid (**41**) was performed using **40** as the starting material (*Scheme 3.1.11*). Pd-C (500 mg, 10% on carbon) was added to a solution of **40** (3.3 g, 4.55 mmol, 1 equiv.) in THF (50 mL) and the reaction mixture was stirred under H_2 at 23 °C. After 20 h, the catalyst was filtered and the filtrate was evaporated under pressure to obtain **41** (2.78 g, 96%). ^{1}H NMR (250 MHz, $CDCl_3$): δ 7.58 (t, 1H, *t*-BocNH), 5.10 (s, 1H, 12β-CH), 4.92 (br s, 1H, 7β-CH), 4.63 (br s, 1H, 3β-CH), 3.98 (d, 2H, *t*-BocNHCH_2), 2.15 (s, 3H, 12α-MeCO), 2.10 (s, 3H, 7α-MeCO), 1.45 [s, 9H, $(CH_3)_3C$], 0.93 (br s, 6H, 19-Me and 21-Me), 0.78 (s, 3H, 18-Me). FABMS: *m/z* 635 $[M + H]^+$ (Tian *et al.*, 2002).

Compound **41** was utilized for the synthesis of pentafluorophenyl 7α,12α-diacetoxy-3α-(*N*-[(*tert*-butoxy)carbonyl]-glycyl)-24-nor-5β-cholan-23-oate (**42**) (*Scheme 3.1.12*). DCC (3.8 g, 18.5 mmol, 1.5 equiv.) was added to a stirred solution of **41** (7.8 g, 12.3 mmol, 1 equiv.) in dry DCM (150mL) and C_6F_5OH (3.4 g, 18.4 mmol, 1.5 equiv.) at 0 °C under N_2 and the mixture was stirred overnight, after which, the reaction mixture was filtered. The filtrate was extracted, successively, with aqueous $NaHCO_3$, 1M HCl and brine, dried (Na_2SO_4), the solvent was evaporated to dryness to provide **42** in a quantitative yield. This product was used without further purification for the next step. ^{1}H NMR (250 MHz, $CDCl_3$): δ 7.51 (t, 1H, *tert*-BocNHCH_2CO_2), 5.44 (s, 1H, 12β-CH), 5.11 (br s, 1H, 7β-CH), 4.92 (br s, 1H, 3β-CH), 4.00 (d, 2H, NHCH_2CO_2), 2.15 (s, 3H, 12α-MeCO), 2.10 (s, 3H, 7α-MeCO), 1.15 (d, 3H, 21-Me), 0.94 (s, 3H, 19-Me), 0.81 (s, 3H, 18-Me). FABMS: *m/z* 824 $[M + Na]^+$ (Tian *et al.*, 2002).

In the next reaction, pentafluorophenyl 7α,12α-diacetoxy-3α-(*N*-[(*tert*-butoxy)carbonyl]-glycyl)-24-nor-5β-cholan-23-oate (**42**) was converted to **43** as follows (*Scheme 3.1.12*). TFA (15 mL) was added to a stirred solution of **42** (12.3 mmol, 1 equiv.) in DCM (100 mL) and stirred at 23 °C. After 60 min, the solvent was evaporated under pressure and the solid was dissolved in DCM (200 mL), washed with saturated aqueous $NaHCO_3$ followed by brine, dried (Na_2SO_4) and rotary evaporated to dryness. A portion of

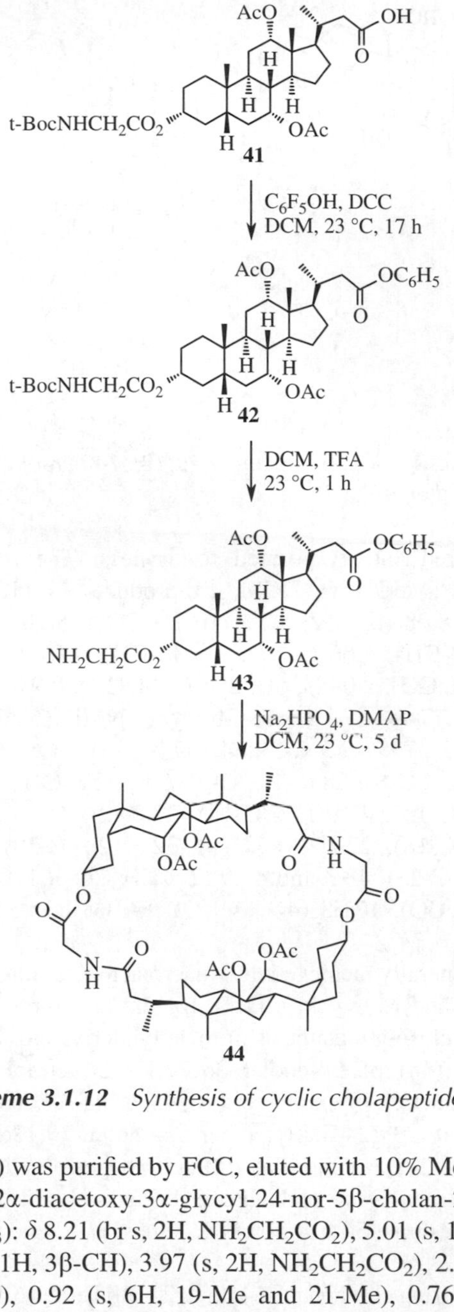

Scheme 3.1.12 *Synthesis of cyclic cholapeptide* ***44***

the crude solid (100 mg) was purified by FCC, eluted with 10% MeOH in $CHCl_3$, to yield pentafluorophenyl 7α,12α-diacetoxy-3α-glycyl-24-nor-5β-cholan-23-oate (**43**, 70 mg). ^1H NMR (250 MHz, $CDCl_3$): δ 8.21 (br s, 2H, $NH_2CH_2CO_2$), 5.01 (s, 1H, 12β-CH), 4.81 (br s, 1H, 7β-CH), 4.64 (br s, 1H, 3β-CH), 3.97 (s, 2H, $NH_2CH_2CO_2$), 2.07 (s, 3H, 12α-MeCO), 2.02 (s, 3H, 7α-MeCO), 0.92 (s, 6H, 19-Me and 21-Me), 0.76 (s, 3H, 18-Me) (Tian *et al.*, 2002).

Finally, pentafluorophenyl 7α,12α-diacetoxy-3α-glycyl-24-nor-5β-cholan-23-oate (**43**) was employed for the synthesis of cyclic cholapeptide **44** (*Scheme 3.1.12*). To a stirred solution of **43** (12.3 mmol) in DCM (3075 mL), DMAP (3.0 g, 24.6 mmol, 2 equiv.) and Na_2HPO_4 (7.0 g, 49.3 mmol, 4 equiv.) were added at 23 °C. After 5 days,

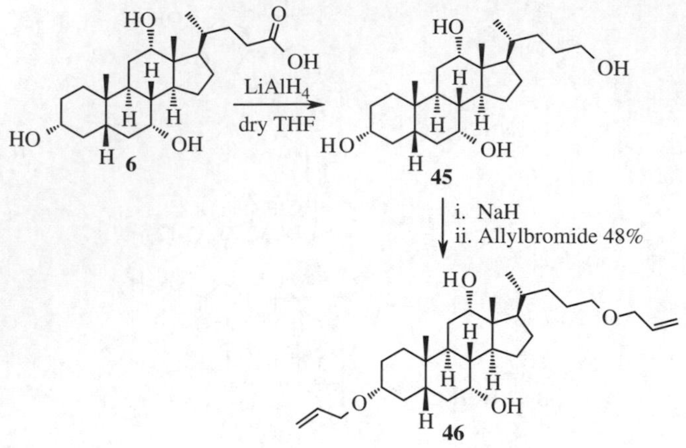

Scheme 3.1.13 *Synthesis of 5β-cholan-3α,7α,12α,24-tetraol (**45**) and 3α,24-diallyl ether-5β-cholan-7α,12α-diol (**46**)*

the mixture was filtered and evaporated to dryness. The residue was purified by CC, eluted with 10% MeOH in DCM, to produce **44** (1.07 g, 17%). ^{1}H NMR (250 MHz, $CDCl_3$): δ 8.08 (br s, 2H, 2 × $NHCH_2CO_2$), 5.10 (br s, 2H, 2 × 12β-CH), 4.91 (br s, 2H, 2 × 7β-CH), 4.66 (s, 2H, 2 × 3β-CH), 4.07 (s, 4H, 2 × $NHCH_2CO_2$), 2.13 (s, 6H, 2 × 12α-MeCO), 2.09 (s, 6H, 2 × 7α-MeCO), 0.94 (s, 6H, 2 × 19-Me), 0.85 (d, 6H, 2 × 21-Me), 0.77 (s, 6H, 2 × 18-Me). ^{13}C NMR (62.5 MHz, $CDCl_3$): δ 34.4 (2 × C-1), 26.7 (2 × C-2), 75.3 (2 × C-3), 34.2 (2 × C-4), 41.6 (2 × C-5), 31.2 (2 × C-6), 70.7 (2 × C-7), 37.7 (2 × C-8), 28.8 (2 × C-9), 34.2 (2 × C-10), 25.3 (2 × C-11), 75.8 (2 × C-12), 45.1 (2 × C-13), 43.9 (2 × C-14), 22.8 (2 × C-15), 27.3 (2 × C-16), 48.2 (2 × C-17), 12.3 (2 × C-18), 22.4 (2 × C-19), 32.3 (2 × C-20), 18.4 (2 × C-21), 40.8 (2 × C-22), 167.7 (2 × C-23), 43.3 and 43.9 (2 × CH_2CO_2), 170.6 (2 × CO_2CH_2), 21.0, 21.3 and 21.5 (4 × MeCO), 171.8 (4 × MeCO). FABMS: *m/z* 1056 $[M + Na - H]^+$ (Tian *et al.*, 2002).

Cholaphanes are generally achieved by macrolactonization or macrolactamization. However, Czajkowska and Morzycki (2007), for the first time, described the synthesis of cholaphanes by ring closing metathesis of the allyl derivatives *e.g.*, 3α,24-diallyl ether-5β-cholan-7α,12α-diol (**46**) of 5β-cholan-3α,7α,12α,24-tetraol (**45**), which could be prepared by $LiAlH_4$ reduction of cholic acid (**6**) (*Scheme 3.1.13*).

While the reactions of 3α,24-diallyl ether-5β-cholan-7α,12α-diol (**46**) to form the cholaphanes **47** and **48** were somewhat difficult to perform (*Scheme 3.1.14*), diallyl derivatives of 3,3'-disteroidal *ortho*-phthalate **53** reacted smoothly affording predominantly the *E* isomer of the cyclic dimer **54** as follows (Schemes *3.1.15* and *3.1.16*) (Czajkowska and Morzycki, 2007).

In the process of preparing cholaphane **54**, first, 24-allyl ether-5β-cholan-3α,7α,12α-triol (**52**) was synthesized from cholic acid (**6**) in four steps through the formation of several intermediates **49-51** (*Scheme 3.1.15*) (Czajkowska and Morzycki, 2007). 24-Allyl ether-5β-cholan-3α,7α,12α-triol (**52**) was treated with DCC and *ortho*-phthalic anhydride for regioselective esterification at the 3α-OH to furnish 3,3'-disteroidal *ortho*-phthalate **53**,

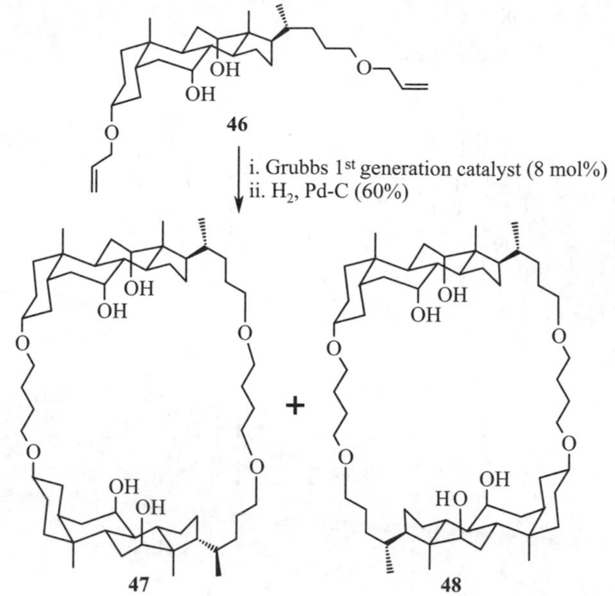

46

i. Grubbs 1st generation catalyst (8 mol%)
ii. H_2, Pd-C (60%)

47 + 48

Scheme 3.1.14 *Synthesis of cholaphanes* ***47*** *and* ***48***

which upon treatment with a solution of a Grubbs 1st-generation catalyst yielded cholaphane **54** (*Scheme 3.1.16*) (Czajkowska and Morzycki, 2007).

The synthesis of methyl cholate (**49**) was accomplished in high yield from cholic acid (**6**), reacted with MeOH and conc. HCl. Selective protection of the 3α-hydroxy group was carried out with TBDMSCl-imidazole in DMF at room temperature to afford methyl 3α-*tert*-butyl dimethylsilyloxy-7α,12α-dihydroxy-5β-cholan-24-oate (**50**) (*Scheme 3.1.15*). Reduction of **50** with $LiAlH_4$ provided 3α-*tert*-butyldimethylsilyloxy-5β-cholan-3α,7α,12α-triol (**51**, 87%).

The allylation of **51** with allyl bromide and NaH in refluxing THF followed by removal of the TBDMS by TBAF (1M in THF) furnished the corresponding 24-allyl ether-5β-cholan-3α,7α,12α-triol (**52**, 62%) (*Scheme 3.1.15*) (Czajkowska and Morzycki, 2007).

To avoid the formation of monophthalate, a portionwise addition of *ortho*-phthalic anhydride to the reaction mixture containing 24-allyl ether-5β-cholan-3α,7α,12α-triol (**52**) was utilized and esterification with DCC, *ortho*-phthalic anhydride and catalytic amount of DMAP was carried out (*Scheme 3.1.16*). The reaction was continued for 5 days in DCM at room temperature. The observed regioselectivity at position 3α-OH group was good enough to give dimer **53** (66%). IR (KBr): ν_{max} cm^{-1} 3434br (O–H), 2941m (C–H), 2862m (C–H), 1715s (C=O), 1655s (C=C), 1526m (C–H), 1292m (C–O), 1075m (C–O). ^1H NMR (300 MHz, $CDCl_3$): δ 7.66 (m, 2H, 2 × Ph-CH), 7.47 (m, 2H, 2 × Ph-CH), 5.87 (m, 2H, 2 × 12β-CH), 5.22 (m, 4H, 2 × =CH_2), 4.78 (m, 2H, 2 × 7β-CH), 3.95 (m, 8H, 4 × OCH_2), 3.82 (s, 6H), 3.39 (t, 4H, $J = 6.5$ Hz, 2 × OCH_2), 0.97 (d, 6H, $J = 6.2$ Hz, 2 × 21-Me), 0.87 (s, 6H, 2 × 19-Me), 0.67 (s, 6H, 2 × 18-Me) (Czajkowska and Morzycki, 2007).

The final ring-closing metathesis (RCM) of dimer **53** with Grubbs 1st-generation catalyst (25 mol%) in toluene produced an isomeric mixture of cholaphane **54** as follows

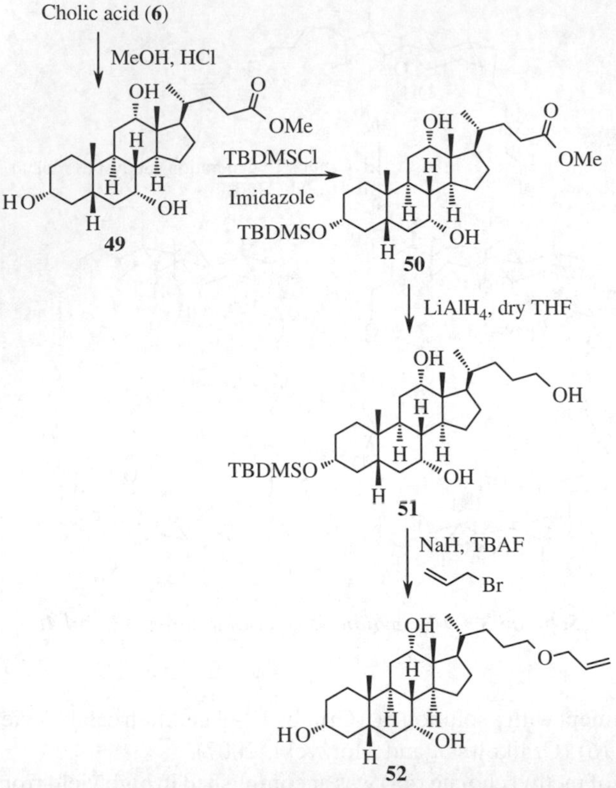

Scheme 3.1.15 *Synthesis of 24-allyl ether-5β-cholan-3α,7α,12α-triol (**52**)*

(*Scheme 3.1.16*). To a stirred solution of **53** (0.008 mM) in dry toluene (10 mL), a solution of Grubbs 1st-generation catalyst (0.002 mM, 25 mol%) in dry toluene (10 mL) was added dropwise over 30 min under Ar. The resulting mixture (4 mM/L) was stirred at 30 °C for another 60 min under Ar. The solvent was rotary evaporated and the crude product was purified by gradient CC (mobile phase: 2% EtOAc in *n*-hexane and 5% EtOAc in *n*-hexane) to obtain cholaphane **54** (72%, *E*:*Z* = 8:1). IR (KBr): ν_{max} cm^{-1} 3454br (O–H), 2945m (C–H), 2869m (C–H), 1713s (C=O), 1294m (C–O), 1133m (C–O), 968m. ^{1}H NMR (300 MHz, $CDCl_3$): δ 7.66 (m, 2H, 2 × Ph-CH), 7.48 (m, 2H, 2 × Ph-CH), 5.77 (m, 2H, 2 × 12β-CH), 4.75 (m, 2H, 2 × 7β-CH), 4.05 (m, 6H), 3.80 (s, 2H, 2 × OCH), 3.41 (m, 4H, 2 × OCH_2), 0.97 (d, 6H, J = 6.2 Hz, 2 × 21-Me), 0.92 (s, 6H, 2 × 19-Me), 0.70 (s, 6H, 2 × 18-Me). ^{13}C NMR (75 MHz, $CDCl_3$): δ 12.5, 17.6, 22.5, 23.0, 23.8, 23.8, 26.5, 26.6, 26.7, 27.6, 27.8, 27.9, 29.4, 29.5, 32.8, 34.5, 34.6, 34.8, 35.0, 35.8, 39.6, 41.2, 42.1, 46.3, 47.4, 67.4, 67.6, 67.7, 70.2, 70.3, 73.1, 76.0, 106.3, 107.6, 129.1, 130.6, 133.2, 167.5. EIMS: *m/z* 993 [M + Na]$^+$. HREIMS: *m/z* 993.6393 calculated for $C_{60}H_{90}O_{10}Na$, found: 993.6426 (Czajkowska and Morzycki, 2007).

Using exactly the same methodology, cholaphane **56** could be obtained from 3α,7α,12α-triacetoxy-24-nor-5β-cholan-23-oic acid (**55**), which was initially prepared from cholic acid (**6**) by the Barbier–Wieland procedure (*Scheme 3.1.17*). $LiAlH_4$ reduction of **55** yielded

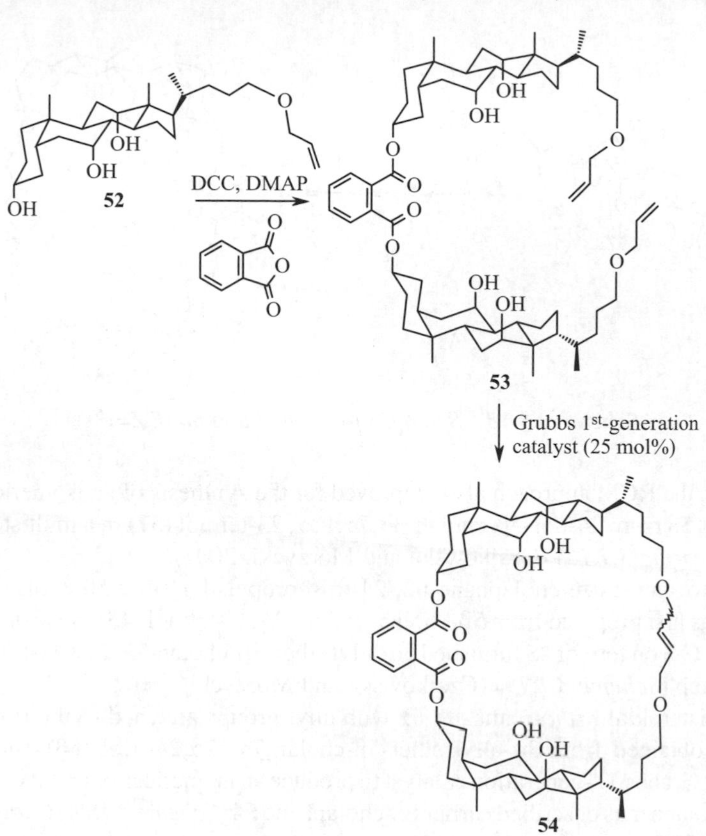

Scheme 3.1.16 *Synthesis of cholaphane **54** (E:Z = 8:1)*

24-nor-5β-cholan-3α,7α,12α,24-tetraol (**57**), which was allylated regioselectively at 23-OH. The corresponding 3,3'-disteroidal *ortho*-phthalate was synthesized and subjected to RCM with the Grubbs 1st-generation catalyst to give an isomeric mixture of cholaphane **56** (*E:Z* = 3:1) (*Scheme 3.1.17*) (Czajkowska and Morzycki, 2007).

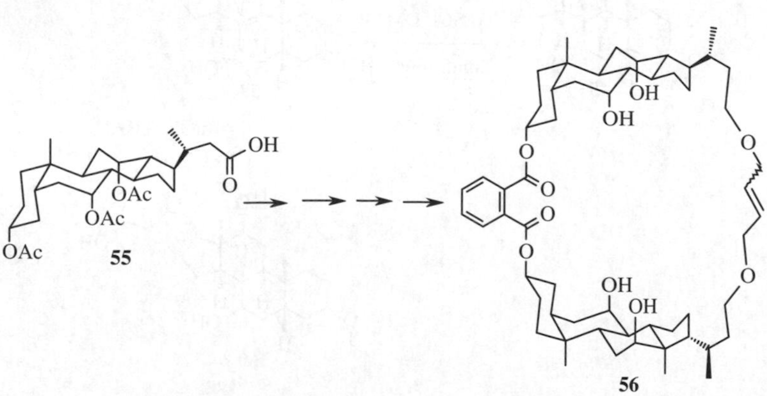

Scheme 3.1.17 *Synthesis of cholaphane **56** (E:Z = 3:1)*

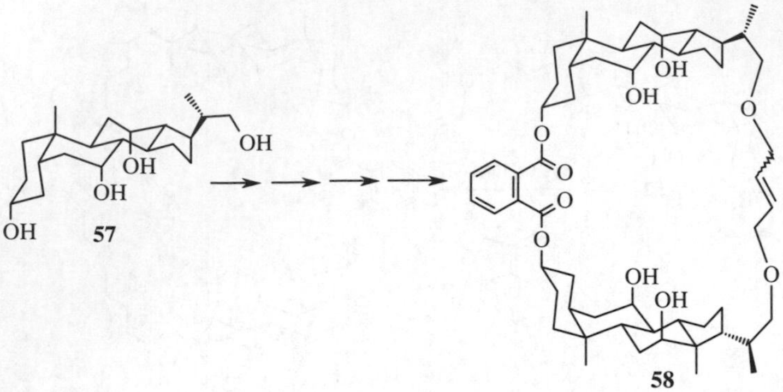

Scheme 3.1.18 *Synthesis of cholaphane* **58** *(E:Z = 3:1)*

Similarly, the RCM approach also employed for the synthesis of an isomeric mixture of cholaphanes **58** from 24-nor-5β-cholan-3α,7α,12α, 23-tetraol (**57**) in a multistep sequence as above (*Scheme 3.1.18*) (Czajkowska and Morzycki, 2007).

In order to synthesize cholaphane **62**, 24-triisopropylsilyl ether-5β-cholan-3α,7α,12α-triol (**59**) was first prepared from 5β-cholan-3α,7α,12α,24-tetraol (**45**) (*Scheme 3.1.19*), and allylation at C-3 on tetraol **45** furnished 3α-allyl ether-5β-cholan-7α,12α,24-triol (**60**) in the following step (*Scheme 3.1.19*) (Czajkowska and Morzycki, 2007).

24,24'-Disteroidal *ortho*-phthalate **61** with allyl groups attached to the ring A oxygen atoms was obtained from 3α-allyl ether-5β-cholan-7α,12α,24-triol (**60**), finally **61** was treated with Grubbs 1st-generation catalyst to produce an isomeric mixture of cholaphane **62** in the same manner as described earlier for cholaphane **54** (*Scheme 3.1.20*) (Czajkowska and Morzycki, 2007).

Several 'head-to-head' cholaphanes were formed with aromatic and diamino spacer groups, respectively, between two A rings and between side chains of two cholic acid molecules

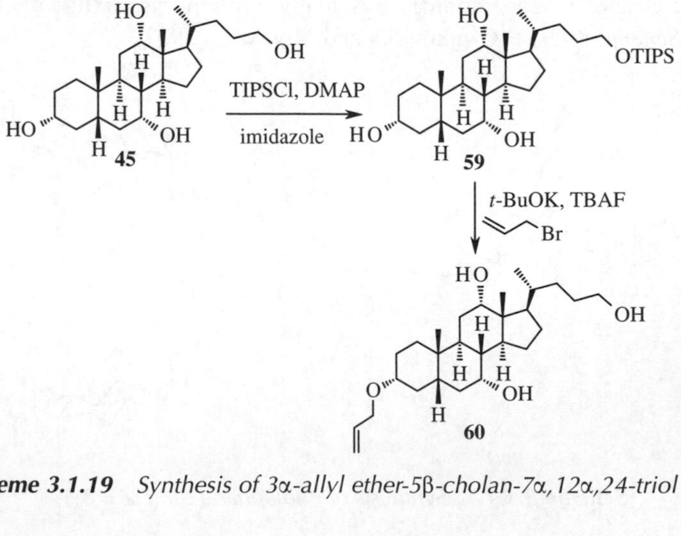

Scheme 3.1.19 *Synthesis of 3α-allyl ether-5β-cholan-7α,12α,24-triol* (**60**)

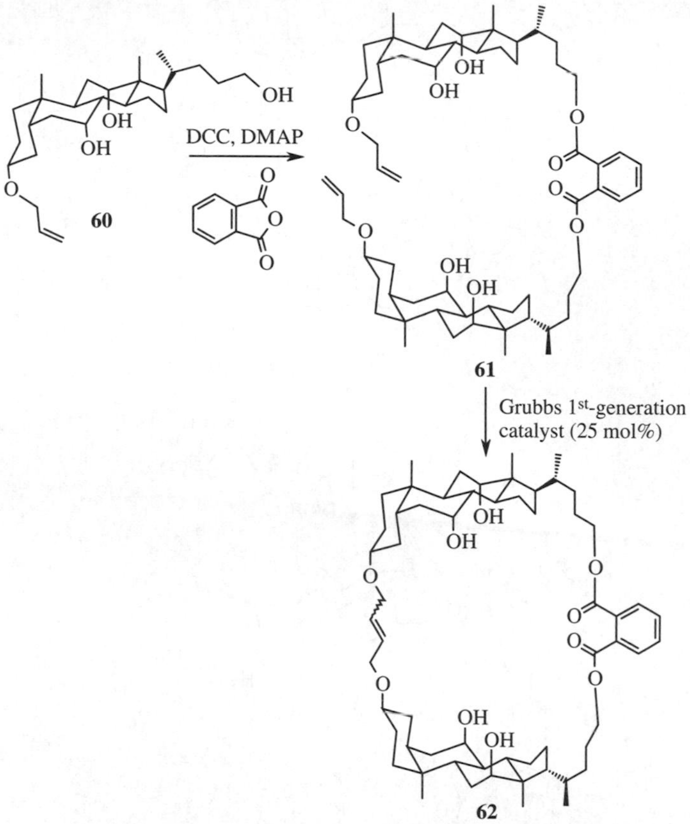

Scheme 3.1.20 *Synthesis of cholaphane **62** (E:Z = 5:3)*

(Pandey *et al.*, 2002; Kolehmainen *et al.*, 1996; Tamminen *et al.*, 2000; Ra *et al.*, 2000; Łotowski *et al.*, 2008; Valkonen *et al.*, 2007; Chattopadhyay and Pandey, 2006).

Three 'head-to-head' cholaphanes **66–68** were prepared from bile acid-based acyclic dimers **63-65** involving Cs-salt methodology (Pandey *et al.*, 2002). The synthetic steps of these acyclic dimers **63–65** were discussed earlier (see *Chapter 2*; *Scheme 2.4.22*). In the final step, the treatment of a bisbromoacetate **63–65** with freshly prepared bis-cesium terephthalate in DMF (5 mL) led to the formation of the cholaphanes **66–68** in excellent yields (*Scheme 3.1.21*).

Bis-caesium terephthalate was synthesized from a stirred solution of terephthalic acid (830 mg, 5 mmol) in dry DMF (8 ml) and caesium carbonate (1.62 g, 5 mmol). After 3 h, the product was filtered off, washed with Me_2CO, and dried *in vacuo* to yield bis-caesium terephthalate (1.85 g, 95%). IR (KBr): ν_{max} cm^{-1} 1580 (C=O). ^1H NMR (300 MHz, DMSO-d_6): δ 7.85 (br s, Ph-CH) (Pandey *et al.*, 2002).

Bis(3α-*O*-hydroxyacetylcholyl)ethylenediamine cyclic terephthalate (**66**) was produced from *N,N'*-bis(3α-*O*-bromoacetylcholyl)ethylenediamine (**63**) as follows (*Scheme 3.1.21*). To a stirred solution of acyclic dimer **63** (217 mg, 0.20 mmol) in dry DMF (6 mL), bis-caesium terephthalate (89 mg, 0.20 mmol, 1 equiv.) was added at room temperature. After

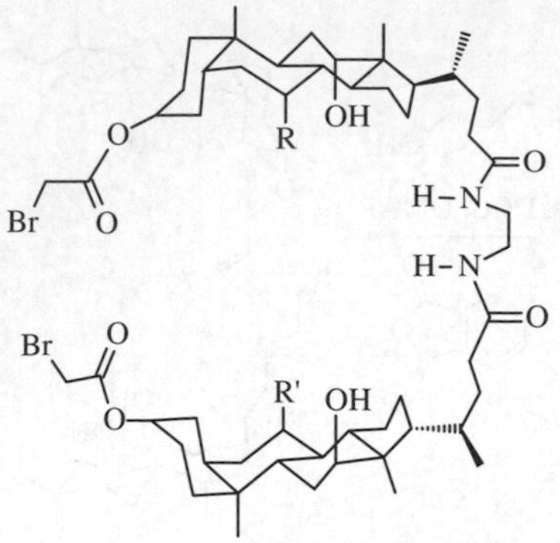

63: R = R' = OH
64: R = R' = H
65: R = OH; R' = H

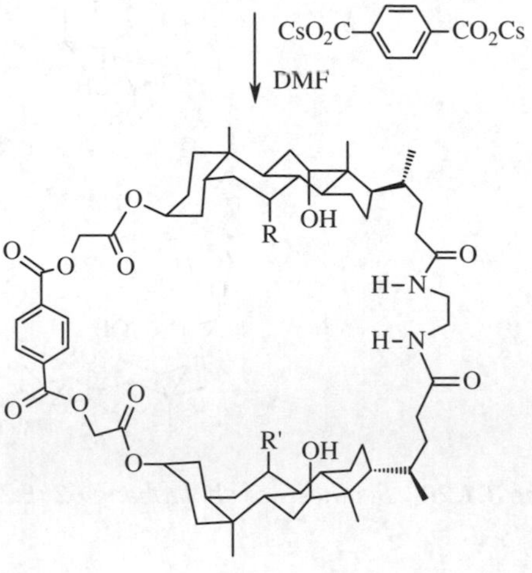

66: R = R' = OH
67: R = R' = H
68: R = OH; R' = H

Scheme 3.1.21 *Synthesis of 'head-to-head' cholaphanes **66–68***

12 h, the reaction mixture was filtered and ice-cold brine (20 mL) was added to the filtrate. The resulting solid was filtered off, dried *in vacuo* and purified by CC, eluted with 0-5% MeOH in $CHCl_3$, to produce the cholaphane **66** (205 mg, 95%, mp: 185–188 °C). IR (KBr): ν_{max} cm^{-1} 3408m, 1734s (C=O), 1652m. ^1H NMR (300 MHz, $CDCl_3$-CD_3OD): δ 8.09 (br s, 4H, Ph-CH), 4.74 (br s, 4H, 2 × $COCH_2O$), 4.56 (m, 2H, 2 × 3β-CH), 3.93 (br s, 2H, 2 × 12β-CH), 3.73 (br s, 2H, 2 × 7β-CH), 3.29 (br s, 4H, 2 × CH_2-NHCO), 2.50–1.00 (m, 48H, steroidal CH and CH_2), 0.92 (br s, 6H, 2 × 21-Me), 0.83 (s, 6H, 2 × 19-Me), 0.60 (s, 6H, 2 × 18-Me). ^{13}C NMR (75 MHz, $CDCl_3$-CD_3OD): δ 175.8 (C-24), 167.2 (OCO-Ar), 165.1 ($OCOCH_2$), 133.3 (Ar-C), 129.8 (CH, Ar-CH), 76.6 (C-3), 72.8 (C-12), 67.9 (C-7), 61.6 ($OCOCH_2$), 39.3 (CH_2NHCO), 34.7 (C-23), 32.9 (C-22). FABMS: *m/z* 1109 [M + Na]$^+$ (Pandey *et al*., 2002).

Using the same reaction protocol as outlined above (*Scheme 3.1.21*), both cholaphanes, bis(3α-*O*-hydroxyacetyldeoxycholyl)ethylenediamine cyclic terephthalate (**67**, 207 mg, 85%, mp: 196–199 °C) and 3α-*O*-hydroxyacetylcholyl-3'-*O*-hydroxyacetyl-deoxycholyl-ethylenediamine cyclic terephthalate (**68**, 165 mg, 87%, mp: 215 °C) could be prepared respectively, from acyclic dimers **64** (244 mg, 0.23 mmol) and **65** (190 mg, 0.17 mmol), by treating with equivalent amounts of bis-caesium terephthalate at room temperature (Pandey *et al.*, 2002).

Cholaphane **67**: IR (KBr): ν_{max} cm^{-1} 3410m, 1735s (C=O), 1648m. ^{1}H NMR (300 MHz, $CDCl_3$-CD_3OD): δ 8.17 (br s, 4H, Ph-CH), 4.83 (br s, 6H, 2 × $COCH_2O$ and 2 × 3β-CH), 3.98 (br s, 2H, 2 × 12β-CH), 3.34 (br s, 4H, 2 × CH_2NHCO), 2.40–1.03 (m, 50H, steroidal CH and CH_2), 0.98 (br s, 6H, 2 × 21-Me), 0.91 (s, 6H, 2 × 19-Me), 0.67 (s, 6H, 2 × 18-Me); ^{13}C NMR (75 MHz, $CDCl_3$-CD_3OD): δ 76.0 (2 × C-3), 72.8 (2 × C-12), 31.6 (2 × C-22), 32.0 (2 × C-23), 175.4 (2 × C-24), 167.1 (2 × OCO-Ar), 165.2 (2 × $OCOCH_2$), 133.4 (2 × Ph-C), 129.9 (2 × Ph-CH), 61.7 (2 × $OCOCH_2$), 39.5 (2 × CH_2NHCO). FABMS: *m/z* 1078 (M + Na]$^+$ (Pandey *et al.*, 2002).

Cholaphane **68**: (IR (KBr): ν_{max} cm^{-1} 3410m, 1735s (C=O), 1650m. ^{1}H NMR (300 MHz, $CDCl_3$-CD_3OD): δ 8.17 (br s, 4H, Ph-CH), 4.83 (br s, 4H, 2 × $COCH_2O$), 4.60 (m, 2H, 2 × 3β-CH), 3.97 (br s, 2H, 2 × 12β-CH), 3.83 (br s, 2H, 2 × 7β-CH), 3.36 (br s, 4H, 2 × CH_2NHCO), 2.25–1.02 (m, 49H, steroidal CH and CH_2), 0.99 (br s, 6H, 2 × 21-Me), 0.91 (br s, 6H, 2 × 19-Me), 0.67 (s, 6H, 2 × 18-Me). ^{13}C NMR (75 MHz, $CDCl_3$-CD_3OD): δ 76.1 (2 × C-3), 68.0 (2 × C-7), 72.8 (2 × C-12), 31.9 (C-23), 175.4 (2 × C-24), 167.1 (2 × OCO-Ph), 165.2 (2 × $OCOCH_2$), 133.4 (2 × Ph-C), 129.9 (2 × Ph-CH), 61.7 (2 × $OCOCH_2$), 39.6 (CH_2NHCO). FABMS: *m/z* 1071 (M + H]$^+$ (Pandey *et al.*, 2002).

Similar cholaphanes **71** and **72**, using spacers *m*-xylenediamine, were synthesized from bile acid-based acyclic dimers, *N,N'*-bis(3α-*O*-bromoacetylcholyl)-*m*-xylylenediamine (**69**) and *N,N'*-bis(3α-*O*-bromo-acetyldeoxycholyl)-*m*-xylylenediamine (**70**), respectively (Pandey *et al.*, 2002). The synthesis of bile acid-based acyclic dimers **69** and **70** has been discussed before (see *Chapter 2*; *Scheme 2.4.24).*

Bis(3α-*O*-hydroxyacetylcholyl)-*m*-xylylenediamine cyclic terephthalate (**71**, 248 mg, 81%, mp: 165–168 °C) was synthesized from acyclic dimer **69** (300 mg, 0.25 mmol) and bis-caesium terephthalate (113 mg, 0.25 mmol) (*Scheme 3.1.22*), following the same protocol as outlined above. IR (KBr): ν_{max} cm^{-1} 3419m, 1733s (C=O), 1653m. ^{1}H NMR (300 MHz, $CDCl_3$-CD_3OD): δ 8.14 (br s, 4H, Ph-CH), 7.19 (br s, 4H, Ph-CH), 4.82 (br s, 4H, 2 × $COCH_2O$), 4.68 (m, 2H, 2 × 3β-CH), 4.36 (m, 4H, 2 × CH_2N), 3.90 (br s, 2H, 2 × 12β-CH), 3.80 (br s, 2H, 2 × 7β-CH), 2.50–1.05 (m, 48H, steroidal CH and CH_2), 0.95 (br s, 6H, 2 × 21-Me), 0.89 (br s, 6H, 2 × 19-Me), 0.65 (s, 6H, 2 × 18-Me). ^{13}C NMR (75 MHz, $CDCl_3$-CD_3OD): δ 76.0 (2 × C-3), 67.8 (2 × C-7), 72.6 (2 × C-12), 31.5 (C-22), 32.5 (2 × C-23), 174.6 (2 × C-24), 167.0 (2 × OCO-Ph), 164.9 (2 × $OCOCH_2$), 138.7 (Ph-C), 133.1 (Ph-C), 129.7 (Ph-CH), 128.6 (Ph-CH), 126.4 (CH, Ph-CH), 61.5 (2 × $OCOCH_2$), 43.0 (2 × CH_2NHCO). FABMS: *m/z* 1165 [M + H]$^+$ (Pandey *et al.*, 2002).

Similarly, another cholaphane, bis(3α-*O*-hydroxyacetyldeoxycholyl)-*m*-xylylenediamine cyclic terephthalate (**72,** 183 mg, 75%, mp: 178–180 °C), was prepared from bis(bromoacetyldeoxycholyl)-*m*xylylenediamine **70** (245 mg, 0.21 mmol) by reacting it with an equivalent amount of bis-caesium terephthalate (97 mg, 0.21 mmol) (*Scheme 3.1.22*). IR (KBr): ν_{max} cm^{-1} 3460m, 1735s (C=O), 1653m. ^{1}H NMR (300 MHz, $CDCl_3$-CD_3OD): δ 8.15 (br s, 4H, Ph-CH), 7.32 (m, 4H, Ph-CH), 4.83 (m, 6H, 2 × $COCH_2O$ and 2 × 3β-H), 4.37

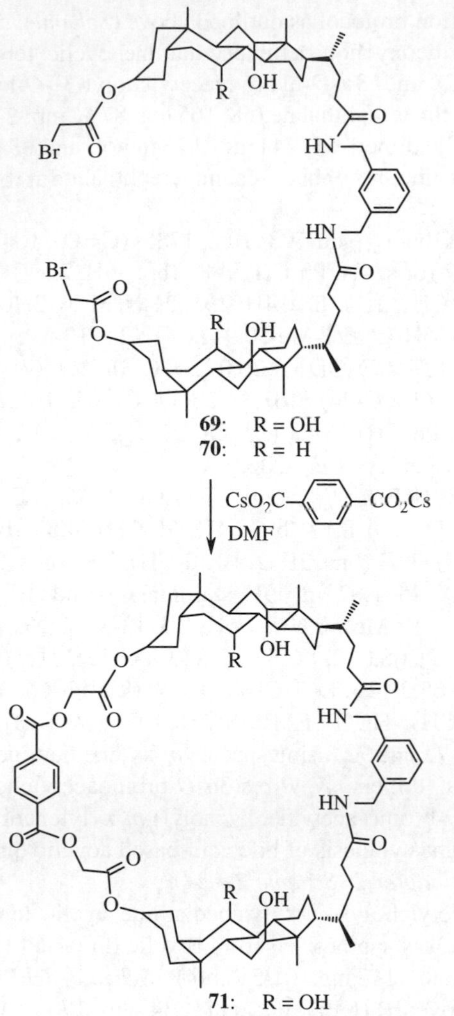

Scheme 3.1.22 *Synthesis of 'head-to-head' cholaphanes* ***71*** *and* ***72***

(m, 4H, 2 × CH_2N), 3.96 (br s, 2H, 2 × 12β-CH), 2.40-1.10 (m, 50H, steroidal CH and CH_2), 0.98 (br s, 6H, 2 × 21-Me), 0.89 (s, 6H, 2 × 19-Me), 0.65 (s, 6H, 2 × 18-Me). ^{13}C NMR (75 MHz, $CDCl_3$-CD_3OD): δ 76.6 (2 × C-3), 26.9 (2 × C-7), 73.1 (2 × C-12), 31.4 (2 × C-22), 32.0 (2 × C-23), 173.6 (2 × C-24), 167.1 (2 × OCO-Ph), 165.2 (2 × $OCOCH_2$), 139.0 (Ph-C), 133.4 (Ph-C), 129.9 (Ph-CH), 129.1 (Ph-CH), 127.0 (Ph-CH), 61.7 (2 × $OCOCH_2$), 43.5 (2 × CH_2NHCO). FABMS: *m/z* 1131 $[M + H]^+$ (Pandey *et al.*, 2002).

The synthesis of a 'head-to-head' cholaphane, 3α-hydroxy-5β-cholan-24-oic acid ethane-1,2-diol diester terephthalate (**75**), was achieved from 3α-hydroxy-5β-cholan-24-oic acid ethane-1,2-diol diester (**74**) (Kolehmainen *et al.*, 1996). The synthetic steps of **74** from lithocholic acid (**73**) have been discussed earlier (see *Chapter 2; Schemes 2.4.31* and *2.4.32*).

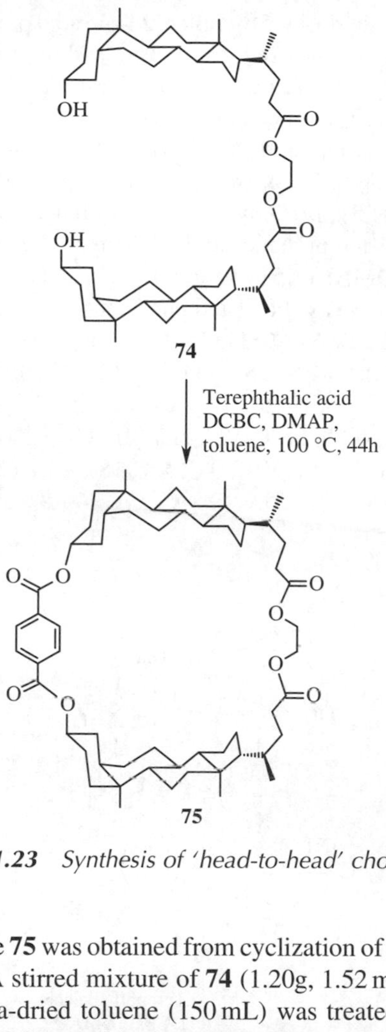

74

Terephthalic acid
DCBC, DMAP,
toluene, 100 °C, 44h

75

Scheme 3.1.23 *Synthesis of 'head-to-head' cholaphane **75***

In the final step, cholaphane **75** was obtained from cyclization of **74** with terephthalic acid as follows (*Scheme 3.1.23*). A stirred mixture of **74** (1.20g, 1.52 mmol) and terephthalic acid (250 mg, 1.52 mmol) in Na-dried toluene (150 mL) was treated with DMAP (1.50 g, 4.0 equiv.) and heated to 100 °C in an oil bath, DCBC (700 mg, 1.1 equiv) was added and the mixture was stirred for 40 h at the same temperature. The solvent was rotary evaporated and the crude solid was dissolved in $CHC1_3$ (75 mL), extracted with 2M HC1 (2 × 60 mL) followed by saturated aqueous $NaHCO_3$ (2 × 60 mL) to remove unreacted reagents. The solution was washed with H_2O (60 mL), dried (MgSO4) and evaporated under pressure. The crude solid was purified by step gradient CC (mobile phase: 12% EtOAc in DCM and 4% EtOAc in DCM) to afford cholaphane 75 (150 mg, 11%, mp: 143–145 °C). ^{13}C NMR (67.7 MHz, $CDCl_3$): δ 35.1 (2 × C-1), 26.9 (2 × C-2), 75.7 (2 × C-3), 32.4 (2 × C-4), 42.0 (2 × C-5), 27.1 (2 × C-6), 26.4 (2 × C-7), 35.9 (2 × C-8), 40.6 (2 × C-9), 34.8 (2 × C-10), 21.0 (2 × C-11), 42.2 (2 × C-12), 42.9 (2 × C-13), 56.5 (2 × C-14), 24.5 (2 × C-15), 28.3 (2 × C-16), 56.1 (2 × C-17), 12.2 (2 × C-18), 23.4 (2 × C-19), 35.4 (2 × C-20), 18.4 (2 × C-21), 31.2 (2 × C-22), 31.0 (2 × C-23), 174.0 (2 × C-24), ethane-1,2-diol: 62.1 (C-1 and C-2), terephthalate: 165.5 (2 × CO), 134.6 (2 × C-1), 129.5 (2 × C-2). CIMS: *m/z* 909 $[M]^+$, 743 $[M - C_6H_4(CO_2H)_2]^+$ (Kolehmainen *et al*., 1996).

The cholaphane, 3α,3'α-dihydroxy-5β-cholan-24-oic acid piperazine diamide terephthalic acid diester (**77**), was synthesized from 3α,3'α-dihydroxy-5β-cholan-24-oic acid piperazine diamide terephthalic acid diester (**76**) (Tamminen *et al*., 2000). The synthesis of **76** from lithocholic acid (**73**) has been discussed before (see *Chapter 2; Schemes 2.4.31* and *2.4.32*). In the final step, cholaphane **77** was achieved from cyclization of **76** with terephthalic acid (*Scheme 3.1.24*), following the same protocol as described above for cholaphane **75**. A mixture of 3α,3'α-dihydroxy-5β-cholan-24-oic acid piperazine diamide (**76**, 1.2 g, 1.49 mmol) and terephthalic acid (250 mg, 1.50 mmol) in Na-dried toluene (150 mL) was treated with DMAP (1.5 g, 4 equiv.) and DCBC (700 mg, 1.1 equiv.) at 100 °C for 44 h. After the usual workup as above and purification by step gradient CC, eluted with 10% Me_2CO in $CHC1_3$ and 20% Me_2CO in DCM, furnished cholaphane **77** (120 mg, 9%). ^{13}C NMR (125 MHz, $CDCl_3$): δ 34.8 (2 × C-1), 26.5 (2 × C-2), 74.4 (2 × C-3), 31.9 (2 × C-4), 41.5 (2 × C-5), 26.6 (2 × C-6), 26.2 (2 × C-7), 35.5 (2 × C-8), 40.7 (2 × C-9), 34.3 (2 × C-10), 20.8 (2 × C-11), 40.4 (2 × C-12), 42.7 (2 × C-13), 57.1 (2 × C-14), 24.0 (2 × C-15), 27.6 (2 × C-16), 54.1 (2 × C-17), 12.0 (2 × C-18), 23.1 (2 × C-19), 33.1 (2 × C-20), 18.7 (2 × C-21), 29.8 (2 × C-22), 25.9 (2 × C-23), 171.8 (2 × C-24), piperazine: 44.8

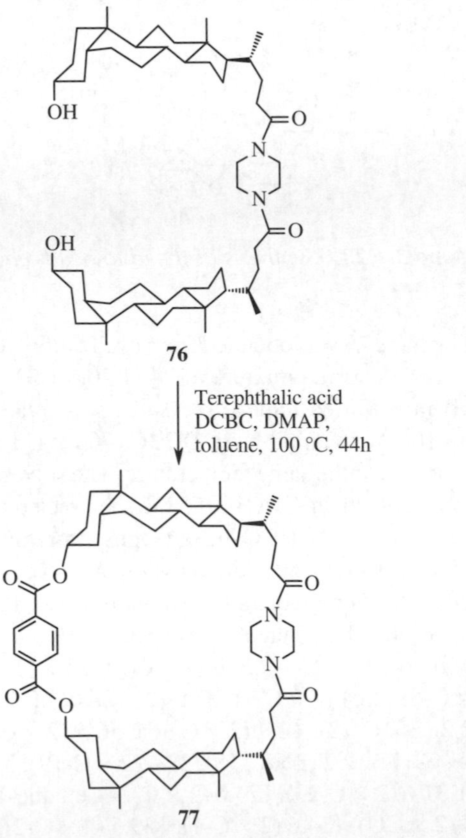

76

Terephthalic acid
DCBC, DMAP,
toluene, 100 °C, 44h

77

Scheme 3.1.24 *Synthesis of 'head-to-head' cholaphane 77*

($2 \times$ C-1), 41.3 ($2 \times$ C-2), terephthalate: 164.8 ($2 \times$ CO), 134.5 ($2 \times$ C-1), 129.3 ($2 \times$ C-2). MS (MALDI–TOF): *m/z* 934.18 $[M + H]^+$ (Tamminen *et al.*, 2000).

A 'head-to-head' cholaphane **79**, with an ethane-1,2-diol bridge, was prepared from acyclic lithocholic acid dimer **78** (Ra *et al.*, 2000). The synthetic steps of acyclic lithocholic acid dimer **78** from lithocholic acid (**73**) have been discussed earlier (see *Chapter 2; Schemes 2.4.45* and *2.4.46*). A mixture of 60% NaH (4.0 mg, 0.10 μmol) and acyclic dimer **78** (48 mg, 0.05 mmol) and refluxed for 72 h, and the oily residue was purified by prep-TLC, eluted with 6% EtOAc in *n*-hexane, to give cholaphane **79** (18.5 mg, 47%, mp: 155 °C) (*Scheme 3.1.25*). ^{1}H NMR (300 MHz, $CDCl_3$): δ 3.64–3.56 (series of m, 6H), 3.53-3.48 (series of m, 2H), 3.34 (t, $J = 7.94$ Hz, 4H), 3.19 (m, 2H), 1.94–0.93 (series of m, 52H), 0.90-0.86 (series of m, 16H), 0.59 (s, 6H); ^{13}C NMR (75 MHz, $CDCl_3$): δ 80.2, 72.9,70.5, 69.0, 56.8, 53.7, 42.4, 42.2, 40.2, 40.1, 35.7, 35.5, 34.7, 34.6, 32.9, 30.8, 28.3, 27.2, 26.9, 26.1, 24.1, 23.5, 23.1, 20.0, 18.7, 11.8 (Ra *et al.*, 2000).

Several other interesting cholaphanes **83–86** containing oxamide spacers were synthesized from cholic acid derivative, methyl 7α,12α-diacetoxy-3α-hydroxy-5β-cholan-24-oate. The spacers bind two identical steroidal subunits 'head-to-head' through 3α,3'α positions and 'head-to-tail' through 3α,24' positions (Łotowski *et al.*, 2008). The synthetic steps of cholic acid-based acyclic dimers, *N,N'*-di(24-amino-7α,12α-diacetoxy-5β-cholan)3α,3'α-oxamide (**80**) and *N*-(24-amino-7α,12α-diacetoxy-5β-cholane)-Scheme *N'*-(3'α-amino-7'α,12'α-diacetoxy-5'β-cholan)-3α,24'-oxamide (**81**) from methyl 7α,

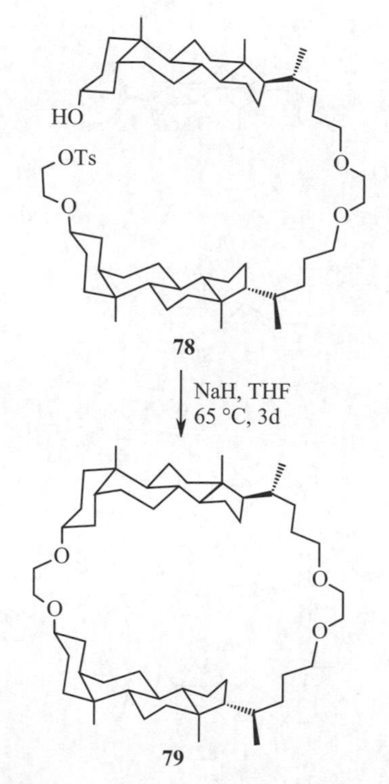

Scheme 3.1.25 *Synthesis of 'head-to-head' cholaphane* ***79***

12α-diacetoxy-3α-hydroxy-5β-cholan-24-oate have been discussed before (see *Chapter 2; Schemes 2.1.54–2.1.57* and *2.5.11–2.5.13*).

N,N'-Di(24-amino-7α,12α-diacetoxy-5β-cholan)3α,3'α-oxamide (**80**) was utilized for the synthesis of *N,N':N',N"*-di(7α,12α-diactoxy-5β-cholane)-3α,3'α:24,24'-dioxamide (**82**) (*Scheme 3.1.26*). A solution of $(COCl)_2$ (43 μL, 0.50 mmol) and dry C_5H_5N (109 μL, 1.37 mmol) was added to a stirred solution of acyclic dimer **82** (343 mg, 0.34 mmol) in dry toluene (200 mL, dried over CaH_2). The reaction mixture was heated at 75 °C for 60 min, after which MeOH (few drops) was added and the solvent was evaporated under vacuum. The crude residue was purified by CC (mobile phase: 4% MeOH in $CHCl_3$) and crystallized from *n*-hexane-DCM to obtain the cholaphane **82** (100 mg, 26%, mp: >300 °C). IR (KBr): ν_{max} cm^{-1} 3396m (N–H), 1724s (C=O), 1674s (C=O), 1514m (N–H), 1256m (C–O), 1022m. ^1H NMR (200 MHz, $CDCl_3$): δ 7.61 (dd, 2H, J = 3.47, 8.95 Hz, CH_2NH), 7.49 (d, 2H, J = 8.15 Hz, 2 × 3α-NH), 5.01 (m, 2H, 2 × 12β-CH), 4.86 (m, 2H, 2 × 7β-CH), 3.61 (m, 2H, 2 × 3β-CH), 3.30 (m, 2H, CH_2NH), 2.95 (m, 2H, CH_2NH), 2.16 (s, 6H, 2 × 12α-MeCO), 2.00 (s, 6H, 2 × 7α-MeCO), 0.86 (s, 6H, 2 × 19-Me), 0.82 (d, 6H, J = 6.4 Hz, 2 × 21-Me), 0.72 (s, 6H, 2 × 18-Me). ^{13}C NMR (50 MHz, $CDCl_3$): δ 12.1, 17.7, 21.5, 21.6, 22.3, 22.5, 23.8, 25.5, 25.6, 27.2, 29.3, 30.9, 32.4, 33.3, 33.6, 33.9, 35.5, 37.5, 40.6, 41.2, 44.3, 44.9, 45.2, 50.6, 70.7, 75.7, 159.3, 159.8, 170.1, 170.5. FABMS: *m/z* 1083.8 $[M + Na]^+$ (Łotowski *et al.*, 2008).

Alkaline hydrolysis of *N,N':N',N"*-di(7α,12α-diactoxy-5β-cholane)-3α,3'α:24,24'-dioxamide (**82**) furnished another cholaphane, *N,N':N',N"*-di(5β-cholane-7α,12α-diol)-3α,3'α:24,24'-dioxamide (**83**) (*Scheme 3.1.27*). To a stirred solution of **82** (40 mg,

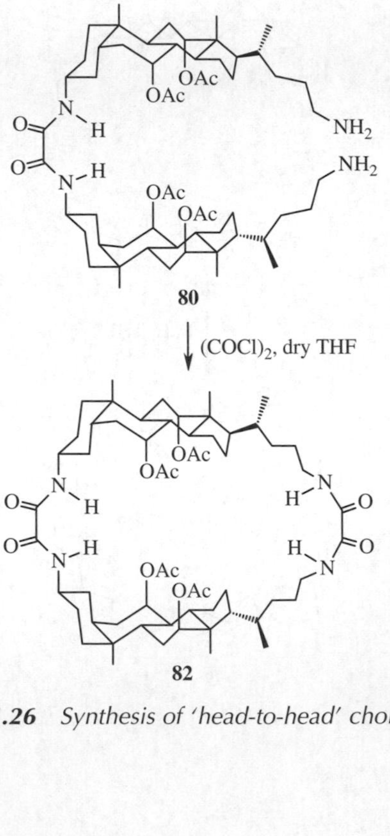

80

$(COCl)_2$, dry THF

82

Scheme 3.1.26 *Synthesis of 'head-to-head' cholaphane **82***

0.038 mmol) in THF (8 mL), 1M NaOH (2 mL) in MeOH (2 mL) was added at room temperature. After 4 days, acidified H_2O (50 mL) was and the resulting mixture was extracted twice with $CHCl_3$. The organic layers were dried ($MgSO_4$), evaporated to dryness and the crude residue was purified by CC, eluted with 10% MeOH in $CHCl_3$, and crystallized from *n*-hexane-DCM to provide cholaphane **83** (16 mg, 48%, mp: 264–266 °C). IR (KBr): ν_{max} cm^{-1} 3397br (O–H and N–H), 1674s (C=O), 1514m (N–H), 1242m (C–O). ^1H NMR (200 MHz, $CDCl_3$): δ 7.58 (m, 4H, 2 × CH_2NH, 2 × 3α-NH), 3.95 (m, 2H, 2 × 12β-CH), 3.85 (m, 2H, 2 × 7β-CH), 3.51 (m, 4H, 2 × 3β-CH, CH_2NH), 3.00 (m, 2H, CH_2NH), 1.02 (d, 6H, J = 6.5 Hz, 2 × 21-Me), 0.89 (s, 6H, 2 × 19-Me), 0.70 (s, 6H, 2 × Scheme 18-Me). ^{13}C NMR (50 MHz, $CDCl_3$): δ 12.4, 17.7, 22.3, 23.1, 24.0, 25.2, 26.8, 27.4, 28.2, 29.7, 32.6, 34.0, 34.1, 34.6, 35.8, 39.4, 40.7, 41.5, 42.3, 45.0, 46.3, 50.7, 68.2, 72.8, 159.4, 159.8. FABMS: *m/z* 915 $[M + Na]^+$. HRFABMS: 915.6187calculated for $C_{52}H_{84}N_4NaO_8$, found 915.6146 (Łotowski *et al.*, 2008).

N,N':N',N"-di(7α,12α-diactoxy-5β-cholane)-3α,24':3'α,24-dioxamide (**84**) was prepared from *N*-(24-amino-7α,12α-diacetoxy-5β-cholane)-*N'*-(3'α-amino-7'α,12'α-diacetoxy-5'β-cholan)-3α,24'-oxamide (**81**) (*Scheme 3.1.28*), following the same method as described for cholaphane **82**. After the usual workup as above, and crystallization from MeOH-EtOAC–heptane yielded cholaphane **84** (51 mg, 24%, mp: 305 °C). IR (KBr): ν_{max} cm^{-1} 3400m (N–H), 1724s (C=O), 1674s (C=O), 1508m (N–H), 1253m (C–O), 1027m. ^1H NMR (200 MHz, $CDCl_3$): δ 7.58 (dd, 2H, J = 3.46, 8.88 Hz, CH_2NH), 7.46 (d, 2H, J = 8.12 Hz, 2 × 3α-NH), 5.04 (m, 2H, 2 × 12β-CH), 4.92 (m, 2H, 2 × 7β-CH), 3.46 (m, 4H,

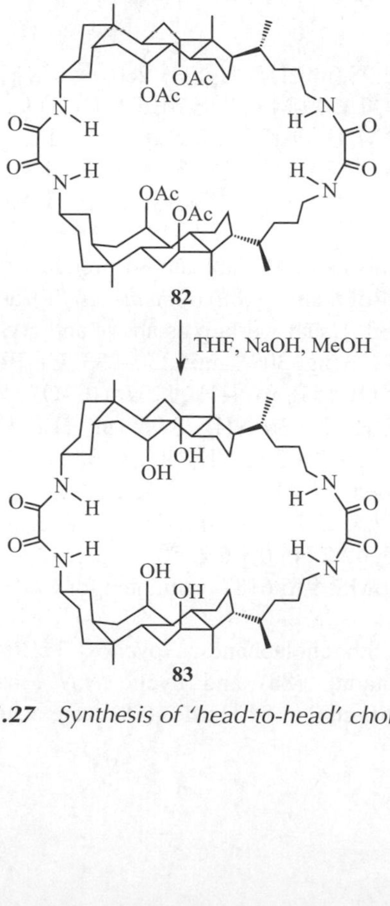

82

THF, NaOH, MeOH

83

Scheme 3.1.27 *Synthesis of 'head-to-head' cholaphane* **83**

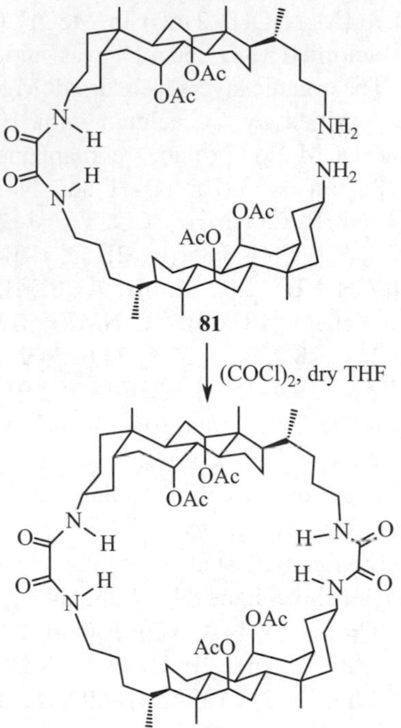

Scheme 3.1.28 *Synthesis of 'head-to-tail' cholaphane* ***84***

2 × 3β-CH, 2 × CH_2NH), 2.98 (m, 2H, CH_2NH), 2.16 (s, 6H, 2 × 12α-MeCO), 2.09 (s, 6H, 2 × 7α-MeCO), 0.91 (s, 6H, 2 × 19-Me), 0.85 (d, 6H, J = 6.69 Hz, 2 × 21-Me), 0.76 (s, 6H, 2 × 18-Me). ^{13}C NMR (50 MHz, $CDCl_3$): δ 12.0, 17.3, 21.3, 22.1, 22.5, 22.6, 22.9, 25.9, 26.7, 27.9, 29.7, 31.0, 31.5, 33.8, 34.3, 34.4, 34.8, 37.3, 41.1, 41.3, 43.9, 44.6, 45.0, 50.1, 70.6, 75.4, 158.9, 160.5, 170.1, 170.5. FABMS: *m/z* 1083.8 $[M + Na]^+$ (Łotowski *et al.*, 2008).

Similarly, alkaline hydrolysis of cholaphane **84** produced *N,N':N',N"*-di(5β-cholane-7α,12α-diol)-3α,24':3'α,24-dioxamide (**85**) (*Scheme 3.1.29*), according to the procedure as described for cholaphane **83**. Usual workup as above and crystallization from *n*-hexane-DCM gave cholaphane **85** (13 mg, 30%, mp: 230–233 °C). IR (KBr): ν_{max} cm^{-1} 3396br (O–H and N–H), 1674s (C=O), 1518m (N–H), 1237m (C–O). ^1H NMR (200 MHz, $CDCl_3$): δ 7.53 (m, 4H, 2 × CH_2NH, 2 × 3α-NH), 3.96 (m, 2H, 2 × 12β-CH), 3.84 (m, 2H, 2 × 7β-CH), 3.51 (m, 4H, 2 × 3β-CH, CH_2NH), 3.01 (m, 2H, CH_2NH), 1.02 (d, 6H, J = 6.43 Hz, 2 × 21-Me), 0.90 (s, 6H, 2 × 19-Me), 0.72 (s, 6H, 2 × 18-Me). ^{13}C NMR (50 MHz, $CDCl_3$): δ 12.3, 17.7, 22.2, 23.1, 24.0, 25.1, 26.8, 27.5, 28.2, 29.8, 32.4, 34.0, 34.1, 34.5, 35.8, 39.6, 40.6, 41.5, 42.2, 45.0, 46.2, 50.5, 68.2, 72.4, 159.3, 159.8. FABMS: *m/z* 915.6 $[M + Na]^+$. HRFABMS: 915.6187 calculated for $C_{52}H_{84}N_4NaO_8$, found 915.6153 (Łotowski *et al.*, 2008).

Two 'head-to-head' lithocholaphanes, cyclic 3α,3'α-(isophthaloyloxy)-bis(5β-cholan-24-yloxy)-3-oxaglutarate (**88**) and cyclic *N,N'*-(3-oxa-1,5-pentanediyl) 3α,3' α-(terephthaloyl oxy)-bis(5β-cholan-24-amide) (**89**) were achieved from lithocholic acid

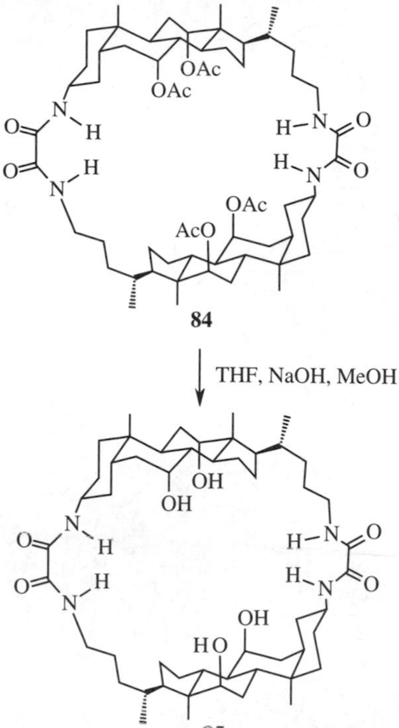

84

THF, NaOH, MeOH

85

Scheme 3.1.29 *Synthesis of 'head-to-tail' cholaphane* ***85***

(**73**) in several steps (Valkonen *et al.*, 2007). The synthesis of lithocholic acid-based acyclic dimers, 3α,3'α-(isophthaloyloxy)-bis(5β-cholan-24-ol) (**86**) and bis(pentafluorophenyl) 3α,3'α-(tere phthaloyloxy)-bis(5β-cholan-24-oate) (**87**) from **73** has been discussed earlier (see *Chapter 2; Schemes 2.1.50* and *2.1.51–2.1.53*). 3α,3'α-(Isophthaloyloxy)-bis(5β-cholan-24-ol) (**86**) was converted to cholaphane **88** as follows (*Scheme 3.1.30*). DMAP (3 mg, 0.02 mmol), 2,2'-oxydiacetylchloride (14 mg, 0.08 mmol), and Et_3N (500 μL) were added to a stirred solution of **86** (68 mg, 0.08 mmol) in toluene (8 mL) and the mixture was heated at 100 °C (oil bath). After 20 h, the solvent was rotary evaporated and the crude solid was purified by CC, eluted with DCM, to yield cyclic 3α,3'α-(isophthaloyloxy)-bis(5β-cholan-24-yloxy)-3-oxaglutarate (**88**, 14 mg, 15%, mp: 314.6 °C). ^1H NMR (500 MHz, $CDCl_3$): δ 8.45 (t, 2H, J=1.6 Hz, Ph-CH), 8.26 (dd, H, J= 1.7, 7.8 Hz, Ph-CH), 7.54 (t, 1H, J=7.8 Hz, Ph-CH), 4.95 (m, 2H, J=4.3, 7.3 Hz, 2 × 3β-CH), 4.25-4.12 (m, 4H, 2 × 26-CH_2), 4.20 and 4.24 (d, 4H, J =16.5 Hz, 2 × 24-CH_2), 0.99 (s, 6H, 2 × 19-Me), 0.93 (d, 6H, J=6.5 Hz, 2 × 21-Me), 0.68 (s, 6H, 2 × 18-Me). ^{13}C NMR (125 MHz, $CDCl_3$): δ 35.2 (2 × C-1), 26.8 (2 × C-2), 76.4 (2 × C-3), 32.4 (2 × C-4), 42.3 (2 × C-5), 27.2 (2 × C-6), 26.5 (2 × C-7), 35.9 (2 × C-8), 40.6 (2 × C-9), 34.8 (2 × C-10), 20.9 (2 × C-11), 40.2 (2 × C-12), 42.8 (2 × C-13), 56.5 (2 × C-14), 24.3 (2 × C-15), 28.3 (2 × C-16), 56.4 (2 × C-17), 12.0 (2 × C-18), 23.4 (2 × C-19), 35.1 (2 × C-20), 18.6 (2 × C-21), 32.1 (2 × C-22), 25.3 (2 × C-23), 65.3 (2 × C-24), 68.5 (2 × C-26), 169.9 (2 × $OCH_2CO_2CH_2$), 165.8 (2 × $PhCO_2$), 134.3 (2 × Ph-C), 131.3 (2 × Ph-CH), 128.8 (Ph-CH), 128.6 (Ph-CH). ESIMS: *m/z* 975.8177 $[M + Na]^+$ (Valkonen *et al.*, 2007).

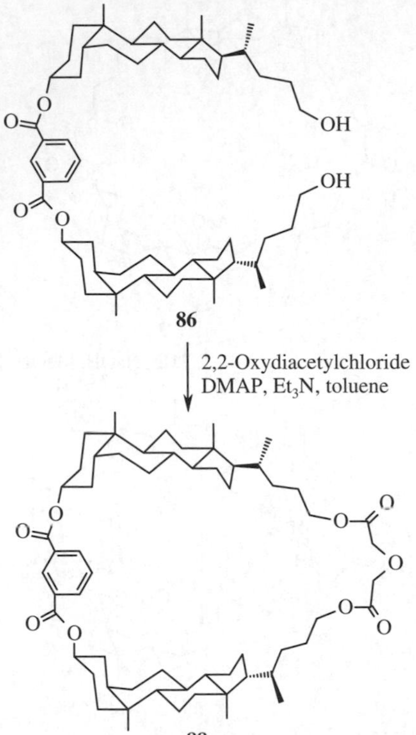

86

2,2-Oxydiacetylchloride
DMAP, Et_3N, toluene

88

Scheme 3.1.30 *Synthesis of lithocholaphane **88***

Cholaphane **89** was synthesized from bis(pentafluorophenyl)3α,3'α-(terephthaloyloxy)-bis(5β-cholan-24-oate) (**87**) in the following manner (*Scheme 3.1.31*). DMAP (22 mg, 0.18 mmol), Et_3N (25 μL, 0.18 mmol) and 2,2'-oxybis(ethylamine) (7.0 mg, 0.07 mmol) were added to a stirred solution of **87** (73 mg, 0.06 mmol) in dry DCM (6 mL) and stirred at room temperature. After 15 h, DCM was added and the mixture was washed with 10% KOH followed by H_2O. The organic layer was dried ($MgSO_4$), evaporated under pressure and the crude product was purified by CC (mobile phase: DCM:MeOH = 30:1) to produce cyclic *N,N'*-(3-oxa-1,5-pentanediyl) 3α,3'α-(terephthaloyloxy)-bis(5β-cholan-24-amide) (**89**, 21 mg, 37%, mp: 352.8 °C). ^{1}H NMR (500 MHz, $CDCl_3$): δ 8.10 (s, 4H, Ph-CH), 5.74 (t, 2H, 2 × NH), 5.06 (m, 2H, J = 4.9, 5.3, 5.8 Hz, 2 × 3β-CH), 3.62-3.34 (m, 8H, 2 × 25-CH and 2 × 26-CH), 0.96 (s, 6H, 2 × 19-Me), 0.95 (d, 6H, J = 6.7 Hz, 2 × 21-Me), 0.68 (s, 6H, 2 × 18-Me). ^{13}C NMR (125 MHz, $CDCl_3$): δ 34.9 (2 × C-1), 26.6 (2 × C-2), 74.7 (2 × C-3), 32.1 (2 × C-4), 41.7 (2 × C-5), 26.8 (2 × C-6), 26.2 (2 × C-7), 35.6 (2 × C-8), 40.7 (2 × C-9), 34.4 (2 × C-10), 21.0 (2 × C-11), 40.5 (2 × C-12), 42.7 (2 × C-13), 57.1 (2 × C-14), 24.0 (2 × C-15), 28.3 (2 × C-16), 55.2 (2 × C-17), 12.0 (2 × C-18), 23.2 (2 × C-19), 35.1 (2 × C-20), 18.4 (2 × C-21), 31.8 (2 × C-22), 32.2 (2 × C-23), 173.8 (2 × C-24), 39.0 (2 × CH_2NH), 69.7 (2 × CH_2O), 165.1 (2 × Ph-CO), 134.6 (2 × Ph-C), 129.4 (2 × Ph-C). ESIMS: *m/z* 973.8069 [M + Na]$^+$ (Valkonen *et al.*, 2007).

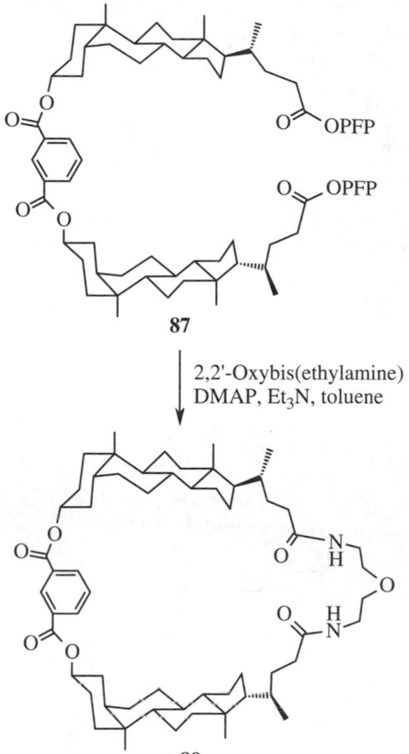

87

2,2'-Oxybis(ethylamine)
DMAP, Et_3N, toluene

89

Scheme 3.1.31 *Synthesis of lithocholaphane* ***89***

The synthesis of cholaphanes, bis(3α-*O*-hydroxyacetyllithocholyl)-pyridine-2,6-diamine cyclic 3,5-pyridine dicarboxylate (**93**) and bis(3α-*O*-hydroxyacetyldeoxycholyl)-pyridine-2,6-diamine cyclic 3,5-pyridine dicarboxylate (**94**) was performed utilizing lithocholic acid (**73**, 3α-hydroxy-5β-cholan-24-oic acid) and deoxycholic acid (**90**, 3α,12α-dihydroxy-5β-cholan-24-oic acid) using a multistep sequence involving Cs-salt methodology of macrocyclization (Chattopadhyay and Pandey, 2006). The synthetic steps leading to the formation of acyclic dimers, *N,N'*-bis(3α-*O*-bromoacetyllithocholyl)-pyridine-2,6-diamine (**91**) and *N,N'*-bis(3α-*O*-bromoacetyldeoxycholyl)-pyridine-2,6-diamine (**92**) (precursor for the cyclic dimer **93** and **94**, respectively) have been discussed previously (see *Chapter 2; Scheme 2.4.24*). Both cholaphanes **93** and **94** showed flavin binding ability and can be used as molecular receptors for recognition of flavin analogues.

An equivalent amount of bis-caesium 3,5-pyridine dicarboxylate (390 mg, 0.92 mmol) was added to a stirred solution of lithocholic acid-based acyclic dimer **91** (990 mg, 0.92 mmol) in dry DMF (250 mL) at room temperature (*Scheme 3.1.32*). After 12 h, DMF was evaporated to dryness, the crude solid was dissolved in $CHCl_3$ (30 mL) and washed with brine (10 mL), dried (Na_2SO_4), concentrated under vacuum and purified by FCC, eluted with 25% EtOAc in *n*-hexane, to afford lithocholaphane **93** (750 mg, 69%, mp: 175–177 °C). IR (KBr): ν_{max} cm^{-1} 3369br (N–H), 2936m (C–H), 1741s (C=O), 1702s (C=O), 1586m,

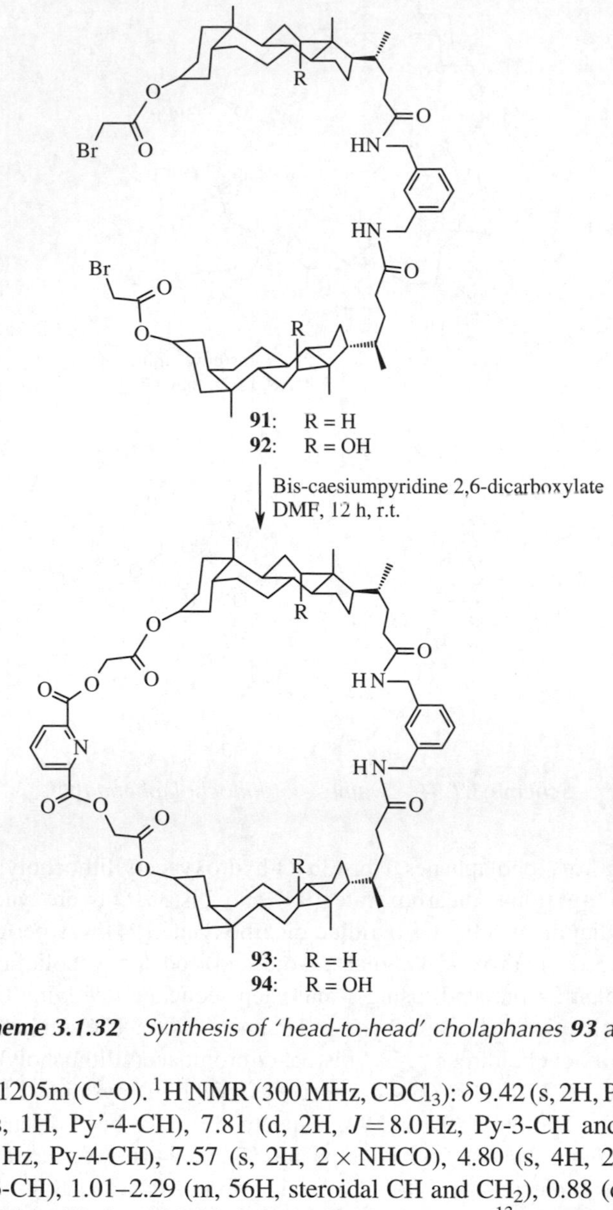

91: R = H
92: R = OH

Bis-caesiumpyridine 2,6-dicarboxylate
DMF, 12 h, r.t.

93: R = H
94: R = OH

Scheme 3.1.32 *Synthesis of 'head-to-head' cholaphanes* ***93*** *and* ***94***

1287m (C–O), 1205m (C–O). ^{1}H NMR (300 MHz, $CDCl_3$): δ 9.42 (s, 2H, Py'-2-CH and Py'-6-CH), 8.91 (s, 1H, Py'-4-CH), 7.81 (d, 2H, $J = 8.0$ Hz, Py-3-CH and Py-5-CH), 7.62 (t, 1H, $J = 8.0$ Hz, Py-4-CH), 7.57 (s, 2H, 2 × NHCO), 4.80 (s, 4H, 2 × $COCH_2$), 4.72 (m, 2H, 2 × 3β-CH), 1.01–2.29 (m, 56H, steroidal CH and CH_2), 0.88 (d, 6H, J =6.2 Hz, 2 × 21-Me), 0.85 (s, 6H, 2 × 19-Me), 0.59 (s, 6H, 2 × 18-Me). ^{13}C NMR (75 MHz, $CDCl_3$): δ 12.0, 18.5, 20.8, 23.2, 24.1, 26.2, 26.4, 26.8, 28.2, 31.0, 31.9, 33.4, 34.5, 34.9, 35.7, 40.2, 40.5, 41.8, 42.7, 54.9, 56.5, 56.7, 62.3, 76.6, 109.2, 125.4, 138.7, 140.9, 149.4, 154.8, 163.8, 163.7, 172.0. HRMS: *m/z* 1073.6579 calculated for $C_{64}H_{89}N_4O_{10}$, found: 1073.6625 (Chattopadhyay and Pandey, 2006).

The synthesis of deoxycholaphane, bis(3α-*O*-hydroxyacetyldeoxycholyl)-pyridine-2,6-diamine cyclic 3,5-pyridine dicarboxylate (**94**, 540 mg, 68%, mp: 170–171 °C) was accomplished from the deoxycholic acid-based acyclic dimer **92** (800 mg, 0.72 mmol) (*Scheme 3.1.32*), following the same protocol as outlined above. IR (KBr): ν_{max} cm^{-1} 3423br (N–H), 2936m (C–H), 1740s (C=O), 1702s (C=O), 1586m (N–H), 1288m (C–O), 1207m (C–O). ^{1}H NMR (300 MHz, $CDCl_3$): δ 9.50 (s, 2H, Py'-2-CH and Py-6-CH), 8.99 (s, 1H, Py'-4-CH), 7.79–7.84 (m, 4H, 2 × NHCO, Py-3-CH and Py-5-CH), 7.69 (t, 1H, J = 7.9 Hz, Py-4-CH), 4.88 (s, 4H, 2 × $COCH_2$), 4.83 (m, 2H, 2 × 3β-CH), 3.98 (s, 2H, 2 × 12β-CH), 1.08–2.40 (m, 52H, steroidal CH and CH_2), 1.01 (d, 6H, J = 5.6 Hz, 2 × 21-Me), 0.91 (s, 6H, 2 × 19-Me), 0.69 (s, 6H, 2 × 18-Me). ^{13}C NMR (75 MHz, $CDCl_3$): δ 11.7, 16.6, 22.0, 22.5, 24.9, 25.3, 25.7, 26.4, 27.9, 29.8, 30.8, 32.1, 32.6, 33.0, 33.5, 33.7, 34.8, 40.7, 44.8, 45.4, 47.6, 61.3, 72.0, 75.3, 108.5, 124.4, 137.7, 140.0, 148.5, 153.9, 162.9, 165.7, 171.4. HRMS: *m/z* 1104.6399 calculated for $C_{64}H_{88}N_4O_{12}$, found: 1104.6387 (Chattopadhyay and Pandey, 2006).

Several 'head-to-tail' cholic acid cyclopeptides (**95–97**) were obtained by joining two molecules of 3α-aminocholic acid or its acetyloxy derivatives to two phenylalanine monomers or dimers (Tamminen and Kolehmainen, 2001).

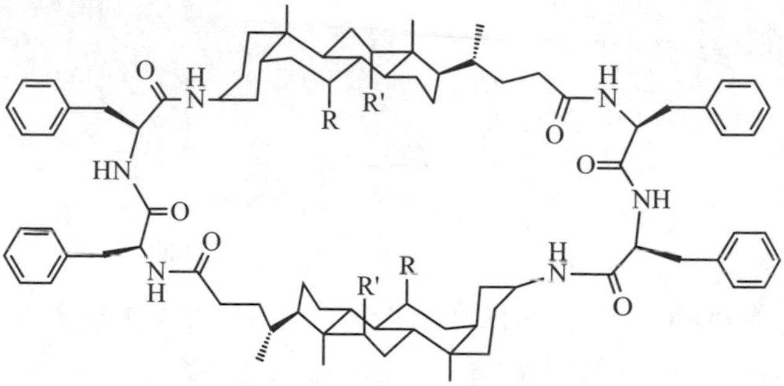

Compounds	R	R'
95	OH	OH
96	OH	OAc
97	OAc	OAc

The 'head-to-tail' cholaphanes **105** and **106** were achieved in a multistep sequence from cholic acid (**6**) (Davis *et al*, 1996; Davis 2007). The secondary hydroxyls of cholic acid (**6**) were protected as triformate **8**, which was transformed in two steps to 23-methyl ester **10** following the method described earlier by Bhattarai *et al.* (1997) (*Scheme 3.1.2*). The 23-methyl ester **10** was degraded to ketone **98** though a multistep sequence described in the literature (Barton *et al.*, 1989) (*Scheme 3.1.33*). The Baeyer–Villiger oxidation of ketone **98** yielded acetate **99** (Dias and Ramachandra, 1977) (*Scheme 3.1.33*), which was selectively deprotected both at 7α and 12α positions to form diol **100**, which was oxidized to diketone **101** (*Scheme 3.1.33*).

A silyl-modified Sakurai reaction introduced both spacer and solubilizing groups, with excellent regio- and stereoselectivity (Mekhalfia and Markó, 1991). Hydrolysis of both ester and acetate on diketone **101**, followed by selective tosylation at position 17β, yielded **102** (*Scheme 3.1.34*) (Davis *et al*, 1996; Davis 2007). Displacement of tosyl with azide,

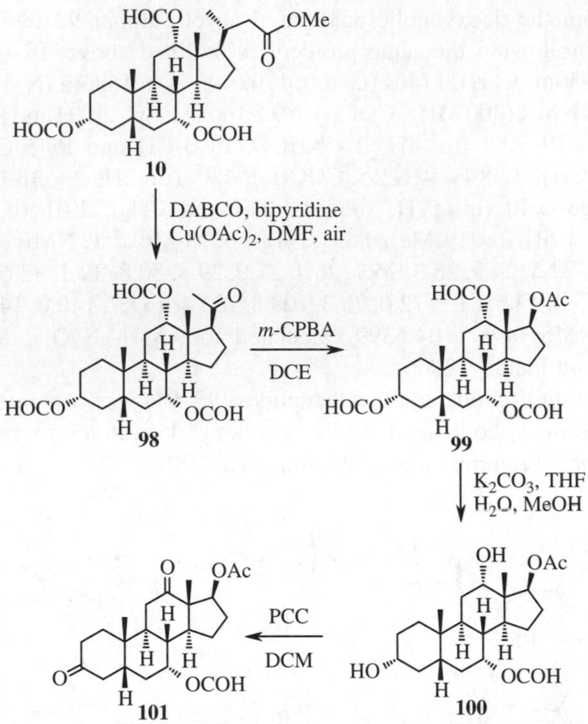
HOCO
OMe
O
H
H
H
HOCO
OCOH
H
10
DABCO, bipyridine
Cu(OAc)$_2$, DMF, air
HOCO
O
H
H
H
HOCO
OCOH
H
98
m-CPBA
DCE
HOCO
OAc
H
H
H
HOCO
OCOH
H
99
K$_2$CO$_3$, THF
H$_2$O, MeOH
OH
OAc
H
H
H
HO
OCOH
H
100
PCC
DCM
O
OAc
H
H
H
O
OCOH
H
101

Scheme 3.1.33 *Synthesis of steroidal derivatives* ***98–101***

reduction to amine, and then protection as Boc afforded **103** (*Scheme 3.1.34*) (Davis *et al*, 1996; Davis 2007).

Oxidative cleavage of the alkene and activation of the ester on **103** provided **104** (*Scheme 3.1.35*) (Davis *et al.*, 1996; Davis 2007). Finally, cyclization using TFA furnished the cholaphane **105**, which was converted to the cholaphane **106** by catalytic hydrogenation (*Scheme 3.1.35*) (Davis *et al.*, 1996; Davis 2007). Methodologies for the synthesis of several other 'head-to-tail' cholaphanes **106–108** are available in the literature (Nahar *et al.*, 2007).

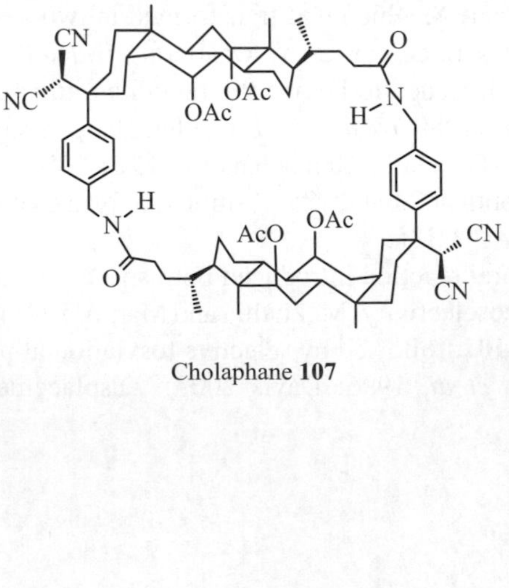
CN
NC
O
OAc
OAc
N
H
H
N
AcO
OAc
CN
O
CN

Cholaphane **107**

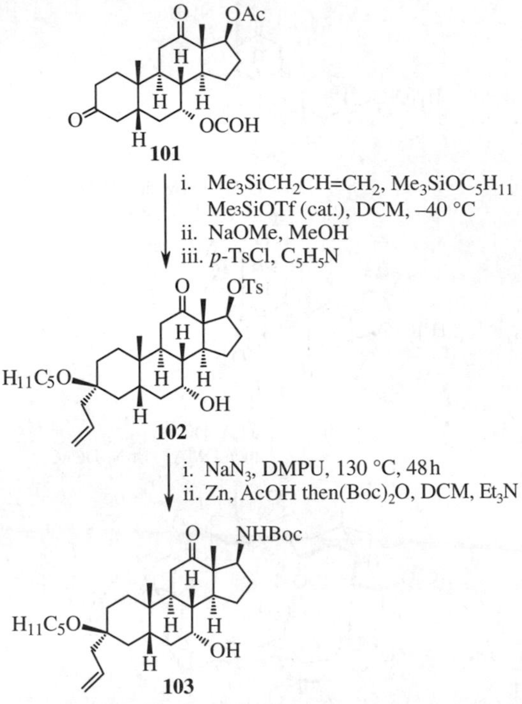

O OAc
H
H H
O OCOH
H
101

i. $Me_3SiCH_2CH{=}CH_2$, $Me_3SiOC_5H_{11}$
Me_3SiOTf (cat.), DCM, –40 °C
ii. NaOMe, MeOH
iii. *p*-TsCl, C_5H_5N

O OTs
H
$H_{11}C_5O$ H H
OH
H
102

i. NaN_3, DMPU, 130 °C, 48 h
ii. Zn, AcOH then$(Boc)_2O$, DCM, Et_3N

O NHBoc
H
$H_{11}C_5O$ H H
OH
H
103

Scheme 3.1.34 *Synthesis of steroidal derivatives **102** and **103***

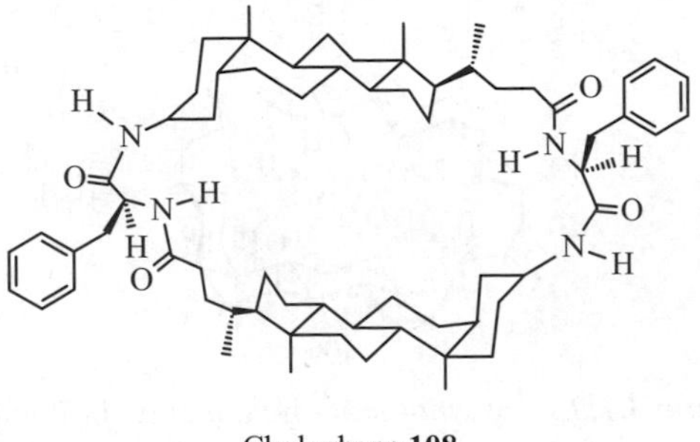

Cholaphane **108**

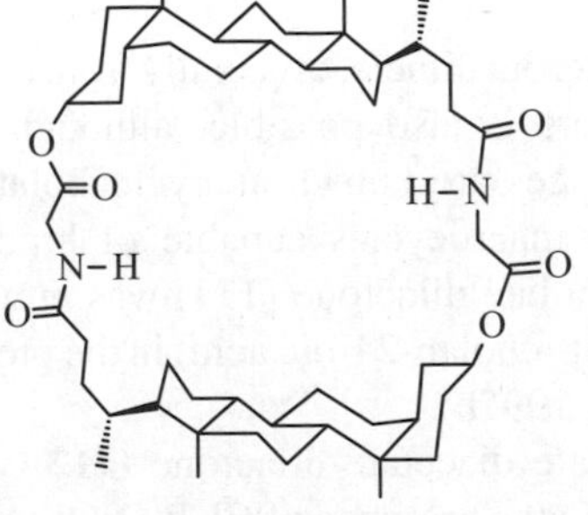

Cholaphane **109**

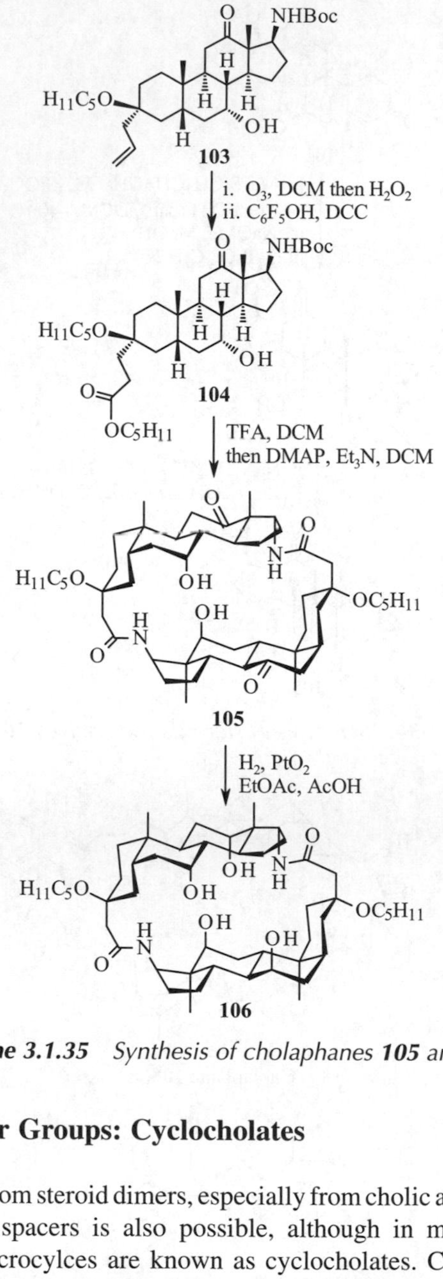

Scheme 3.1.35 *Synthesis of cholaphanes* ***105*** *and* ***106***

3.2 Without Spacer Groups: Cyclocholates

Macrocyles formation from steroid dimers, especially from cholic acids and their derivative, without involving any spacers is also possible, although in much less numbers than cholaphanes. These macrocylces are known as cyclocholates. Cyclocholates are two to six units of bile acid-based macrocycles capable of binding other biomolecules. For example, cyclochenodeoxycholate dilactone (**111**) was prepared from chenodeoxycholic acid (**110**, 3α,7α-dihydroxy-5β-cholan-24-oic acid) in the presence of $PhSO_2Cl$ and C_5H_5N (*Scheme 3.2.1*) (Li and Dias, 1997b).

Similarly, cyclodeoxycholate diacetate dilactone (**112**) was synthesized from deoxycholic acid (**90**) in DCM and in the presence of $Cl_2P(O)OEt$ at room temperature over 88 h.

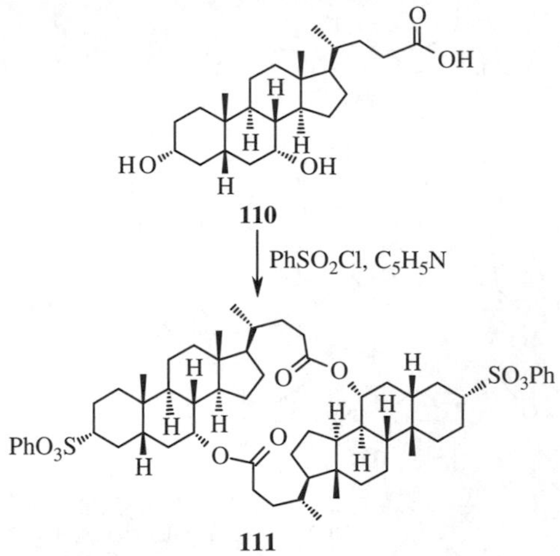

110

$PhSO_2Cl$, C_5H_5N

111

Scheme 3.2.1 *Synthesis of cyclochenodeoxycholate dilactone (**111**)*

Hydrolysis of cyclodeoxycholate dilactone diacetate (**112**) was carried out using NaOMe in MeOH to generate cyclodeoxycholate dilactone diol (**113**).

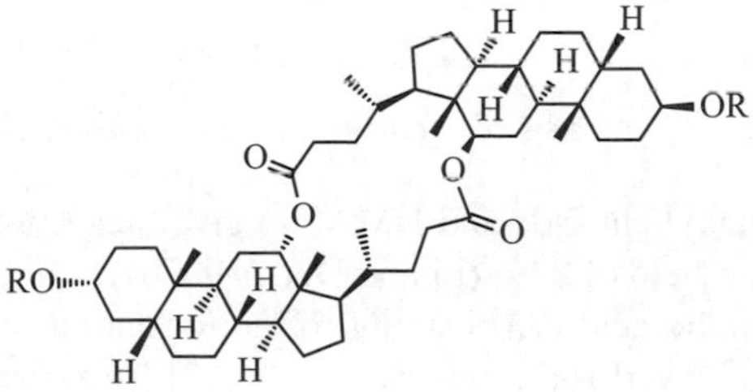

Compound	R	Compound	R
112	Ac	**113**	H

The 'head-to-tail' cyclodilithocholate (**115**) was produced from lithocholic acid (**73**) with an overall yield of 23% (Li and Dias, 1997b). It was observed that cyclotetramerization of lithocholic acid (**73**) was favoured over cyclodimerization under the reaction conditions used. Functionalized cylocholates resulted from 'head-to-tail' cyclization of cholic acid derivatives were reported by Bonar-Law and Sanders (1992). A few other functionalized cyclocholates were reported by other groups (Li and Dias, 1997a; Nahar *et al.*, 2007).

The macrolactonization of lithocholic acid (**73**) using DCC and DMAP was performed in one step to furnish of 'head-to-tail' cyclodilithocholate (**115**) with a poor yield of 2.1% (Gao and Dias, 1998a; Li and Dias, 1996). This dimer was also prepared from **73** in multisteps sequence, and the synthesis of acyclic lithocholic acid dimer (**114**) from **73** has been discussed earlier (see *Chapter 2; Scheme 2.5.5*). In the final step, the acyclic dilithocholate (**114**) was

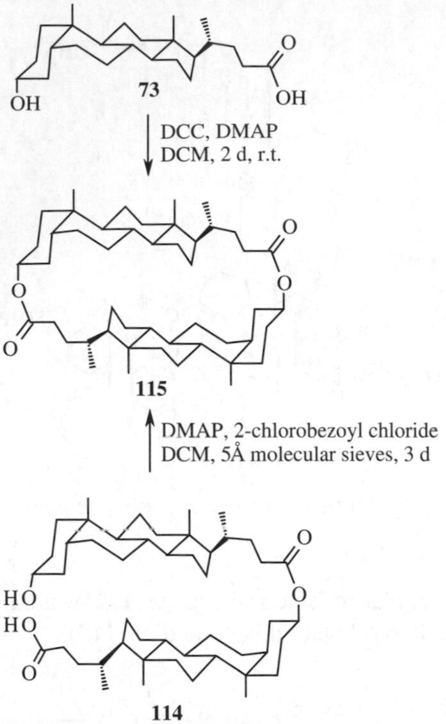

Scheme 3.2.2 *Synthesis of 'head-to-tail' cyclodilithocholate (**115**)*

treated with 2-chlorobenzoyl chloride and DMAP to give 'head-to-tail' cyclodilithocholate (**115**) with a much higher yield of 23% (Li and Dias, 1997b).

Cyclization of lithocholic acid (**73**) leading to the formation of 'head-to-tail' cyclodilithocholate (**115**) was achieved as follows (*Scheme 3.2.2*). To a stirred solution of **73** (1.5 g, 3.98 mmol) in dry DCM (125 mL), DMAP (1.72 g, 14.1 mmol) and DCC (2.72 g, 13.2 mmol) were added at room temperature. After 48 h, the precipitate was filtered and the filtrate was washed, successively, with 1M HCl, saturated $NaHCO_3$ and brine. The dried (Na_2SO_4) solution was concentrated and purified by CC, eluted with EtOAc:*n*-hexane = 1:5, to obtain 'head-to-tail' cyclodilithocholate (**115**, 2.1%, mp: 309–311 °C). ^{1}H NMR (250 MHz, $CDCl_3$): δ 4.75 (m, 1H, 3β-CH), 2.24 (t, 2H, 23-CH_2), 0.92–0.93 (m, 6H, 19-Me and 21-Me), 0.64 (s, 3H, 18-Me). ^{13}C NMR (62.5 MHz, $CDCl_3$): δ 35.2 (2 × C-1), 28.1 (2 × C-2), 74.1 (2 × C-3), 32.0 (2 × C-4), 41.8 (2 × C-5), 26.8 (2 × C-6), 26.4 (2 × C-7), 36.0 (2 × C-8), 40.3 (2 × C-9), 34.7 (2 × C-10), 20.7 (2 × C-11), 39.8 (2 × C-12), 42.8 (2 × C-13), 56.5 (2 × C-14), 24.1 (2 × C-15), 27.2 (2 × C-16), 52.6 (2 × C-17), 11.9 (2 × C-18), 23.3 (2 × C-19), 34.7 (2 × C-20), 19.5 (2 × C-21), 31.1 (2 × C-22), 28.6 (2 × C-23), 175.3 (2 × C-24). FABMS: *m/z* 717.7 $[M + H]^+$ (Gao and Dias, 1998a).

Cyclization of acyclic dilithocholate (**113**) was accomplished in the following manner (*Scheme 3.2.2*). DMAP (0.64 mmol), 5 Å molecular sieves and 2-chlorobezoyl chloride (0.20 mmol) were added to a stirred solution of **113** (0.16 mmol) in dry DCM (80 mL) at room temperature. After 3 days, the mixture was filtered, concentrated under pressure, sequentially, washed with 5% HCl (50 mL), saturated aqueous $NaHCO_3$ (50 mL) and H_2O

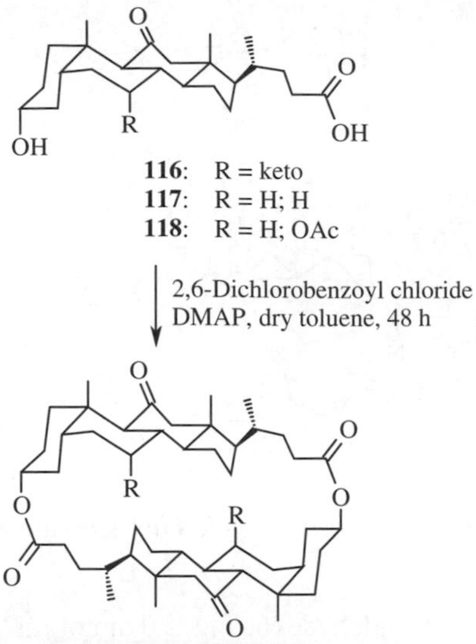

116: R = keto
117: R = H; H
118: R = H; OAc

2,6-Dichlorobenzoyl chloride
DMAP, dry toluene, 48 h

119: R = keto
120: R = H; H
121: R = H; OAc

Scheme 3.2.3 *Synthesis of 'head-to-tail' cyclodichenodeoxycholates **119–121***

(100 mL), the solvent was dried ($MgSO_4$), evaporated to dryness, purified by CC and crystallized from *n*-hexane-EtOAc to provide 'head-to-tail' cyclodilithocholate (**115**, 23%, mp: 298–300 °C). FABMS: *m/z* 717.7 $[M + H]^+$. The ^1H and ^{13}C NMR data were also same as above (Li and Dias, 1997b).

Three 'head-to-tail' cyclodichenodeoxycholates (**119–121**) were synthesized from respective chenodeoxycholic acid derivatives (**116–118**) (Gao and Dias, 1998b). A mixture of cholic acid derivative (**116**, **117** or **118**, 3.4 mmol), 2,6-dichlorobenzoyl chloride (0.53 mL, 3.7 mmol), DMAP (1.6 g, 13 mmol) in Na-dried toluene (800 mL) was refluxed (*Scheme 3.2.3*). After 48 h, the solvent was removed under pressure, the crude residue was purified by FCC and the product was further recrystallized to afford respective 'head-to-tail' cyclodichenodeoxycholates, **119–121**.

Cyclodichenodeoxycholate (**119**, 5%, mp: 305–307 °C). ^1H NMR ($CDCl_3$): δ 4.80 (m, 2H, 2 × 3β-CH), 2.33 (m, 4H, 2 × 11-CH_2), 2.88 (m, 2H, 2 × 8-CH), 1.01 (s, 12H, 2 × 18-Me and 2 × 19-Me), 0.87 (d, 6H, 2 × 21-Me), 2.74 (dd, 4H, 2 × 6-CH_2). FABMS: *m/z* 773 $[M + H]^+$ (Gao and Dias, 1998b).

Cyclodichenodeoxycholate (**120**, 18%, mp: 145–147 °C). ^1H NMR ($CDCl_3$): δ 4.72 (m, 2H, 2 × 3β-CH), 2.49 (m, 4H, 2 × 11-CH_2), 1.05 (s, 12H, 2 × 18-Me and 2 × 19-Me), 0.93 (d, 6H, 2 × 21-Me). FABMS: *m/z* 745 $[M + H]^+$ (Gao and Dias, 1998b).

Cyclodichenodeoxycholate (**121**, 9%, mp: 220–222 °C). ^1H NMR ($CDCl_3$): δ 5.03 (s, 2H, 2 × 7β-CH), 4.60 (m, 2H, 2 × 3β-CH), 2.40 (dd, 4H, 2 × 11-CH_2), 2.24 (m, 4H, 2 × 23-CH_2), 2.10 (s, 6H, 2 × 7α-CH_3CO_2), 0.97 (s, 12H, 2 × 18-me and 2 × 19-Me), 0.81 (d, 6H, 2 × 21-Me). FABMS: *m/z* 861 $[M + H]^+$ (Gao and Dias, 1998b).

The synthesis of 'head-to-tail' cyclodilithocholate amide (**123**), utilizing a unidirectional Ugi-MiB approach, was achieved from 3α-aminolithocholic acid (**122**) (Daniel *et al.*, 2007).

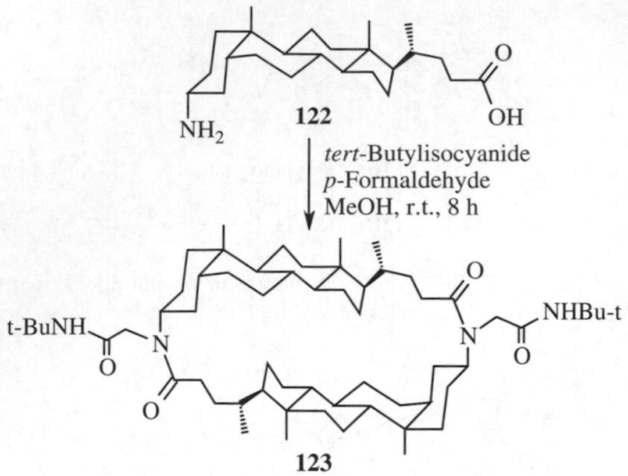

Scheme 3.2.4 *Synthesis of cyclodilithocholate amide (**123**)*

To a stirred suspension of *p*-formaldehyde (60 mg, 2.0 mmol) in MeOH (100 mL), a solution of **122** (375 mg, 1 mmol) in MeOH (20 mL) and a solution of *tert*-butylisocyanide (230 μL, 2 mmol) in MeOH (20 mL) were added at room temperature (*Scheme 3.2.4*). The reaction mixture was stirred for 8 h and the solvent was evaporated under vacuum to produce a crude product, which was subsequently purified by FCC (mobile phase: *n*-hexane:EtOAc = 10:1) and recrystallized from DCM-MeOH to yield 'head-to-tail' cyclodilithocholate amide (**123**, 155 mg, 33%, mp: 288–290 °C) (Rivera and Wessjohann, 2007). ^{1}H NMR (300 MHz, $CDCl_3$): δ 6.60-6.54 (m, 2H, 2 × NH), 4.16–4.12 (m, 2H, CH_2), 3.86–3.81 (m, 2H, CH_2), 3.47 (br m, 2H, 2 × 3β-CH), 1.31 [s, 9H, 2 × $(CH_3)_3C$], 1.30 [s, 9H, 2 × $(CH_3)_3C$], 0.93–0.92 (m, 6H, 2 × 21-Me), 0.89 (s, 6H, 2 × 19-Me), 0.65 (s, 6H, 2 × 18-Me). ^{13}C NMR (75 MHz, $CDCl_3$): δ 12.1, 19.7, 20.6, 23.2, 24.4, 26.7, 26.8, 27.4, 28.0, 28.1, 28.7, 31.4, 32.2, 34.6, 34.8, 35.4, 35.9, 39.3, 40.8, 41.4, 42.5, 45.1, 45.6, 52.5, 54.5, 56.3, 70.6, 168.5, 169.4. HRESIMS: *m/z* 963.7651 calculated for $C_{60}H_{100}O_4NaN_4$, found: 963.7646 (Daniel *et al.*, 2007).

The synthesis of 'head-to-tail' cyclodichenodeoxycholate diol (**124**) and 'head-to-tail' cyclodichenodeoxycholate diacetate (**125**) using the DCC/DMAP method as well as the Yamaguchi method for cyclodichenodeoxycholate diol (**124**) was accomplished from chenodeoxycholic acid (**110**) (Karabulut *et al.*, 2007). In the DCC/DMAP method, to a stirred solution of **110** (1.56 g, 4.0 mmol) in dry DCM (125 mL), DCC (2.74 g, 13.2 mmol) and DMAP (1.73 g, 14 mmol) were added and the mixture was stirred at room temperature. After 48 h, the reaction mixture was filtered and the filtrate was washed, successively, with 1N HCl, saturated aqueous $NaHCO_3$, saturated aqueous NaCl and H_2O. The solvent was dried (Na_2SO_4), rotary evaporated and the crude solid was purified by CC, eluted with 25% EtOAc in *n*-hexane, to furnish **124** (340 mg, 23%).

This 'head-to-tail' cyclodichenodeoxycholate diol (**124**) could also be prepared by the Yamaguchi method as follows (*Scheme 3.2.5*). A mixture of 2,6-dichlorobenzoyl chloride (126 mg, 0.6 mmol), DMAP (300 mg, 2.46 mmol) was added to a stirred solution **110** (235 mg, 0.6 mmol) in toluene (20 mL) and refluxed for 24 h. The mixture was extracted with Et_2O and ethereal layer was dried (Na_2SO_4), evaporated to dryness, and the crude product was purified as above to give **124** (122 mg, 54%). ^{1}H NMR (250 MHz, $CDCl_3$): δ 4.62 (m, 2H, 2 × 3α-CH), 3.88 (s, 2H, 2 × 7α-CH), 0.92 (d, 6H, 2 × 21-Me), 0.91 (s, 6H,

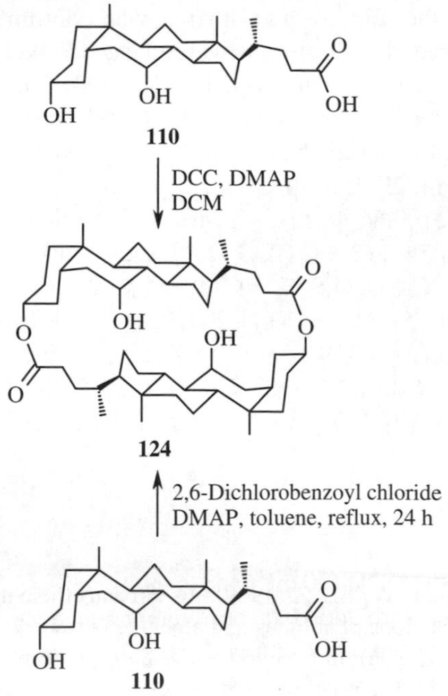
OH
OH
OH
110
DCC, DMAP
DCM
O
OH
OH
O
O
O
124
2,6-Dichlorobenzoyl chloride
DMAP, toluene, reflux, 24 h
OH
OH
OH
110

Scheme 3.2.5 *Synthesis of 'head-to-tail' cyclodichenodeoxycholate diol (**124**)*

2 × 19-Me), 0.64 (s, 6H, 2 × 19-Me). FABMS (NaI): *m/z* 771.5 [cyclodimer + Na]$^+$ (Karabulut *et al.*, 2007).

The 'head-to-tail' cyclodichenodeoxycholate diacetate (**125**) was obtained from chenodeoxycholic acid (**110**) (*Scheme 3.2.6*). DCC (2.74 g, 13.2 mmol) and DMAP (1.73 g, 14 mmol) were added to a stirred solution of **110** (1.56 g, 4.0 mmol) in dry DCM (125 mL) and the reaction mixture was stirred at room temperature. After 48 h, Ac_2O (0.22 mL,

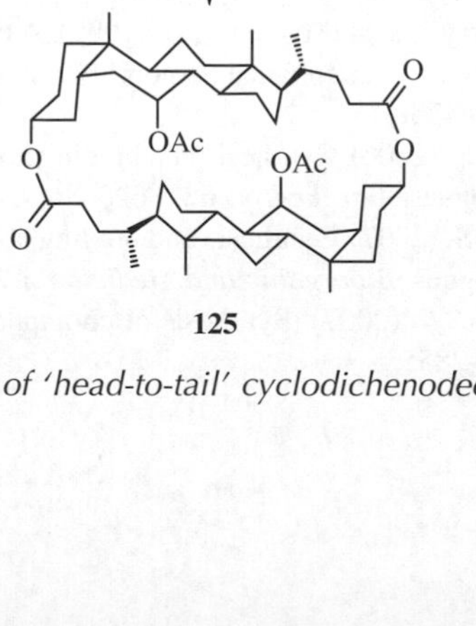
OH
OH
OH
110
DCC, DMAP
DCM then Ac2O
O
OAc
OAc
O
O
O
125

Scheme 3.2.6 *Synthesis of 'head-to-tail' cyclodichenodeoxycholate diacetate (**125**)*

4.0 mmol) was added to the mixture and stirring was continued for another 24 h. The reaction mixture was filtered, the filtrate was washed, successively, with 1N HCl, saturated $NaHCO_3$, brine, and H_2O. The solvent was dried (MgSO4), rotary evaporated to dryness. The crude solid was purified by step gradient CC (mobile phase: EtOAc:*n*-hexane = 1:9, 2:8, 3:7, 4:6 and 5:5) to obtain **125** (52 mg, 3.1%). ^{1}H NMR (250 MHz, $CDCl_3$): δ 4.84 (s, 2H, 2 × 7β-CH), 4.62 (m, 2H, 2 × 3β-CH), 2.06 (s, 6H, 2 × 7α-MeCO), 0.95 (br s, 12H, 2 × 19-Me and 2 × 21-Me), 0.66 (s, 6H, 18-Me). ^{13}C NMR (62.5 MHz, $CDCl_3$): δ 34.6 (2 × C-1), 27.8 (2 × C-2), 74.2 (2 × C-3), 34.5 (2 × C-4), 40.9 (2 × C-5), 32.9 (2 × C-6), 71.1 (2 × C-7), 38.0 (2 × C-8), 34.5 (2 × C-9), 34.6 (2 × C-10), 20.5 (2 × C-11), 39.7 (2 × C-12), 42.3 (2 × C-13), 51.3 (2 × C-14), 23.6 (2 × C-15), 26.7 (2 × C-16), 53.1 (2 × C-17), 11.7 (2 × C-18), 22.6 (2 × C-19), 34.9 (2 × C-20), 17.5 (2 × C-21), 31.4 (2 × C-22), 28.8 (2 × C-23), 174.2 (2 × C-24), 170.6 (2 × 7α-CO), 21.5 (2 × 7α-MeCO). FABMS (LiI): (*m/z*) 839.8 [cyclodimer + Li]$^+$ (Karabulut *et al.*, 2007).

References

Barton, D. H. R., Wozniak, J. and Zard, S. Z. (1989). A short and efficient degradation of the bile-acid side-chain: Some novel reactions of sulfines and alpha-ketoesters. *Tetrahedron* **45**, 3741–3754.

Bell, D. A. and Anslyn, E. V. (1996). In *Comprehensive Supramolecular Chemistry.* Atwood, J. L., Davis, J. E. D., Macnicol, D. D., Vögtle F. Eds., Elsevier, Oxford, Vol. **2**, pp. 439–475.

Bonar-Law, R. P. and Davis, A. P. (1989). Synthesis of steroidal cyclodimers from cholic acid: A molecular framework with potential for recognition and catalysis. *Journal of Chemical Society, Chemical Communications*, 1050–1052.

Bonar-Law, R. P. and Davis, A. P. (1993a). Cholic acid as an architectural component in molecular recognition chemistry. *Synthesis of the first cholaphanes. Tetrahedron* **49**, 9829–9844.

Bonar-Law, R. P. and Davis, A. P. (1993b). Cholic-acid as an architectural component in biomimetic molecular recognition chemistry - NMR and molecular mechanics study of a tetra-acetoxycholaphane. *Tetrahedron* **49**, 9845–9854.

Bonar-Law, R. P., Davis, A. P. and Dorgan, B. J. (1993). Cholic-acid as an architectural component in biomimetic molecular recognition chemistry - synthesis of cholaphanes with facial differentiation of functionality. *Tetrahedron* **49**, 9855–9866.

Bonar-Law, R. P., Davis, A. P. and Murray, B. A. (1990). Artificial receptors for carbohydrate-derivatives. *Angewandte Chemie International Edition* **29**, 1407–1408.

Bonar-Law, R. P. and Sanders, J. K. M. (1992). Cyclocholates: synthesis and ion binding. *Tetrahedron Letters* **33**, 2071–2074.

Bhattarai, K. M., Bonar-Law, R. P., Davis, A. P. and Murray, B. A. (1992). Diastereoselective and enantio-selective binding of octyl glucosides by an artificial receptor. *Journal of Chemical Society, Chemical Communication* **10**, 752–754.

Bhattarai, K. M., Davis, A. P., Perry, J. J. and Walter, C. J. (1997). A new generation of 'cholaphanes': Steroid-derived macrocyclic hosts with enhanced solubility and controlled flexibility. *Journal of Organic Chemistry* **62**, 8463–8473.

Chattopadhyay, P and Pandey, P. S. (2006). Synthesis and binding ability of bile acid-based receptors for recognition of flavin analogues. *Tetrahedron* **62**, 8620–8624.

Chattopadhyay, P and Pandey, P. S. (2007). Synthesis and binding ability of bile acid-based receptors for recognition of flavin analogues. *Bioorganic and Medicinal Chemistry Letters* **17**, 1553–1557.

Czajkowska, D. and Morzycki, J. W. (2007). Synthesis of cholaphanes by ring closing metathesis. *Tetrahedron Letters* **48**, 2851–2855.

Davis, A. P. and Bhattarai, K. M. (1995). Antimony-based 'forcing knoevenagel' methodology for the conversion of ketones into alkylidenemalonates. *Tetrahedron* **51**, 8033–8042.

Davis, A. P., Gilmer, J. F. and Perry, J. J. (1996). A steroid-based cryptand for halide anions. *Angewandte Chemie International Edition* **35**, 1312–1314.

Davis, A. P. and Wareham, R. S. (1999). Carbohydrate recognition through noncovalent interactions: A challenge for biomimetic and supramolecular chemistry *Angewandte Chemie International Edition* **38**, 2978–2996.

Dias, J. R. and Ramachandra, R. (1977). Studies directed toward synthesis of quassinoids. 4. D-Ring cleavage of cholic-acid derivatives. *Organic Preparations and Procedures International* **9**, 109–115.

Gao, H. and Dias, J. R. (1998a). Highly selective cyclotrimerization of lithocholic acid by DCC and DMAP reagent. *Croatica Chemica Acta* **71**, 827–831.

Gao, H. and Dias, J. R. (1998b). Synthesis of cyclocholates and derivatives, III. Cyclocholates with 12-oxo and 7, 12-oxo groups. *European Journal of Organic Chemistry* 719–724.

Karabulut, H. R., Rashdan, S. A. and Dias, J. R. (2007). Notable chenodeoxycholic acid oligomers synthesis, characterization, and 7α-OR steric hindrance evaluation. *Tetrahedron* **63**, 5030–5035.

Kolehmainen, E., Tamminen, J., Lappalainen, K., Torkkel, T. and Seppala, R. (1996). Substituted methyl 5β-cholan-24-oates. Part III: Synthesis of a novel cholaphane ethylene glycol diester of lithocholic acid by cyclization with terephthalic acid. *Synthesis*, 1082–1084.

Li, Y. and Dias, J. R. (1996). Synthesis of linear dimeric and cyclic oligomeric cholate ester derivatives. *Synthesis*, 425–430.

Li, Y. X. and Dias, J. R. (1997a). Dimeric and oligomeric steroids. *Chemical Review* **97**, 283–304.

Li, Y. X. and Dias, J. R. (1997b) Syntheses of linear dimeric and cyclic oligomeric cholate ester derivatives. *Synthesis* 425–430.

Łotowski, Z., Piechowska, J. and Jarocki, L. (2008). Dimeric cholaphanes with oxamide spacers. *Monatshefte für Chemie* **139**, 213–222.

Mekhalfia, A. and Markó, I. E. (1991). The silyl modified Sakurai (SMS) reaction. An efficient and versatile one-pot synthesis of homoallylic ethers. *Tetrahedron Letters* **32**, 4779–4782.

Nahar, L., Sarker, S. D. and Turner, A. B. (2007). A review on synthetic and natural steroid dimers: 1997-2006. *Current Medicinal Chemistry* **14**, 1349–1370.

Pandey, P. S. and Singh, R. B. (1997). Synthesis of a head-to-head cholaphane. *Tetrahedron Letters* **38**, 5045–5046.

Pandey, P. S., Rai, R. and Singh, R. B. (2002). Synthesis of cholic acid-based molecular receptors: head-to-head cholaphanes. *Journal of Chemical Society, Perkin Trans.* **1**, 918–923.

Ra, C. S., Cho, S. W. and Choi, J. W. (2000). A steroidal cyclic dimer with ethylene glycol bridges. *Bulletin of Korean Chemical Society* **21**, 342–344.

Rivera, D. G. and Wessjohann, L. A. (2007). Synthesis of novel steroid-peptoid hybrid macrocycles by multiple multicomponent macrocyclizations including bifunctional building blocks (MiBs). *Molecules* **12**, 1890–1899.

Schteingart, C. D. and Hofmann, A. F. (1988). Convenient and efficient one-carbon degradation of the side chain of natural bile acids. *Journal of Lipid Research* **29**, 1387–1395.

Tamminen, J., Kolehmainen, E., Linnanto, J., Salo, H. and Manttari, P. (2000). 3α,3'α bis(n-acetoxyphenylcarboxy)-5β-cholan-24-oic acid ethane-1,2-diol diesters (n = 2-4): ^{13}C NMR chemical shifts, variable-temperature and NOE ^{1}H NMR measurements and MO calculations of novel bile acid-based dimers. *Magnetic Resonance in Chemistry* **38**, 877–882.

Tamminen, J. and Kolehmainen, E. (2001). Bile acids as building blocks of supramolecular hosts. *Molecules* **6**, 21–46.

Tian, Z., Cui, H. and Wang, Y. (2002). Synthesis of linear and cyclic dimeric ester derivatives from bile acids and glycine. *Synthetic Communication* **32**, 3821–3829.

Tserng, K.-Y. and Klein, P. D. (1977). Formylated bile acids: improved synthesis, properties, and partial deformylation. *Steroids* **29**, 635–648.

Valkonen, A., Sievänen, E., Ikonen, S., Lukashev, N. V., Donez, P. A., Averin, A. D., Lahtinen, M. and Kolehmainen, E. (2007). Novel lithocholaphanes: Syntheses, NMR, MS, and molecular modeling studies. *Journal of Molecular Structure* **846**, 65–73.

4

Naturally Occurring Steroid Dimers

Since the discovery of the first naturally occurring sulphur-containing dimeric steroid japindine (**1**) from the root-bark of the Indian medicinal plant *Chonemorpha macrophylla* (family: Apocynaceae) (Banerji and Chatterjee, 1973), a number of other steroid dimers, mainly belonging to the classes of cephalostatins, crellastatins and ritterazines, have been reported from various marine organisms (Li and Dias, 1997; Nahar *et al.*, 2007). Because of different pharmacological properties, particularly, cytotoxicity towards cancer cell lines, these compounds have been utilized as possible templates for anticancer drug discovery and design. Cephalostatins and ritterazines exhibit potent cytotoxicity towards the murine P388 lymphocytic leukaemia cell line. Cephalostatin 1 (**2**, ED_{50} 0-1-0-001 pM) was found to be one of the most powerful cancer cell growth inhibitors ever tested by the US National Cancer Institute (NCI) (Moser, 2008). The structure of cephalostatin 1 (**2**) is characterized by an asymmetric union of two highly oxygenated steroidal spiroketal subunits with a central pyrazine ring. Its unique structure, extremely powerful cytotoxicity, and natural scarcity have prompted several synthetic endeavours, and as a result, a number of synthetic analogues of these naturally occurring steroid dimers have been designed and synthesized by synthetic chemists over the last few decades (see *Chapter 5*). In this chapter, however, the discussion will primarily be confined to various aspects of the naturally occurring steroidal dimers, *e.g.*, cephalostatins, crellastatins and ritterazines.

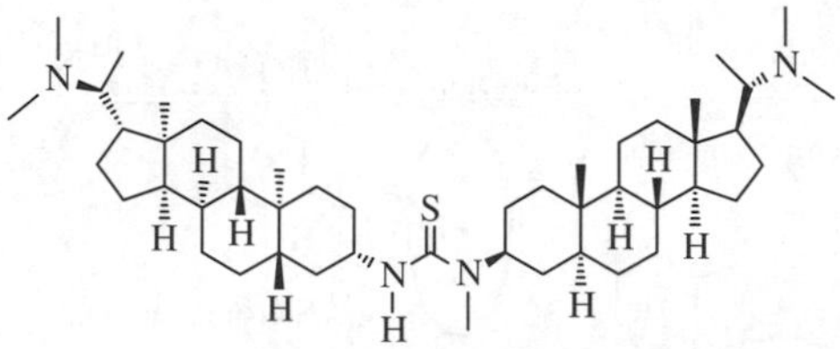

Japindine (**1**)

Steroid Dimers: Chemistry and Applications in Drug Design and Delivery, First Edition. Lutfun Nahar and Satyajit D. Sarker. Published 2012 by John Wiley & Sons, Ltd.

4.1 Cephalostatins

Cephalostatins are a group of naturally occurring highly cytotoxic steroid dimers containing a pyrazine moiety isolated from marine worms. All these cephalostatins contain the unique structural framework of two polyoxygenated steroidal units with side chains forming spiroketals linked through a pyrazine core involving C-2 and C-3 of each monomeric unit (*Figure 4.1.1*).

Compounds of this group were first isolated from the Indian Ocean (South African) invertebrate marine chordates, *Cephalodiscus gilchristi* (family: Cephalodiscidae) (*Figure 4.1.2*), by Pettit's group in 1988 (Nahar *et al*., 2007), and most of the cephallostatins, *e.g*., cephalostatins 1–19 (**2**–**20**), were reported by this group during 1988–1998. Of these cephalostatins, the most potent cytotoxins are cephalostatin 1 (**2**) and cephalostatin 7 (**11**). Cephalostatin 1 (**2**) is probably biosynthesized partly from a condensation of two 2-amino-3-oxo steroid units.

The natural habitat of the small, *ca*. 5 mm long, colonial marine worm *C. gilchristi* is in the shark-infested waters of the Indian Ocean off the coast of southeastern Africa (Moser, 2008). In 1972, Pettit and coworkers collected specimens of *C. gilchristi* by scuba (depth of 20 m), but it took fifteen years of painstaking research directed at structure elucidation of the active constituents of the MeOH and aqueous extracts of *C. gilchristi*, including re-collection efforts in 1981, to finally isolate (yield: 8.36×10^{-4} wt%) and establish the structure of cephalostatin 1 (**2**) and report in 1988 (Pettit *et al*., 1988a).

Pettit *et al*. (1988a) tried a number of extraction protocols, until they successfully isolated **2** using the following method. A DCM-MeOH extract of *C. gilchristi* was partitioned, successively, by using the system MeOH-H_2O against *n*-hexane, CCl_4 and DCM. The active DCM fraction was subjected to bioassay-guided purification using a combination of gel permeation and partition gradient CC (Sephadex LH20 and LH60, SiO_2, *e.g*., with *n*-hexane:EtOAc:MeOH = 10:10:1 to *n*-hexane:DCM:MeOH = 10:10:1), to HPLC on Partisil M9 using MeOH-H_2O and *n*-hexane-EtOAc-MeOH gradients, followed by crystallization from EtOAc-MeOH to afford cephalostatin 1 (**2**, 138.8 mg, 8.36×10^{-4}%, mp: 236 °C). IR (KBr): ν_{max} cm^{-1} 3430br (O–H), 3050m (CH=CH), 2970m (C–H), 2930m (C–H), 2880m (C–H), 2860m (C–H), 1708s (C=O), 1650s (C=C), 1615m (C=C), 1445m

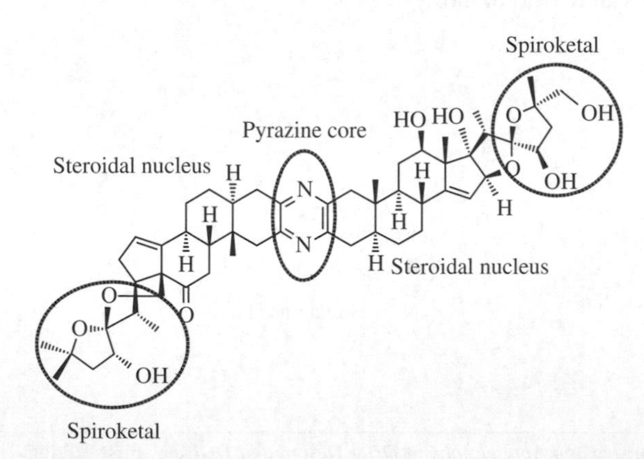

Figure 4.1.1 Unique structural features of cephalostatins (e.g. cephalostatin 1, **2**)

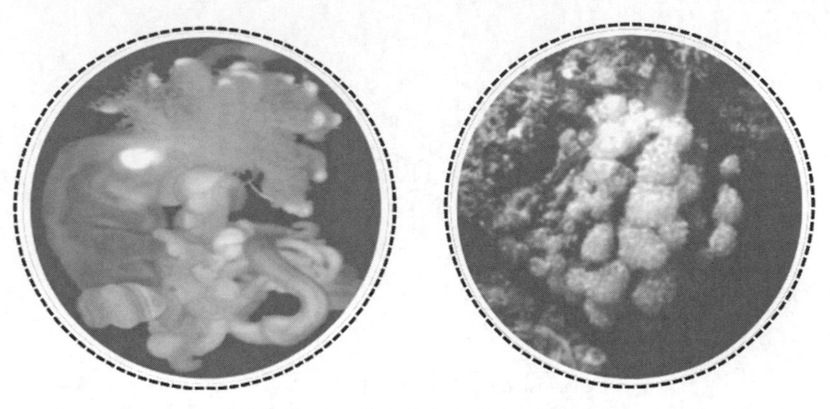

Figure 4.1.2 *Cephalodiscus gilchristi (left) and Ritterella tokioka (right)*

(C–H), 1400s (pyrazine), 1152m (C–O), 1115m (C–O), 1090m (C–O), 1045m (C–O), 950m, 892m. UV (EtOH): λ_{max} nm 289, 309. ^{1}H NMR (400 MHz, C_5D_5N): δ 8.06 (br s, 1H, 23-OH), 7.19 (br s, 1H, 23'-OH), 6.54 (br s, 1H, 26-OH), 6.23 (s, 1H, 17-OH), 5.64 (s, 1H, 15-CH), 5.44 (s, 1H, 15'-CH), 5.24 (s, 1H, 16-CH), 4.80 (m, 2H, 23-CH and 23'-CH), 4.70 (s, 1H, 12-OH), 4.06 (d, 1H, J = 12.2 Hz, 18'-CHb), 4.05 (m, 1H, 12-CH), 4.02 (d, 1H, J = 12.2 Hz, 18'-CHa), 3.78 (d, 1H, J = 11.2 Hz, 26-CHb), 3.72 (d, 1H, J = 11.2 Hz, 26-CHa), 3.17 (dq, 1H, J = 7.0 Hz, 20'-CH), 3.07 (d, 1H, J = 17.0 Hz, 1-CHa), 3.04 (d, 1H, J = 17.0 Hz, 1'-CHa), 2.93 (dd, 1H, J = 5.5, 17.9 Hz, 4-CHa), 2.91 (dd,1H, J = 5.5, 17.9 Hz, 4'-CHa), 2.87 (dt, 1H, J = 4.0, 11.8 Hz, 16'-CHa), 2.86 (q, 1H, J = 7.0 Hz, 20-CH), 2.78 (t, 1H, J = 14.0 Hz, 11'-CHa), 2.77 (dd, 1H, J = 7.9, 10.5 Hz, 24-CHa), 2.72 (dd, 1H, J = 7.0, 12.0 Hz, 17'-CH), 2.56 (d, 2H, J = 12.5, 17.0 Hz, 4-CHb and 4'-CHb), 2.36 (m, 1H, 24-CHb), 2.35 (m, 1H, 24'-CHa), 2.33 (m, 1H, 16'-CHb), 2.13 (m, 1H, 8'-CH), 2.07 (m, 1H, 8-CH), 2.04 (m, 1H, 11-CHa), 1.99 (m, 1H, 7'-CHa), 1.95 (dd, 1H, J = 6.2, 12.0 Hz, 24'-CHb), 1.77 (dt, 1H, J = 10.0, 14.0 Hz, 11-CHb), 1.69 (m, 1H, 7-CHa), 1.65 (s, 3H, 27-Me), 1.61 (m, 1H, 5-CH), 1.60 (m, 1H, 5'-CH), 1.59 (m, 1H, 6-CHa), 1.53 (m, 1H, 6'-CHa), 1.47 (d, 3H, J = 7.0 Hz, 21'-Me), 1.47 (s, 3H, 26'-Me), 1.39 (s, 3H, 27'-Me), 1.35 (d, 3H, J = 7.0 Hz, 21-Me), 1.35 (m, 1H, 7b-CH), 1.34 (m, 1H, 6-CHb), 1.33 (s, 3H, 18-Me), 1.30 (m, 1H, 7'-CHb), 1.28 (m, 1H, 6'-CHb), 1.26 (m, 1H, 9'-CH), 0.88 (dt, 1H, J = 4.5, 13.8 Hz, 9-CH), 0.75 (s, 3H, 19-Me), 0.72 (s, 3H, 19'-Me). ^{13}C NMR (100 MHz, C_5D_5N): δ 46.0 (C-l), 45.8 (C-1'), 148.4, 148.4, 148.7, 149.0, and 149.5 (C-2, C-3, C-2', C-3', and C-14'), 35.7 (C-4), 35.8 (C-4'), 41.8 (C-5), 41.2 (C-5'), 28.2 (C-6'), 28.7 (C-7), 29.5 (C-7'), 33.8 (C-8), 35.6 (C-8'), 53.2 (C-9), 52.2 (C-9'), 36.3 (C-10), 36.3 (C-10'), 28.9 (C-11), 38.8 (C-11'), 75.6 (C-12), 211.8 (C-12'), 55.4 (C-13), 61.8 (C-13'), 152.7 (C-14), 122.3 (C-15), 123.2 (C-15'), 93.2 (C-16), 32.4 (C-16'), 91.7 (C-17), 44.2 (C-17'), 12.6 (C-18), 64.2 (C-18'), 11.7 (C-19), 11.3 (C-19'), 44.5 (C-20), 32.9 (C-20'), 9.0 (C-21), 15.5 (C-21'), 117.2 (C-22), 111.0 (C-22'), 71.5 (C-23), 81.5 (C-23'), 39.5 (C-24), 47.3 (C-24'), 82.8 (C-25), 81.1 (C-25'), 69.3 (C-26), 26.3 (C-27), 29.4 and 29.8 (C-26' and C-27'). SPHRSIMS: *m/z* 911.5423 calculated for $C_{54}H_{74}N_2O_{10}$, found: 911.5442 (Pettit *et al.*, 1988a).

In the same year Pettit *et al.* (1988b) reported the bioassay-monitored isolation of three more steroidal alkaloids, cephalostatins 2–4 (**5**, **6** and **8**, respectively), with similar cytotoxic potential as that of **2**, from the DCM-MeOH extract of the same source materials. The isolation and purification protocol was quite similar to that described earlier for cephalostatin 1 (**2**).

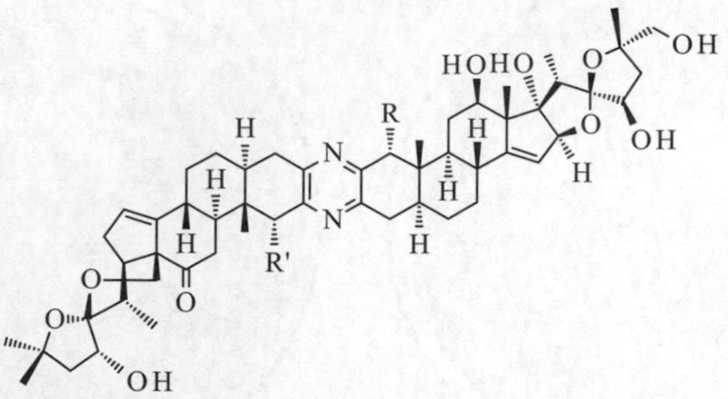

Cephalostatin 1 (**2**)	R = H	R' = H
Cephalostatin 18 (**3**)	R = OMe	R' = H
Cephalostatin 19 (**4**)	R = H	R' = OMe

Cephalostatin 2 (**5**, 242.8 mg, 14.6×10^{-4}%, mp: > 350 °C). IR (KBr): ν_{max} cm^{-1} 3430br (O–H), 3055m (CH=CH), 2975m (C–H), 2930m (C–H), 2880m (C–H), 1710s (C=O), 1655s (C=C),1625m (C=C), 1448m (C–H), 1400s (pyrazine), 1385m (C–H), 1092m (C–O), 1045m (C–O), 950m. UV (EtOH): λ_{max} nm 290, 308. ^{13}C NMR (100 MHz, C_5D_5N): δ 46.0 (C-1), 39.5 (C-1'), 148.6 (C-2, C-3 and C-3'), 148.7 (C-2'), 35.8 (C-4), 36.2 (C-4'), 41.8 (C-5), 34.2 (C-5'), 28.2 (C-6), 28.3 (C-6'), 28.7 (C-7), 24.6 (C-7'), 33.8 (C-8), 39.0 (C-8'), 53.2 (C-9), 78.7 (C-9'), 36.4 (C-10), 41.2 (C-10'), 29.0 (C-11), 45.6 (C-11'), 75.6 (C-12), 211.1 (C-12'), 55.4 (C-13), 61.6 (C-13'), 152.7 (C-14), 148.3 (C-14'), 122.3 (C-15), 124.5 (C-15'), 93.2 (C-16), 32.6 (C-16'), 91.7 (C-17), 44.2 (C-17'), 12.6 (C-18), 64.1 (C-18'), 11.8 (C-19), 15.1 (C-19'), 44.5 (C-20), 32.9 (C-20'), 9.0 (C-21), 15.5 (C-21'), 117.2 (C-22), 111.0 (C-22'), 71.5 (C-23), 81.6 (C-23'), 39.5 (C-24), 47.4 (C-24'), 82.8 (C-25), 81.1 (C-25'), 69.3 (C-26), 29.8 (C-26'), 26.5 (C-27), 29.5 (C-27'). SPHRSIMS: *m/z* 927.5372 calculated for $C_{54}H_{75}N_2O_{11}$, found: 927.5230 (Pettit *et al.*, 1988b).

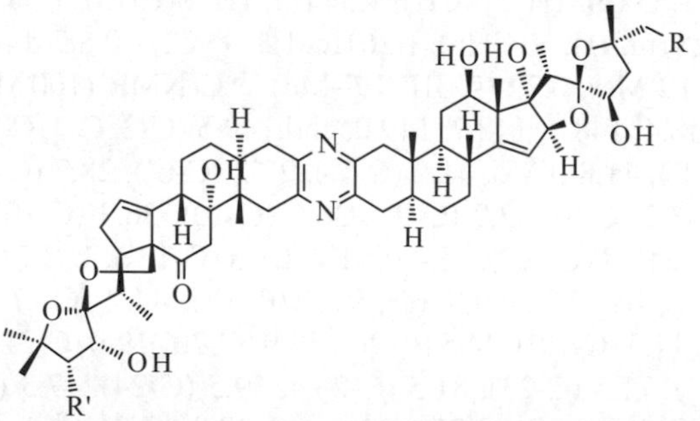

Cephalostatins	R	R'
Cephalostatin 2 (**5**)	OH	H
Cephalostatin 3 (**6**)	OH	Me
Cephalostatin 17 (**7**)	H	H

Cephalostatin 3 (**6**, 21.2 mg, 1.2×10^{-4}%, mp: >350 °C). IR (KBr): ν_{max} cm^{-1} 3430br (O–H), 3050m (CH=CH), 2967m (C–H), 2928m (C–H), 2872m (C–H), 1707s (C=O), 1645s (C=C), 1615m (C=C), 1446m (C–H), 1400s (pyrazine), 1383m (C–H), 1040m (C–O), 977m, 952m. UV (EtOH): λ_{max} nm 290, 308. ^{13}C NMR (100 MHz, C_5D_5N): δ 46.0 (C-1), 39.5 (C-1'), 148.7 (C-2, C-2' and C-3), 148.5 (C-3'), 35.8 (C-4), 36.2 (C-4'), 41.8 (C-5), 34.2 (C-5'), 28.2 (C-6), 28.3 (C-6'), 28.7 (C-7), 24.6 (C-7'), 33.8 (C-8), 39.0 (C-8'), 53.2 (C-9), 78.7 (C-9'), 36.3 (C-10), 41.2 (C-10'), 29.0 (C-11), 45.5 (C-11'), 75.6 (C-12), 211.1 (C-12'), 55.4 (C-13), 61.2 (C-13'), 152.7 (C-14), 148.3 (C-14'), 122.3 (C-15), 124.5 (C-15'), 93.2 (C-16), 32.4 (C-16'), 91.7 (C-17), 43.9 (C-17'), 12.6 (C-18), 65.0 (C-18'), 11.7 (C-19), 15.0 (C-19'), 44.5 (C-20), 32.6 (C-20'), 9.0 (C-21), 15.2 (C-21'), 117.2 (C-22), 109.1 (C-22'), 71.5 (C-23), 87.2 (C-23'), 39.5 (C-24), 51.7 (C-24'), 82.8 (C-25), 81.3 (C-25'), 69.3 (C-26), 28.0 (C-26'), 26.4 (C-27), 29.3 (C-27'), 12.7 (C-28'). SPHRSIMS: *m/z* 941.5528 calculated for $C_{55}H_{77}N_2O_{11}$, found: 941.5546 (Pettit *et al.*, 1988b).

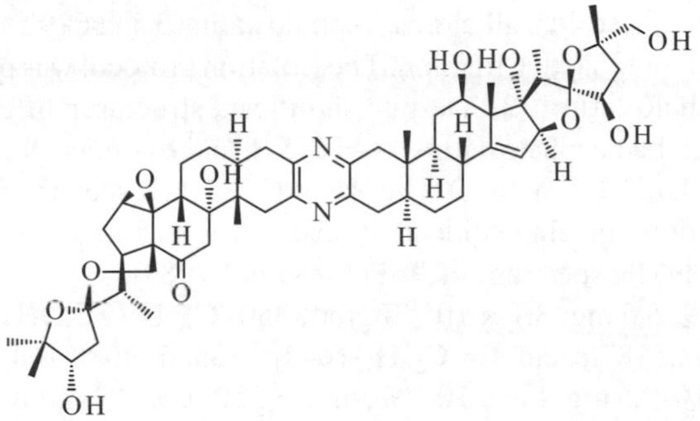

Cephalostatin 4 (**8**)

Cephalostatin 4 (**8**, 8.0 mg, 5.5×10^{-5}%, mp: >350 °C). IR (KBr): ν_{max} cm^{-1} 3430br (O–H), 2970m (C–H), 2928m (C–H), 2875m (C–H), 1711s (C–H), 1660s (C=C), 1600m (C=C), 1447m (C–H), 1400s (pyrazine), 1383m (C–H), 1089s (C–O), 948m, 904m. UV (EtOH): λ_{max} nm 290, 308. ^{13}C NMR (100 MHz, C_5D_5N): δ 46.0 (C-1), 39.0 (C-1'), 148.7 (C-2), 149.2 (C-2'), 148.6 (C-3), 148.1 (C-3'), 35.8 (C-4), 36.1 (C-4'), 41.8 (C-5), 33.9 (C-5'), 28.2 (C-6), 27.6 (C-6'), 28.7 (C-7), 20.1 (C-7'), 33.8 (C-8), 34.6 (C-8'), 53.2 (C-9), 81.0 (C-9'), 36.3 (C-10), 41.5 (C-10'), 29.0 (C-11), 45.6 (C-11'), 75.6 (C-12), 209.4 (C-12'), 55.4 (C-13), 56.3 (C-13'), 152.7 (C-14), 72.8 (C-14'), 122.3 (C-15), 54.1 (C-15'), 93.2 (C-16), 27.7 (C-16'), 91.7 (C-17), 33.1 (C-17'), 12.6 (C-18), 62.4 (C-18'), 11.7 (C-19), 14.9 (C-19'), 44.5 (C-20), 32.0 (C-20'), 9.0 (C-21), 15.0 (C-21'), 117.2 (C-22), 110.2 (C-22'), 71.5 (C-23), 81.5 (C-23'), 39.5 (C-24), 47.3 (C-24'), 82.8 (C-25), 81.1 (C-25'), 69.3 (C-26), 29.6 (C-26'), 26.4 (C-27), 29.3 (C-27'). SPHRSIMS: *m/z* 943.5343 calculated for $C_{54}H_{75}N_2O_{12}$, found: 943.5309 (Pettit *et al.*, 1988b).

Cephalostatins are rather rare marine natural products and available in only small amounts. For example, 166 kg of *C. gilchristi* (5 mm long tube worms) yielded only 139 mg of cephalostatin 1 (**2**) and a total of 272 mg of other cephalostatins (Li and Dias, 1997). Cephalostatins 5 and 6 (**9** and **10**) are unusual in the sense that they both contain an aromatic C ring. While naturally occurring and synthetic steroids with aromatic A rings are well documented, steroids bearing an aromatic C ring are quite rare.

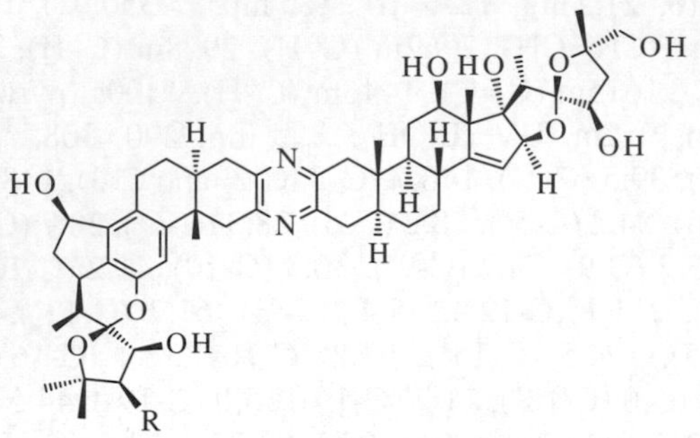

Cephalostatin 5 (**9**) R = Me Cephalostatin 6 (**10**) R = H

In continuation of their quest for more cephalostatins, Pettit *et al.* (1989) reported the isolation of these rare disteroidal alkaloids, cephalostatins 5 and 6 (**9** and **10**) from the same source, but from a minor bioactive fraction. The isolation protocol was pretty much the same as described for cephalostatin 1 (**2**). Despite significant structural differences as compared with previously described cephalostatins 1–4 (**2**, **5**, **6** and **8**) dimers **9** and **10** retained their cytotoxic potency (ED_{50} 10^{-2} mg/mL). The ^{1}H and ^{13}C NMR spectra of cephalostatins 5 (**9**) and 6 (**10**) were found to be almost identical except for an additional methyl doublet at δ 1.07 ppm in the ^{1}H NMR spectrum of **9** (Pettit *et al.*, 1989).

Cephalostatin 5 (**9**, 5.5 mg, 3.1×10^{-5}%, mp: 350 °C). UV (EtOH): λ_{max} nm 290, 310. SPHRSIMS: *m/z* 908.5187 cacld. for $C_{54}H_{72}N_2O_{10}$, found: 908.5164 (Pettit *et al.*, 1989).

Cephalostatin 6 (**10**, 2.5 mg, 1.8×10^{-5}%, mp: > 350 °C). UV (EtOH): λ_{max} nm 289, 310. ^{1}H NMR (400 MHz, C_5D_5N): δ 8.08 (d, 1H, $J = 7.4$ Hz, 23-OH), 6.93 (s, 1H, 11'-CH), 6.55 (s, 1H, 26-OH), 6.54 (s, 1H, 15'-OH), 6.51 (s, 1H, 23'-OH), 6.23 (s, 1H, 17-OH), 5.55 (t, 1H, $J = 5.5$ Hz, 15'-CH), 5.24 (s, 1H, 16-CH), 4.80 (m, 2H, 23-CH_2), 4.70 (d, 1H, $J = 1.1$ Hz, 12-OH), 5.63 (s, 1H, 15-CH), 4.56 (q, 1H, $J = 8.7$ Hz, 23'-CH), 4.44 (dt, 1H, $J = 6.0$, 10.0 Hz, 17'-CH), 4.05 (ddd, 1H, $J = 1.1$, 4.9, 12.4 Hz, 12-CH), 3.81 (dd, 1H, $J = 5.3$, 10.9 Hz, 26-CHa), 3.71 (dd, 1H, $J = 4.6$, 10.9 Hz, 26-CHb), 3.56 (d, 1H, $J = 16.9$ Hz, 1'-CH), 3.09 (br t, 2H, 7'-CH_2), 3.08 (d, 1H, $J = 17.0$ Hz, 1-CHa), 2.96 (m, 1H, 1'-CH and 4'-CHa), 2.90 (m, 1H, 4-CHa), 2.86 (m, 1H, 20-CH), 2.80 (m, 1H, 4'-CHb), 2.73 (dd, 1H, $J = 8.0$ and 11.2 Hz, 24-CHa), 2.65 (d, 1H, $J = 16.8$ Hz, 4-CHb), 2.63 (d, 1H, $J = 16.8$ Hz, 1-CHb), 2.58 (t, 1H, $J = 11.2$ Hz, 24'-CHa), 2.46 (dd, 1H, $J = 8.0$, 11.7 Hz, 24'-CH), 2.38 (dd, 1H, $J = 6.0$, 12.5 Hz, 16'-CHa), 2.35 (t, 1H, $J = 11.1$ Hz, 24-CHb), 2.27 (quint, 1H, $J = 7.0$ Hz, 20'-CH), 2.18 (ddd, 1H, $J = 5.5$, 10.5, 12.5 Hz, 16'-CHb), 2.05 (m, 2H, 8-CH and 11-CH), 1.95 (m, 1H, 5'-CH), 1.71 (m, 2H, 6'-CH_2), 1.69 (m, 1H, 7-CHa), 1.64 (s, 3H, 27-Me), 1.57 (m, 1H, 5-CH), 1.49 (m, 1H, 6-CHa), 1.43 (s, 3H, 27'-Me), 1.35 (d, 3H, $J = 7.1$ Hz, 21-Me), 1.34 (s, 3H, 18-Me), 1.31 (m, 1H, 6-CHb), 1.27 (s, 3H, 26'-Me), 1.24 (m, 1H, 7-CHb), 1.09 (s, 3H, 19'-Me), 1.07 (d, 3H, $J = 7.1$ Hz, 21'-Me), 0.87 (m, 1H, 9-CH), 0.71 (s, 3H, 19-Me) (Pettit *et al.*, 1989). ^{13}C NMR (100 MHz, C_5D_5N): δ 46.0 (C-1), 47.2 (C-1'), 149.6 (C-2), 149.5 (C-2'), 149.3 (C-3), 148.9 (C-3'), 35.8 (C-4), 35.8 (C-4'), 41.7 (C-5), 39.1 (C-5'), 28.2 (C-6), 25.3 (C-6'), 28.7 (C-7), 26.4 (C-7'), 33.8 (C-8), 124.6 (C-8'), 53.2 (C-9), 146.2 (C-9'), 36.3 (C-10), 37.1 (C-10'), 30.0 (C-11), 111.5 (C-11'), 75.6 (C-12), 149.6 (C-12'), 55.4 (C-13), 127.9 (C-13'), 152.7 (C-14), 144.3 (C-14'), 122.3 (C-15), 74.5

(C-15'), 93.2 (C-16), 39.8 (C-16'), 91.6 (C-17), 36.9 (C-17'), 12.6 (C-18), 11.7 (C-19), 22.6 (C-19'), 44.5 (C-20), 37.0 (C-20'), 9.0 (C-21), 8.7 (C-21'), 117.1 (C-22), 107.4 (C-22'), 71.5 (C-23), 77.2 (C-23'), 39.5 (C-24), 46.2 (C-24'), 82.8 (C-25), 78.5 (C-25'), 69.3 (C-26), 30.1 (C-26'), 26.4 (C-27), 31.3 (C-27'). SPHRSIMS: *m/z* 994.5031 cacld. for $C_{53}H_{70}N_2O_{10}$, found: 894.4985 (Pettit *et al.*, 1989).

About a couple of years later, the same group isolated three more of these cytotoxic compounds, cephalostatins 7–9 (**11–13**), from *C. gilchirsti* (Pettit *et al.*, 1992) using similar extraction and isolation protocols as described for previously isolated cephalostatins 1–6 (**2**, **5**, **6**, **8–10**) (Pettit *et al.*, 1988a,b, 1989).

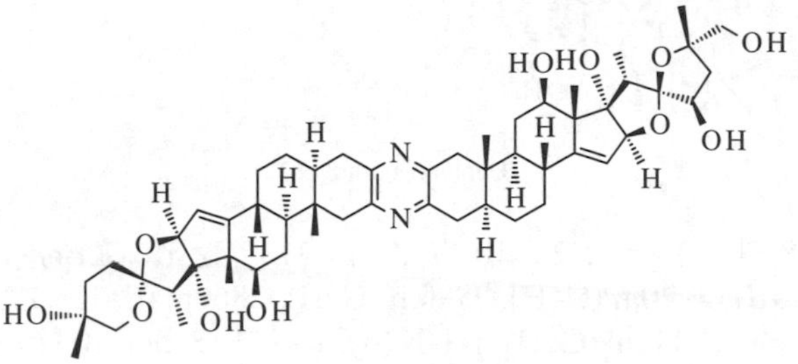

Cephalostatin 7 (**11**)

Cephalostatin 7 (**11**, 18.7 mg, 1.1×10^{-6}%, mp: 315 °C). IR (KBr): ν_{max} cm^{-1} 3430br (O–H), 2960m (C–H), 2920m (C–H), 2880m (C–H), 2860m (C–H), 1715s (C=O), 1650s (C=C), 1615m (C=C), 1450m (C–H), 1400s (pyrazine), 1385m (C–H), 1056m (C–O), 1043m (C–O), 950m. UV (MeOH): λ_{max} nm 286, 310. ^{1}H NMR (400 MHz, C_5D_5N): 'Right half' δ 5.64 (s, 1H, 15-CH), 5.24 (s, 1H, 16-CH), 4.79 (m, 2H, 23-CH), 4.05 (ddd, 1H, J = 1.0, 5.2, 12 Hz, 12-CH), 3.79 (d, 1H J = 11.2 Hz, 26-CHa), 3.70 (d, 1H, J = 11.3 Hz, 26-CHb), 3.08 (d, 1H, J = 16.7 Hz, 1-CHa), 2.89 (dd, 1H, J = 4.0, 17.6 Hz, 4-CHa), 2.83 (q, 1H, J = 7.0 Hz, 20-CH), 2.71 (dd, 1H, J = 8.0, 11.5 Hz, 24-CHa), 2.66 (m, 1H, 4-CHb), 2.61 (m, 1H, 1-CHb), 2.34 (t, 1H, J = 11.2 Hz, 24-CHb), 2.04 (m, 2H, 8-CH and 11-CHa), 1.72 (q, 1H, J = 12.0 Hz, 11-CHb), 1.63 (m, 1H, 7-CHa), 1.61 (s, 3H, 27-Me), 1.56 (m, 1H, 5-CH), 1.48 (m, 1H, 6-CHa), 1.42 (m, 1H, 7-CHb), 1.33 (d, 3H, J = 7.0 Hz, 21-Me), 1.31 (s, 3H, 18-Me), 1.20 (m, 1H, 6-CHb), 0.88 (dt, 1H, J = 3.5, 11.2 Hz, 9-CH), 0.75 (s, 3H, 19-Me); 'Left half' δ 5.60 (s, 1H, 15-CH), 5.17 (s, 1H, 16-CH), 4.17 (ddd, 1H, J = 1.0, 5.0, 11.9 Hz, 12-CH), 3.98 (d, 1H, J = 11.4 Hz, 26-CHa), 3.57 (dd, 1H, J = 1.0, 11.2 Hz, 26-CHb), 3.08 (d, 1H, J = 18.0 Hz, 1-CHa), 2.89 (dd, 1H, J = 4.0, 17.6 Hz, 4-CHa), 2.61 (m, 1H, 1-CHb), 2.60 (m, 1H, 4-CHb), 2.53 (dt, 1H, J = 4.9, 13.3, 23-CHa), 2.19 (m, 1H, 20-CH), 2.11 (m, 1H, 24-CHa), 2.08 (m, 1H, 11-CHa), 2.06 (m, 1H, 8-CH), 1.85 (m, 1H, 24-CHb), 1.75 (m, 1H, 11-CHb), 1.56 (m, 1H, 5-CH), 1.52 (m, 1H, 23-CHb), 1.48 (m, 1H, 6-CHa), 1.55 (m, 1H, 7-CHa), 1.31 (s, 3H, 18-Me), 1.31 (m, 1H, 7-CHb), 1.26 (d, 3H, J = 6.8 Hz, 21-Me), 1.23 (s, 3H, 27-Me), 1.20 (m, 1H, 6-CHb), 0.98 (dt, 1H, J = 4.0, 11.8 Hz, 9-CH), 0.76 (s, 3H, 19-Me). ^{13}C NMR (100 MHz, C_5D_5N): 'Right half' δ 46.0 (C-1), 148.9 (C-2), 148.6 (C-3), 35.8 (C-4), 41.8 (C-5), 28.3 (C-6), 28.7 (C-7), 33.8 (C-8), 53.2 (C-9), 36.3 (C-10), 29.0 (C-11), 75.6 (C-12), 55.4 (C-13), 152.7 (C-14), 122.3 (C-15), 93.2 (C-16), 91.7 (C-17), 12.6 (C-18), 11.7 (C-19), 44.5 (C-20), 9.0 (C-21), 117.2 (C-22), 71.5 (C-23), 39.5 (C-24), 82.8 (C-25), 69.3 (C-26), 26.4 (C-27); 'Left half' δ 46.0 (C-1), 148.9 (C-2), 148.6 (C-3), 35.8 (C-4), 41.8 (C-5), 28.3 (C-6), 29.0 (C-7), 34.0 (C-8), 52.9

(C-9), 36.3 (C-l0), 29.2 (C-11), 75.7 (C-12), 56.0 (C-13), 154.9 (C-14), 120.0 (C-15), 93.7 (C-16), 93.3 (C-17), 13.0 (C-18), 11.7 (C-19), 44.5 (C-20), 8.1 (C-21), 107.9 (C-22), 27.7 (C-23), 33.3 (C-24), 65.8 (C-25), 70.2 (C-26), 27.0 (C-27). SPHRSIMS: *m/z* 967.6413 calculated for $C_{54}H_{76}N_2O_{11}$, found: 967.5067 (Pettit *et al.*, 1992).

Cephalostatin 8 (**12**)

Cephalostatin 8 (**12**, 3.4 mg, 2×10^{-7}%, mp: 313 °C). IR (KBr): ν_{max} cm^{-1} 3400br (O–H), 2970m (C–H), 2920m (C–H), 2880m (C–H), 2860m (C–H), 1715s (C=O), 1650s (C=C), 1600m (C=C), 1445m (C–H), 1400s (pyrazine), 1385m, 1037m (C–O), 965m. UV (MeOH): λ_{max} nm 286, 310. SPHRSIMS: *m/z* 965.6640 calculated for $C_{55}H_{78}N_2O_{10}$, found: 965.5261 (Pettit *et al.*, 1992).

Cephalostatin 9 (**13**)

Cephalostatin 9 (**13**, mp: 307 °C). IR (KBr): ν_{max} cm^{-1} 3400br (O–H), 2990m (C–H), 2930m (C–H), 2880m (C–H), 2860m (C–H), 1710s (C=O), 1645s (C=C), 1630m (C=C), 1445m (C–H), 1400s (pyrazine), 1385m (C–H), 1055m (C–O), 950m. UV (MeOH): λ_{max} nm 286, 310. SPHRSIMS: *m/z* 929.5527 calculated for $C_{54}H_{76}N_2O_{11}$, found: 929.5527 (Pettit *et al.*, 1992).

In 1994, cephalostatins 10 and 11 (**14** and **15**) were reported (Pettit *et al.*, 1994a). These steroidal dimers were isolated from a 1990 scale-up 450 kg re-collection of *C. gilchristi* from the Indian Ocean (Southeast Africa). Again, these compounds also showed *in vitro* cytotoxicity towards a number of human cancer cell lines. Initial extraction and isolation methods for these dimers were similar to that described earlier, but in the final purification step, RP-HPLC using a C_8 reversed-phase column (mobile phase: MeCN-MeOH-H_2O =10:10:12) was employed to obtain cephalostatins 10 (**14**, 14.6 mg, 3.24×10^{-6}%, mp: > 300 °C) and 11 (**15**, 6.1 mg, 1.36×10^{-6}%, mp: > 300 °C). The ^{1}H and ^{13}C NMR and FABMS spectra of cephalostatins 10 (**14**) and 11 (**15**) clearly established that the structures of **14** and **15** were similar to that of cephalostatin 2 (**5**) with an additional OMe. The relative stereochemistry at the major chiral centres within **14** and **15** was established by studying nOe interactions observed in their ^{1}H-^{1}H ROSEY spectra.

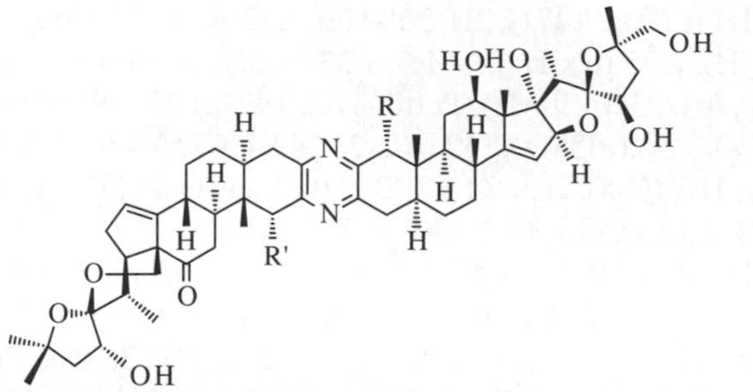

Cephalostatin 10 (**14**) R = OMe; R' = H Cephalostatin 11 (**15**) R = H; R' = OMe

Cephalostatin 10 (**14**): UV (MeOH): λ_{max} nm 289, 304. ^{1}H NMR (500 MHz, C_5D_5N): δ 6.23 (s, 1H, 17-OH), 5.97 (s, 1H, 9'-OH), 5.62 (br s, 1H, 15-CH), 5.56 (br s, 1H, 15'-CH), 5.24 (br s, 1H, 16-CH), 4.81 (m, 1H, 23-CH), 4.79 (s, 1H, 12-OH), 4.83 (m, 1H, 23'-CH), 4.24 (m, 1H, 12-CH), 4.14 (s, 1H, 1-CH), 4.14 (dd, 1H, 18'-CH), 3.80 (d, 1H, 26-CHa), 3.35 (s, 3H, OMe), 3.74 (m, 1H, 1'-CHb), 3.72 (d, 1H, 26-CHb), 3.31 (m, 1H, 11'-CHb), 3.18 (m, 1H, 20'-CH), 3.06 (m, 1H, 4'-CHb), 3.00 (m, 1H, 4-CHa), 2.99 (m, 1H, 1'-CHa), 2.92 (m, 1H, 16'-CHb), 2.96 (m, 1H, 11'-CHa), 2.87 (m, 1H, 20-CH), 2.80 (m, 1H, 17'-CH), 2.75 (m, 1H, 8'-CH), 2.73 (m, 1H, 4-CHa), 2.70 (m, 1H, 24-CHa), 2.61 (m, 1H, 4-CHb), 2.55 (m, 1H, 5'-CH), 2.35 (m, 1H, 24-CHb and 24'-CHb), 2.27 (m, 2H, 5-CH and 16'-CHa), 2.13 (m, 1H, 8-CH), 2.12 (m, 1H, 11-CHb), 1.94 (m, 1H, 24'-CHa), 1.90 (m, 1H, 7'-CH), 1.85 (m, 1H, 11-CHa), 1.64 (m, 1H, 7-CHb), 1.63 (s, 3H, 27-Me), 1.52 (m, 1H, 6-CHb), 1.48 (m, 1H, 6'-CH), 1.47 (s, 3H, 26'-Me), 1.40 (s, 3H, 27'-Me), 1.39 (d, 3H, 21'-Me), 1.35 (d, 3H, 21-Me), 1.34 (s, 3H, 18-Me), 1.34 (m, 1H, 7-CHa), 1.28 (m, 1H, 6-CHa), 0.98 (s, 3H, 19'-Me), 0.85 (m, 1H, 9-CH), 0.71 (s, 3H, 19-Me) (Pettit *et al.*, 1994a). ^{13}C NMR (125 MHz, C_5D_5N): δ 83.4 (C-1), 57.3 (OMe), 39.8 (C-1'), 149.9 (C-2), 150.0 (C-2'), 147.5 (C-3), 147.7 (C-3'), 35.8 (C-4), 36.2 (C-4'), 35.1 (C-5), 34.2 (C-5'), 28.3 (C-6), 28.1 (C-6'), 28.6 (C-7), 24.6 (C-7'), 33.7 (C-8), 39.0 (C-8'), 45.9 (C-9), 78.6 (C-9'), 40.7 (C-10), 41.3 (C-10'), 28.8 (C-11), 45.6 (C-11'), 75.7 (C-12), 211.0 (C-12'), 55.5 (C-13), 61.6 (C-13'), 153.1 (C-14), 148.5 (C-14'), 122.2 (C-15), 123.8 (C-15'), 93.2 (C-16), 32.6 (C-16'), 91.7 (C-17), 44.2 (C-17'), 12.7 (C-18), 64.0 (C-18'), 11.1 (C-19), 15.1 (C-19'), 44.5 (C-20), 32.9 (C-20'), 9.0 (C-21), 15.5 (C-21'), 117.2 (C-22), 110.9 (C-22'), 71.6 (C-23), 81.6 (C-23'), 39.5 (C-24), 47.3 (C-24'), 82.8 (C-25), 81.1 (C-25'), 69.3 (C-26), 29.8 (C-26'), 26.4 (C-27), 29.5 (C-27'). HRFABMS: *m/z* 957.5444 calculated for $C_{55}H_{77}N_2O_{12}$, found: 957.5492 (Pettit *et al.*, 1994a).

Cephalostatin 11 (**15**): UV (MeOH): λ_{max} nm 288, 305. ^{1}H NMR (500 MHz, C_5D_5N): δ 6.23 (s, 1H, 17-OH), 5.95 (s, 1H, 9'-OH), 5.64 (br s, 1H, 15-CH), 5.56 (br s, 1H, 15'-CH), 5.24 (br s, 1H, 16-CH), 4.81 (m, 1H, 23-CH), 4.81 (m, 1H, 23'-CH), 4.58 (m, 1H, 1'-CH), 4.15 (dd, 1H, 18'-CH), 4.05 (dd, 1H, 12-CH), 3.80 (d, 1H, 26-CHa), 3.42 (m, 1H, 11'-CHb), 3.72 (d, 1H, 26-CHb), 3.35 (s, 3H, OMe), 3.25 (m, 1H, 20'-CH), 3.20 (dd, 1H, 4-CHb), 3.10 (d, 1H, 1-CHb), 2.95 (m, 1H, 4-CHa), 2.90 (m, 1H, 17'-CH), 2.87 (m, 1H, 11'-CHa), 2.85 (m, 1H,

16'-CHb), 2.82 (m, 1H, 20-CH), 2.75 (m, 1H, 24-CHb), 2.65 (m, 2H, 4-CHa' and 8'-CH), 2.62 (m, 1H, 4-CHb), 2.60 (m, 1H, 1-CHa), 2.36 (m, 1H, 24-CHa), 2.35 (m, 1H, 24'-CHb), 2.31 (m, 1H, 16'-CHa), 2.07 (m, 2H, 8-CH and 7'-CHb), 2.05 (m, 1H, 11-CHb), 2.00 (m, 1H, 5'-CH), 1.93 (dd, 1H, 24'-CHa), 1.85 (m, 1H, 7'-CHa), 1.77 (m, 1H, 11-CHa), 1.64 (s, 3H, 27-Me), 1.59 (m, 1H, 5-CH), 1.47 (s, 3H, 26'-Me), 1.43 (d, 3H, 21'-Me), 1.40 (s, 3H, 27'-Me), 1.39 (m, 1H, 6'-CH), 1.35 (d, 3H, 21-Me), 1.33 (s, 3H, 18-Me), 0.86 (m, 1H, 9-CH), 0.81 (s, 3H, 19'-Me), 0.76 (s, 3H, 19-Me) (Pettit *et al.*, 1994a). ^{13}C NMR (125 MHz, C_5D_5N): δ 46.1 (C-1), 85.8 (C-1'), 56.6 (OMe), 152.0 (C-2), 149.4 (C-2'), 148.4 (C-3), 145.9 (C-3'), 35.7 (C-4), 35.8 (C-4'), 41.7 (C-5), 28.9 (C-5'), 28.2 (C-6 and C-7), 30.0 (C-6'), 24.4 (C-7'), 33.8 (C-8), 39.5 (C-8'), 53.1 (C-9), 80.2 (C-9'), 36.3 (C-10), 43.1 (C-10'), 28.6 (C-11), 46.6 (C-11'), 75.5 (C-12), 210.7 (C-12'), 55.4 (C-13), 61.6 (C-13'), 152.6 (C-14), 148.0 (C-14'), 122.3 (C-15), 123.8 (C-15'), 93.1 (C-16), 32.5 (C-16'), 91.6 (C-17), 44.2 (C-17'), 12.6 (C-18), 64.6 (C-18'), 11.8 (C-19), 14.4 (C-19'), 44.5 (C-20), 33.0 (C-20'), 9.0 (C-21), 15.9 (C-21'), 117.2 (C-22), 110.9 (C-22'), 71.5 (C-23), 81.6 (C-23'), 39.5 (C-24), 47.3 (C-24'), 82.8 (C-25), 81.1 (C-25'), 69.3 (C-26), 29.7 (C-26'), 26.4 (C-27), 29.5 (C-27'). HRFABMS: *m/z* 957.5444 calculated for $C_{55}H_{77}N_2O_{12}$, found: 957.5474 (Pettit *et al.*, 1994a).

In the same year, Pettit *et al.* (1994b,c) published the isolation, structure determination and assessment of cytotoxicity of four further members of this new class of dimeric steroidal alkaloids, cephalostatins 12–15 (**16–19**). Cephalostatins 12 (**16**) is the first symmetrical cephalosatatin. Isolation of **16** and **17** compounds was achieved from an active butanol subfraction of the DCM fraction of the initial extract by a combination of steric exclusion and partition type chromatographic separations followed by a RP-semipreparative HPLC on Phenomenex Prepex C_8 column eluting with MeOH-H_2O. The ^{1}H and ^{13}C NMR data of **16** were identical to those for the 'right-side' of cephalostatin 1 (**2**) as described earlier, and because of the symmetry in **16**, the data for both halves were exactly the same.

Similar initial purification of the extract and fractions, followed by RP-HPLC on a C_8 column (mobile phase: MeCN:MeOH:H_2O = 5:5:6) afforded cephalostatins 14 (**18**) and 15 (**19**). A detailed schematic protocol for the isolation of these compounds can be found in Pettit *et al.* (1994c). Like previously isolated cephalostatins, all these compounds (**16–19**) also displayed growth inhibitory activity against many of the human tumour cell lines tested, but the activity was much less potent than cephalostatin 1 (**2**).

Cephalostatin 12 (**16**, 47.3 mg, 1.1×10^{-5}%, mp: > 300 °C). IR (KBr): ν_{max} cm^{-1} 3416br (O–H), 2977m (C–H), 2924m (C–H), 2882m (C–H), 1645s (C=C), 1448m (C–H), 1400s (pyrazine), 1219m (C–O), 1117m (C–O), 1049m (C–O), 951m, 888m. UV (MeOH): λ_{max} nm 288, 308. HRFABMS: *m/z* 945.5444 calculated for $C_{54}H_{76}N_2O_{12}$, found: 945.5444 (Pettit *et al.*, 1994b).

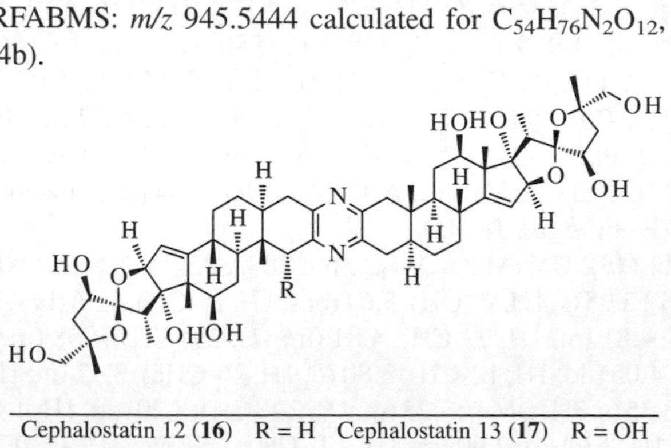

Cephalostatin 12 (**16**) R = H Cephalostatin 13 (**17**) R = OH

Cephalostatin 13 (**17**, 3.7 mg, 8.2×10^{-7}%, mp: >300 °C). IR (KBr): ν_{max} cm^{-1} 3426br (O–H), 2978m (C–H), 2926m (C–H), 2859m (C–H), 1643s (C=C), 1449m (C–H) 1398s (pyrazine), 1385m (C–H), 1221s (C–O), 1119m (C–O), 1105m (C–O), 1044m (C–O), 953m. UV (MeOH): λ_{max} nm 287, 308. ^{1}H NMR (400 MHz, C_5D_5N): δ 'Right half' δ 8.07 (d, 1H, $J = 6.8$, 23-OH), 6.55 (br s, 1H, 26-OH), 6.21 (s, 1H, 17-OH), 5.62 (s, 1H, 15-CH), 5.23 (s, 1H, 16-CH), 4.79 (m, 1H, 23-CH), 4.70 (s, 1H, 12-OH), 4.03 (dd, 1H, $J = 4.8$, 11.4 Hz, 12-CH), 3.81 (br d, 1H, $J = 12.2$ Hz, 26-CHb), 3.71 (br d, 1H, $J = 12.2$ Hz, 26-CHa), 3.08 (d, 1H, $J = 17.4$ Hz, 1-CHb), 2.87 (m, 1H, 20-CH), 2.86 (m, 1H, 4-CHb), 2.62 (m, 1H, 4-CHa), 2.71 (m, 1H, 24-CHb), 2.54 (d, 1H, $J = 17.4$ Hz, 1-CHa), 2.33 (t, 1H, $J = 11.0$ Hz, 24-CHa), 2.07 (m, 1H, 8-CH), 2.03 (m, 1H, 11-CHb), 1.76 (q, 1H, $J = 11.4$ Hz, 11-CHa), 1.64 (m, 1H, 7-CHb), 1.63 (s, 3H, 27-Me), 1.49 (m, 1H, 5-CH), 1.47 (m, 1H, 6-CHb), 1.33 (s, 3H, 18-Me), 1.35 (d, 3H, $J = 6.9$ Hz, 21-Me), 1.26 (m, 1H, 7-CHa), 1.23 (m, 1H, 6-CHa), 0.85 (m, 1H, 9-CH), 0.73 (s, 3H, 19-Me); 'Left half' δ 8.07 (d, 1H, $J = 6.8$ Hz, 23-OH), 7.34 (d, 1H, $J = 4.5$ Hz, 1-OH), 6.55 (br s, 1H, 26-OH), 6.22 (s, 1H, 17-OH), 5.66 (s, 1H, 15-CH), 5.24 (s, 1H, 16-CH), 4.85 (d, $J = 4$ Hz, 5-CH), 4.79 (m, 1H, 23-CH), 4.72 (s, 1H, 12-OH), 4.24 (dd, 1H, $J = 5.0$, 11.9 Hz, 12-CH), 3.81 (br d, 1H, $J = 12.2$ Hz, 26-CHb), 3.71 (br d, 1H, $J = 12.2$ Hz, 26-CHa), 3.02 (dd, 1H, $J = 5.3$, 18.0 Hz, 4-CHb), 2.87 (m, 1H, 20-CH), 2.65 (m, 1H, 4-CHa), 2.71 (m, 1H, 24-CHb), 2.50 (m, 1H, 5-CH), 2.47 (m, 1H, 11-CHb), 2.34 (t, 1H, $J = 11.0$ Hz, 24-CHa), 2.18 (m, 1H, 8-CH), 2.09 (m, 1H, 9-CH), 1.86 (q, 1H, $J = 11.9$ Hz, 11-CHa), 1.70 (m, 1H, 7-CHb), 1.64 (s, 3H, 27-Me), 1.61 (m, 1H, 6-CHb), 1.41 (m, 1H, 7-CHa), 1.37 (m, 1H, 6-CHa), 1.38 (s, 3H, 18-Me), 1.35 (d, 3H, $J = 6.9$ Hz, 21-Me), 0.77 (s, 3H, 19-Me). ^{13}C NMR (100 MHz, C_5D_5N): 'Right half' δ 46.1 (C-1), 150.6 (C-2), 148.9 (C-3), 35.7 (C-4), 41.8 (C-5), 28.2 (C-6), 28.7 (C-7), 33.8 (C-8), 53.2 (C-9), 36.3 (C-10), 28.9 (C-11), 75.6 (C-12), 55.4 (C-13), 152.7 (C-14), 122.3 (C-15), 93.2 (C-16), 91.7 (C-17), 12.6 (C-18), 11.7 (C-19), 44.5 (C-20), 9.0 (C-21), 117.2 (C-22), 71.5 (C-23), 39.5 (C-24), 82.8 (C-25), 69.3 (C-26), 26.4 (C-27); 'Left half' δ 74.4 (C-1), 151.2 (C-2), 149.1 (C-3), 36.0 (C-4), 34.7 (C-5), 28.5 (C-6), 28.8 (C-7), 33.8 (C-8), 46.2 (C-9), 40.4 (C-10), 28.4 (C-11), 75.9 (C-12), 55.5 (C-13), 153.4 (C-14), 122.1 (C-15), 93.3 (C-16), 91.7 (C-17), 12.6 (C-18), 11.2 (C-19), 44.5 (C-20), 9.0 (C-21), 117.2 (C-22), 71.5 (C-23), 39.5 (C-24), 82.8 (C-25), 69.3 (C-26), 26.4 (C-27). HRFABMS: *m/z* 961.5451 calculated for $C_{54}H_{76}N_2O_{13}$, found: 961.5451 (Pettit *et al.*, 1994b).

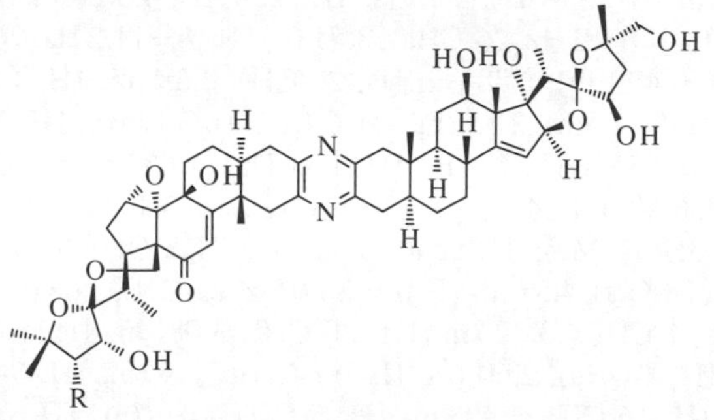

Cephalostatin 14 (**18**) R = H	Cephalostatin 15 (**19**) R = Me

Cephalostatin 14 (**18**, 5.2 mg, 2.5 × 10^{-2}%, mp: >300 °C). IR (KBr): ν_{max} cm^{-1} 3418br (O–H), 2928m (C–H), 1674s (C=O and C=C), 1449m (C–H), 1400s (pyrazine), 1233m (C–O), 1101m (C–O), 1049m (C–O), 941m, 889m. UV (MeOH): λ_{max} nm 230, 288, 305. ^{1}H NMR (500 MHz, C_5D_5N): 'Right half' δ 6.26 (s, 1H, 17-OH), 5.66 (br s, 1H, 15-CH), 5.28 (br s, 1H, 16-CH), 4.82 (dd, 1H, J = 8.4, 10.2 Hz, 23-CH), 4.70 (s, 1H, 12-OH), 4.07 (dd, 1H, J = 4.5, 11.0 Hz, 12-CH), 3.77 (d, 1H, J = 11.0 Hz, 26-CHa), 3.72 (d, 1H, J = 11.0 Hz, 26-CHb), 3.11 (m, 1H, 1-CHb), 2.96 (m, 1H, 4-CHb), 2.80 (m, 1H, 20-CH), 2.76 (m, 1H, 4-CHa), 2.66 (m, 1H, 1-CHa), 2.75 (m, 1H, 24-CHb), 2.40 (m, 1H, 24-CHa), 2.08 (m, 1H, 8-CH), 2.07 (m, 1H, 11-CHb), 1.78 (m, 1H, 11-CHa), 1.72 (m, 1H, 7-CHb), 1.67 (s, 3H, 27-Me), 1.62 (m, 1H, 5-CH), 1.55 (m, 1H, 6-CHb), 1.45 (m, 1H, 7-CHa), 1.36 (s, 3H, 18-Me), 1.36 (d, 3H, J = 6.8 Hz, 21-Me), 1.32 (m, 1H, 6-CHa), 0.92 (dt, 1H, J = 4.0, 11.5 Hz, 9-CH), 0.78 (s, 3H, 19-Me); 'Left half' δ 6.92 (s, 1H, 8-OH), 6.47 (s, 1H, 11-CH), 4.88 (t, 1H, J = 7.0 Hz, 23-CH), 4.68 (dd, 1H, J = 13.0 Hz, 18-CH), 3.75 (br s, 1H, 15-CH), 3.36 (d, 1H, J = 16.4 Hz, 1-CH), 3.24 (dq, 1H, J = 6.0, 13.0 Hz, 20-CH), 3.06 (dd, 1H, J = 6.0, 17.0 Hz, 4-CHb), 2.73 (m, 1H, 4-CHa), 2.56 (m, 1H, 15-CH), 2.38 (m, 1H, 16-CHb), 2.35 (m, 1H, 24-CHb), 2.18 (m, 1H, 6-CHb), 2.17 (m, 1H, 16-CIIa), 2.02 (m, 1H, 7-CHb), 1.95 (m, 1H, 24-CIIa), 1.90 (m, 1H, 5-CH), 1.82 (m, 1H, 7-CHa), 1.57 (m, 1H, 6-CHa), 1.44 (s, 3H, 26-Me), 1.43 (s, 3H, 19-Me), 1.38 (d, 3H, J = 6.6 Hz, 21-Me), 1.33 (s, 3H, 27-Me). ^{13}C NMR (125 MHz, C_5D_5N): 'Right half' δ 45.7 (C-1), 148.7 (C-2 and C-3), 35.5 (C-4), 41.4 (C-5), 27.9 (C-6), 28.4 (C-7), 33.5 (C-8), 51.9 (C-9), 36.0 (C-10), 28.7 (C-11), 75.3 (C-12), 55.1 (C-13), 152.4 (C-14), 122.0 (C-15), 92.9 (C-16), 91.4 (C-17), 12.3 (C-18), 11.5 (C-19), 44.2 (C-20), 8.7 (C-21), 116.2 (C-22), 71.2 (C-23), 39.2 (C-24), 82.5 (C-25), 69.0 (C-26), 26.1 (C-27); 'Left half' δ 45.1 (C-1), 152.0 (C-2), 147.7 (C-3), 36.2 (C-4), 34.2 (C-5), 28.1 (C-6), 24.6 (C-7), 70.2 (C-8), 165.8 (C-9), 39.7 (C-10), 123.6 (C-11), 201.3 (C-12), 52.1 (C-13), 71.7 (C-14), 57.6 (C-15), 28.3 (C-16), 36.0 (C-17), 64.4 (C-18), 20.5 (C-19), 32.3 (C-20), 14.8 (C-21), 109.5 (C-22), 81.3 (C-23), 47.1 (C-24), 81.1 (C-25), 29.4 (C-26), 28.9 (C-27). HRFABMS: *m/z* 963.4983 calculated for $C_{54}H_{72}N_2O_{12}Na$, found: 963.4974 (Pettit *et al.*, 1994c).

Cephalostatin 15 (**19**, 32.0 mg, 3.3 × 10^{-3}%, mp: >300 °C). IR (KBr): ν_{max} cm^{-1} 3439br (O–H), 2924m (C–H), 1674s (C=C), 1451m (C–H), 1400s (pyrazine), 1171m (C–O), 1101m (C–O), 1049m (C–O), 941m, 889m. UV (MeOH): λ_{max} nm 228, 288, 305. ^{1}H NMR (500 MHz, C_5D_5N): 'Right half' δ 6.24 (s, 1H, 17-OH), 5.64 (br s, 1H, 15-CH), 5.24 (br s, 1H, 16-CH), 4.80 (m, 1H, 23-CH), 4.70 (s, 1H, 12-OH), 4.06 (dd, 1H, J = 4.0, 11.0 Hz, 12-CH), 3.81 (d, 1H, J = 11.2 Hz, 26-CHa), 3.71 (d, 1H, J = 11.2 Hz, 26CHb), 3.07 (m, 1H, 1-CHb), 2.92 (m, 1H, 4-CHb), 2.88 (m, 1H, 20-CH), 2.68 (m, 1H, 4-CHa), 2.62 (m, 1H, 1-CHa), 2.78 (m, 1H, 24-CHb), 2.35 (m, 1H, 24-CHa), 2.12 (m, 1H, 7-CH), 2.05 (m, 1H, 11-CHb), 2.03 (m, 1H, 8-CH), 1.77 (m, 1H, 11-CHa), 1.65 (s, 3H, 27-Me), 1.62 (m, 1H, 5-CH), 1.54 (m, 1H, 6-CH), 1.34 (s, 3H, 18-Me), 1.35 (d, 3H, J = 6.4 Hz, 21-Me), 0.89 (m, 1H, 9-CH), 0.76 (s, 3H, 19-Me); 'Left half' δ 6.88 (s, 1H, 8-OH), 6.46 (s, 1H, 11-CH), 4.72 (d, 1H, J = 11.0 Hz, 18-CH), 4.37 (d, 1H, J = 10.0 Hz, 23-CH), 3.68 (br s, 1H, 15-CH), 3.33 (d, 1H, J = 16.5 Hz, 1-CHb), 3.22 (m, 1H, 20-CH), 3.08 (m, 1H, 1-CHa), 3.02 (m, 1H, 4-CHb), 2.88 (m, 1H, 4-CHa), 2.60 (m, 1H, 17-CH), 2.15 (m, 1H, 6-CHb), 2.03 (m, 1H, 24-CH), 2.02 (m, 1H, 16-CH), 1.90 (m, 1H, 5-CH), 1.82 (m, 1H, 7-CH), 1.53 (m, 1H, 6-CHa), 1.36 (s, 3H, 19-Me), 1.32 (s, 3H, 27-Me), 1.24 (d, 3H, J = 6.8 Hz, 21-Me), 1.18 (s, 3H, 26-Me), 1.08 (d, 3H, J = 6.7 Hz, 28-Me). ^{13}C NMR (125 MHz, C_5D_5N): 'Right half' δ 46.0 (C-1), 149.2 (C-2), 148.2 (C-3), 35.6 (C-4), 41.7 (C-5), 28.2 (C-6), 28.9 (C-7), 33.8

(C-8), 53.2 (C-9), 36.3 (C-10), 28.7 (C-11), 75.6 (C-12), 55.4 (C-13), 152.7 (C-14), 122.3 (C-15), 93.2 (C-16), 91.6 (C-17), 12.6 (C-18), 11.7 (C-19), 44.5 (C-20), 9.0 (C-21), 117.2 (C-22), 71.5 (C-23), 39.5 (C-24), 82.8 (C-25), 69.3 (C-26), 26.4 (C-27); 'Left half' δ 45.4 (C-1), 148.9 (C-2), 148.1 (C-3), 35.8 (C-4), 41.2 (C-5), 23.7 (C-6), 24.4 (C-7), 71.0 (C-8), 165.9 (C-9), 39.9 (C-10), 123.9 (C-11), 201.7 (C-12), 52.2 (C-13), 78.0 (C-14), 58.0 (C-15), 28.5 (C-16), 36.1 (C-17), 63.5 (C-18), 20.8 (C-19), 32.4 (C-20), 14.9 (C-21), 108.0 (C-22), 86.9 (C-23), 51.6 (C-24), 81.3 (C-25), 23.7 (C-26), 28.1 (C-27), 12.6 (C-28). HRFABMS: *m/z* 955.5320 calculated for $C_{55}H_{75}N_2O_{12}$, found: 955.5318 (Pettit *et al.*, 1994c).

Two more highly potent human cancer cell growth inhibitors, cephalostatins 16 (**20**), and 17 (**7**), were reported from *C. gilchristi* in 1995 (Pettit *et al.*, 1995). Because of structural similarities among cephalostatins 2, 3 and 17 (**5**–**7**), the structure of cephalostatin 17 (**7**) could be easily deduced by comparison of its ^{1}H and ^{13}C NMR data with those of compounds **5** and **6** as well as of cephalostatin 1 (**2**), as discussed earlier.

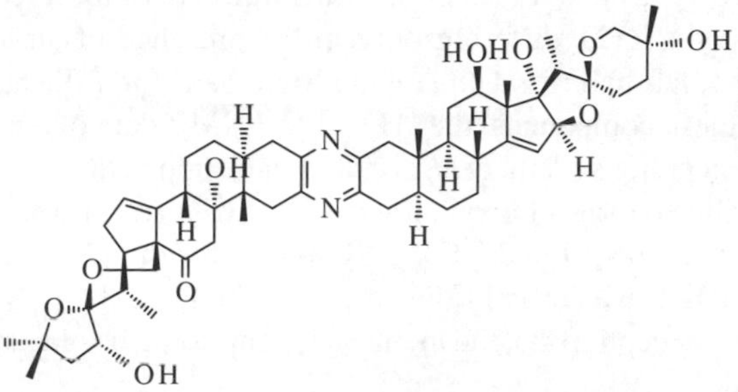

Cephalostatin 16 (**20**)

Cephalostatin 16 (**20**, 4.1 mg, $9 \times 10^{-7}\%$, mp: >300 °C). IR (KBr): ν_{max} cm^{-1} 3453br (O–H), 2928m (C–H), 1711s (C=O), 1632s (C=C), 1449m (C–H), 1400s (pyrazine), 1045m (C–O), 889m. UV (MeOH): λ_{max} nm 288, 305. ^{1}H NMR (500 MHz, C_5D_5N): 'Right half' δ 5.59 (br s, 1H, 15-CH), 5.18 (br s, 1H, 16-CH), 5.09 (s, 1H, 17-OH), 4.71 (s, 1H, 12-OH), 4.20 (m, 1H, 12-CH), 4.01 (d, 1H, J = 14.5 Hz, 26-CHb), 3.61 (d, 1H, J = 14.5 Hz, 26-CHa), 3.12 (m, 1H, 1-CHb), 2.90 (m, 1H, 4-CHb), 2.68 (m, 1H, 4-CHa), 2.60 (m, 1H, 1-CHa), 2.55 (m, 1H, 23-CHa), 2.22 (m, 1H, 20-CH), 2.16 (m, 1H, 24-CHb), 2.08 (m, 1H, 8-CH), 1.87 (m, 1H, 24-CHa), 1.74 (m, 2H, 7-CHb and 11-CHb), 1.62 (m, 1H, 5-CH), 1.57 (m, 1H, 23-CHb), 1.36 (m, 1H, 7-CHa and 11-CHa), 1.35 (s, 3H, 18-Me), 1.27 (d, 3H, J = 7.0 Hz, 21-Me), 1.23 (s, 3H, 27-Me), 0.85 (m, 1H, 9-CH), 0.78 (s, 3H, 19-Me); 'Left half' δ 6.47 (s, 1H, 11-CH), 6.02 (s, 1H, 9-OH), 5.57 (br s, 1H, 15-CH), 4.82 (t, 1H, J = 7.0 Hz, 23-CH), 3.77 (m, 1H, 1-CHb), 4.16 (dd, 1H, J = 13.0 Hz, 18-CH), 3.20 (m, 1H, 20-CH), 3.08 (m, 1H, 4-CHb), 2.95 (m, 1H, 1-CHa), 2.90 (m, 1H, 16-CHb), 2.81 (m, 1H, 17-CH), 2.77 (m, 1H, 8-CH), 2.72 (m, 1H, 4-CHa), 2.36 (dd, 1H, J = 7.0, 12.2 Hz, 24-CHb), 2.05 (m, 1H, 5-CH), 2.12 (m, 1H, 16-CHa), 2.01 (m, 1H, 7-CHb), 1.93 (dd, 1H, J = 12.2 Hz, 24-CHa), 1.92 (m, 1H, 7-CHa), 1.46 (s, 3H, 26-Me), 1.40 (d, 3H, J = 6.8 Hz, 21-Me), 1.40 (s, 3H, 27-Me), 0.98 (s, 3H, 19-Me). ^{13}C NMR (125 MHz, C_5D_5N): 'Right half' δ 46.0 (C-1), 149.7 (C-2), 148.3 (C-3), 35.8 (C-4), 41.8 (C-5), 28.8 (C-6), 29.0 (C-7 and C-11), 34.0 (C-8), 52.9 (C-9), 36.3 (C-10), 75.7 (C-12), 55.8 (C-13), 154.8 (C-14), 119.9 (C-15), 93.7 (C-16), 91.3 (C-17), 13.0 (C-18), 11.8 (C-19), 48.5 (C-20), 8.2 (C-21), 107.9 (C-22), 27.7 (C-23), 33.3 (C-24), 65.8 (C-25), 70.2 (C-26), 27.0 (C-27); 'Left half' δ 39.5 (C-1), 150.2

(C-2), 148.6 (C-3), 36.2 (C-4), 34.2 (C-5), 28.1 (C-6), 24.6 (C-7), 39.0 (C-8), 78.7 (C-9), 42.2 (C-10), 123.6 (C-11), 211.1 (C-12), 61.4 (C-13), 148.6 (C-14), 124.4 (C-15), 32.5 (C-16), 36.0 (C-17), 64.0 (C-18), 15.0 (C-19), 32.9 (C-20), 15.5 (C-21), 110.9 (C-22), 81.6 (C-23), 47.3 (C-24), 81.1 (C-25), 29.8 (C-26), 28.9 (C-27). HRFABMS: *m/z* 911.5422 calculated for $C_{54}H_{75}N_2O_{10}$, found: 911.5422 (Pettit *et al.*, 1995).

Cephalostatin 17 (**7**, 3.8 mg, 8.4×10^{-7}%, mp: >300 °C). IR (KBr): ν_{max} cm^{-1} 3439br (O–H), 2928m (C–H), 1713s (C=O), 1464m (C–H), 1400s (pyrazine), 1090m (C–O), 1051m (C–O), 952m, 891m. UV (MeOH): λ_{max} nm 288, 305. HRFABMS: *m/z* 911.5422 calculated for $C_{54}H_{75}N_2O_{10}$, found: 911.5402 (Pettit *et al.*, 1995).

A few years later, Pettit's group reported the isolation and structure elucidation of the last two of the cephalostatin series of cytotoxic steroidal alkaloids, cephalostatins 18 (**3**) and 19 (**4**), which are structural isomers (Pettit *et al.*, 1998). Reinvestigation of the active fraction, which yielded cephalostatins 1 (**2**), 2 (**5**), 14 (**18**) and 15 (**19**) as outlined before, also afforded dimers **3** (6.1 mg) and **4** (1.3 mg). Structurally, both these compounds are quite similar to cephalostatin 1 (**2**), differing only in the presence of an additional methoxyl moiety at C-1, on the 'left half' for **3** and on the 'right half' for **4**. Because of the structural similarities among these compounds, the ^{1}H and ^{13}C NMR data of **3** and **4** were similar to that of **2** with a few exceptions (Pettit *et al.*, 1998). For example, in the ^{1}H and ^{13}C NMR of **3**, two new signals attributed to a methoxyl moiety (δ_C 57.6, δ_H 3.53) and to an oxymethine (δ_C 82.9, δ_H 4.13), replaced the ^{1}H and ^{13}C NMR signals for the C-1 methylene group ('left half') in **2**. The situation was exactly the same (δ_C 57.4, δ_H 3.53; δ_C 83.3, δ_H 4.14) for compound **4**, with the exception that the methoxyl group could be placed on C-1 of the 'right half' of the molecule.

Cephalostatin 18 (**3**, 3.8 mg, 1.3×10^{-6}%, mp: >320 °C). IR (KBr): ν_{max} cm^{-1} 3426br (O–H), 2925m (C–H), 1707s (C=O), 1398m (pyrazine), 1088m (C–O). UV (MeOH): λ_{max} nm 288, 308. HRFABMS: *m/z* 947.5632 calculated for $C_{55}H_{76}N_2O_{11}Li$, found: 947.5609 (Pettit *et al.*, 1998).

Cephalostatin 19 (**4**, 1.3 mg, 2.9×10^{-7}%, mp: >320 °C). IR (KBr): ν_{max} cm^{-1} 3426br (O–H), 2925m (C–H), 1711s (C=O), 1398m (pyrazine), 1088m (C–O). UV (MeOH): λ_{max} nm 288, 305. HRFABMS: *m/z* 941.5545 calculated for $C_{55}H_{77}N_2O_{11}$, found: 941.5527 (Pettit *et al.*, 1998).

4.2 Crellastatins

Crellastatin A (**21**), an unsymmetrical cytotoxic sulphur-containing steroid dimer, was first isolated from Vanuatu Island marine sponge *Crella* sp. (family: Crellidae, order: Poecilo-scleridae), a sample of which was collected in 1996 (D'Auria *et al.*, 1998). This was the first example of a naturally occurring unsymmetrical steroid dimer connected through side chains. This dimer exhibited *in vitro* cytotoxicity against NSCLC–N6 cells ($IC_{50} = 1.5$ µg/mL). Structure elucidation of **21** was accomplished by extensive 1D and 2D NMR experiments, and the relative stereochemistry at the major chiral centres was established by the ROESY experiment as well as by molecular mechanics and dynamics calculations. The lyophilized sponge (80 g) was extracted with MeOH (3 × 2 L) using a modified Kupchan partitioning method (Kupchan *et al.*, 1973). Briefly, the MeOH extract (20 g) was dissolved in a mixture of 90% MeOH in H_2O, and partitioned against *n*-hexane. The water content of the MeOH extract was adjusted to 20–40% and partitioned, successively, against CCl_4 and $CHCl_3$. The aqueous phase was concentrated to remove MeOH and then extracted with *n*-BuOH. The cytotoxic $CHCl_3$ fraction (1.6 g) was subjected to DCCC using the solvent system, $CHCl_3$:MeOH:H_2O (7:13:8), followed by purification by RP-HPLC on a μ-Bondapak C_{18} column, eluting with MeOH:H_2O = 17:8, to obtain crellastatin A (**21**, 231 mg, 0.29%). IR (KBr): ν_{max} cm^{-1} 3420br (O–H), 2924m (C–H), 1651m (C=C), 1540m, 1458m (C–H), 1200m (C–O), 1060m (C–O). UV (MeOH): λ_{max} nm 208 (D'Auria *et al.*, 1998).

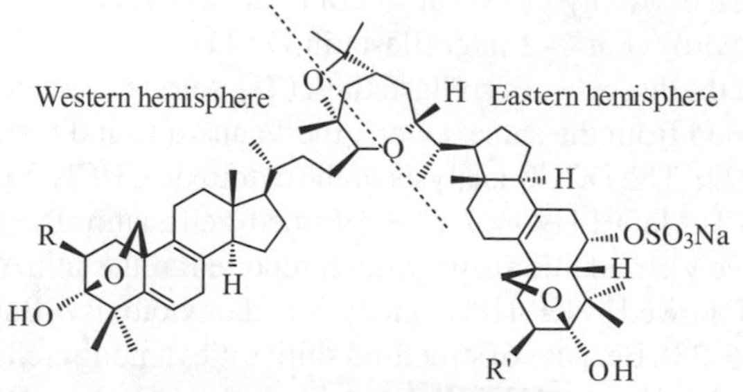

Crellastatin A (**21**)	R = R' = OH	Crellastatin B (**22**)	R = H; R' = OH
Crellastatin C (**23**)	R = OH; R' = H	Crellastatin D (**24**)	R = R' = H

^{1}H NMR (500 MHz, CD_3OD): 'Western hemisphere' δ 5.60 (t, 1H, J = 3.1 Hz, 6-CH), 4.30 (d, 1H, J = 8.6 Hz, 19*R*-CH), 4.08 (dd, 1H, J = 3.3, 8.8 Hz, 2-CH), 3.66 (dd, 1H, J = 2.7, 8.6 Hz, 19*S*-CH), 3.22 (br t, 1H, J = 4.0 Hz, 24-CH), 2.65 (br s, 1H, 7-CH), 2.26* (m, 1H, 11α-CH), 2.25* (1H, 26-CHa), 2.22* (m, 1H, 14-CH), 2.20* (1H, 1β-CH), 2.17* (2H, 1α-CH and 16α-CH), 2.08* (m, 1H, 12α-CH), 2.07* (1H, 26-CHb), 2.04* (m, 1H, 11β-CH), 1.67* (m, 1H, 15α-CH), 1.66* (2H, 20-CH and 23-CHa), 1.58* (1H, 22-CHb), 1.46* (m, 1H, 12β-CH), 1.45* (1H, 22-CHa), 1.40* (1H, 23-CHb), 1.39* (m, 1H, 16β-CH), 1.36* (m, 1H, 15β-CH), 1.33* (1H, 17-CH), 1.22 (s, 3H, 28-Me), 1.19 (s, 3H, 27-Me), 1.16 (s, 3H, 29-Me), 1.01 (d, 3H, J = 6.6 Hz, 21-Me), 0.72 (s, 3H, 18-Me); 'Eastern hemisphere' δ 4.65 (dt, 1H, J = 1.0, 10.5 Hz, 6-CH), 4.04 (dd, 1H, J = 4.8, 9.7 Hz, 2-CH), 3.94 (dd, 1H, J = 1.0, 8.8 Hz, 19*R*-CH), 3.83 (dd, 1H, J = 1.8, 8.8 Hz, 19*S*-CH), 3.72 (dd, 1H, J = 4.2, 11.4 Hz, 22-CH), 2.24* (m, 1H, 14-CH), 2.17* (2H, 11α-CH and 24-CH), 2.06* (3H, 11β-CH, 12β-CH and 23-CHb), 2.04* (2H, 1α-CH and 7α-CH), 1.94* (1H, 1β-CH), 1.88* (1H, 16α-CH), 1.74* (1H, 15α-CH), 1.58 (dd, 1H, J = 1.0, 10.5 Hz, 5-CH), 1.58 (1H, 17-CH), 1.47* (1H, 16β-CH), 1.42* (1H, 12α-CH), 1.38* (m, 1H, 20-CH), 1.38 (s, 3H, 28-Me), 1.36 (s, 3H, 26-Me), 1.32* (2H, 15β-CH and 23-CHa), 1.26 (s, 3H, 27-Me), 1.19 (s, 3H, 29-Me), 1.02 (d, 3H, J = 6.6 Hz, 21-Me), 0.68 (s, 3H, 18-Me); *overlapped peaks. ^{13}C NMR (125 MHz, CD_3OD): 'Western hemisphere' δ 42.8 (C-1), 68.2 (C-2), 99.7 (C-3), 42.6 (C-4), 147.8 (C-5), 115.7 (C-6), 30.0 (C-7), 130.3 (C-8), 126.6 (C-9), 38.4 (C-10), 23.4 (C-11), 38.2 (C-12), 42.5 (C-13), 53.3 (C-14), 23.9 (C-15), 30.5 (C-16), 54.5 (C-17), 11.6 (C-18), 72.8 (C-19), 36.2 (C-20), 19.2 (C-21), 36.1 (C-22), 25.1 (C-23), 85.7 (C-24), 87.9 (C-25), 36.4 (C-26), 27.1 (C-27), 25.6 (C-28 and C-29); 'Eastern hemisphere' δ 44.7 (C-1), 67.2 (C-2), 99.5 (C-3), 40.5 (C-4), 52.4 (C-5), 75.3 (C-6), 35.2 (C-7), 127.7 (C-8), 131.5 (C-9), 40.7 (C-10), 23.2 (C-11), 38.2 (C-12), 42.6 (C-13), 53.8 (C-14), 24.9 (C-15), 29.5 (C-16), 52.8 (C-17), 11.2 (C-18), 69.8 (C-19), 43.8 (C-20), 13.2 (C-21), 71.4 (C-22), 34.0 (C-23), 45.6

(C-24), 85.4 (C-25), 25.5 (C-26), 32.4 (C-27), 17.7 (C-28), 28.4 (C-29). HRFABMS: *m/z* 933.6154 calculated for $C_{58}H_{86}O_8$, found: 933.6154 (D'Auria *et al.*, 1998).

Later, a series of potential cytotoxic crellastatins B–M (**22–33**) were isolated from *Crella* sp. (Nahar *et al.*, 2007). All these new compounds possess the same unique junction between the monomeric units through their side chains and they differ from crellastatin A (**21**) mainly in the hydroxylation patterns of the two tetracyclic cores. For example, crellastatins B–D (**22–24**) are quite similar to crellastatin A (**21**) with the exceptions that in crellastatins B and C (**22** and **23**), instead of two hydroxyls at C-2 of both monomeric units, there was only one, and there was no hydroxyl at C-2 in crellastatin D (**24**).

Just after a year of the discovery of crellastatin A (**21**), seven new dimers, creallastatins B–H (**22–28**), were reported from the same source, the Vanuatu Island marine sponge *Crella* sp. (Zampella *et al.*, 1999). The DCCC analysis of the cytotoxic $CHCl_3$ fraction (1.6 g) using the solvent system, $CHCl_3$:MeOH:H_2O = 7:13:8, afforded crellastatins B-E (**22-25**, 19.0, 15.0, 4.5 and 3.0 mg, respectively), in addition to previously reported crellastatin A (**21**, 231 mg). Similar DCCC purification followed by RP-HPLC analysis of the cytotoxic *n*-BuOH fraction produced crellastatins F–H (**26–28**). Because of structural similarities among crellastatins A–D (**21–24**), as mentioned above, the ^{1}H and ^{13}C NMR data (obtained in CD_3OD) of these compounds were quite similar with a few minor exceptions in signals arising from the presence/absence of hydroxyl functionalities at C-2 position of both monomeric units (Zampella *et al.*, 1999). The FABMS data of crellastatins B–D (**22–24**) are presented in *Table 4.2.1*.

Crellastatin E (**25**) was significantly different from **21–24** because of the presence of a ketone functionality (instead of a hydroxyl or methylene) on C-3 of the 'Eastern hemisphere' of the molecule. Crellastatin E (**25**): ^{1}H NMR (500 MHz, CD_3OD): 'Western hemisphere' δ 5.60 (t, 1H, J = 2.9 Hz, 6-CH), 4.30 (d, 1H, J = 8.3 Hz, 19*R*-CH), 4.05 (dd, 1H, J = 3.0, 9.2 Hz, 2-CH), 3.66 (dd, 1H, J = 1.9, 8.3 Hz, 19*S*-CH), 2.63 (br s, 1H, 7-CH), 2.20* (1H, 1β-CH), 2.15* (1H, 1α-CH), 1.22 (s, 3H, 28-Me), 1.13 (s, 3H, 29-Me), 0.72 (s, 3H, 18-Me); 'Eastern hemisphere' δ 4.66 (dd, 1H, J = 5.0, 9.5 Hz, 2-CH), 4.65 (dd, 1H, J = 4.9, 11.0 Hz, 6-CH), 2.92 (dd, 1H, J = 4.9, 17.2 Hz, 7-CHa), 2.50* (1H, 1-CHa), 2.26* (m, 1H, 5-CH), 2.05* (1H, 7-CHb), 1.60* (1H, 1-CHb), 1.45 (s, 3H, 28-Me), 1.38 (s, 3H, 29-Me), 0.94 (s, 3H, 19-Me), 0.71 (s, 3H, 18-Me); *overlapped peaks (Zampella *et al.*, 1999).

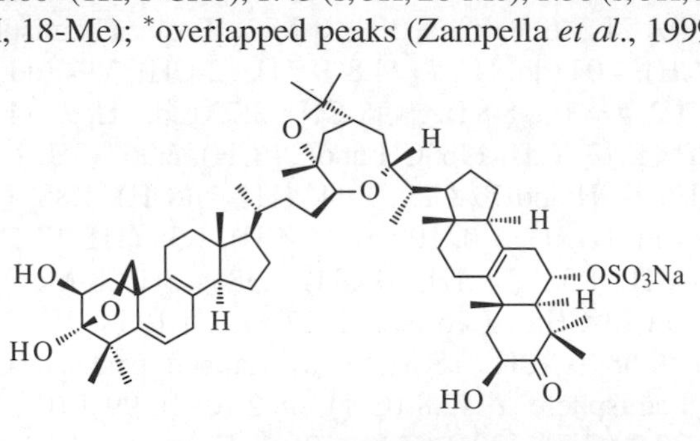

Crellastatin E (**25**)

Table 4.2.1 *FABMS (negative ion mode) data of crellastatins B–D (**22–24**)*

	Crellastatin B (**22**)	Crellastatin C (**23**)	Crellastatin D (**24**)
m/z: [M-H]$^-$	991	991	975

^{13}C NMR (125 MHz, CD_3OD): 'Western hemisphere' δ 42.8 (C-1), 68.4 (C-2), 99.7 (C-3), 42.6 (C-4), 147.8 (C-5), 116.0 (C-6), 30.0 (C-7), 130.5 (C-8), 126.8 (C-9), 38.4 (C-10), 23.6 (C-11), 38.2 (C-12), 42.5 (C-13), 53.3 (C-14), 24.2 (C-15), 30.6 (C-16), 54.5 (C-17), 11.8 (C-18), 73.0 (C-19), 37.2 (C-20), 19.4 (C-21), 36.4 (C-22), 25.4 (C-23), 85.8 (C-24), 87.8 (C-25), 36.2 (C-26), 27.3 (C-27), 25.1 (C-28), 25.6 (C-29); 'Eastern hemisphere' δ 48.2 (C-1), 70.0 (C-2), 221.5 (C-3), 47.8 (C-4), 53.8 (C-5), 76.0 (C-6), 35.5 (C-7), 128.3 (C-8), 134.8 (C-9), 41.7 (C-10), 23.3 (C-11), 38.2 (C-12), 43.2 (C-13), 53.0 (C-14), 25.2 (C-15), 29.7 (C-16), 52.8 (C-17), 11.5 (C-18), 22.5 (C-19), 44.2 (C-20), 13.4 (C-21), 71.8 (C-22), 33.0 (C-23), 46.0 (C-24), 85.6 (C-25), 25.5 (C-26), 32.5 (C-27), 19.5 (C-28), 33.7 (C-29) (Zampella *et al.*, 1999).

Unlike crellastatins A–E (**21**–**25**), crellastatins F–H (**26**–**28**), and M (**33**) have sulphur-containing functionality on both steroidal units.

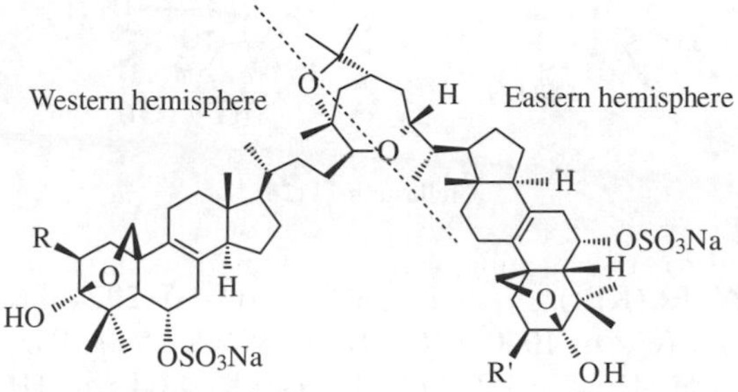

Crellastatins	R	R'
Crellastatin F (**26**)	OH	OH
Crellastatin G (**27**)	OH	H

Crellastatin F (**26**): ^{1}H NMR (500 MHz, CD_3OD): 'Western hemisphere' δ 4.69 (m, 1H, 6-CH), 4.24 (dd, 1H, $J = 5.1$, 8.8 Hz, 2-CH), 3.90 (br dd, 1H, $J = 8.8$ Hz, 19*R*-CH), 3.83 (d, 1H, $J = 8.8$ Hz, 19*S*-CH), 2.94* (1H, 7-CHa), 2.40 (t, 1H, $J = 8.8$ Hz, 1-CHb), 2.04* (1H, 7-CHb), 1.95* (1H, 1-CHa), 1.83 (d, 1H, $J = 11.7$ Hz, 5-CH), 1.37 (s, 3H, 28-Me), 1.23 (s, 3H, 29-Me), 0.69 (s, 3H, 18-Me); 'Eastern hemisphere' δ 4.69 (m, 1H, 6-CH), 4.30 (dd, 1H, $J = 5.1$, 9.6 Hz, 2-CH), 3.90 (br d, 1H, $J = 8.7$ Hz, 19*R*-CH), 3.83 (dd, 1H, $J = 2.5$, 8.7 Hz, 19*S*-CH), 2.93 (m, 1H, 7-CHa), 2.21* (1H, 1-CHb), 2.07* (1H, 7-CHb), 1.97* (1H, 1-CHa), 1.81 (d, 1H, $J = 11.0$ Hz, 5-CH), 1.38 (s, 3H, 28-Me), 1.30 (s, 3H, 29-Me), 0.65 (s, 3H, 18-Me); *overlapped peaks (Zampella *et al.*, 1999). ^{13}C NMR (125 MHz, CD_3OD): 'Western hemisphere' δ 44.5 (C-1), 67.0 (C-2), 100.0 (C-3), 41.0 (C-4), 51.5 (C-5), 75.8 (C-6), 35.5 (C-7), 127.8 (C-8), 131.0 (C-9), 41.0 (C-10), 23.4 (C-11), 38.6 (C-12), 42.4 (C-13), 53.1 (C-14), 25.2 (C-15), 30.9 (C-16), 54.1 (C-17), 11.3 (C-18), 70.1 (C-19), 36.2 (C-20), 19.5 (C-21), 36.9 (C-22), 24.7 (C-23), 86.3 (C-24), 88.3 (C-25), 36.5 (C-26), 27.1 (C-27), 18.0 (C-28), 28.3 (C-29); 'Eastern hemisphere' δ 44.9 (C-1), 67.2 (C-2), 100.1 (C-3), 41.0 (C-4), 52.0 (C-5), 76.0 (C-6), 35.8 (C-7), 126.2 (C-8), 131.1 (C-9), 40.8 (C-10), 23.4 (C-11), 38.7 (C-12), 42.7 (C-13), 53.4 (C-14), 25.5 (C-15), 29.6 (C-16), 53.1 (C-17), 11.7 (C-18), 70.1 (C-19), 44.0 (C-20), 13.0 (C-21), 71.7 (C-22), 34.0 (C-23), 45.8 (C-24), 85.7 (C-25), 25.7 (C-26), 32.7 (C-27), 18.0 (C-28), 28.9 (C-29). FABMS: *m/z* 1105 $[M-H]^-$ (Zampella *et al.*, 1999).

The ^{1}H and ^{13}C NMR data of crellastatin G (**27**, FABMS: *m/z* 1105 $[M - H]^-$) were similar to that of crellastatin F (**26**) with the only notable difference being the absence of the signals associated with the oxymethine on C-2 of the 'Eastern hemisphere', and the presence of methylene signals instead. Although crellastatin H (**28**) has sulphur-containing functionality on both steroidal units, the 'Western hemisphere' is significantly different from dimers **26** and **27**.

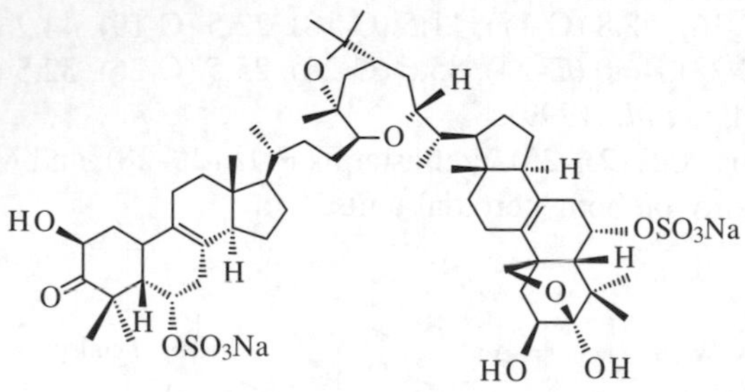

Crellastatin H (**28**)

Crellastatin H (**28**): IR (KBr): ν_{max} cm^{-1} 3420br (O–H), 2924m (C–H), 1704s (C=O), 1458m (C–H), 1200m (C–O), 1060m (C–O). ^{1}H NMR (500 MHz, CD_3OD): 'Western hemisphere' δ 5.04 (dd, 1H, $J = 5.2$, 11.7 Hz, 2-CH), 4.64 (m, 1H, 6-CH), 2.96* (1H, 7-CHa), 2.61 (t, 1H, $J = 11.7$ Hz, 1-CHb), 2.45 (d, 1H, $J = 11.0$ Hz, 5-CH), 2.16* (1H, 7-CHb), 1.55* (1H, 1-CHa), 1.51 (s, 3H, 29-Me), 1.45 (s, 3H, 28-Me), 0.93 (s, 3H, 19-Me), 0.72 (s, 3H, 18-Me); 'Eastern hemisphere' δ 4.70 (m, 1H, 6-CH), 4.25 (dd, 1H, $J = 5.0$, 9.5 Hz, 2-CH), 3.93 (d, 1H, $J = 8.8$ Hz, 19*R*-CH), 3.84 (dd, 1H, $J = 2.7$, 8.9 Hz, 19*S*-CH), 2.98 (m, 1H, 7-CHa), 2.27* (1H, 7-CHb), 2.39* (1H, 1-CHb), 1.95* (1H, 1-CHa), 1.89 (d, 1H, $J = 11.7$ Hz, 5-CH), 1.35 (s, 3H, 28-Me), 1.18 (s, 3H, 29-Me), 0.66 (s, 3H, 18-Me); *overlapped peaks (Zampella *et al.*, 1999). ^{13}C NMR (125 MHz, CD_3OD): 'Western hemisphere' δ 48.0 (C-1), 69.0 (C-2), 221.4 (C-3), 47.4 (C-4), 53.0 (C-5), 75.9 (C-6), 36.3 (C-7), 128.0 (C-8), 134.1 (C-9), 40.7 (C-10), 22.5 (C-11), 38.2 (C-12), 42.4 (C-13), 52.4 (C-14), 24.5 (C-15), 30.5 (C-16), 53.7 (C-17), 10.9 (C-18), 22.2 (C-19), 36.5 (C-20), 19.1 (C-21), 36.7 (C-22), 25.2 (C-23), 85.8 (C-24), 87.8 (C-25), 36.2 (C-26), 26.7 (C-27), 19.0 (C-28), 33.1 (C-29); 'Eastern hemisphere' δ 44.2 (C-1), 66.9 (C-2), 99.5 (C-3), 40.5 (C-4), 51.0 (C-5), 75.8 (C-6), 35.4 (C-7), 126.4 (C-8), 130.9 (C-9), 41.2 (C-10), 22.5 (C-11), 38.2 (C-12), 42.3 (C-13), 54.0 (C-14), 24.3 (C-15), 29.6 (C-16), 52.8 (C-17), 11.4 (C-18), 69.8 (C-19), 43.7 (C-20), 12.8 (C-21), 71.5 (C-22), 33.9 (C-23), 45.5 (C-24), 85.3 (C-25), 25.3 (C-26), 32.3 (C-27), 17.7 (C-28), 27.7 (C-29). FABMS: *m/z* 1089 $[M - H]^-$ (Zampella *et al.*, 1999).

In the same year, the same group reported further cytotoxic bis-steroidal derivatives, crellastatins I–M (**29**–**33**) from the same source (Giannini *et al.*, 1999). However, these compounds were isolated in minute amounts, and clearly were not the major components of this sponge. The initial extraction, solvent partitioning, and DCCC methodologies for analyzing the active $CHCl_3$ and *n*-BuOH fractions were pretty much the same as applied for the isolation of crellastatins A–H (**21**–**28**) as outlined earlier, but the amount of the starting material was much higher, *i.e.*, 300 g of *Crella* spp. In the final step of purification, RP-HPLC on C_8 column (mobile phase: 52–65% aqueous MeOH) was used.

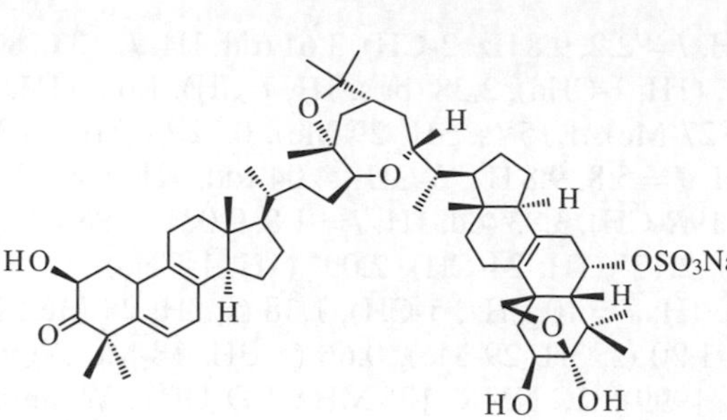

Crellastatin I (**29**)

Crellastatin I (**29**, 12 mg). IR (KBr): ν_{max} cm^{-1} 3410br (O–H), 2920m (C–H), 1710s (C=O), 1200m (C–O), 1060m (C–O). ^{1}H NMR (500 MHz, CD_3OD): 'Western hemisphere' δ 5.84 (t, 1H, $J = 3.1$ Hz, 6-CH), 4.55 (dd, 1H, $J = 6.6$, 9.6 Hz, 2-CH), 3.22 (t, 1H, $J = 4.4$ Hz, 24-CH), 2.70* (1H, 1-CHa), 2.65* (1H, 7-CH), 1.85* (1H, 1-CHb), 1.35 (s, 3H, 28-Me), 1.33 (s, 3H, 29-Me), 1.19 (s, 3H, 27-Me), 1.10 (s, 3H, 19-Me), 0.73 (s, 3H, 18-Me); 'Eastern hemisphere' δ 4.65 (dt, 1H, $J = 5.1$, 11.0 Hz, 6-CH), 4.05 (dd, 1H, $J = 5.1$, 10.3 Hz, 2-CH), 3.94 (dd, 1H, $J = 1.0$, 8.8 Hz, 19*R*-CH), 3.83 (dd, 1H, $J = 1.5$, 8.8 Hz, 19*S*-CH), 3.73* (1H, 22-CH), 2.92 (dd, 1H, $J = 5.1$, 16.2 Hz, 7-CHa), 2.17* (1H, 24-CH), 2.05* (2H, 1CH-a and 7-CHb), 2.02* (1H, 1-CHb), 1.59 (d, $J = 11.0$ Hz, H-5), 1.36 (s, 6H, 26-Me and 28-Me), 1.25 (s, 3H, 27-Me), 1.17 (s, 3H, 29-Me), 0.67 (s, 3H, 18-Me); *overlapped/unresolved peaks (Giannini *et al.*, 1999). ^{13}C NMR (125 MHz, CD_3OD): 'Western hemisphere' δ 45.5 (C-1), 70.6 (C-2), 219.3 (C-3), 49.1 (C-4), 147.7 (C-5), 120.4 (C-6), 30.0 (C-7), 130.3 (C-8), 132.5 (C-9), 39.4 (C-10), 23.5 (C-11), 23.5 (C-12), 42.9 (C-13), 53.8 (C-14), 24.2 (C-15), 30.5 (C-16), 54.6 (C-17), 11.6 (C-18), 26.7 (C-19), 36.4 (C-20), 19.3 (C-21), 36.4 (C-22), 24.9 (C-23), 85.7 (C-24), 88.4 (C-25), 36.5 (C-26), 27.3 (C-27), 29.5 (C-28), 28.6 (C-29); 'Eastern hemisphere' δ 44.8 (C-1), 66.9 (C-2), 99.8 (C-3), 40.9 (C-4), 52.8 (C-5), 73.3 (C-6), 35.5 (C-7), 127.9 (C-8), 131.8 (C-9), 40.8 (C-10), 23.5 (C-11), 38.5 (C-12), 42.6 (C-13), 53.4 (C-14), 25.0 (C-15), 29.7 (C-16), 52.8 (C-17), 11.2 (C-18), 69.7 (C-19), 44.1 (C-20), 13.6 (C-21), 71.6 (C-22), 34.0 (C-23), 45.9 (C-24), 85.7 (C-25), 25.6 (C-26), 32.5 (C-27), 18.1 (C-28), 28.6 (C-29). HRFABMS: *m/z* 917.6271 calculated for $C_{58}H_{86}O_7Na$, found: 917.6252 (Giannini *et al.*, 1999).

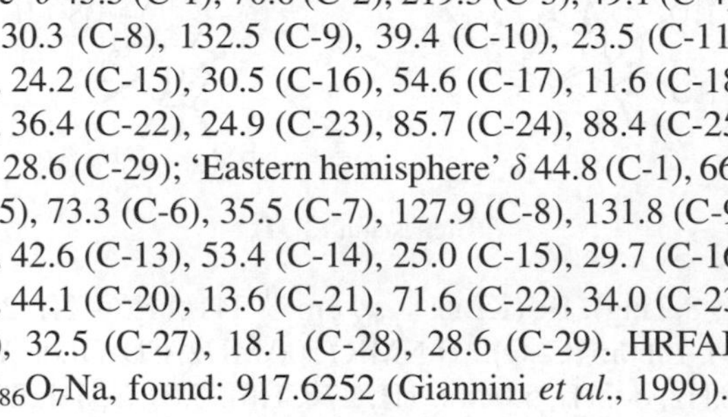

Crellastatin J (**30**)

Crellastatin J (**30**, 6.2 mg). ^{1}H NMR (500 MHz, CD_3OD): 'Western hemisphere' δ 5.69 (d, 1H, $J = 6.2$ Hz, 11-CH), 5.46 (dd, 1H, $J = 2.4$, 5.4 Hz, 6-CH), 4.41 (d, 1H, $J = 8.0$ Hz,

19*R*-CH), 3.97 (dd, 1H, $J = 2.2, 9.8$ Hz, 2-CH), 3.61 (dd, 1H, $J = 3.0, 8.0$ Hz, 19*S*-CH), 3.29 (m, 1H, 24-CH), 2.90* (1H, 1-CHa), 2.28 (br s, 1H, 7-CH), 1.62* (1H, 1-CHb), 1.22 (s, 3H, 28-Me), 1.19 (s, 3H, 27-Me), 1.15 (s, 3H, 29-Me), 0.72 (s, 3H, 18-Me); 'Eastern hemisphere' δ 4.68 (dt, 1H, $J = 5.8, 9.8$ Hz, 6-CH), 4.04 (dd, 1H, $J = 5.2, 9.8$ Hz, 2-CH), 3.93 (br d, 1H, $J = 9.0$ Hz, 19*R*-CH), 3.83 (dd, 1H, $J = 1.8, 9.0$ Hz, 19*S*-CH), 3.77* (1H, 22-CH), 2.95 (dd, 1H, 7-CHa), 2.17* (1H, 24-CH), 2.09* (1H, 1-CHa), 2.08* (1H, 7-CHb), 2.00* (1H, 1-CHb), 1.71 (d, 1H, $J = 10.5$ Hz, 5-CH), 1.38 (s, 3H, 28-Me), 1.35 (s, 3H, 26-Me), 1.25 (s, 3H, 27-Me), 1.20 (s, 3H, 29-Me), 0.66 (s, 3H, 18-Me); *overlapped/unresolved peaks (Giannini *et al.*, 1999). ^{13}C NMR (125 MHz, CD_3OD): 'Western hemisphere' δ 44.6 (C-1), 68.5 (C-2), 99.7 (C-3), 42.8 (C-4), 148.1 (C-5), 115.9 (C-6), 31.0 (C-7), 83.7 (C-8), 138.8 (C-9), 39.2 (C-10), 126.5 (C-11), 42.9 (C-12), 44.9 (C-13), 52.9 (C-14), 22.2 (C-15), 30.4 (C-16), 57.6 (C-17), 13.2 (C-18), 74.9 (C-19), 36.2 (C-20), 18.7 (C-21), 35.1 (C-22), 25.0 (C-23), 84.8 (C-24), 87.6 (C-25), 36.6 (C-26), 27.8 (C-27), 25.0 (C-28), 26.2 (C-29); 'Eastern hemisphere' δ 44.4 (C-1), 67.5 (C-2), 100.0 (C-3), 40.9 (C-4), 52.2 (C-5), 75.9 (C-6), 35.4 (C-7), 127.6 (C-8), 132.0 (C-9), 40.8 (C-10), 23.5 (C-11), 38.4 (C-12), 43.4 (C-13), 54.0 (C-14), 25.0 (C-15), 29.2 (C-16), 52.9 (C-17), 11.6 (C-18), 70.1 (C-19), 44.0 (C-20), 13.2 (C-21), 71.3 (C-22), 34.3 (C-23), 45.8 (C-24), 85.5 (C-25), 25.7 (C-26), 32.5 (C-27), 17.9 (C-28), 28.2 (C-29). FABMS: *m/z* 1023 $[M - H]^-$ (Giannini *et al.*, 1999).

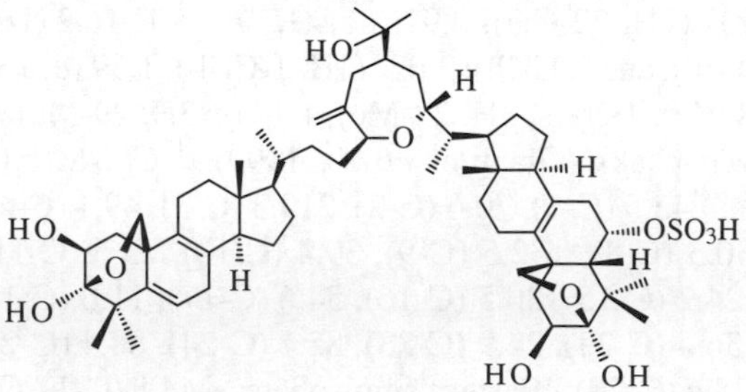

Crellastatin K (**31**)

Crellastatin K (**31**, 8.2 mg). IR (KBr): ν_{max} cm^{-1} 3410br (O–H), 2920m (C–H), 1630s (C=C), 1200m (C–O), 1060m (C–O). ^{1}H NMR (500 MHz, CD_3OD): 'Western hemisphere' δ 5.61 (t, 1H, $J = 2.8$ Hz, 6-CH), 5.04 (s, 1H, 27-CHb), 5.01 (s, 1H, 27-CHa), 4.33 (d, 1H, $J = 8.0$ Hz, 19*R*-CH), 4.02 (dd, 1H, $J = 2.2, 9.8$ Hz, 2-CH), 4.02 (dd, 1H, 24-CH), 3.66 (dd, 1H, $J = 2.4, 8.0$ Hz, 19*S*-CH), 2.68* (1H, 26-CHb), 2.65 (br s, 1H, 7-CH), 2.20* (1H, 1-CHa), 2.13* (1H, 1-CHb), 1.93* (1H, 26-CHa), 1.23 (s, 3H, 28-Me), 1.13 (s, 3H, 29-Me), 0.73 (s, 3H, 18-Me); 'Eastern hemisphere' δ 4.68 (dt, 1H, $J = 5.5, 10.4$ Hz, 6-CH), 4.05 (dd, 1H, $J = 5.6, 8.8$ Hz, 2-CH), 3.93 (dd, 1H, $J = 1.6, 9.6$ Hz, 19*R*-CH), 3.84 (dd, 1H, $J = 0.8, 9.6$ Hz, 19*S*-CH), 3.48* (1H, 22-CH), 2.94 (dd, 1H, $J = 5.6, 17.6$ Hz, 7-CHa), 2.08* (1H, 7-CHb), 2.00* (2H, 1-CH and 23-CHb), 1.64 (d, 1H, $J = 10.4$ Hz, 5-CH), 1.55* (1H, 23-CHa), 1.52* (1H, 24-CH), 1.38 (s, 3H, 28-Me), 1.20 (s, 3H, 29-Me), 1.19 (s, 6H, 26-Me and 27-Me), 0.65 (s, 3H, 18-Me); *overlapped or unresolved peaks (Giannini *et al.*, 1999). ^{13}C NMR (125 MHz, CD_3OD): 'Western hemisphere' δ 44.4 (C-1), 68.0 (C-2), 99.9 (C-3), 42.5 (C-4), 147.6 (C-5), 116.0 (C-6), 30.8 (C-7), 130.0 (C-8), 127.4 (C-9), 38.1 (C-10), 23.4 (C-11), 37.8 (C-12), 42.8 (C-13), 53.3 (C-14), 24.5 (C-15), 30.8 (C-16), 56.0 (C-17), 11.5 (C-18), 72.5 (C-19), 36.7 (C-20), 19.2 (C-21), 36.2 (C-22), 25.5 (C-23), 80.4 (C-24), 148.0

(C-25), 37.0 (C-26), 112.5 (C-27), 24.9 (C-28), 25.3 (C-29); 'Eastern hemisphere' δ 44.0 (C-1), 66.9 (C-2), 99.9 (C-3), 40.6 (C-4), 52.2 (C-5), 75.4 (C-6), 35.1 (C-7), 128.4 (C-8), 131.3 (C-9), 41.0 (C-10), 23.2 (C-11), 37.8 (C-12), 42.8 (C-13), 53.3 (C-14), 24.9 (C-15), 29.6 (C-16), 53.1 (C-17), 11.5 (C-18), 69.6 (C-19), 49.7 (C-20), 13.3 (C-21), 75.8 (C-22), 34.0 (C-23), 54.0 (C-24), 73.4 (C-25), 26.8 (C-26 and C-27), 17.4 (C-28), 27.9 (C-29). HRFABMS: *m/z* 933.6220 calculated for $C_{58}H_{86}O_8Na$, found: 933.6210 (Giannini *et al.*, 1999).

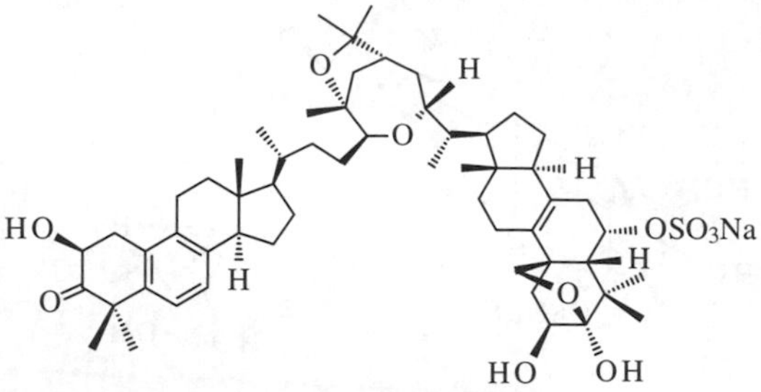

Crellastatin L (**32**)

Crellastatin L (**32**, 5.0 mg). IR (KBr): ν_{max} cm^{-1} 3410br (O–H), 2920m (C–H), 1720s (C=O), 1640s (C=C), 1200m (C–O), 1060m (C–O). ^{1}H NMR (500 MHz, CD_3OD): 'Western hemisphere' δ 7.19 (d, 1H, J = 8.1 Hz, 7-CH), 6.98 (d, 1H, J = 8.1 Hz, 6-CH), 4.77 (dd, 1H, J = 6.8, 12.3 Hz, 2-CH), 3.56 (dd, 1H, J = 6.8, 15.6 Hz, 1-CHa), 3.22 (br t, 1H, J = 3.8, 24-CH), 2.78* (1H, 11-CH), 2.60* (1H, 1-CHb), 2.24* (1H, 26-CHa), 2.06* (1H, 26-CHb), 1.55 (s, 3H, 28-Me), 1.36 (s, 3H, 29-Me), 1.20* (1H, 27-CH), 0.57 (s, 3H, 18-Me); 'Eastern hemisphere' δ 4.57 (dd, 1H, J = 4.9, 11.0 Hz, 6-CH), 3.98 (dd, 1H, J = 4.8, 9.9 Hz, 2-CH), 3.90 (dd, 1H, J = 1.4, 9.3 Hz, 19*R*-CH), 3.78 (dd, 1H, J = 1.6, 9.3 Hz, 19*S*-CH), 3.76 (dd, 1H, J = 2.6, 10.0 Hz, 22-CH), 2.80 (dd, 1H, J = 3.6, 15.0 Hz, 7-CHa), 2.15* (1H, 24-CH), 2.05* (1H, 7-CHb), 2.04* (1H, 23-CHb), 1.98* (1H, 1-CHb), 1.88* (1H, 5-CH), 1.82* (1H, 1-CHa), 1.38* (1H, 26-CH), 1.32 (s, 3H, 28-Me), 1.26* (1H, 27-CH), 1.30* (1H, 23-CHa), 1.09 (s, 3H, 29-Me), 0.65 (s, 3H, 18-Me); *overlapped/unresolved peaks. ^{13}C NMR (125 MHz, CD_3OD): 'Western hemisphere' δ 71.6 (C-2), 216.5 (C-3), 50.0 (C-4), 142.0 (C-5), 38.5 (C-12), 43.0 (C-13), 52.8 (C-14), 24.0 (C-15), 53.5 (C-17), 11.5 (C-18), 36.5 (C-20), 19.3 (C-21), 36.3 (C-22), 86.0 (C-24), 88.0 (C-25), 36.2 (C-26), 27.1 (C-27), 30.0 (C-28), 25.5 (C-29); 'Eastern hemisphere' δ 45.9 (C-1), 67.5 (C-2), 99.8 (C-3), 40.5 (C-4), 52.0 (C-5), 74.8 (C-6), 35.4 (C-7), 127.9 (C-8), 131.8 (C-9), 38.2 (C-12), 42.5 (C-13), 54.0 (C-14), 53.5 (C-17), 11.5 (C-18), 70.0 (C-19), 43.9 (C-20), 13.2 (C-21), 71.4 (C-22), 46.4 (C-24), 85.4 (C-25), 25.5 (C-26), 32.5 (C-27), 17.7 (C-28), 28.5 (C-29). HRFABMS: *m/z* 901.5933 calculated for $C_{57}H_{82}O_7Na$, found: 901.5958 (Giannini *et al.*, 1999).

Crellastatin M (**33**, 3.5 mg). IR (KBr): ν_{max} cm^{-1} 3410br (O–H), 2920m (C–H), 1200m (C–O), 1060m (C–O). ^{1}H NMR (500 MHz, CD_3OD): 'Western hemisphere' δ δ 4.69 (m, 1H, 6-CH), 4.24 (dd, 1H, J = 4.8, 9.6 Hz, 2-CH), 3.90 (br d, 1H, J = 8.8 Hz, 19*R*-CH), 3.82 (dd, 1H, J = 1.6, 8.8 Hz, 19*S*-CH), 3.12 (br d, 1H, J = 6.4 Hz, 24-CH), 2.94* (1H, 7-CHa), 2.42* (1H, 26-CHa), 2.39* (t, 1H, 1-CHb), 2.18* (1H, 7-CHb), 1.94* (1H, 1-CHa), 1.84 (d, 1H, J = 12.0 Hz, 5-CH), 1.58* (1H, 26-CHa), 1.38 (s, 3H, 28-Me), 1.23 (s, 3H, 29-Me), 0.97 (s, 3H, 27-Me), 0.67 (s, 3H, 18-Me); 'Eastern hemisphere' δ 4.94 (s, 1H, 26-CH),

4.68 (m, 1H, 6-CH), 4.30 (dd, 1H, $J=5.6, 9.6$ Hz, 2-CH), 3.90 (d, 1H, $J=8.8$ Hz, 19*R*-CH), 3.84 (dd, 1H, $J=3.2, 11.2$, 22-CH), 3.82 (dd, 1H, $J=1.6, 9.0$ Hz, 19*S*-CH), 2.94* (1H, 7-CHa), 2.10* (1H, 24-CH), 2.18* (1H, 1-CHb), 2.06* (1H, 7-CHb), 1.96* (1H, 1-CHa), 1.81 (d, 1H, $J=12.0$ Hz, 5-CH), 1.81* (1H, 23-CHb), 1.74 (s, 3H, 27-Me), 1.38 (s, 3H, 28-Me), 1.34* (1H, 23-CHa), 1.30 (s, 3H, 29-Me), 0.64 (s, 3H, 18-Me); *overlapped/unresolved peaks (Giannini *et al.*, 1999).

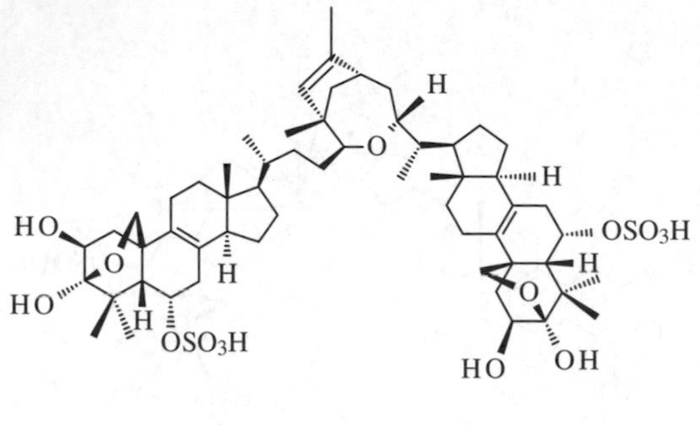

Crellastatin M (**33**)

^{13}C NMR (125 MHz, CD_3OD): 'Western hemisphere' δ 44.4 (C-1), 67.1 (C-2), 99.4 (C-3), 40.4 (C-4), 50.7 (C-5), 75.9 (C-6), 35.5 (C-7), 128.1 (C-8), 133.5 (C-9), 41.0 (C-10), 22.5 (C-11), 38.1 (C-12), 41.9 (C-13), 52.9 (C-14), 24.5 (C-15), 30.7 (C-16), 53.8 (C-17), 11.5 (C-18), 71.0 (C-19), 36.1 (C-20), 19.1 (C-21), 36.1 (C-22), 25.2 (C-23), 84.8 (C-24), 54.4 (C-25), 38.3 (C-26), 24.0 (C-27), 17.4 (C-28), 28.0 (C-29); 'Eastern hemisphere' δ 44.5 (C-1), 67.3 (C-2), 99.4 (C-3), 40.4 (C-4), 51.4 (C-5), 75.9 (C-6), 35.6 (C-7), 126.3 (C-8), 130.9 (C-9), 40.9 (C-10), 22.7 (C-11), 38.1 (C-12), 42.2 (C-13), 52.9 (C-14), 25.0 (C-15), 29.5 (C-16), 52.8 (C-17), 11.5 (C-18), 70.0 (C-19), 43.6 (C-20), 13.0 (C-21), 72.6 (C-22), 34.5 (C-23), 45.5 (C-24), 145.7 (C-25), 132,5 (C-26), 13.9 (C-27), 17.4 (C-28), 27.9 (C-29). HRFABMS: *m/z* 1087.5486 calculated for $C_{58}H_{87}O_{15}S_2$, found: 1087.5463 (Giannini *et al.*, 1999).

While crellastatins B–H (**22**–**28**) displayed *in vitro* antitumour activity against human bronchopulmonary nonsmall-cell lung carcinoma cell lines (NSCLC) with IC_{50} values in the range of 2-10 μg/mL, the IC_{50} values of crellastatins I–M (**29**–**33**) were within the range of 1-5 μg/mL (Giannini *et al.*, 1999; Zampella *et al.*, 1999).

4.3 Ritterazines

Dimeric steroidal alkaloids, ritterazines A–Z (**34**–**59**) were isolated from the Japanese tunicate, *Ritterella tokioka* (family: Polyclinidae) (*Figure 4.1.2*), collected from the Izu Peninsula, 100 km southwest of Tokyo, by Fukuzawa and coworkers during 1994–1997 (Li and Dias, 1997; Moser, 2008). The ritterazines and cephalostatins share several common structural features in which two highly oxygenated steroidal units with side chains forming either 5/5 or 5/6 spiroketals are fused *via* a pyrazine core. It was postulated that ritterazine A (**34**) might follow the biosynthetic route starting from a 12-hydroxy-Δ^{14}-steroid skeleton (Fukuzawa *et al.*, 1994; Nahar *et al.*, 2007). To date, the naturally occurring cephalostatins and ritterazines comprise a family of 45 structurally unique marine natural products with extreme cytotoxic activity towards various human tumour cell lines. Ritterazine B (**36**) is the most cytotoxic among the ritterazines to P388 murine leukaemia cells.

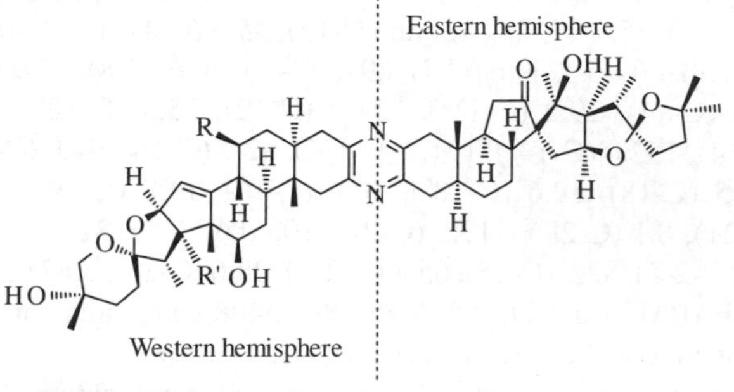

Ritterazine A (**34**)	R = OH	R' = OH
Ritterazine T (**35**)	R = H	R' = H

Colonies of *Ritterella tokioka* (5.5 kg) were extracted with EtOH followed by Me_2CO, and concentrated extracts were initially partitioned between H_2O and EtOAc. The organic phase was partitioned by a series of solvent mixtures, notably, *n*-hexane-90% aqueous MeOH, and DCM-60% aqueous MeOH (Fukuzawa *et al*., 1994). The active DCM fraction was subjected to repeated CC on ODS and LH-20 stationary phases to give ritterazine A (**34**, 2.9 mg, 5.3×10^{-5}%). The structure of **34** was determined unequivocally by spectroscopic means, especially by a series of 1D and 2D NMR spectroscopic techniques including COSY, HMBC and NOESY. IR (KBr): ν_{max} cm^{-1} 3420br (O–H), 2970m (C–H), 2940m (C–H), 2880m (C–H), 2360m, 1770s (C=O), 1730s (C=O), 1600s (C=C), 1450m (C–H), 1400s (pyrazine), 1260m (C–O), 1230m (C–O), 1160m (C–O), 1130m (C–O), 1080m (C–O), 1040m (C–O), 990m, 970m, 960m, 940m, 890w, 870w, 820w. ^1H NMR (500 MHz, C_5D_5N): δ 6.13 (dd, 1H, J = 1.9, 1.8 Hz, 15'-CH), 6.02 (br s, 1H, 7'-OH), 5.28 (br s, 1H, 13-OH), 5.25 (d, 1H, J = 1.9 Hz, 16'-CH), 5.20 (ddd, 1H, J = 2.9, 3.9, 9.3 Hz, 16-CH), 4.89 (br s, 1H, 17'-OH), 4.66 (br s, 1H, 12'-OH), 4.22 (ddd, 1H, J = 1.7, 4.8, 11.3 Hz, 12'-CH), 4.07 (ddd, 1H, J = 5.6, 9.8, 10.6 Hz, 7'-CH), 4.01 (d, 1H, J = 11.5 Hz, 26'β-CH), 3.64 (br s, 1H, 25'-OH), 3.61 (dd, 1H, J = 2.6, 11.5 Hz, 26'α-CH), 3.17 (d, 1H, J = 16.8 Hz, l'β-CH), 3.02 (dd, 1H, J = 5.1, 17.7 Hz, 4'α-CH), 2.97 (dd, 1H, J = 17.8, 5.5 Hz, 4α-CH), 2.91 (d, 1H, J = 16.4 Hz, 1β-CH), 2.85 (dd, 1H, J = 1.5, 9.3 Hz, 17-CH), 2.81 (dd, 1H, J = 13.0, 17.7 Hz, 4'β-CH), 2.76 (d,1H, J = 16.4 Hz, 1-CHa), 2.70 (d, 1H, J = 16.8 Hz, 1' α-CH), 2.69 (dd, 1H, J = 12.7, 17.8 Hz, 4β-CH), 2.54 (dq, 1H, J = 1.5, 7.3 Hz, 20-CH), 2.50 (ddd, 1H, J = 4.9, 13.1, 13.3 Hz, 23'α-CH), 2.43 (dd, 1H, J = 10.6, 10.9 Hz, 8'-CH), 2.40 (ddd, 1H, J = 3.6, 11.9, 12.1 Hz, 8-CH), 2.35 (dd, 1H, J = 6.3, 16.5 Hz, 11α-CH), 2.27 (m, 1H, 7β-CH), 2.24 (m, 1H, 15β-CH), 2.22 (m,1H, 6α-CH), 2.22 (q, 1H, J = 6.9 Hz, 20'-CH), 2.17 (m, 1H, 11'α-CH), 2.15 (m, 1H, 24'β-CH), 2.14 (m, 1H, 11β-CH), 2.08 (m, 1H, 15α-CH), 2.01 (dd, 1H, J = 3.5, 11.3 Hz, 24α-CH), 1.94 (m, 1H, 23α-CH), 1.90 (m, 1H, 23β-CH), 1.88 (m, 1H, 5'-CH), 1.87 (m, 1H, 11'β-CH), 1.86 (m, 1H, 24'α-CH), 1.77 (dd, 1H, J = 10.3, 12.7 Hz, 6'β-CH), 1.74 (m, 1H, 5-CH), 1.68 (m, 1H, 24β-CH), 1.65 (m, 1H, 6α-CH), 1.62 (m, 1H, 9-CH), 1.45 (m, 1H, 23'β-CH), 1.45 (m, 1H, 6β-CH), 1.42 (s, 3H, 27-Me), 1.36 (m, 1H, 7-CHa), 1.34 (s, 3H, 18'-Me), 1.30 (s, 3H, 18-Me), 1.27 (d, 3H,

$J = 6.9$ Hz, 21'-Me), 1.23 (s, 3H, 27'-Me), 1.19 (ddd, 1H, $J = 4.0$, 10.9, 12.2 Hz, 9'-CH), 1.12 (s, 3H, 26-Me), 0.98 (d, 3H, $J = 7.3$ Hz, 21-Me), 0.88 (s, 3H, 19'-Me), 0.81 (s, 3H, 19-Me). ^{13}C NMR (125 MHz, C_5D_5N): δ 46.7 (C-1), 45.9 (C-1'), 149.0 (C-2), 149.2 (C-2' and C-3), 148.7 (C-3'), 35.9 (C-4, C-10' and C-15), 35.6 (C-4' and C-10), 41.9 (C-5), 40.0 (C-5'), 29.1 (C-6), 38.4 (C-6'), 30.6 (C-7), 69.4 (C-7'), 40.6 (C-8), 42.9 (C-8'), 50.1 (C-9), 51.2 (C-9'), 40.9 (C-11), 29.2 (C-11'), 221.2 (C-12), 75.8 (C-12'), 80.9 (C-13), 56.1 (C-13'), 69.3 (C-14), 152.0 (C-14'), 121.2 (C-15'), 82.9 (C-16), 94.0 (C-16'), 61.7 (C-17), 93.3 (C-17'), 23.5 (C-18), 12.6 (C-18'), 10.9 (C-19), 11.8 (C-19'), 40.8 (C-20), 48.3 (C-20'), 19.0 (C-21), 8.1 (C-21'), 119.8 (C-22), 108.0 (C-22'), 32.7 (C-23), 27.5 (C-23'), 37.4 (C-24), 33.2 (C-24'), 82.7 (C-25), 65.8 (C-25'), 28.6 (C-26), 70.2 (C-26'), 30.4 (C-27), 27.0 (C-27'). HRFABMS: *m/z* 913.5576 and 895.5408 calculated for $C_{54}H_{77}N_2O_{10}$ and $C_{54}H_{75}N_2O_9$, respectively (Fukuzawa *et al.*, 1994).

1n 1995, the same group first reported the isolation and identification of ritterazines B (**36**) and C (**38**) (Fukuzawa *et al.*, 1995a), and then ten more dimers, ritterazines D–M (**37**, **39–47**) (Fukuzawa *et al.*, 1995b) from the same source material using the same protocol as outlined earlier for ritterazine A (**34**).

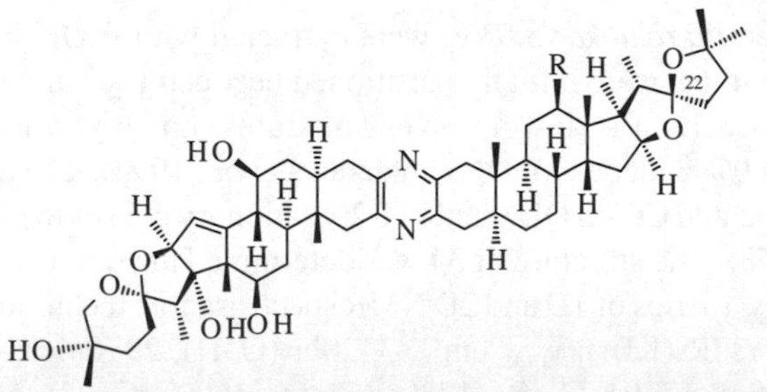

Ritterazine B (**36**) R = OH 22*R* Ritterazine H (**37**) R = Oxo 22*R*

Retterazine B (**36**, 13.4 mg). IR (film): ν_{max} cm^{-1} 3480br (O–H), 2960m (C–H), 2920m (C–H), 2870m (C–H), 2850m (C–H), 1680m (C=O), 1610s (C=C) 1460m (C–H), 1400s (pyrazine), 1140m (C–O), 1120m (C–O), 1060m (C–O), 1040m (C–O), 1000m (C–O), 980m, 940m, 880w, 850w. UV (MeOH): λ_{max} nm 288, 308. ^{1}H NMR (600 MHz, C_5D_5N): δ 5.80 (br s, 1H, 12-OH), 6.13 (d, 1H, $J = 2.0$ Hz, 15'-CH), 5.25 (d, 1H, $J = 2.0$ Hz, 16'-CH), 5.00 (s, 1H, 17'-OH), 4.78 (dd, 1H, $J = 7.0$, 9.7 Hz, 16-CH), 4.67 (d, 1H, $J = 2.0$ Hz, 12'-OH), 4.20 (ddd, 1H, $J = 2.0$, 4.6, 11.2 Hz, 12'-CH), 4.06 (dddd, 1H, $J = 2.2$, 4.4, 10.5, 10.6 Hz, 7'-CH), 4.02 (d, 1H, $J = 11.6$ Hz, 26'β-CH), 3.69 (br s, 1H, 25'-OH), 3.64 (dd, 1H, $J = 3.7$, 11.7 Hz, 12-CH), 3.63 (d, 1H, $J = 2.2$ Hz, 7'-OH), 3.61 (dd, 1H, $J = 2.7$, 11.6 Hz, 26'α-CH), 3.17 (d, 1H, $J = 18.0$ Hz, 1β-CH), 3.15 (dd, 1H, $J = 9.6$, 9.7 Hz, 17-CH), 3.15 (d, 1H, $J = 18.3$ Hz, 1'β-CH), 2.98 (dd, 1H, $J = 5.2$, 17.7 Hz, 4'α-CH), 2.94 (dd, 1H, $J = 5.2$, 17.8 Hz, 4α-CH), 2.77 (dd, 1H, $J = 12.5$, 17.7 Hz, 4'β-CH), 2.71 (d, 1H, $J = 18.0$ Hz, 1α-CH), 2.68 (dd, 1H, $J = 13.1$, 17.5 Hz, 4β-CH), 2.68 (d, 1H, $J = 18.3$ Hz, 1'α-CH), 2.50 (ddd, 1H, $J = 4.8$, 13.4, 13.4 Hz, 23'α-CH), 2.41 (dd, 1H, $J = 10.0$, 10.6 Hz, 8'-CH), 2.21 (d, 1H, $J = 6.9$ Hz, 20'-CH), 2.18 (dd, 1H, $J = 4.4$, 12.4 Hz, 6'α-CH), 2.17 (m, 1H, 11'α-CH), 2.16 (m, 1H, 24'β-CH), 2.12 (m, 1H, 23β-CH), 2.08 (m, 1H, 14-CH), 2.04 (m, 1H, 24β-CH), 2.04 (m, 1H, 11α-CH), 2.01 (dq, 1H, $J = 6.7$, 9.6 Hz, 20-CH), 1.88 (m, 1H, 11'β-CH), 1.87

(m, 1H, 24'α-CH), 1.85 (m, 1H, 23α-CH), 1.84 (m, 1H, 5'-CH), 1.83 (dd, 1H, $J = 7.0$, 15.1 Hz, 15β-CH), 1.80 (dd, 1H, $J = 13.8$, 15.1 Hz, 15α-CH), 1.76 (ddd, 1H, $J = 10.5$, 12.3, 12.4 Hz, 6'β-CH), 1.68 (m, 1H, 24α-CH), 1.68 (m, 1H, 8-CH), 1.67 (m, 1H, 11β-CH), 1.57 (m, 1H, 5-CH), 1.49 (m, 1H, 7β-CH), 1.49 (m, 1H, 23β-CH), 1.48 (m, 1H, 6α-CH), 1.43 (s, 3H, 27-Me), 1.36 (m, 1H, 9-CH), 1.33 (s, 3H, 18'-Me), 1.28 (m, 1H, 6β-CH), 1.26 (d, 3H, $J = 6.9$ Hz, 21'-Me), 1.26 (s, 3H, 18-Me), 1.22 (s, 3H, 27'-Me), 1.18 (d, 3H, $J = 6.7$ Hz, 21-Me), 1.18 (s, 3H, 26-Me), 1.16 (m, 1H, 9'-CH), 1.10 (m, 1H, 7α-CH), 0.85 (s, 3H, 19'-Me), 0.75 (s, 3H, 19-Me). ^{13}C NMR (150 MHz, C_5D_5N): δ 46.3 (C-1), 45.9 (C-1'), 149.3 (C-2), 148.6 (C-2'), 149.0 (C-3), 148.1 (C-3'), 35.9 (C-4), 35.6 (C-4'), 41.5 (C-5), 40.0 (C-5'), 29.0 (C-6), 38.4 (C-6'), 31.7 (C-7), 69.4 (C-7'), 32.6 (C-8), 42.8 (C-8'), 45.5 (C-9), 51.2 (C-9'), 35.9 (C-10 and C-10'), 30.7 (C-11), 29.2 (C-11'), 71.8 (C-12), 75.6 (C-12'), 48.6 (C-13), 55.1 (C-13'), 47.8 (C-14), 151.6 (C-14'), 32.8 (C-15), 121.1 (C-15'), 80.0 (C-16), 94.0 (C-16'), 57.5 (C-17), 93.3 (C-17'), 13.7 (C-18), 12.5 (C-18'), 11.9 (C-19), 11.5 (C-19'), 42.0 (C-20), 48.2 (C-20'), 14.7 (C-21), 8.1 (C-21'), 117.0 (C-22), 107.9 (C-22'), 33.2 (C-23), 27.5 (C-23'), 37.3 (C-24), 33.2 (C-24'), 81.4 (C-25), 65.8 (C-25'), 28.8 (C-26), 70.2 (C-26'), 30.8 (C-27), 27.0 (C-27'). HRFABMS: *m/z* 899.5873 calculated for $C_{54}H_{79}N_2O_9$ (Fukuzawa *et al.*, 1995a).

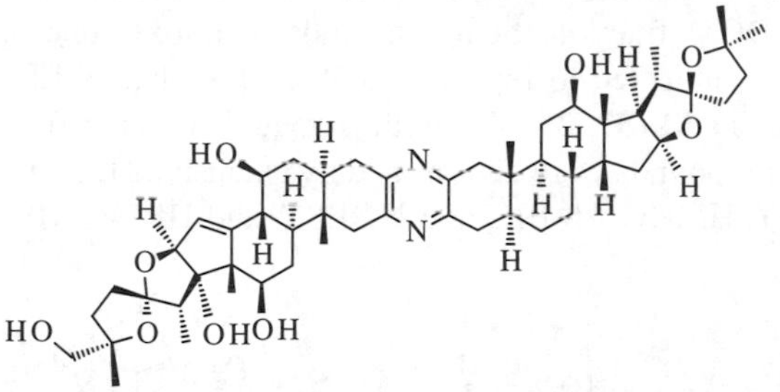

Ritterazine C (**38**)

Retterazine C (**38**, 7.8 mg). IR (film): ν_{max} cm^{-1} 3400br (O–H), 2970m (C–H), 2940m (C–H), 2880m (C–H), 1780s (C=O), 1680m (C=C), 1610s (C=C), 1510m, 1460m (C–H), 1390m (pyrazine), 1200m (C–O), 1140m (C–O), 1030m (C–O), 1000m (C–O), 980m, 950m, 870m, 800w, 720w, 700w. UV (MeOH): λ_{max} nm 285, 303. ^{1}H NMR (600 MHz, C_5D_5N): δ 5.98 (dd, 1H, $J = 1.6$, 1.7 Hz, 15'-CH), 5.25 (d, 1H, $J = 1.7$ Hz, 16'-CH), 4.77 (dd, 1H, $J = 7.0$, 7.0 Hz, 16-CH), 4.16 (dd, 1H, $J = 4.8$, 11.3 Hz, 12'-CH), 4.00 (dddd, 1H, $J = 4.6$, 10.1, 10.6 Hz, 7'-CH), 3.80 (d, 1H, $J = 10.8$ Hz, 26'α-CH), 3.76 (d, 1H, $J = 10.8$ Hz, 26'β-CH), 3.64 (dd, 1H, $J = 4.4$, 9.5 Hz, 12-CH), 3.63 (br s, 12-OH), 3.62 (br s, 1H, 7'-OH), 3.16 (d, 1H, $J = 16.6$ Hz, 1β-CH), 3.15 (m, 1H, 17-CH), 3.11 (d, 1H, $J = 17.1$ Hz, 1'β-CH), 2.96 (dd, 1H, $J = 5.2$, 17.7 Hz, 4'α-CH), 2.95 (dd, 1H, $J = 3.6$, 18.1 Hz, 4α-CH), 2.75 (dd, 1H, $J = 12.1$, 17.7 Hz, 4'β-CH), 2.72 (d, 1H, $J = 16.6$ Hz, 1α-CH), 2.66 (dd, 1H, $J = 12.9$, 18.1 Hz, 4β-CH), 2.64 (d, 1H, $J = 17.1$ Hz, 1'α-CH), 2.40 (dd, 1H, $J = 10.2$, 10.6 Hz, 8'-CH), 2.36 (m, 1H, 23'α-CH, 2.34 (m, 1H, 20'-CH), 2.17 (m, 1H, 6'α-CH), 2.16 (m, 1H, 11'α-CH), 2.12 (m, 1H, 23β-CH), 2.08 (m, 1H, 14-CH), 2.04 (m, 1H, 11α-CH), 2.03 (m, 1H, 20-CH), 2.02 (m, 1H, 24β-CH), 2.02 (m, 1H, 24'α-CH), 1.87 (m, 1H, 11'β-CH), 1.84 (m, 1H, 15β-CH), 1.81 (m, 1H, 5'-CH), 1.80 (m, 1H, 15α-CH), 1.72 (m, 1H, 6'β-CH), 1.70 (m, 1H, 23α-CH), 1.67 (m, 1H, 24α-CH), 1.68 (m, 1H, 11β-CH), 1.67 (m, 1H, 24'β-CH),

1.65 (m, 1H, 8-CH), 1.65 (m, 1H, 23'β-CH), 1.56 (m, 1H, 5-CH), 1.49 (m, 1H, 7β-CH), 1.47 (m, 1H, 6α-CH), 1.43 (s, 3H, 27-Me), 1.36 (m, 1H, 9-CH), 1.35 (s, 3H, 18'-Me), 1.29 (m, 1H, 6β-CH), 1.29 (s, 3H, 27'-Me), 1.26 (s, 3H, 18-Me), 1.19 (s, 3H, 26-Me), 1.19 (d, 3H, $J = 7.0$ Hz, 21'-Me), 1.17 (d, 3H, $J = 6.6$ Hz, 21-Me), 1.15 (m, 1H, 9'-CH), 1.11 (m, 1H, 7α-CH), 0.83 (s, 3H, 19'-Me), 0.75 (s, 3H, 19-Me). ^{13}C NMR (150 MHz, C_5D_5N): δ 46.3 (C-1), 45.8 (C-1'), 149.4 (C-2), 148.5 (C-2'), 148.8 (C-3), 148.3 (C-3'), 35.8 (C-4), 35.6 (C-4'), 41.5 (C-5), 40.0 (C-5'), 29.0 (C-6), 38.4 (C-6'), 31.8 (C-7), 69.4 (C-7'), 32.6 (C-8), 42.8 (C-8'), 45.5 (C-9), 51.3 (C-9'), 35.8 (C-10 and C-10'), 30.8 (C-11), 29.4 (C-11'), 71.8 (C-12), 75.7 (C-12'), 48.6 (C-13), 55.9 (C-13'), 47.8 (C-14), 151.5 (C-14'), 32.8 (C-15), 121.3 (C-15'), 80.0 (C-16), 94.4 (C-16'), 57.5 (C-17), 92.9 (C-17'), 13.7 (C-18), 12.6 (C-18'), 11.9 (C-19), 11.8 (C-19'), 42.0 (C-20), 45.0 (C-20'), 14.7 (C-21), 8.2 (C-21'), 117.0 (C-22), 118.1 (C-22'), 33.2 (C-23), 32.1 (C-23'), 37.8 (C-24), 33.5 (C-24'), 81.4 (C-25), 86.1 (C-25'), 28.8 (C-26), 69.7 (C-26'), 30.3 (C-27), 23.7 (C-27'). HRFABMS: *m/z* 899.5861 corresponding to $C_{54}H_{79}N_2O_9$ (Fukuzawa *et al.*, 1995a).

For the isolation of ritterazines D–M (**37**, **39–47**), homogenized colonies of turnicate (8.2 kg) were extracted with EtOH (4 × 10 L) (Fuzukawa *et al.*, 1995b). Pooled extracts were concentrated and partitioned between H_2O and EtOAc. The EtOAc layer was portioned again using the modified Kupchan partitioning with various solvents (Kupchan *et al.*, 1973). The DCM fraction, being the most cytotoxic one against P388 murine leukaemia cells, was subjected to repeated ODS and Sephadex LH-20 chromatography to isolate ritterazines D–M (**37**, **39–47**), with the yields ranging from 1.1 to 6.2 mg. The structures of these compounds were determined by extensive 1D and 2D NMR techniques, notably, DQF-COSY, HMQC, HMBC and NOESY, and HRFABMS.

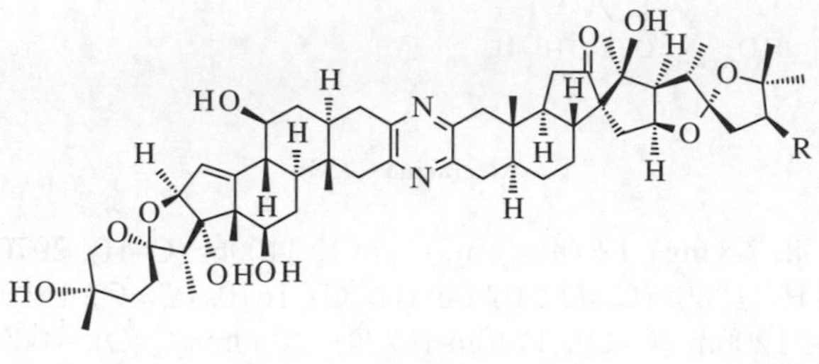

Ritterazine D (**39**)	R = H
Ritterazine E (**40**)	R = Me

Ritterazine D (**39**) had the same molecular formula of $C_{54}H_{76}N_2O_{10}$ as ritterazine A (**34**) as determined by the HRFABMS experiment. In general, the gross structure of ritterazine D (**39**) determined by the interpretation of DQF-COSY, HMQC and HMBC data was almost identical with that of ritterazine A (**34**), with the significant differences in the ^{1}H and ^{13}C NMR chemical shift values between the two compounds were observed for signals for C-20 and C-22 in the 'Eastern hemisphere' (δ_C 38.4 and 120.7 for **39** as opposed to 40.8 and 119.8 for **34**) (Fukuzawa *et al.*, 1995a,b). However, the NMR data for the 'Western hemisphere' of both **34** and **39** were superimposable on each other. In fact, ritterazine D (**39**) is the 22*S* isomer of ritterazine A (**34**). Retterazine D (**39**, 4.0 mg): IR (film): ν_{max} cm^{-1} 3460br (O–H), 2980m (C–H), 2940m (C–H), 2890m (C–H), 2370m, 2340m, 1740s (C=O), 1460m (C–H),

1400s (pyrazine), 1130m (C–O), 1040m (C–O), 1000m (C–O), 900m. UV (MeOH): λ_{max} nm 286, 304. HRFABMS: *m/z* 913.5579 calculated for $C_{54}H_{77}N_2O_{10}$, found: 913.5633 (Fukuzawa *et al.*, 1995b).

Structurally ritterazine E (**40**) differs from ritterazine D (**39**) only by the presence of an additional methyl group at C-24. Retterazine E (**40**, 2.8 mg). IR (film): ν_{max} cm^{-1} 3440br (O–H), 2960m (C–H), 2920m (C–H), 2870m (C–H), 1730s (C=O), 1710s (C=O), 1600s (C=C), 1460m (C–H), 1400s (pyrazine), 1140m (C–O), 1030 (C–O), 960m, 880m. UV (MeOH): λ_{max} nm 288, 307. ^{1}H NMR (500 MHz, C_5D_5N): δ 6.25 (s, 1H, 15'-CH), 6.00 (s, 1H, 13-OH), 5.26 (s, 1H, 16'-CH), 5.30 (ddd, 1H, J = 2.9, 3.9, 9.3 Hz, 16-CH), 4.21 (dd, 1H, J = 4.0, 11.0 Hz, 12'-CH), 4.06 (m, 1H, 7'-CH), 4.01 (d, 1H, J = 11.5 Hz, 26'β-CH), 3.61 (dd, 1H, J = 2.6, 11.5 Hz, 26'α-CH), 3.15 (d, 1H, J = 16.5 Hz, l'β-CH), 3.00 (dd, 1H, J = 5.0, 16.0 Hz, 4'α-CH), 3.00 (m, 1H, 20-CH), 2.98 (m, 1H, 17-CH), 2.90 (dd, 1H, J = 5.5, 17.8 Hz, 4α-CH), 2.87 (d, 1H, J = 16.5 Hz, 1β-CH), 2.79 (dd, 1H, J = 12.5, 16.0 Hz, 4'β-CH), 2.72 (d, 1H, J = 16.5 Hz, lα-CH), 2.68 (d, 1H, J = 16.5 Hz, 1'α-CH), 2.57 (dd, 1H, J = 12.7, 17.8 Hz, 4β-CH), 2.51 (ddd, 1H, J = 4.9, 13.1, 13.3 Hz, 23'α-CH), 2.44 (m, 1H, 24-CH), 2.42 (dd, 1H, J = 10.5, 11.5 Hz, 8'-CH), 2.37 (m, 1H, 15β-CH), 2.37 (m, 1H, 11α-CH), 2.35 (m, 1H, 8-CH), 2.34 (m, 1H, 7β-CH), 2.21 (m, 1H, 6'α-CH), 2.22 (q, 1H, J = 6.8 Hz, 20'-CH), 2.18 (m, 1H, 11'α-CH), 2.15 (m, 1H, 23β-CH), 2.15 (m, 1H, 11β-CH), 2.10 (m, 1H, 15α-CH), 2.16 (m, 1H, 24'β-CH), 1.88 (m, 1H, 5'-CH), 1.88 (m, 1H, 11'β-CH), 1.88 (m, 1H, 24'α-CH), 1.76 (m, 1H, 6'β-CH), 1.74 (m, 1H, 23α-CH), 1.64 (m, 1H, 5-CH), 1.48 (m, 1H, 6α-CH), 1.45 (m, 1H, 23'β-CH), 1.45 (s, 3H, 18-Me), 1.39 (s, 3H, 27-Me), 1.37 (m, 1H, 7α-CH), 1.33 (s, 3H, 18'-Me), 1.32 (m, 1H, 9-CH), 1.26 (d, 3H, J = 6.7 Hz, 21'-Me), 1.22 (3H, s, 27'-Me), 1.17 (m, 1H, 9'-CII), 1.15 (m, 1H, 6β-CH), 1.14 (d, 3H, J = 6.0 Hz, 21-Me), 1.02 (s, 3H, 26-Me), 0.86 (s, 3H, 19'-Me), 0.82 (d, 3H, J = 7.0 Hz, 28-Me), 0.73 (s, 3H, 19-Me). ^{13}C NMR (125 MHz, C_5D_5N): δ 46.5 (C-1), 46.8 (C-1'), 149.0 (C-2 and C-3'), 148.9 (C-2' and C-3), 35.5 (C-4 and C-4'), 41.7 (C-5), 40.0 (C-5'), 28.6 (C-6), 38.7 (C-6'), 32.5 (C-7), 69.6 (C-7'), 40.7 (C-8), 43.8 (C-8'), 50.0 (C-9), 51.1 (C-9'), 35.6 (C-10), 35.8 (C-10'), 40.8 (C-11), 29.0 (C-11'), 220.7 (C-12), 75.8 (C-12'), 80.0 (C-13), 56.0 (C-13'), 70.8 (C-14), 151.3 (C-14'), 33.8 (C-15), 120.2 (C-15'), 80.2 (C-16), 94.0 (C-16'), 60.5 (C-17), 93.2 (C-17'), 23.7 (C-18), 12.4 (C-18'), 10.8 (C-19), 11.7 (C-19'), 37.7 (C-20), 48.1 (C-20'), 14.7 (C-21), 7.8 (C-2l'), 118.6 (C-22), 108.0 (C-22'), 41.0 (C-23), 27.4 (C-23'), 41.6 (C-24), 33.2 (C-24'), 83.5 (C-25), 65.6 (C-25'), 23.3 (C-26), 70.2 (C-26'), 29.6 (C-27), 26.7 (C-27'), 14.0 (C-28). HRFABMS: *m/z* 927.5635 calculated for $C_{55}H_{79}N_2O_{10}$, found: 927.5724 (Fukuzawa *et al.*, 1995b).

Structurally, in general, ritterazine F (**41**) and ritterazine B (**36**) are the same with the only exception in the stereochemistry at C-22; it is 22*S* in **41** and 22*R* in **36**. As a consequence of this structural similarity, the ^{1}H and ^{13}NMR spectra of these compounds are almost identical with a few differences around C-22 in the 'Eastern hemisphere' of the molecules; ritterazine F (**41**)/ ritterazine B (**36**): δ_C 40.7/42.1 (C-20), 16.3/14.7 (C-21), 30.1/33.2 (C-23); δ_H 2.84/3.15 (H-17); 2.29/2.01 (H-20); 1.08/1.18 (H-21), 1.87/2.12 (H-23β) (Fukuzawa *et al.*, 1995b). Retterazine F (**41**, 2.6 mg). IR (film): ν_{max} cm^{-1} 3470br (O–H), 2960m (C–H), 2920m (C–H), 2860m (C–H), 1700s (C=O), 1600m (C=C), 1460m (C–H), 1400s (pyrazine), 1220m (C–O), 1140m (C–O), 1040m (C–O), 880m. UV (MeOH): λ_{max} nm 288, 306. HRFABMS: *m/z* 8995786 calculated for $C_{54}H_{79}N_2O_9$, found: 899.5764 (Fukuzawa *et al.*, 1995b).

Ritterazine G (**42**) is the 14/15 dehydro derivative of ritterazine F (**41**). The ^{1}H and ^{13}C NMR data for the 'Western hemisphere' of both these molecules as well as ritterazine B (**36**) of are identical, but the signals, especially around C-14 in the 'Eastern hemisphere', because of the presence of the double bond ($\Delta^{14,15}$) in ritterazine G (**42**) changed noticeably. The NMR data for the 'Eastern hemisphere' of **42** are presented below.

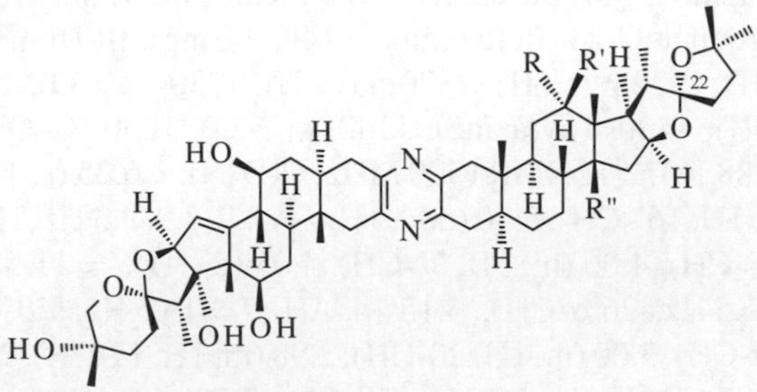

Ritterazine F (**41**)	R = OH	R' = H	R" = H 22 *S*
Ritterazine G (**42**)	R = OH	R' = H	R" = $\Delta^{14,15}$ 22*S*
Ritterazine I (**43**)	R, R' = Oxo		R" = OH 22 *S*

Retterazine G (**42**, 2.2 mg). IR (film): ν_{max} cm^{-1} 3450br (O–H), 2960m (C–H), 2930m (C–O), 2870m (C–O), 1730s (C=O), 1600s (C=C), 1450m (C–H), 1400s (pyrazine), 1230m (C–O), 1140m (C–O), 1040m (C–O), 1000m (C–O), 880m. UV (MeOH): λ_{max} nm 288, 306. ^{1}H NMR (500 MHz, C_5D_5N): 'Eastern hemisphere' δ 5.55 (s, 1H, 15-CH), 5.27 (br d, 1H, $J = 8.0$ Hz, 16-CH), 3.48 (dd, 1H, $J = 4.5$, 11.5 Hz, 12-CH), 3.10 (d, 1H, $J = 16.5$ Hz, 1β-CH), 3.09 (q, 1H, $J = 8.0$ Hz, 17-CH), 2.92 (dd, 1H, $J = 5.0$, 18.0 Hz, 4α-CH), 2.67 (dd, 1H, $J = 11.0$, 18.0 Hz, 4β-CH), 2.22 (q, 1H, $J = 6.8$ Hz, 20-CH), 2.12 (m, 1H, 23β-CH), 2.10 (m, 1H, 11α-CH), 2.16 (m, 1H, 8-CH), 2.03 (m, 1H, 24β-CH, 1.88 (m, 1H, 11β-CH), 1.81 (m, 1H, 7β-CH), 1.68 (m, 1H, 23α-CH), 1.67 (m, 1H, 24α-CH), 1.61 (m, 1H, 5-CH), 1.54 (m, 1H, 6α-CH), 1.46 (s, 3H, 27-Me), 1.33 (m, 1H, 7α-CH), 1.32 (s, 3H, 18-Me), 1.27 (m, 1H, 6β-CH), 1.25 (d, 3H, $J = 6.6$ Hz, 21-Me), 1.19 (s, 3H, 26-Me), 0.92 (m, 1H, 9-CH), 0.76 (s, 3H, 19-Me). ^{13}C NMR (125 MHz, C_5D_5N): 'Eastern hemisphere' δ 46.2 (C-1), 149.0 (C-2), 148.3 (C-3), 35.8 (C-4), 41.8 (C-5), 28.4 (C-6), 29.7 (C-7), 33.8 (C-8), 52.7 (C-9), 36.2 (C-10), 30.9 (C-11), 78.9 (C-12), 53.5 (C-13), 152.0 (C-14), 120.9 (C-15), 85.0 (C-16), 56.2 (C-17), 13.8 (C-18), 11.8 (C-19), 42.1 (C-20), 14.4 (C-21), 117.9 (C-22), 33.8 (C-23), 37.8 (C-24), 82.6 (C-25), 28.7 (C-26), 30.3 (C-27). HRFABMS: *m/z* 897.5629 calculated for $C_{54}H_{77}N_2O_{10}$, found: 897.5598 (Fukuzawa *et al.*, 1995b).

Ritterazine H (**37**) is the 12-oxo derivative of ritterazine B (**36**), *i.e.*, the 12-hydroxyl group in the 'Eastern hemisphere' of **36**, is replaced by an oxo functionality in **37**. The ^{1}H and ^{13}C NMR data for the 'Western hemisphere' of both ritterazines F and H (**36** and **37**) are identical, but the signals, especially around C-12 in the 'Eastern hemisphere', in ritterazine H (**37**) are significantly different from those in **36**. The NMR data for the 'Eastern hemisphere' of **37** are presented below. Retterazine H (**37**, 1.2 mg). IR (film): ν_{max} cm^{-1} 3460br (O–H), 2960m (C–H), 2920m (C–H), 2860m (C–H), 1700s (C=O), 1590s (C=C), 1450m (C–H), 1400s (pyrazine), 1360m, 1230m (C–O), 1130m (C–O), 1040m, 880m. UV (MeOH): λ_{max} nm 287, 306. ^{1}H NMR (500 MHz, C_5D_5N): 'Eastern hemisphere' δ 4.31 (dd, 1H, $J = 7.0$, 7.0 Hz, 16-CH), 3.22 (dd, 1H, $J = 7.0$, 7.5 Hz, 17-CH), 2.97 (d, 1H, $J = 16.0$ Hz, 1β-CH), 2.89 (dd, 1H, $J = 5.5$, 17.5 Hz, 4α-CH), 2.61 (dd, 1H, $J = 11.0$, 17.5 Hz, 4β-CH),

2.60 (d, 1H, $J=16.0$ Hz, 1α-CH), 2.55 (dd, 1H, $J=12.0$, 13.0 Hz, 11β-CH), 2.52 (d, 1H, $J=13.5$ Hz, 14-CH), 2.45 (dd, 1H, $J=4.5$, 13.0 Hz, 11α-CH), 2.08 (m, 1H, 20-CH), 1.97 (m, 1H, 24β-CH), 1.95 (m, 1H, 8-CH), 1.85 (m, 1H, 23β-CH, 1.73 (m, 1H, 15β-CH, 1.75 (m, 1H, 23α-CH, 1.67 (m, 1H, 24α-CH, 1.62 (m, 1H, 9-CH), 1.48 (m, 1H, 5-CH), 1.42 (m, 1H, 6α-CH, 1.40 (m, 1H, 7β-CH), 1.36 (s, 3H, 27-Me), 1.24 (m, 1H, 15α-CH, 1.21 (s, 3H, 18-Me), 1.20 (m, 1H, 6β-CH), 1.14 (s, 3H, 26-Me), 1.00 (m, 1H, 6α-CH), 0.94 (d, 3H, $J=7.0$ Hz, 21-Me), 0.70 (s, 3H, 19-Me). ^{13}C NMR (125 MHz, C_5D_5N): 'Eastern hemisphere' δ 45.8 (C-1), 148.8 (C-2 and C-3), 36.0 (C-4), 41.1 (C-5), 29.0 (C-6), 31.4 (C-7), 48.1 (C-8), 47.9 (C-9), 35.8 (C-10), 37.7 (C-11), 214.5 (C-12), 57.8 (C-13), 49.2 (C-14), 32.5 (C-15), 77.7 (C-16), 52.2 (C-17), 19.4 (C-18), 11.9 (C-19), 41.1 (C-20), 17.0 (C-21), 117.5 (C-22), 33.1 (C-23), 37.4 (C-24), 81.2 (C-25), 28.9 (C-26), 30.3 (C-27). HRFABMS: *m/z* 897.5629 calculated for $C_{54}H_{77}N_2O_9$, found: 897.5591 (Fukuzawa *et al.*, 1995b).

Ritterazine I (**43**) is a 12-oxo, 14-hydroxyl derivative of ritterazine F (**41**), *i.e.* the 12-hydroxyl group of **41** is replaced by an oxo functionality, and an additional hydroxyl group is present at C-14 ('Eastern hemisphere') of **43**. The 'Western hemisphere' of ritterazine B (**36**), F (**41**) and I (**43**) are exactly the same, and the differences are only in the ('Eastern hemisphere'), and because of this fact, the ^{1}H and ^{13}C NMR data corresponding to the 'Western hemisphere' of these molecules are identical. Therefore, the NMR data for the 'Eastern hemisphere' of **43** are presented below.

Retterazine I (**43**, 4.4 mg). IR (film): ν_{max} cm^{-1} 3460br (O–H), 2960m (C–H), 2920m (C–H), 2860m (C–H), 1700s (C=O), 1450m (C–H), 1400s (pyrazine), 1300m (C–H), 1230m (C–O), 1110m (C–O), 1040m (C–O), 880m. UV (MeOH): λ_{max} nm 286, 306. ^{1}H NMR (500 MHz, C_5D_5N): 'Eastern hemisphere' δ 4.79 (dd, 1H, $J=7.5$, 8.0 Hz, 16-CH), 3.74 (dd, 1H, $J=8.0$, 9.0 Hz, 17-CH), 3.05 (d, 1H, $J=16.0$ Hz, 1β-CH), 2.94 (dd, 1H, $J=5.5$, 17.5 Hz, 4α-CH), 2.90 (dq, 1H, $J=7.0$, 9.0 Hz, 20-CH), 2.73 (dd, 1H, $J=12.0$, 13.0 Hz, 11β-CH), 2.66 (dd, 1H, $J=11.0$, 17.5 Hz, 4β-CH), 2.67 (d, 1H, $J=16.0$ Hz, 1α-CH), 2.54 (dd, 1H, $J=4.5$, 13.0 Hz, 11α-CH), 2.42 (m, 1H, 7β-CH), 2.17 (m, 1H, 24β-CH), 2.09 (m, 1H, 8-CH), 2.05 (d, $J=15.0$ Hz, H-15β), 1.99 (m, H-23β), 1.88 (m, H-24α), 1.67 (dd, $J=7.7$, 8.0 Hz, 1H, 15α-CH), 1.62 (m, 1H, 23α-CH), 1.64 (m, 1H, 5-CH), 1.56 (m, 1H, 6α-CH), 1.50 (s, 3H, 18-Me), 1.43 (m, 1H, 9-CH), 1.41 (s, 3H, 27-Me), 1.26 (m, 1H, 6β-CH), 1.17 (m, 1H, 7α-CH), 1.17 (s, 3H, 26-Me), 1.09 (d, 3H, $J=7.0$ Hz, 21-Me), 0.76 (s, 3H, 19-Me) (Fukuzawa *et al.*, 1995b). ^{13}C NMR (125 MHz, C_5D_5N): 'Eastern hemisphere' δ 45.6 (C-1), 148.9 (C-2 and C-3), 35.5 (C-4), 41.3 (C-5), 28.1 (C-6), 27.2 (C-7), 40.5 (C-8), 46.7 (C-9), 36.5 (C-10), 37.2 (C-11), 213.5 (C-12), 62.5 (C-13), 87.0 (C-14), 38.6 (C-15), 79.2 (C-16), 53.1 (C-17), 15.1 (C-18), 11.3 (C-19), 42.1 (C-20), 15.0 (C-21), 117.8 (C-22), 32.9 (C-23), 37.3 (C-24), 81.7 (C-25), 28.3 (C-26), 30.0 (C-27). HRFABMS: *m/z* 913.5578 calculated for $C_{54}H_{77}N_2O_{10}$, found: 913.5663 (Fukuzawa *et al.*, 1995b).

Ritterazines J–L (**44**–**46**) have 22*R* stereochemistry as ritterazine B (**36**), and they differ structurally only on the basis of presence/absence of hydroxyl functionalities at C-7 ('Western hemisphere') and C-17 ('Eastern hemisphere'). Ritterazine J (**44**) is the most oxygenated of all ritterazines, and has hydroxyl groups at both C-7 ('Western hemisphere') and C-17 ('Eastern hemisphere') positions. The structure including stereochemistry of the 'Western hemisphere' of **44** is the same as that of **36**, and this results in identical ^{1}H and ^{13}C NMR signals associated with the 'Western hemisphere' of both molecules. The NMR data for the 'Western hemisphere' of **44** are shown below.

Retterazine J (**44**, 2.8 mg). IR (film): ν_{max} cm^{-1} 3450br (O–H), 2960m (C–H), 2920m (C–H), 2880m (C–H), 1730s (C=O), 1700s (C=O), 1450m (C–H), 1400s (pyrazine), 1300m (C–H), 1230m (C–H), 1110m (C–H), 1060m (C–H), 1040m (C–H), 990m, 970m, 940w, 880w, 850w. UV (MeOH): λ_{max} nm 289, 308. ^{1}H NMR (500 MHz, C_5D_5N): 'Eastern hemisphere' δ 5.60 (s, 1H, 15-CH), 5.18 (s, 1H, 16-CH), 4.19 (dd, 1H, J=4.5, 11.5 Hz, 12-CH), 4.01 (d, 1H, J= 11.5 Hz, 26β-CH), 3.13 (d, 1H, J= 17.0 Hz, 1β-CH), 2.93 (dd, 1H, J=5.5, 17.5 Hz, 4α-CH), 2.67 (dd, 1H, J=11.0, 17.5 Hz, =4β-CH), 2.62 (br d, 1H, J=11.5 Hz, 26α-CH), 2.66 (d, 1H, J=17.0 Hz, 1α-CH), 3.56 (ddd, 1H, J=4.5, 13.0, 14.0 Hz, 23α-CH), 2.22 (q, 1H, J=6.5 Hz, 20-CH), 2.17 (m, 1H, 24β-CH), 2.09 (m, 1H, 8-CH), 2.15 (m, 1H, 11α-CH), 1.87 (m, 1H, 24α-CH), 1.81 (m, 1H, 11β-CH), 1.75 (m, 1H, 7β-CH), 1.59 (m, 1H, 23β-CH), 1.63 (m, 1H, 5-CH), 1.54 (m, 1H, 6α-CH), 1.36 (m, 1H, 7α-CH), 1.33 (s, 3H, 18-Me), 1.29 (d, 3H, J=6.9 Hz, 21-Me), 1.27 (m, 1H, 6β-CH), 1.23 (s, 3H, 27-Me), 0.98 (m, 1H, 9-CH), 0.77 (s, 3H, 19-Me) (Fukuzawa *et al.*, 1995b).

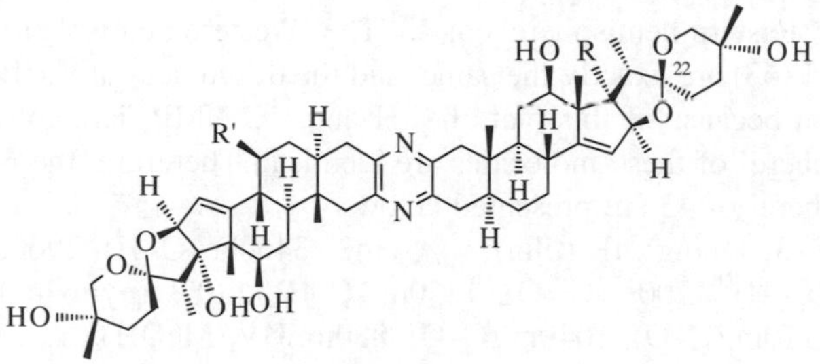

Ritterazine J (**44**)	R = OH	R' = OH 22*R*
Ritterazine K (**45**)	R = OH	R' = H 22*R*
Ritterazine L (**46**)	R = H	R' = H 22*R*

^{13}C NMR (125 MHz, C_5D_5N): 'Eastern hemisphere' δ 46.0 (C-1), 148.8 (C-2 and C-3), 35.7 (C-4), 41.8 (C-5), 28.3 (C-6), 29.0 (C-7), 34.0 (C-8), 53.0 (C-9), 36.1 (C-10), 29.2 (C-11), 75.8 (C-12), 56.0 (C-13), 155.0 (C-14), 120.2 (C-15), 93.9 (C-16), 93.3 (C-17), 12.7 (C-18), 11.6 (C-19), 48.2 (C-20), 8.0 (C-21), 108.0 (C-22), 27.4 (C-23), 33.3 (C-24), 65.8 (C-25), 70.4 (C-26), 26.9 (C-27). HRFABMS: *m/z* 929.5528 calculated for $C_{54}H_{77}N_2O_{11}$, found: 929.5471 (Fukuzawa *et al.*, 1995b).

Ritterazine K (**45**) is a symmetrical dimer, *i.e.*, its 'Western hemisphere' and 'Eastern hemisphere' are exactly the same, and identical to the 'Eastern hemisphere' of ritterazine J (**44**). The ^{1}H and ^{13}C NMR data for both 'hemispheres' of **45** were found to be identical to those for the 'Eastern hemisphere' of **44** as outlined earlier. Retterazine K (**45**, 6.2 mg). IR (film): ν_{max} cm^{-1} 3480br (O–H), 2960m (C–H), 2930m (C–H), 2870m (C–H), 1700s (C=O), 1450m (C–H), 1400s (pyrazine), 1300m (C–H), 1230m (C–O), 1200m (C–O), 1140m (C–O), 1110m (C–O), 1060m (C–O), 1040m (C–O), 990m, 960m, 940m, 880w, 850w. UV (MeOH): λ_{max} nm 288, 306. HRFABMS: *m/z* 913.5578 calculated for $C_{54}H_{77}N_2O_{10}$, found: 913.5532 (Fukuzawa *et al.*, 1995b).

Ritterazine L (**46**) is, in fact, 17-deoxyritterazine K, *i.e.*, it has one less oxygen compared to **45**. The ^{1}H and ^{13}C NMR data arising from the 'Western hemisphere' of **46** were identical to those of the 'Eastern hemisphere' of ritterazine J (**44**). The signals for the 'Eastern hemisphere' of ritterazine L (**46**) were slightly different from those of **44** or **45** because of the lack of oxygenation at C-17 in **46**. However, the notable differences were pretty much around C-17; ritterazine L (**46**)/ritterazine J (**44**) or ritterazine K (**45**): δ_C 157.9/155.0 (C-14), 56.6/93.3 (C-17), 13.8/12.7 (C-18), 44.9/48.2 (C-20), 14.2/8.0 (C-21), and δ_H 3.18/ none (17-CH), 2.11/2.22 (20-CH), 1.38/1.29 (21-CH) (Fukuzawa *et al.*, 1995b). Retterazine L (**46**, 1.1 mg). IR (film): ν_{max} cm^{-1} 3440br (O–H), 2960m (C–H), 2920m (C–H), 2860m (C–H), 1600m (C=C), 1450m (C–H), 1400s (pyrazine), 1230m (C–O), 1040m (C–O), 940m, 880w, 850w. UV (MeOH): λ_{max} nm 288, 307. HRFABMS: *m/z* 897.5629 calculated for $C_{54}H_{77}N_2O_9$, found: 897.5598 (Fukuzawa *et al.*, 1995b).

Ritterazine M (**47**) is a 22*S* stereoisomer of ritterazine L (**46**). Its NMR data were different from those of **46**, particularly, in signals around C-22, but the signals associated with the rest of the molecule were superimposable (Fukuzawa *et al.*, 1995b). Ritterazine M (**47**, 1.9 mg). IR (film): ν_{max} cm^{-1} 3460br (O–H), 2960m (C–H), 2920m (C–H), 2860m (C–H), 1700s (C=O), 1460m (C–H), 1400s (pyrazine), 1040m (C–O), 940m, 880w, 850w. UV (MeOH): λ_{max} nm 289, 306. HRFABMS: *m/z* 897.5629 calculated for $C_{54}H_{77}N_2O_9$, found: 897.5591 (Fukuzawa *et al.*, 1995b).

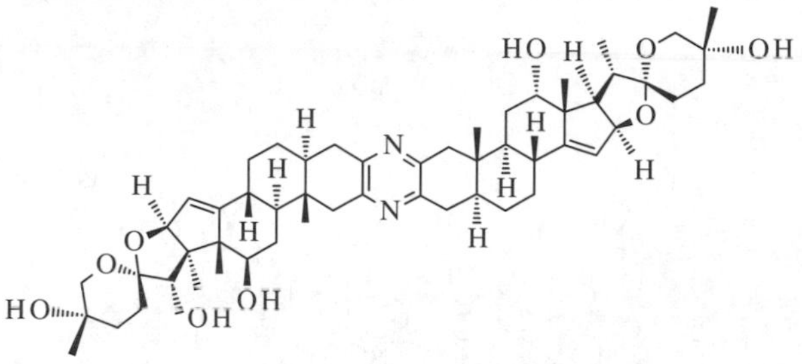

Ritterazine M (**47** 22*S*)

A couple of years later, the same group reported the isolation and structure elucidation of yet another thirteen ritterazines, ritterazines N–Z (**35**, **48–59**) from *Ritterella tokioka* (9 kg) using a combination of solvent partitioning and chromatographic techniques as outlined for the isolation of the earlier members of this class of compounds (Fuluzawa *et al.*, 1994, 1995a,b,1997). While ritterazines N (**48**), O (**49**), P (**50**), Q (**51**), R (**52**), and S (**53**) (yields: 1.5, 3.8, 0.8, 0.6, 0.3 and 0.5 mg, respectively) were obtained from the MeOH fraction of the initial ODS flash chromatography, ritterazines T (**35**), U (**54**), V (**55**), W (**56**), × (**57**), Y (**58**) and Z (**59**) (yields: 2.3, 3.0, 0.9, 0.7, 0.7, 3.5, and 1.7 mg, respectively) were purified from the MeCN–H_2O (7:3) fraction together with ritterazines A–C (**34, 36** and **38**) (Fukuzawa *et al.*, 1997). While the UV absorption maxima and HRFABMS data of these compounds are summarized in *Table 4.3.1*, ^{13}C NMR data are presented in Tables 4.3.2–4.3.5.

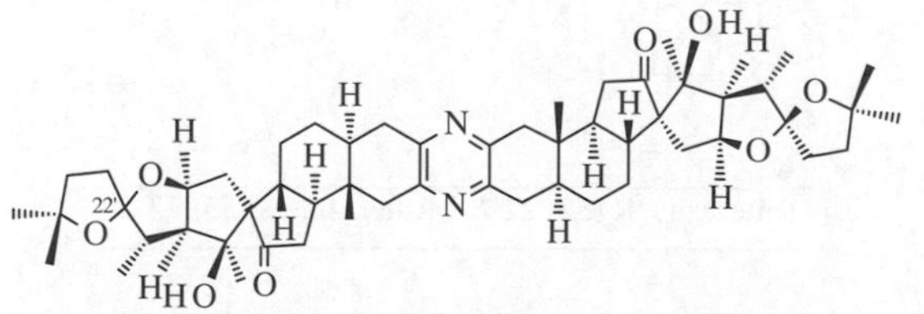

Ritterazine N (**48**) 22'*R* Ritterazine O (**49**) 22'*S*

Table 4.3.1 *The UV absorption maxima and MS data of ritterazines N-Z (**35, 48–59**)*

Ritterazines	UV (MeOH) λ_{max} in nm	HRFABMS (*m/z*)
N (**48**)	289, 309	881.5629 corresponding to $C_{54}H_{77}N_2O_8$
O (**49**)	289, 308	881.5619 corresponding to $C_{54}H_{77}N_2O_8$
P (**50**)	289, 309	867.5815 corresponding to $C_{54}H_{79}N_2O_7$
Q (**51**)	289, 309	867.5938 corresponding to $C_{54}H_{79}N_2O_7$
R (**52**)	288, 309	853.6071 corresponding to $C_{54}H_{81}N_2O_6$
S (**53**)	290, 309	853.6045 corresponding to $C_{54}H_{81}N_2O_6$
T (**35**)	290, 308	881.5719 corresponding to $C_{54}H_{77}N_2O_8$
U (**54**)	290, 308	897.5646 corresponding to $C_{54}H_{77}N_2O_9$
V (**55**)	290, 308	897.5627 corresponding to $C_{54}H_{77}N_2O_9$
W (**56**)	290, 308	881.5691 corresponding to $C_{54}H_{77}N_2O_8$
X (**57**)	290, 308	881.5632 corresponding to $C_{54}H_{77}N_2O_8$
Y (**58**)	289, 308	867.5938 corresponding to $C_{54}H_{79}N_2O_7$
Z (**59**)	289, 308	911.5773 corresponding to $C_{55}H_{79}N_2O_9$

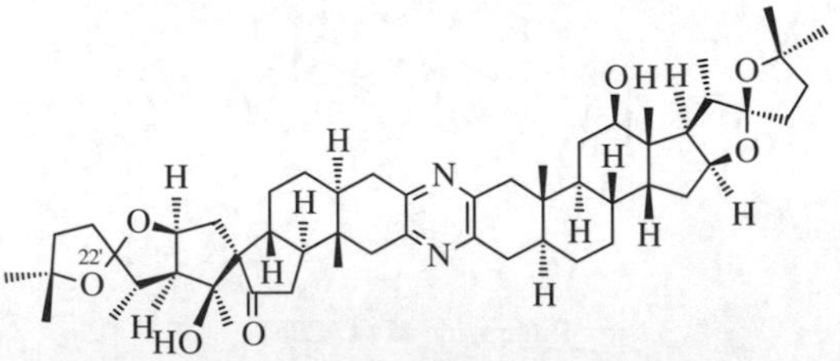

Ritterazine P (**50**) 22'*R* Ritterazine Q (**51**) 22'*S*

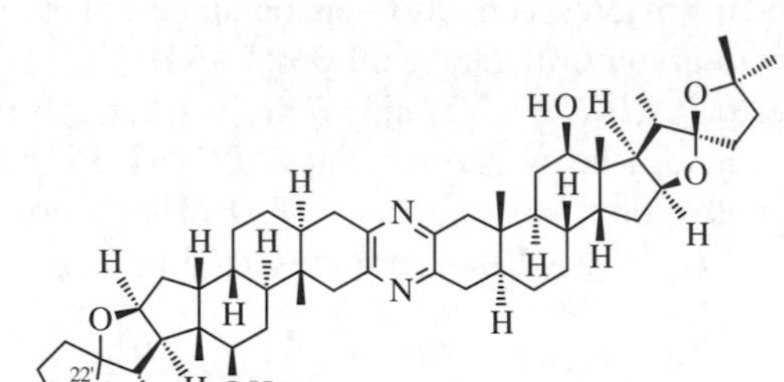

Ritterazine R (**52**) 22'*R* Ritterazine S (**53**) 22'*S*

Table 4.3.2 *^{13}C NMR data (500 MHz, C_5D_5N) of the 'Eastern hemisphere' of ritterazines N–S (**48–53**)*

Carbon	Chemical shifts (δ) in ppm					
	48	49	50	51	52	53
1	46.7	46.8	46.7	46.7	46.5	46.5
2	149.0	149.0	149.0	149.0	149.1	149.0
3	149.0	149.0	149.0	149.0	149.1	149.2
4	35.8	35.9	35.9	35.9	36.0	36.1
5	42.0	42.0	41.5	41.5	41.5	41.5
6	29.2	29.0	28.7	29.0	28.8	29.0
7	30.6	30.8	31.7	31.7	31.7	31.5
8	40.5	40.5	32.7	32.5	32.6	32.5
9	50.3	50.3	45.3	45.1	45.2	44.9
10	35.4	35.3	35.5	35.6	35.7	35.5
11	40.8	40.9	30.6	30.4	30.5	30.5
12	220.2	220.2	72.3	72.2	72.2	72.1
13	81.0	82.0	48.0	48.0	48.0	48.0
14	69.1	69.2	47.9	48.0	47.9	48.0
15	36.0	36.0	52.7	32.7	32.6	32.6
16	83.3	83.3	79.9	79.7	79.7	79.8
17	61.9	61.8	57.5	57.4	57.5	57.5
18	23.5	23.8	13.2	13.2	13.2	13.2
19	10.8	11.1	11.9	11.8	11.8	11.8
20	41.0	41.0	42.0	42.0	42.4	42.3
21	19.1	19.4	14.6	14.6	14.7	14.6
22	119.0	119.6	116.7	116.7	116.9	116.8
23	33.0	33.1	33.3	33.3	33.8	33.3
24	37.4	37.7	38.0	38.0	37.7	38.0
25	82.4	82.4	81.6	81.6	81.7	81.5
26	28.7	28.7	28.8	28.8	28.8	28.8
27	30.4	30.4	30.4	30.3	30.7	30.4

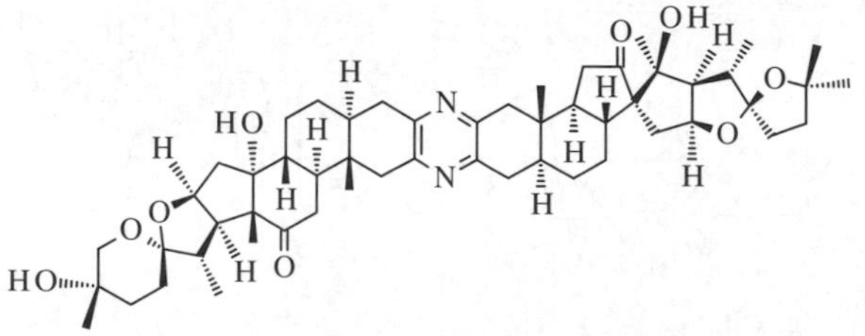

Ritterazine U (**54**)

Table 4.3.3 *^{13}C NMR data (500 MHz, C_5D_5N) of the 'Western hemisphere' of ritterazines N–S (**48–53**) (*Identical to the 'Eastern hemisphere')*

Carbon	Chemical shifts (δ) in ppm					
	48	49	50	51	52	53
1'	∗	46.6	46.7	46.6	∗	46.6
2'	∗	148.8	148.9	148.8	∗	149.0
3'	∗	148.8	148.9	148.8	∗	149.2
4'	∗	35.9	35.9	35.9	∗	36.1
5'	∗	41.6	41.7	41.6	∗	41.5
6'	∗	29.0	29.0	29.0	∗	29.0
7'	∗	30.8	30.5	30.4	∗	31.5
8'	∗	40.4	40.5	40.6	∗	32.6
9'	∗	49.6	50.3	50.0	∗	44.9
10'	∗	35.3	35.5	35.3	∗	35.9
11'	∗	40.6	40.8	40.3	∗	30.5
12'	∗	218.6	220,0	218.3	∗	72.8
13'	∗	82.4	81.0	82.0	∗	48.0
14'	∗	69.6	69.2	69.4	∗	48.0
15'	∗	36.0	33.3	33.1	∗	33.7
16'	∗	80.9	83.4	80.3	∗	79.2
17'	∗	60.4	61.9	60.5	∗	57.8
18'	∗	21.9	23.5	21.6	∗	13.8
19'	∗	11.1	10.8	10.8	∗	11.8
20'	∗	38.5	41.0	38.0	∗	41.4
21'	∗	15.1	19.2	15.1	∗	17.1
22'	∗	119.6	119.5	119.5	∗	117.5
23'	∗	33.1	33.4	34.3	∗	33.3
24'	∗	37.7	37.5	37.6	∗	37.5
25'	∗	82.1	82.4	82.0	∗	81.1
26'	∗	28.8	28.7	28.7	∗	28.8
27'	∗	30.5	30.4	30.4	∗	30.1

Ritterazine V (**55**)

Table 4.3.4 *^{13}C NMR data (500 MHz, C_5D_5N) of the 'Eastern hemisphere' of ritterazines T–Z (**37, 54–59**)*

Carbon	Chemical shifts (δ) in ppm						
	37	54	55	56	57	58	59
1	46.8	46.6	46.6	46.7	46.4	46.2	46.4
2	149.0	148.8	148.7	148.7	148.7	148.6	148.0
3	149.0	148.8	148.7	148.7	148.7	149.1	148.5
4	35.5	35.9	35.5	35.7	35.7	35.8	35.1
5	41.7	41.9	42.0	41.8	41.6	41.7	41.5
6	29.2	29.1	29.1	29.2	28.9	29.0	28.7
7	30.7	30.6	30.7	30.6	30.6	31.8	30.2
8	40.5	40.5	40.7	40.5	41.0	32.6	40.2
9	50.1	50.2	50.1	50.0	50.0	45.5	49.7
10	35.6	35.5	35.6	35.8	35.4	35.9	35.6
11	40.9	40.9	41.0	41.0	41.0	30.7	40.5
12	221.6	221.2	221.2	221.2	220.5	71.7	220.5
13	80.9	80.9	80.4	80.9	79.8	48.5	80.6
14	69.3	69.3	69.3	69.7	70.8	47.7	68.9
15	35.9	35.8	35.7	35.9	34.0	32.9	35.5
16	82.8	82.8	82.8	81.9	80.4	80.0	82.5
17	61.7	61.7	61.6	61.7	60.7	57.4	61.4
18	23.5	23.5	23.5	23.4	23.8	23.7	23.1
19	10.9	10.9	11.0	10.9	10.9	11.9	10.5
20	40.8	40.7	40.8	40.7	38.4	42.0	40.4
21	18.9	19.0	19.0	19.0	14.7	14.7	18.6
22	119.9	119.8	119.9	119.6	120.6	119.0	119.2
23	32.7	32.7	32.7	32.8	33.9	33.3	32.4
24	37.4	37.4	37.3	37.4	37.8	37.8	37.0
25	82.3	82.5	81.7	81.6	81.6	81.4	82.3
26	28.7	28.5	28.6	28.6	28.8	28.7	28.2
27	30.5	30.4	30.4	30.4	30.3	30.3	30.1

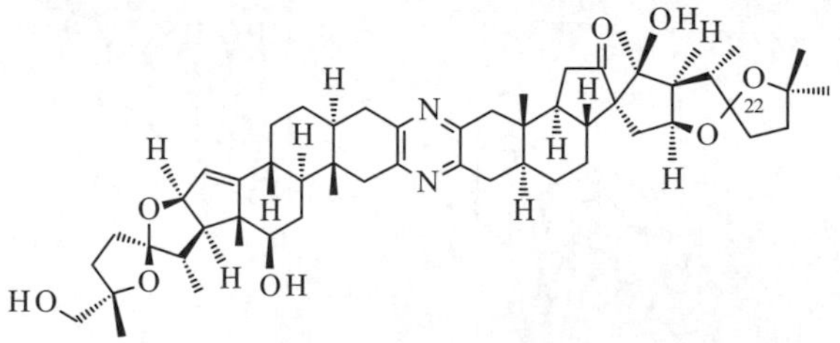

Ritterazine W (**56**) 22*R* Ritterazine X (**57**) 22*S*

Table 4.3.5 *^{13}C NMR data (500 MHz, C_5D_5N) of the 'Western hemisphere' of ritterazines T–Z (**37, 54–59**)*

Carbon	Chemical shifts (δ) in ppm						
	37	54	55	56	57	58	59
1'	46.3	46.0	46.5	46.2	46.1	46.2	45.4
2'	149.0	148.8	148.7	148.7	148.7	148.8	148.5
3'	149.0	148.8	148.7	148.7	148.7	148.7	148.0
4'	35.8	35.8	35.5	35.8	35.7	36.0	35.4
5'	42.0	42.1	41.7	41.9	42.0	41.6	41.3
6'	28.3	28.6	29.0	28.3	28.3	28.3	27.9
7'	29.9	27.2	30.7	29.9	29.8	29.7	30.6
8'	35.1	38.6	40.7	34.7	34.7	33.1	33.9
9'	49.5	48.4	49.8	49.6	49.1	52.7	54.3
10'	36,2	36.4	35.4	36.1	36.0	36.2	35.9
11'	29.8	37.5	41.0	29.8	29.7	30.9	39.4
12'	76.1	212.6	220.6	76.1	76.1	78.7	208.3
13'	52.5	60.3	80.0	52.6	52.6	52.6	60.0
14'	151.9	90.5	70.9	153.8	153.5	157.7	53.9
15'	120.9	39.5	34.0	121.5	121.5	120.5	36.2
16'	86.4	81.0	81.0	86.3	86.3	85.4	79.4
17'	54.6	53.1	60.8	54.6	54.5	56.4	49.6
18'	18.9	20.0	23.7	19.0	19.0	14.0	59.5
19'	11.9	11.6	11.0	11.8	11.8	11.8	11.3
20'	45.1	43.0	41.5	41.8	41.8	45.2	37.3
21'	14.8	14.4	14.7	14.6	14.7	14.5	14.0
22'	107.1	110.0	110.3	118.0	118.0	107.1	111.2
23'	27.8	28.1	27.7	33.3	33.3	27.7	70.4
24'	33.8	33.8	33.9	32.8	32.7	33.8	44.8
25'	66.0	65.9	66.0	85.4	85.5	66.0	72.6
26'	70.3	69.8	70.1	69.8	69.7	70.2	26.5
27'	26.9	27.0	27.0	24.2	24.2	27.0	29.7
28'	-	-	-	-	-	-	9.3

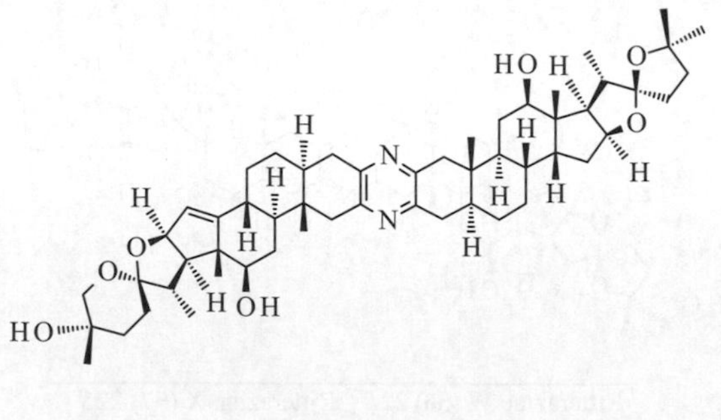

Ritterazine Y (**58**)

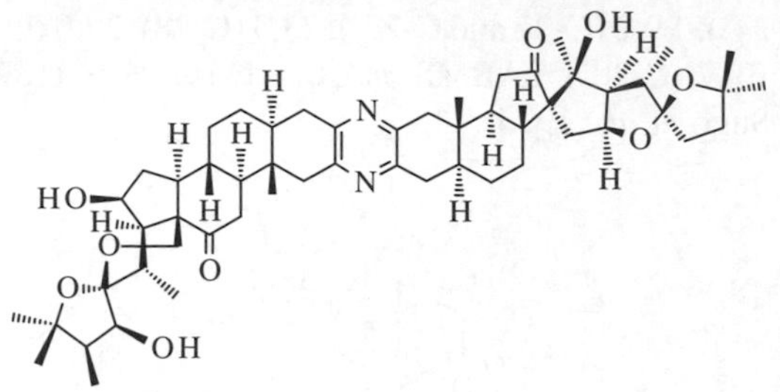

Ritterazine Z (**59**)

4.4 Others

A few other natural steroid dimers, *e.g*., bistheonellasterone (**60**) and bisconicasterone (**61**), which are the representative of Diels–Alder adducts of the corresponding 3-keto-4-methylene sterols, were isolated from the Okinawan marine sponge *Theonella swinhoei*, collected from the Nazumado bay on Hachijo Island (Kobayashi *et al*., 1992; Inouye *et al*., 1994; Sugo *et al*., 1995). Bistheonellasterone (**60**) resulted through a Diels–Alder-type cycloaddition of theonellasterone to an isomeric $\Delta^{4(5)}$-theonellasterone, and similarly, compound **61** could be produced *in vitro* by a Diels–Alder-type dimerization.

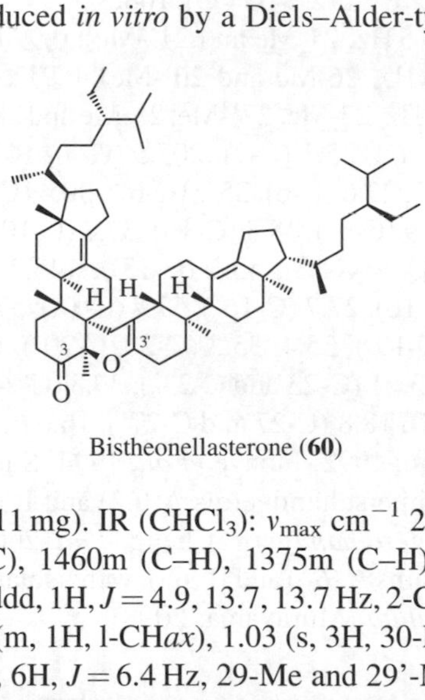

Bistheonellasterone (**60**)

Bistheonellasterone (**60**, 11 mg). IR ($CHCl_3$): ν_{max} cm^{-1} 2950m (C–H), 2870m (C–H), 1720s (C=O), 1685s (C=C), 1460m (C–H), 1375m (C–H), 1160m, 1050m. ^1H NMR (500 MHz, $CDCl_3$): δ 2.69 (ddd, 1H, $J = 4.9, 13.7, 13.7$ Hz, 2-CH*ax*), 2.25 (m, 1H, 2-CH*eq*), 1.99 (m, 1H, l-CH*eq*), 1.56 (m, 1H, l-CH*ax*), 1.03 (s, 3H, 30-Me), 0.94 (d, 6H, $J = 6.1$ Hz, 21-Me and 21'-Me), 0.86 (t, 6H, $J = 6.4$ Hz, 29-Me and 29'-Me), 0.84 (d, 6H, $J = 7.4$ Hz, 26-Me and 26'-Me), 0.84 (s, 6H, 18-Me and 18'-Me), 0.81 (d, 6H, $J = 7.1$ Hz, 27-Me and 27'-Me), 0.65 (s, 6H, 19-Me and 19'-Me). ^{13}C–NMR (125 MHz, $CDC1_3$): δ 33.9 (C-1 and C-1'), 209.8 (C-3), 146.1 (C-3'), 83.9 (C-4), 102.9 (C-4'), 38.4 (C-5 and C-10), 5.13 (C-5'), 27.2 (C-6), 27.1 (C-6'), 29.8 (C-7) 29.7 (C-7'), 126.4 (C-8), 125.3 (C-8'), 48.2 (C-9 and C-9'), 37.1 (C-10'), 20.3 (C-11 and C-11'), 37.6 (C-12 and C-12'), 42.9 (C-13 and C-13'), 143.6 (C-14), 142.3 (C-14'), 25.0 (C-15), 24.9 (C-15'), 26.6 (C-16 and C-16'), 57.3 (C-17), 57.0 (C-17'), 18.6 (C-18), 18.4 (C-18'), 13.0 (C-19), 12.4 (C-l9'), 19.4 (C-26 and C-26'), 35.0 (C-20 and C-20'), 19.7(C-21 and C-21'), 37.3 (C-22 and C-22'), 26.0 (C-23 and C-23'),

46.3 (C-24), 46.1 (C-24'), 1 9.2 (C-27 and C-27'), 23.3 (C-28), 22.0 (C-28'), 14.8 (C-29 and C-29'), 29.2 (C-30), 39.7 (C-30'). FABMS: *m/z* 849 $[M + H]^+$ (Kobayashi *et al.*, 1992; Inouye *et al.*, 1994; Sugo *et al.*, 1995).

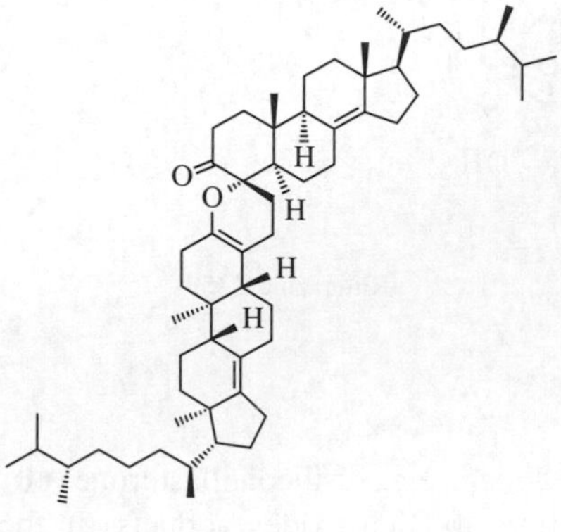

Bisconicasterone (**61**)

Chromatographic separation of the EtOH extract of *Theonella swinhoei* gave the dimer bisconicasterone (**61**, mp: 213–216 °C) together with its monomer conicasterol (Inouye *et al.*, 1994). IR (CCl_4): ν_{max} cm^{-1} 1721s (C=O), 1682s (C=C). ^1H NMR (500 MHz, C_6D_6): δ 1.07 and 1.05 (d, 6H, J = 6.5 Hz, 21-Me and 21'-Me), 0.99 (s, 3H, 18'-Me), 0.97 (s, 3H, 18-Me), 0.95 (d, 6H, J = 6.5 Hz, 26-Me and 26'-Me), 0.91 (s, 3H, 19'-Me), 0.90 (s, 3H, 19-Me), 0.91 (d, 12H, J = 6.5 Hz, 27-Me, 27'-Me, 28-Me and 28'-Me). ^{13}C NMR (125 MHz, C_6D_6): δ 40.0 (C-1), 34.5 (C-1'), 35.3 (C-2), 207.5 (C-3), 147.3 (C-3'), 84.3 (C-4), 103.2 (C-4'), 57.6 (C-5), 46.9 (C-5'), 22.6 (C-6), 25.9 (C-6'), 30.4 (C-7), 30.6 (C-7'), 126.2 (C-8), 127.23 (C-8'), 51.8 (C-9), 48.9 (C-9'), 38.8 (C-10), 37.7 (C-10'), 20.4 (C-11), 20.9 (C-11'), 37.9 (C-12), 38.3 (C-12'), 43.4 (C-13), 43.5 (C-13'), 143.7 (C-14), 142.7 (C-14'), 26.6 (C-15), 26.5 (C-15'), 27.9 (C-16), 27.7 (C-16'), 57.7 (C-17), 57.6 (C-17'), 19.0 (C-18), 19.2 (C-18'), 15.1 (C-19), 13.7 (C-19'), 35.4, 35.3 (C-20, C-20'), 19.8, 19.8 (C-21 and C-21'), 34.4, 34.3 (C-22 and C-22'), 31.1 (C-23 and C-23'), 39.8 (C-24 and C-24'), 33.1 (C-25 and C-25'), 20.8 (C-26 and C-26'), 18.8 (C-27 and C-27'), 16.0 (C-28 and C-28'), 26.5 (C-29), 19.6 (C-29') (Kobayashi *et al.*, 1992; Inouye *et al.*, 1994; Sugo *et al.*, 1995).

Two polysulphate sterol dimers, hamigerols A (**62**) and B (**63**), were reported from the Mediterranean sponge *Hamigera hamigera* (Cheng *et al.*, 2007). Very recently, two other similar dimers, shishicrellastains A (**64**) and B (**65**), were isolated from the Japanese marine sponge *Crella (Yvesia) spinulata* (Murayama, 2008).

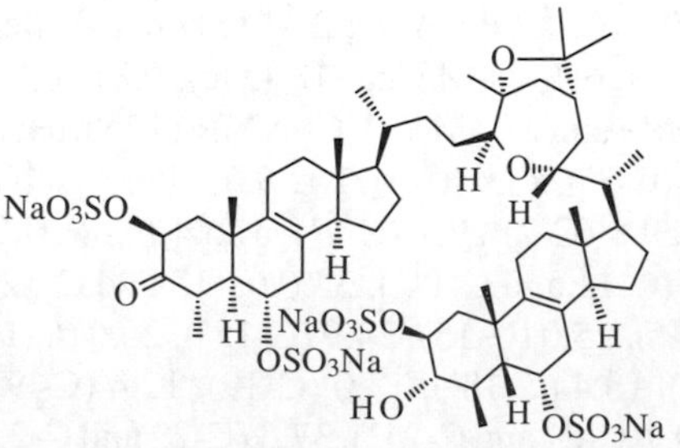

Hamigerol A (**62**)

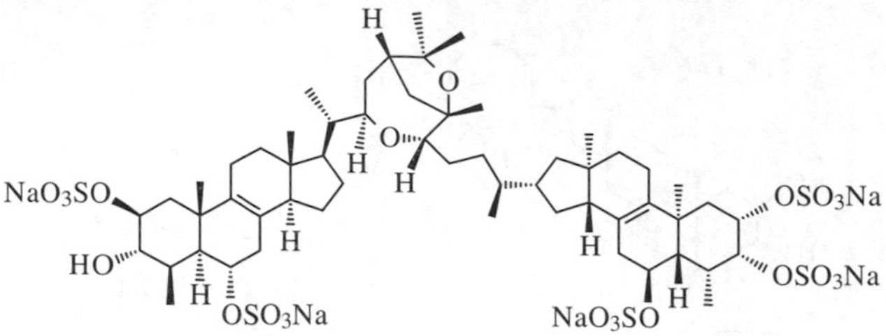

Hamigerol B (**63**)

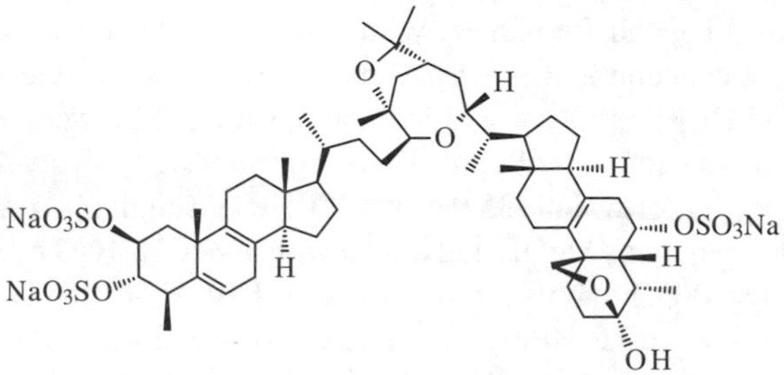

Shishicrellastatin A (**64**)

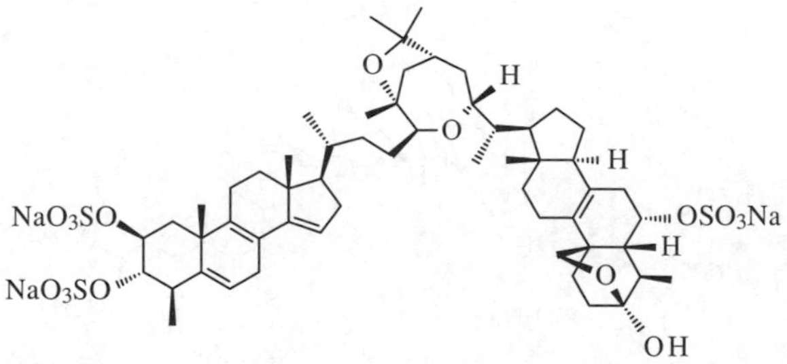

Shishicrellastatin B (**65**)

Further polysulphate sterol dimers, amaroxocanes A (**66**) and B (**67**), were isolated from *Phorbas amaranthus* collected on shallow coral reefs off Key Largo, Florida (Morinaka *et al.*, 2009). Both dimers are composed of two sulphated sterol cores bridged by an oxocane formed by different oxidative side-chain fusions. The crellastatins (see *Section 4.2*) and hamigerols (**62** and **63**) are unique among the dimeric steroids possessing a 3,8-dioxabicyclo-[4.2.1]nonane system formed by oxidative fusion of unsaturated sterol side chains (Morinaka *et al.*, 2009). Amaroxocane B (**67**) possesses the same heterocycle found in the crellastatins and hamigerols, and amaroxocane A (**66**) is similar to crellastatin M (**33**) because the dimer formed by oxidative cyclization of the side chains contains C_5 carbocycles by variations in their biosynthetic pathways. It was suggested that amaroxocane B (**67**) might have some ecological role as a chemical defense against fish predators (Morinaka *et al.*, 2009).

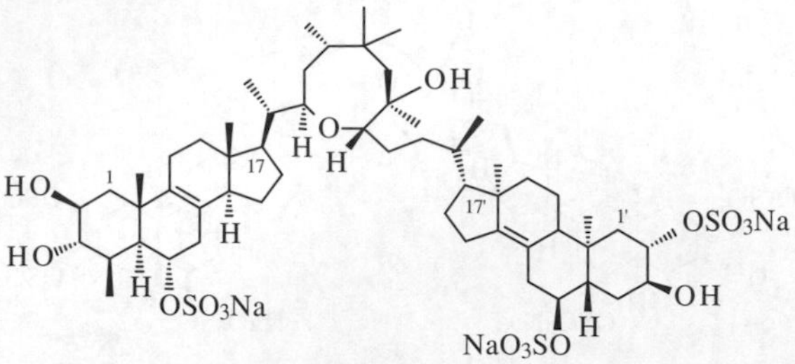

Amaroxocane A (**66**)

A freeze-dried and lyophilized sample of *P. amaranthus* (131.2 g) was extracted by 50% aqueous MeOH (2 × 1 L), and then finally with 100% MeOH (Morinaka *et al*., 2009). The pooled extracts were concentrated *in vacuo*, and the resulting residue was extracted with DCM (2 × 1 L). The DCM extract was dried and partitioned between *n*-hexane and 90% aqueous MeOH. The aqueous MeOH layer was concentrated *in vacuo* to afford a crude extract (48.6 g), a portion of which (33.9 g) was filtered through C_{18} reversed-phase SiO_2, conditioned with 90% aqueous MeOH and eluted with a MeOH–H_2O mixture of decreasing polarity, followed by *i*-PrOH to obtain three fractions I-III. A portion (1.16 g) of fraction II was analyzed by preparative RP-HPLC on C_{18} column (25 × 200 mm, flow rate 25 mL/min) using a gradient elution by H_2O-MeCN containing 1.5 M $NaClO_4$ (detection at 240 nm) to provide seven fractions. Further RP-HPLC (using a slightly modified solvent system) of a portion of the sixth fraction (72.5 mg) afforded amaroxocane A (**66**, 17.9 mg, 0.46%) and amaroxocane B (**67**, 6.5 mg, 0.17%).

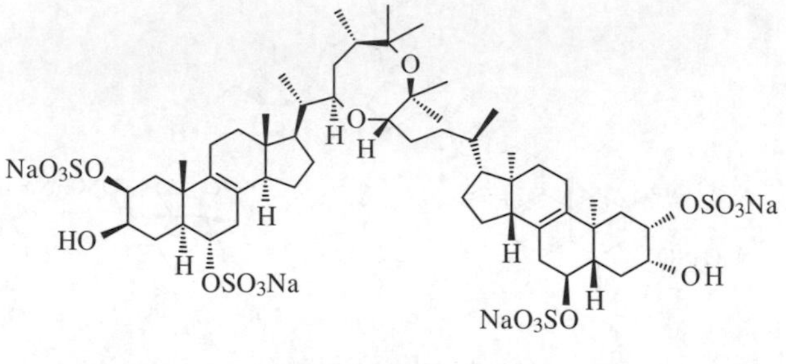

Amaroxocane B (**67**)

Amaroxocane A (**66**): IR (ZnSe): ν_{max} cm^{-1} 3453br (O–H), 2949m (C–H), 1637m (C=C), 1457m (C–H), 1383m (C–H), 1216m (C–O), 1061m (C–O), 959m, 914m. UV (MeOH): λ_{max} nm 218. ^{1}H NMR (600 MHz, CD_3OD): δ 4.87 (dt, 1H, J = 7.6, 12.0 Hz, 6-CH), 4.72 (br q, 1H, J = 2.6 Hz, 2'-CH), 4.16 (br q, 1H, J = 3.0 Hz, 2-CH), 4.09 (td, 1H, J = 5.2, 10.7 Hz, 6'-CH), 3.96 (dd, 1H, J = 3.5, 6.1 Hz, 3-CH), 3.80 (m, 1H, 3'-CH), 3.63 (dd, 1H, J = 4.6, 11.8 Hz, 22-CH), 3.26 (d, 1H, J = 7.9 Hz, 24'-CH), 3.02 (dd, 1H, J = 5.2, 13.6 Hz, 7'-CHb), 2.91 (dd, 1H, J = 6.8, 18.2 Hz, 7-CHb), 2.50 (m, 1H, 4-CH), 2.35 (m, 1H, 15'-CHb), 2.33 (m, 1H, 7-CHa), 2.32 (dd, 1H, J = 3.2, 14.6 Hz, 1'-CHb), 2.26 (m, 1H, 15'-CHa), 2.20 (m, 1H, 11-CHb), 2.13 (m, 1H, 16-CHb), 2.10 (m, 1H, 11-CHa), 2.05 (m, 2H, 4'-CHb and 26'-CHb), 2.02 (dd, 1H, J = 4.1, 12.0 Hz, 5-CH), 2.00 (m, 1H, 12-CHb), 1.97 (m, 1H, 1-CHb and 12'-CHb), 1.94 (m, 1H, 14-CH), 1.93 (m, 2H, 23-CHb and 26'-CHa), 1.92 (m, 1H,

7'-CHa), 1.87 (m, 1H, 9'-CH), 1.85 (dd, 1H, $J = 3.0$, 14.6 Hz, 1-CHa), 1.83 (m, 1H, 16'-CHb), 1.81 (m, 1H, 24-CH), 1.79 (m, 1H, 23'-CHb), 1.76 (m, 1H, 27'-CHb), 1.69 (m, 1H, 11'-CHb), 1.68 (m, 1H, 20'-CH), 1.63 (m, 1H, 1'-CHa), 1.62 (m, 2H, 15-CHb and 27'-CHa), 1.61 (m, 1H, 5'-CH), 1.60 (m, 1H, 4'-CHa), 1.56 (m, 1H, 11'-CHa), 1.41 (m, 1H, 16'-CHa), 1.39 (m, 1H, 23'-CHa), 1.38 (m, 1H, 15-CHa), 1.37 (m, 1H, 17-CH), 1.36 (m, 1H, 12-CHa), 1.29 (m, 2H, 16-CHa and 17'-CH), 1.28 (d, 3H, $J = 6.6$ Hz, 28-Me), 1.27 (s, 3H, 19-Me), 1.24 (m, 1H, 20-CH), 1.16 (m, 1H, 23-CHa), 1.14 (s, 3H, 27-Me), 1.13 (m, 1H, 12'-CHa), 1.03 (s, 3H, 26-Me), 1.00 (d, 3H, $J = 6.6$ Hz, 21'-Me),0.95 (d, 3H, $J = 6.7$ Hz, 21-Me), 0.91 (s, 3H, 19'-Me), 0.86 (s, 3H, 18'-Me), 0.64 (s, 3H, 18-Me). ^{13}C NMR (150 MHz, CD_3OD): δ 42.3 (C-1), 41.8 (C-1'), 72.8 (C-2), 79.4 (C-2'), 73.4 (C-3), 71.8 (C-3'), 34.6 (C-4), 28.5 (C-4'), 48.7 (C-5), 50.5 (C-5'), 75.1 (C-6), 78.7 (C-6'), 37.2 (C-7), 37.8 (C-7'), 126.8 (C-8), 125.9 (C-8'), 138.1 (C-9), 51.0 (C-9'), 38.8 (C-10), 38.2 (C-10'), 23.0 (C-11), 20.8 (C-11'), 38.2 (C-12), 39.3 (C-12'), 43.2 (C-13), 43.7 (C-13'), 53.0 (C-14), 146.1 (C-14'), 25.2 (C-15), 27.0 (C-15'), 30.8 (C-16), 28.5 (C-16'), 52.4 (C-17), 56.1 (C-17'), 11.3 (C-18), 18.5 (C-18'), 23.4 (C-19), 16.5 (C-19'), 44.1 (C-20), 34.9 (C-20'), 13.4 (C-21), 19.7 (C-21'), 72.0 (C-22), 37.1 (C-22'), 35.0 (C-23), 25.4 (C-23'), 46.3 (C-24), 87.1 (C-24'), 41.5 (C-25), 85.1 (C-25'), 27.3 (C-26), 37.8 (C-26'), 35.5 (C-27), 57.1 (C-27'), 10.4 (C-28). HRESMS (MALDI–TOF): m/z 1159.4764 calculated for $C_{55}H_{85}Na_2O_{17}S_3$, found: 1159.4744 (Morinaka *et al.*, 2009).

Amaroxocane B (**67**): IR (ZnSe): ν_{max} cm^{-1} 3480br (O–H), 2945m (C–H), 1652m (C=C), 1456m (C–H), 1376m (C–H), 1220m (C–O), 1062m (C–O), 968m, 913m. UV (MeOH): λ_{max} nm 218. ^{1}H NMR (600 MHz, CD_3OD): δ 4.74 (br q, 1H, $J = 3.1$ Hz, 2'-CH), 4.72 (br q, 1H, $J = 3.1$ Hz, 2-CH), 4.48 (m, 2H, 6-CH and 6'-CH), 4.01 (dd, 1H, $J = 4.2$, 11.5 Hz, 3-CH), 3.94 (m, 1H, 3'-CH), 3.67 (dd, 1H, $J = 4.3$, 11.4 Hz, 22-CH), 3.16 (d, 1H, $J = 6.6$ Hz, 24'-CH), 2.70 (m, 2H, 7-CHb and 7'-CHb), 2.51 (m, 1H, 4-CHb), 2.45 (dd, 1H, $J = 3.1$, 14.6 Hz, 1-CHb), 2.36 (dd, 1H, $J = 3.1$, 14.6 Hz, 1'-CHb), 2.26 (m, 2H, 14-CH and 26'-CHb), 2.23 (m, 1H, 14'-CH), 2.22 (m, 2H, 11-CHb and 11'-CHb), 2.17 (m, 2H, 11-CHa and 11'-CHa), 2.16 (m, 1H, 4'-CHb), 2.15 (m, 2H, 16-CHb and 24-CH), 2.07 (m, 1H, 7'-CHa), 2.05 (m, 1H, 26'-CH), 2.04 (m, 2H, 12-CHb and 12'-CHb), 2.03 (m, 2H, 23-CHb and 22'-CHb), 2.00 (m, 1H, 7-CHa), 1.90 (m, 1H, 16'-CHb), 1.75 (td, 1H, $J = 2.2$, 12.4 Hz, 5'-CH), 1.69 (m, 1H, 1'-CHa), 1.68 (m, 1H, 5-CH), 1.67 (m, 1H, 15'-CHb), 1.65 (m, 1H, 20'-CH), 1.64 (m, 1H, 4-CHa), 1.61 (m, 1H, 4'-CHa), 1.58 (m, 1H, 15-CHb), 1.51 (m, 3H, 17-CH, 22'-CHa and 23'-CHb), 1.46 (m, 1H, 1-CHa), 1.42 (m, 1H, 23'-CHa), 1.40 (m, 2H, 15'-CHa and 16'-CHa), 1.37 (m, 2H, 12-CHa and 12'-CHa), 1.36 (m, 1H, 15-CHa), 1.33 (m, 1H, 20-CH), 1.32 (s, 3H, 26-Me), 1.31 (m, 1H, 16-CHa), 1.29 (m, 1H, 17'-CH), 1.27 (m, 1H, 23-CHa), 1.24 (s, 3H, 19-Me), 1.23 (s, 3H, 27-Me), 1.22 (s, 3H, 19'-Me), 1.15 (s, 3H, 27'-Me), 0.99 (d, 3H, $J = 6.6$ Hz, 21-Me), 0.98 (d, 3H, $J = 7.4$ Hz, 21'-Me), 0.68 (s, 3H, 18'-Me), 0.65 (s, 3H, 18-Me). ^{13}C NMR (150 MHz, CD_3OD): δ 40.8 (C-1), 40.3 (C-1'), 79.1 (C-2), 79.4 (C-2'), 71.0 (C-3), 71.4 (C-3'), 28.0 (C-4 and C-4'), 47.6 (C-5), 46.9 (C-5'), 76.2 (C-6 and C-6'), 36.7 (C-7 and C-7'), 137.5 (C-9), 137.2 (C-9'), 39.2 (C-10), 39.0 (C-10'), 23.5 (C-11 and C-23'), 38.3 (C-12 and C-12'), 42.6 (C-13), 42.4 (C-13'), 52.6 (C-14), 53.4 (C-14'), 24.5 (C-15), 25.0 (C-15'), 29.4 (C-16), 30.6 (C-16'), 52.7 (C-17), 53.9 (C-17'), 11.1 (C-18), 10.8 (C-18'), 22.1 (C-19), 22.1 (C-19'), 43.7 (C-20), 36.0 (C-20'), 12.9 (C-21), 19.2 (C-21'), 71.7 (C-22), 36.9 (C-22'), 33.8 (C-23), 25.4 (C-23'), 45.5 (C-24), 86.2 (C-24'), 85.5 (C-25), 88.0 (C-25'), 25.4 (C-26), 36.2 (C-26'), 32.5 (C-27), 27.0 (C-27'). HRESIMS (MALDI–TOF): m/z 1247.3965 calculated for $C_{54}H_{82}Na_3O_{20}S_4$, found: 1247.3975 (Morinaka *et al.*, 2009).

Whitson *et al.* (2009) reported the isolation and identification of three PKCζ inhibitors, fibrosterol sulphates A–C (**68–70**), from the sponge *Lissodendoryx* (*Acanthodoryx*) *fibrosa*, collected in the Philippines.

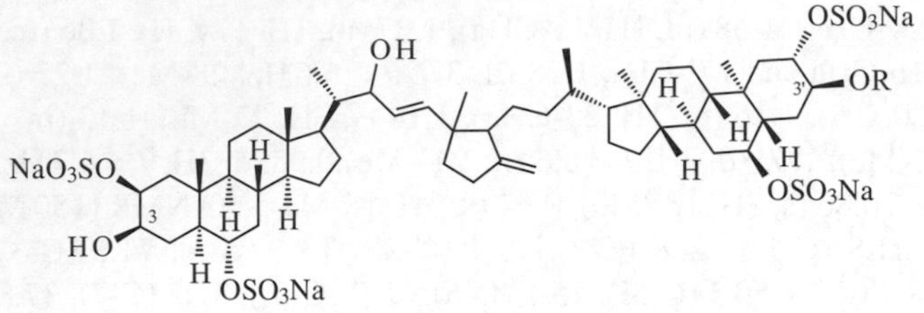

Fibrosterol A (**68**)	R = H	Fibrosterol B (**69**)	SO_3Na

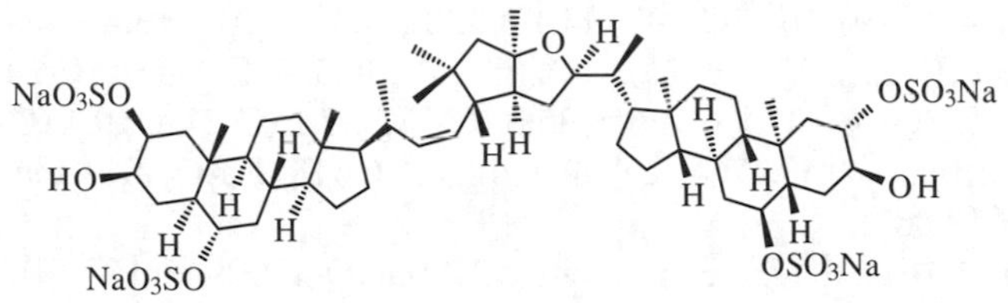

Fibrosterol C (**70**)

The sponge sample was exhaustively extracted with MeOH and the crude extract was subjected to CC using the HP20SS resin eluting with a step gradient of H_2O to *i*-PrOH (25% steps, 5 fractions) (Whitson *et al.*, 2009). Bioassay-guided fractionation of the second (mobile phase: H_2O: *i*-PrOH = 75:25) and the third fractions (mobile phase: H_2O: *i*-PrOH = 50:50), using a combination of RP-CC and RP-HPLC, afforded fibrosterol sulphates A–C (**68–70**), which appear to be composed of two cholestene monomers, with differing configuration at C-3, and oxygenation at C-22 in only one monomer. The structures of these dimers **68–70** were elucidated by spectroscopic means, particularly, by the use of extensive 1D and 2D NMR techniques.

Fibrosterol sulphate A (**68**, 6.4 mg). IR (film, NaCl): ν_{max} cm^{-1} 337br (O–H), 1660m (C=C), 1641m (C=C), 1444m (C–H), 1221m (C–O), 1063m (C–O), 966m, 708w. UV (MeOH): λ_{max} nm 206. ^1H NMR (600 MHz, CD_3OD): δ 5.54 (d, 1H, J = 15.7 Hz, 24-CH), 5.35 (dd, 1H, J = 8.7, 15.7 Hz, 23-CH), 4.85 (m, 1H, 26'-CHa), 4.81 (m, 1H, 26'-CHb), 4.64 (m, 1H, 2-CH), 4.46 (m, 1H, 2'-CH), 4.19 (m, 2H, 6-CH and 6'-CH), 4.06 (m, 1H, 3'-CH), 3.98 (dd, 1H, J = 3.4, 8.7 Hz, 22-CH), 3.58 (ddd, 1H, J = 4.0, 4.0, 11.8 Hz, 3-CH), 2.43 (m, 1H, 27'-CH), 2.42 (m, 1H, 1β-CH), 2.36 (m, 2H, 7β-CH and 7β'-CH), 2.10 (m, 1H, 4α-CH, 2.06 (m, 1H, 1β'-CH), 2.04 (m, 1H, 4α'-CH), 2.01 (m, 1H, 12β-CH), 2.02 (m, 1H, 12β'-CH), 1.87 (m, 1H, 16β'-CH),1.84 (m, 1H, 24'-CH), 1.74 (m, 1H, 16β-CH), 1.72 (m, 1H, 4β'-CH), 1.69 (m, 1H, 22'-CHa), 1.68 (m, 1H, 27-CHa), 1.66 (m, 1H, 5'-CH), 1.69 (m, 1H, 20-CH), 1.62 (m, 2H, 4β-CH and 15β'-CH), 1.578 (m, 1H, 27-CHb), 1.56 (m, 2H, 15β-CH and 11α'-CH), 1.55 (m, 1H, 11α-CH), 1.53 (m, 2H, 8-CH and 8'-CH), 1.47 (dd, 1H, J = 3.5, 14.8 Hz, 1α'-CH), 1.42 (m, 2H, 20'-CH and 23'-CHa), 1.38 (m, 1H, 11β-CH), 1.33 (m, 1H, 11β'-CH), 1.30 (m, 1H, 16α-CH and 16'α-CH), 1.19 (m, 1H, 5-CH), 1.18 (m, 1H, 12α'-CH),

1.16 (m, 1H, 12α-CH), 1.15 (m, 1H, 17'-CH), 1.15 (s, 3H, 26-Me), 1.14 (m, 1H, 1α-CH), 1.12 (m, 3H, 14'-CH, 15α-CH and 15α'-CH), 1.08 (s, 3H, 19-Me), 1.07 (m, 2H, 14-CH and 23'-CHb), 1.06 (m, 1H, 7α'-CH, 1.05 (s, 3H, 19'-Me), 1.03 (m, 1H, 17-CH), 0.99 (m, 1H, 22'-CHb), 0.99 (d, 3H, $J = 6.5$ Hz, 21'-Me), 0.98 (m, 1H, 7α-CH), 0.97 (d, 3H, $J = 6.6$ Hz, 21-Me), 0.75 (m, 1H, 9'-CH), 0.71 (s, 3H, 18-Me), 0.69 (m, 1H, 9-CH), 0.72 (s, 3H, 18'-Me). ^{13}C NMR (150 MHz, CD_3OD): δ 42.8 (C-1), 38.8 (C-1'), 78.2 (C-2), 77.8 (C-2'), 72.2 (C-3), 68.4 (C-3'), 27.7 (C-4), 26.3 (C-4'), 51.7 (C-5), 44.5 (C-5'), 78.0 (C-6), 78.6 (C-6'), 39.8 (C-7 and C-7'), 35.0 (C-8 and C-8'), 55.7 (C-9 and C-9'), 37.8 (C-10), 38.1 (C-10'), 22.1 (C-11), 21.7 (C-11'), 41.0 (C-12), 41.3 (C-12'), 43.9 (C-13), 43.8 (C-13'), 56.9 (C-14), 57.4 (C-14'), 25.0 (C-15 and C-15'), 28.1 (C-16), 29.1 (C-16'), 54.3 (C-17), 57.3 (C-17'), 12.4 (C-18), 13.1 (C-18'), 15.6 (C-19), 15.1 (C-19'), 43.1 (C-20), 37.7 (C-20'), 12.7 (C-21), 19.8 (C-21'), 76.3 (C-22), 36.9 (C-22'), 126.2 (C-23), 26.6 (C-23'), 138.6 (C-24), 56.8 (C-24'), 47.8 (C-25), 157.5 (C-25'), 24.0 (C-26), 105.1 (C-26'), 38.5 (C-27), 29.8 (C-27'). HRESIMS: *m/z* 605.1480 calculated for $C_{54}H_{84}Na_2O_{19}S_4$, found: 605.1459 (Whitson *et al.*, 2009).

Fibrosterol sulphate B (**69**, 8.0 mg). IR (film): ν_{max} cm^{-1} 3224br (O–H), 1662m (C=C), 1639m (C=C), 1444m (C–H), 1221m (C–O), 968m. UV (MeOH): λ_{max} nm 206. ^{1}H NMR (600 MHz, CD_3OD): δ 5.57 (d, 1H, $J = 15.7$ Hz, 24-CH), 5.37 (dd, 1H, $J = 8.7, 15.7$ Hz, 23-CH), 4.87 (m, 1H, 26'-CHa), 4.83 (m, 1H, 2'-CH), 4.82 (m, 1H, 26'-CHb), 4.76 (m, 1H, 3'-CH), 4.67 (m, 1H, 2-CH), 4.20 (m, 2H, 6-CH and 6'-CH), 3.99 (dd, 1H, $J = 3.4, 8.7$ Hz, 22-CH), 3.59 (ddd, 1H, $J = 4.0, 4.0, 11.8$ Hz, 3-CH), 2.44 (m, 1H, 1β-CH), 2.43 (m, 1H, 27'-CH), 2.37 (m, 2H, 7β-CH and 7β'-CH), 2.28 (m, 1H, 4β'-CH), 2.12 (m, 1H, 4β-CH), 2.06 (m, 1H, 1β'-CH), 2.04 (m, 1H, 12β'-CH), 2.02 (m, 1H, 12β-CH), 1.88 (m, 1H, 16β'-CH), 1.85 (m, 1H, 24'-CH), 1.80 (m, 1H, 4α'-CH), 1.74 (m, 1H, 16β-CH), 1.71 (m, 1H, 22'-CHa), 1.70 (m, 1H, 20-CH), 1.68 (m, 1H, 27-CHa), 1.64 (m, 1H, 15β'-CH, 1.63 (m, 1H, 4α-CH), 1.62 (m, 1H, 5'-CH), 1.58 (m, 1H, 15β-CH), 1.57 (m, 1H, 27-CHb), 1.55 (m, 2H, 11β-CH and 11β'-CH), 1.53 (m, 2H, 8-CH and 8'-CH), 1.47 (dd, 1H, $J = 3.5, 14.8$ Hz, 1α'-CH, 1.43 (m, 1H, 23'-CHa), 1.42 (m, 1H, 20'-CH), 1.38 (m, 1H, 11α-CH), 1.36 (m, 1H, 11α'-CH), 1.31 (m, 2H, 16α-CH and 16'α-CH), 1.19 (m, 1H, 5-CH and 12α'-CH), 1.16 (m, 2H, 1α-CH and 12α-CH), 1.16 (s, 3H, 26-Me), 1.15 (m, 1H, 14'-CH), 1.13 (m, 3H, 15α-CH, 15α'-CH and 17'-CH), 1.08 (m, 2H, 14-CH and 23'-CHb), 1.08 (s, 3H, 19-Me), 1.05 (m, 2H, 17-CH and 7α'-CH), 1.05 (s, 3H, 19'-Me), 0.99 (m, 1H, 22'-CHb), 0.99 (d, 3H, $J = 7.0$ Hz, 21'-Me), 0.98 (m, 1H, 7α-CH), 0.98 (d, 3H, $J = 6.8$ Hz, 21-Me), 0.74 (m, 1H, 9'-CH), 0.73 (s, 3H, 18'-Me), 0.71 (s, 3H, 18-Me), 0.68 (m, 1H, 9-CH). ^{13}C NMR (150 MHz, CD_3OD): δ 43.0 (C-1), 39.5 (C-1'), 78.3 (C-2), 75.4 (C-2'), 72.3 (C-3), 75.4 (C-3'), 27.9 (C-4), 24.9 (C-4'), 51.7 (C-5), 45.2 (C-5'), 78.4 (C-6 and C-6'), 40.0 (C-7 and C-7'), 35.0 (C-8 and C-8'), 55.9 (C-9), 55.8 (C-9'), 38.0 (C-10), 37.5 (C-10'), 22.2 (C-11), 22.1 (C-11'), 41.1 (C-12), 41.3 (C-12'), 44.0 (C-13), 43.9 (C-13'), 57.0 (C-14), 57.3 (C-14'), 25.2 (C-15), 25.1 (C-15'), 28.2 (C-16), 29.3 (C-16'), 54.5 (C-17), 57.5 (C-17'), 12.7 (C-18), 13.2 (C-18'), 15.8 (C-19), 15.2 (C-19'), 43.1 (C-20), 37.9 (C-20'), 13.0 (C-21), 19.8 (C-21'), 76.5 (C-22), 36.9 (C-22'), 126.2 (C-23), 26.7 (C-23'), 138.9 (C-24), 56.9 (C-24'), 47.9 (C-25), 157.5 (C-25'), 24.2 (C-26), 105.3 (C-26'), 38.8 (C-27), 30.0 (C-27'). HRESIMS: *m/z* 656.18418 calculated for $C_{54}H_{83}Na_3O_{22}S_5$, found: 656.18432 (Whitson *et al.*, 2009).

Fibrosterol sulphate C (**70**, 0.3 mg). IR (film): ν_{max} cm^{-1} 3346br (O–H), 2953m (C–H), 1662m (C=C), 1641m (C=C), 1446m (C–H), 1219m (C–O), 1063m (C–O), 962m, 708w. UV (MeOH): λ_{max} nm 204. ^{1}H NMR (600 MHz, CD_3OD): δ 5.29 (dd, $J = 10.8$ Hz, H-22),

5.02 (dd, 1H, $J = 10.8$ Hz, 23-CH), 4.76 (m, 1H, 2-CH), 4.42 (m, 1H, 2'-CH), 4.19 (m, 2H, 6-CH and 6'-CH), 4.16 (m, 1H, 22'-CH), 4.08 (m, 1H, 3'-CH), 3.71 (ddd, 1H, $J = 4.0$, 4.0, 11.8 Hz, 3-CH), 2.45 (m, 1H, 20-CH), 2.39 (m, 1H, 24-CH), 2.34 (m, 1H, 7β'-CH), 2.38 (m, 1H, 7β-CH), 2.24 (ddd, 1H, $J = 3.3$, 10.1 Hz, 24'-CH), 2.08 (m, 1H, 4β-CH), 2.07 (m, 1H, 1β'-CH), 2.04 (m, 1H, 4β'-CH), 2.02 (m, 1H, 12β'-CH), 2.01 (m, 1H, 12β-CH), 1.88 (d, 1H, $J = 13.9$ Hz, 27β'-CH), 1.84 (m, 2H, 20'-CH and 23α'-CH), 1.76 (m, 2H, 4α'-CH and 16β'-CH), 1.72 (m, 1H, 16β-CH), 1.66 (d, 1H, $J = 13.9$ Hz, 27α'-CH), 1.65 (m, 1H, 5'-CH), 1.63 (m, 2H, 15β-CH and 15β'-CH), 1.59 (m, 1H, 4α-CH), 1.58 (m, 1H, 11β'-CH), 1.56 (m, 1H, 11β-CH), 1.50 (m, 2H, 8-CH and 8'-CH), 1.47 (m, 1H, 23β'-CH), 1.46 (m, 1H, 1α'-CH), 1.41 (m, 1H, 1α-CH), 1.39 (m, 1H, 16α'-CH), 1.35 (m, 1H, 11α'-CH), 1.33 (m, 1H, 5-CH and 11α-CH), 1.31 (m, 3H, 26'-Me), 1.29 (m, 1H, 16α-CH), 1.22 (m, 1H, 17-CH), 1.20 (m, 1H, 12α-CH), 1.13 (m, 1H, 12α'-CH), 1.14 (m, 1H, 14-CH), 1.11 (m, 2H, 15α-CH and 15α'-CH), 1.07 (s, 3H, 19-Me), 1.05 (s, 3H, 19'-Me), 1.04 (m, 3H, 7α-CH, 7α'-CH and 14'-CH), 1.00 (d, 6H, $J = 6.8$ Hz, 21-Me and 21'-Me), 1.00 (m, 1H, 17'-CH), 0.94 (s, 3H, 26-Me), 0.88 (s, 3H, 27-Me), 0.75 (m, 1H, 9'-CH), 0.74 (m, 1H, 9-CH), 0.73 (s, 3H, 18-Me), 0.71 (s, 3H, 18'-Me). ^{13}C NMR (150 MHz, CD_3OD): δ 42.4 (C-1), 38.8 (C-1'), 79.1 (C-2), 77.8 (C-2'), 71.9 (C-3), 68.4 (C-3'), 27.9 (C-4), 26.5 (C-4'), 51.0 (C-5), 44.6 (C-5'), 78.4 (C-6 and C-6'), 39.9 (C-7), 39.7 (C-7'), 35.0 (C-8 and C-8'), 55.3 (C-9), 55.7 (C-9'), 37.7 (C-10), 37.5 (C-10'), 21.7 (C-11), 22.0 (C-11'), 41.0 (C-12), 41.4 (C-12'), 43.4 (C-13), 43.9 (C-13'), 57.6 (C-14), 56.9 (C-14'), 25.1 (C-15 and C-15'), 29.8 (C-16), 28.2 (C-16'), 57.5 (C-17), 55.2 (C-17'), 12.9 (C-18), 12.3 (C-18'), 15.8 (C-19), 15.1 (C-19'), 35.8 (C-20), 38.6 (C-20'), 21.6 (C-21), 13.1 (C-21'), 140.1 (C-22), 82.4 (C-22'), 126.7 (C-23), 30.1 (C-23'), 56.6 (C-24), 56.4 (C-24'), 46.4 (C-25), 91.0 (C-25'), 29.3 (C-26), 26.2 (C-26'), 23.2 (C-27), 56.2 (C-27'). HRESIMS: *m/z* 1233.4183 calculated for $C_{54}H_{84}Na_3O_{19}S_4$, found: 1233.4167 (Whitson *et al.*, 2009).

References

Banerji, J. and Chatterjee, A. (1973). New steroid alkaloids from *Chonemorpha macrophylla*. *Indian Journal of Chemistry* **11**, 1056–1057.

D'Auria, M. V., Giannini, C., Zampella, A., Minale, L., Debitus, C. and Roussakis, C. (1998). Crellastatin A: a cytotoxic bis-steroid sulfate from the Vanuatu marine sponge *Crella* sp. *Journal of Organic Chemistry* **63**, 7382–7388.

Cheng, J-F., Lee, J-S., Sun, F., Jares-Erijman, E. A., Cross, S. and Rinehart, K. L. (2007). Hamigerols A and B, unprecedented polysulfate sterol dimers from the Mediterranean Sponge *Hamigera hamigera*. *Journal of Natural Porducts* **70**, 1195–1199.

Fuzukawa, S., Matsunaga, S. and Fusetani, N. (1994). Ritterazine A, a highly cytotoxic dimeric steroidal alkaloid, from the Tunicate *Ritterella tokioka*. *Journal of Organic Chemistry* **59**, 6164–6166.

Fuzukawa, S., Matsunaga, S. and Fusetani, N. (1995a). Isolation and structure elucidation of ritterazines B and C, highly cytotoxic dimeric steroidal alkaloids, from the tunicate *Ritterella tokioka*. *Journal of Organic Chemistry* **60**, 608–614.

Fuzukawa, S., Matsunaga, S. and Fusetani, N. (1995b). Ten more ritterazines, cytotoxic steroidal alkaloids from the turnicate *Ritterella tokioka*. *Tetrahedron* **51**, 6707–6716.

Fuzukawa, S., Matsunaga, S. and Fusetani, N. (1997). Isolation of 13 new ritterazines from the tunicate *Ritterella tokioka* and chemical transformation of ritterazine B. *Journal of Organic Chemistry* **62**, 4484–4491.

Giannini, C., Zampella, A., Debitus, C., Menou, J. L., Roussakis, C. and D'Auria, M. V. (1999). Isolation and structural elucidation of the crellastatins I-M: cytotoxic bis-steroid derivatives from the Vanuatu marine sponge *Crella* sp. *Tetrahedron* **55**, 13749–13756.

Inouye, Y., Sugo, Y., Kusumi, T. and Fusetani, N. (1994). Structure and absolute stereochemistry of bisconicasterone from the marine sponge *Theonella swinhoei*. *Chemistry Letters*, 419–420.

Kobayashi, M., Kawazoe, K., Katori, T. and Kitagawa, I. (1992). Marine natural products.30. 2 new 3-keto-4-methylene steroids, theonellasterone and conicasterone, and a Diels–Alder type dimeric steroid bistheonellasterone, from the Okinawan marine sponge *Theonella swinhoei*. *Chemical and Pharmaceutical Bulletin* **40**, 1773–1778.

Kupchan, S. M., Britton, R. W., Ziegler, M. F. and Sigel, C. W. (1973). Tumor inhibitors.82. bruceantin, a new potent antileukemic simaroubolide from *Brucea antidysenterica*. *Journal of Organic Chemistry* **38**, 178–179.

Li, Y. X. and Dias, J. R. (1997). Dimeric and oligomeric steroids. *Chemical Review* **97**, 283–304.

Morinaka, B. I., Pawlik, J. R. and Molinski, T. F. (2009). Amaroxocanes A and B: Sulfated Dimeric Sterols Defend the Caribbean Coral Reef Sponge *Phorbas amaranthus* from Fish Predators. *Journal of Natural Products* **72**, 259–264.

Moser, B. R. (2008). Review of cytotoxic cephalostatins and ritterazines: isolation and synthesis. *Journal of Natural Products* **71**, 487–491.

Murayama, S. (2008). *PhD thesis: Studies on inhibitors of enzymes involved in tumor metastasis from marine invertebrates*, University of Tokyo, Japan.

Nahar, L., Sarker, S. D. and Turner, A. B. (2007). A review on synthetic and natural steroid dimers: 1997-2006. *Current Medicinal Chemistry* **14**, 1349–1370.

Pettit, G. R., Inoue, M., Kamano, Y., Herald, D. L., Arm, C., Dufresne, C., Christie, N. D., Schmidt, J. M., Doubek, D. L. and Krupa, T. S. (1988a). Isolation and structure of powerful cell growth inhibitor cephalostatin 1. *Journal of the Americal Chemical Society* **110**, 2006–2007.

Pettit, G. R., Inoue, M., Kamano, Y., Dufresne, C., Christie, N., Niven, M. L. and Herald, D. L. (1988b). Isolation and structure of the hemichordate cell growth inhibitors cephalostatins 2, 3, and 4. *Journal of the Chemical Society Chemical Communications* 865–867.

Pettit, G. R., Kamano, Y., Dufresne, C., Inoue, M., Christie, N., Schmidt, J. M. and Doubek, D. L. (1989). Isolation and structure of the unusual Indian Ocean *Cephalodiscus gilchristi* components, cephalostatins 5 and 6. *Canadian Journal of Chemistry* **67**, 1509–1513.

Pettit, G. R., Kamano, Y., Inoue, M., Dufresne, C., Boyd, M. R., Herald, C. L., Schmidt, J. M., Doubek, D. L. and Christie, N. (1992). Antineoplastic agents. 214. isolation and structure of cephalostatins 7-9. *Journal of Organic Chemistry* **57**, 429–431.

Pettit, G. R., Xu, J-P., Williams, M. D., Christie, N. D., Doubek, D. L. and Schmidt, J. M. (1994a). Isolation and structure of cephalostatins 10 an 11. *Journal of Natural Products* **57**, 52–63.

Pettit, G. R., Ichihara, Y., Xu, J-P., Boyd, M. R. and Williams, M. D. (1994b). Isolation and structure of the symmetrical disteroidal alkaloids cephalostatins 12 an 13. *Bioorganic and Medicinal Chemistry Letters* **4**, 1507–1512.

Pettit, G. R., Xu, J-P., Ichihara, Y., Williams, M. D. and Boyd, M. R. (1994c). Antineoplastic agents.285. isolation and structures of cephalostatin -14 and cephalostatin-15. *Canadian Journal of Chemistry-Revue Canadienne de Chimie* **72**, 2260–2267.

Pettit, G. R., Xu, J-P. and Schmidt, J. M. (1995). Isolation and structure of the exceptional pterobranchia human canver inhibitors cephalostatins 16 and 17. *Bioorganic and Medicinal Chemistry* **5**, 2027–2032.

Pettit, G. R., Tan, R., Xu, J-P., Ichihara, Y., Williams, M. D. and Boyd, M. R. (1998). Antineoplastic agents. 398. Isolation and structure elucidation of cephalostatins 18 and 19. *Journal of Natural Products* **61**, 955–958.

Sugo, Y., Inouye, Y. and Nakayama, N. (1995). Structures of 9 oxygenated 4-methylene sterols from Hachijo marine sponge *Theonella swinhoei*. *Steroids* **60**, 738–742.

Whitson, E. L., Bugni, T. S., Chockalingam, P. S., Concepcion, G. P., Feng, X., Jin, G., Harper, M. K., Mangalindan, G. C., McDonald, L. A. and Ireland, C. M. (2009). Fibrosterol Sulfates from the Philippine Sponge *Lissodendoryx (Acanthodoryx) fibrosa*: Sterol Dimers that Inhibit PKCζ. *Journal of Organic Chemistry* **74**, 5902–5908.

Zampella, A., Giannini, C., Debitus, C., Roussakis, C. and D'Auria, M. V. (1999). Isolation and structural elucidation of crellastatins B-H: cytotoxic bis(steroid) derivatives from the Vanuatu marine sponge *Crella* sp. *European Journal of Organic Chemistry* 949–953.

5

Synthesis of Cephalostatin and Ritterazine Analogues

5.1 Introduction

Following the discovery of cytotoxic naturally occurring dimeric steroids, cephalostatins and ritterazines, from marine sponges, *Cephalodiscus gilchristi* (family: Cephalodiscidae) and *Ritterella tokioka* (family: Polyclinidae), respectively (see *Chapter 4* for further details), a significant body of research has been carried out in this area, resulting in the successful protocols for the total synthesis of some of these compounds as well as for the synthesis of a number of their analogues (Nahar *et al.*, 2007). The quest for developing successful synthetic methodologies for these compounds was primarily driven by the fact that potent cytotoxic group of compounds cephalostatins [*e.g.*, cephalostatin 1 (**1**)] and ritterazines [*e.g.*, ritterazine A, (**2**)] were only found in two specific sponge species, and the yields were generally extremely poor. For example, only 139 mg of cephalostatin 1 (**1**) could be isolated from 166 kg of crude marine worms (*C. gilchristi*).

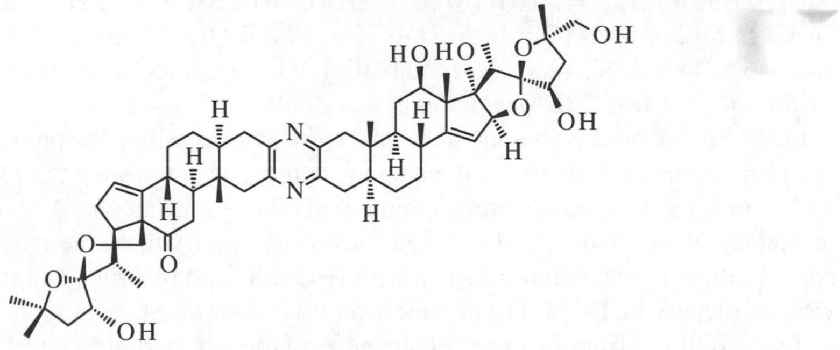

Cephalostatin 1 (**1**)

Steroid Dimers: Chemistry and Applications in Drug Design and Delivery, First Edition. Lutfun Nahar and Satyajit D. Sarker. © 2012 John Wiley & Sons, Ltd. Published 2012 by John Wiley & Sons, Ltd.

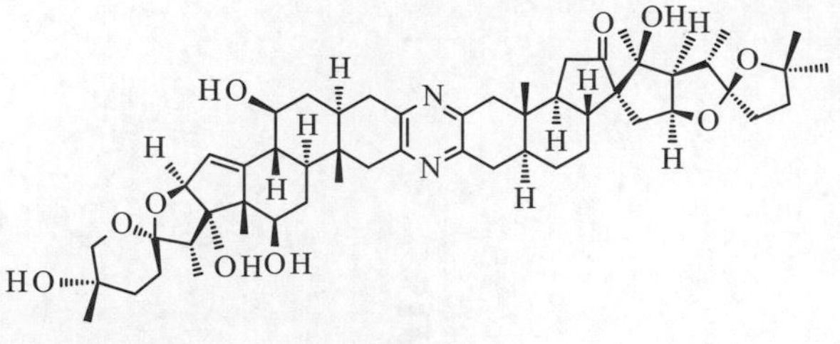

Ritterazine A (**2**)

It is noteworthy that pyrazine-linked steroid dimers were synthesized even before the discovery of natural cephalostatins (Li and Dias, 1997), but the major cephalostatin analogues were prepared after the discovery.

5.2 Synthesis of Cephalostatin and Ritterazine Analogues

The synthesis of several cephalostatin analogues, *e.g.*, **9–14** and **23**, was accomplished from (25*R*)-3β-acetyloxy-5α-spirostan-12-one (**3**, hecogenin acetate) in a multistep sequence (Drögemüller *et al.*, 1998). Initially, the synthesis of (25*R*)-3β-acetyloxy-5α-spirostan-12α-ol (**4**) was carried out from a cooled (0 °C) solution of **3** (1.0 g) in dry toluene (6 mL) and $BF_3.Et_2O$ (310 μL, 2.54 mmol) complex in dry toluene (15 mL) by stirring the mixture for 25 min maintaining the same temperature (*Scheme 5.2.1*). After 40 min, 0.5M $NaHCO_3$ (10 mL, 5 mmol) was added and the mixture was extracted with EtOAc, the organic layer was washed with saturated aqueous NaCl. The organic layer was dried (Na_2SO_4), evaporated under pressure and crystallized from Et_2O-petroleum ether to afford homoallylic alcohol **4** (773 mg, 80%, mp: 224 °C). IR (KBr): ν_{max} cm^{-1} 3499br (O–H), 3060w (C–H), 2947s (C–H), 2873s (C–H), 1733s (C=O), 1650m (C=C), 1244s (C–O), 1063s (C–O). ^1H NMR (200 MHz, $CDC1_3$): δ 5.54 (br s, 1H), 4.87 (dd, 1H, J = 1.6, 8.0 Hz), 4.67 (m, 1H), 3.69 (m, 1H), 3.46 (m, 2H), 2.66 (dd, 1H, J = 8.0, 10.0 Hz), 2.15 (m, 1H), 2.02 (s, 3H), 1.11 (s, 3H), 1.01 (d, 3H, J = 7.0 Hz), 0.89 (s, 3H), 0.80 (d, 3H, J = 6.0 Hz). ^{13}C NMR (50 MHz, $CDC1_3$): δ 12.0, 14.2, 17.2, 18.6, 21.4, 27.3, 28.2, 28.6, 28.7, 29.5, 30.4, 31.1, 33.9, 34.4, 35.8, 36.4, 44.5, 44.7, 49.8, 52.31, 53.6, 67.1, 73.3, 76.2, 85.0, 106.4, 121.1, 153.9, 170.6. EIMS: *m/z* 472 $[M]^+$ (Drögemüller *et al.*, 1998).

(25*R*)-3β-Acetyloxy-5α-spirost-14-en-12α-ol (**4**) was utilized to prepare (25*R*)-5α-spirost-14-ene-3α,12α-diol (**5**) by alkaline hydrolysis (*Scheme 5.2.1*). KOH (2.51 g, 44.77 mmol) was added to a stirred solution of homoallylic alcohol **4** (7.05 g, 14.92 mmol) in MeOH-DCM (70 mL) and refluxed for 40 min, after which, the reaction mixture was cooled to room temperature, washed with H_2O followed by brine, and the aqueous layers were extracted with DCM. The organic layer was dried ($MgSO_4$), evaporated under vacuum and recrystallized from Et_2O-petroleum ether to furnish enediol **5** (mp: 175 °C). IR (KBr): ν_{max} cm^{-1} 3436br (O–H), 3058w (CH=CH), 2928s (C–H), 2860s (C–H), 1650w (C=C), 1460w (C–H), 1377m (C–H), 1243m (C–O). ^1H NMR (200 MHz, $CDCl_3$): δ 5.54 (br s, 1H, 15-CH), 4.87 (dd, 1H, J = 1.5, 8.0 Hz, 16α-CH), 3.70 (m, 1H, 12H), 3.58 (tt, 1H, J = 5.5, 10.5 Hz, 3α-CH), 3.49 (br, 1H, J = 4.5 Hz, 11.0 Hz, 26-CHb), 3.43 (t, 1H, J = 11.0 Hz, 26-CHa), 2.64 (dd, 1H, J = 8.0, 9.5 Hz, 17α-CH), 2.15 (m, 1H, 8-CH), 1.12 (s, 3H, 18-Me), 1.00

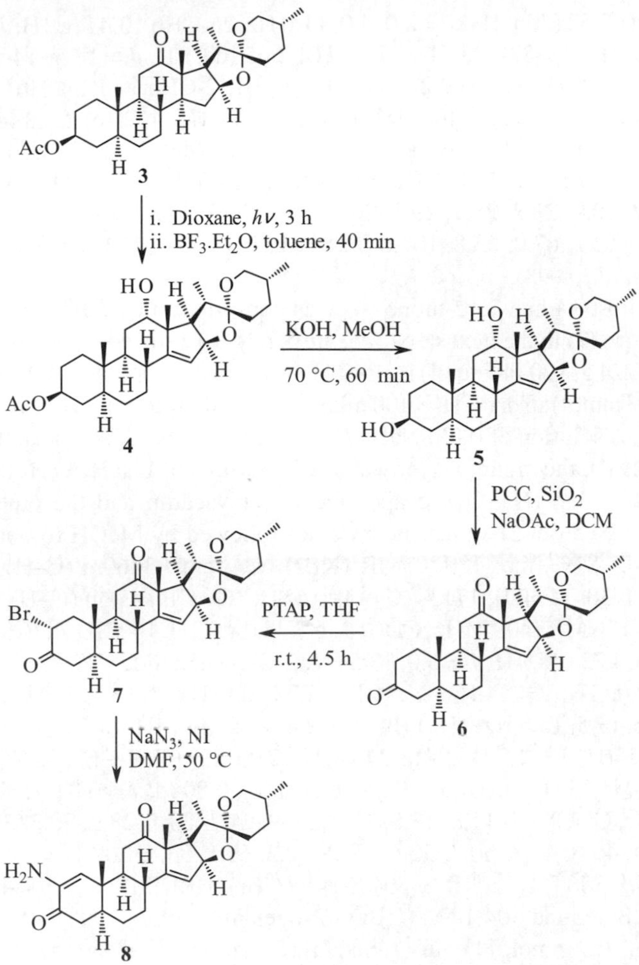

i. Dioxane, $h\nu$, 3 h
ii. $BF_3.Et_2O$, toluene, 40 min

KOH, MeOH
70 °C, 60 min

PCC, SiO_2
NaOAc, DCM

PTAP, THF
r.t., 4.5 h

NaN_3, NI
DMF, 50 °C

Scheme 5.2.1 *Synthesis of hecogenin derivatives **4–8***

(d, 3H, $J = 6.5$ Hz, 21-Me), 0.87 (s, 3H, 19-Me), 0.80 (d, 3H, $J = 6.0$ Hz, 27-Me). ^{13}C NMR (50 MHz, DMSO-d_6): δ 11.8, 14.1, 17.1, 18.3, 28.1, 28.4, 28.7, 29.4, 30.0, 30.8, 31.2, 33.9, 35.3, 36.3, 38.1, 44.1, 44.3, 48.9, 51.2, 52.8, 66.1, 69.3, 74.7, 85.0, 105.5, 119.9, 153.1. EIMS: *m*/*z* 430 $[M]^+$ (Drögemüller *et al.*, 1998).

(25*R*)-5α-Spirost-14-ene-3,12-dione (**6**) was synthesized from (25*R*)-5α-spirost-14-ene-3α,12α-diol (**5**) by PCC oxidation (*Scheme 5.2.1*). A mixture PCC (7.98 g, 37.03 mmol) and SiO_2 (8.0 g) in DCM (100 mL) was stirred vigorously at room temperature. After 30 min, NaOAc (200 mg, 2.44 mmol) and enediol **5** (7.39 g) were added and stirred for another 16 h. The mixture was filtered and the solvent was evaporated under vacuum. The resulting crude product was purified by FCC (mobile phase: *n*-hexane:EtOAc = 2:1) and crystallized from Et_2O to give enedione **6** (5.98 g, 95%, mp: 215 °C). IR (KBr): ν_{max} cm^{-1} 3070w (CH=CH), 2929s (C–H), 2872m (C–H), 1713s (C=O), 1645w (C=C), 1459m (C–H), 1376m (C–H), 1243m (C–O). ^1H NMR (400 MHz, $CDCl_3$): δ 5.47 (br s, 1H, 15-CH), 4.77 (dd, 1H, $J = 2.0$,

8.0 Hz, 16α-CH), 3.52 (ddd, 1H, $J=2.0, 4.0, 11.0$ Hz, 26-CHb), 3.41 (t, 1H, $J=11.0$ Hz, 26-CHa), 3.34 (dd, 1H, $J=8.0, 9.0$ Hz, 17α-CH), 2.65 (dd, 1H, $J=13.5, 14.5$ Hz, 11-CHb), 2.52 (t, 1H, $J=10.5$ Hz, 8-CH), 2.34–2.43 (m, 2H, 2-CHa and 2-CHb), 2.38 (dd, 1H, $J=5.0, 14.5$ Hz, 11-CHa), 2.28 (dd, 1H, $J=13.5, 15.0$ Hz, 4-CHb), 2.18 (dd, 1H, $J=4.0, 15.0$ Hz, 4a-CH), 2.03 (m, 1H, 7-CHb), 1.32 (s, 3H, 18-Me), 1.13 (s, 3H, 19-Me), 1.05 (d, 3H, $J=7.0$ Hz, 21-Me), 0.80 (d, 3H, $J=6.5$ Hz, 27-Me). ^{13}C NMR (100 MHz, $CDCl_3$): δ 11.0, 13.7, 17.1, 20.9, 28.2, 28.7, 29.1, 30.2, 31.2, 34.0, 36.3, 37.3, 37.8, 37.7, 44.1, 44.3, 45.7, 49.7, 53.0, 62.2, 67.0, 83.8, 107.0, 121.3, 154.1, 210.4, 210.6. EIMS: *m/z* 426 $[M]^+$ (Drögemüller *et al.*, 1998).

(25*R*)-5α-Spirost-14-ene-3,12-dione (**6**) was converted to (25*R*)-2α-bromo-5α-spirost-14-ene-3,12-dione (**7**) in the next step (*Scheme 5.2.1*). To a stirred cooled (0 °C) solution of enedione **6** (51.224 g, 120.08 mmol) in dry THF (800 mL), a cooled (0 °C) solution of PTAP (48.86 g, 126.08 mmol) in dry THF (400 mL) was added dropwise over 3 h, then stirred for another 40 min. A solution of 0.5M $NaHCO_3$ (1000 mL) was added, the aqueous layer was extracted with Et_2O, the organic layer was washed with 0.5M $NaHCO_3$ followed by brine. The solvent was dried ($NaSO_4$), evaporated under vacuum and the crude residue was recrystallized twice from Et_2O-petroleum ether followed by MeOH to obtain 2α-bromoketone **7** (49.442, 81%, mp: 189 °C). IR (KBr): ν_{max} cm^{-1} 3060m (C–H), 2928s (C–H), 2872m (C–H), 1730s (C=O), 1713s (C=O), 1651w (C=C), 1456m (C–H), 1376m (C–H), 1243m (C–O). ^{1}H NMR (400 MHz, $CDCl_3$): δ 5.48 (br s, 1H, 15-CH), 4.76 (dd, 1H, $J=2.0, 8.0$ Hz, 16-CH), 4.72 (dd, 1H, $J=6.0, 13.5$ Hz, 2-CH), 3.52 (ddd, 1H, $J=2.0, 4.0, 11.0$ Hz, 26b-CH), 3.40 (t, 1H, $J=11$ Hz, 26a-CH), 3.33 (dd, 1H, $J=8.0, 9.0$ Hz, 17-CH), 2.65 (dd, 1H, $J=9.5, 13.5, 14.5$ Hz, 1H, 11b-CH), 2.46–2.54 (m, 4H, 1a-CH, 4a-CH, 4b-CH and 8-CH), 2.38 (dd, 1H, $J=4.5, 14.5$ Hz, 11a-CH), 2.04 (m, 1H, 9-CH), 1.31 (s, 3H, 18-Me), 1.21 (s, 3H, 19-Me), 1.04 (d, $J=7.0$ Hz, 3H, 21-Me), 0.80 (d, $J=6.0$ Hz, 3H, 27-Me). ^{13}C NMR (100 MHz, $CDCl_3$): δ 11.8, 13.8, 17.1, 20.9, 27.7, 28.7, 28.9, 30.3, 31.2, 33.6, 37.3, 39.3, 43.6, 44.1, 46.6, 49.8, 50.7, 52.6, 53.2, 62.3, 67.1, 83.8, 107.1, 121.8, 153.5, 200.0, 211.5. EIMS: *m/z* 506 $[M]^+$ (^{81}Br); 504 $[M]^+$ (^{79}Br). HREIMS: *m/z* 504.1875 calculated for $C_{27}H_{38}O_4{}^{79}Br$, found 504.1872 (^{79}Br) (Drögemüller *et al.*, 1998).

NaN_3 (5.18 g, 79.7 mmol, 11 equiv.) and NaI (few mg) were added to a stirred solution of (25*R*)-5α-spirost-14-en-3,12-dione (**7**, 3.70 g, 7.32 mmol) in DMF (200 mL) at 50 °C and stirred under Ar (*Scheme 5.2.1*). After 60 min, H_2O (100 mL) was added and mixture was extracted with *n*-hexane-MTBE (1:2). The organic layer was washed with brine and dried (Na_2SO_4), evaporated to dryness and recrystallized from *n*-hexane-Et_2O to produce (25*R*)-2-enamino-5α-spirost-14-ene-3,12-dione (**8**, 2.91 g, 91%). IR (KBr): ν_{max} cm^{-1} 3452m (N–H), 3368m (N–H), 3060s (C–H), 1708s (C=O), 1676s (C=O), 1628m (C=C). UV (MeOH): λ_{max} nm 214, 290. ^{1}H NMR (400 MHz, $CDCl_3$): δ 5.86 (s, 1H, 15-CH), 5.41 (t, 1H, $J=2.0$ Hz, 16-CH), 4.71 (dd, 1H, $J=2.0, 8.0$ Hz, 2-CH), 3.34 (m, 4H, 1a-CH, 4a-CH, 4b-CH and 8-CH), 2.50 (m, 5H), 1.30 (s, 3H, 18-Me), 1.10 (s, 3H, 19-Me), 1.01 (d, 3H, $J=7.0$ Hz, 21-Me), 0.79 (d, 3H, $J=6.0$ Hz, 27-Me). ^{13}C NMR (100 MHz, $CDCl_3$): δ 13.4, 13.8, 17.1, 21.2, 26.8, 28.7, 29.0, 30.3, 31.2, 34.4, 37.4, 38.5, 40.0, 44.1, 44.2, 49.9, 50.6, 62.3, 67.1, 83.8, 107.1, 121.3, 123.2, 138.3, 154.2, 195.0, 210.5. EIMS: *m/z* 439 $[M]^+$ (Drögemüller *et al.*, 1998).

(25*R*)-2-Enamino-5α-spirost-14-ene-3,12-dione (**8**) was employed to achieve cephalostatin analogue **9** by catalytic hydrogenation reaction (*Scheme 5.2.2*). To a solution of enaminoketone **8** (1.38 g, 3.14 mmol) in MeOH (22 mL), a suspension of Pd-$BaSO_4$

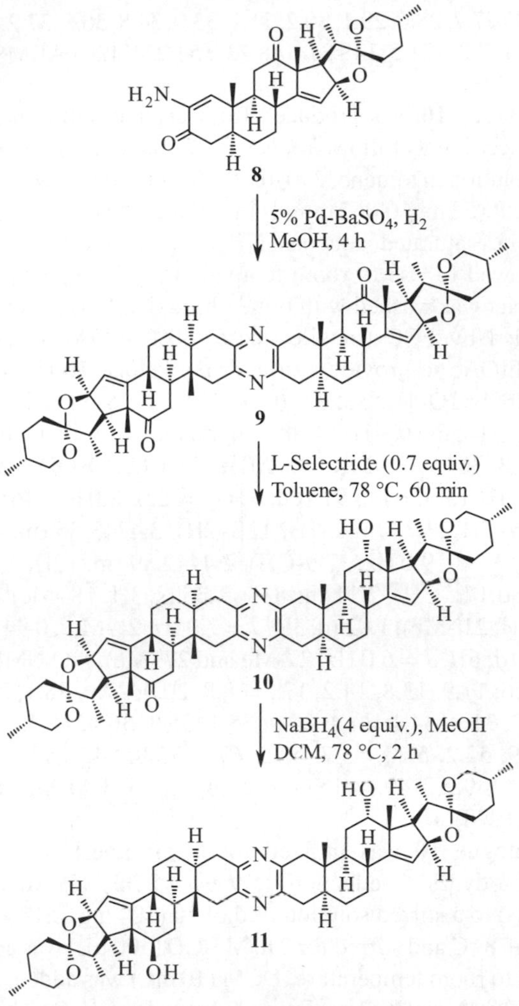

8

5% Pd-$BaSO_4$, H_2
MeOH, 4 h

9

L-Selectride (0.7 equiv.)
Toluene, 78 °C, 60 min

10

$NaBH_4$(4 equiv.), MeOH
DCM, 78 °C, 2 h

11

Scheme 5.2.2 *Synthesis of cephalostatin analogues* ***9–11***

(588 mg, 0.27 mmol, 5% on $BaSO_4$) in MeOH (5 mL) was added and the mixture was saturated with H_2 gas. The mixture was hydrogenated under a slight positive hydrogen pressure at room temperature. After 4 h, the catalyst was filtered through SiO_2 (mobile phase: 10% MeOH in DCM), and the eluate was evaporated under pressure to yield crude solid, which was recrystallized from Et_2O to yield diketone **9** (73%, mp: >300 °C). IR (KBr): ν_{max} cm^{-1} 2928s (C–H), 2873m (C–H), 1714s (C=O), 1646w (C=C), 1456m (C–H), 1399m (pyrazine), 1376m (C–H), 1243m (C–O). UV (MeOH): λ_{max} nm 288, 305. ^{1}H NMR (400 MHz, $CDCl_3$:CD_3OD = 5:1): δ 5.49 (br s, 2H, 2 × 15-CH), 4.79 (dd, 2H, J = 2.0, 8.0 Hz, 2 × 16α-CH), 3.53 (br d, 2H, J = 11.0 Hz, 2 × 26-CHb), 3.42 (t, 2H, J = 11.0 Hz, 2 × 26a-H), 3.36 (dd, 2H, J = 8.0, 9.0 Hz, 2 × 17α-CH), 2.44–2.91 (8H, 2 × 8-CH_2 and 2 × 11-CH_2), 1.35 (s, 6H, 2 × 18-Me), 1.05 (d, 6H, J = 7.0 Hz, 2 × 21-Me), 0.95 (s, 6H, 2 × 19-Me), 0.82 (d, 6H, J = 6.5 Hz, 2 × 27-Me). ^{13}C NMR (100 MHz, $CDCl_3$:CD_3OD = 5:1):

δ 11.5, 13.6, 17.0, 20.7, 27.7, 28.6, 29.1, 30.2, 31.1, 33.9, 34.8, 36.3, 37.2, 41.0, 44.1, 44.8, 49.7, 53.1, 62.3, 67.1, 84.0, 107.2, 121.5, 148.37, 148.43, 154.2, 211.3. FABMS: *m/z* 846 $[M + H]^+$ (Drögemüller *et al.*, 1998).

Cephalostatin analogue **10** was produced from cephalostatin analogue **9** by selective reduction using L-selectride as follows (*Scheme 5.2.2*). A dry toluene solution containing L-selectride (0.7M solution in toluene, 240 μL, 0.168 mmol) was added to a cooled solution (−78 °C) of diketone **9** (52 mg, 0.062 mmol) in dry toluene (5 mL) was added dropwise at −78 °C. After 60 min, saturated aqueous NH_4Cl (2 mL) was added and the reaction temperature was allowed to rise to room temperature. The aqueous layer was extracted with EtOAc, the extract was washed with brine, dried (Na_2SO_4), rotary evaporated and the crude product purified by FCC (mobile phase: 40% EtOAc in petroleum ether) and recrystallized from EtOAc to provide 12α-hydroxy ketone **10** (82%, mp: >300 °C). IR (KBr): ν_{max} cm^{-1} 3460br (O–H), 3056w (CH=CH), 2928s (C–H), 2872s (C–H), 1712s (C=O), 1652w (C=C), 1456m (C–H), 1398m (pyrazine), 1376m (C–H), 1240m (C–O). UV (MeOH): λ_{max} nm 288, 304. ^1H NMR (400 MHz, $CDCl_3$:CD_3OD = 5:1): δ 5.53 (br s, 1H, 15-CH), 5.49 (br s, 1H, 15′-CH), 4.91 (dd, 1H, *J* = 2.0, 8.0 Hz, 16α-CH), 4.79 (dd, 1H, *J* = 8.0 Hz, 2Hz, 16′α-CH), 3.77 (br s, 1H, 12β-CH), 3.42–3.55 (m, 4H, 26-CH_2 and 26′-CH_2), 3.36 (dd, 1H, *J* = 8.0, 9.0 Hz, 17′α-CH), 2.44–2.89 (m, 12H, 4 × CH_2-C=N, 8′-CH, 11′-CHa, 11′-CHb and 17α-CH), 2.11 (m, 1H), 1.35 (s, 3H, 18′-Me), 1.14 (s, 3H, 18-Me), 1.06 (d, *J* = 7.0 Hz, 3H, 21′-Me), 1.02 (d, 3H, *J* = 7.0 Hz, 21-Me), 0.94 (s, 3H, 19′-Me), 0.87 (s, 3H, 19-Me), 0.82 (d, 6H, *J* = 6.0 Hz, 27-Me and 27′-Me). ^{13}C NMR (100 MHz, $CDCl_3$: CD_3OD = 5:1): δ 11.6, 11.9, 13.8, 14.2, 17.2, 18.8, 20.9, 27.9, 28.1, 28.8, 28.8, 29.0, 29.3, 29.3, 30.4, 30.5, 31.3, 31.3, 34.1, 34.4, 34.9, 35.1, 35.9, 36.4, 37.3, 41.2, 41.6, 44.3, 44.9, 44.9, 45.2, 49.2, 49.9, 52.2, 53.3, 53.7, 62.5, 67.3, 75.8, 84.2, 85.6, 107.0, 107.4, 121.1, 121.7, 148.2, 148.3, 149.2, 149.3, 154.0, 154.4, 211.5. FABMS: *m/z* 848 $[M + H]^+$ (Drögemüller *et al.*, 1998).

Cephalostatin analogue **10** was utilized for the synthesis of another cephalostatin analogue **11** by borodydride reduction (*Scheme 5.2.2*). Powdered $NaBH_4$ (18 mg, 0.48 mmol) was added to a stirred solution of diol **10** (104 mg, 0.123 mmol) in 50% DCM in MeOH (2 mL) at −78 °C and stirred for 2 h. Me_2CO (500 μL) was added and the reaction mixture was warmed to room temperature, DCM (10 mL) was added, and the organic layer was washed with a 0.5 M $NaHCO_3$ solution followed by H_2O. The aqueous layer was extracted with EtOAc. The solvent was dried (Na_2SO_4), evaporated under vacuum and the crude residue was recrystallized from EtOAc to afford 12α,12′β-diol **11** (96%, mp: > 300 °C). IR (KBr): ν_{max} cm^{-1} 3472br (O–H), 3056w (CH=CH), 2928s (C–H), 2872s (C–H), 1648w (C=C), 1456m (C–H), 1398m (pyrazine), 1372m (C–H), 1240m (C–O). UV (MeOH): λ_{max} nm 288, 305. ^1H NMR (400 MHz, $CDCl_3$:CD_3OD = 5:1): δ 5.53 (br s, 1H, 15′-CH), 5.42 (br s, 1H, 15-CH), 4.91 (dd, 1H, *J* = 1.5, 8.0 Hz, 16′α-CH), 4.88 (dd, 1H, *J* = 1.5, 8.0 Hz, 16α-CH), 3.77 (br s, 1H, 12′β-CH), 3.42–3.52 (m, 4H, 26-CH_2 and 26′-CH_2), 3.22 (dd, 1H, *J* = 4.5, 11.0 Hz, 12-CH), 2.43–2.91 (10H, 4 × CH_2-C=N, 17α-CH and 17′α-CH), 2.81–2.91 (m, 4H), 2.43–2.70 (m, 6H), 2.19 (m, 1H, 8′-CH), 2.08 (m, 1H, 8-CH), 1.14 (s, 3H, 18′-Me), 1.05 (d, 3H, *J* = 7.0 Hz, 21-Me), 1.04 (s, 3H, 18-Me), 1.02 (d, 3H, *J* = 7.0 Hz, 21′-Me), 0.88 (s, 6H, 19-Me and 19′-Me), 0.83 (d, 3H, *J* = 6.0 Hz, 27-Me), 0.82 (d, 3H, *J* = 6.0 Hz, 27′-Me). ^{13}C NMR (100 MHz, $CDCl_3$:CD_3OD = 5:1): δ 11.9, 12.0, 13.4, 13.6, 14.2, 17.1, 17.2, 18.8, 28.1, 28.2, 28.8, 29.1, 29.3, 29.9, 30.5, 30.5, 31.2, 31.3, 33.8, 34.4, 35.0, 35.1, 35.9, 36.1, 41.5, 41.6, 45.2, 45.3, 49.1, 52.1, 52.3, 52.9, 53.7, 55.7, 67.27, 67.33,

75.6, 78.5, 84.9, 85.6, 106.9, 107.1, 119.5, 121.1, 148.6, 148.8, 148.9, 149.0, 153.9, 157.7. FABMS: *m/z* 850 [M + H]$^+$ (Drögemüller *et al.*, 1998).

The same procedure, as outlined for the synthesis of cephalostatin analogue **10** but using an excess of L-selectride, was applied for the synthesis of cephalostatin analogue **12** from diketone **9** (50 mg, 0.059 mmol) by treating it with an excess of L-selectride (0.7M solution, 240 μL, 0.168 mmol, 2.9 equiv.) in dry toluene (*Scheme 5.2.3*). The resulting crude solid was recrystallized from EtOAc to furnish 12α,12′α-diol **12** (98%, mp: >300 °C). IR (KBr): ν_{max} cm^{-1} 3444br (O–H), 3056w (CH=CH), 2928s (C–H), 2872s (C–H), 1644w (C=C), 1456m (C–H), 1400m (pyrazine), 1370m (C–H), 1240m (C–O). UV (MeOH): λ_{max} nm 288, 305. ^{1}H NMR (400 MHz, $CDCl_3$:CD_3OD = 5:1): δ 5.53 (br s, 2H, 2 × 15-CH), 4.91 (br d, 2H, *J* = 8.0 Hz, 2 × 16α-CH), 3.77 (br s, 2H, 2 × 12β-CH), 3.42–3.52 (m, 4H, 2 × 26-CH_2), 2.79–2.85 (m, 4H, 2 × CH_2-C=N), 2.47–2.70 (m, 6H, 2 × CH_2-C=N, and 2 × 17α-CH), 2.18 (br t, 4H, *J* = 10.5 Hz, 2 × 8-CH_2), 1.14 (s, 6H, 2 × 18-Me), 1.02 (d, 6H, *J* = 7.0 Hz, 2 × 21-Me), 0.87 (s, 6H, 2 × 19-Me), 0.82 (d, 6H, *J* = 6.5 Hz, 2 × 27-Me). ^{13}C NMR (100 MHz, $CDCl_3$:CD_3OD = 5:1): δ 11.9, 14.2, 17.2, 18.8, 28.2, 28.8, 29.0, 29.3, 30.5, 31.3, 34.4, 35.1, 35.9, 41.6, 44.9, 45.1, 49.1, 52.2, 53.7, 67.3, 75.7, 85.6, 106.9, 121.1, 148.8, 149.0, 154.0. FABMS: *m/z* 850 [M + H]$^+$ (Drögemüller *et al.*, 1998).

Cephalostatin analogue **9** was transformed to cephalostatin analogue **13** using borohydride reduction procedure as before (*Scheme 5.2.4*). Powdered $NaBH_4$ (20 mg, 0.53 mmol) was added to a cooled (−78 °C) solution of diketone **9** (100 mg, 0.118 mmol) in 50% DCM in MeOH (3 mL). After 30 min, acetaldehyde (200 μL, 3.58 mmol) was added and the reaction temperature was allowed to warm to room temperature. The solution was washed with saturated aqueous NH_4Cl followed by H_2O, and the aqueous layer was extracted with DCM. The organic solvent was dried (Na_2SO_4), evaporated under vacuum, purified by FCC (mobile phase: DCM:MeOH = 30:1) and recrystallized from EtOAc to obtain 12β-hydroxy ketone **13** (81%, mp: >300 °C). IR (KBr): ν_{max} cm^{-1} 3444br (O–H), 2928s (C–H), 2875m (C–H), 1715m (C=O), 1645w (C=C), 1456m (C–H), 1399m (pyrazine), 1376m (C–H),

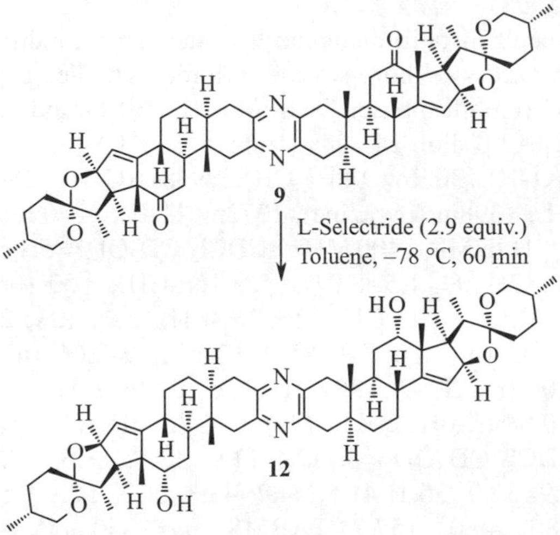

Scheme 5.2.3 *Synthesis of cephalostatin analogue* ***12***

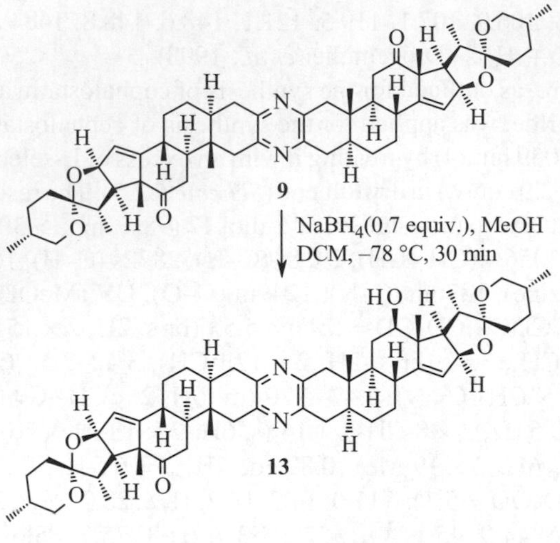

9

$NaBH_4$(0.7 equiv.), MeOH
DCM, –78 °C, 30 min

13

Scheme 5.2.4 *Synthesis of cephalostatin analogue **13***

1240m (C–O). UV (MeOH): λ_{max} nm 288, 305. ^{1}H NMR (400 MHz, $CDCl_3$:CD_3OD = 5:1): δ 5.49 (br s, 1H, 15′-CH), 5.42 (br s, 1H, 15-CH), 4.88 (dd, 1H, J = 2.0, 8.0 Hz, 16α-CH), 4.79 (dd, 1H, J = 2.0, 8.0 Hz, 16′α-CH), 3.34–3.54 (m, 5H, 17′α-CH, 26-CH_2 and 26′-CH_2), 3.21 (dd, 1H, J = 4.5, 11.0 Hz, 12α-CH), 2.43–2.92 (12H, 4 × CH_2-C=N, and 2 × 11-CH_2), 2.04–2.11 (m, 2H, 2 × 8β-CH), 1.35 (s, 3H, 18′-Me), 1.06 (d, 3H, J = 7.0 Hz, 21′-Me), 1.05 (d, 3H, J = 7.0 Hz, 21-Me), 1.03 (s, 3H, 18-Me), 0.94 (s, 3H, 19′-Me), 0.87 (s, 3H, 19-Me), 0.83 (d, 3H, J = 6.0 Hz, 27-Me), 0.82 (d, 3H, J = 6.0 Hz, 27′-Me). ^{13}C NMR (100 MHz, $CDCl_3$:CD_3OD = 5:1): δ 11.6, 11.9, 13.4, 13.6, 13.8, 17.2, 20.9, 27.9, 28.1, 28.7, 28.8, 29.2, 29.3, 29.8, 30.4, 30.5, 31.2, 31.3, 33.8, 34.1, 34.9, 35.0, 36.1, 36.5, 37.4, 41.2, 41.5, 44.3, 44.6, 44.9, 45.3, 49.8, 52.3, 53.0, 53.2, 55.6, 62.5, 67.4, 67.5, 78.6, 84.2, 84.9, 107.2, 107.4, 119.5, 121.7, 148.3, 148.4, 149.1, 149.2, 154.5, 157.7, 211.6. FABMS: m/z 848 $[M + H]^+$ (Drögemüller *et al.*, 1998).

The treatment of cephalostatin analogue **9** in the same conditions as stated above, but utilizing a large excess of borohydride, afforded another cephalostatin analogue **14**, which after FCC (mobile phase: DCM:MeOH = 30:1) and recrystallization from EtOAc produced 12β,12′β-diol **14** (98%, mp: >300 °C) (*Scheme 5.2.5*). IR (KBr): ν_{max} cm^{-1} 3448br (O–H), 3056w (CH=CH), 2928s (C–H), 2872 s (C–H), 1640w (C=C), 1456m (C–H), 1398m (pyrazine), 1372m (C–H), 1240m (C–O). UV (MeOH): λ_{max} nm 288, 305. ^{1}H NMR (400 MHz, $CDCl_3$:CD_3OD = 5:1): δ 5.42 (br s, 2H, 2 × 15-CH), 4.88 (dd, 2H, J = 1.5, 8.0 Hz, 2 × 16α-CH), 3.52 (br d, 2H, J = 11.0 Hz, 2 × 26b-CH), 3.45 (t, 2H, J = 11.0 Hz, 2 × 26a-CH), 3.21 (dd, 2H, J = 4.5, 11.0 Hz, 2 × 12α-CH), 2.79–2.93 (m, 4H, 2 × CH_2-C=N), 2.49–2.66 (m, 6H, 2 × CH_2-C=N and 2 × 17α-CH), 2.08 (m, 2H, 2 × 8β-CH), 1.05 (d, J = 7.0 Hz, 6H, 2 × 21-Me), 1.03 (s, 6H, 2 × 18-Me), 0.88 (s, 6H, 2 × 19-Me), 0.82 (d, 6H, J = 6.5 Hz, 2 × 27-Me). ^{13}C NMR (100 MHz, $CDCl_3$:CD_3OD = 5:1): δ 11.9, 13.4, 13.6, 17.2, 28.1, 28.8, 29.3, 29.9, 30.5, 31.2, 33.9, 35.0, 36.1, 41.5, 44.7, 45.3, 52.3, 52.9, 55.7, 67.4, 78.6, 84.9, 107.1, 119.6, 148.8, 149.0, 157.7. FABMS: m/z 850 $[M + H]^+$ (Drögemüller *et al.*, 1998).

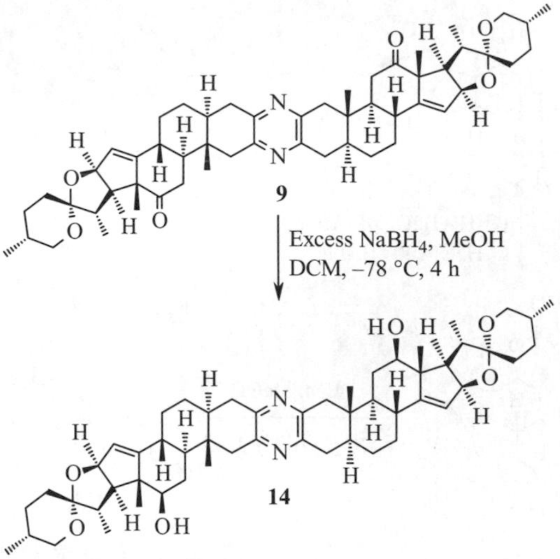

9

Excess $NaBH_4$, MeOH
DCM, –78 °C, 4 h

14

Scheme 5.2.5 *Synthesis of cephalostatin analogue* ***14***

Propionic anhydride (32.84 mL, 254.84 mmol) and DMAP (200 mg) were added to a stirred solution of (25*R*)-3β-acetyloxy-5α-spirostan-12α-ol (**4**, 40.18 g, 84.95 mmol) in dry C_5H_5N (500 mL), heated at 100 °C for 4 h, and after cooling to room temperature, washed with 2N HCl (4 × 200 mL) (*Scheme 5.2.6*). The resulting crude product was dissolve in MeOH (200 mL) and sequentially washed with 2N HCl (100 mL), saturated aqueous $NaHCO_3$ and brine. The organic layer was dried ($MgSO_4$), concentrated *in vacuo*, the residue was filtered through SiO_2 and recrystallized from DCM-MeOH to obtain (25*R*)-3β-acetyloxy-5α-spirostan-12α-propionate (**15**, 38.6 g, 86%). IR ($CHCl_3$): ν_{max} cm^{-1} 2980m (C–H), 2952m (C–H), 2872m (C–H), 1724s (C=O), 1460w (C–H). ^1H NMR (400 MHz, $CDCl_3$): δ 5.45 (m, 1H, 15-CH), 4.88 (t, 1H, *J* = 2.6 Hz, 12β-CH), 4.83 (dd, 1H, *J* = 1.8, 8.1 Hz, 16α-CH), 4.66 (m, 1H, 3α-CH), 3.38–3.53 (m, 2H, 26-CH_2), 2.27–2.35 (m, 3H, 17α-CH, prop-CH_2), 2.16 (m, 1H, 8α-CH), 2.01 (s, 3H, 3β-MeCO), 0.77–1.88 (m, 35H), 1.43 (m, 3H, prop-Me), 1.13 (s, 3H, 18-Me), 0.96 (d, 3H, *J* = 6.8 Hz, 21-Me), 0.87 (s, 3H, 19-Me), 0.80 (d, 3H, *J* = 6.3 Hz, 27-Me). ^{13}C NMR (100 MHz, $CDCl_3$): δ 9.3, 11.9, 14.1, 17.2, 18.7, 21.4, 26.1, 27.3, 28.1, 28.2, 28.7, 29.5, 30.4, 31.2, 33.9, 34.4, 35.6, 36.6, 44.5, 44.5, 50.1, 50.2, 53.7, 67.2, 73.3, 78.1, 85.2, 106.6, 120.6, 153.6, 170.6, 174.0 (Drögemüller *et al.*, 1998).

(25*R*)-3β-Hydroxy-5α-spirostan-12α-propionate (**16**) could be achieved from a solution of **15** (43.3 g, 81.85 mmol) in MeOH (350 mL) and DCM (20 mL), treated with a solution of Na_2CO_3 (9.5 g, 90 mmol) in H_2O (80 mL) with constant stirring for 2 h (*Scheme 5.2.6*). A further solution of Na_2CO_3 (10 g, 94.7 mmol) in H_2O (80 mL) was added, stirred for another 20 h. The mixture was extracted with DCM, the organic layer was dried ($MgSO_4$), concentrated under vacuum and the solid was filtered through SiO_2 to produce 2α-hydroxy **16** (36.6 g, 92%). IR ($CHCl_3$): ν_{max} cm^{-1} 3608br (O–H), 2980m (C–H), 2932s (C–H), 2860m (C–H), 1724s (C=O), 1652w (C=C), 1460m (C–H), 1376m (C–H), 1240m (C–O). ^1H NMR (400 MHz, $CDCl_3$): δ 5.45 (m, 1H, 15-CH), 4.88 (t, 1H, *J* = 2.7 Hz, 12β-CH), 4.82 (dd, 1H, *J* = 1.9, 8.2 Hz, 16α-CH), 3.57 (m, 1H, 3α-CH),

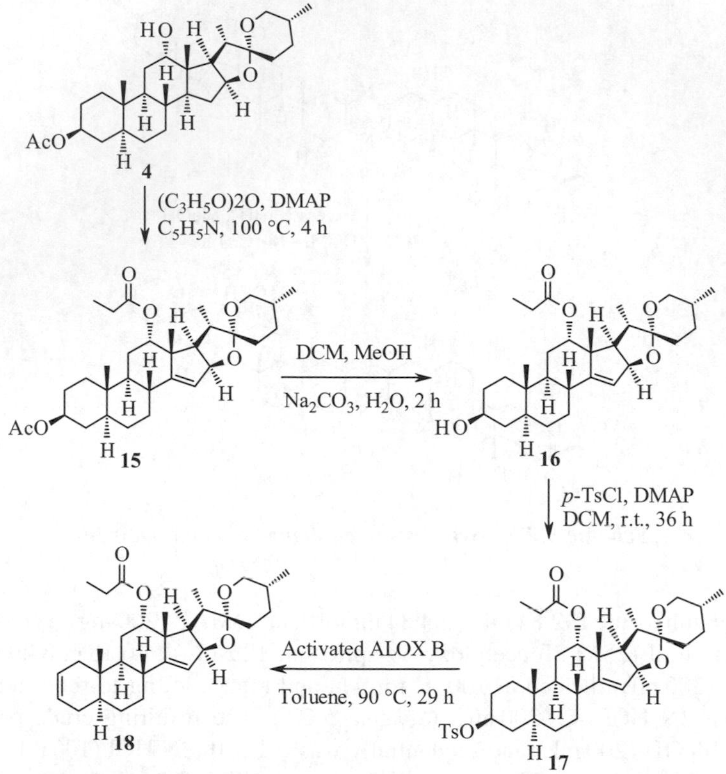

Scheme 5.2.6 *Synthesis of hecogenin derivatives **15–18***

3.38–3.52 (m, 2H, 26-CH_2), 2.26–2.35 (m, 3H, 17α-CH and CH_2), 2.16 (m, 1H, 8β-CH), 1.13 (s, 3H, 18-Me), 1.11 (t, 3H, $J = 7.6$ Hz, prop-Me), 0.97 (d, 3H, $J = 6.8$ Hz, 21-Me), 0.85 (s, 3H, 19-Me), 0.80 (d, 3H, $J = 6.2$ Hz, 27-Me). ^{13}C NMR (100 MHz, $CDCl_3$): δ 9.3, 12.0, 14.1, 17.1, 18.7, 26.1, 28.2, 28.2, 28.7, 29.5, 30.4, 31.2, 31.3, 34.4, 35.6, 36.8, 38.0, 44.5, 44.7, 50.1, 50.3, 53.6, 67.1, 71.0, 78.2, 85.2, 106.6, 120.5, 153.7, 174.0. EIMS: *m/z* 486 $[M]^+$ (Drögemüller *et al.*, 1998).

Simple tosylation of (25*R*)-3β-hydroxy-5α-spirostan-12α-propionate (**16**) provided the (25*R*)-3β-tosyloxy-5α-spirostan-12α-propionate (**17**) (*Scheme 5.2.6*). To a stirred solution of **16** (35.5 g, 73 mmol) in DCM (400 mL), *p*-TsCl (16.7 g, 87.6 mmol) and DMAP (10.7 g, 87.6 mmol) were added at room temperature. After 24 h, an additional portion of DMAP (7.0 g, 57.4 mmol) was added to the reaction mixture and stirring was continued for another 36 h. H_2O (200 mL) was added to the reaction mixture, the aqueous layer was extracted three times with DCM. The pooled organic layers were successively washed with 2N aqueous HCl, saturated aqueous $NaHCO_3$ and brine. The organic layer was dried ($MgSO_4$), concentrated under vacuum and chromatographed to yield 2α-tosylate **17** (33.7 g, 72%). IR ($CHCl_3$): ν_{max} cm^{-1} 2980m (C–H), 2952m (C–H), 2932m (C–H), 2876m (C–H), 1724m (C=O), 1460m (C–H), 1376m (C–H), 1240m (C–O), 1172s (C–O). ^1H NMR (400 MHz, $CDCl_3$): δ 7.76–7.80 (m, 2H, Ts-CH), 7.30–7.34 (m, 2H, Ts-CH), 5.43 (m, 1H, 15-CH), 4.85 (t, 1H, $J = 2.4$ Hz, 12β-CH), 4.81 (dd, 1H, $J = 1.8$, 8.1 Hz, 16α-CH), 4.38 (m, 1H, 3α-CH),

3.38–3.52 (m, 2H, 26-CH_2), 2.44 (s, 3H, Ts-Me), 2.12 (m, 1H, 8β-CH), 1.11 (s, 3H, 18-Me), 1.10 (t, 3H, $J = 7.5$ Hz, prop-Me), 0.96 d, 3H, $J = 6.8$ Hz, 21-Me), 0.82 (s, 3H, 19-Me), 0.80 (d, 3H, $J = 6.3$ Hz, 27-Me). ^{13}C NMR (100 MHz, $CDCl_3$): δ 9.3, 11.8, 14.1, 17.1, 18.7, 21.6, 26.0, 28.0, 28.2, 28.7, 29.4, 30.4, 31.2, 34.2, 34.7, 35.4, 36.6, 44.5, 44.6, 50.0, 50.1, 53.6, 67.2, 78.0, 81.9, 85.2, 106.7, 120.7, 127.6, 129.7, 134.7, 144.4, 153.3, 173.9. ESIMS: *m/z* 641 $[M + H]^+$ (Drögemüller *et al.*, 1998).

(25*R*)-5α-Spirosta-2,14-diene-12α-propionate (**18**) was prepared from (25*R*)-3β-tosyloxy-5α-spirostan-12α-propionate (**17**) (*Scheme 5.2.6*). Activated ALOX B (50 g) was added, to a stirred solution of **17** (28.6 g, 44.69 mmol) in dry toluene (350 mL) and the mixture was heated at 90 °C for 45 min, after which, an additional portion of ALOX B (50 g) was added and stirred for another 3 h. The reaction mixture was cooled, filtered, the filtrate was concentrated *in vacuo*, chromatographed and recrystallized from DCM-MeOH to provide **18** (17.8 g, 85%, mp: 105 °C). IR ($CHCl_3$): ν_{max} cm^{-1} 2956s (C–H), 2928s (C–H), 2876m (C–H), 1724s (C=O), 1652w (C=C), 1460m (C–H), 1380m (C–H), 1240m (C–O). ^{1}H NMR (400 MHz, $CDCl_3$): δ 5.50–5.64 (m, 2H, 2-CH and 3-CH), 5.45 (m, 1H, 15-CH), 4.89 (t, 1H, $J = 2.6$ Hz, 12β-CH), 4.83 (dd, 1H, $J = 1.5$, 8.1 Hz, 16α-CH), 3.4–3.53 (m, 2H, 26-CH_2), 2.33 (m, 1H, 17α-CH), 2.13 (m, 1H, 8β-CH), 1.15 (s, 3H, 18-Me), 1.11 (t, 3H, $J = 7.5$ Hz, prop-Me), 0.97 (d, 3H, $J = 6.8$ Hz, 21-Me), 0.79–0.81 (m, 6H, 27-Me and 19-Me). ^{13}C NMR (100 MHz, $CDCl_3$): δ 9.3, 11.4, 14.1, 17.2, 18.6, 25.8, 28.2, 28.3, 28.7, 29.3, 30.1, 30.4, 31.2, 34.5, 34.8, 39.6, 41.3, 44.5, 50.0, 50.1, 53.7, 67.1, 78.3, 85.2, 106.6, 120.6, 125.4, 125.9, 153.8, 174.0. FABMS: *m/z* 491 $[M + Na]^+$ (Drögemüller *et al.*, 1998).

(25*R*)-5α-Spirosta-2,14-diene-12α-propionate (**18**) was employed for the synthesis of several other hecogenin derivatives **19–22**, as the precursors for the preparation of cephalostatin analogue **23**. DMDO (0.09–0.1 M solution, 125 mL) in Me_2CO was added dropwise to a cooled (0 °C) solution of **18** (4.8 g, 10.26 mmol) in DCM (100 mL) (*Scheme 5.2.7*). After 90 min at the same temperature, saturated aqueous $NaHSO_3$ (50 mL) was added to the reaction mixture, the aqueous layer was separated and extracted with DCM. The organic solvent was dried ($MgSO_4$), evaporated to dryness and chromatographed to afford (25*R*)-2α,3α-epoxy-5α-spirostan-12α-propionate (**19**, 4.38 g, 88%). IR ($CHCl_3$): ν_{max} cm^{-1} 2996m (C–H), 2956s (C–H), 2876m (C–H), 1724s (C=O), 1460m (C–H). ^{1}H NMR (400 MHz, $CDCl_3$): δ 5.44 (m, 1H, 15-CH), 4.87 (t, 1H, $J = 2.8$ Hz, 12β-CH), 4.81 (dd, 1H, $J = 1.5$, 7.9 Hz, 16α-CH), 3.38–3.52 (m, 2H, 26-CH_2), 3.16 (m, 1H, 3β-CH), 3.09 (m, 1H, 2β-CH), 2.01 (m, 1H, 8β-CH), 1.12 (s, 3H, 18-Me), 1.11 (t, 3H, $J = 7.4$ Hz, prop-Me), 0.97 (d, 3H, $J = 6.8$ Hz, 21-Me), 0.78–0.82 (m, 6H, 19-Me and 27-Me). ^{13}C NMR ($CDCl_3$, 100 MHz): δ 9.3, 12.7, 14.1, 17.2, 18.6, 25.8, 28.0, 28.2, 28.7, 28.9, 29.2, 30.4, 31.2, 33.8, 34.5, 36.1, 38.1, 44.5, 49.8, 50.0, 50.6, 52.3, 53.7, 67.1, 78.0, 85.1, 106.6, 120.9, 153.3, 174.0. EIMS: *m/z* 484 $[M]^+$. HRMS: *m/z* 484.3189 calculated for $C_{30}H_{44}O_5$, found 484.3199 (Drögemüller *et al.*, 1998).

(25*R*)-2β-Chloro-3α-hydroxy-5α-spirostan-12α-propionate (**20**) was synthesized from (25*R*)-2α,3α-epoxy-5α-spirostan-12α-propionate (**19**) by epoxidation (*Scheme 5.2.7*). A solution of PPh_3Cl (1.4 g, 1.2 equiv.) in DCM (10 mL) was added to a stirred cooled (−15 °C) solution of **19** (1.8 g, 3.72 mmol) in DCM (20 mL). After 60 min, saturated aqueous $NaHSO_3$ (30 mL) was added to the reaction mixture and the aqueous layer was extracted with DCM. The solvent was dried ($MgSO_4$), rotary evaporated, chromatographed and recrystallized from DCM-MeOH to furnish **20** (1.42 g, 73%, mp: 198 °C). IR ($CHCl_3$):

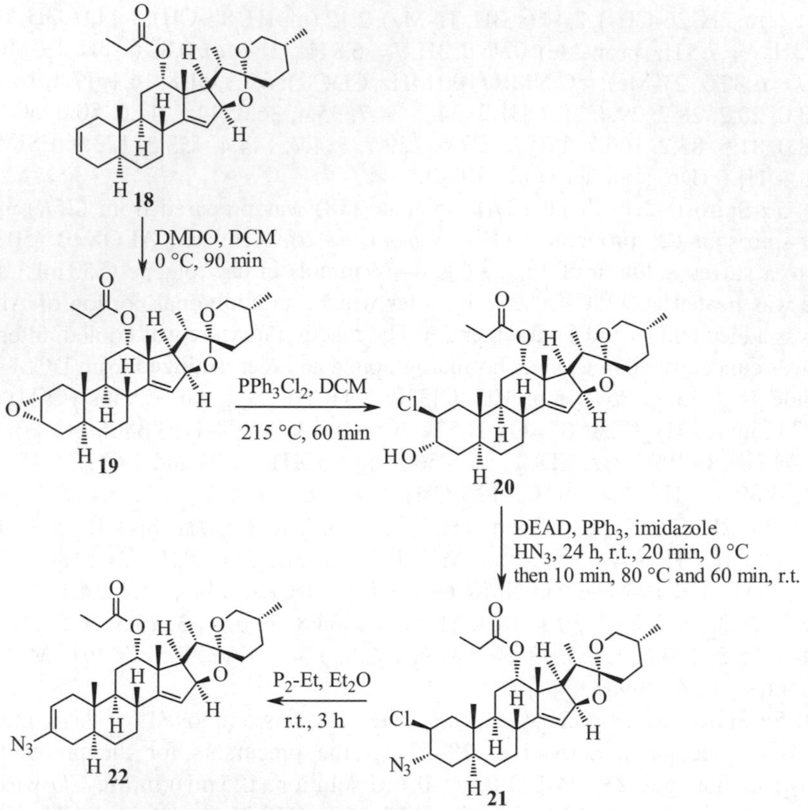

Scheme 5.2.7 *Synthesis of hecogenin derivatives* ***19–22***

ν_{max} cm^{-1} 3608w (O–H), 2980m (C–H), 2932s (C–H), 2880m (C–H), 1724s (C=O), 1460m (C–H). ^{1}H NMR (200 MHz, $CDCl_3$): δ 5.45 (m, 1H, 15-CH), 4.88 (m, 1H, 12β-CH), 4.82 (dd, 1H, J = 1.7, 7.9 Hz, 16α-CH), 4.04–4.13 (m, 2H, 2α-CH and 3β-CH), 3.38–3.53 (m, 2H, 26-CH_2), 1.13 (s, 3H, 18-Me), 1.11 (t, 3H, J = 7.6 Hz, prop-Me), 1.09 (s, 3H, 19-Me), 0.97 (d, 3H, J = 6.7 Hz, 21-Me), 0.80 (d, 3H, J = 6.0 Hz, 27-Me). ^{13}C NMR (50 MHz, $CDCl_3$): δ 9.2, 14.1, 14.4, 17.2, 18.7, 25.8, 27.6, 28.1, 28.7, 29.3, 30.4, 30.7, 31.2, 33.7, 36.2, 38.4, 40.1, 44.5, 50.1, 51.1, 53.6, 59.2, 67.2, 70.9, 78.1, 85.2, 106.7, 120.6, 153.6, 174.0. HRMS: m/z 520.2956 calculated for $C_{30}H_{45}ClO_5$, found 520.2933 (Drögemüller *et al.*, 1998).

(25*R*)-2β-Chloro-3α-hydroxy-5α-spirostan-12α-propionate (**20**) could be converted to (25*R*)-3α-azido-2β-chloro-5α-spirostan-12α-propionate (**21**) (*Scheme 5.2.7*). DEAD (675 mg, 3.9 mmol) was added slowly by a syringe to a solution of **20** (675 mg, 1.30 mmol) and PPh_3 (1.02 g, 3.9 mmol) in dry toluene (20 mL) at 0 °C, followed by the addition of 1.4M HN_3 in toluene (4.6 mL, 6.5 mmol). After 15 min at this temperature, the reaction mixture was heated rapidly to 80 °C for 10 min. The reaction temperature was allowed to cool and subsequently stirred at room temperature for 90 min. The solvent was evaporated under pressure and the product was chromatographed to give **21** (283 mg, 40%) and a mixture of regio- and diastereoisomeric allylic azides (264 mg, 40%). IR ($CHCl_3$): ν_{max} cm^{-1} 2980w (C–H), 2952m (C–H), 2880w (C–H), 2864w (C–H), 2104s (N_3), 1724s (C=O), 1460m (C–H), 1240m (C–O). ^{1}H NMR (400 MHz, $CDCl_3$): δ 5.49 (m, 1H, 15-CH), 4.88 (t, 1H, $J=2.8$ Hz, 12β-CH), 4.81 (dd, 1H, $J=1.7, 8.1$ Hz, 16α-H), 4.43 (m, 1H, 2α-CH), 3.37–3.52 (m, 3H, 3-CH and 26-CH_2), 1.15 (s, 3H, 18-Me), 1.14 (s, 3H, 19-Me), 1.11 (t, 3H, $J=7.6$ Hz, prop-Me), 0.97 (d, 3H, $J=6.7$ Hz, 21-Me), 0.80 (d, 3H, $J=6.0$ Hz, 27-Me). ^{13}C NMR (100 MHz, $CDCl_3$): δ 9.3, 14.1, 15.2, 17.2, 18.8, 26.2, 27.6, 28.2, 28.2, 28.7, 29.3, 30.4, 31.2, 33.6, 36.1, 44.5, 44.7, 46.6, 50.2, 51.2, 53.6, 60.5, 62.5, 67.2, 77.9, 85.1, 106.7, 120.9, 153.2, 173.8. HRMS: *m/z* 545.3020 calculated for $C_{30}H_{44}ClN_3O_4$, found 545.3031 (Drögemüller *et al.*, 1998).

(25*R*)-3α-Azido-2β-chloro-5α-spirostan-12α-propionate (**21**) was transformed to (25*R*)-3-azido-5α-spirost-2-ene-12α-propionate (**22**) (*Scheme 5.2.7*). Phosphazene base, P_2-Et (262 μL, 0.79 mmol) was added slowly at room temperature to the solution of **21** (358 mg, 0.66 mmol) in Et_2O (3 mL), and stirred for 5 h, after which, the solvent was rotary evaporated and the crude residue was purified on SiO_2 to obtain **22** (92%). IR ($CHCl_3$): ν_{max} cm^{-1} 3012m (CH=CH), 2956s (C–H), 2928s (C–H), 2876m (C–H), 2100s (N_3), 1724s (C=O), 1460w (C–H), 1224s (C–O). ^{1}H NMR (400 MHz, $CDCl_3$): δ 5.46 (m, 1H, 15-CH), 5.17 (m, 1H, 2-CH), 4.89 (t, 1H, $J=2.8$ Hz, 12β-CH), 4.83 (dd, 1H, $J=1.8, 8.3$ Hz, 16α-CH), 3.39–3.53 (m, 2H, 26-CH_2), 2.13 (m, 1H, 8β-H), 1.14 (s, 3H, 18-Me), 1.11 (t, 3H, $J=7.6$ Hz, prop-Me), 0.98 (d, 3H, $J=6.8$ Hz, 21-Me), 0.80 (d, 3H, $J=6.2$ Hz, 27-Me), 0.80 (s, 3H, 19-Me). ^{13}C NMR (100 MHz, $CDCl_3$): δ 9.3, 11.5, 14.1, 17.2, 18.6, 26.6, 28.0, 28.2, 28.7, 29.1, 30.3, 30.4, 31.2, 34.3, 34.7, 38.8, 41.5, 44.5, 49.7, 50.0, 53.7, 67.2, 78.1, 85.2, 106.7, 110.3, 120.9, 133.8, 153.3, 174.0. HRMS: *m/z* 509.3254 calculated for $C_{30}H_{43}N_3O_4$, found 509.3229 (Drögemüller *et al.*, 1998).

(25*R*)-3-Azido-5α-spirost-2-ene-12α-propionate (**22**) and (25*R*)-2-enamino-5α-spirost-14-ene-3,12-dione (**8**) were utilized as the building blocks to synthesize another cephalostatin analogue **23** (*Scheme 5.2.8*). A catalytic amount of PPTS (1.0 mg) was added to a stirred mixture of compounds **8** (50 mg), **22** (30 mg) and 4 Å activated molecular sieves (30 mg) in dry toluene (2 mL) under Ar. The reaction mixture was refluxed for 2.5 h. The solvent was evaporated under vacuum and the crude solid was chromatographed to produce **23** (31 mg, 36%). IR ($CHCl_3$): ν_{max} cm^{-1} 2956s (C–H), 2932s (C–H), 2876m (C–H), 1712s (ester C=O), 1460m (C–H), 1400s (pyrazine). ^{1}H NMR (400 MHz, $CDCl_3$): δ 5.50 (m, 1H, 15-CH), 5.48 (m, 1H, 15′-CH), 4.95 (m, 1H, 12β′-CH), 4.84 (dd, 1H, $J=1.8, 7.9$ Hz, 16α-CH), 4.78 (dd, 1H, $J=2.0, 8.1$ Hz, 16′α-CH), 3.34–3.56 (m, 5H, 17α-CH, 26-CH_2 and 26′-CH_2), 2.18 (m, 2H, 2 × 8β-CH), 1.33 (s, 3H, 18-Me), 1.17 (s, 3H, 18′-Me), 1.11 (t, 3H, $J=7.7$ Hz, prop-Me), 1.05 (d, 3H, $J=6.8$ Hz, 21-Me), 0.99 (d, 3H, $J=6.8$ Hz, 21′-Me), 0.85 (s, 3H, 19-Me), 0.92 (s, 3H, 19′-Me), 0.81 (d, 6H, $J=6.2$ Hz, 27-Me and 27′-Me). ^{13}C NMR (100 MHz, $CDCl_3$): δ 9.3, 11.5, 11.7, 13.8, 14.1, 17.2, 18.7, 20.8, 26.3, 27.9, 27.9, 28.1, 28.7, 28.8, 29.0, 29.1, 30.3, 30.4, 31.2, 31.3, 34.0, 34.2, 35.2, 35.2, 35.7, 36.4, 37.2, 41.2, 41.6, 44.2, 44.5, 45.2, 45.8, 49.8, 49.8, 50.0, 53.1, 53.7, 62.3, 67.1, 67.2, 77.9, 83.9, 85.1, 106.7, 107.0, 121.1, 121.6, 148.0, 148.1, 148.6, 148.7, 153.2, 154.3, 173.7, 210.7. HRMS: *m/z* 903.5887 calculated for $C_{57}H_{78}N_2O_7$, found 903.5999 (Drögemüller *et al.*, 1998).

A series of cholic acid derivatives **25**–**28** were synthesized as the precursors for the synthesis of cephalostatin analogue **29** (*Scheme 5.2.9*). Initially, methyl 3α,12α-dihydroxy-5β,

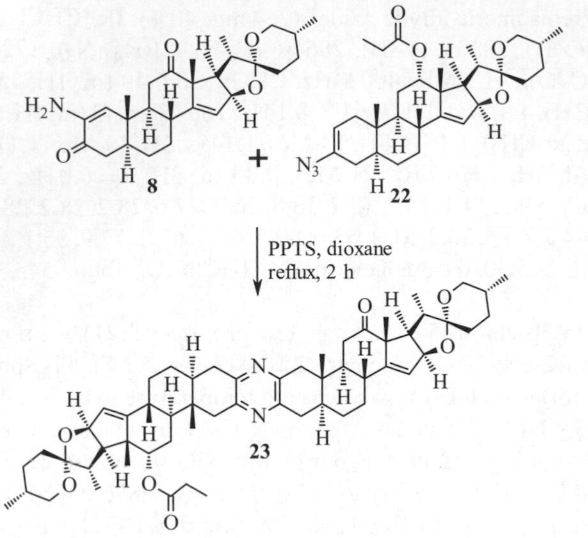

Scheme 5.2.8 Synthesis of cephalostatin analogue **23**

14α-chol-14-en-24-oate (**25**) was prepared from cholic acid (**24**) in the following manner. $ZnCl_2$ (30 g, 220.6 mmol, 3 equiv.) was added to a stirred solution of **24** (30 g, 73.4 mmol) in freshly distilled Me_2CO (300 mL), and the mixture was heated at 80 °C for 2 h. The solvent was slowly distilled off until total conversion of the starting material (monitored on TLC). The solution was cooled to room temperature, 0.5% aqueous AcOH (300 mL) was added resulting in the formation of white precipitates, which were collected by filtration and the solid was dried under vacuum. The dried solid was redissolved in MeOH and treated with the ion-exchange resin Amberlyst 15 (12 g). The mixture was stirred for 12 h, filtered, and the filtrate was evaporated under vacuum to yield a white solid, which was again dissolved in freshly distilled $CHCl_3$ (250 mL) and the solution was cooled (−78 °C). A dry stream of HCl gas was passed through the solution for 2 h, followed by a stream of N_2. A solution of 0.5M $NaHCO_3$ (100 mL) was added and the mixture was allowed to warm to room temperature. The organic layer was separated, washed with H_2O, dried ($MgSO_4$), concentrated under pressure and subjected to FCC (mobile phase: 50–75% EtOAc in *n*-hexane) to provide **25** (14.8 g, 50%). IR ($CHCl_3$): ν_{max} cm^{-1} 3420br (O–H), 2928m (C–H), 2804m (C–H), 1740m (C=O), 1632m (C=C). ^{1}H NMR (400 MHz, $CDCl_3$): δ 5.25 (d, 1H, J = 2.0 Hz, 15-CH), 3.76 (t, 1H, J = 2.0 Hz, 12β-CH), 3.63 (s, 3H, 24-OMe), 3.56 (m, 1H, 3β-CH), 0.88–0.96 (m, 9H, 18-Me, 19-Me and 21-Me). ^{13}C NMR (100 MHz, $CDCl_3$): δ 16.8, 17.8, 23.0, 24.0, 26.9, 29.3, 30.6, 30.9, 31.2, 32.0, 33.6, 34.3, 34.8, 35.0, 35.1, 36.3, 42.0, 47.0, 51.6, 51.8, 71.7, 73.3, 120.2, 151.5, 174.9. EIMS: m/z 404 $[M]^+$. HREIMS: m/z 404.2929 calculated for $C_{25}H_{40}O_4$, found 404.2926 (Drögemüller *et al.*, 1998).

Methyl 3α,12α-dihydroxy-5β-chol-14-en-24-oate (**25**) was employed for the synthesis of methyl 3,12-dioxo-5β-chol-14-en-24-oate (**26**) (*Scheme 5.2.9*). A mixture of PCC (2.13 g, 9.9 mmol, 4 equiv.), SiO_2 (2.13 g) and NaOAc (176 mg, 2.5 mmol, 1 equiv.) was added to a solution of **25** (1.0 g, 2.5 mmol) in DCM (30 mL), stirred at room temperature for 4 h, filtered through a plug of SiO_2, the eluate was washed once with brine, dried

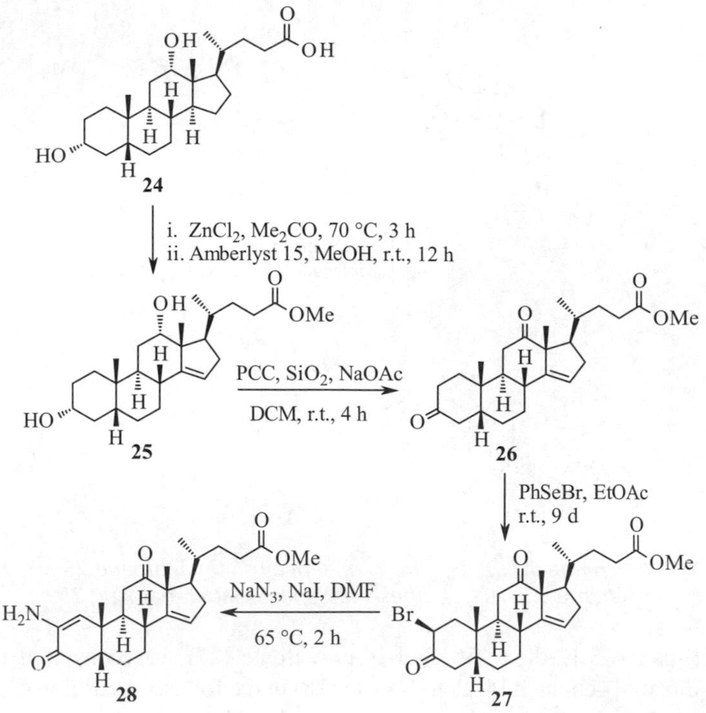

Scheme 5.2.9 *Synthesis of cholic acid derivatives* ***25–28***

($MgSO_4$), evaporated to dryness and the crude product was purified by FCC (mobile phase: 50% EtOAc in *n*-hexane) to afford diketone **26** (810 mg, 81%). IR ($CHCl_3$): ν_{max} cm^{-1} 2936m (C–H), 2872m (C–H), 1708m (C=O). ^{1}H NMR (400 MHz, $CDCl_3$): δ 5.28 (d, 1H, J = 2.0 Hz, 15-CH), 3.63 (s, 3H, 24-OMe), 1.21 (s, 3H, 19-Me), 1.06 (s, 3H, 18-Me), 0.90 (d, 3H, J = 7.0 Hz, 21-Me). ^{13}C NMR (100 MHz, $CDCl_3$): δ 17.6, 19.2, 22.0, 23.5, 26.1, 30.6, 31.4, 33.6, 34.4, 34.9, 35.5, 36.5, 37.1, 38.5, 41.0, 42.1, 43.7, 47.1, 51.6, 62.8, 120.9, 151.7, 174.7, 212.1, 213.3. EIMS: *m/z* 400 $[M]^+$. HREIMS: *m/z* 400.2600 calculated for $C_{25}H_{36}O_4$, found 400.2600 (Drögemüller *et al.*, 1998).

Methyl 2β-bromo-3,12-dioxo-5β-chol-14-en-24-oate (**27**) was achieved from methyl 3,12-dioxo-5β-chol-14-en-24-oate (**26**) (*Scheme 5.2.9*). PhSeBr (69 mg, 0.3 mmol, 1.2 equiv.) was added to a stirred solution of **26** (98 mg, 0.24 mmol) in EtOAc (2 mL) at room temperature. After 96 h, the solvent was evaporated under pressure and the crude residue was purified by FCC, using the same solvent system as mentioned for **25**, to furnish 2β-bromoketone **27** (34 mg, 29%). IR ($CHCl_3$): ν_{max} cm^{-1} 2932m (C–H), 2872m (C–H), 1732m (C=O), 1708m (C=O), 1648m (C=C). ^{1}H NMR (400 MHz, $CDCl_3$): δ 5.34 (m, 1H, 15-CH), 4.65 (dd, 1H, J = 5.6, 13.8 Hz, 2-CH), 3.68 (s, 3H, 24-OMe), 1.25 (s, 3H, 19-Me), 1.12 (s, 3H, 18-Me), 0.95 (d, 3H, J = 6.6 Hz, 21-Me). ^{13}C NMR (100 MHz, $CDCl_3$): δ 17.7, 19.2, 21.7, 23.4, 26.0, 30.7, 31.7, 33.6, 34.5, 34.9, 38.5, 39.1, 41.3, 41.6, 44.5, 47.3, 48.8, 51.6, 52.3, 62.9, 121.5, 151.1, 176.7, 201.5, 212.6. EIMS: *m/z* 478 $[M]^+$. HREIMS: *m/z* 478.1719 calculated for $C_{25}H_{35}BrO_4$, found: 478.1718 (Drögemüller *et al.*, 1998).

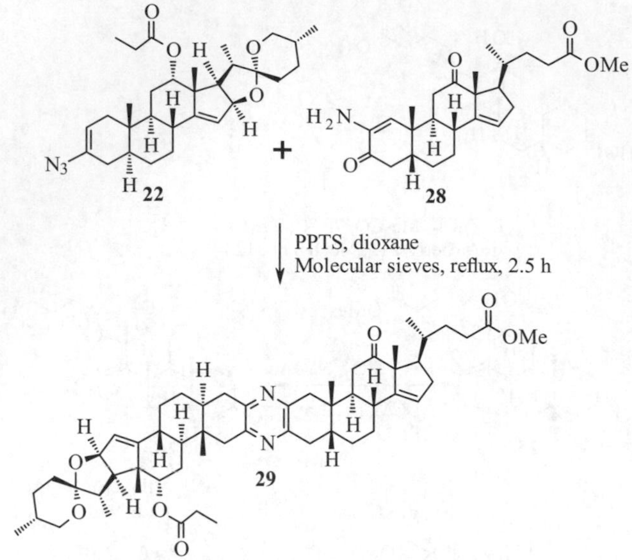

PPTS, dioxane
Molecular sieves, reflux, 2.5 h

Scheme 5.2.10 *Synthesis of cephalostatin analogue* ***29***

Methyl 2β-bromo-3,12-dioxo-5β-chol-14-en-24-oate (**27**) was converted to methyl 2-amino-3,12-dioxo-5β-chola-1,14-dien-24-oate (**28**) in the following fashion (*Scheme 5.2.9*). NaN_3 (71.5 mg, 1.1 mmol, 11 equiv.) and one crystal of NaI were added to a stirred solution of **27** (45 mg, 0.10 mmol) in degassed DMF (2.5 mL) under Ar, and the mixture was heated to 65 °C for 2 h. The reaction mixture was allowed to cool to room temperature, H_2O (1 mL) was added and extracted three times with MTBE:*n*-hexane (2:1). The pooled organic solvents were washed with brine, dried ($MgSO_4$) and evaporated under vacuum to give aminoketone **28** (34 mg, 81%). IR (KBr): ν_{max} cm^{-1} 3484m (N–H), 3456m (N–H), 1724s (C=O), 1672s (C=C), 1640s (C=C). UV: λ_{max} nm 209, 286. ^1H NMR (400 MHz, $CDCl_3$): δ 5.60 (s, 1H, 1-CH), 5.28 (m, 1H, 15-CH), 3.64 (s, 3H, 24-OMe), 1.21 (m, 6H, 18-Me and 19-Me), 0.92 (d, 3H, $J = 6.6$ Hz). ^{13}C NMR (100 MHz, $CDCl_3$): δ 14.4, 17.7, 19.2, 27.4, 29.5, 30.7, 31.7, 32.4, 33.6, 33.8, 34.6, 36.9, 35.4, 38.2, 38.8, 47.1, 53.0, 60.4, 121.8, 124.9, 150.0, 150.2, 174.29, 198.9, 212.5. EIMS: *m/z* 413 $[M]^+$. HREIMS: *m/z* 413.2513 calculated for $C_{25}H_{35}NO_4$, found 413.2512 (Drögemüller *et al.*, 1998).

For the synthesis of the cephalostatin analogue **29**, a stirred mixture of methyl 2-amino-3,12-dioxo-5β-chola-1,14-dien-24-oate (**28**) (40 mg, 0.10 mmol) and (25*R*)-3-azido-5α-spirost-2-ene-12α-propionate (**22**, 51 mg, 0.10 mmol) in dry dioxane (1 mL) was treated with PPTS (~1.0 mg) (*Scheme 5.2.10*). The solution was degassed with Ar and 4 Å activated molecular sieves (30 mg) were added, the suspension was refluxed for 2.5 h under Ar. After cooling, the mixture was filtered through a bed of SiO_2 using Me_2CO as eluent. The solvent was rotary evaporated and the crude solid was subjected to FCC (mobile phase: *n*-hexane-EtOAc) to obtain **29** (26 mg, 30%). IR ($CHCl_3$): ν_{max} cm^{-1} 1724s (C=O), 1672s (C=C), 1460m (C–H), 1398m (pyrazine). UV: λ_{max} nm 288, 305. ^1H NMR (400 MHz, $CDCl_3$): δ 5.60 (s, 1H, 15′-CH), 5.28 (m, 1H, 15-CH), 4.90 (s, 1H, 12′β-CH), 4.81 (d, 1H, 16′α-CH), 3.62 (s, 3H, 24-OMe), 3.45 (m, 2H, 26′-CH). ^{13}C NMR (100 MHz, $CDCl_3$): δ 9.3, 11.7, 14.1, 17.2, 17.6, 18.7, 19.1, 21.9, 28.1, 28.7, 29.7, 30.4, 30.5, 31.2, 31.3, 31.7, 31.9, 33.5, 33.6,

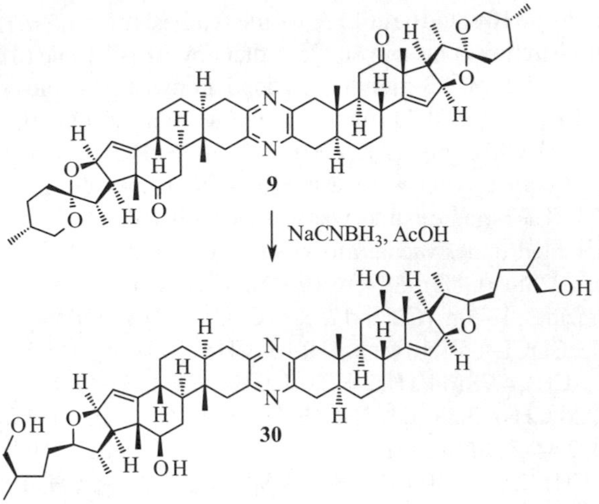

$NaCNBH_3$, AcOH

Scheme 5.2.11 *Synthesis of (5α,5′α,12β,12β,22β,22′β,25R,25′R)-difurosta-2,14-dieno-pyrazine-12,12′,26,26′-tetrol (**30**)*

33.8, 34.2, 34.4, 34.6, 34.9, 35.5, 35.8, 38.3, 38.7, 38.9, 41.4, 41.7, 42.4, 44.5, 45.6, 46.9, 47.0, 49.7, 50.0, 51.5, 52.9, 53.6, 62.7, 67.3, 78.0, 85.2, 106.8, 120.5, 121.1, 147.3, 148.5, 148.6, 149.0, 151.6, 153.4, 173.9, 176.8, 213.4. HRFABMS: *m/z* 877.5731 calculated for $C_{55}H_{76}N_2O_7$, found: 877.5838 (Drögemüller *et al.*, 1998).

Several cephalostatin analogues **30–37** with a modified spiroketal skeleton were prepared *via* successful introduction of a 17-*O*-functionality into hecogenin derivatives with a closed spiroketal moiety. Different remote-oxidation procedures resulted in the synthesis of these tetradecacyclic cephalostatin analogues with improved tumour-inhibiting properties (Bäsler *et al.*, 2000).

To accomplish the synthesis of the dimer, (5α,5′α,12β,12β,22β,22′β,25*R*,25′*R*)-difurosta-2,14-dieno-pyrazine-12,12′,26,26′-tetrol (**30**), a suspension of diketone **9** (170 mg, 0.2 mmol, 1 equiv.) in AcOH (2 mL) was treated with $NaCNBH_3$ (71 mg, 1.1 mmol, 5.5 equiv.) at room temperature (*Scheme 5.2.11*). After 60 min, DCM and saturated aqueous Na_2CO_3 were added and the aqueous layer was extracted with DCM. The organic solvent was separated and dried ($MgSO_4$) and evaporated under vaccum to produce **30** (164 mg, 96%). IR ($CHCl_3$): ν_{max} cm^{-1} 3616br (O–H), 2956s (C–H), 2932s (C–H), 2872s (C–H), 1648w (C=C), 1600w (C=C), 1460m (C–H), 1400s (pyrazine), 1328m (C–H), 1232m (C–O), 1252w (C–O), 1100m (C–O), 1032m (C–O), 964m. ^1H NMR (400 MHz, $CDCl_3$): δ 5.43 (br s, 2H, 2 × 15-CH), 4.76 (dd, 2H, $J = 2.0, 8.5$ Hz, 26-CHa and 26′-CHa), 3.48 (dd, 1H, $J = 6.0, 10.5$ Hz, 26′-CHb), 3.42 (dd, 1H, $J = 6.0, 10.5$ Hz, 26-CHb), 3.31 (t d, 2H, $J = 6.0, 8.5$ Hz, 2 × 22α-CH), 3.23 (dd, 2H, $J = 4.5, 10.5$ Hz, 2 × 12α-CH), 2.76–2.93 (m, 4H, 2 × 1-CHa and 2 × 4-CHa), 2.44–2.64 (m, 4H, 2 × 1-CHb and 2 × 4-CHb), 2.31 (t, 2H, $J = 8$ Hz, 2 × 17α-CH), 0.85 (s, 6H, 2 × 19-Me), 1.04 (d, 6H, $J = 6.5$ Hz, 2 × 21-Me), 1.03 (s, 6H, 2 × 18-Me), 0.89 (d, 6H, $J = 7.0$ Hz, 2 × 27-Me). ^{13}C NMR (100 MHz, $CDCl_3$): δ 11.8, 13.9, 16.6, 16.8, 28.0, 29.2, 30.0, 30.1, 30.3, 33.8, 35.2, 35.8, 35.9, 41.2, 41.4, 45.6, 51.9, 52.9, 59.2, 68.0, 79.2, 86.0, 87.1, 119.8, 148.4, 148.5, 157.6. FABMS: *m/z* 853 $[M + H]^+$ (Bäsler *et al.*, 2000).

Cephalostatin analogue **10** could be utilized for the synthesis of (5α,5′α,12′β,25*R*,25′*R*)-12′-([(*tert*-butyl)dimethylsilyl]oxy)-bisspirosta-2,14-dienopyrazin-12-one (**31**) (*Scheme 5.2.12*). TBSOTf (380 μL, 1.633 mmol, 2.3 equiv.) was added dropwise to a cold (−20 °C) solution of α-hydroxy ketone **10** (600 mg, 0.708 mmol, 1 equiv.) and 2,6-dimethylpyridine (250 μL, 2.13 mmol, 3 equiv.) in DCM (3 mL). After 90 min, 0.5M $NaHCO_3$ (5 mL) and DCM (10 mL) were added and the reaction mixture was warmed to room temperature. The aqueous layer was extracted with $CHCl_3$. The organic extract was washed with 0.3M HC1 followed by brine, dried (Na_2SO_4) and evaporated under vacuum to yield **31** (749 mg, 110%). IR (KBr): ν_{max} cm^{-1} 3052w (CH=CH), 2928m (C–H), 2876m (C–H), 1716m (C=O), 1648w (C=C), 1460m (C–H), 1398m (pyrazine), 1370m (C–H), 1252m (C–O). UV (MeOH): λ_{max} nm 208, 288, 304. ^1H NMR (400 MHz, $CDCl_3$): δ 5.48 (br s, 1H, 15′-CH), 5.43 (br s, 1H, 15-CH), 4.85 (dd, 1H, *J* = 2.0, 8.0 Hz, 16′α-CH), 4.78 (dd, 1H, *J* = 2.0, 8.0 Hz, 16α-CH), 3.50–3.55 (m, 2H, 26′-CH_2), 3.39–3.45 (m, 2H, 26-CH_2), 3.37 (dd, 1H, *J* = 8.0, 9.0 Hz, 17α-CH), 3.22 (dd, 1H, *J* = 5.0, 10.5 Hz, 12′β-CH), 2.44–2.90 (m, 11H, 2 × 1-CHa, 2 × 4-CHa, 2 × 1-CHb and 2 × 4-CHb, 11′-CH_2 and 17′α-CH), 2.02–2.10 (m, 4H, 2 × 8β-CH), 1.33 (s, 3H, 18-Me), 1.05 (d, 3H, *J* = 7.0 Hz, 21-Me), 1.02 (d, 3H, *J* = 7.0 Hz, 21′-Me), 1.01 (s, 3H, 18′-Me), 0.92 (s, 3H, 19-Me), 0.91 (s, 3H, *t*-BuSi), 0.86 (s, 3H, 19′-Me), 0.81 (br d, 6H, *J* = 6.0 Hz, 2 × 27-Me), 0.09 (s, 3H, MeSi), 0.08 (s, 3H, MeSi). FABMS: *m*/*z* 962 $[M + H]^+$ (Bäsler *et al.*, 2000).

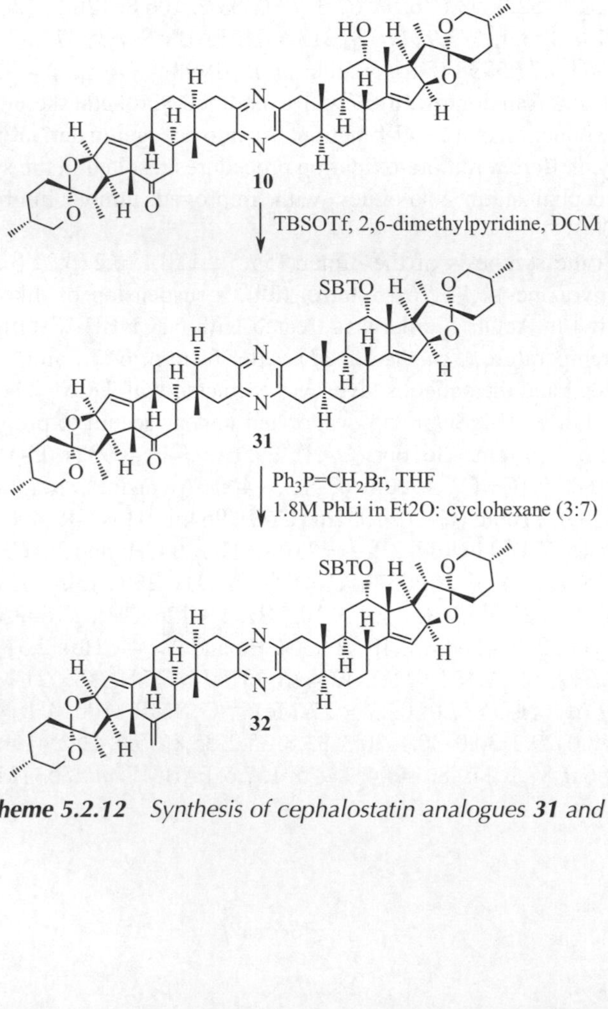

10

TBSOTf, 2,6-dimethylpyridine, DCM

31

$Ph_3P{=}CH_2Br$, THF
1.8M PhLi in Et2O: cyclohexane (3:7)

32

Scheme 5.2.12 *Synthesis of cephalostatin analogues* ***31*** *and* ***32***

Cephalostatin analogue **31** was converted to another analogue, (5α,5′α,12′β,25*R*,25′*R*)-12′-([(*tert*-butyl)dimethylsilyl]oxy)-12-methylene-bisspirosta-2,14-dienopyrazine (**32**) in the following manner (*Scheme 5.2.12*). PhLi in 30% Et_2O in cyclohexane (1.8M 2.83 mL, 5.1 mmol) was added slowly to a cooled (0 °C) suspension of $Ph_3P{=}CH_2Br$ (2.0 g, 5.6 mmol) in THF (12 mL), and the mixture was stirred at 0 °C for 60 min. A part of this mixture (6 mL) was decanted into a solution of **31** (700 mg, 0.662 mmol) in THF (1 mL) and stirred for 5 h. DCM (20 mL) was added and the mixture was washed with half-saturated brine (5 mL). The organic layer was dried (Na_2SO_4), evaporated to dryness and the crude product was subjected to CC (mobile phase: 30% EtOAc in petroleum ether) followed by crystallization from $CHCl_3$ to provide **32** (538 mg, 85%, mp: 236 °C). IR (KBr): ν_{max} cm^{-1} 2928m (C–H), 2876m (C–H), 1632w (C=C), 1460m (C–H), 1400s (pyrazine), 1376m (C–H), 1240m (C–H). UV (MeOH): λ_{max} nm 209, 288, 305. ^{1}H NMR (400 MHz, $CDCl_3$): δ 5.43 (br s, 1H, 15′-CH), 5.35 (br s, 1H, 15-CH), 4.83–4.86 (m, 2H, 2 × 16α-CH), 4.70–4.80 (m, 2H, $CH_2{=}C$), 3.40–3.54 (m, 4H, 2 × 26-CH_2), 3.22 (dd, 1H, *J* = 5.0, 10.5 Hz, 12′β-CH), 2.79–2.99 (m, 5H, 2 × 1-CHa, 2 × 4-CHa, 17α-CH), 2.37–2.67 (m, 7H, 2 × 1-CHb, 2 × 4-CHb, 11-CH_2 and 17′α-CH), 2.28 (m, 1H, 8′β-CH), 2.05 (m, 1H, 8β-CH), 1.22 (s, 3H, 18-Me), 1.09 (d, 3H, *J* = 7.0 Hz, 21-Me), 1.02 (d, 3H, *J* = 7.0 Hz, 21′-Me), 1.01 (s, 3H, 18′-Me), 0.91 (s, 3H, *t*-BuSi), 0.88 (s, 3H, 19-Me), 0.86 (s, 3H, 19′-Me), 0.81 (d, 3H, *J* = 6.0 Hz, 27-Me), 0.80 (d, 3H, *J* = 6.0 Hz, 27′-Me), 0.09 (s, 3H, MeSi), 0.08 (s, 3H, MeSi). ^{13}C NMR (100 MHz, $CDCl_3$): δ −4.2, −3.6, 11.6, 11.8, 13.6, 14.0, 14.1, 17.2, 18.0, 23.7, 26.0, 27.9, 28.2, 28.7, 28.8, 29.2, 29.4, 30.4, 30.5, 30.7, 31.2, 31.5, 32.3, 33.8, 34.6, 35.3, 35.4, 36.0, 36.1, 41.3, 41.5, 44.2, 44.8, 45.5, 45.6, 52.0, 53.3, 53.4, 54.0, 54.8, 55.5, 67.0, 67.1, 79.7, 84.3, 84.5, 106.1, 106.5, 107.4, 117.8, 119.7, 148.3, 148.6, 148.6, 148.7, 155.2, 157.2, 158.8. FABMS: *m/z* 960 $[M + H]^+$ (Bäsler *et al.*, 2000).

Cephalostatin analogue **32** yielded (5α,5′α,12α,12β,12′β,25*R*,25′*R*)-12′-[([(*tert*-butyl) dimethysilyl]oxy)bisspirosta-2,14-dieno] pyrazin-12-methanol (**33**), following the protocol outlined as follows (Scheme 5.2.13). BH_3 in THF (1M, 1.2 ml, 1.2 mmol, 5.8 equiv.) was added to a cold (0 °C) solution of **32** (200 mg, 0.209 mmol, 1 equiv.) in THF (2 mL) and the reaction mixture was stirred at 0 °C for 6 h. NaOH (6N, 1.5 mL) was added slowly to the reaction mixture, followed by the sequential addition of EtOH (200 μL), H_2O (1 mL), and Et_2O (1 mL). After another 30 min, the mixture was warmed to room temperature, and 35% H_2O_2 (400 μL) was added and stirred for further 60 min. Solid NH_4Cl was added to the mixture, the aqueous layer separated and was extracted with DCM. The organic layer was dried (Na_2SO_4), evaporated under pressure and the crude residue was purified by CC (mobile phase: DCM:MeOH = 30:1) to afford inseparable isomeric mixture (α:β = 1:2) of **33** (184 mg, 90%). IR ($CHCl_3$): ν_{max} cm^{-1} 3040w (C–H), 2956s (C–H), 2928s (C–H), 2876s (C–H), 2860s (C–H), 1648w (C=C), 1460m (C–H), 1400s (pyrazine), 1376m (C–H), 1240m (C–H). UV (MeOH): λ_{max} nm 206, 288, 305. ^{1}H NMR (400 MHz, $CDCl_3$): δ 5.43 (br s, 1H, 15′-CH), 5.38 (br s, 1H, 15-CH), 4.87 (dd, 1H, *J* = 1.5, 8.0 Hz, 16α-CH), 4.79 (dd, 1H, *J* = 1.5, 8.0 Hz, 16′α-CH), 3.77 (dd, 1H, *J* = 10 Hz, CH_2O), 3.71–3.74 (m, 1H, CH_2O), 3.47–3.54 (m, 2H, 26′-CH), 3.35–3.47 (m, 3H, 26-CH and CH_2O), 3.22 (dd, 1H, *J* = 4.5, 10.5 Hz, 12′α-CH), 2.78–3.00 (m, 6H, 2 × 1-CHa, 2 × 4-CHa, 17′α-CH and 17α-CH), 2.45–2.66 (m, 6H, 2 × 1-CHb, 2 × 4-CHb, 17′α-CH and 17α-CH), 2.02–2.16 (m, 2H, 2 × 8α-CH), 1.25 (s, 4H, 18-Me and 12α-CH), 1.04 (d, 4H, *J* = 7.0 Hz, 21-Me and 12β-CH), 1.02 (d, 4H, *J* = 7.0 Hz, 21′-Me and 12β-CH), 1.00 (s, 3H, 18′-Me), 0.91 (s, 3H, *t*-BuSi),

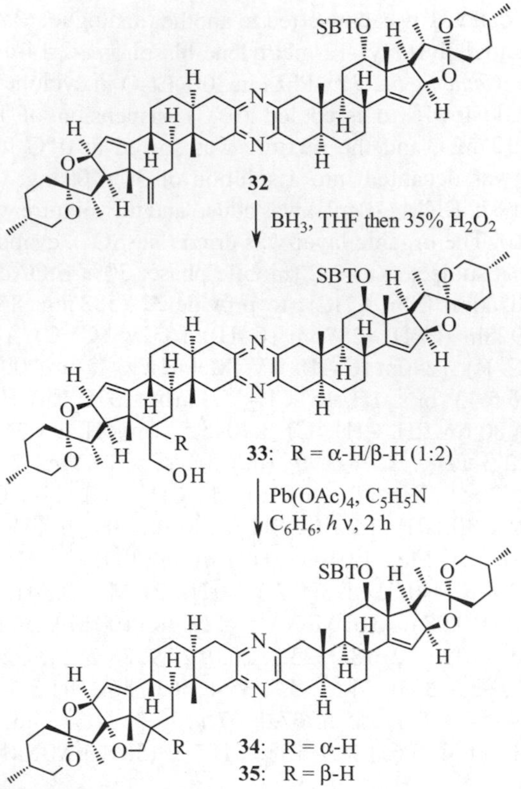

Scheme 5.1.13 *Synthesis of cephalostatin analogues* ***33–35***

0.87 (s, 4H, 19-Me and 12α-CH), 0.86 (s, 3H, 19′-Me), 0.81 (br d, 6H, 2 × 27-Me), 0.09 (s, 3H, MeSi), 0.08 (s, 3H, MeSi). FABMS: *m*/*z* 978 $[M + H]^+$ (Bäsler *et al.*, 2000).

Two further isomeric cephalostatin analogues, (5α,5′α,12α,12′β,25*R*,25′*R*)-12′-[([(*tert*-butyl)dimethylsilyl]oxy)-17,12-epoxymethano]bis-spirosta-2,14-dienopyrazine (**34**) and (5α,5′α,12β,12β,25*R*,25′*R*)-12′-[([(*tert*-butyl)dimethylsilyl]oxy)-17,12-epoxy methano]-bisspirosta-2,14-dienopyrazine (**35**) were synthesized from an isomeric mixture of **33** (Scheme 5.2.13). A solution of Pb(OAc)$_4$ (200 mg, 0.45 mmol, 3 equiv.) in C_5H_5N (50 μL, 0.60 mmol, 4 equiv.) was added to a stirred cold (10 °C) suspension of **33** (150 mg, 0.154 mmol, 1 equiv.) in C_6H_6 (3 mL). The mixture was irradiated for 2 h at the same temperature, the solid were removed, the organic layer was rotary evaporated and the crude solid was purified by gradient CC (mobile phase: petroleum ether:EtOAc = 2:1 followed by DCM:MeOH = 30:1) to furnish **34** (20 mg, 14%) and **35** (50 mg, 34%) (Bäsler *et al.*, 2000).

Cephalostatin analogue **34**: IR (CHCl$_3$): ν_{max} cm^{-1} 3040w (CH=CH), 2956s (C–H), 2928s (C–H), 2872s (C–H), 2860s (C–H), 1644w (C=C), 1460m (C–H), 1400s (pyrazine), 1376m (C–H), 1244m (C–H). UV (MeOH): λ_{max} nm 207, 288, 304. ^{1}H NMR (400 MHz, CDCl$_3$): δ 5.57 (br s, 1H, 15-CH), 5.43 (br s, 1H, 15′-CH′), 4.85 (dd, 1H, *J* = 2.0, 8.0 Hz, 16′α-CH), 4.47 (d, 1H, *J* = 2.5 Hz, 16α-CH), 3.81 (dd, 1H, *J* = 7.0, 8.0 Hz, 17-CH$_2$O), 3.62 (br d, 1H, *J* = 11.0 Hz, 26-CHa), 3.51 (br d, 1H, *J* = 11.0 Hz, 26′-CHa), 3.38–3.46 (m, 2H,

26′-CHb and 26-CHb), 3.22 (dd, 1H, $J = 5.0$, 10.5 Hz, 12′α-CH), 2.77–2.90 (m, 4H, 2 × 1-CHa and 2 × 4-CHa), 2.41–2.67 (m, 5H, 2 × 1-CHb, 2 × 4-CHb and 17′α-CH), 1.15 (s, 3H, 21-Me), 1.02 (d, 3H, $J = 7.0$ Hz, 21′-Me), 1.00 (s, 3H, 18′-Me), 0.93 (d, 3H, $J = 7.0$ Hz, 21-Me), 0.91 (s, 3H, *t*-BuSi), 0.86 (s, 3H, 19′-Me), 0.85 (s, 3H, 19-Me), 0.80 (br d, 6H, 2 × 27-Me), 0.09 (s, 3H, MeSi), 0.08 (s, 3H, MeSi). ^{13}C NMR (100 MHz, $CDCl_3$): δ −4.2, −3.6, 8.4, 11.6, 11.8, 13.6, 14.1, 17.2, 18.0, 20.7, 21.7, 26.0, 27.7, 28.2, 28.4, 28.7, 28.8, 29.2, 30.0, 30.5, 30.7, 31.2, 31.8, 33.8, 35.3, 35.4, 35.4, 36.0, 36.1, 41.5, 41.6, 44.8, 45.5, 45.6, 45.7, 45.9, 50.4, 52.0, 53.3, 55.5, 67.1, 67.2, 68.8, 79.8, 84.5, 90.7, 98.4, 106.5, 107.9, 118.5, 119.7, 148.2, 148.5, 148.6, 148.7, 157.2, 157.5. FABMS: *m/z* 976 $[M + H]^+$ (Bäsler *et al.*, 2000).

Cephalostatin analogue **35**: IR ($CHCl_3$): ν_{max} cm^{-1} 2956s (C–H), 2928s (C–H), 2876s (C–H), 2860s (C–H), 1645w (C=C), 1620w (C=C), 1460m (C–H), 1400s (pyrazine), 1384m (C–H), 1240m (C–H). UV (MeOH): λ_{max} nm 205, 287, 303. ^{1}H NMR (400 MHz, $CDCl_3$): δ 5.43 (br s, 1H, 15-CH), 5.13 (br s, 1H, 15′-CH′), 4.85 (dd, 1H, $J = 2.0$, 8.0 Hz, 16′α-CH), 4.76 (d, 1H, $J = 2.0$ Hz, 16α-CH), 3.96 (dd, 1H, $J = 5.5$, 7.5 Hz, 17-CH_2O), 3.64 (br d, 1H, $J = 11.0$ Hz, 26-CHb), 3.55 (dd, 1H, $J = 7.5$, 12.0 Hz, 17-CH_2O), 3.50 (br d, 1H, $J = 11.0$ Hz, 26′-CHb), 3.45 (t, 1H, $J = 11.0$ Hz, 26-CHa), 3.43 (t, 1H, $J = 11.0$ Hz, 26′-CHa), 3.22 (dd, 1H, $J = 4.5$, 10.5 Hz, 12′α-CH), 2.79–2.91 (m, 4H, 2 × 1-CHa and 2 × 4-CHa), 2.48–2.65 (m, 5H, 2 × 1-CHb, 2 × 4-CHb and 17′α-CH), 2.40 (m, 1H, 8β-CH), 2.05 (m, 1H, 8′β-CH), 1.02 (d, 3H, $J = 7.0$ Hz, 21′-Me), 1.01 (d, 3H, $J = 7$ Hz, 21-Me), 0.97 (s, 3H, 18′-Me), 0.93 (d, 3H, 18-Me), 0.91 (s, 3H, *t*-BuSi), 0.86 (s, 3H, 19′-Me), 0.85 (s, 3H, 19-Me), 0.81 (br d, 6H, 2 × 27-Me), 0.09 (s, 3H, MeSi), 0.08 (s, 3H, MeSi). ^{13}C NMR (100 MHz, $CDCl_3$): δ −4.2, −3.6, 7.8, 11.0, 11.8, 12.2, 13.6, 14.1, 17.1, 17.2, 18.0, 21.2, 26.0, 28.1, 28.4, 28.5, 28.8, 29.0, 29.2, 30.1, 30.4, 30.7, 31.2, 31.6, 33.8, 35.4, 36.0, 36.1, 36.7, 41.5, 42.2, 44.8, 45.6, 45.9, 47.3, 52.0, 53.3, 55.5, 56.8, 57.0, 67.2, 67.4, 71.2, 79.7, 84.5, 93.2, 99.3, 106.5, 106.7, 113.3, 119.7, 148.3, 148.5, 148.6, 148.8, 157.2, 158.7. FABMS: *m/z* 976 $[M + H]^+$ (Bäsler *et al.*, 2000).

The synthesis of cephalostatin analogue, (5α,5′α,12α,12′β,25*R*,25′*R*)-17,12-(epoxymethano)bisspirosta-2,14-dienopyrazin-12′-ol (**36**), was achieved from a mixture of cephalostatin analogue **34** (11.5 mg, 0.012 mmol) in THF (1 mL) and TBAF (20 mg) by constant stirring for 13 days in the dark (*Scheme 5.2.14*). DCM and H_2O were added to the reaction mixture, the aqueous layer was extracted with DCM, the organic layer was separated and dried (Na_2SO_4). The solvent was evaporated under vacuum and the crude residue was subjected to CC (mobile phase: DCM:MeOH = 30:1) to yield **36** (9.1 mg, 90%). IR ($CHCl_3$): ν_{max} cm^{-1} 3605w (O–H), 3065w (C–H), 2956s (C–H), 2928s (C–H), 2872s (C–H), 1456m (C–H), 1400s (pyrazine), 1376m (C–H), 1240m (C–O). UV (MeCN): λ_{max} nm 289, 306. ^{1}H NMR (400 MHz, $CDCl_3$:CD_3OD = 5:1): δ 5.57 (br s, 1H, 15-CH), 5.42 (br s, 1H, 15′-CH), 4.88 (dd, 1H, $J = 1.5$, 8.0 Hz, 16′-CH), 4.46 (d, 1H, $J = 2.5$ Hz, 16α-CH), 3.83 (dd, 1H, $J = 7.0$, 8.5 Hz, 17-CH_2O), 3.59 (br d, 1H, $J = 11.0$ Hz, 26-CHb), 3.52 (br d, $J = 11.0$ Hz, 26′-CHb), 3.45 (t, 1H, $J = 11.0$ Hz, 26-CHa), 3.43 (t, 1H, $J = 11.0$ Hz, 26′-CHa), 3.41 (dd, 1H, $J = 8.5$, 11.5 Hz, 17-CH_2O), 3.21 (dd, 1H, $J = 4.5$, 11.0 Hz, 12′α-CH), 2.79–2.92 (m, 4H, 2 × 1-CHb, 2 × 4-CHb,), 2.42–2.67 (m, 5H, 2 × 1-CHa, 2 × 4-CHa, 17′α-CH), 1.17 (s, 3H, 18-Me), 1.05 (d, 3H, $J = 7.0$ Hz, 21′-Me), 1.03 (s, 3H, 18′-Me), 0.93 (d, 3H, $J = 7.0$ Hz, 21-Me), 0.88 (s, 3H, 19′-Me), 0.87 (s, 3H, 19-Me), 0.83 (d, 3H, $J = 6.0$ Hz, 27′-Me), 0.82 (d, 3H, $J = 6.0$ Hz, 27-Me). ^{13}C NMR (100 MHz, $CDCl_3$:

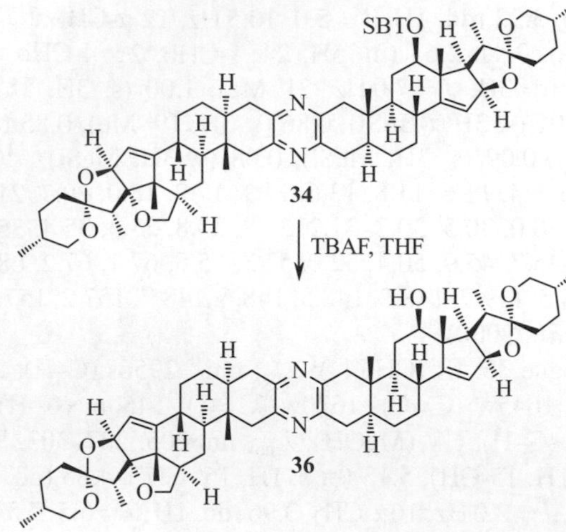

SBTO H O N N H H H H H H H H H O O O O H H H 34

TBAF, THF

HO H O N N H H H H H H H H H O O O O H H H 36

Scheme 5.2.14 *Synthesis of (5α,5′α,12α,12′β,25R,25′R)-17,12-(epoxymethano) bisspirosta-2,14-dienopyrazin-12′-ol (**36**)*

$CD_3OD = 5{:}1$): δ 8.4, 11.8, 12.0, 13.4, 13.6, 17.1, 17.2, 20.8, 21.8, 27.8, 28.2, 28.5, 28.8, 29.0, 29.4, 29.9, 30.2, 30.5, 31.3, 31.8, 33.9, 35.0, 35.1, 35.6, 36.1, 36.2, 41.5, 41.6, 44.7, 45.3, 45.4, 45.7, 46.0, 50.5, 52.4, 53.0, 55.6, 55.7, 67.2, 67.4, 69.0, 78.7, 85.0, 90.9, 98.8, 107.2, 108.3, 118.5, 119.6, 148.6, 148.8, 148.9, 149.1, 157.8, 157.9. HRMS: *m/z* 861.5782 calculated for $C_{55}H_{77}N_2O_6$, found: 861.5811 (Bäsler *et al.*, 2000).

Using the similar protocol to the one outlined above for **36**, (5α,5′α,12β,12′β,25*R*,25′*R*)-17,12-(epoxymethano)bisspirosta-2,14-dienopyrazin-12′-ol (**37**) could be prepared from cephalostatin analogue **35** (*Scheme 5.2.15*). MeCN (1.5 mL) and 48% HF (200 μL) solutions were added to a stirred solution of **35** (20.4 mg, 0.021 mmol) in DCM (0.5 mL), and the mixture was kept in a *Nalgene* vessel for 16 h. Saturated aqueous $NaHCO_3$ was added to the reaction mixture and the aqueous layer was extracted with DCM. The organic layer was washed with saturated aqueous $NaHCO_3$, dried (Na_2SO_4), evaporated under vacuum and the crude product was purified by CC (mobile phase: DCM:MeOH = 30:1) to obtain **37** (13 mg, 72%) (Bäsler *et al.*, 2000).

IR ($CHC1_3$): ν_{max} cm^{-1} 3610w (O–H), 2956s (C–H), 2932s (C–H), 2872s (C–H), 1448m (C–H), 1400s (pyrazine), 1384m (C–H), 1224s (C–O). UV (MeCN): λ_{max} nm 290, 308. ^1H NMR (400 MHz, $CDCl_3$:$CD_3OD = 5{:}1$): δ 5.42 (br s, 1H, 15′-CH), 5.12 (br s, 1H, 15-CH), 4.88 (dd, 1H, *J* = 1.5, 8.0 Hz, 16′-CH), 4.75 (d, 1H, *J* = 2.0 Hz, 16α-CH), 3.96 (dd, 1H, *J* = 5.5, 7.5 Hz, 17-CH_2O), 3.61 (br d, 1H, *J* = 11.0 Hz, 26-CHb), 3.58 (dd, 1H, *J* = 7.5, 12.0 Hz, 17-CH_2O), 3.53 (br d, *J* = 11.0 Hz, 26′-CHb), 3.52 (t, 1H, *J* = 11.0 Hz, 26-CHa), 3.43 (t, 1H, *J* = 11.0 Hz, 26′-CHa), 3.21 (dd, 1H, *J* = 4.5, 11.0 Hz, 12′α-CH), 2.80–2.93 (m, 4H, 2 × 1-CHb, 2 × 4-CHb,), 2.49–2.66 (m, 5H, 2 × 1-CHa, 2 × 4-CHa, 17′α-CH), 2.40 (br t, 1H, *J* = 10.0 Hz, 8β-CH), 2.08 (m, 1H, 8′β-CH) 1.05 (d, 3H, *J* = 7.0 Hz, 21′-Me), 1.03 (s, 3H, 18′-Me), 1.02 (d, 3H, *J* = 7.0 Hz, 21-Me), 0.99 (s, 3H, 18-Me), 0.88 (s, 3H, 19-Me), 0.82 (d, 3H, *J* = 6.0 Hz, 27′-Me), 0.81 (d, 3H, *J* = 6.0 Hz, 27-Me). ^{13}C NMR (100 MHz, $CDCl_3$:$CD_3OD = 5{:}1$): δ 7.8, 11.1, 12.0, 12.4, 13.4, 13.6, 17.1, 17.2, 21.4, 28.2, 28.5, 28.6, 28.8, 29.2, 29.4, 29.9, 30.2, 30.5, 31.3, 31.5, 33.9, 35.0, 36.1, 36.9, 41.6, 42.3, 44.7, 45.4, 45.6,

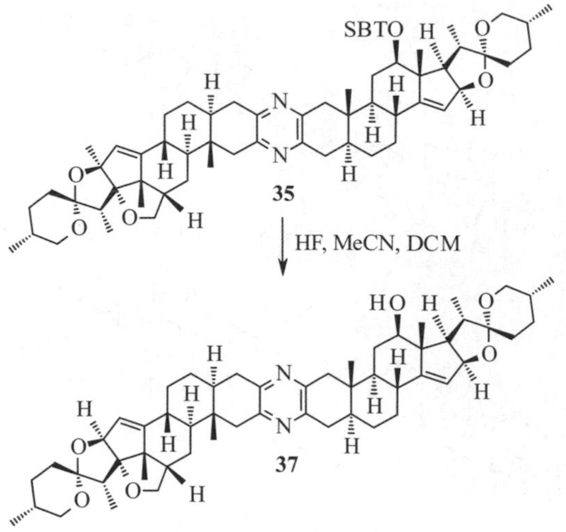

Scheme 5.2.15 *Synthesis of (5α,5′α,12β,12′β,25R,25′R)-17,12-(epoxymethano)bisspirosta-2,14-dienopyrazin-12′-ol (**37**)*

47.4, 52.4, 53.0, 55.7, 57.0, 57.2, 58.7, 67.4, 71.3, 78.7, 85.0, 93.4, 99.8, 107.0, 107.2, 113.5, 119.6, 148.8, 148.90, 148.95, 149.1, 157.7, 158.9. HRMS: *m/z* 861.5782 calculated for $C_{55}H_{77}N_2O_6$, found: 861.5838 (Bäsler *et al.*, 2000).

Highly oxygenated cephalostatin analogues such as spiroketal pyrazines **38–40**, were synthesized and the importance of the $\Delta^{14,15}$ double bond for their biological activity was established (Jautelat *et al.*, 1999). Pyrazine triketal ketone **38** was prepared from the reaction between a cold (0 °C) suspension of cephalostatin analogue **10** (105 mg, 0.124 mmol, 1 equiv.) in a mixture of EtOAc-CH_3CN (6.5 mL) and a solution of $RuCl_3.H_2O$ (20 mg, 0.08 mmol, 0.6 equiv.) and $NaIO_4$ (65 mg, 0.304 mmol, 2.4 equiv.) in H_2O (650 μL) under vigorous stirring (*Scheme 5.2.16*). After 48 h, $Na_2S_2O_3$ (220 mg, 1.40 mmol, 4.5 equiv. to $NaIO_4$) was added to the reaction mixture and stirred for another 30 min. The solution was warmed to room temperature, SiO_2 (1 g) and EtOAc (10 mL) were added and solid was filtered off. The filtrate was evaporated under vacuum and the crude residue was subjected to FCC (mobile phase: DCM:MeOH = 30:1) to produce triketal ketone **38** (34 mg, 31%). IR ($CHCl_3$): ν_{max} cm^{-1} 3586br (O–H), 2956m (C–H), 2932m (C–H), 2876m (C–H), 1708s (C=O), 1645w (alkene C=C), 1456m (C–H), 1400s (pyrazine), 1360m (C–H), 1228m (C–O). UV (MeCN): λ_{max} nm 289, 307. ^{1}H NMR (400 MHz, $CDCl_3$): δ 5.48 (s, 1H, 15′-CH), 5.20 (d, 1H, *J* = 3 Hz, 15-CH), 4.78 (dd, 1H, *J* = 2.0, 8.0 Hz, 16′α-CH), 4.11 (dd, 1H, *J* = 9.0 Hz, 3 Hz, 16α-CH), 3.76 (s, 1H, 12β-CH), 3.49–3.55 (m, 2H, 26-CHb and 26′-CHb), 3.43 (t, 1H, *J* = 11.0 Hz, 26′-CHa), 3.42 (t, 1H, *J* = 11.0 Hz, 26′-CHa), 3.37 (1H, dd, 1H, *J* = 8.0, 9.0 Hz, 17′α-CH), 2.44–2.95 (m, 14H, 4 × CH_2-C=N and 14-OH, 2 × 20-CH, 8′β-CH, 11′-CH_2,), 1.33 (s, 3H, 18′-Me), 1.06 (s, 3H, 18-Me), 1.05 (d br, 6H, *J* = 7.0 Hz, 2 × 21-Me), 0.92(s, 3H, 19′-Me), 0.82 (s, 3H, 19-Me), 0.80 (d br, 6H, *J* = 6 Hz, 2 × 27-Me). ^{13}C NMR (100 MHz, $CDCl_3$): δ 11.5, 12.1, 13.8, 15.6, 15.8, 17.1, 20.8, 22.9, 25.6, 27.9, 28.0, 28.8, 29.2, 30.2, 30.3, 31.3, 31.6, 34.0, 35.2, 35.3, 35.4, 36.4, 36.7, 37.2, 40.4, 41.2, 43.3, 44.0, 44.2, 45.2, 45.3, 49.8, 52.2, 53.1, 62.3, 67.0, 67.1, 74.0, 79.0, 83.9, 90.5, 100.3, 106.4,

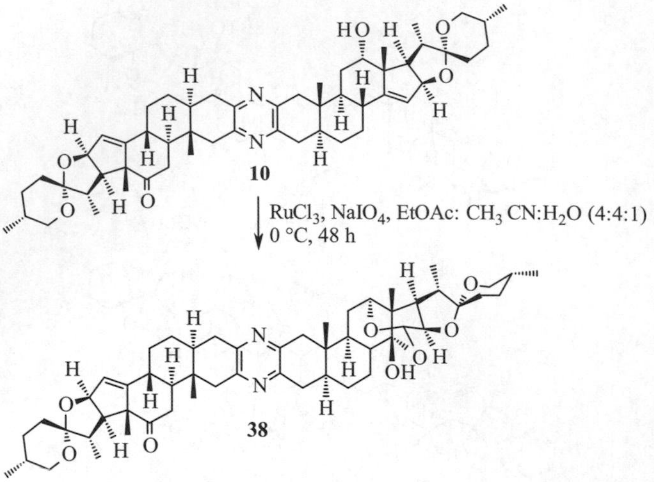

$RuCl_3$, $NaIO_4$, EtOAc: CH_3CN:H_2O (4:4:1)
0 °C, 48 h

Scheme 5.2.16 *Synthesis of pyrazine triketal ketone **38***

107.1, 121.6, 148.0, 148.8, 149.0, 154.3, 210.7. HRMS: *m/z* 879.5523 calculated for $C_{54}H_{75}N_2O_8$, found: 879.5490 (Jautelat *et al.*, 1999).

Pyrazines glycol-triketal **39** and tetraketal **40** could be synthesized from cephalostatin analogue **12**, following the same method as described for pyrazine triketal ketone **38** (*Scheme 5.2.17*). A solution of $RuCl_3.H_2O$ (40 mg, 0.16 mmol, 1 equiv.) and $NaIO_4$ (150 mg, 0.702 mmol, 4.5 equiv.) in H_2O (1.0 mL) was added to a cold (0 °C) suspension of **12** (132 mg, 0.155 mmol, 1 equiv.) in a mixture of EtOAc-CH_3CN (8 mL) and the mixture was stirred for 12 h. Usual workup and purification as above afforded **39** (37 mg, 26%) and the FCC of the concentrated second eluent, using the same mobile phase as above, produced **40** (41 mg, 29%) (Jautelat *et al.*, 1999).

Pyrazine glycol-triketal **39**: IR ($CHC1_3$): ν_{max} cm^{-1} 3592br (O–H), 3460br (O–H), 2932s (C–H), 2876m (C–H), 1620w (C=C), 1456m (C–H), 1400s (pyrazine), 1264s (C–O), 1240m (C–O), 1048s (C–O). UV (MeCN): λ_{max} nm 290, 306. ^{1}H NMR (400 MHz, $CDCl_3$): δ 5.20 (d, 1H, $J = 3.0$ Hz, 15-CH), 4.49 (t, 1H, $J = 7.5$ Hz, 16′α-CH), 4.28 (dd, 1H, $J = 6.0$, 7.5 Hz, 15′-CH), 4.11 (dd, 1H, $J = 3.0$, 9.5 Hz, 16α-CH), 3.76 (s, 1H, 12β-CH), 3.58 (s, 1H, 12′β-CH), 3.49–3.53 (m, 2H, 26-CHb and 26′-CHb), 3.40–3.46 (m, 3H, 15-OH, 26-CHa and 26′-CHa), 3.07 (s, 1H, 14′-OH), 2.94 (q, 1H, $J = 7.0$ Hz, 20β-CH), 2.79–2.90 (m, 4H, 2 × CH_2-C=N), 2.54–2.63 (m, 4H, 2 × CH_2-C=N), 2.51 (s, 1H, 14-OH), 2.45 (d q, 1H, $J = 7.0$, 9.5 Hz, 20′β-CH), 2.26 (dd, 1H, $J = 8.0$, 9.5 Hz, 17′α-CH), 1.08 (s, 3H, 18′-Me), 1.06 (s, 3H, 18-Me), 1.05 (d, 3H, $J = 7.0$ Hz, 21-Me), 0.95 (d, 3H, $J = 7.0$ Hz, 21′-Me), 0.79–0.82 (m, 12H, 2 x19-Me and 2 × 27-Me). ^{13}C NMR (100 MHz, $CDCl_3$): δ 12.06, 12.13, 14.8, 15.5, 15.6, 15.8, 17.1, 17.2, 22.8, 25.6, 25.8, 28.0, 28.4, 28.7, 28.8, 29.1, 30.2, 30.3, 31.5, 31.7, 35.1, 35.3, 35.4, 35.5, 36.7, 40.4, 41.0, 41.1, 41.2, 41.8, 43.3, 44.0, 45.1, 45.3, 45.8, 48.0, 52.2, 57.6, 67.0, 72.4, 74.0, 75.7, 79.0, 79.2, 82.6, 90.4, 100.3, 106.4, 107.1, 148.3, 148.5, 148.7, 148.9. HRMS: *m/z* 915.5735 calculated for $C_{54}H_{79}N_2O_{10}$, found: 915.5731 (Jautelat *et al.*, 1999).

Pyrazine tetraketal **40**: IR ($CHC1_3$): ν_{max} cm^{-1} 3588br (O–H), 3428br (O–H), 3056w (C–H), 2932s (C–H), 2876m (C–H), 1652w (C=C), 1616w (C=C), 1456m (C–H), 1400s

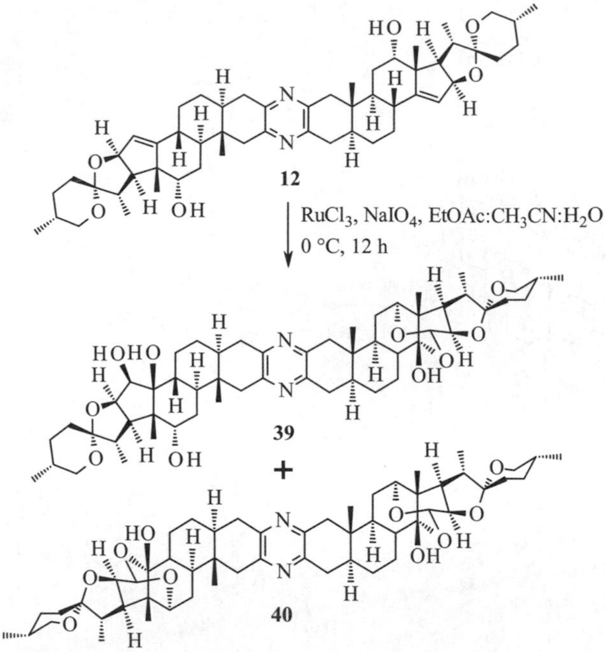

12

$RuCl_3$, $NaIO_4$, EtOAc:CH_3CN:H_2O
0 °C, 12 h

39 + **40**

Scheme 5.2.17 *Synthesis of pyrazines glycol-triketal **39** and tetraketal **40***

(pyrazine), 1372m (C–H), 1240m (C–O), 1044s (C–O). UV (MeCN): λ_{max} nm 288, 306. ^1H NMR (400 MHz, $CDCl_3$): δ 5.20 (d, 2H, J = 3.0 Hz, 2 × 15-CH), 4.11 (dd, 2H, J = 3.0, 9.0 Hz, 2 × 16α-CH), 3.76 (s, 2H, 2 × 12β-CH), 3.51 (d br, 2H, J = 11.0 Hz, 2 × 26-CHb), 3.43 (t, 2H, J = 11.0 Hz, 2 × 26-CHa), 2.94 (q, 2H, J = 7.0 Hz, 2 × 20-CH), 2.80–2.87 (m, 4H, 2 × CH_2-C=N), 2.55–2.64 (m, 4H, 2 × CH_2-C=N), 1.06 (s, 6H, 2 × 18-Me), 2.52 (s, 2H, 2 × 14-OH), 1.05 (d, 6H, J = 7.0 Hz, 2 × 21-Me), 0.81 (d br, 6H, J = 6.0 Hz, 2 × 27-Me), 0.80 (s, 6H, 2 × 19-Me). ^{13}C NMR (100 MHz, $CDCl_3$): δ 12.1, 15.6, 15.8, 17.2, 22.9, 25.6, 28.1, 28.8, 30.2, 31.6, 35.3, 35.4, 36.8, 40.5, 41.2, 43.3, 44.0, 45.3, 52.2, 67.0, 74.0, 79.0, 90.4, 100.3, 106.4, 148.4, 148.7. HRMS: m/z 913.5578 calculated for $C_{54}H_{77}N_2O_{10}$, found: 913.5748 (Jautelat *et al*., 1999).

The synthesis of di(17β-hydroxy-5α-androstano)pyrazine (**44**) from androstanolone (**41**, 17β-hydroxy-5α-androstan-3-one) was accomplished in three steps as follows (Smith and Hicks, 1971). First, 2-oximino-17β-hydroxy-5α-androstan-3-one (**42**) was prepared from **41** by nitrosation with *t*-BuOH and *t*-BuOK using 2-octyl nitrite under N_2 (*Scheme 5.2.18*). Acidified H_2O was added and the resulting gelatinous solid was extracted into Et_2O, the ethereal layer was evaporated to dryness, the crude residue was thoroughly washed with H_2O and recrystallized twice from MeOH, to provide **42** (12%, mp: 265–268 °C). IR (KBr): ν_{max} cm^{-1} 3150br (O–H), 3480br (O–H), 3620br (O–H), 1720s (C=O), 1620s (C=N) (Smith and Hicks, 1971).

Reduction of 2-oximino-17β-hydroxy-5α-androstan-3-one (**42**) in ethanolic HCl with H_2 over Pd-C (10% on carbon), followed by recrystallization from *i*-PrOH-Et_2O (Norit) and then from MeOH-Et_2O yielded 2α-amino-17β-hydroxy-5α-androstan-3-one hydrochloride

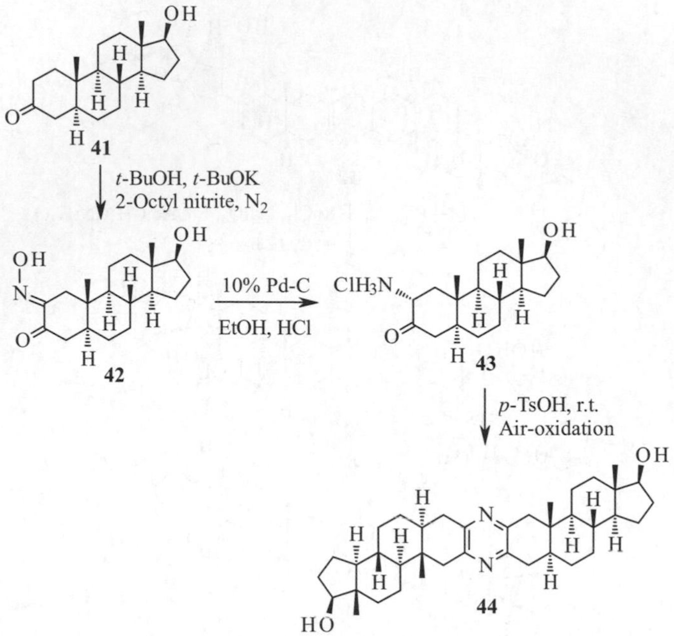

Scheme 5.2.18 *Synthesis of di(17β-hydroxy-5α-androstano)pyrazine (**44**)*

(**43**, 14%, mp: >300 °C). IR (KBr): ν_{max} cm^{-1} 3450br (O–H), 3245br (O–H), 1720s (C=O). ^1H NMR (60 MHz, CD_3OD): δ 4.78 (s, 4H, NH_3 and OH), 4.18 (m, 1H, 2-CH), 1.20 (s, 3H, 19-Me), 0.77 (s, 3H, 18-Me) (Smith and Hicks, 1971).

Di(17β-hydroxy-5α-androstano)pyrazine (**44**) was achieved from 2α-amino-17β-hydroxy-5α-androstan-3-one hydrochloride (**43**) (*Scheme 5.2.18*). Saturated aqueous Na_2CO_3 was added to an aqueous solution of 2α-amino-17β-hydroxy-5α-androstan-3-one hydrochloride (**43**, 431 mg, 1.26 mmol) and the mixture was extracted with Et_2O. The organic layer was dried (Na_2SO_4) and evaporated under pressure. The dried solid was dissolved in absolute EtOH, *p*-TsOH (27 mg, 0.16 mmol) was added and the mixture was stirred for 30 min to form precipitates, which were separated by filtration, washed with absolute EtOH, and dried under vacuum. The crude product was purified by recrystallization from absolute EtOH-$CHCl_3$ to furnish **44** (196 mg, 54%, mp: >300 °C). IR (KBr): ν_{max} cm^{-1} 3460br (O–H), 3440br (O–H), 1445s (C–H), 1400s (pyrazine). ^1H NMR (60 MHz, $CDCl_3$): δ 0.82 (s, 6H, 2 × 19-Me), 0.78 (s, 6H, 2 × 18-Me) (Smith and Hicks, 1971).

The synthesis of bis-acryloylated derivative **45** was achieved from di(17β-hydroxy-5α-androstano)pyrazine (**44**) in a single step reaction (Tanaka *et al.*, 2005). Acryloyl chloride (3.12 mL, 38.4 mmol) was added slowly to a stirred mixture of **44** (10 mg, 17.5 mmol) and *i*-Pr_2NEt (6.08 mL, 34.9 mmol) in DCM (2 mL) at room temperature (*Scheme 5.2.19*). After 3.5 h, H_2O and saturated aqueous NH_4Cl were added and the resulting mixture was extracted with $CHCl_3$. The organic solvent was washed with brine, dried (Na_2SO_4), rotary evaporated and to crude residue was subjected to CC (mobile phase: 17–25% EtOAc in *n*-hexane) to give **45** (9.1 mg, 77%).

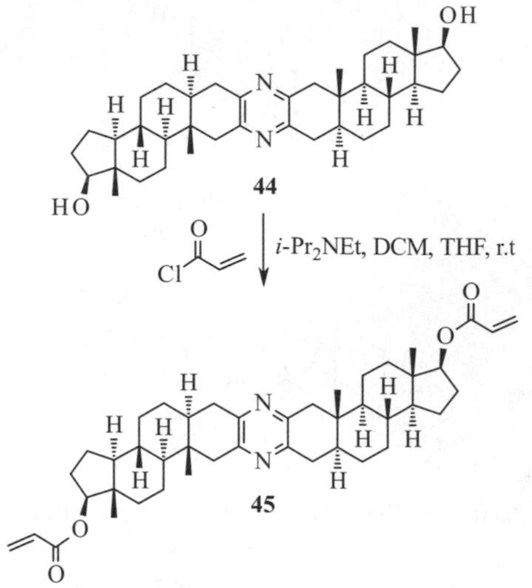

Scheme 5.2.19 Synthesis of bis-acryloylated derivative **45**

^{1}H NMR (300 MHz, $CDCl_3$): δ 6.39 (dd, 2H, $J = 1.6$, 17.3 Hz), 6.12 (dd, 2H, $J = 10.4$, 17.3 Hz), 5.81 (dd, 2H, $J = 1.6$, 10.4 Hz), 4.70 (dd, 2H, $J = 8.0$ Hz), 2.91 (d, 2H, $J = 16.7$ Hz), 2.78 (dd, 2H, $J = 5.2$, 17.8 Hz), 2.50–2.63 (m, 4H), 2.15–2.27 (m, 2H), 0.88–1.84 (m, 30H), 0.84 (s, 6H), 0.81 (s, 6H). HRMS: *m/z* 681.4632 calculated for $C_{44}H_{61}O_4N_2$, found: 681.4631 (Tanaka *et al.*, 2005).

Symmetrical pyrazine dimer could be synthesized *via* classical dimerization of 2α-azidoketones or 2α-bromoketones. Cholestane pyrazines, *trans*-di(5α-cholestano)pyrazine (**47**) was obtained in one step, but *cis*-di(5α-cholestano)pyrazine (**54**) was prepared from 2α-azido-5α-cholestan-3-one (**46**) in several steps (Heathcock and Smith, 1994). Ph_3P (184 mg, 0.702 mmol, 3 equiv.) was added to a stirred solution of **46** (100 mg, 0.234 mmol) in THF (2 mL) under N_2 (*Scheme 5.2.20*). When gas started evolving from the solution, H_2O (100 μL, 5.6 mmol) was added by syringe, and stirred at room temperature for 24 h to form precipitates. The reaction mixture was concentrated under vacuum and the solid was azeotroped with toluene to remove H_2O. Absolute EtOH (8 mL) and *p*-TsOH (5.0 mg) were added to the mixture was stirred at room temperature for 16 h, the resulting solid was filtered through a pad of Celite and the filtrate was discarded. The solid was dissolved in $CHCl_3$ (20 mL) and absolute EtOH (5 mL), and concentrated until 3–4 mL of solvent remained. The resulting suspension was filtered over Celite, and the solid was redissolved in $CHC1_3$, evaporated under vacuum and recrystallized from hot toluene to obtain **47** (78.6 mg, 87%, mp: >265 °C). ^{1}H NMR (500 MHz, $CDCl_3$): δ 2.90 (d, 2H, $J = 16.5$ Hz), 2.76 (dd, 2H, $J = 5.2$, 17.7 Hz), 2.58 (dd, 2H, $J = 13.2$, 17.6 Hz), 2.50 (d, 2H, $J = 16.5$ Hz), 2.03 (m, 2H, $J = 2.5$, 12.9 Hz), 1.79–1.85 (m, 2H), 1.72–1.75 (m, 2H, $J = 12.8$ Hz), 1.58–1.67 (m, 8H), 1.52 (m, 2H, $J = 6.6$ Hz), 1.43 (dq, 2H, $J = 3.6$, 13.1 Hz), 0.92 (d, 6H, $J = 6.5$ Hz), 0.85–1.38 (m, 32H), 0.87 (d, 6H, $J = 6.5$ Hz), 0.86 (d, 6H, $J = 6.6$ Hz), 0.79 (s, 6H), 0.68 (s, 6H). ^{13}C NMR (125 MHz, $CDCl_3$): δ 12.0, 18.7, 21.2, 22.6, 22.8, 23.9, 24.3, 28.0, 28.2, 28.5, 31.6, 35.4, 35.6, 35.6, 35.8, 36.2, 39.5, 39.9, 41.8, 42.5, 46.0,53.7, 56.31, 56.4, 148.5, 149.0. EIMS: *m/z* 764 $[M]^+$ (Heathcock and Smith, 1994).

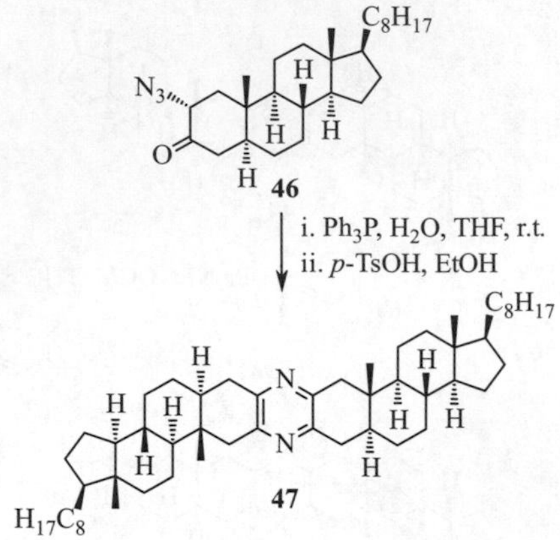

Scheme 5.2.20 *Synthesis of* trans-*di(5α-cholestano)pyrazine* (**47**)

2α-Azido-5α-cholestan-3-one (**46**) was employed to produce 2α-azido-3-hydroxyimino-5α-cholestane (**48**), which in turn was converted to 2α-amino-3-hydroxyimino-5α-cholestane (**49**) (*Scheme 5.2.21*). To a solution of **46** (87.7 mg, 0.205 mmol) in C_5H_5N (8 mL), $HONH_2$.HCl (143 mg, 2.05 mmol) was added and stirred at room temperature for 30 min. The reaction mixture was filtered through a pad of SiO_2 (mobile phase: EtOAc) and the filtrate was evaporated under vacuum to yield **48** (91 mg, 100%, mp: 68–72 °C). ^{1}H NMR (500 MHz, $CDCl_3$): δ 4.02 (dd, 1H, $J = 5.4$, 12.3 Hz), 3.12 (dd, 1H, $J = 3.0$, 15.1 Hz), 2.21 (dd, 1H, $J = 5.5$, 12.5 Hz), 2.00 (dt, 1H, $J = 12.7$ Hz), 1.78–1.84 (m, 1H), 1.64–1.70 (m, 2H), 1.43–1.58 (m, 4H), 0.96–1.37 (m, 19H), 0.93 (s, 3H), 0.90 (d, 3H, $J = 6.5$ Hz), 0.87 (d, 3H, $J = 6.6$ Hz), 0.86 (d, 3H, $J = 6.6$ Hz), 0.72–0.77 (m, 1H, CH), 0.66 (s, 3H, Me). ^{13}C NMR (125 MHz, $CDCl_3$): δ 12.1, 12.4, 18.6, 21.3, 22.5, 22.8, 23.8, 24.2, 26.9, 28.0, 28.1, 28.2, 31.5, 34.9, 35.8, 36.1, 36.6, 39.5, 39.8, 42.5, 44.8, 45.0, 53.8, 56.2, 58.4, 156.7. FABMS: *m*/*z* 443 $[M + H]^+$ (Heathcock and Smith, 1994).

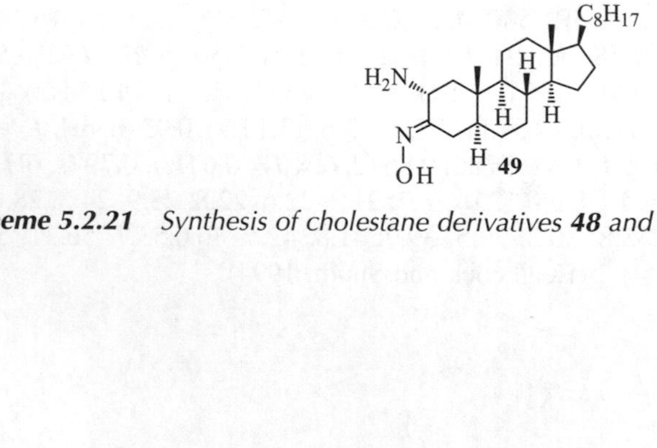

Scheme 5.2.21 *Synthesis of cholestane derivatives* **48** *and* **49**

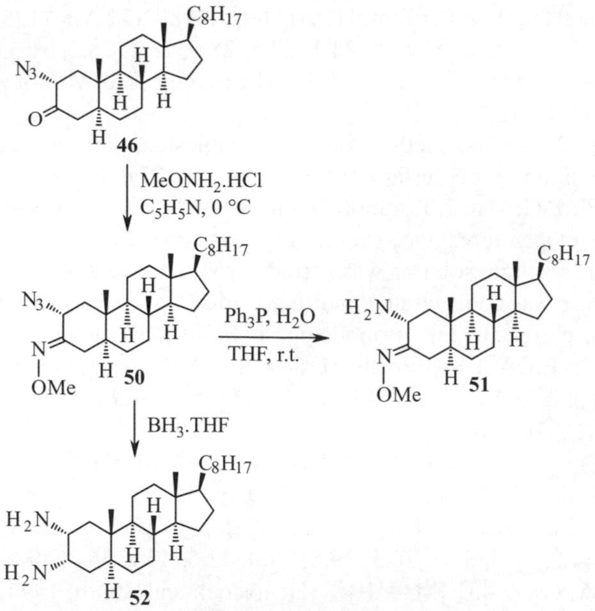

46

$MeONH_2.HCl$
C_5H_5N, 0 °C

50

Ph_3P, H_2O
THF, r.t.

51

$BH_3.THF$

52

Scheme 5.2.22 *Synthesis of cholestane derivatives **50–52***

To a mixture of **48** (91 mg, 0.205 mmol) and Ph_3P (161 mg, 0.615 mmol, 3 equiv.), THF (1 mL) followed by H_2O (80 mL) were added by syringe under N_2 at room temperature (*Scheme 5.2.21*). After 24 h, the reaction mixture was concentrated under pressure, the residue was azeotroped with toluene to remove H_2O, and the crude solid was purified by FCC (mobile phase: MeOH:DCM:NH_4OH = 10:90:0.5) followed by crystallization from THF to provide 2α-amino-3-hydroxyimino-5α-cholestane (**49**, 78.6 mg, 92%). ^1H NMR (500 MHz, $CDCl_3$): δ 3.53 (dd, 1H, J = 5.1, 12.3 Hz), 3.07 (dd, 1H, J = 2.4, 15.1 Hz), 2.17 (dd, 1H, J = 5.1, 12.5 Hz), 1.97 (dt, 1H, J = 3.3, 12.7 Hz), 1.77–1.83 (m, 1H), 0.95–1.69 (m, 27H), 0.93 (s, 3H), 0.89 (d, 3H, J = 6.6 Hz), 0.86 (d, 3H, J = 6.6 Hz), 0.85 (d, 3H, J = 6.6 Hz), 0.68–0.71 (m, 1H), 0.66 (s, 3H). ^{13}C NMR (125 MHz, $CDCl_3$): δ 12.1, 12.5, 18.7, 21.3, 22.5, 22.8, 23.8, 24.2, 26.8, 28.0, 28.2, 28.3, 31.7, 34.9, 35.8, 36.1, 36.5, 39.5, 39.9, 42.6, 45.9, 48.3, 49.6, 54.0, 56.3, 110.1, 160.8. FABMS: m/z 417 $[M + H]^+$ (Heathcock and Smith, 1994).

2α-Azido-5α-cholestan-3-one (**46**) was utilized for the synthesis of other cholestane derivatives **50–52** (*Scheme 5.2.22*). To a stirring solution of **46** (6.0 g, 14.0 mmol) in C_5H_5N (300 mL) at 0 °C, $MeONH_2.HCl$ (3.51 g, 42.0 mmol, 3 equiv.) was added. Once a fine suspension was formed, EtOAc (300 mL) was added and the mixture was passed through a pad of SiO_2, eluting with EtOAc. The filtrate was evaporated to dryness to afford 2α-azido-3-methoxyimino-5α-cholestane (**50**, 6.39 g, 100%, mp: 132–135 °C). ^1H NMR (500 MHz, $CDCl_3$): δ 3.98 (dd, 1H, J = 5.5, 12.0 Hz), 3.91 (s, 3H), 3.00 (dd, 1H, J = 3.3, 15.2 Hz), 2.18 (dd, 1H, J = 5.3, 12.5 Hz), 1.98 (dt, 1H, J = 3.1, 12.9 Hz), 1.78–1.83 (m, 1H), 1.58–1.69 (m, 2H), 0.95–1.52 (m, 22H), 0.90 (d, 3H, J = 3.0, 6.2 Hz), 0.86 (d, 3H, J = 6.6 Hz),

0.85 (d, 3H, $J = 6.6$ Hz), 0.71–0.76 (m, 1H), 0.66 (s, 3H). ^{13}C NMR (125 MHz, $CDCl_3$): δ 12.0, 12.4, 18.6, 21.2, 22.5, 22.8, 23.8, 24.1, 27.5, 28.0, 28.1, 28.2, 31.5, 34.9, 35.7, 36.1, 36.5, 39.5, 39.8, 42.5, 44.9, 53.8, 56.2, 57.8, 61.9, 155.7. FABMS: *m/z* 457 $[M + H]^+$ (Heathcock and Smith, 1994).

In the next step, 2α-azido-3-methoxyimino-5α-cholestane (**50**) was transformed to 2α-amino-3-methoxyimino-5α-cholestane (**51**) (*Scheme 5.2.22*). To a mixture of **50** (1.07 g, 2.34 mmol) and Ph_3P (1.84 g, 7.02 mmol, 3 equiv.), THF (30 mL) was added by a syringe under N_2 at room temperature. Once evolution of gas had started, H_2O (900 μL, 50 mmol) was added, and the resulting solution was stirred at room temperature for 48 h, after which, the reaction mixture was concentrated and the residue was azeotroped with toluene. The resulting solid was purified by FCC (mobile phase: EtOAc to remove Ph_3P and Ph_3PO, then 10–30% MeOH in EtOAc) to furnish **51** (900 mg, 89%, mp: 113–115 °C). ^1H NMR (500 MHz, $CDCl_3$): δ 3.84 (s, 3H), 3.47 (dd, 1H, $J = 5.0$, 12.2 Hz), 2.95 (dd, 1H, $J = 3.3$, 15 Hz), 2.35 (br s, 2H), 2.16 (dd, 1H, $J = 5.2$, 12.5 Hz), 1.97 (dt, 1H, $J = 3.4$, 12.6 Hz), 1.77–1.83 (m, 1H), 0.92–1.68 (m, 24H), 0.91(s, 3H), 0.89 (d, 3H, $J = 6.5$ Hz), 0.86 (d, 3H, $J = 6.6$ Hz), 0.85 (d, 3H, $J = 6.6$ Hz), 0.65–0.70 (s, 3H, and also overlapping m, 1H). ^{13}C NMR (125 MHz, $CDCl_3$): δ 12.1, 12.5, 18.7, 21.2, 22.5, 22.8, 23.8, 24.2, 27.4, 28.0, 28.2, 28.3, 31.7, 34.9, 35.8, 36.2, 36.5, 39.5, 39.9, 42.6, 45.9, 48.7, 49.6, 54.0, 56.2, 56.3, 61.4, 160.7. FABMS: *m/z* 431 $[M + H]^+$ (Heathcock and Smith, 1994).

2α,3α-Diamino-5α-cholestane (**52**) was also prepared from 2α-azido-3-methoxyimino-5α-cholestane (**50**) in the following fashion (*Scheme 5.2.22*). To a stirred cooled (−5 °C) solution of **50** (64.6 mg, 0.15 mmol) in THF (3 mL), BH_3.THF complex (4.25 mL of a 1M solution in THF, 4.25 mmol) was added dropwise. The reaction mixture was allowed to warm to room temperature and stirred for 1.5 h and then refluxed for another 2 h. The reaction mixture was cooled (0 °C) again and H_2O (250 mL) followed by 20% KOH (250 mL) were added and refluxed for 90 min. Further H_2O (20 mL) was added and the mixture was extracted with DCM (5 × 20 mL). The pooled organic extracts were dried (Na_2SO_4), evaporated under pressure and the crude product was purified by FCC (mobile phase: 40% MeOH in DCM containing 2% NH_4OH solution) to obtain **52** (50 mg, 82%). ^1H NMR (500 MHz, $CDCl_3$): δ 3.07 (br s, 1H), 2.99 (br dt, 1H, $J = 3.5$, 12.2 Hz), 2.22 (br s, 4H), 1.96 (br d, 1H, $J = 12.4$ Hz), 1.76–1.81 (m, 1H), 0.89 (d, 3H, $J = 6.4$ Hz), 0.86 (dd, 6H, $J = 6.6$ Hz), 0.80 (s, 3H), 0.66 (s, 3H), 0.73–1.68 (m, 27H). ^{13}C NMR (125 MHz, $CDCl_3$): δ 12.1, 12.5, 18.6, 20.9, 22.5, 22.8, 23.8, 24.1, 28.0, 28.2, 31.9, 35.0, 35.6, 35.8, 36.1, 36.8, 38.5, 39.5, 39.9, 41.2, 42.6, 48.9, 50.5, 54.3, 56.2, 56.4. FABMS: *m/z* 403 $[M + H]^+$ (Heathcock and Smith, 1994).

cis-Di(5α-cholestano)pyrazine (**54**) was accomplished from 3β-acetyloxy-5α-cholestan-2-one (**53**) and 2α-amino-3-methoxyimino-5α-cholestane (**51**) (*Scheme 5.2.23*). A degassed stirred mixture of **53** (105 mg, 0.236 mmol) and **51** (105 mg, 0.236 mmol) in toluene (500 μL) was heated to 90 °C for 120 h. After cooling, the mixture was passed through SiO_2, eluted with 5% EtOAc in DCM, to produce **54** (24 mg, 20%, mp: 152 °C). ^1H NMR (500 MHz, $CDCl_3$): δ 2.89 (d, 2H, $J = 16.9$ Hz), 2.77 (dd, 2H, $J = 5.2$, 18 Hz), 2.57 (dd, 2H, $J = 12.7$, 18.0 Hz), 2.50 (d, 2H, $J = 16.9$ Hz), 2.03 (dt, 2H, $J = 3.31$, 12.8 Hz), 1.80–1.85 (m, 2H), 1.72–1.79 (m, 2H), 1.57–1.68 (m, 12H), 1.51 (hept, 2H, $J = 6.6$ Hz), 1.44 (dq, 2H, $J = 3.6$, 13.0 Hz), 0.96–1.40 (m, 28H), 0.92 (d, 6H, $J = 6.5$ Hz), 0.86 (d, 6H, $J = 6.5$ Hz), 0.85 (d, 6H, $J = 6.6$ Hz), 0.79 (s, 6H), 0.68 (s, 6H). ^{13}C NMR (125 MHz, $CDCl_3$): δ 11.9, 12.0, 18.7, 21.2, 22.6, 22.8, 23.9, 24.3, 28.0, 28.2, 28.5, 31.4, 35.4, 35.6, 35.8, 36.2, 39.5,

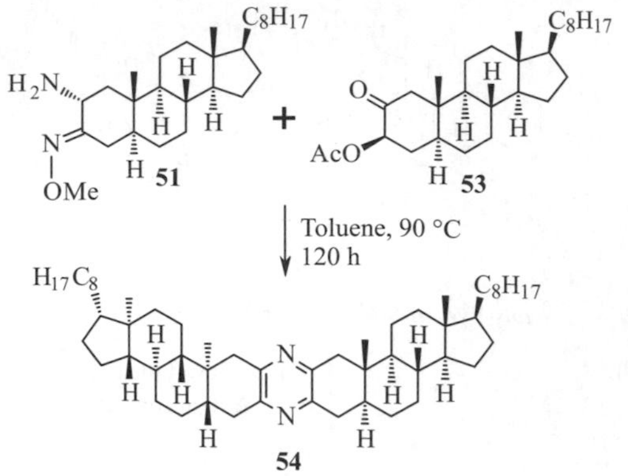

C_8H_{17} H_2N N OMe H **51** + C_8H_{17} O AcO H **53**
Toluene, 90 °C
120 h
$H_{17}C_8$ C_8H_{17} N N **54**

Scheme 5.2.23 *Synthesis of* cis-*di(5α-cholestano)pyrazine (**54**)*

40.0, 41.8, 42.5,46.0, 53.7, 56.3, 56.4, 148.4, 149.1. FABMS: *m/z* 765 $[M + H]^+$ (Heathcock and Smith, 1994).

trans-Di(5α-cholestano)pyrazine (**47**) and *cis*-di(5α-cholestano)pyrazine (**54**) could be synthesized from a one-pot reaction from 2α,3α-diamino-5α-cholestane (**52**) and 5α-cholestan-2,3-dione (**55**) (Heathcock and Smith, 1994). Later, these dimers were achieved also from 2α-bromo-5α-cholestan-3-one (**56**) (Łotowski *et al.*, 1999).

A degassed stirred mixture of **52** (76 mg, 0.189 mmol) and **55** (83.2 mg, 0.21 mmol) in toluene (1.5 mL) was heated in a sealed tube to 110 °C for 24 h, and then cooled to room temperature (*Scheme 5.2.24*). The solvent was rotary evaporated, the solid was dissolved in EtOH, and stirred for 3 h. EtOH was evaporated under vacuum to yield crude residue, which was dissolved in $CHCl_3$-EtOH (6 mL), and the solvent was evaporated to induce precipitation. Precipitates were collected by filtration and repeated precipitation from $CHCl_3$-EtOH afforded *trans*-di(5α-cholestano)pyrazine (**47**, 40 mg, 28%). The combined filtrates were evaporated to dryness and the solid was purified by FCC (mobile phase: 5% EtOAc in DCM) to yield *cis*-di(5α-cholestano)pyrazine (**54**, 44.2 mg, 31%) (Heathcock and Smith, 1994).

To prepare *trans*-di(5α-cholestano)pyrazine (**47**) and *cis*-di(5α-cholestano)pyrazine (**54**) from 2α-bromo-5α-cholestan-3-one (**56**), the following protocol was applied (Łotowski *et al.*, 1999). A mixture of **56** (2.35 g, 5.05 mmol) in C_6H_6 (10 mL) and *n*-BuOH (50 mL) was stirred at room temperature, and then treated with NH_3 under pressure (4–5 atm) at room temperature overnight (*Scheme 5.2.24*). H_2O (12 mL) was added and the reaction mixture was refluxed for 24 h, and stirring was continued at room temperature for another 2 h. The solvent was evaporated under pressure and the crude product was recrystallized from toluene to provide *trans*-dimer **47** (750 mg, 39%). The mother liquor was concentrated under vacuum and the crude residue was subjected to CC, eluted with C_6H_6, to give *cis*-dimer **54** (450 mg, 23%).

An equal mixture of *cis*- and *trans*-isomers of di-[(25*R*)-12-oxo-5α-spirostano]pyrazines (**58** and **59**) was prepared from (25*R*)-2α-bromo-5α-spirostan-3,12-dione (**57**) using the same reaction protocol as outlined above (Łotowski *et al.*, 1999). A stirred solution of **57** (500 mg, 1 mmol) in *n*-BuOH (70 mL) was treated with NH_3 under pressure at 30 °C for 168 h (*Scheme 5.2.25*). After addition of H_2O (6 mL), the reaction mixture was refluxed

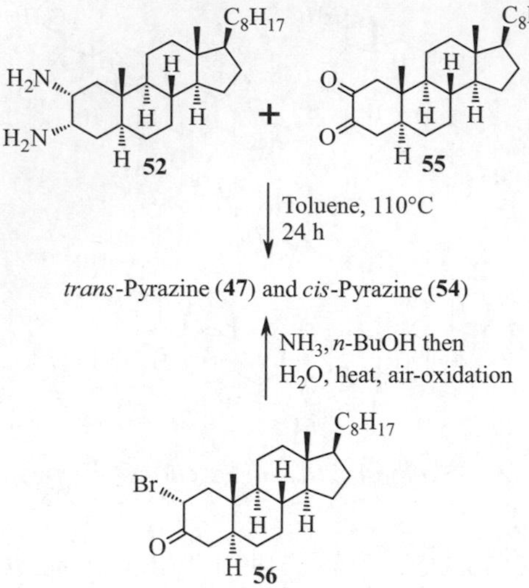

Toluene, 110°C
24 h

trans-Pyrazine (**47**) and *cis*-Pyrazine (**54**)

NH_3, *n*-BuOH then
H_2O, heat, air-oxidation

Scheme 5.2.24 *Synthesis of* trans-*di(5α-cholestano)pyrazine (**47**) and cis-di(5α-cholestano) pyrazine (**54**)*

overnight, then stirred at room temperature for another 2 h. The solvent was rotary evaporated and the crude solid was purified by CC (mobile phase: C_6H_6:$CHCl_3$:EtOAc = 66:20:14) to furnish *cis*-dimer **58** (47 mg), a mixture of *cis*-dimer **58** and *trans*-dimer **59** (80 mg), and *trans*-dimer **59** (46 mg). Total yield of dimeric pyrazines **58** and **59** was 41% (mp: >300 °C).

cis-Di(25*R*-12-oxo-5α-spirostano)pyrazine (**58**): IR ($CHCl_3$): ν_{max} cm^{-1} 2959s (C–H), 2931s (C–H), 2875s (C–H), 2861s (C–H), 1706s (C=O), 1455m (C–H), 1400s (pyrazine), 1159m (C–H), 1076m (C–H), 1055m (C–H), 1040m (C–H), 980w, 898w. ^1H NMR (400 MHz, $CDCl_3$): δ 4.36 (m, 2H, 2 × 16α-CH), 3.48 (m, 2H, 2 × 26-CHa), 3.36 (t, 2H, J = 10.6 Hz, 2 × 26-CHb), 2.95–2.42 (m, 8H, 2 × 1-CH_2 and 2 × 4-CH_2), 1.08 (s, 6H, 2 × 19-Me), 1.07 (d, 6H, J = 5.8 Hz, 2 × 21-Me), 0.89 (s, 6H, 2 × 18-Me), 0.79 (d, 6H, J = 6.1 Hz, 2 × 27-Me). ^{13}C NMR (100 MHz, $CDCl_3$): δ 213.0, 148.5, 148.2, 109.3, 79.1, 66.9, 55.6, 55.0, 54.8, 53.5, 45.2, 42.2, 41.6, 37.6, 36.1, 35.2, 34.2, 31.4, 31.2, 30.2, 28.8, 28.1, 17.1, 15.9, 13.2, 11.7. The pyrazine carbon signals for *trans*-dimer **59** merged at δ 148.3, other signals were the same as *cis*-dimer **58**. LSIMS: *m/z* 849 $[M + H]^+$. Compounds **58** and **59** gave identical IR and MS spectra, nearly identical ^1H NMR spectra and quite similar ^{13}C NMR spectra (Łotowski *et al.*, 1999).

Several symmetrical pyrazines **60–65** were prepared from *trans*-di(5α-cholestano)pyrazine (**47**) and *cis*-di(5α-cholestano)pyrazine (**54**) (Łotowski *et al.*, 2000). Both *trans*-1α-bromo-di(5α-cholestano)pyrazine (**60**) and *trans*-4α-bromo-di(5α-cholestano) pyrazine (**61**) could be carried out from *trans*-di(5α-cholestano)pyrazine (**47**) in the following manner (*Scheme 5.2.26*). A solution of *trans*-dimer **47** (100 mg, 0.13 mmol) in dry CCl_4 (15 mL) was treated with NBS (24 mg, 0.13 mmol) and AIBN (18 mg, 0.11 mmol), and the reaction mixture was refluxed for 60 min. The solvent was evaporated under vacuum, and the crude product was purified by CC (mobile phase: 40% *n*-hexane in C_6H_6) to give a

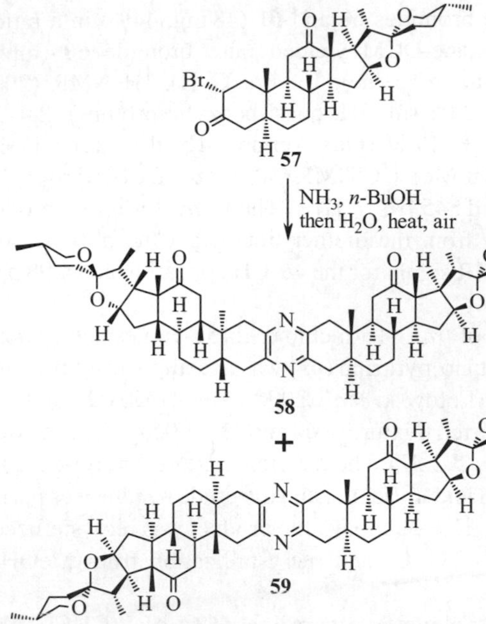

57

NH_3, *n*-BuOH
then H_2O, heat, air

58

+

59

Scheme 5.2.25 *Synthesis of* cis*-di-[(25R)-12-oxo-5α-spirostano]pyrazine* (***58***) *and* trans*-di-[(25R)-12-oxo-5α-spirostano]pyrazine* (***59***)

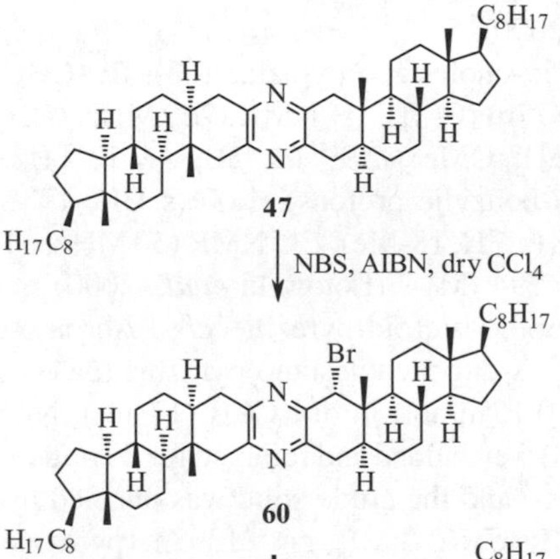

47

NBS, AIBN, dry CCl_4

60

+

61

Scheme 5.2.26 *Synthesis of* trans*-1α-bromo-di(5α-cholestano)pyrazine* (***60***) *and* trans*-4α-bromo-di(5α-cholestano)pyrazine* (***61***) mixture of *trans*-pyrazine bromides **60** and **61** (48 mg, 44%) in a ratio of 2:1, which on crystallization from *n*-hexane-DCM yielded pure bromide **60** (mp: 314–318 °C). IR ($CHCl_3$): ν_{max} cm^{-1} 1398m (pyrazine), 1384m (C–H). ^1H NMR (200 MHz, $CDCl_3$): δ 5.17 (s, 1H, 1β-CH), 2.8–3.05 (m, 3H, quasi-benzylic protons), 2.4–2.7 (m, 3H, quasi-benzylic protons), 0.89 (s, 3H, 19-Me plus overlapped by the side chain signals), 0.76 (s, 3H, 19′-Me), 0.69 (s, 6H, 2 × 18-Me). ^{13}C NMR (50 MHz, $CDCl_3$): δ 64.1, 147.6, 148.2, 149.4, 151.9. ESIMS: *m/z* 843 and 845 $[M + H]^+$. The *trans*-4α-bromo product **61** could not be obtained in its pure form from the mother liquor, but the presence of bromide **61** was confirmed from the ^1H NMR signal for the 4α-CH appearing at δ 4.98 ppm (d, $J = 10.4$ Hz) (Łotowski *et al.*, 2000).

Similarly, the synthesis of *trans*-1α-methoxy-di(5α-cholestano)pyrazine (**62**) and *trans*-4β-methoxy-di(5α-cholestano)pyrazine (**63**) was accomplished utilizing a mixture of both bromo products **60** and **61** (Łotowski *et al.*, 2000). MeOH (20 mL) and a catalytic amount of *p*-TsOH were added to a stirred mixture of **60** and **61** (100 mg, 0.12 mmol) in dioxane (1 mL) and $CHCl_3$ (1 mL) (*Scheme 5.2.27*). The reaction mixture was refluxed for 6 h, after which, the solvent was evaporated under vacuum and the crude residue was purified by CC (mobile phase: 30% *n*-hexane in C_6H_6). After CC, the product was recrystallized from *n*-hexane to give **62** (16 mg, mp: 281–284 °C) and recrystallization from MeOH-DCM yielded **63** (28 mg, mp: 306–309 °C).

trans-1α-Methoxy-di(5α-cholestano)pyrazine (**62**): IR ($CHCl_3$): ν_{max} cm^{-1} 1398m (pyrazine), 1078m (C–H), 909w. ^1H NMR (200 MHz, $CDCl_3$): δ 3.93 (s, 1H, 1β-CH), 3.41 (s, 3H, OMe), 2.78–3.04 (m, 3H, quasi-benzylic protons), 2.28–2.70 (m, 3H, quasi-benzylic protons), 0.82 (s, 3H, 19′-Me), 0.69 (s, 3H, 19-Me), 0.68 (s, 3H, 18′-Me), 0.67 (s, 3H, 18-Me). ^{13}C NMR (50 MHz, $CDCl_3$): δ 83.4, 147.6, 147.8, 149.5, 151.1. EIMS: *m/z* 794 $[M]^+$ (Łotowski *et al.*, 2000).

trans-4β-Methoxy-di(5α-cholestano)pyrazine (**63**): IR ($CHCl_3$): ν_{max} cm^{-1} 1401s (pyrazine), 1113m (C–H), 1077m (C–H). ^1H NMR (200 MHz, $CDCl_3$): δ 4.08 (d, 1H, $J = 4.2$ Hz, 4α-CH), 3.66 (s, 3H, OMe), 2.99 (d, 2H, $J = 17.4$ Hz, quasi-benzylic protons), 2.36–2.88 (m, 4H, quasi-benzylic protons), 1.00 (s, 3H, 19′-Me), 0.83 (s, 3H, 19-Me), 0.70 (s, 3H, 18′-Me), 0.68 (s, 3H, 18-Me). ^{13}C NMR (50 MHz, $CDCl_3$): δ 81.8, 148.2, 148.4, 149.6, 150.7. EIMS: *m/z* 794 $[M]^+$ (Łotowski *et al.*, 2000).

The synthesis of yet another steroid-pyrazine, *cis*-1α-bromo-di(5α-cholestano)pyrazine (**64**), was achieved from *cis*-di(5α-cholestano)pyrazine (**54**) (Łotowski *et al.*, 2000). To a solution of **53** (115 mg, 0.15 mmol) in dry CCl_4 (17 mL), NBS (27 mg, 0.15 mmol) and AIBN (20 mg, 0.12 mmol) were added and refluxed for 60 min (*Scheme 5.2.28*). The solvent was evaporated to dryness, and the crude solid was purified by CC, eluted with C_6H_6, to obtain **64** (36 mg, 28%). IR ($CHCl_3$): ν_{max} cm^{-1} 1398m (pyrazine), 1383m (C–H). ^1H NMR (200 MHz, $CDCl_3$): δ 5.15 (s, 1H, 1β-CH), 2.3–3.1 (m, 8H, quasi-benzylic protons), 0.88 (s, 3H, 19-Me), 0.84 (s, 3H, 19′-Me), 0.69 (s, 6H, 2 × 18-Me). ^{13}C NMR (50 MHz, $CDCl_3$): δ 63.9, 147.6, 148.3, 149.9, 151.3. ESIMS: *m/z* 762 $[M - HBr]^+$ (Łotowski *et al.*, 2000).

In the next step, *cis*-1α-methoxy-di(5α-cholestano)pyrazine (**65**) was prepared from *cis*-1α-bromo-di(5α-cholestano)pyrazine (**64**) (*Scheme 5.2.28*). A mixture of MeOH (10 mL) and a catalytic amount of *p*-TsOH was added to a stirred solution of **64** (20 mg, 0.02 mmol) in dioxane (1 mL) and $CHCl_3$ (1 mL). The reaction mixture was refluxed for six days, the solvents were evaporated under pressure and the crude product was purified by CC (mobile phase: C_6H_6-$CHCl_3$) followed by recrystallization from MeOH-DCM to produce **65**

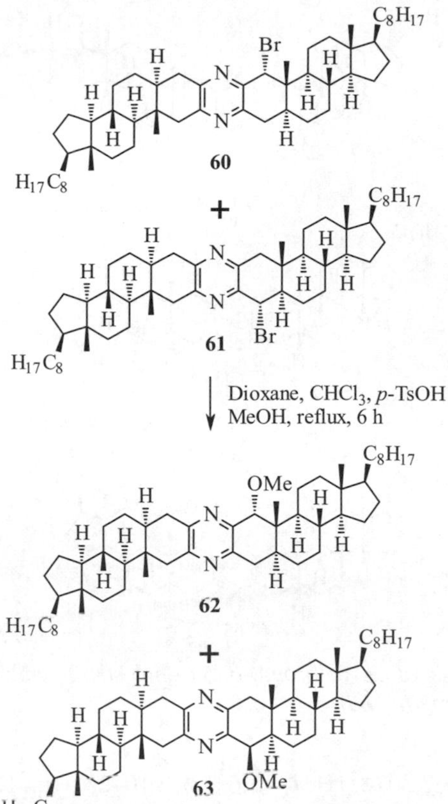

Scheme 5.2.27 *Synthesis of* trans-1α*-methoxy-di(5α-cholestano)pyrazine (***62***) and* trans-4β-*methoxy-di(5α-cholestano)pyrazine (***63***)*

(11 mg, 58%, mp: 205–208 °C). IR ($CHCl_3$): ν_{max} cm^{-1} 1400s (pyrazine), 1383m (C–H), 1087m (C–H), 1078m (C–H). ^1H NMR (200 MHz, $CDCl_3$): δ 3.94 (s, 1H, 1β-CH), 3.41 (s, 3H, OMe), 2.72–3.04 (m, 4H, quasi-benzylic protons), 2.38–2.70 (m, 4H, quasi-benzylic protons), 0.81 (s, 3H, 19′-Me), 0.69(s, 3H, 19-Me), 0.68 (s, 3H, 18-Me), 0.66 (s, 3H, 18-Me). ^{13}C NMR (50 MHz, $CDCl_3$): δ 83.3, 147.7, 148.4, 149.3, 150.5. EIMS: *m*/*z* 794 $[M]^+$ (Łotowski *et al.*, 2000).

The synthesis of (17β-acetyloxy-5α-androstano)-(5α-cholestano)pyrazine (**70**) could be accomplished from 5α-androst-2-en-3,17β-yl diacetate (**66**) in several steps (Heathcock and Smith, 1994). In order to synthesize cephalostain analogue **70**, initially 2α,3α-epoxy-5α-androstan-3β,17β-yl diacetate (**67**) was accomplished from **66** (*Scheme 5.2.29*). A freshly prepared cold (0 °C) solution of DMDO (0.041M solution in Me_2CO, 180 mL, 7.38 mmol, 1.2 equiv.) was added to **66** (2.30 g, 6.14 mmol) under N_2. The reaction mixture was allowed to warm to room temperature and stirred for 4 h. Additional amount of DMDO solution (40 mL, 1.64 mmol) was added and stirred for another 2 h. DMDO solution (20 mL, 0.82 mmol) was added again, and the solution was refrigerated for 12 h. The solvent was evaporated under vacuum and the crude residue was crystallized from Et_2O at room temperature to yield **67** (2.16 g, 90%, mp: 142–144 °C). ^1H NMR (500 MHz, $CDCl_3$): δ 4.56

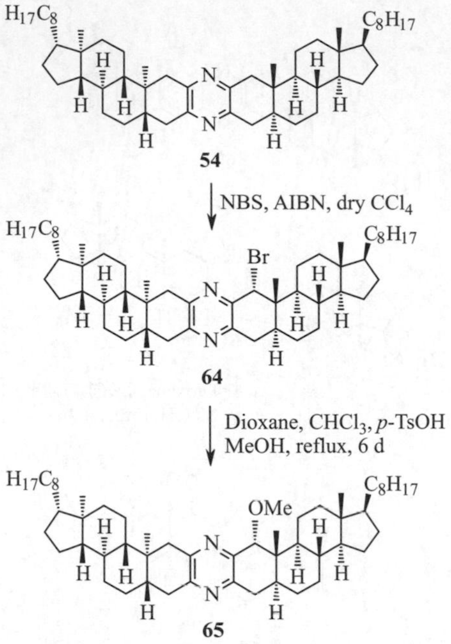

Scheme 5.2.28 *Synthesis of* cis-*1α-bromo-di(5α-cholestano)pyrazine (**64**) and* cis-*1α-methoxy-di(5α-cholestano)pyrazine (**65**)*

(dd, 1H, $J=7.0$, 8.0 Hz), 3.30 (d, 1H, $J=5.6$ Hz), 2.10–2.16 (m, 1H), 2.07 (dd, 1H, $J=4.4$, 14.2 Hz), 2.07 (dd, 1H, $J=4.4$, 14.2 Hz), 2.04 (s, 3H), 2.00 (s, 3H, overlapping dd, 1H, $J=5.7$ and 15.4 Hz), 1.91 (dd, 1H, $J=11.6$, 14.2 Hz), 1.71 (dt, 1H, $J=3.3$, 12.5 Hz), 1.03 (m, 1H), 1.57–1.66 (m, 2H), 1.41–1.50 (m, 2H), 1.37 (dm, 1H, $J=14.8$ Hz), 1.19–1.33 (m, 6H), 1.11 (dt, 1H, $J=4.3$, 12.9 Hz), 0.95–1.00 (m, 1H, overlapping 0.95, s, 3H), 0.81 (dq, 1H, $J=4.2$, 13.0 Hz), 0.76 (s, 3H), 0.62 (dt, 1H, $J=6.3$, 12.2 Hz). ^{13}C NMR (125 MHz, $CDCl_3$): δ 12.0, 12.8, 20.5, 21.0, 21.1, 23.4, 27.4, 27.9, 30.7, 31.0, 34.4, 35.3, 36.8, 38.7, 38.8, 42.4, 50.5, 53.3, 58.3, 82.7, 83.0, 169.4, 171.1 (Heathcock and Smith, 1994).

In the next reaction, 2α,3α-epoxy-5α-androstan-3β,17β-yl diacetate (**67**) was converted to 3-oxo-5α-androstan-2β,17β-yl diacetate (**68**) in the following fashion (*Scheme 5.2.29*). A mixture of **67** (2.16 g, 5.53 mmol) in toluene (100 mL) and C_5H_5N (10 mL) was refluxed for 24 h. The solution was warmed to room temperature, the solvent was rotary evaporated and the crude solid was recrystallized from Et_2O to provide **68** (1.79 g, 83%, mp: 169–171 °C). ^{1}H NMR (500 MHz, $CDCl_3$): δ 5.34 (dd, 1H, $J=7.1$, 10.1 Hz), 4.56 (t, 1H, $J=8.9$ Hz), 2.36 (dd, 1H, $J=6.2$ and 17.8 Hz), 2.20 (dd, 1H, $J=12.2$, 17.8 Hz), 2.10 (s, 3H), 2.09–2.16 (m, 2H), 1.95–2.02 (m, 1H, overlapping 2.00, s, 3H), 0.83–1.90 (m, 15H), 0.83 (s, 3H), 0.76 (s, 3H). ^{13}C NMR (125 MHz, $CDCl_3$): δ 12.0, 14.3, 20.7, 20.9, 21.1, 23.3, 27.4, 28.1, 30.7, 34.9, 36.1, 36.7, 41.6, 42.1, 42.6, 43.4, 50.4, 54.6, 74.2, 82.5, 169.8, 171.0, 206.7. EIMS: 390 $[M]^+$ (Heathcock and Smith, 1994).

HBr (48% aqueous, 5 drops) was added to a stirred solution of 3-oxo-5α-androstan-2β,17β-yl diacetate (**68**, 270 mg, 0.691 mmol) in AcOH (20 mL) at room temperature

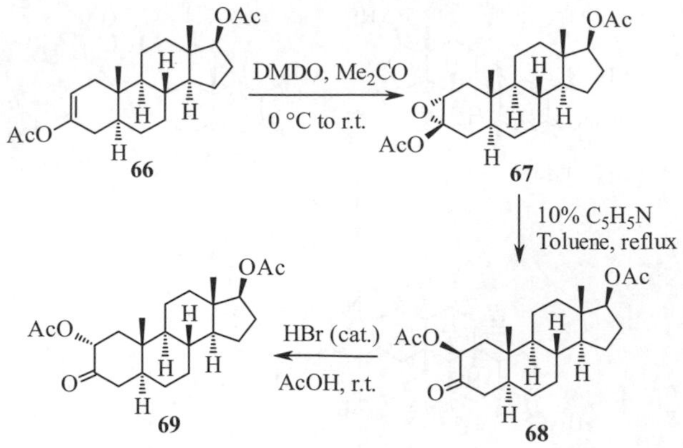

Scheme 5.2.29 *Synthesis of androstane derivatives* ***67–69***

(*Scheme 5.2.29*). After 24 h, the solvent was rotary evaporated, and azeotroped with toluene to remove the AcOH and H_2O. The crude product was subjected to CC (mobile phase: 20% EtOAc in *n*-hexane) and the resulting solid was further purified by crystallization from Et_2O to afford 3-oxo-5α-androstan-2α,17β-yl diacetate (**69**, 160 mg, 59%, mp: 189–193 °C). ^1H NMR (500 MHz, $CDCl_3$): δ 5.27 (dd, 1H, J= 6.7, 12.9 Hz), 4.57 (dd, 1H, J= 7.9, 9.1 Hz), 2.41 (t, 1H, J= 14.3 Hz), 2.25 (dd, 1H, J= 6.7, 12.4 Hz), 2.19 (dd, 1H, J= 3.6, 14.1 Hz), 2.13 (s, 3H, overlapping m, 1H, CH), 2.02 (s, 3H), 1.70–1.75 (m, 2H), 1.27–1.57 (m, 1H, CH), 1.15 (dt, 1H, J= 4.1, 12.9 Hz), 1.12 (s, 3H), 0.91 (m, 1H), 0.76–0.82 (m, 1H, overlapping 0.79, s, 3H). ^{13}C NMR (125 MHz, $CDCl_3$): δ 12.1, 12.7, 20.7, 21.0, 21.1, 23.4, 27.4, 28.2, 31.0, 34.5, 36.7, 37.2, 42.6, 43.4, 44.7, 47.7, 50.4, 53.7, 74.3, 82.5, 170.1, 171.1, 203.9 (Heathcock and Smith, 1994).

Heating 2α-amino-3-methoxyimino-5α-cholestane (**51**) and 3-oxo-5α-androstan-2α,17β-yl diacetate (**69**) in toluene at 145 °C resulted in the synthesis of (17β-acetyloxy-5α-androstano)-(5α-cholestano)pyrazine (**70**), which upon alkaline hydrolysis yielded (17β-hydroxy-5α-androstano)-(5α-cholestano)pyrazine (**71**) (Heathcock and Smith, 1994). A stirred solution of **69** (70.3 mg, 0.18 mmol) and **51** (64.6 mg, 0.15 mmol) in toluene (1 mL) was degassed by cooling (–78 °C), and the reaction temperature was allowed to rise to room temperature before introducing N_2 (*Scheme 5.2.30*). This freeze/thaw procedure was repeated twice. The reaction mixture was heated in a sealed tube for 24 h at 90 °C and then for further 24 h at 145 °C. Slow cooling to room temperature yielded precipitation. The slurry was removed from the reaction vessel by dissolving in $CHC1_3$. The solvent was evaporated under vacuum yielding a residue, which was dissolved in absolute EtOH (6 mL), filtered through Celite and the filtrate was discarded. The solid was washed with 50% $CHC1_3$ and then dissolved in EtOH (3 mL), which was concentrated to a smaller volume (2 mL) to afford **70** (45.9 mg, 43%; mp: > 260 °C). ^1H NMR (500 MHz, $CDCl_3$): δ 4.61 (t, 1H, J= 8.5 Hz), 2.89 (d, 2H, J= 16.9 Hz), 2.76 (dt, 2H, J= 6.3, 18.4 Hz), 2.53–2.60 (m, 2H), 2.50 (dd, 2H, J= 6.0, 16.9 Hz), 2.12–2.20 (m, 1H), 2.01–2.09 (m, 1H, overlapping 2.03, s, 3H), 0.91 (d, 3H, J= 6.4 Hz), 0.86 (d, 3H, J= 6.5 Hz), 0.85 (d, 3H, J= 6.6 Hz), 0.80 (s, 6H), 0.78 (s, 3H), 0.67 (s, 3H). ^{13}C NMR (125 MHz, $CDCl_3$): δ 11.9, 12.01, 18.7, 20.7, 21.2, 21.2, 22.5, 22.8, 23.5, 23.8, 24.2, 27.5, 28.0, 28.2, 28.3, 28.4, 31.1, 31.6, 35.1, 35.3,

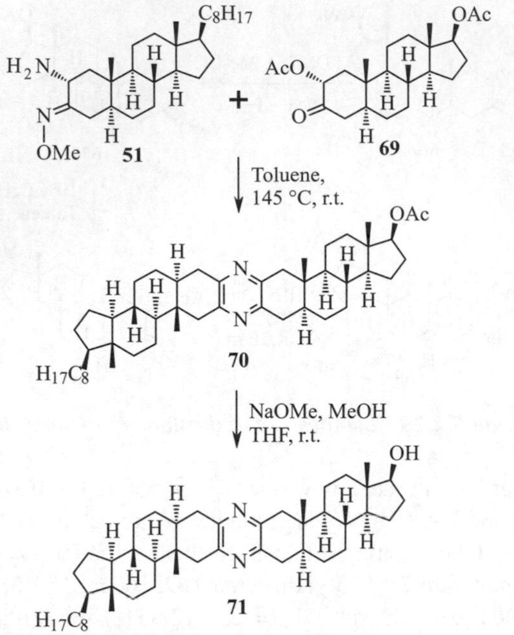

Scheme 5.2.30 *Synthesis of (17β-acetyloxy-5α-androstano)-(5α-cholestano)pyrazine (**70**) and (17β-hydroxy-5α-androstano)-(5α-cholestano)pyrazine (**71**)*

35.4, 35.52, 35.5, 35.6, 35.8, 36.2, 36.9, 39.5, 39.9, 41.7, 42.4, 42.5, 45.9, 46.0, 50.6, 53.6, 53.7, 56.2, 56.3, 82.7, 148.3, 148.6, 148.7, 149.0, 171.2. FABMS: 711 $[M + H]^+$ (Heathcock and Smith, 1994).

In the next reaction, (17β-hydroxy-5α-androstano)-(5α-cholestano)pyrazine (**71**) was obtained from alkaline hydrolysis of (17β-acetyloxy-5α-androstano)-(5α-cholestano) pyrazine (**70**) (*Scheme 5.2.30*). To a suspension of **70** (130 mg, 0.183 mmol) in THF (10 mL) at room temperature, NaOMe (2.5 mL of a 0.36M solution in MeOH, 0.92 mmol, 5 equiv.) was added. After 24 h, once most of the solid had been dissolved, the reaction mixture was poured into H_2O (20 mL) and extracted with DCM (3 × 30 mL). Pooled organic solvents were dried (Na_2SO_4), concentrated and purified by FCC (mobile phase: 25–40% EtOAc in *n*-hexane) to provide **71** (101 mg, 82%, mp: > 270 °C). ^{1}H NMR (500 MHz, $CDCl_3$): δ 3.64 (br t, 1H, $J = 8.4$ Hz), 2.89 (dd, 2H, $J = 7.1$, 17.0 Hz), 2.75 (dt, 2H, $J = 4.9$, 17.9 Hz), 2.54–2.60 (m, 2H), 2.49 (dd, 2H, $J = 6.9$, 16.8 Hz), 2.01–2.06 (m, 2H), 0.80–1.86 (m, 40H), 0.91 (d, 3H, $J = 6.4$ Hz), 0.85 (d, 3H, $J = 6.6$ Hz), 0.84 (d, 3H, $J = 6.6$ Hz), 0.80 (s, 3H), 0.78 (s, 3H), 0.76 (s, 3H), 0.69 (s, 3H). ^{13}C NMR (125 MHz, $CDCl_3$): δ 11.1, 11.9, 12.0, 18.7, 20.8, 21.2, 22.5, 22.8, 23.4, 23.9, 24.2, 28.0, 28.2, 28.3, 28.4, 30.5, 31.2, 31.6, 35.3, 35.4, 35.5, 35.6, 35.7, 35.8, 36.2, 36.7, 39.5, 39.9, 41.7, 41.8, 42.5, 42.9, 46.0, 50.9, 53.7, 53.8, 56.3, 56.3, 81.8, 148.4, 148.6, 148.7, 149.0 (Heathcock and Smith, 1994).

Two other cephalostatin analogues, nonsymmetric trisdecacyclic pyrazines **75** and **76**, were synthesized from hecogenin derivative, (25*R*)-5α,14α-spirostan-3α,12α-diol (**72**), using the similar synthetic protocol as discussed earlier (Heathcock and Smith, 1994). First, (25*R*)-12α-hydroxy-5α,14α-spirostan-3-one (**73**) was prepared from **72**

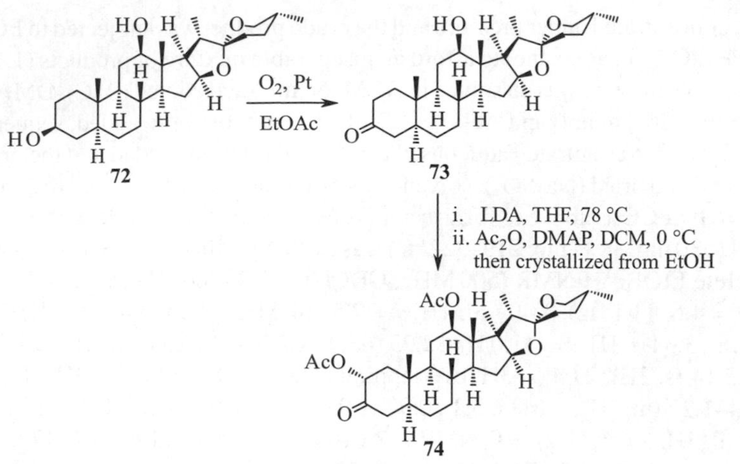

Scheme 5.2.31 *Synthesis of (25R)-12α-hydroxy-5α,14α-spirostan-3-one (**73**) and (25R)-2α,12β-diacetoxy-5α,14α-spirostan-3-one (**74**)*

(*Scheme 5.2.31*). To a solution of **72** (2.72 g, 6.29 mmol) in EtOAc (60 mL), PtO_2 (1.0 g, 4.40 mmol, 0.7 equiv.) was added and the mixture was degassed under aspirator vacuum and refilled five times with N_2, and H_2 was introduced by degassing and refilling three times with H_2 from a balloon. The mixture was hydrogenated at 1 atm for 30 min, then degassing and refilling three times with N_2 was followed by further degassing and refilling three times with O_2. The reaction mixture was stirred under O_2 at 1 atm for 30 h. The catalyst was removed by filtration through Celite, washing with EtOAc (1500 mL) and Me_2CO (500 mL). Concentration of the filtrate under pressure and FCC of the resulting solid (mobile phase: 30–40% EtOAc in *n*-hexane) afforded **73** (2.32 g, 86%, mp: 239–241 °C). ^1H NMR (500 MHz, $CDCl_3$): δ 3.45 (ddd, 1H, $J = 4.3$, 10.8 Hz), 3.31–3.37 (m, 1H, overlapping t, 1H, $J = 3.3$, 10.9 Hz), 2.36 (dt, 1H, $J = 6.6$, 15.5 Hz), 2.21–2.31 (m, 1H, overlapping 2.24, t, 1H, $J = 14.3$ Hz), 2.08 (ddd, 1H, $J = 3.8$, 15.1 Hz), 2.01, 4.39 (dd, 1H, $J = 7.5$, 15.0 Hz), 1.95–2.03 (m, 2H), 1.81–1.90 (m, 2H), 1.29–1.72 (m, 14H), 1.00–1.08 (m, 2H), 1.02 (s, 3H), 1.01 (d, 3H, $J = 6.8$ Hz), 0.80–0.91 (m, 2H), 0.77 (s, 3H, overlapping d, 3H, $J = 6.1$ Hz). ^{13}C NMR (125 MHz, $CDCl_3$): δ 10.5, 11.4, 13.9, 17.1, 28.7, 28.8, 30.2, 30.6, 31.3, 31.4, 31.5, 33.9, 35.7, 38.0, 38.4, 42.1, 44.5, 46.0, 46.5, 52.7, 54.5, 61.8, 66.9, 79.6, 80.6, 109.4, 211.5 (Heathcock and Smith, 1994).

(25*R*)-2α,12β-Diacetoxy-5α,14α-spirostan-3-one (**74**) was transformed from (25*R*)-12α-hydroxy-5α,14α-spirostan-3-one (**73**) in two steps (*Scheme 5.2.31*). A solution of LDA was prepared *in situ* by adding *n*-BuLi (2M solution in *n*-hexane, 5.0 mL, 10.0 mmol, 2.1 equiv.) in DIPA (1.40 mL, 10.0 mmol, 2.1 equiv.) and THF (40 mL) and stirring at −20 °C for 20 min. To a cooled (−78 °C) LDA solution, a solution of **73** (2.05 g, 4.76 mmol) in THF (40 mL + 5 mL washing) was added by cannula and stirred at the same temperature for 30 min. The solution was warmed to −40 °C, and oxodiperoxomolybdenum-pyridine-dimethylpyrrolidinone (2.74 g, 7.14 mmol, 1.5 equiv.) was added and stirred for another 60 min. Saturated aqueous $Na_2S_2O_3$ (5 mL) and Et_2O (20 mL) were added, and the mixture was stirred for 10 min. The aqueous layer was extracted with Et_2O, the ethereal layer was washed with H_2O (3 × 20 mL) followed by brine (20 mL), the organic layer was dried

(Na_2SO_4), concentrated under pressure and the crude product was subjected to FCC (mobile phase: 40% EtOAc in *n*-hexane) to afford an inseparable mixture of products (1.18 g, 55%). This inseparable mixture was dissolved in DCM (50 mL) and cooled (0 °C), DMAP (30 mg), Et_3N (3.68 mL, 26.4 mmol) and Ac_2O (1.25 mL, 13.2 mmol) were added, sequentially, and stirred for 5 h at 0 °C. Saturated aqueous $NaHCO_3$ (50 mL) was added and the organic layer was separated and dried (Na_2SO_4), solvent was evaporated to dryness and the crude residue was purified by FCC as above to furnish a mixture of compounds (1.35 g, 96%). The major product **74** (530 mg, 38%, mp: 219–222 °C) was further purified by repeated crystallizations from absolute EtOH. ^1H NMR (500 MHz, $CDCl_3$): δ 5.27 (dd, 1H, J = 6.6, 12.8 Hz), 4.52 (dd, 1H, J = 4.6, 11.1 Hz), 4.40 (dd, 1H, J = 7.5, 14.6 Hz), 3.45–3.47 (m, lH), 3.40 (t, 1H, J = 14.2 Hz), 3.34 (t, 1H, J = 11.0 Hz), 2.23 (dd, 1H, J = 3.5, 14.3 Hz), 2.19 (dd, 1H, J = 5.8, 12.5 Hz), 2.14 (s, 3H), 2.04 (s, 3H, overlapping m, 1H), 1.83–1.90 (m, 3H), 1.33–1.76 (m, 12H), 1.14–1.22 (m, 2H), 1.14 (s, 3H), 0.92–1.03 (m, 2H), 0.90 (d, 3H, J = 6.7 Hz), 0.88 (s, 3H), 0.78 (d, 3H, J = 6.3 Hz). ^{13}C NMR (125 MHz, $CDCl_3$): δ 11.6, 12.6, 13.6, 17.0, 20.7, 21.4, 27.0, 28.1, 28.7, 30.2, 31.1, 31.2, 31.3, 33.4, 37.2, 42.1, 43.3, 44.5, 47.4, 52.2, 54.4, 61.2, 66.8, 74.0, 80.3, 81.2, 109.2, 170.0, 170.4, 203.5 (Heathcock and Smith, 1994).

Nonsymmetric trisdecacyclic pyrazine **75** was prepared from (25*R*)-2α,12β-diacetoxy-5α,14α-spirostan-3-one (**74**) and 2α-amino-3-methoxyimino-5α-cholestane (**51**). In the next reaction, another trisdecacyclic pyrazine **76** was achieved from trisdecacyclic pyrazine **75** by alkaline hydrolysis A degassed solution of **74** (502 mg, 0.946 mmol) and **51** (340 mg, 0.788 mmol) in toluene (5 mL) was heated at 90 °C for 24 h, then at 145 °C for a further 24 h (*Scheme 5.2.32*), and finally cooled, the solvent was removed and the solid was suspended in absolute EtOH. The suspension was filtered through Celite to give **75**. Partial purification of the filtrate by FCC (mobile phase: 25% EtOAc in *n*-hexane), followed by precipitation from $CHC1_3$-EtOH yielded more of **75** (combined yield was 197 mg, 29%, mp: 228 °C). ^1H NMR (500 MHz, $CDCl_3$): δ 4.57 (dd, 1H, J = 4.7, 11.2 Hz), 4.41 (q, 1H, J = 7.6 Hz), 3.44 (m, 1H), 3.34(t, 1H, J = 11.0 Hz), 2.89 (d, 1H, J = 16.8 Hz), 2.82 (d, 1H, J = 16.8 Hz), 2.77 (t,1H, J = 17.7 Hz), 2.76 (t, 1H, J = 17.7 Hz), 2.01–2.59 (m, 2H), 2.04 (s, 3H), 1.95 (dt, 1H, J = 4.3, 12.5 Hz), 0.70–1.91 (m, 45H), 0.91 (dd, 6H, J = 6.5, 6.8 Hz), 0.88 (s, 3H), 0.86 (d, 3H, J = 6.6 Hz), 0.85 (d, 3H, J = 6.6 Hz), 0.81(s, 3H), 0.78 (br s, 6H), 0.67 (s, 3H). ^{13}C NMR (125 MHz, $CDCl_3$): δ 11.5, 12.0, 13.6, 17.1, 18.7, 21.2, 21.4, 22.5, 22.8, 23.8, 24.2, 26.8, 28.0, 28.2, 28.4, 28.8, 30.2, 31.2, 31.3, 31.4, 31.6, 33.9, 35.3, 35.5, 35.6, 35.65, 35.8, 36.2, 39.5, 39.9, 41.6, 41.7, 42.2, 42.5, 44.5, 45.8, 46.0, 52.2, 53.7, 54.8, 56.3, 56.3, 61.2, 66.8, 80.5, 81.4, 109.3, 148.0, 148.3,1 48.8, 149.1, 170.2. FABMS: m/z 851 $[M + H]^+$ (Heathcock and Smith, 1994).

To a stirred solution of **75** (58.0 mg, 68.1 mL) in THF (5 mL), NaOMe (0.51M solution in MeOH, 670 mL) was added by syringe at room temperature under N_2 (*Scheme 5.2.32*). After 16 h, H_2O (10 mL) was added and the mixture was extracted with DCM (5 × 20 mL). The pooled organic layers were dried (Na_2SO_4), concentrated under vacuum and the crude residue was subjected to FCC, eluted with 40% EtOAc in *n*-hexane, to obtain trisdecacyclic pyrazine **76** (47.0 mg, 85%, mp: > 270 °C). ^1H NMR (500 MHz, $CDCl_3$): δ 4.42 (q, 1H, J = 7.3 Hz), 3.46–3.48 (m, 1H), 3.37 (t, 1H, J = 10.9 Hz, overlapping m, 1H), 2.89 (dd, 2H, J = 16.9, 17.0 Hz), 2.77 (ddd, 2H, J = 5.0, 11.0, 16.9 Hz), 2.57 (dd, 1H, J = 1.0, 2.5, 17.3 Hz), 2.51 (dd, 2H, J = 16.8, 17.0 Hz), 2.02–2.08 (m, 2H), 1.92 (dd, 1H, J = 7.7, 8.3 Hz), 0.80–1.89 (m, 44H), 1.04 (d, 3H, J = 6.8 Hz), 0.91 (d, 3H, J = 6.5 Hz), 0.86 (dd, 6H,

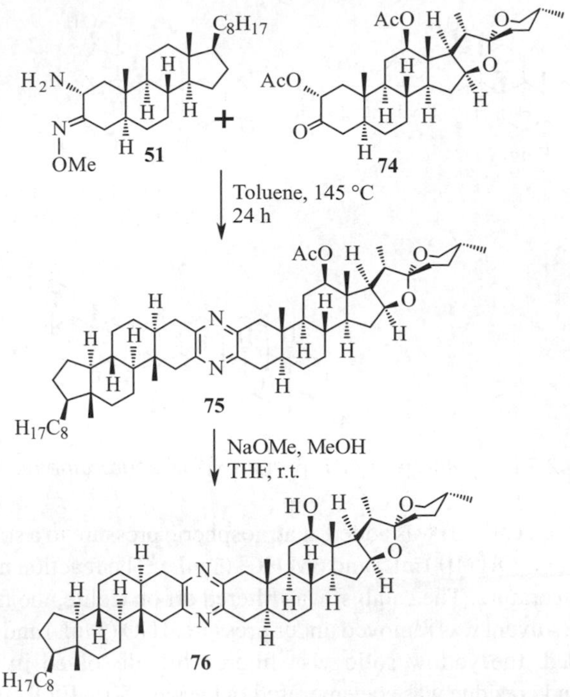

Scheme 5.2.32 Synthesis of nonsymmetric trisdecacyclic pyrazines 75 and 76

$J = 6.6$ Hz), 0.82 (s, 3H), 0.79 (br s, 9H), 0.68 (s, 3H). ^{13}C NMR (125 MHz, $CDCl_3$): δ 10.4, 12.0, 13.9, 17.1, 18.7, 21.2, 22.6, 22.8, 23.9, 24.3, 28.0, 28.2, 28.3, 28.5, 28.8, 30.3, 30.6, 31.3, 31.4, 31.6, 33.9, 35.3, 35.4, 35.6, 35.7, 35.8, 36.2, 39.5, 40.0, 41.7, 41.8, 42.2, 42.5, 45.9, 46.0, 52.7, 53.7, 54.7, 56.3, 56.4, 61.8, 66.9, 79.6, 80.7, 109.5, 148.2, 148.4, 148.8, 149.2 (Heathcock and Smith, 1994).

After almost 30 years, the synthesis of di(17β-hydroxy-5α-androstano)pyrazine (**44**) was achieved from 2α-bromo-17β-hydroxy-5α-androstan-3-one (**77**) in two steps (Černý *et al.*, 2000). Initially, 2α-azido-17β-hydroxy-5α-androstan-3-one (**78**) was accomplished from 2α-bromo-17β-hydroxy-5α-androstan-3-one (**77**) (*Scheme 5.2.33*). AcOH (250 μL, 4.37 mmol) and NaN_3 (1.0 g, 15.38 mmol) were added, sequentially, to a stirred solution of **77** (1.5 g, 4.06 mmol) in DMF (25 mL) at room temperature. After 20 min, a mixture ice–H_2O was added to the reaction mixture and the crystalline material was collected by filtration, washed with H_2O and dried under vacuum to produce the crude product that after purification by CC, eluted isocratically with C_6H_6-Me_2CO = 20:1), was crystallized from hot Me_2CO to yield **78** (1.03 g, 77%, mp: 170–173 °C). IR (KBr): ν_{max} cm^{-1} 3614br (O–H), 2107 (N_3), 1723 (C=O), 1054 (C–O). ^1H NMR (200 MHz, $CDCl_3$): δ 3.99 (dd, 1H, $J = 6.3$, 13.0 Hz, 2β-CH), 3.65 (t, 1H, $J = 8.2$ Hz, 17α-CH), 1.10 (s, 3H, 19-Me), 0.76 (s, 3H 18-Me) (Černý *et al.*, 2000).

Cephalostatin analogue, di(17β-hydroxy-5α-androstano)pyrazine (**44**) was prepared from 2α-azido-17β-hydroxy-5α-androstan-3-one (**78**) in the final step (*Scheme 5.2.33*).

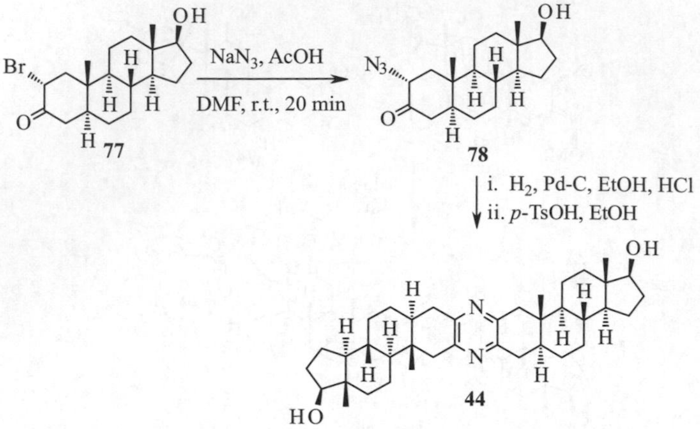

Scheme 5.2.33 *Synthesis of di(17β-hydroxy-5α-androstano)pyrazine (**44**)*

Pd-C (100 mg, 10% on carbon) was added at atmospheric pressure to a stirred mixture of 77 (1.0 g, 3.02 mmol) in EtOH (100 mL) and 6M HCl (1 mL). The reaction mixture was stirred for 2 h at room temperature. The catalyst was filtered off on Celite, the filtrate was washed with EtOH, and the solvent was removed under pressure. H_2O (5 mL) and saturated aqueous $KHCO_3$ were added, the yellow solid was filtered off, dissolved in EtOH and rotary evaporated. The crude residue was coevaporated twice with 50% EtOH in C_6H_6 (20 mL) to dryness. The crude material was dissolved in EtOH (20 mL), *p*-TsOH.H_2O (50 mg) was added and the mixture was boiled to dissolve all solids. The stirred solution was left at room temperature for 24 h, precipitate was formed, which was filtered off and washed with aqueous EtOH to provide **44** (480 mg, 55%, mp: 340 °C). IR (KBr): ν_{max} cm^{-1} 3394br (O–H), 1398s (pyrazine), 1059m (C–O), 1046m (C–O). ^{1}H NMR (500 MHz, $CDCl_3$): δ 3.66 (t, 2H, *J* = 8.7 Hz, 2 × 17α-CH), 2.92 (d, 2H, *J* = 16.5 Hz, 2 × 1β-CH), 2.78 (dd, 2H, *J* = 5.1, 17.5 Hz, 2 × 4α-CH), 2.59 (bdd, 2H, *J* = 12.5, 17.5 Hz, 2 × 4β-CH), 2.52 (bd, 2H, *J* = 16.5 Hz, 2 × 1α-CH), 2.07 (ddt, 2H, *J* = 5.8, 9.2, 13.5 Hz, 2 × 16β-CH), 1.86 (ddd, 2H, *J* = 2.8, 3.9, 12.5 Hz, 2 × 12β-CH), 1.75 (dq, 2H, *J* = 3.5, 12.8 Hz, 2 × 6α-CH), 1.70 (m, 2H, 2 × 11α-CH), 1.66 (m, 2H, 2 × 5-CH), 1.64 (m, 2H, 7β-CH), 1.62 (m, 2H, 2 × 15α-CH), 1.46 (m, 2H, 2 × 16α-CH), 1.45 (m, 2H, 2 × 11β-CH), 1.40 (m, 2H, 2 × 8-CH), 1.35 (m, 2H, 2 × 7α-CH), 1.28 (dq, 2H, *J* = 5.9, 12.2 Hz, 2 × 15β-CH), 1.12 (dt, 2H, *J* = 4.1, 12.8 Hz, 2 × 12α-CH), 0.99 (ddd, 2H, *J* = 7.2, 11.0, 12.2 Hz, 2 × 14-CH), 0.94 (m, 2H, 2 × 6β-CH), 0.88 (ddd, 2H, *J* = 4.0, 10.6, 12.2 Hz, 2 × 9-CH), 0.81 (s, 6H, 2 × 19-Me), 0.77 (s, 6H, 18-Me). ^{13}C NMR (125 MHz, $CDCl_3$): δ 46.0 (2 × C-1), 148.9 (2 × C-2), 148.5 (2 × C-3), 35.5 (2 × C-4), 41.8 (2 × C-5), 28.3 (2 × C-6), 31.2 (2 × C-7), 35.4 (2 × C-8),53.8 (2 × C-9), 35.7 (2 × C-10), 20.8 (2 × C-11), 36.7 (2 × C-12), 42.9 (2 × C-13), 50.9 (2 × C-14), 23.4 (2 × C-15), 30.5 (2 × C-16), 81.9 (2 × C-17), 11.1 (2 × C-18), 12.0 (2 × C-19). FABMS: *m*/*z* 573 $[M + H]^+$.

A series of other cephalostatin analogues **79–91** were synthesized from di(17β-hydroxy-5α-androstano)pyrazine (**44**) (Černý *et al.*, 2000). In the first reaction, acetylation of 5′α-androstano(2′,3′-5″,6″)pyrazino(2″,3″-2,3)-5α-androstane-17β,17′β-diol (**44**) yielded a mixture of 5′α-androstano(2′,3′-5″,6″)pyrazino(2″,3″-2,3)-5α-androstan-17β,17′β-yl diacetate (**79**) and 5′α-androstano(2′,3′-5″,6″)pyrazino(2″,3″-2,3)-5α-androstan-17β-ol-17′

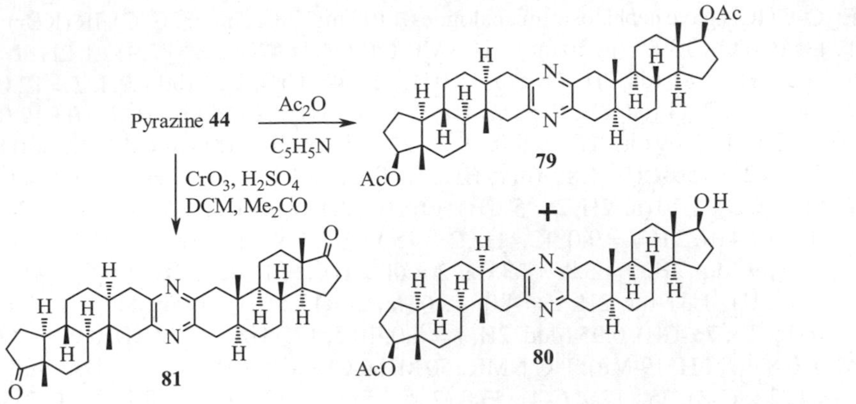

Pyrazine **44** $\xrightarrow[C_5H_5N]{Ac_2O}$ **79** + **80**

Pyrazine **44** $\xrightarrow{CrO_3, H_2SO_4, DCM, Me_2CO}$ **81**

Scheme 5.2.34 *Synthesis of cephalostatin analogues **79–81***

β-yl acetate (**80**) (*Scheme 5.2.34*). Ac_2O (50% solution in C_5H_5N, 600 μL, 3.18 mmol) was added dropwise to a stirred solution of **44** (1.0 g, 1.75 mmol) in C_5H_5N (25 mL), and stirring was continued at room temperature. After 48 h, a mixture of ice–H_2O was added to the reaction mixture and the solid was filtered off, washed with H_2O and dried under vacuum to afford the crude solid, which was subjected to CC (mobile phase: $CHCl_3$: EtOAc = 100:1 and $CHCl_3$: EtOAc = 50:1) to furnish, sequentially, the diacetate **79** (50 mg, 4%), the monoacetate **80** (460 mg, 43%) and the recovered diol **44** (372 mg, 37%).

5′α-Androstano(2′,3′-5″,6″)pyrazino(2″,3″-2,3)-5α-androstan-17β,17′β-yl diacetate (**79**, mp: 330 °C). IR (KBr): ν_{max} cm^{-1} 1724s (C=O), 1399s (pyrazine), 1258m (C–O), 1032m. ^{1}H NMR (200 MHz, $CDCl_3$): δ 4.62 (t, 2H, J = 8.3 Hz, 2 × 17α-CH), 2.91 (d, 2H, J = 16.4 Hz, 2 × 1β-CH), 2.79 (dd, 2H, J = 5.8, 18.0 Hz, 2 × 4α-CH), 2.57 (dd, 2H, J = 11.9, 18.0 Hz, 4β-CH), 2.52 (bd, 2H, J = 16.4 Hz, 2 × lα-CH), 2.04 (s, 6H, 17β-MeCO and 17β′-MeCO), 0.81 (s, 12H, 2 × 18-Me and 2 × 19-Me). FABMS: *m*/*z*: 657 $[M + H]^+$ (Černý *et al.*, 2000).

5′α-Androstano(2′,3′-5″,6″)pyrazino(2″,3″-2,3)-5α-androstan-17β-ol-17′β-yl acetate (**80**, mp: 345 °C). IR (KBr): ν_{max} cm^{-1} 3450br (O–H), 1738s (C=O), 1397s (pyrazine), 1247m (C–O), 1239m (C–O), 1030m (C–O), 1045m (C–O). ^{1}H NMR (200 MHz, $CDCl_3$): δ 4.62 (t, 1H, J = 8.3 Hz, 17α-CH), 3.66 (bt, 1H, J = 8.3 Hz, 17′α-CH), 2.95 (d, 2H, J = 17.4 Hz, 2 × 1β-CH), 2.82 (dd, 2H, J = 4.3, 18.2 Hz, 2 × 4α-CH), 2.60 (dd, 2H, J = 12.2, 18.2 Hz, 2 × 4β-CH), 2.53 (d, 2H, J = 17.4 Hz, 2 × lα-CH), 2.04 (s, 3H, 17β-MeCO), 0.81 (s, 9H, 2 × 19-Me, 18-Me), 0.77 (s, 3H, 18′-Me). FABMS: *m*/*z* 615 $[M + H]^+$ (Černý *et al.*, 2000).

Oxidation of 5′α-androstano(2′,3′-5′,6′)pyrazino(2′,3′-2,3)-5α-androstane-17β,17′β-diol (**44**) yielded 5′α-androstano(2′,3′-5′,6′)pyrazino(2′,3′-2,3)-5α-androstan-17,17′-dione (**81**) (*Scheme 5.2.34*). Jones′ reagent (300 μL, 0.54 mmol) in Me_2CO (5 mL) was added dropwise to a solution of **44** (120 mg, 0.21 mmol) in DCM (5 mL), and stirred for 10 min, the excess reagent was destroyed by 2-PrOH (300 μL), then saturated aqueous $KHCO_3$ (10 mL) was added and organic solvents were removed under pressure. The reaction mixture was extracted with $CHCl_3$ (3 × 5 mL), the pooled extracts were washed with saturated aqueous $KHCO_3$ followed by H_2O, dried ($MgSO_4$) and the solvent was evaporated to dryness. The solid was boiled with Me_2CO to give the crude residue, which after recrystallization from

Et_2O-$CHCl_3$ gave cephalostatin analogue **81** (93 mg, 78%, mp: 320 °C). IR (KBr): ν_{max} cm^{-1} 1743s (C=O), 1399s (pyrazine). ^{1}H NMR (200 MHz, $CDCl_3$): δ 2.93 (d, 2H, *J* = 16.7 Hz, 2 × 1β-CH), 2.81 (dd, 2H, *J* = 5.0, 18.0 Hz, 2 × 4α-CH), 2.61 (bdd, 2H, *J* = 12.0, 18.0 Hz, 2 × 4β-CH), 2.54 (bd, 2H, *J* = 16.7 Hz, 2 × 1α-CH), 2.46 (ddd, 2H, *J* = 0.8, 9.0, 19.2 Hz, 2 × 16β-CH); 2.09 (dt, 2H, *J* = 9.0, 19.2 Hz, 2 × 16α-CH), 1.98 (dddd, 2H, *J* = 0.8, 5.8, 9.0, 12.2 Hz, 2 × 15α-CH), 1.89 (m, 2H, 2 × 7β-CH), 1.78 (ddd, 2H, *J* = 3.0, 4.0, 13.0 Hz, 2 × 11α-CH), 1.70 (m, 2H, 2 × 5-CH), 1.68 (m, 2H, 6α-CH), 1.57 (dq, 2H, *J* = 3.7, 11.0 Hz, 8-CH), 1.54 (tt, 2 H, *J* = 9.0, 12.4 Hz, 2 × 15β-CH), 1.49 (dq, 2H, *J* = 4.0. 13.0 Hz, 2 × 11β-CH), 1.39 (dq, 2H, *J* = 3.8, 13.3 Hz, 2 × 6β-CH), 1.31 (ddd, 2H, *J* = 5.8, 11.0, 12.6 Hz, 2 × 14-CH), 1.31 (dt, 2H, *J* = 4.0, 13.0 Hz, 2 × 12α-CH), 1.06 (ddt, 2H, *J* = 3.6, 12.0, 13.0 Hz, 2 × 7α-CH), 0.95 (ddd, 2H, *J* = 4.0, 10.5, 12.5 Hz, 2 × 9-CH), 0.90 (s, 6H, 2 × 18-Me), 0.83 (s, 6H, 19-Me). ^{13}C NMR (50 MHz, $CDCl_3$): δ 45.9 (2 × C-1), 148.7 (2 × C-2), 148.4 (2 × C-3), 35.8 (2 × C-4), 53.8 (2 × C-5), 30.5 (2 × C-6), 28.1 (2 × C-7), 34.9 (2 × C-8), 41.8 (2 × C-9), 35.4 (2 × C-10), 20.5 (2 × C-11), 31.5 (2 × C-12), 47.6 (2 × C-13), 51.3 (2 × C-14), 21.8 (2 × C-15), 35.8 (2 × C-16), 220.9 (2 × C-17), 13.7 (2 × C-18), 12.0 (2 × C-19). FABMS: *m/z* 569 $[M + H]^+$ (Černý *et al*., 2000).

Cephalostatin analogue **81** was oxidized to 5′α-androstano(2′,3′-5″,6″)pyrazino(2″,3″-2,3)-5α-androstan-17-on-17′β-yl acetate (**82**) by the Jones′ oxidation. The method was exactly the same as that was described for **81** (*Scheme 5.2.34*). Jones′ reagent (400 μL, 0.72 mmol) in Me_2CO (10 mL) was added dropwise to a stirred solution of **81** (400 mg, 0.65 mmol) in DCM (10 mL) (*Scheme 5.2.35*). After the usual workup, the crude product was purified by prep-TLC (mobile phase: $CHCl_3$:MeOH = 50:1) and then crystallized from MeOH to obtain cephalostatin analogue **82** (310 mg, 78%, mp: 330 °C). IR (KBr): ν_{max} cm^{-1} 1740s (C=O), 1 398s (pyrazine), 1246m (C–O), 1032m (C–O). ^{1}H NMR (200 MHz, $CDCl_3$): δ 4.62 (t, 1H, *J* = 8.3 Hz, 17α-CH), 2.04 (s, 3H, 17β-MeCO), 0.89 (s, 3H, 18′-Me), 0.83 (s, 3H, 19′-Me), 0.81 (s, 6H, 19-Me and 18-Me). FABMS: *m/z* 613 $[M + H]^+$ (Černý *et al*., 2000).

Deacetylation of 5′α-androstano(2′,3′-5″,6″)pyrazino(2″,3″-2,3)-5α-androstan-17-on-17′β-yl acetate (**82**) afforded another cephalostatin analogue, 5′α-androstano(2′,3′-5″,6″) pyrazino(2″,3″-2,3)-5α-androstan-17′β-ol-17-one (**83**) (*Scheme 5.2.35*). NaOMe (600 μL, 0.52 mmol) in MeOH (10 mL) was added to a stirred solution **82** (200 mg, 0.32 mmol) in C_6H_6 (10 mL) at room temperature. After 24 h, solid CO_2 (500 μL) was added and the mixture was evaporated under pressure. The residue was extracted with $CHCl_3$ (3 × 10 mL), the pooled extracts were evaporated to dryness and the crude solid was purified by prep-TLC (mobile phase: $CHCl_3$:MeOH = 50:1) and crystallized from MeOH:H_2O (100:1) to produce cephalostatin analogue **83** (136 mg, 73%, mp: 335 °C). IR (KBr): ν_{max} cm^{-1} 3461br (O–H), 1742s (C=O), 1729s (C=O), 1397s (pyrazine), 1048 (C–O). ^{1}H NMR (200 MHz, $CDCl_3$): δ 3.65 (t, 1H, *J* = 8.0 Hz, 17′α-CH), 0.90 (s, 3H, 18-Me), 0.83 (s, 3H, 19-Me), 0.82 (s, 3H, 19′-Me), 0.77 (s, 3H, 18′-Me). FABMS: *m/z* 571 $[M + H]^+$ (Černý *et al*., 2000).

The synthesis of 5′α-androstano(2′,3′-5″,6″)pyrazino(2″,3″-2,3)-5α-androstan-16-oxim-17′β-ol-17-one (**84**) was achieved by utilizing cephalostatin analogue **83** by nitrosation with *t*-BuOK, *t*-BuOH and DMF using isopentyl nitrile (*Scheme 5.2.35*), following the procedure as described earlier for **42** (*Scheme 5.2.18*). ^{1}H NMR (200 MHz, $CDCl_3$): δ 10.05 (br s, NOH), 3.66 (t, 1H, *J* = 9.3 Hz, 17α-CH), 2.97 (dd, 1H, *J* = 6.6, 17.5 Hz, 15′α-CH), 2.94 (d, 1H, *J* = 17.0 Hz, 1′β-CH), 2.92 (d, 1H, *J* = 17.0 Hz, 1β-CH), 2.83 (bdd, 1H, *J* = 5.2,

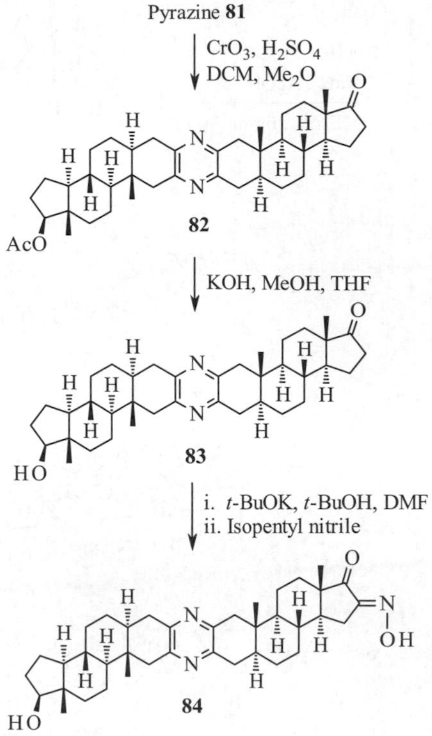

Pyrazine **81**

CrO_3, H_2SO_4
DCM, Me_2O

82

KOH, MeOH, THF

83

i. *t*-BuOK, *t*-BuOH, DMF
ii. Isopentyl nitrile

84

Scheme 5.2.35 *Synthesis of cephalostatin analogues **82–84***

17.5 Hz, 4′α-CH); 2.79 (bdd, 1H, J = 5.2, 17.5 Hz, 4α-CH), 2.62 (bdd, 1H, J = 11.5, 17.5 Hz, 4′β-CH), 2.60 (bdd, 2H, J = 11.5, 17.5 Hz, 4β-CH), 2.55 (bd, 1H, J = 17.0 Hz, 1′α-CH), 2.53 (bd, 1H, J = 17.0 Hz, 1α-CH), 2.16 (dd, 1H, J = 13.2, 17.5 Hz, 15′β-CH), 2.07 (dddd, 1H, J = 6.0, 9.0, 9.3, 13.6 Hz, 16β-CH), 1.68 (m, 2H, 5α-CH and 5′α-CH), 1.62 (m, 1H, 15α-CH), 1.46 (m, 1H, 16α-CH), 1.45 (m, 1H, 14′α-CH), 1.27 (m, 1H, 15β-CH), 0.99 (s, 3H, 18′-Me), 0.85 (s, 3H, 19′-Me), 0.81 (s, 3H, 19-Me), 0.77 (s, 3H, 18′-Me). ^{13}C NMR (50 MHz, $CDCl_3$): δ 45.9 (C-1), 45.6 (C-1′), 148.5 (C-2), 149.2 (C-2′), 148.7 (C-3), 148.3 (C-3′), 35.4 (C-4), 35.2 (C-4′), 41.8 (C-5), 41.7 (C-5′), 28.3 (C-6), 28.0 (C-6′), 31.1 (C-7), 30.5 (C-7′), 35.4 (C-8), 34.3 (C-8′), 53.8 (C-9), 53.6 (C-9′), 35.7 (C-10), 35.8 (C-10′), 20.8 (C-11), 20.34 (C-11′), 36.7 (C-12), 25.5 (C-12′), 42.9 (C-13), 48.8 (C-13′), 50.9 (C-14), 46.4 (C-14′), 23.4 (C-15), 31.1 (C-15′), 30.5 (C-16), 156.6 (C-16′), 81.9 (C-17), 205.1 (C-17′), 11.0 (C-18), 14.1 (C-18′), 12.0 (C-19), 12.0 (C-19′) (Černý *et al.*, 2000).

The synthesis of 16α-azido-3β-hydroxy-5α-androstan-l7-one (**86**) could be carried out from 16α-bromo-3β-hydroxy-5α-androstan-l7-one (**85**) (*Scheme 5.2.36*), utilizing the method as described for **77** (*Scheme 5.2.33*). NaN_3 (1.8 g, 27.69 mmol) in AcOH (500 μL, 8.74 mmol) was added to a stirred solution of **85** (3.0 g, 8.12 mmol) in DMF (60 mL). After the usual workup, the crude product was crystallized from hot Me_2CO affording **86** (2.65 g, 98%, mp: 184–185 °C). IR (KBr): ν_{max} cm^{-1} 3610br (O–H), 2102 (N_3), 1749s (C=O), 1034m (C–O). ^1H NMR (200 MHz, $CDCl_3$): δ 3.71 (t, 1H, J = 8 Hz,

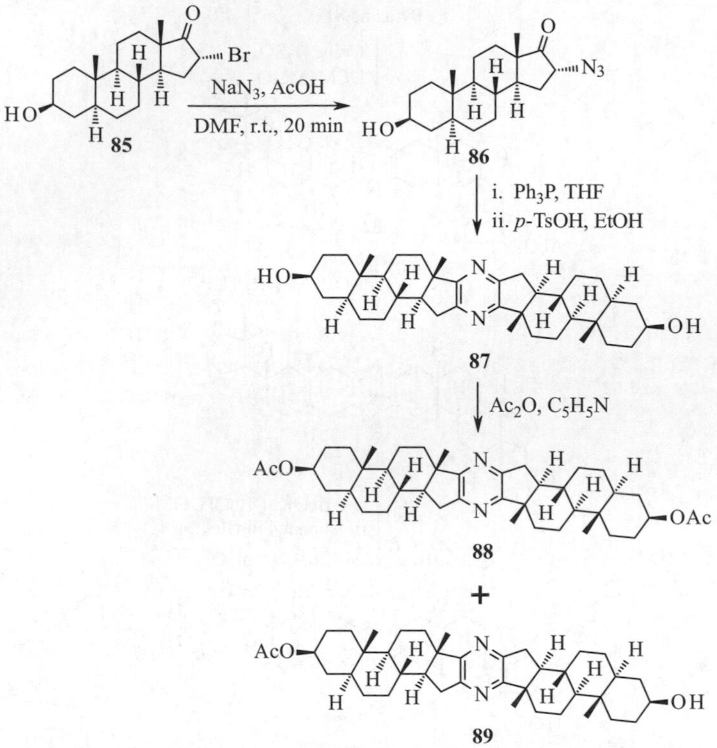

Scheme 5.2.36 *Synthesis of cephalostatin analogues **87–89***

16α-CH), 3.58 (m, 1H, 3α-CH), 2.32 (ddd, 1H, $J = 4.6, 8.4, 11.6$ Hz, 15α-CH), 0.92 (s, 3H, 19-Me), 0.83 (s, 1H, 18-Me) (Černý *et al.*, 2000).

5′α-Androstano(16′,17′-5″,6″)pyrazino(2″,3″-16,17)-5α-androstane-3β,3′β-diol (**87**) was achieved from 16α-azido-3β-hydroxy-5α-androstan-l7-one (**86**) (*Scheme 5.2.36*). A dry mixture of Ph_3P (2.0 g, 7.62 mmol) and **86** (1.0 g, 3.02 mmol) was stirred under Ar. After 10 min, dry THF (20 mL) was added through septum and stirring was continued for another 60 min. H_2O (1 mL, 55.56 mmol) was added and the mixture was stirred for another 24 h. The reaction mixture was transferred to a round flask (250 mL), the reaction vessel was washed with THF and the organic solvent was evaporated to dryness. The solid was coevaporated twice with toluene (20 mL). Then EtOH (10 mL) and *p*-TsOH.H_2O (50 mg) were added, and boiled to dissolve all the solids and stirred at room temperature for 24 h. The organic solvents were rotary evaporated and the residue was coevaporated twice with toluene (20 mL), the crude product was purified by CC (mobile phase: $CHCl_3$:MeOH = 100:1) and crystallized from EtOH to yield cephalostatin analogue **87** (160 mg, 18%, > 340 °C). IR (KBr): ν_{max} cm^{-1} 3415br (O–H), 1041m (C–O). ^1H NMR (200 MHz, $CDCl_3$): δ 3.61 (m, 2H, 2 × 3β-CH), 2.78 (dd, 2H, $J = 5.6, 15.8$ Hz, 2 × 15α-CH), 2.59 (dd, 2H, $J = 10.2, 15.8$ Hz, 2 × 15β-CH), 2.21 (m, 2H, $J = 12$ Hz, 12β-CH), 1.01 (s, 6H, 2 × 19-Me), 0.89 (s, 6H, 2 × 18-Me). ^{13}C NMR (50 MHz, $CDCl_3$): δ 36.7 (2 × C-1), 31.6 (2 × C-2), 71.2 (2 × C-3), 38.1 (2 × C-4), 45.0 (2 × C-5), 28.5 (2 × C-6), 31.5 (2 × C-7), 34.1 (2 × C-8), 54.8 (2 × C-9), 35.8 (2 × C-10), 20.7 (C-11), 33.6 (2 × C-12), 44.1 (2 × C-13), 54.7 (2 ×

C-14), 31.9 (2 × C-15), 155.5 (2 × C-16), 163.8 (2 × C-17), 17.56 (C-18), 12.32 (C-19). FABMS: *m/z* 573 $[M + H]^+$ (Černý *et al.*, 2000).

5′α-Androstano(16′,17′-5″,6″)pyrazino(2″,3″-16,17)-5α-androstane-3β,3′β-diol (**87**) was acetylated to a mixture of analogues, 5′α-androstano(16′,17′-5″,6″)pyrazino(2″,3″-16,17)-5α-androstan-3β,3′β-yl diacetate (**88**) and 5′α-androstano(16′,17′-5″,6″)pyrazino(2″,3″-16,17)-5α-androstan-3′β-ol 3β-yl acetate (**89**) (*Scheme 5.2.36*). Ac_2O (50% solution in C_5H_5N, 600 μL, 3.18 mmol) was added dropwise to a stirred solution of **87** (1.0 g, 1.75 mmol) in C_5H_5N (25 mL) at room temperature. After 48 h, a mixture of ice-H_2O was added to the reaction mixture and the solid was filtered off, washed with H_2O and dried *in vacuo*. The crude product was chromatographed (mobile phase: $CHCl_3$:EtOAc = 50:1 and then $CHCl_3$:EtOAc = 30:1) to provide the diacetate **88** (70 mg, 6%), the monoacetate **89** (527 mg, 49%) and the recovered diol **87** (278 mg, 27%) (Černý *et al.*, 2000).

5′α-Androstano(16′,17′-5″,6″)pyrazino(2″,3″-16,17)-5α-androstan-3β,3′β-yl diacetate (**88**, mp: 342–343 °C). IR (KBr): ν_{max} cm^{-1} 1724s (C=O), 1261m (C–O), 1254m (C–O), 1025m. ^{1}H NMR (500 MHz, $CDCl_3$): δ 4.77 (tt, 2H, *J* = 5.0, 11.4 Hz, 2 × 3α-CH), 2.77 (dd, 2H, *J* = 6.4, 15.0 Hz, 2 × 15α-CH), 2.59 (m, 2H, 2 × 15β-CH), 2.21 (m, 2H, 2 × 12β-CH), 2.03 (s, 6H, 2 × 3β-MeCO), 1.85 (m, 2H, 2 × 2α-CH), 1.78 (m, 2H, 2 × 7α-CH), 1.75 (m, 2H, 2 × 1β-CH), 1.73 (m, 2H, 2 × 11α-CH), 1.70 (m, 2H, 2 × 8-CH), 1.67 (m, 2H, 2 × 14-CH), 1.62 (m, 2H, 2 × 4α-CH), 1.52 (m, 4H, 2 × 11β-CH and 2 × 12α-CH), 1.50 (m, 2H, 2 × 2β-CH), 1.37 (m, 2H, 2 × 4β-CH), 1.35 (m, 4H, 2 × 6α-CH and 2 × 6β-CH), 1.23 (m, 2H, 2 × 5-CH), 1.05 (m, 2H, 2 × 1α-CH), 1.03 (m, 2H, 2 × 7β-CH), 1.01 (s, 6H, 2 × 18-Me), 0.90 (s, 6H, 2 × 19-Me), 0.85 (m, 2H, 2 × 9-CH). ^{13}C NMR (125 MHz, $CDCl_3$): δ 36.5 (2 × C-1), 27.4 (2 × C-2), 73.5 (2 × C-3), 170.7 (2 × 3β-CO), 21.4 (2 × 3β-MeCO), 34.0 (2 × C-4), 44.8 (2 × C-5), 28.3 (2 × C-6), 31.5 (2 × C-7), 34.1 (2 × C-8), 54.7 (2 × C-9), 35.8 (2 × C-10), 20.7 (2 × C-11), 33.5 (2 × C-12), 44.1 (2 × C-13), 54.6 (2 × C-14), 31.9 (2 × C-15), 155.5 (2 × C-16), 163.8 (2 × C-17), 17.6 (2 × C-18), 12.2 (2 × C-19). FABMS: *m/z* 657 $[M + H]^+$ (Černý *et al.*, 2000).

5′α-Androstano(16′,17′-5′,6′)pyrazino(2′,3′-16,17)-5α-androstan-3′β-ol-3β-yl acetate (**89**, mp: 312–313 °C). IR (KBr): ν_{max} cm^{-1} 3437br (O–H), 1738s (C=O), 1247m (C–O), 1030m (C–O). ^{1}H NMR (500 MHz, $CDCl_3$): δ 4.70 (m, 1H, 3α-CH), 3.61 (m, 1H, *J* = 32.0 Hz, 3′α-CH), 2.77 (dd, 2H, *J* = 6.5, 15.0 Hz, 2 × 15α-CH), 2.60 (m, 2H, 2 × 15β-CH), 2.21 (m, 2H, 2 × 12β-CH), 2.03 (s, 3H, 3β-MeCO), 1.01 (s, 6H, 2 × 18-Me), 0.90 (s, 3H, 19-Me), 0.89 (s, 3H, 19′-Me). FABMS: *m/z* 615 $[M + H]^+$ (Černý *et al.*, 2000).

Jones′ oxidation of 5′α-androstano(16′,17′-5″,6″)pyrazino(2″,3″-16,17)-5α-androstan-3′β-ol-3β-yl acetate (**89**) produced its keto analogue, 5′α-androstano(16′,17′-5″,6″) pyrazino(2″,3″-16,17)-5α-androstan-3′-on-3β-yl acetate (**90**) (*Scheme 5.2.37*). Jones′ reagent (200 μL, 0.36 mmol) in and Me_2CO (3 mL) was added dropwise to a stirred solution of the monoacetate **89** (80 mg, 0.13 mmol) in DCM (3 mL). After the usual work up, as described earlier, the crude solid was crystallized from EtOH to afford cephalostatin analogue **90** (55 mg, 69%, mp: 330–331 °C) (Černý *et al.*, 2000).

IR (KBr): ν_{max} cm^{-1} 1738s (C=O), 1713s (C=O), 1250m (C–O), 1027m (C–O). ^{1}H NMR (500 MHz, $CDCl_3$): δ 4.70 (m, 1H, 3α-CH), 2.78 (m, 2H, 2 × 15α-CH), 2.60 (m, 2H, 2 × 15β-CH), 2.02 (s, 3H, 3β-MeCO), 1.09 (s, 3H, 19′-Me), 1.04 (s, 3H, 18′-Me), 1.01 (s, 3H, 18-Me), 0.91 (s, 3H, 19-Me). ^{13}C NMR (125 MHz, $CDCl_3$): δ 36.5 (C-1), 38.2 (C-1′), 27.4 (C-2), 38.1 (C-2′), 73.6 (C-3), 170.7 (3β-CO), 21.4 (3β-MeCO), 211.7 (C-3′), 44.6 (C-4), 34.0 (C-4), 44.8 (C-5), 46.8 (C-5′), 28.3 (C-6), 28.7 (C-6′), 31.5 (C-7), 31.2 (C-7′),

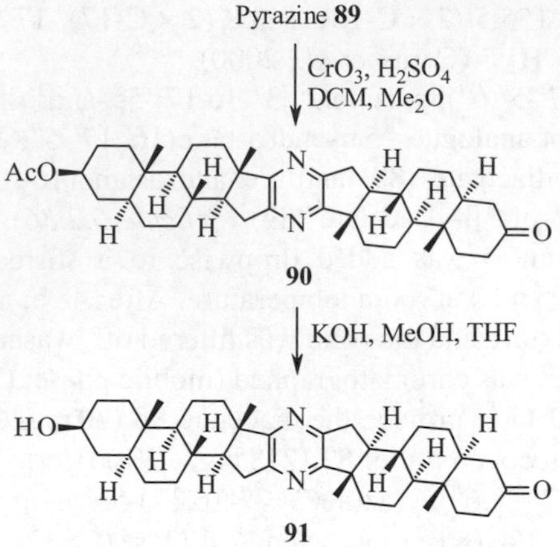

Pyrazine **89**

CrO_3, H_2SO_4
DCM, Me_2O

90

KOH, MeOH, THF

91

Scheme 5.2.37 *Synthesis of cephalostatin analogues **90** and **91***

34.1 and 34.0 (C-8 and C-8′), 54.7 and 54.6 (C-9, C-9), 36.0 and 35.8 (C-10 and C-10′), 21.0 and 20.7 (C-11 and C-11′), 33.5 (2 × C-12), 44.1 and 44.0 (C-13 and C-13′), 54.5 and 54.3 (C-14 and C-14′), 31.9 (2 × C-15), 155.6 and 155.3 (C-16 and C-16′), 164.0 and 163.6 (C-17 and C-17), 17.6 (2 × C-18), 12.2 (C-19), 11.5 (C-19′). FABMS: *m/z* 613 $[M + H]^+$ (Černý *et al.*, 2000).

In the next step, 5′α-androstano(16′,17′-5′,6′)pyrazino(2′,3′-16,17)-5α-androstan-3′-on-3β-yl acetate (**90**) was deacetylated to 5′α-androstano(16′,17′-5′,6′)pyrazino(2′,3′-16,17)-5α-androstan-3β-ol-3′-one (**91**) (*Scheme 5.2.37*), following the procedure as described for **78** and **79** (*Scheme 5.2.34*). NaOMe (300 μL, 0.26 mmol) in MeOH (10 mL) was added to a stirred solution of **90** (100 mg, 0.16 mmol) in C_6H_6 (10 mL) at room temperature. After the usual workup and purification, as described earlier for **83**, yielded cephalostatin analogue **91** (50 mg, 54%, 320 °C). IR (KBr): ν_{max} cm^{-1} 3446br (O–H), 1713s (C=O), 1045m (C–O). ^1H NMR (200 MHz, $CDCl_3$): δ 3.60 (m, 1H, 3α-CH), 2.78 (m, 2H, 2 × 15α-CH), 2.60 (m, 2H, 2 × 15β-CH), 1.09 (s, 3H, 19-Me), 1.03 (s, 3H, 18-Me), 1.01 (s, 3H, 18′-Me), 0.88 (s, 3H, 19′-Me). FABMS: *m/z* 571 $[M + H]^+$ (Černý *et al.*, 2000).

Cephalostatin/ritterazine analogues **101** and **102**, having higher cytotoxic activity than the natural cytotoxic 22-*epi*-hippuristanol (**92**) against MDA-MB-231, A-549 and HT-29 cultured tumour cell lines, were prepared *via* condensation reaction of the derivatives of commercially available compound (25*R*)-3β-hydroxy-5α-spirostan-12-one (**96**, hecogenin) and the natural product 22-*epi*-hippuristanol (**92**) (Poza *et al.*, 2010). Initially, compound **92** was converted to 22-*epi*-hippuristan-3,11-dione (**93**) by selective oxidation using the Corey–Schmidt reagent (*Scheme 5.2.38*). A mixture of 22-*epi*-hippuristanol (**92**, 100 mg, 0.22 mmol) and PDC (200 mg, 0.5 mmol) in dry DMF (2 mL) in the presence of 4 Å activated molecular sieves was stirred at room temperature. After 6 h, the reaction mixture was filtered off through a Celite pad and the filtrate was evaporated under pressure, the crude product was subjected to CC (mobile phase: 20% EtOAc in *n*-hexane) to give diketone **93** (95 mg, 96%, mp: 220 °C). IR (neat): ν_{max} cm^{-1} 3434r (O–H), 2924s (C–H), 1701s (C=O), 1050m (C–O), 1031m (C–O), 1009m (C–O), 970m, 923w, 874w. ^1H NMR (300 MHz,

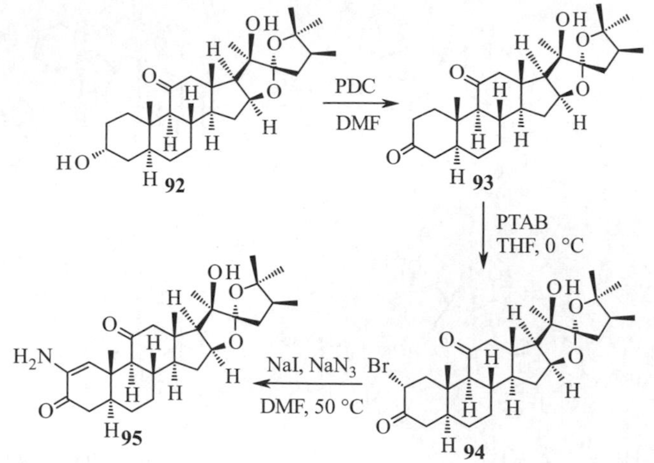

Scheme 5.2.38 *Synthesis of 22-epi-hippuristanol derivatives **93–95***

$CDCl_3$): δ 4.48 (m, 1H, 16α-CH), 1.30 (s, 3H, 18-Me), 1.27 (s, 3H, 27-Me), 1.21 (s, 3H, 21-Me), 1.07 (s, 3H, 19-Me), 0.97 (s, 3H, 26-Me), 0.94 (d, 3H, $J = 6.5$ Hz, 28-Me). ^{13}C NMR (75 MHz, $CDCl_3$): δ 37.9 (C-1), 37.1 (C-2), 210.4 (C-3), 39.6 (C-4), 46.9 (C-5), 28.2 (C-6), 32.3 (C-7), 35.7 (C-8), 63.3 (C-9), 35.2 (C-10), 211.4 (C-11), 58.1 (C-12), 45.8 (C-13), 55.7 (C-14), 31.5 (C-15), 79.0 (C-16), 64.0 (C-17), 26.0 (C-18), 11.0 (C-19), 81.8 (C-20), 17.5 (C-21), 118.5 (C-22), 44.3 (C-23), 40.9 (C-24), 84.3 (C-25), 29.0 (C-26), 22.9 (C-27), 14.0 (C-28). EIMS: *m/z* 458 $[M]^+$. FABMS: *m/z* 459 $[M+H]^+$. HRESIMS: *m/z* 459.3105 calculated for $C_{28}H_{43}O_5$, found: 459.3104 (Poza *et al.*, 2010).

2α-Bromo-22-*epi*-hippuristan-3,11-dione (**94**) was synthesized from 22-*epi*-hippuristan-3,11-dione (**93**) (*Scheme 5.2.38*). To a solution of **93** (95 mg, 0.21 mmol) in dry THF (2 mL), a cold (0 °C) solution of PTAB (80 mg, 0.2 mmol) in dry THF (2 mL) was added dropwise over a period of 3 h, and stirred for another 45 min. Saturated aqueous $NaHCO_3$ (5 mL) was added, the mixture was extracted with EtOAc, and the organic layer was washed with brine, dried ($MgSO_4$), concentrated under vacuum and purified by CC (mobile phase: 10% EtOAc in *n*-hexane) to obtain 2α-bromoketone **94** (76 mg, 68%, mp: 115–121 °C). IR (neat): ν_{max} cm^{-1} 3494br (O–H), 2925m (C–H), 1702s (C=O), 1025s (C–O), 970m, 923m, 871w. ^1H NMR (300 MHz, $CDCl_3$): δ 4.78 (dd, 1H, $J = 6.4$, 13.2 Hz, 2β-CH), 4.48 (m, 1H, 16α-CH), 1.30 (s, 3H, 18-Me), 1.28 (s, 3H, 27-Me), 1.24 (s, 3H, 21-Me), 1.07 (s, 3H, 19-Me); 0.98 (s, 3H, 26-Me), 0.94 (d, 3H, $J = 6.5$ Hz, 28-Me). ^{13}C NMR (75 MHz, $CDCl_3$): δ 37.4 (C-1), 56.1 (C-2), 210.4 (C-3), 39.9 (C-4), 45.7 (C-5), 28.2 (C-6), 32.6 (C-7), 35.8 (C-8), 63.3 (C-9), 35.6 (C-10), 211.4 (C-11), 57.1 (C-12), 45.3 (C-13), 54.7 (C-14), 31.8 (C-15), 79.0 (C-16), 64.0 (C-17), 25.9 (C-18), 11.4 (C-19), 81.8 (C-20), 17.4 (C-21), 118.5 (C-22), 44.8 (C-23), 40.6 (C-24), 84.3 (C-25), 28.7 (C-26), 22.5 (C-27), 14.2 (C-28). FABMS: *m/z* 539 $[M + H]^+$ (^{81}Br); 537 $[M + H]^+$ (^{79}Br). HRESIMS: *m/z* 537.2210 calculated for $C_{28}H_{42}O_5{}^{79}Br$, found: 537.2212 (^{79}Br) (Poza *et al.*, 2010).

2α-Bromo-22-*epi*-hippuristan-3,11-dione (**94**) was converted to 2-amino-22-*epi*-hippurist-1-ene-3,11-dione (**95**) (*Scheme 5.2.38*). Catalytic amounts of NaI and NaN_3 (0.1 g, 8.0 mmol) were added to a solution of **94** (76 mg, 0.14 mmol) in DMF (5 mL). The

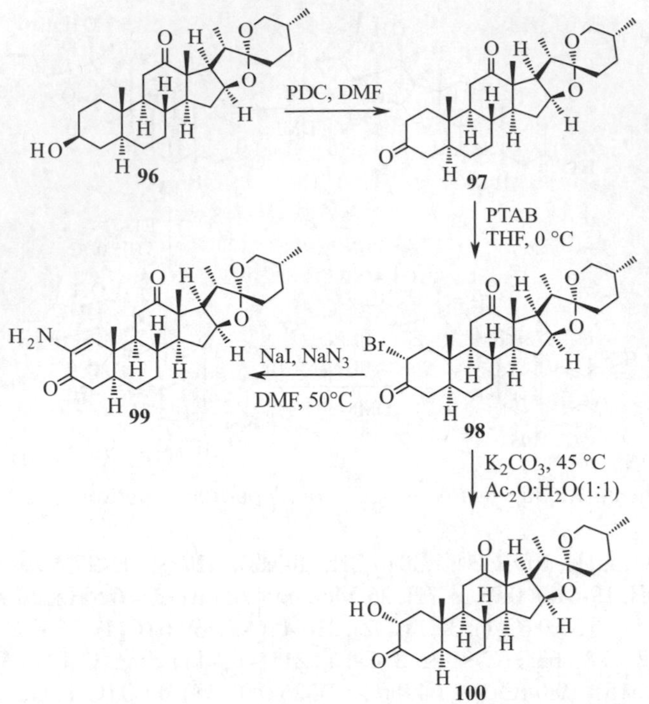

Scheme 5.2.39 *Synthesis of hecogenin derivatives* ***97–100***

suspension was stirred under Ar at 50 °C for 60 min. H_2O (50 mL) was added and the mixture was extracted with Et_2O, the ethereal extract was washed with brine, dried (Na_2SO_4), evaporated to dryness and the crude solid was subjected to CC, eluted with 30% EtOAc in *n*-hexane, to produce enaminoketone **95** (65 mg, 97%). ^{1}H NMR (300 MHz, $CDCl_3$): δ 6.57 (br, 1H, 1-CH), 4.53 (m, 1H, 16α-CH), 1.36 (s, 3H, 18-Me), 1.31 (s, 3H, 27-Me), 1.26 (s, 3H, 21-Me), 1.12 (s, 3H, 19-Me), 1.04 (s, 3H, 26-Me), 0.99 (d, 3H, J = 6.5 Hz, 28-Me). ^{13}C NMR (75 MHz, $CDCl_3$): δ 127.5 (C-1), 137.8 (C-2), 195.3 (C-3), 39.5 (C-4), 45.8 (C-5), 28.2 (C-6), 32.1 (C-7), 35.7 (C-8), 61.4 (C-9), 35.5 (C-10), 210.2 (C-11), 57.1 (C-12), 45.2 (C-13), 54.7 (C-14), 31.0 (C-15), 79.1 (C-16), 63.2 (C-17), 25.9 (C-18), 11.4 (C-19), 81.3 (C-20), 17.7 (C-21), 118.6 (C-22), 44.5 (C-23), 40.4 (C-24), 84.5 (C-25), 29.0 (C-26), 23.0 (C-27), 14.7 (C-28). EIMS: *m/z* 471 $[M]^+$. FABMS: *m/z* 472 $[M + H]^+$. HRESIMS: *m/z* 472.3057 calculated for $C_{27}H_{42}NO_5$, found: 472.3063 (Poza *et al.*, 2010).

The synthesis of (25*R*)-5α-spirostan-3,12-dione (**97**) was accomplished using the Corey–Schmidt oxidation, in a similar way as described earlier for **92**, from hecogenin (**96**, 1.0 g, 2.3 mmol) by treating it with PDC (2.0 g, 5.0 mmol) in dry DMF (20 mL) (*Scheme 5.2.39*), and the crude product was purified by CC (mobile phase: 20% EtOAc in *n*-hexane) to yield hecogenin-3-one (**97**, 990 mg, 99%). ^{1}H NMR (300 MHz, $CDCl_3$): δ 4.34 (m, 1H, 16α-CH), 3.49 (ddd, 1H, J = 2, 4.4 and 10.9 Hz, 26-CHb), 3.34 (t, 1H, J = 10.9 Hz, 26-CHa), 2.54 (dd, 1H, J = 8.0, 9.0 Hz, 1H, 17α-CH), 2.48 (dd, 1H, J = 10.5, 13.5 Hz, 11-CHb), 2.38 (br s, 1H, 8β-CH), 2.43–2.32 (m, 2H, 2-CH_2), 2.17 (dd, 1H, J = 5.0, 10.5 Hz, 11-CHa), 2.20–1.90 (m, 4H, 2 × CH_2), 1.25 (s, 3H, 18-Me), 1.09 (d, 3H, J = 7.0 Hz, 3H,

21-Me), 1.06 (s, 3H, 19-Me), 0.79 (d, 3H, $J = 6.5$ Hz, 27-Me). ^{13}C NMR (75 MHz, $CDCl_3$): δ 11.2, 13.2, 16.0, 17.1, 28.7, 28.5, 30.2, 31.1, 31.2, 31.4, 31.5, 34.2, 36.2, 37.7, 37.8, 42.2, 44.4, 46.2, 53.5, 54.8, 55.1, 55.5, 66.8, 79.1, 109.2, 210.7, 212.8. EIMS: *m/z* 428 $[M]^+$. FABMS: *m/z* 429 $[M + H]^+$. HRESIMS: *m/z* 429.2999 calculated for $C_{27}H_{41}O_4$, found: 429.2998 (Poza *et al.*, 2010).

(25*R*)-5α-Spirostan-3,12-dione (**97**) was converted to (25*R*)-2α-bromo-5α-spirostan-3,12-dione (**98**), following the same method as outlined for **94** (*Scheme 5.2.39*). A solution of **97** (6.0 g, 14.0 mmol) in dry THF (20 mL) was reacted with PTAB (5.0 g, 12.5 mmol) in dry THF (20 mL) to provide the crude product, which was purified by CC, eluted with 10% EtOAc in *n*-hexane to afford 2α-bromoketone **98** (4.5 g, 63%). ^{1}H NMR (300 MHz, $CDCl_3$): δ 4.70 (dd, 1H, $J = 6.4$, 13.2 Hz, 2β-CH), 4.32 (m, 1H, 16α-CH), 3.46 (m, 1H, 26-CHb), 3.32 (t, 1H, $J = 11.0$ Hz, 26-CHa), 2.23 (dd, 1H, $J = 5.0$, 10.5 Hz, 11-CHa), 2.20–1.90 (m, 2H, 4-CH_2), 1.22 (s, 3H, 18-Me), 1.05 (d, 3H, $J = 6.8$ Hz, 21-Me), 1.05 (s, 3H, 19-Me), 0.78 (d, 3H, $J = 6.5$ Hz, 27-Me). ^{13}C NMR (75 MHz, $CDCl_3$): δ 11.8, 13.2, 15.9, 17.0, 27.9, 28.7, 30.1, 30.8, 31.0, 31.3, 33.7, 37.6, 39.2, 42.1, 43.5, 46.9, 50.8, 53.4, 53.5, 54.4, 55.2, 61.8, 66.8, 78.9, 109.2, 200.2, 211.8. EIMS: *m/z* 508 $[M]^+$ (^{81}Br) and 506 $[M]^+$ (^{79}Br). FABMS: *m/z* 509 $[M + H]^+$ (^{81}Br) and 507 $[M + H]^+$ (^{79}Br). HRESIMS: *m/z* 507.2110 calculated for $C_{27}H_{40}O_4{}^{79}Br$, found: 507.2114 (^{79}Br) (Poza *et al.*, 2010).

NaN_3 (5.2 g, 80 mmol) and a catalytic amount of NaI were added to a stirred solution of (25*R*)-2α-bromo-5α-spirostan-3,12-dione (**98**, 4.0 g, 7.9 mmol) in DMF (100 ml). Usual workup as described for **95**, afforded a crude residue, which was purified by CC (mobile phase: 30% EtOAc in *n*-hexane) followed by crystallization from Et_2O-EtOAc to give (25*R*)-2-amino-5α-spirost-1-ene-3,12-dione (**99**, 3.2 g, 92%, mp: 235–240 °C) (*Scheme 5.2.39*). IR (neat): ν_{max} cm^{-1} 3447m (N–H), 3349m (N–H), 2929m (C–H), 1703s (C=O), 1667s (C=C), 1633s (C=C), 1038s (C–O), 1007m, 954w, 868w. ^{1}H NMR (300 MHz, $CDCl_3$): δ 5.87 (br s, 1H, 1-CH), 4.34 (m, 1H, 16α-CH), 3.47 (m, 1H, 26-CHb), 3.34 (t, 1H, $J = 10.8$ Hz, 26-CHa), 1.25 (s, 3H, 18-Me), 1.08 (s, 3H, 19-Me); 1.07 (d, 3H, $J = 6.8$ Hz, 21-Me); 0.79 (d, 3H, $J = 6.6$ Hz, 27-Me). ^{13}C NMR (75 MHz, $CDCl_3$): δ 12.8, 13.4, 15.6, 16.6, 26.6, 28.3, 29.7, 30.5, 30.6, 31.0, 34.0, 37.3, 38.0, 39.6, 41.8, 44.1, 51.8, 53.2, 54.8, 55.1, 66.4, 78.6, 108.8, 123.5, 137.7, 194.7, 211.9. EIMS: *m/z* 441 $[M]^+$. FABMS: *m/z* 442 $[M + H]^+$. HRESIMS: *m/z* 442.2951 calculated for $C_{27}H_{40}NO_4$, found: 442.2959 (Poza *et al.*, 2010).

(25*R*)-2α-bromo-5α-spirostan-3,12-dione (**98**) was also transformed to (25*R*)-2α-hydroxy-5α-spirostan-3,12-dione (**100**) (*Scheme 5.2.39*). A solution of K_2CO_3 (1.0 g, 7.2 mmol) in H_2O (5 mL) was added to a stirred solution of **98** (200 mg, 0.39 mmol) in Me_2CO (15 mL), and the reaction mixture was heated at 45 °C. After 15 h, H_2O was added and the mixture was partly evaporated under pressure. The solid was extracted with $CHCl_3$, the organic extract was dried (Na_2SO_4) and rotary evaporated to obtain 2α-hydroxyketone **100** (140 mg, 80%). ^{1}H NMR (300 MHz, $CDCl_3$): δ 5.80 (br s, 1H, OH), 4.39 (m, 1H, 16α-CH), 4.28 (dd, 1H, $J = 7.0$, 12.0 Hz, 2β-CH), 3.53 (m, 1H, 26-CHb), 3.38 (t, 1H, $J = 10.8$ Hz, 26-CHa), 1.28 (s, 3H, 18-Me), 1.11 (d, 3H, $J = 6.8$ Hz, 21-Me); 1.10 (s, 3H, 19-Me), 0.83 (d, 3H, $J = 6.8$ Hz, 27-Me). ^{13}C NMR (75 MHz, $CDCl_3$): δ 12.6, 13.3, 16.0, 17.7, 28.2, 28.8, 29.7, 30.2, 31.1, 31.2, 31.4, 33.6, 37.5, 37.9, 42.3, 42.2, 47.7, 48.0, 54.6, 55.1, 55.2, 66.9, 72.5, 79.1, 109.3, 210.2, 212.4. EIMS: *m/z* 444 $[M]^+$. FABMS: *m/z* 445 $[M + H]^+$. HRESIMS: *m/z* 445.2948 calculated for $C_{27}H_{41}O_5$, found: 445.2961 (Poza *et al.*, 2010).

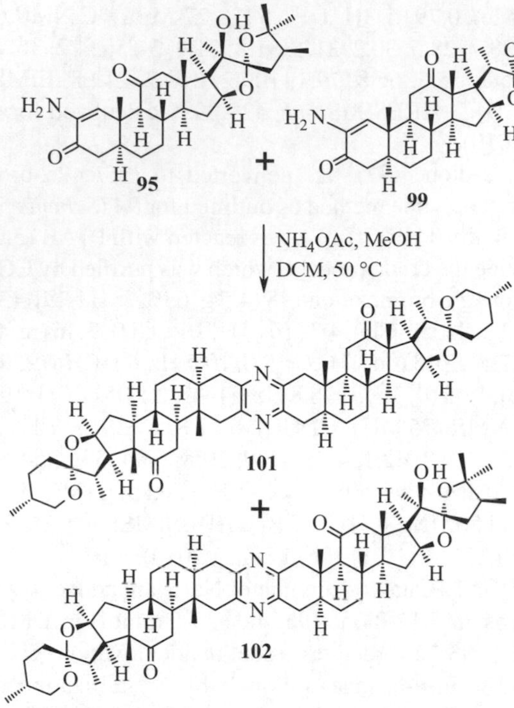

NH$_4$OAc, MeOH
DCM, 50 °C

Scheme 5.2.40 *Synthesis of cephalostatin/ritterazine analogues **101** and **102***

A mixture of cephalostatin/ritterazine analogues **101** and **102** was obtained simply *via* condensation reaction of 2-amino-22-*epi*-hippurist-1-ene-3,11-dione (**95**) and (25*R*)-2-amino-5α-spirost-1-ene-3,12-dione (**99**) (Poza *et al.*, 2010). Interestingly, analogue **102** could be prepared solely *via* condensation reaction of 2-amino-22-*epi*-hippurist-1-ene-3,11-dione (**95**) and (25*R*)-2α-hydroxy-5α-spirostan-3,12-dione (**100**) (Poza *et al.*, 2010). MeOH (100 μL) and two drops of AcOH was added to a stirred mixture of **95** (12 mg, 0.025 mmol) and **99** (12 mg, 0.025 mmol) in EtOAc (3 mL). The mixture was hydrogenated with Pd-C (65 mg, 5% on carbon) at room temperature overnight (*Scheme 5.2.40*). The mixture was filtered and the filtrate was washed with saturated aqueous $NaHCO_3$ followed by saturated aqueous NaCl, dried (Na_2SO_4), concentrated and purified by HPLC (Scharlau C_{18}, flow rate 1.5 mL/min, eluted with 50% EtOAc in *n*-hexane) to produce cephalostatin/ritterazine analogues **101** and **102** (Poza *et al.*, 2010).

Cephalostatin/ritterazine analogue **101** (3.0 mg). ^{1}H NMR (500 MHz, $CDCl_3$): δ 4.40 (m, 1H, 16α-CH), 3.54 (m, 1H, 26-CHb), 3.39 (d, 1H, $J = 11.0$ Hz, 26-CHa), 1.29 (s, 3H, 18-Me); 1.10 (d, 3H, $J = 6.3$ Hz, 21-Me), 1.08 (s, 3H, 19-Me), 0.84 (d, 3H, $J = 6.5$ Hz, 27-Me). ^{13}C NMR (125 MHz, $CDCl_3$): δ 11.8, 13.3, 16.2, 17.5, 28.2, 28.8, 29.7, 30.5, 31.2, 31.5, 34.0, 35.3, 36.2, 37.6, 42.8, 43.1, 45.4, 49.7, 51.8, 54.9, 55.4, 66.9, 78.8, 109.2, 148.4, 213.3. EIMS: *m/z* 848 $[M]^+$. FABMS: *m/z* 849 $[M + H]^+$. HRESIMS: *m/z* 849.5776 calculated for $C_{54}H_{77}N_2O_6$, found: 849.5775 (Poza *et al.*, 2010).

Cephalostatin/ritterazine analogue **102** (5.0 mg, 22%). ^{1}H NMR (500 MHz, $CDCl_3$): δ 4.48 (m, 1H, 16α-CH), 4.40 (m, 1H, 16′α-CH), 3.54 (m, 1H, 26′-CHb), 3.41 (m, 1H,

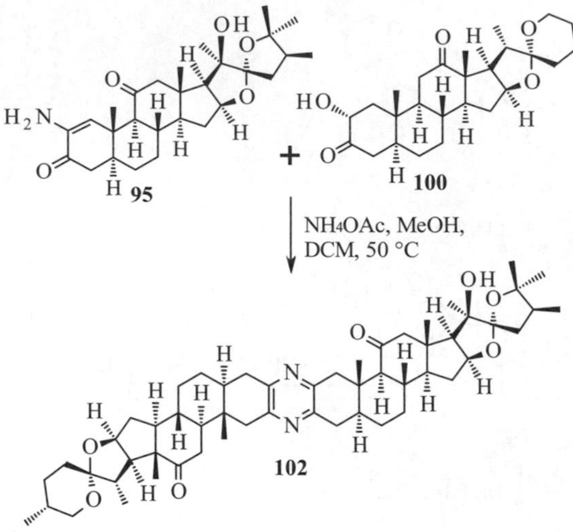

Scheme 5.2.41 *Synthesis of cephalostatin/ritterazine analogue* ***102***

26′-CHa), 2.60 (m, 2H, 15′-CH_2), 2.19 (m, 2H, 15-CH_2), 1.65 (m, 1H, 25-CH), 1.30 (s, 6H, 18-Me and 27-Me), 1.11 (s, 3H, 18′-Me); 1.26 (s, 3H, 21-Me), 1.13 (d, 3H, $J = 6.5$ Hz, 21′-Me); 1.12 (s, 3H, 19-Me), 1.07 (s, 3H, 19′-Me), 1.03 (s, 3H, 26-Me), 0.99 (d, 3H, $J = 6.8$ Hz, 26′-Me), 0.84 (d, 3H, $J = 6.4$ Hz, 27′-Me). ^{13}C NMR (125 MHz, $CDCl_3$): δ 11.7, 13.1, 13.8, 14.2, 16.0, 17.2, 19.8, 22.7, 26.4, 27.1, 28.1, 28.9, 29.4, 29.6, 29.7, 29.8, 30.1, 30.3, 30.6, 31.2, 31.4, 31.9, 32.2, 32.8, 33.8, 34.2, 35.3, 36.1, 37.1, 37.7, 41.0, 41.6, 42.5, 45.3, 45.4, 53.5, 54.8, 55.0, 55.7, 57.0, 63.6, 65.6, 66.9, 79.2, 79.8, 81.9, 84.5, 109.4, 117.2, 148.3, 148.5, 204.5, 213.0. FABMS: *m/z* 879 $[M + H]^+$. HREIMS: *m/z* 878.5804 calculated for $C_{55}H_{78}N_2O_7$, found: 878.5771. HRESIMS: *m/z* 879.5882 calculated for $C_{55}H_{79}N_2O_7$, found: 879.5888 (Poza *et al.*, 2010).

A mixture of NH_4OAc (9.0 mg) in moist MeOH (5 mL, <0.1% H_2O) and 2-amino-22-*epi*-hippurist-1-ene-3,11-dione (**95**, 10 mg, 0.021 mmol) was refluxed for 30 min. Then, a solution of (25*R*)-2α-hydroxy-5α-spirostan-3,12-dione (**100**, 10 mg, 0.022 mmol) in DCM (60 μL) was added and refluxed for another 3 h (*Scheme 5.2.41*). H_2O was added and the mixture was extracted with DCM, the organic layer was washed with brine, dried (Na_2SO_4), concentrated under vacuum and purified by HPLC as described above to afford cephalostatin/ritterazine analogue **102** (3.0 mg, 16%) (Poza *et al.*, 2010).

An efficient synthesis of cephalostatin analogues, diosgenin pyrazine dimers **120–126**, was performed starting from (25*R*)-5α-spirost-5-en-3β-ol (**103**, diosgenin) and through its conversion to several diosgenin derivatives **104–119** (Shawakfeh *et al.*, 2008, 2011). These symmetrical cephalostatin analogues were synthesized by the classical condensation of aminoketones. First, (25*R*)-spirost-5-en-3-one (**104**) was accomplished from diosgenin (**103**) by PCC oxidation (*Scheme 5.2.42*). A stirred mixture of powdered $CaCO_3$ (4.0 g, 4.0 mmol) and **103** (2.0 g, 4.8 mmol) in DCM (50 mL) was treated with PCC (1.7 g, 4.0 mmol) at room temperature, After 30 min, Et_2O (50 mL) was added and the solution was filtered through a short column of florisil. The filtrate was evaporated under pressure, and the crude product was purified by CC (mobile phase: 10% EtOAc in *n*-hexane) to furnish **104** (1.6 g, 80%, mp: 193–194 °C). IR (KBr): ν_{max} cm^{-1} 2961m (C–H), 1726s (C–O), 1681s

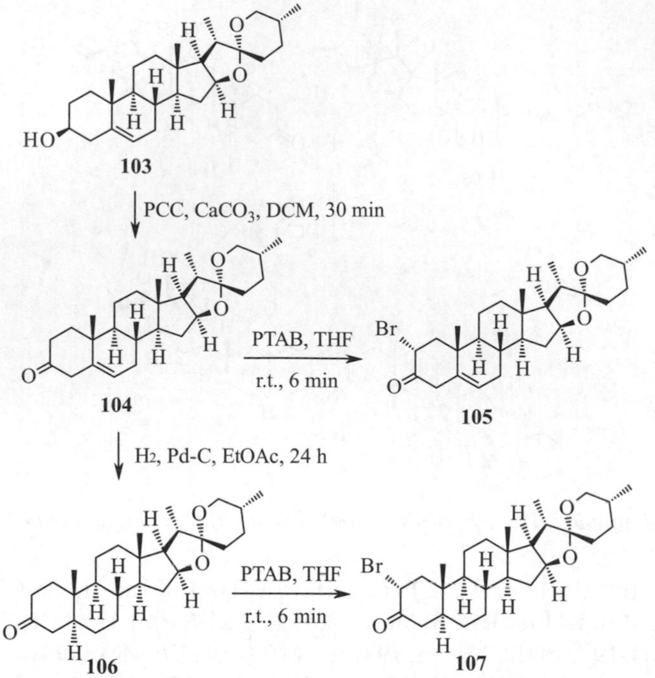

Scheme 5.2.42 *Synthesis of diosgenin derivatives* ***104–107***

(C=C). ^{1}H NMR (400 MHz, $CDCl_3$): δ 5.30 (s, 1H, 6-CH), 4.40 (dt, 1H, 16α-CH), 3.48 (m, 1H, 26-CHb), 3.38 (t, 1H, $J = 10.8$ Hz, 26-CHa). ^{13}C NMR (100 MHz, $CDCl_3$): δ 62.7, 66.7, 80.7, 109.0, 121.0, 140.8, 211.0 (Shawakfeh *et al.*, 2008).

(25*R*)-Spirost-5-en-3-one (**104**) was converted to (25*R*)-2α-bromospirost-5-en-3-one (**105**) in the following way (*Scheme 5.2.42*). PTAB (230 mg, 0.63 mmol, 1.3 equiv.) in THF (7 mL) was added rapidly to a solution of **104** (200 mg, 0.48 mmol) in THF (20 mL) at room temperature. After 6 min, brine (10 mL) was added and the mixture was extracted twice with $CHCl_3$ (20 mL) The solvent was dried (Na_2SO_4), evaporated to dryness and the crude residue was subjected to CC, eluted with 15% EtOAc in *n*-hexane to give 2α-bromoketone **105** (260 mg, 72%, mp: 160–163 °C). IR (KBr): ν_{max} cm^{-1} 2946m (C–H), 1681s (C=O), 1619m (C=C), 797m. ^{1}H NMR (400 MHz, $CDCl_3$): δ 4.50 (dd, 1H, $J = 6.2$, 7.1 Hz, 2β-CH), 4.36 (dt, 1H, $J = 7.2$, 7.6 Hz, 16α-CH), 3.46 (m, 1H, 26-CHb), 3.3 (t, 1H, $J = 10.8$ Hz, 26-CHa). ^{13}C NMR (100 MHz, $CDCl_3$): δ 54.0 (C-2, 204 (C-3), 121 (C-6), 80.6 (C-16), 62.7 (C-17), 109 (C-22), 66.7 (C-26) (Shawakfeh *et al.*, 2008).

(25*R*)-Spirost-5-en-3-one (**104**) was also converted to (25*R*)-5α-spirostan-3-one (**106**) by hydrogenation reaction (*Scheme 5.2.42*). A solution of diosgenin derivative **104** (1.0 g, 2.4 mmol) in EtOAc (30 mL) was treated with Pd-C (10 mg, 10% on carbon) under H_2 at 40 °C. After 24 h, the reaction mixture was filtered through a bed of Celite and the filtrate was rotary evaporated. The resulting crude solid was crystallized from MeOH to obtain **106**

(900 mg, 90%, mp: 155–158 °C). IR (KBr): ν_{max} cm^{-1} 2945m (C–H), 1718s (C=O). ^{1}H NMR (400 MHz, $CDCl_3$): δ 4.37 (dt, 1H, 16α-CH, J = 7.2, 7.6 Hz), 3.46 (m, 1H, 26-CHa), 3.32 (t, 1H, J = 10.8 Hz, 26-CHb). ^{13}C NMR (100 MHz, $CDCl_3$: δ 210.0 (C-3), 80.7 (C-16), 62.7 (C-17), 109.0 (C-22), 66.7 (C-26) (Shawakfeh *et al.*, 2008).

In the next reaction, (25*R*)-5α-spirostan-3-one (**106**) could easily be transformed to (25*R*)-2α-bromo-5α-spirostan-3-one (**107**, 180 mg, 75%, mp: 256 °C), following the same protocol as outlined for **105** (*Scheme 5.2.42*). IR (KBr): ν_{max} cm^{-1} 2950m (C–H), 1697s (C=O), 798m. ^{1}H NMR (400 MHz, $CDCl_3$): δ 4.73 (dd, 1H, 2β-CH, J = 6.0, 6.4, 7.2 Hz), 4.37 (dt, 1H, J = 7.6, 7.2 Hz, 16α-CH), 3.46 (m, 2H, 26-CHb), 3.32 (t, 2H, J = 10.8 Hz, 26-CHa). ^{13}C NMR (100 MHz, $CDCl_3$: δ 54.0 (C-2), 201.0 (C-3), 80.7 (C-16), 62.7 (C-17), 109 (C-22), 67.0 (C-26) (Shawakfeh *et al.*, 2008).

Diosgenin (**103**) was also employed to synthesize (25*R*)-spirost-4-ene-3,6-dione (**108**) by Jones' oxidation (*Scheme 5.2.43*). To a stirred solution of **103** (2 g, 4.82 mmol) in Me_2CO (150 mL) at 10 °C, freshly prepared Jones' reagent was added dropwise and the reaction mixture was stirred for 30 min. The reaction mixture was washed twice with brine (20 mL), the organic solvent was separated and dried (Na_2SO_4), concentrated under vacuum and the crude residue was subjected to CC (mobile phase: 20% EtOAc in *n*-hexane) to yield diketone **108** (1.95 g, 95%, mp: 190–192 °C). IR (KBr): ν_{max} cm^{-1} 2955m (C–H), 1687s

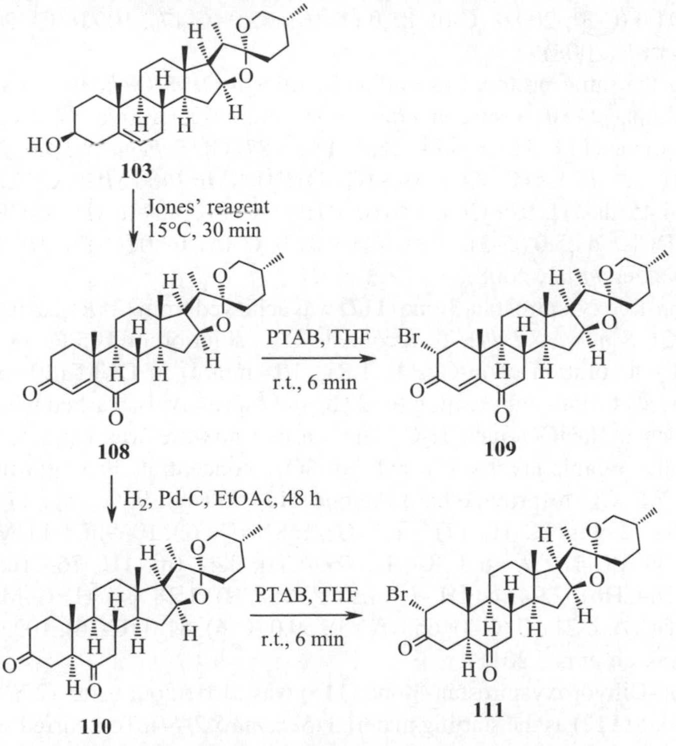

103

Jones' reagent
15°C, 30 min

108

PTAB, THF
r.t., 6 min

109

H₂, Pd-C, EtOAc, 48 h

110

PTAB, THF
r.t., 6 min

111

Scheme 5.2.43 *Synthesis of diosgenin derivatives* ***108–111***

(C=O), 1610s (C=C), 1456m. ^{1}H NMR (400 MHz, $CDCl_3$): δ 6.30 (s, 1H, 4-CH), 4.36 (dt, 1H, 16α-CH), 3.45 (m, 1H, 26-CHb), 3.37 (t, 1H, 26-CHa). ^{13}C NMR (400 MHz, $CDCl_3$): δ 202.0 (C-3), 127.0 (C-4), 163.0 (C-5), 199.0 (C-6), 56.0 (C-14), 80.6 (C-16), 62.7.0 (C-17), 109.0 (C-22), 67.0 (C-26) (Shawakfeh *et al*., 2008).

(25*R*)-2α-Bromospirost-4-ene-3,6-dione (**109**, 160 mg, 67%, mp: 187–190 °C) could be synthesized from (25*R*)-spirost-4-ene-3,6-dione (**108**) (*Scheme 5.2.43*), following the same protocol as outlined earlier for **105** (*Scheme 5.2.42*). IR (KBr): ν_{max} cm^{-1} 2946m (C–H), 1680s (C=O), 1611m (C=C), 983w. ^{1}H NMR (400 MHz, $CDCl_3$): δ 6.30 (s, 1H, 4-CH), 4.47 (dt, 1H, *J* = 6.3, 6.7 Hz, 16α-CH), 4.44 (d, 1H, *J* = 3.6 Hz, 2β-CH), 3.46 (m, 1H, 26-CHb), 3.37 (t, 1H, *J* = 10.8 Hz, 26-CHa). ^{13}C NMR (400 MHz, $CDCl_3$): δ 198.0 (C-3), 129.0 (C-4), 159.0 (C-5), 195.0 (C-6), 57.0 (C-14), 80.6 (C-16), 62.7 (C-17), 109.5 (C-22), 67.0 (C-26) (Shawakfeh *et al*., 2008).

(25*R*)-Spirost-4-ene-3,6-dione (**108**) was utilized for the synthesis of (25*R*)-5α-spirostan-3,6-dione (**110**) (*Scheme 5.2.43*) (Shawakfeh *et al*., 2008, 2011). A solution of **108** (1.0 g, 2.35 mmol) in EtOAc (30 mL) was stirred with Pd-C (10 mg, 10% on carbon) under H_2 at 40 °C for 48 h, the reaction mixture was filtered over a bed of Celite, the solvent was evaporated under vacuum, and the resulting crude solid was recrystallized from MeOH to obtain diketone **110** (800 mg, 80%, mp: 230–231 °C). IR (KBr): ν_{max} cm^{-1} 2960m (C–H), 1708s (C=O), 1706s (C=O). ^{1}H NMR (400 MHz, $CDCl_3$): δ 4.67 (dd, 1H, 2β-CH), 4.36 (dt, 1H, 16α-CH), 3.45 (m, 1H, 26-CHb), 3.37 (t, 1H, 26-CHa). ^{13}C NMR (400 MHz, $CDCl_3$): δ 204.0 (C-3), 200.0 (C-6), 81.0 (C-16), 62.0 (C-17), 109.0 (C-22), 67.0 (C-26) (Shawakfeh *et al*., 2008).

Employing the same protocol as outlined earlier for **105** (*Scheme 5.2.42*), (25*R*)-5α-spirostan-3,6-dione (**110**) was converted to its bromo-derivative, (25*R*)-2α-bromo-5α-spirostan-3,6-dione (**111**, 350 mg, 74%, mp: 184–187 °C) (*Scheme 5.2.43*). IR (KBr): ν_{max} cm^{-1} 2965m (C–H), 1705s (C=O), 1686s (C=O). ^{1}H NMR (400 MHz, $CDCl_3$): δ 4.72 (dd, 1H, 2β-CH), 4.45 (dt, 1H, 16α-CH), 3.46 (m, 1H, 26-CHb), 3.37 (t, 1H, 26-CHa). ^{13}C NMR (400 MHz, $CDCl_3$): δ 25.0 (C-3), 201.0 (C-6), 81.0 (C-16), 63.0 (C-17), 109.0 (C-22), 67.0 (C-26) (Shawakfeh *et al*., 2008).

(25*R*)-5α,6α-Epoxy-spirostan-3-one (**112**) was achieved from (25*R*)-2α-bromospirost-5-en-3-one (**105**) (*Scheme 5.2.44*). To a cooled (0 °C) solution of **105** (1.0 g, 2.4 mmol) in DCM (25 mL), a solution of *m*-CPBA (1.8 g, 10.4 mmol) in DCM (20 mL) was added dropwise, stirred at room temperature for 22 h, and successively washed with 5% Na_2SO_3, saturated aqueous $NaHCO_3$ and H_2O. The washed mixture was extracted with $CHCl_3$ (2 × 50 mL), the organic layer was dried (Na_2SO_4), concentrated and purified by crystallization from Me_2CO to provide epoxyketone **112** (500 mg, 95%, mp: 171–173 °C). IR (KBr): ν_{max} cm^{-1} 2943m (C–H), 1715s (C=O), 1681s (C=C), 1059m. ^{1}H NMR (400 MHz, $CDCl_3$): δ 4.39 (dt, 1H, *J* = 6.8, 8.1 Hz, 16α-CH), 3.41 (m, 1H, 26-CHa), 3.3 (t, 1H, *J* = 10.8 Hz, 26-CHb), 2.94 (d, 1H, *J* = 4.5 Hz, 6α-CH), 1.18 (s, 3H, 19-Me). ^{13}C NMR (100 MHz, $CDCl_3$): δ 211.0 (C-3), 65.0 (C-5), 59.0 (C-6), 81.0 (C-16), 109.0 (C-22), 67.0 (C-26) (Shawakfeh *et al*., 2011).

(25*R*)-5α,6β-Dihydroxyspirostan-3-one (**113**) was carried out using (25*R*)-5α,6α-epoxyspirostan-3-one (**112**) as the starting material (*Scheme 5.2.44*). To a stirred solution of **112** (1.0 g, 2.33 mmol) in Me_2CO (60 mL), 70% $HClO_4$ (400 μL) in H_2O (20 mL) was added and stirred vigorously. After 48 h, the reaction mixture was extracted with $CHCl_3$ (3 × 25 mL), the pooled organic solvents were dried (Na_2SO_4), concentrated *in vacuo* and purified by CC

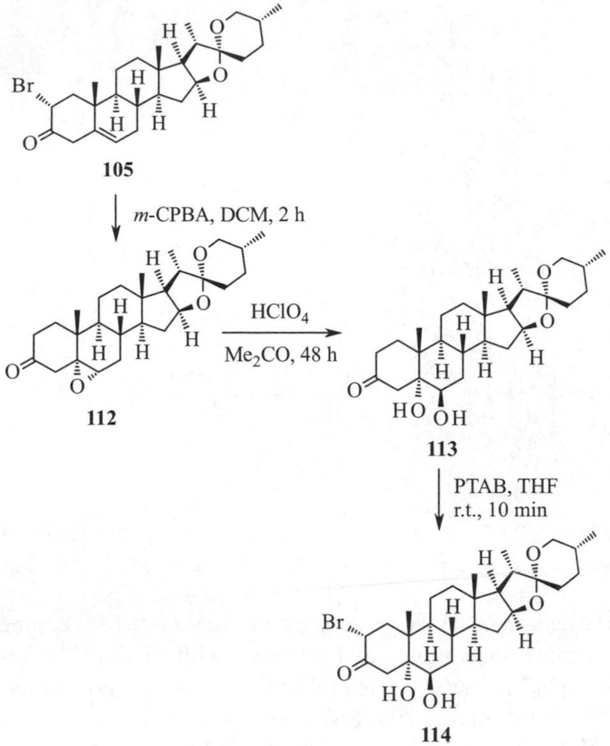

105

m-CPBA, DCM, 2 h

112

$HClO_4$

Me_2CO, 48 h

113

PTAB, THF
r.t., 10 min

114

Scheme 5.2.44 *Synthesis of diosgenin derivatives **112–114***

(mobile phase: 30% EtOAc in *n*-hexane) to afford **113** (550 mg, 50%, mp: 142–144 °C). IR (KBr): ν_{max} cm^{-1} 3327br (O–H), 2957m (C–H), 1712s (C=O), 1681s (C=C), 1058m (C–O). ^1H NMR (400 MHz, $CDCl_3$): δ 4.37 (dt, 1H, $J=6.6$, 8.1 Hz, 16α-CH), 3.5 (t, 1H, $J=2.7$ Hz, 6α-CH). ^{13}C NMR (100 MHz, $CDCl_3$): δ 212.0 (C-3), 75.0 (C-5), 79.0 (C-6), 81.0 (C-16), 109.0 (C-22), 67.0 (C-26) (Shawakfeh *et al.*, 2011).

(25*R*)-5α,6β-Dihydroxyspirostan-3-one (**113**) could easily be converted to its bromo-derivative, (25*R*)-2α-bromo-5α,6β-dihydroxyspirostan-3-one (**114**) (*Scheme 5.2.44*), using the same method as described for **105** (*Scheme 5.2.42*). To a stirred solution of **113** (200 mg, 0.45 mmol) in THF (10 mL), PTAB (220 mg, 0.59 mmol, 1.3 equiv.) in THF (7 mL) was added rapidly at room temperature. The reaction was quenched with brine (10 mL) and extracted with $CHCl_3$ (2 × 20 mL). The solvent was dried (Na_2SO_4), concentrated under vacuum and the crude product was purified by CC, eluted 30% EtOAc in *n*-hexane, to furnish **114** (160 mg, 69%, mp: 147–150 °C). IR (KBr): ν_{max} cm^{-1} 3319br (O–H), 2951m (C–H), 1687s (C=O), 1053m (C–O). ^1H NMR (400 MHz, $CDCl_3$): δ 4.76 (dd, 1H, $J=5.2$, 7.2 Hz, 2β-CH), 4.37 (dt, 1H, $J=6.8$, 7.7 Hz, 16α-CH), 3.57 (t, 1H, $J=2.7$ Hz, 6α-CH). ^{13}C NMR (100 MHz, $CDCl_3$): δ 211.0 (C-3), 76.0 (C-5), 79.0 (C-6), 81.0 (C-16), 109.0 (C-22), 67.0 (C-26) (Shawakfeh *et al.*, 2011).

(25*R*)-4α,5α-Epoxyspirostan-3,6-dione (**115**) was prepared from a solution of (25*R*)-spirost-4-ene-3,6-dione (**108**, 200 mg, 0.47 mmol) in MeOH (20 mL) at 15 °C, by the treatment with 30% H_2O_2 (5 mL), rapidly followed by dropwise addition of 4M NaOH

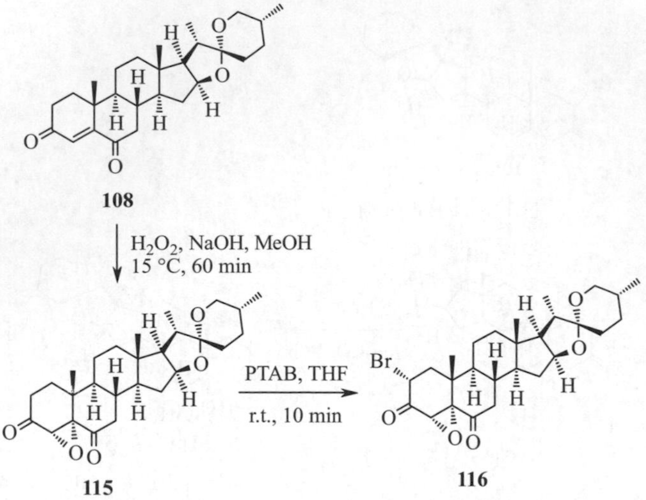

108

H_2O_2, NaOH, MeOH
15 °C, 60 min

PTAB, THF
r.t., 10 min

115

116

Scheme 5.2.45 *Synthesis of diosgenin derivatives **115** and **116***

solution (3 mL) (*Scheme 5.2.45*). The reaction mixture was stirred keeping the temperature below 15 °C for 60 min, concentrated and extracted with $CHCl_3$ (2 × 20 mL). The extract was washed with H_2O (2 × 15 mL), dried (Na_2SO_4), rotary evaporated and the crude solid was purified by CC (mobile phase: 20% EtOAc in *n*-hexane) to give **115** (130 mg, 63%, mp: 167–169 °C). IR (KBr): ν_{max} cm^{-1} 2949m (C–H), 1726s (C=O), 1705s (C=O), 1054m (C–H). ^{1}H NMR (400 MHz, $CDCl_3$): δ 4.37 (dt, 1H, *J* = 6.6, 8.1 Hz, 16α-CH), 2.98 (s, 1H, 4β-CH), 3.48 (m, 1H, 26-CHb), 3.35 (t, 1H, 26-CHa), 1.26 (s, 3H, 19-Me). ^{13}C NMR (100 MHz, $CDCl_3$): δ 203.0 (C-3), 58.0 (C-4), 70.0 (C-5), 201.0 (C-6), 81.0 (C-16), 109.0 (C-22), 67.0 (C-26) (Shawakfeh *et al.*, 2011).

Employing the same method as described for **105** (*Scheme 5.2.42*), (25*R*)-2α-bromo-4α,5α-epoxyspirostan-3,6-dione (**116**, 170 mg, 72%, mp: 170–172 °C) could easily be synthesized from (25*R*)-4α,5α-epoxyspirostan-3,6-dione (**115**) (*Scheme 5.2.45*). IR (KBr): ν_{max} cm^{-1} 2951m (C–H), 1720s (C=O), 1704s (C=O), 1054m, 838w. ^{1}H NMR (400 MHz, $CDCl_3$): δ 4.68 (dd, 1H, *J* = 5.2 and 6.9 Hz, 2β-CH), 4.36 (dt, 1H, *J* = 6.6, 8.1 Hz, 16α-CH), 3.88 (s, 1H, 4β-CH), 3.48 (m, 1H, 26-CHb), 3.35 (t, 1H, 26-CHa), 1.26 (s, 3H, 19-Me). ^{13}C NMR (100 MHz, $CDCl_3$): δ 54.0 (C-2), 209.0 (C-3), 69.0 (C-4), 70.0 (C-5), 206.0 (C-6), 81.0 (C-16), 109.0 (C-22) (Shawakfeh *et al.*, 2011).

The same procedure, as described for the synthesis of **115**, was applied to synthesize (25*R*)-5α,6α-epoxyspirostan-3β-ol (**117**, 950 mg, 92%, mp: 200–202 °C) from diosgenin (**103**) (*Scheme 5.2.46*). IR (KBr): ν_{max} cm^{-1} 3510br (O–H), 3400m (O–H), 2950m (C–H), 1645m (C=C), 1053m (C–O). ^{1}H NMR (400 MHz, $CDCl_3$): δ 4.36 (m, 1H, 16α-CH), 3.90 (m, 1H, 3β-CH), 3.48 (m, 2H, 26-CH_2), 2.90 (d, 1H, *J* = 14.0 Hz, 6β-CH), 1.08 (s, 3H, 19-Me), 0.76 (s, 3H, 18-Me). ^{13}C NMR (100 MHz, $CDCl_3$): δ 70.0 (C-3), 66.0 (C-5), 59.0 (C-6), 81.0 (C-16), 109.0 (C-22), 67.0 (C-26) (Shawakfeh *et al.*, 2011).

Jones' oxidation as described for the synthesis of **108** (*Scheme 5.2.43*), could also be employed to prepare (25*R*)-5α-hydroxyspirostan-3,6-dione (**118**, 1.8 g, 87%, mp:

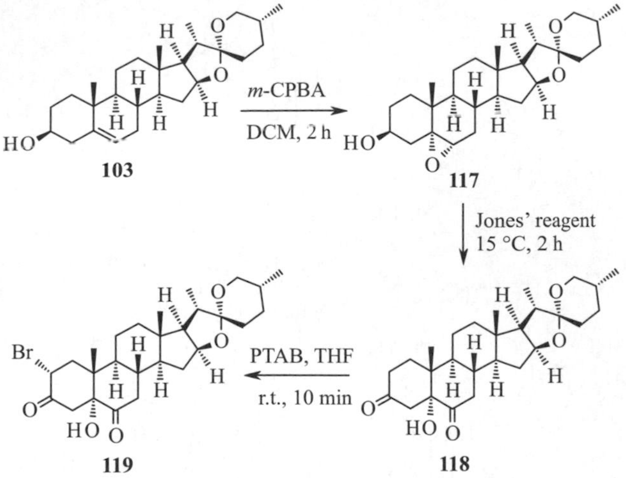

Scheme 5.2.46 *Synthesis of diosgenin derivatives* **117–119**

191–192 °C) from (25*R*)-5α,6α-epoxyspirostan-3β-ol (**117**, 2.0 g, 4.65 mmol) (*Scheme 5.2.46*). IR (KBr): ν_{max} cm^{-1} 3346br (O–H), 2949m (C–H), 1713s (C=O), 1705s (C=O). ^{1}H NMR (400 MHz, $CDCl_3$): δ 4.48 (m, 1H, 16α-CH), 3.48 (t, 2H, 26-CH_2), 2.80–2.90 (s, 2H, 4-CH_2), 2.26 (m, 2H, 2-CH_2). ^{13}C NMR (100 MHz, $CDCl_3$): δ 209.0 (C-3), 82.0 (C-5), 210.0 (C-6), 56.0 (C-14), 63.0 (C-17), 109.0 (C-22), 67.0 (C-26) (Shawakfeh *et al.*, 2011).

Utilizing the same protocol as outlined earlier for **105** (*Scheme 5.2.42*), (25*R*)-2α-bromo-5α-hydroxyspirostan-3,6-dione (**119**, 260 mg, 74%, mp: 195–197 °C) was synthesized from (25*R*)-5α-hydroxyspirostan-3,6-dione (**118**) (*Scheme 5.2.46*). IR (KBr): ν_{max} cm^{-1} 3340br (O–H), 2951m (C–H), 1711s (C=O), 1705s (C=O), 840m. ^{1}H NMR (400 MHz, $CDCl_3$): δ 4.70 (dd, 1H, 2β-CH), 4.45 (dt, 1H, 16α-CH), 3.48 (m, 2H, 26-CH_2). ^{13}C NMR (100 MHz, $CDCl_3$): δ 54.0 (C-2), 209.0 (C-3), 82.0 (C-5), 208.0 (C-6), 109.0 (C-22) (Shawakfeh *et al.*, 2011).

Cephalostatin analogue, di[(25*R*)-5-en-spirostano]pyrazine (**120**), was accomplished from (25*R*)-2α-bromospirost-5-en-3-one (**105**) in the following manner (*Scheme 5.2.47*). To a stirred solution of **105** (400 mg, 0.81 mmol) in DMF (50 mL), KI (a few mg) was added followed by the addition of NaN_3 (520 m g, 8.1 mmol, 10 equiv.), and stirred for 3 h at 50 °C. The solvent was evaporated to dryness, the solid was dissolved in $CHCl_3$ (50 mL), washed with brine (2 × 20 mL), dried (Na_2SO_4) and rotary evaporated to obtain azido ketone intermediate. To a solution of the intermediate azido ketone in THF (10 mL), Ph_3P (700 mg, 2.64 mmol, 3 equiv.) was added under N_2, stirred for 5 h until gas ceased to evolve. THF (5 mL) and H_2O (200 μL, 11 mmol) were added and the reaction mixture was stirred overnight. After concentration, the residue was azeotroped with toluene, and absolute EtOH (10 mL) and *p*-TsOH (catalytic amount) were added, and stirred vigorously at room temperature under atmospheric pressure for further 48 h. The residue was filtered through a bed of Celite and washed with $CHCl_3$ followed by evaporation under pressure to produce **120** (120 mg, 70%, mp: 290 °C). IR (KBr): ν_{max} cm^{-1} 2970m(C–H), 1660s (C=C), 1412m

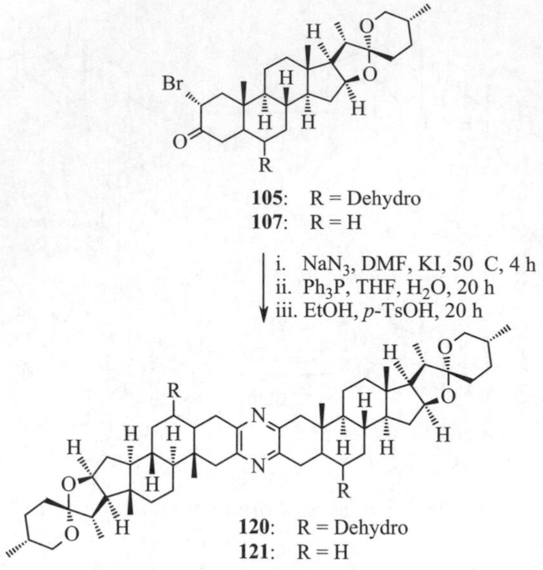

105: R = Dehydro
107: R = H

i. NaN_3, DMF, KI, 50 C, 4 h
ii. Ph_3P, THF, H_2O, 20 h
iii. EtOH, *p*-TsOH, 20 h

120: R = Dehydro
121: R = H

Scheme 5.2.47 *Synthesis of di[(25R)-5-en-spirostano]pyrazine (**120**) and di[(25R)-5α-spirostano]pyrazine (**121**)*

(C–H), 1398m (pyrazine). ^{1}H NMR (400 MHz, $CDCl_3$): δ 5.40 (t, 2H, 2 × 6-CH), 4.40 (dt, 2H, 2 × 16a-CH), 3.40 (m, 4H, 2 × 26-CH_2). ^{13}C NMR (400 MHz, $CDCl_3$): δ 144.0 (pyrazine carbons, 2 × C-2), 143.0 (pyrazine carbons, 2 × C-3), 140.0 (2 × C-6), 81.0 (2 × C-16), 63.0 (2 × C-17), 109.0 (2 × C-22), 67.0 (2 × C-26) (Shawakfeh *et al.*, 2008).

The same procedure, as outlined above for the synthesis of cephalostatin analogue **120**, was applied for the preparation of another cephalostatin analogue, di[(25*R*)-5α-spirostano] pyrazine (**121**, 280 mg, 82%, mp: 250 °C) from (25*R*)-2α-bromo-5α-spirostan-3-one (**107**) (*Scheme 5.2.47*). IR (KBr): ν_{max} cm^{-1} 2949m (C–H), 1412m (C–H), 1398m (pyrazine), 1186m, 695w. ^{1}H NMR (400 MHz, $CDCl_3$): δ 4.40 (dt, 2H, 2 × 16α-CH), 3.50 (m, 4H, 2 × 26-CH_2). ^{13}C NMR (100 MHz, $CDCl_3$): δ 147.0 (pyrazine carbons, 2 × C-2 and 2 × C-3), 80.0 (2 × C-16), 66.0 (2 × C-17), 109.0 (2 × C-22), 67.0 (2 × C-26) (Shawakfeh *et al.*, 2008).

Again, the same procedure, as described earlier for the synthesis of the cephalostatin analogue **120**, was employed for the synthesis of di[(25*R*)-6-oxo-4-en-spirostano]pyrazine (**122**, 110 mg, 68%, mp: 300 °C) from (25*R*)-2α-bromospirost-4-ene-3,6-dione (**109**) (*Scheme 5.2.48*). IR (KBr): ν_{max} cm^{-1} 2922m (C–H), 1684s (C=O), 1607s (C=C), 1416m (C–H), 1398m (pyrazine). ^{1}H NMR (400 MHz, $CDCl_3$): δ 6.00 (s, 2H, 2 × 4-CH), 4.40 (dt, 2H, 2 × 16α-CH), 3.40 (m, 4H, 2 × 26-CH_2). ^{13}C NMR (100 MHz, $CDCl_3$): δ 146.0 (pyrazine carbons, 2 × C-2), 145.0 (pyrazine carbons, 2 × C-3), 129.0 (2 × C-4), 161.0 (2 × C-5), 202.0 (2 × C-6), 81.0 (C-16), 62.0 (2 × C-17), 109.0 (2 × C-22), 67.0 (2 × C-26) (Shawakfeh *et al.*, 2008).

Similarly, another cephalostatin analogue, di[(25*R*)-6-oxo-5α-spirostano]pyrazine (**123**, 200 mg, 78%, mp: 280 °C), could be obtained from (25*R*)-2α-bromo-5α-spirostan-3,

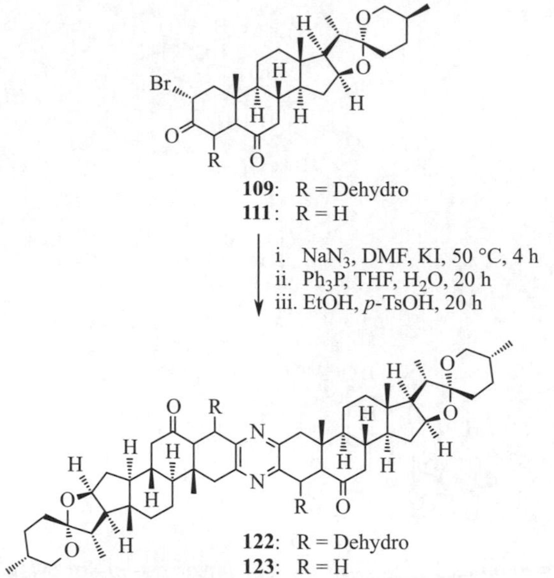

Scheme 5.2.48 *Synthesis of di[(25R)-6-oxo-4-en-spirostano]pyrazine (**122**) and di[(25R)-6-oxo-5α-spirostano]pyrazine (**123**)*

6-dione (**111**) (*Scheme 5.2.48*). IR (KBr): ν_{max} cm^{-1} 2950m (C–H), 1705s (C=O), 1407m (C–H), 1398m (pyrazine). ^1H NMR (400 MHz, $CDCl_3$): δ 4.40 (dt, 2H, 2 × 16α-CH), 3.40 (m, 4H, 2 × 26-CH_2). ^{13}C NMR (100 MHz, $CDCl_3$): δ 142.0 (pyrazine carbons, 2 × C-2), 141.0 (pyrazine carbons, 2 × C-3), 203.0 (2 × C-6), 81.0 (2 × C-16), 63.0 (2 × C-17), 109.0 (2 × C-22), 67.0 (C-26) (Shawakfeh *et al.*, 2008).

The procedure that was applied for the synthesis of the cephalostatin analogue **120** was also applied for the preparation of di[(25*R*)-5α,6β-dihydroxy-5α-spirostano]pyrazine (**124**, 110 mg, 64%, mp: 260 °C) from (25*R*)-2α-bromo-5α,6β-dihydroxyspirostan-3-one (**114**) (*Scheme 5.2.49*). IR (KBr): ν_{max} cm^{-1} 3383br (O–H), 2950m (C–H), 1417m (C–H), 1398m (pyrazine). ^1H NMR (400 MHz, $CDCl_3$): δ 4.40 (dt, 2H, $J = 6.8$, 8.1 Hz, 2 × 16α-CH), 3.46 (t, 2H, $J = 2.7$ Hz, 2 × 6α-CH), 1.09 (s, 6H, 2 × 19-Me). ^{13}C NMR (100 MHz, $CDCl_3$): δ 143.0 (pyrazine carbons, 2 × C-2), 141.0 (pyrazine carbons, 2 × C-3), 76.0 (2 × C-5), 79.0 (2 × C-6), 81.0 (C-16), 109.0 (2 × C-22), 67.0 (2 × C-26) (Shawakfeh *et al.*, 2011).

In the same manner, the synthesis of di[(25*R*)-4α,5α-epoxy-6-oxo-spirostano]pyrazine (**125**, 120 mg, 65%, mp: 300 °C) was accomplished from (25*R*)-2α-bromo-4α,5α-epoxyspirostan-3,6-dione (**116**) (*Scheme 5.2.50*). IR (KBr): ν_{max} cm^{-1} 2922m (C–H), 1719s (C=O), 1417m (C–H), 1398m (pyrazine). ^1H NMR (400 MHz, $CDCl_3$): δ 4.40 (dt, 2H, $J = 6.8$, 8.1 Hz, 2 × 16α-CH), 3.84 (s, 2H, 2 × 4β-CH), 1.07 (s, 6H, 2 × 19-Me). ^{13}C NMR (100 MHz, $CDCl_3$): δ 142.0 (pyrazine carbons, 2 × C-2), 141.0 (pyrazine carbons, 2 × C-3), 67.0 (2 × C-4), 70.0 (2 × C-5), 203.0 (2 × C-6), 81.0 (2 × C-16), 109.0 (2 × C-22) (Shawakfeh *et al.*, 2011).

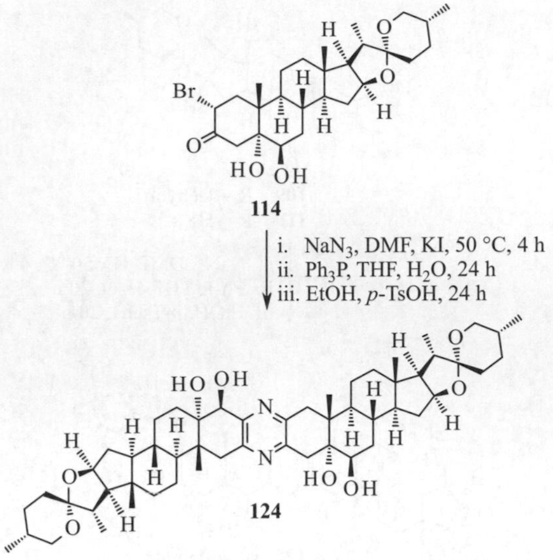

114

i. NaN_3, DMF, KI, 50 °C, 4 h
ii. Ph_3P, THF, H_2O, 24 h
iii. EtOH, *p*-TsOH, 24 h

124

Scheme 5.2.49 *Synthesis of di[(25R)-5α,6β-dihydroxy-5α-spirostano]pyrazine (**124**)*

Again, application of the same process of dimerization as mentioned above yielded di [(25*R*)-3,6-dioxo-5α-hydroxy-5α-spirostano]pyrazine (**126**, 160 mg, 62%, mp: 290 °C) from (25*R*)-2α-bromo-5α-hydroxyspirostan-3,6-dione (**119**) (*Scheme 5.2.51*). IR (KBr): ν_{max} cm^{-1} 3353br (O–H), 2944m (C–H), 1706s (C=O), 1404s (pyrazine). ^1H NMR (400 MHz, $CDCl_3$): δ 4.40 (dt, 2H, 2 × 16α-CH), 3.45 (m, 4H, 2 × 26-CH_2), 1.14 (s,

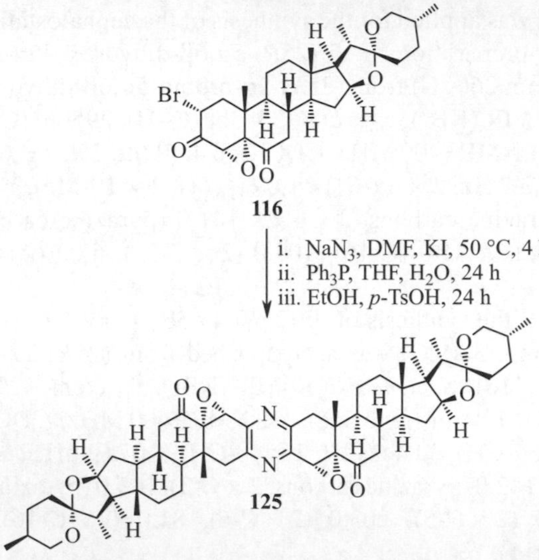

116

i. NaN_3, DMF, KI, 50 °C, 4 h
ii. Ph_3P, THF, H_2O, 24 h
iii. EtOH, *p*-TsOH, 24 h

125

Scheme 5.2.50 *Synthesis of di[(25R)-4α,5α-epoxy-6-oxo-spirostano]pyrazine (**125**)*

6H, 2 × 19-Me). ^{13}C NMR (100 MHz, $CDCl_3$): δ 139.0 (pyrazine carbons, 2 × C-2), 138 (pyrazine carbons, 2 × C-4), 79.0 (2 × C-5), 203.0 (2 × C-6), 108.0 (2 × C-22), 67.0 (2 × C-26) (Shawakfeh *et al.*, 2011).

The synthesis of 20- and 25′-epimers of cephalostatin 7 (**127** and **128**) by directed unsymmetrical pyrazine synthesis, and several other analogues of cephalostatin 1 and ritterazines, *e.g.*, 12β-hydroxycephalostatin 1 (**129**), 7′-deoxyritterazine G (ritterostatin G_N7_S, **130**), 14-*epi*-7′-deoxyritterazine B (**131**), dihydro-ornithostatin O_11_N (**132**) and dihydro-ornithostatin (**133**) were prepared (LaCour *et al.*, 1998, 1999, 2000).

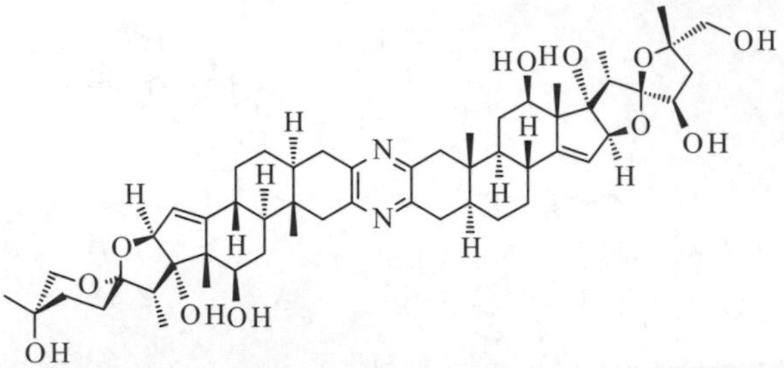

20-*epi*-Cephalostatin 7 (**127**)

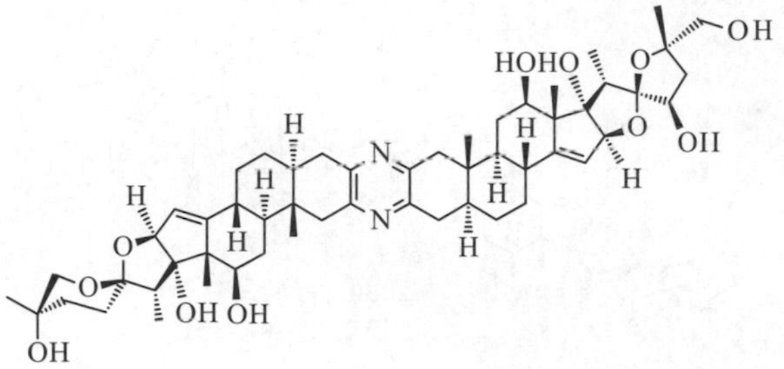

25’-*epi*-Cephalostatin 7 (**128**)

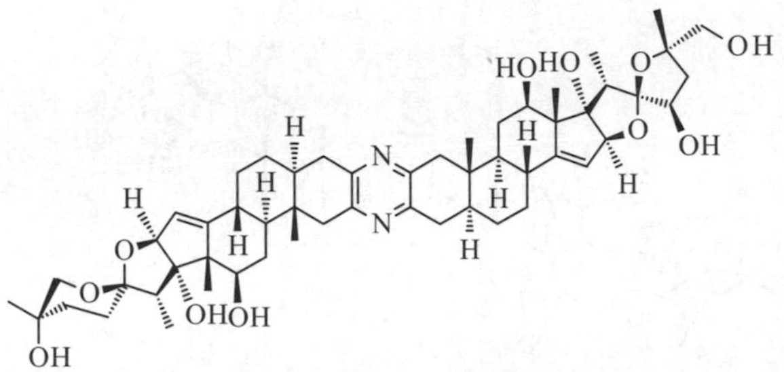

12β-Hydroxycephalostatin 1 (**129**)

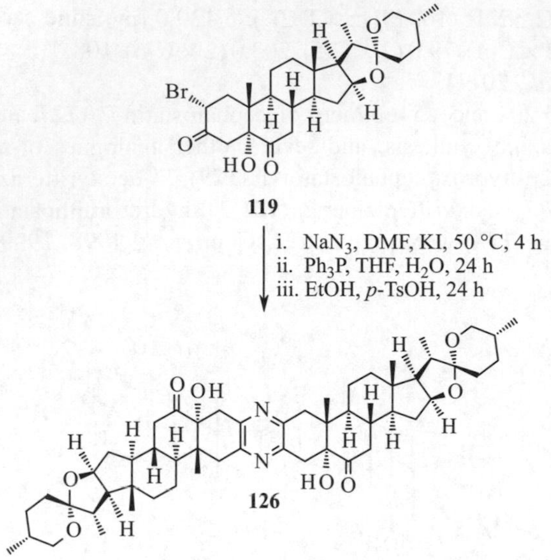

119

i. NaN_3, DMF, KI, 50 °C, 4 h
ii. Ph_3P, THF, H_2O, 24 h
iii. EtOH, *p*-TsOH, 24 h

126

Scheme 5.2.51 *Synthesis of di[(25R)-5α-hydroxy-5α-spirostan-3,6-dione]pyrazine (**126**)*

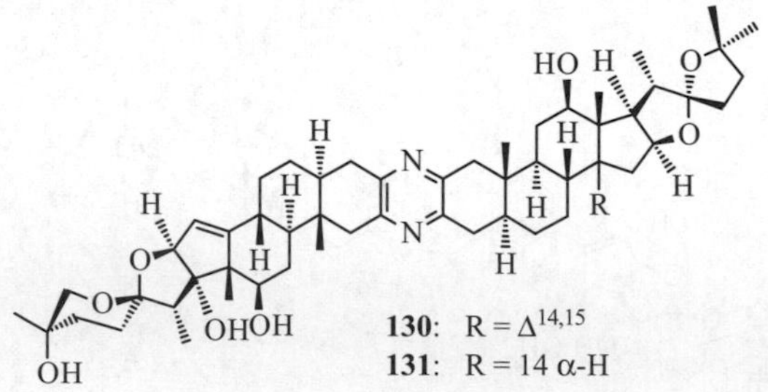

130: R = $\Delta^{14,15}$
131: R = 14 α-H

Ritterostatin G_N7_S (7'-deoxyritterazine G, **130**) and 14-*epi*-7'-deoxyritterazine B (**131**)

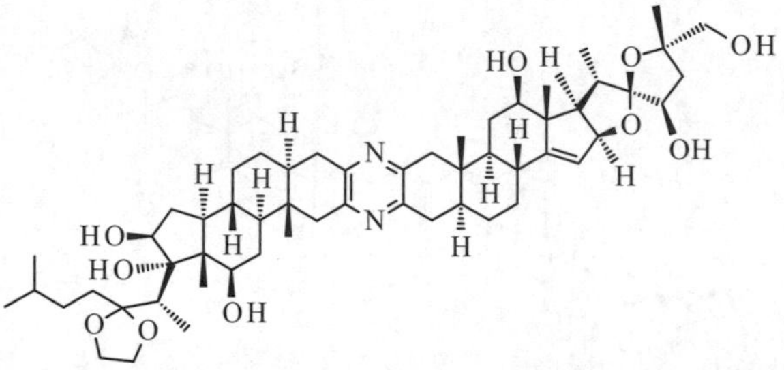

Dihydro-ornithostatin O_11_N (**132**)

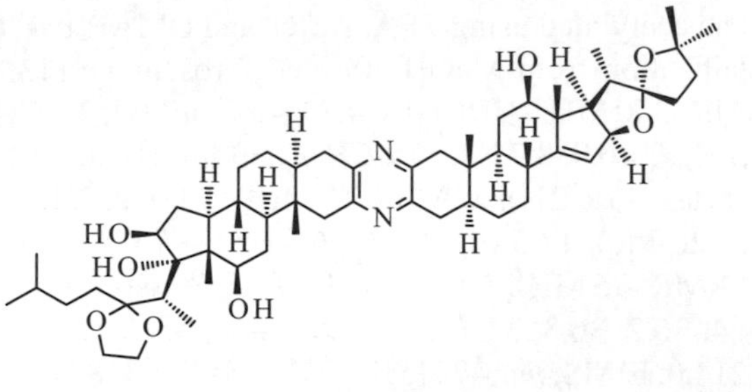

Dihydro-ornithostatin (**133**)

The synthesis of (+)-cephalostatin 1 (**1**), ritterostatin G_N1_N (**152**) and ritterostatin G_N1_S (**154**) was achieved through the synthesis of several steroid monomers using a multistep sequence as outlined as follows (LaCour *et al.*, 1998). In this process, furostanone (**135**) was yielded from a stirred solution of the alkene **134** (890 mg, 1.86 mmol, contained 5% hecogenin) in 75% AcOH by heating at 90–95 °C for 40 h (*Scheme 5.2.52*). H_2O was added to the reaction mixture and the resulting solution was extracted with EtOAc. The organic layer was washed with saturated aqueous $NaHCO_3$ and the original aqueous layer was back-extracted with EtOAc, the pooled organic layers were washed with saturated aqueous $NaHCO_3$, and dried (Na_2SO_4). The solvent was evaporated under vacuum and solid was

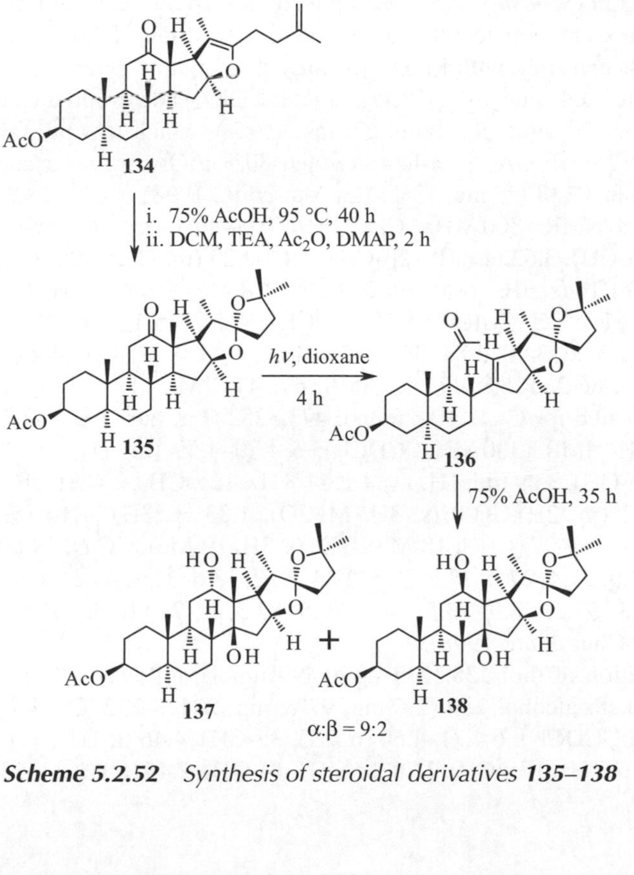

i. 75% AcOH, 95 °C, 40 h
ii. DCM, TEA, Ac_2O, DMAP, 2 h

hν, dioxane, 4 h

75% AcOH, 35 h

α:β = 9:2

Scheme 5.2.52 *Synthesis of steroidal derivatives **135–138***

dissolved in DCM and acetylated using TEA, Ac_2O, and DMAP at 0 °C for 2 h. After usual work up, and crystallization from MeOH afforded furostanone (**135**, 646 mg, 73%, mp: 230–232 °C). ^{1}H NMR (300 MHz, $CDCl_3$): δ 4.73–4.60 (m, 1H, 3α-CH), 4.43–4.35 (m, 1H, 16α-CH), 2.51 (dd, 1H, J = 6.9, 8.7 Hz, 17α-CH), 2.39 (m, 1H, J = 13.9 Hz, 11β-CH), 2.21 (dd, 1H, J = 5.1, 14.2 Hz, 11α-CH), 2.09 (m, 1H, CH), 2.00 (s, 3H, CH_3CO_2), 1.32 (s, 3H, 27-Me), 1.16 (s, 3H, 26-Me), 1.05 (d, 1H, J = 6.4 Hz, 21-Me), 1.03 (s, 3H, 18-Me), 0.91 (s, 3H, 19-Me). ^{13}C NMR (75 MHz, $CDCl_3$): δ 11.9, 13.6, 16.0, 21.5, 27.3, 28.2, 28.5, 30.2, 31.2, 31.5, 33.8, 34.4, 36.2, 36.3, 37.1, 37.8, 39.0, 44.5, 53.3, 55.3, 55.4, 55.6, 73.4, 79.1, 82.6, 120.1, 171.0, 214.0. EIMS: *m/z* 472 $[M]^+$. HRMS: 472.3189 calculated for $C_{29}H_{44}O_5$, found 472.3175 (LaCour *et al.*, 1998).

A stirred solution of furostanone (**135**, 630 mg, 1.33 mmol) in dioxane (70 mL) was deoxygenated with Ar, and the mixture was irradiated for 4 h (*Scheme 5.2.52*). The solvent was rotary evaporated to yield *seco*-aldehyde **136** (36 mg). ^{1}H NMR (300 MHz, $CDCl_3$): δ 9.47 (s,1H, 12-CHO), 4.73–4.60 (m,1H, 3α-CH), 4.55–4.51 (m, 1H, 16α-CH), 2.70–2.64 (t, 1H, J = 7.3 Hz, 17α-CH), 2.45–2.05 (m, 4H), 2.02 (s, 3H, CH_3CO_2), 1.61 (s, 3H, 18-Me), 1.35 (s, 3H, 27-Me), 1.16 (s, 3H, 26-Me), 1.06 (d, 1H, J = 6.9 Hz, 21-CH), 0.84 (s, 3H, 19-Me). ^{13}C NMR (75 MHz, $CDCl_3$): δ 12.2, 12.8, 14.6, 21.5, 28.3, 28.6, 30.2, 30.9, 33.6, 34.0, 36.5, 37.3, 37.4, 37.8, 37.9, 43.4, 43.8, 45.3, 46.3, 61.3, 73.2, 77.4, 81.9, 117.0, 135.3, 135.5, 170.7, 201.4. CIMS: *m/z* 473 $[M + H]^+$. HRMS: *m/z* 473.3267 calculated for $C_{29}H_{45}O_5$, found: 473.3257 (LaCour *et al.*, 1998).

In the next step, a mixture of 12α,14β-diol **137** and 12β,14β-diol **138** was obtained from *seco*-aldehyde **136** (*Scheme 5.2.52*). Compound **136** (14 mg, 1.27 mmol) and 75% AcOH (6 mL) were stirred at room temperature for 35 h, after which, H_2O (90 mL) was added and the mixture was extracted with EtOAc (30 mL). The organic layer washed with H_2O and followed by saturated aqueous $NaHCO_3$, dried (Na_2SO_4) and evaporated under vacuum to provide a mixture of diols (9:2 ratio, 628 mg, 101%), which was purified further by CC (mobile phase: 25% EtOAc in *n*-hexane, then 80% EtOAc in *n*-hexane) to obtain **137** (400 mg, 64%) and **138** (77 mg, 13%) (LaCour *et al.*, 1998).

Diol **137**: ^{1}H NMR (300 MHz, $CDCl_3$): δ 4.71–4.61 (m, 1H, 3α-CH), 4.55 (t, 1H, J = 6.8 Hz, 16α-CH), 3.62 (s, 1H, 12β-CH), 2.44–2.29 (m, 3H), 2.00 (s, 3H, MeCO), 1.35 (s, 3H, 27-Me), 1.16 (s, 3H, 26-Me), 1.02 (s, 3H, 18-Me), 0.95 (d, 3H, J = 6.4 Hz, 21-Me), 0.80 (s, 3H, 19-Me). ^{13}C NMR (75 MHz, $CDCl_3$): δ 12.2, 15.1, 15.4, 21.6, 27.3, 27.4, 28.5, 28.6, 28.7, 30.2, 33.6, 34.0, 35.4, 36.8, 37.3, 40.1, 40.8, 42.8, 43.0, 44.4, 50.9, 57.9, 73.7, 76.1, 81.2, 82.3, 86.2, 116.8, 171.1. CIMS: *m/z* 473 $[M + H - H_2O]^+$. HR CIMS: *m/z* 473.3267 calculated for $C_{29}H_{45}O_5$, found: 473.3252 (LaCour *et al.*, 1998).

Diol **138**: ^{1}H NMR (300 MHz, $CDCl_3$): δ 4.70–4.59 (m, 1H, 3α-CH), 4.55 (t, 1H, J = 6.8 Hz, 16α-CH), 3.09 (dd, 1H, J = 4.1, 11.8 Hz, 12α-CH), 2.48 (t, 1H, J = 7.7 Hz, 17α-CH), 2.37–2.27 (m, 2H), 2.00 (s, 3H, MeCO), 1.33 (s, 3H), 1.16 (s, 3H), 0.97 (3H, J = 6.4 Hz, 21-Me), 0.97 (s, 3H, 18-Me), 0.80 (s, 3H, 19-Me). ^{13}C NMR (75 MHz, $CDCl_3$): δ 12.2, 14.8, 21.5, 27.4, 27.7, 28.5, 28.5, 29.4, 30.2, 33.6, 33.9, 35.8, 36.9, 37.2, 38.8, 39.3, 42.8, 44.6, 46.3, 53.2, 57.0, 73.5, 74.6, 79.5, 82.5, 86.7, 116.8, 170.8. CIMS: *m/z* 491 $[M + H]^+$ (LaCour *et al.*, 1998).

Jones' oxidation of diol **138** (245 mg, 0.499 mmol), followed by crystallization from DCM yielded keto alcohol **139** (237 mg, 97%, mp: 231.5–232 °C) (*Scheme 5.2.53*). ^{1}H NMR (300 MHz, $CDCl_3$): δ 4.71–4.60 (m, 1H, 3α-CH), 4.46 (t, 1H, J = 6.6 Hz, 16α-CH), 3.32 (t, 1H, J = 7.8 Hz, 17α-CH), 2.78 (br s, 1H, 14-OH), 2.45 (t, 1H, J = 13.5 Hz, 11β-CH),

2.31–2.11 (m, 3H), 2.00 (s, 3H, MeCO), 1.33 (s, 3H), 1.23 (s, 3H, 18-Me), 1.16 (s, 3H), 0.96 (d, 1H, $J = 6.7$ Hz, 21-Me), 0.88 (s, 3H, 19-Me). ^{1}H NMR (300 MHz, C_5D_5N): δ 5.46 (br s, 1H, 14-OH), 4.86–4.74 (m, 1H, 3α-CH), 3.74 (t, 1H, $J = 6.6$ Hz, 16α-CH), 2.95–2.86 (m, 1H, 17α-CH), 2.55 (t, 1H, $J = 13.4$ Hz, 11β-CH), 2.38–2.30 (m, 2H), 2.05 (s, 3H, MeCO), 1.48 (s, 3H, 27-Me), 1.44 (s, 3H, 18-Me), 1.20 (s, 3H, 26-Me), 1.10 (d, 1H, $J = 6.7$ Hz, 21-CH), 0.76 (s, 3H, 19-Me). ^{13}C NMR (75 MHz, $CDCl_3$): δ 12.0, 14.8, 14.9, 21.5, 27.3, 27.9, 28.3, 28.5, 30.2, 33.8, 33.9, 36.3, 36.6, 37.2, 37.6, 39.0, 42.5, 44.6, 47.0, 51.3, 62.4, 73.2, 78.8, 82.7, 87.6, 116.7, 170.7, 213.6. CIMS: *m/z* 489 $[M + H]^+$. HRCIMS: *m/z* 489.3216 calculated for $C_{29}H_{45}O_6$ [M + H] +, found: 489.3206 (LaCour *et al.*, 1998).

A solution of $SOCl_2$ (55 μL, 0.886 mmol, 10 equiv.) in toluene (1 mL) was added dropwise to a stirred solution of keto alcohol **139** (216 mg, 0.443 mmol) in toluene (3.4 mL) and C_5H_5N (36 μL, 0.487 mmol, 10 equiv.) over 5 min and the reaction mixture was stirred at room temperature for another 2 h (*Scheme 5.2.53*). A mixture of toluene (15 mL), HCl (27 mL, 0.003M aqueous solution) and ice (5.0 g) were added to the reaction mixture and the aqueous layer was extracted with toluene. The organic layer was washed with brine, dried (Na_2SO_4), evaporated to dryness and the crude residue was subjected to CC (mobile phase: DCM:THF = 50:1 and DCM:THF = 10:1) followed by crystallization from *n*-hexane-DCM furnished (22*R*)-keto olefin **140** (130 mg, 63%, mp: 189–191 °C). ^{1}H NMR (300 MHz, $CDCl_3$): δ 5.40 (br s, 1H, 15-CH), 4.81 (d, 1H, $J = 8.0$ Hz, 16α-CH), 4.70–4.61 (m, 1H, 3α-CH), 3.32 (t, 1H, $J = 8.6$ Hz), 2.54 (t, 1H, $J = 14.2$ Hz, 11β-CH), 2.49–2.38 (m, 1H, CH), 2.33 (dd, 1H, $J = 4.4$, 14.5 Hz, 11α-CH), 2.01 (s, 3H, MeCO), 1.99 (s, 2H), 1.35 (s, 3H), 1.28 (s, 3H, 18-Me), 1.18 (s, 3H), 1.02 (d, 1H, $J = 6.6$ Hz, 21-Me), 0.93 (s, 3H, 19-Me). ^{13}C NMR (75 MHz, $CDCl_3$): δ 11.8, 14.1, 21.2, 21.5, 27.3, 28.0, 28.6, 29.5, 30.1, 33.5, 33.8, 34.2, 36.3, 36.4, 37.3, 37.4, 41.0, 44.1, 49.9, 53.5, 62.4, 73.1, 82.1, 83.7, 117.8, 121.3, 154.6, 170.7, 211.3. CIMS: *m/z* 471 $[M + H]^+$. HRCIMS: *m/z* 471.3111 calculated for $C_{29}H_{43}O_5$, found: 471.3117 (LaCour *et al.*, 1998).

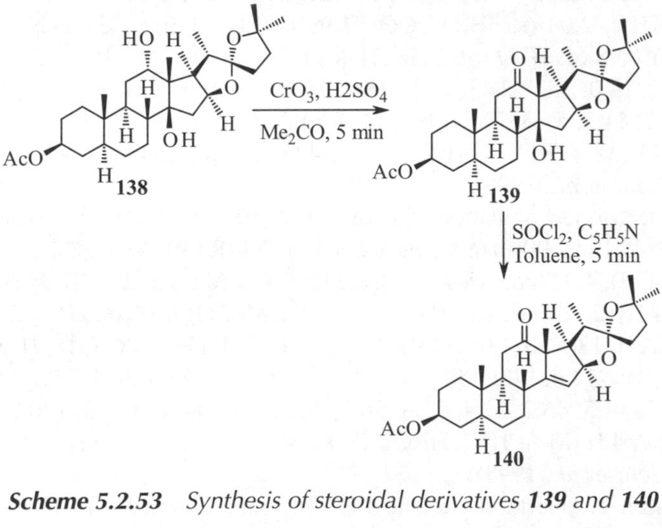

Scheme 5.2.53 *Synthesis of steroidal derivatives **139** and **140***

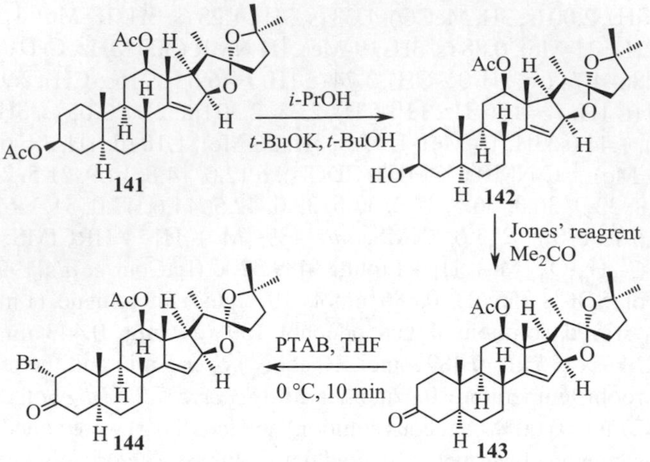

Scheme 5.2.54 *Synthesis of "North G" 2α-bromoketone* ***144***

A cooled (−15 °C) mixture of diacetate **141** (101.6 mg, 0.197 mmol) in *i*-PrOH (4 mL) and *t*-BuOK (217 μL of a 1.0 M solution in *t*-BuOH, 0.217 mmol, 1.1 equiv.) was stirred at −15 °C (*Scheme 5.2.54*). The reaction temperature was allowed to rise to 0 °C, stirring was continued for 8 h and then the reaction mixture was kept in a freezer (−15 °C) for another 14 h. The reaction temperature was allowed to warm slowly to 0 °C during 10 h of stirring. Et_2O and HCl (0.5% aqueous solution) were added to the reaction mixture and the aqueous layer was extracted with Et_2O. The ethereal layer was washed with $NaHCO_3$, dried (Na_2SO_4), evaporated under pressure, and the crude solid was purified by CC (mobile phase: 30% EtOAc in *n*-hexane) to give pure alcohol **142** (63 mg, 68%). ^1H NMR (300 MHz, $CDCl_3$): δ 5.40 (t, 1H, $J = 1.8$ Hz, 15-CH), 4.92 (dd, 1H, $J = 1.8$ and 8.2 Hz), 4.35 (dd, 1H, $J = 4.7$, 11.3 Hz, 12α-CH), 3.64–3.53 (m, 1H, 3α-CH), 2.34 (t, 1H, $J = 8.8$ Hz), 2.04 (s, 3H, MeCO), 1.99 (s, 2H), 1.36 (s, 3H), 1.18 (s, 3H), 1.08 (s, 3H, 18-Me), 0.98 (d, 1H, $J = 6.7$ Hz, 21-Me), 0.86 (s, 3H, Me-19). ^{13}C NMR (75 MHz, $CDCl_3$): δ 12.2, 14.0, 15.0, 21.3, 26.5, 28.3, 28.5, 29.6, 30.1, 31.3, 33.3, 34.1, 36.0, 36.8, 37.3, 37.9, 41.2, 44.5, 51.3, 52.3, 56.1, 71.0, 81.1, 82.1, 84.3, 117.4, 120.2, 156.6, 170.6. CIMS: *m/z* 473 $[M + H]^+$. HRCIMS: *m/z* 473.3267 calculated for $C_{29}H_{45}O_5$, found: 473.3281 (LaCour *et al.*, 1998).

In the next step, Jones' oxidation of alcohol **142** (50.3 mg, 0.106 mmol) afforded "North G" ketone **143** (45.1 mg, 90%) (*Scheme 5.2.54*). ^1H NMR (300 MHz, $CDCl_3$): δ 5.43 (t, 1H, $J = 1.8$ Hz, 15-CH), 4.92 (dd, 1H, $J = 1.8$, 8.3 Hz), 4.37 (dd, 1H, $J = 4.7$, 11.3 Hz, 12α-CH), 2.39–2.22 (m, 4H), 2.16–2.05 (m, 3H), 2.05 (s, 3H, MeCO), 2.00 (s, 2H), 1.37 (s, 3H), 1.18 (s, 3H), 1.11 (s, 3H, 18-Me), 1.06 (s, 3H, 19-Me), 0.99 (d, 1H, $J = 6.7$ Hz, 21-Me). ^{13}C NMR (75 MHz, $CDCl_3$): δ 11.4, 14.0, 15.0, 21.3, 26.7, 28.5, 28.6, 29.3, 30.1, 33.3, 34.0, 36.1, 37.3, 38.0, 38.2, 41.2, 44.5, 46.2, 51.4, 51.8, 56.1, 80.8, 82.2, 84.2, 117.5, 120.8, 155.9, 170.6, 211.2. CIMS: *m/z* 471 $[M + H]^+$. HRCIMS: *m/z* 471.3111 calculated for $C_{29}H_{43}O_5$, found: 471.3134 (LaCour *et al.*, 1998).

The conversion of "North G" ketone **143** (34.8 mg, 73.9 mmol in THF) to its bromo-derivative, "North G" 2α-bromoketone **144**, was achieved employing the same procedure as described earlier for **105** (*Scheme 5.2.42*), with the main exception that the reaction was

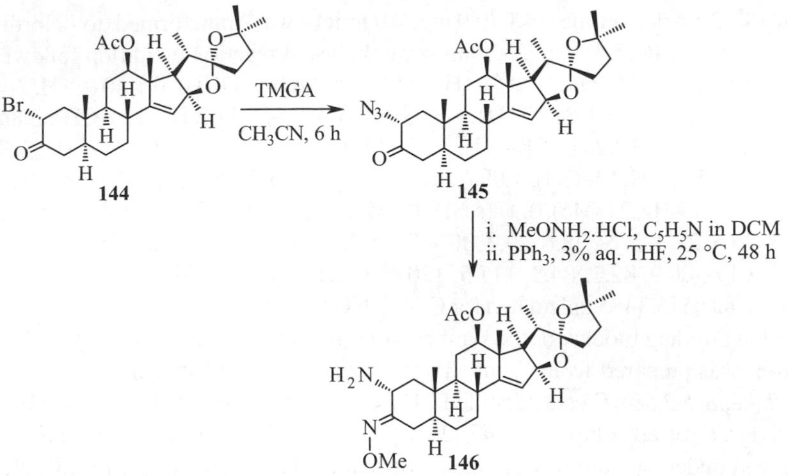

Scheme 5.2.55 Synthesis of "North G" 2α-azidoketone **145** and "North G" 2α-aminomethoxime **146**

prepared at 0 °C instead of room temperature (*Scheme 5.2.54*). After the usual workup, the product was purified by CC (mobile phase: 33% EtOAc in *n*-hexane) to obtain pure **144** (41 mg, 100%). ^{1}H NMR (300 MHz, $CDCl_3$): δ 5.44 (t, 1H, J = 1.9 Hz, 15-CH), 4.92 (dd, 1H, J = 1.9, 8.3 Hz), 4.71 (dd, 1H, J = 6.3, 13.4 Hz, 2β-CH), 4.37 (dd, 1H, J = 4.7, 11.2 Hz, 12α-CH), 2.55 (dd, 1H, J = 6.3, 13.0 Hz, 1β-CH), 2.48–2.42 (m, 2H), 2.37 (t, 1H, J = 8.8 Hz), 2.06 (s, 3H, MeCO), 2.00 (s, 2H), 1.36 (s, 3H), 1.18 (s, 3H), 1.14 (s, 3H, 19-Me), 1.10 (s, 3H), 0.99 (d, 1H, J = 6.7 Hz, 21-Me). ^{13}C NMR (75 MHz, $CDCl_3$): δ 12.0, 14.0, 15.1, 21.3, 26.7, 28.0, 28.6, 29.1, 30.1, 33.3, 33.5, 37.3, 39.3, 41.2, 43.7, 47.0, 51.0, 51.4, 51.4, 53.7, 56.1, 80.4, 82.2, 84.1, 117.5, 121.2, 155.2, 170.6, 200.6. CIMS: *m/z* 549 $[M + H]^+$. HRCIMS: *m/z* 549.2216 calculated for $C_{29}H_{42}BrO_5$, found 549.2226 (LaCour *et al*., 1998).

To a cooled (0 °C) solution of TMGA (45.8 mg, 290 mmol, 5 equiv.) in CH_3CN (2.6 mL) under N_2, a cooled (0 °C) solution of "North G" 2α-bromoketone **144** (31.4 mg, 57 mmol) in CH_3CN (3.2 mL) was added rapidly via cannula, the reaction temperature was allowed to warm slowly to 25 °C during 18 h of stirring. EtOAc was added to the reaction mixture and washed with brine. The organic layer was dried (Na_2SO_4), rotary evaporated and the crude residue was subjected to CC (mobile phase: 33% EtOAc in *n*-hexane) to produce the "North G" 2α-azidoketone **145** (24.7 mg, 85%). ^{1}H NMR (300 MHz, $CDCl_3$): δ 5.43 (t, 1H, J = 1.9 Hz, 15-CH), 4.92 (dd, 1H, J = 1.9, 8.2 Hz), 4.37 (dd, 1H, J = 4.7, 11.2 Hz, 12α-CH), 3.95 (dd, 1H, J = 6.7, 13.0 Hz, 2β-CH), 2.42–2.32 (m, 1H), 2.33 (t, 1H, J = 8.5 Hz), 2.29 (dd, 1H, J = 4.2, 14.4 Hz, 4α-CH), 2.22 (dd, 1H, J = 6.3, 12.6 Hz, 1β-CH), 2.06 (s, 3H), 1.99 (s, 2H), 1.36 (s, 3H), 1.18 (s, 3H), 1.13 (s, 3H, 19-Me), 1.10 (s, 3H), 0.99 (d, 1H, J = 6.7 Hz, 21-Me). ^{13}C NMR (75 MHz, $CDCl_3$): δ 12.4, 14.0, 15.0, 21.3, 26.8, 28.0, 28.6, 29.1, 30.1, 33.3, 33.4, 37.3, 37.3, 41.2, 43.6, 45.1, 47.1, 51.4, 51.5, 56.1, 63.7, 80.5, 82.2, 84.1, 117.5, 121.2, 155.3, 170.6, 204.6. FABMS: *m/z* 512 $[M + H]^+$. HRFABMS: *m/z* 512.3124 calculated for $C_{29}H_{42}N_3O_5$, found 512.3134 (LaCour *et al*., 1998).

'North G' 2α-azidoketone **145** (9.9 mg, 19 μmol) was transformed to 'North G' 2α-aminomethoxime **146** (7.4 mg, 75%), utilizing the usual reaction condition followed by the Staudinger reduction (*Scheme 5.2.55*). ^{1}H NMR (300 MHz, $CDCl_3$): δ 5.40 (t, 1H, J = 1.8 Hz, 15-CH), 4.91 (dd, 1H, J = 1.8, 8.3 Hz), 4.36 (dd, 1H, J = 4.7, 11.2 Hz, 12α-CH), 3.84 (s, 3H), 3.49 (dd, 1H, J = 5.0, 12.2 Hz, 2β-CH), 3.01 (dd, 1H, J = 2.6, 14.3 Hz, 4α-CH), 2.34 (t, 1H, J = 8.8 Hz), 2.05 (s, 3H, MeCO), 1.99 (s, 2H), 1.36 (s, 3H), 1.18 (s, 3H), 1.08 (s, 3H, 18-Me), 0.98 (d, 1H, J = 6.7 Hz, 21-Me), 0.97 (s, 3H, 19-Me). ^{13}C NMR (75 MHz, $CDCl_3$): δ 12.4, 14.0, 15.1, 21.3, 26.5, 27.3, 27.9, 28.6, 29.3, 30.1, 33.3, 33.6, 37.0, 37.3, 41.2, 45.5, 48.8, 49.7, 51.4, 51.9, 56.1, 61.6, 80.9, 82.1, 84.2, 117.5, 120.6, 156.1, 170.5. FABMS: m/z 515 $[M + H]^+$. HRFABMS: m/z 515.3485 calculated for $C_{30}H_{47}NO_5$, found: 515.3468 (LaCour *et al*., 1998).

One of the building blocks for the synthesis of (+)-cephalostain 1 (**1**), 'South 1' 2α-azido ketone **148**, was prepared from 'South 1' 2α-bromoketone **147** in a single step reaction as follows (*Scheme 5.2.56*). Freshly distilled CH_3NO_2 (5.2 mL) and TMGA (33 mg, 0.21 mmol) were added to a stirred solution of **147** (29 mg, 0.05 mmol) at 25 °C. After 2.5 h, the solvent was removed under vacuum to yield the crude solid, which was purified by CC, eluted with 15% EtOAc in n-hexane, to provide **148** (24 mg, 87%). ^{1}H NMR (300 MHz, $CDCl_3$): δ 5.44 (s, 1H, 15-CH), 5.16 (dd, 1H, J = 4.8, 6.3 Hz), 3.98 (dd, 1H, J = 6.9, 12.9 Hz, 2β-CH), 3.83 (d, 1H, J = 12.0 Hz), 3.75 (d, 1H, J = 12.0 Hz), 2.08 (s, 3H, MeCO), 1.29 (s, 3H), 1.27 (s, 3H), 1.12 (s, 3H), 1.06 (d, 3H, J = 6.0 Hz, 21-Me). ^{13}C NMR (75 MHz, $CDCl_3$): δ 11.9, 14.8, 21.4, 27.7, 29.2, 29.5, 29.6, 32.0, 32.2, 35.2, 37.3, 38.7, 43.5, 44.0, 44.4, 44.9, 46.6, 51.5, 61.4, 63.5, 63.6, 82.3, 82.8, 109.3, 123.8, 147.4, 169.9, 204.4, 210.9. FABMS: m/z 526 $[M + H]^+$ (LaCour *et al*., 1998).

Protected cephalostatin 1 (**150**) was accomplished from two cephalostatin building blocks, 'South 1' 2α-azido ketone **148** and 'North 1' 2α-aminomethoxime **149** in the following fashion (*Scheme 5.2.57*). Bu_2SnCl_2 (30 μg, 10 μmol) and PVP (15 mg) were added to a stirred solution of **148** (5 mg, 0.01 mmol) and **149** (9.6 mg, 0.01 mmol) in C_6H_6 (3 mL). The mixture was refluxed for 3 h (2–4 mL of fresh C_6H_6 was added twice to maintain the solvent level in the reaction vessel equipped with a Dean–Stark trap set for azeotropic distillation). The reaction was cooled and the solid was filtered off. The crude product was washed with DCM and purified by CC (mobile phase: 15–20% EtOAc in n-hexane) to afford protected cephalostatin 1 (**150**, 7.6 mg, 59%). ^{1}H NMR (600 MHz, $CDCl_3$): δ 7.86 (d, 2H, J = 6.9 Hz), 7.75 (d, 2H, J = 7.2 Hz), 7.48–7.38 (m, 6H), 5.57 (s, 1H), 5.48 (s, 1H), 5.17 (dd, 1H, J = 4.2, 6.6 Hz), 5.07 (dd, 1H, J = 5.4, 11.4 Hz), 4.95 (s, 1H), 3.97 (s, 1H), 3.90–3.79 (m, 3H), 3.10 (d, 1H, J = 9.6 Hz), 2.98 (d, 1H, J = 10.2 Hz), 2.91–2.81 (m, 4H), 2.09 (s, 3H),

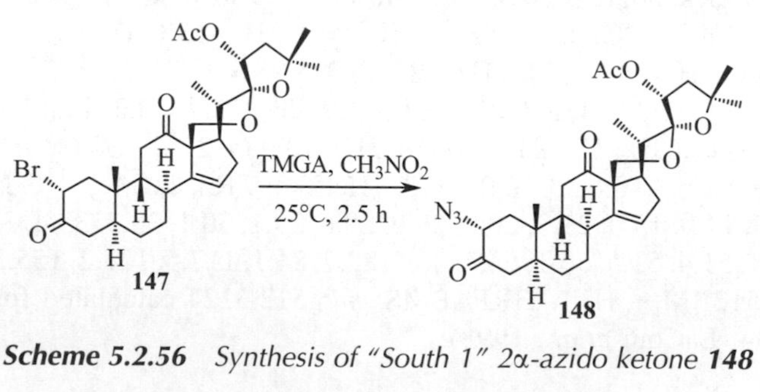

Scheme 5.2.56 *Synthesis of "South 1" 2α-azido ketone* ***148***

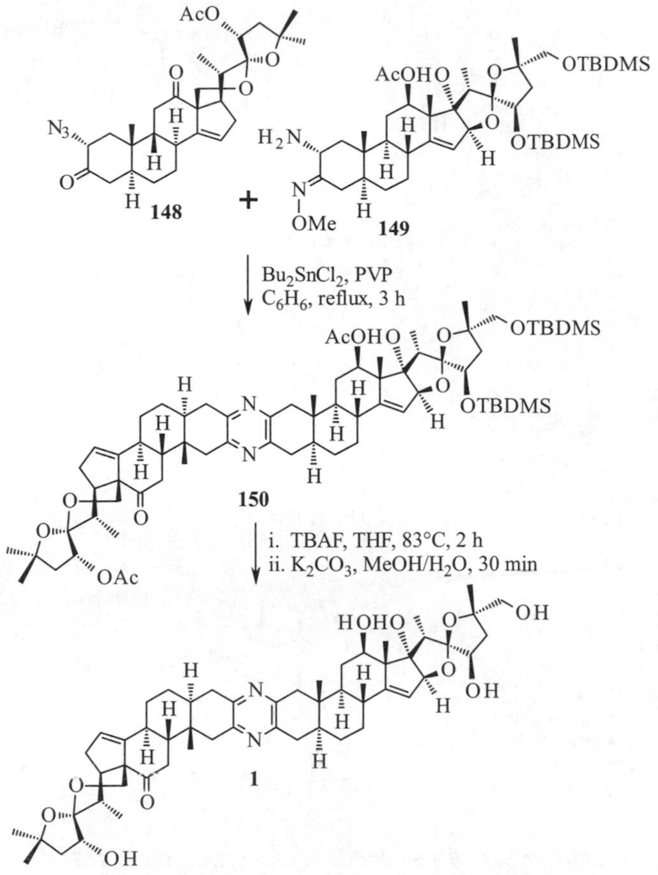

Scheme 5.2.57 *Synthesis of (+)-cephalostain 1 (**1**)*

2.00 (s, 3H), 1.31 (s, 3H), 1.29 (s, 3H), 1.26 (s, 3H), 1.25 (s, 3H), 1.12 (d, 3H), 1.08 (d, 3H, $J = 6$ Hz), 1.01 (s, 9H), 0.86 (s, 3H), 0.85 (s, 3H), 0.76 (s, 9H), –0.13 (s, 3H), –0.14 (s, 3H) (LaCour *et al.*, 1998).

Protected cephalostatin 1 (**150**) was deprotected to furnish (+)-cephalostatin 1 (**1**) in the following manner (*Scheme 5.2.57*). TBAF (1 M solution in THF, 16 mL, 0.016 mmol) was added to a stirred solution of **150** (7 mg, 5 μmol) in THF (2 mLI and the reaction mixture was refluxed for 2 h, cooled and the solvent was rotary evaporated. The solid was dissolved in a mixture of MeOH:H_2O (8:1, 2 mL), K_2CO_3 (7.5 mg, 0.054 mmol) was added and the resulting mixture was refluxed for 30 min. The solution was cooled, concentrated under vacuum, and the resulting crude residue was dissolved in EtOAc, washed with H_2O and dried (Na_2SO_4) and the solvent was evaporated to dryness. CC (mobile phase: 3–5% MeOH in $CHCl_3$) of the crude product yielded cephalostatin 1 (**1**, 3.8 mg, 80%). ^1H NMR (500 MHz, C_5D_5N): δ 8.16 (d, 1H, $J = 7.5$ Hz), 7.27 (d, 1H, $J = 4.0$ Hz), 6.63 (t, 1H, $J = 5.5$ Hz), 6.26 (s, 1H), 5.64 (s, 1H), 5.44 (s, 1H), 5.25 (s, 1H), 4.81 (m, 2H), 4.71 (d, 1H, $J = 1.0$ Hz), 4.08 (d, 1H, $J = 12$ Hz), 4.06 (m, 1H), 4.03 (d, 1H, $J = 12$ Hz), 3.82 (dd, 1H, $J = 5.5, 11.0$ Hz), 3.72 (dd, 1H, $J = 5.0, 11.5$ Hz), 3.18 (dq, 1H, $J = 6.0, 7.0$ Hz), 3.08 (d,

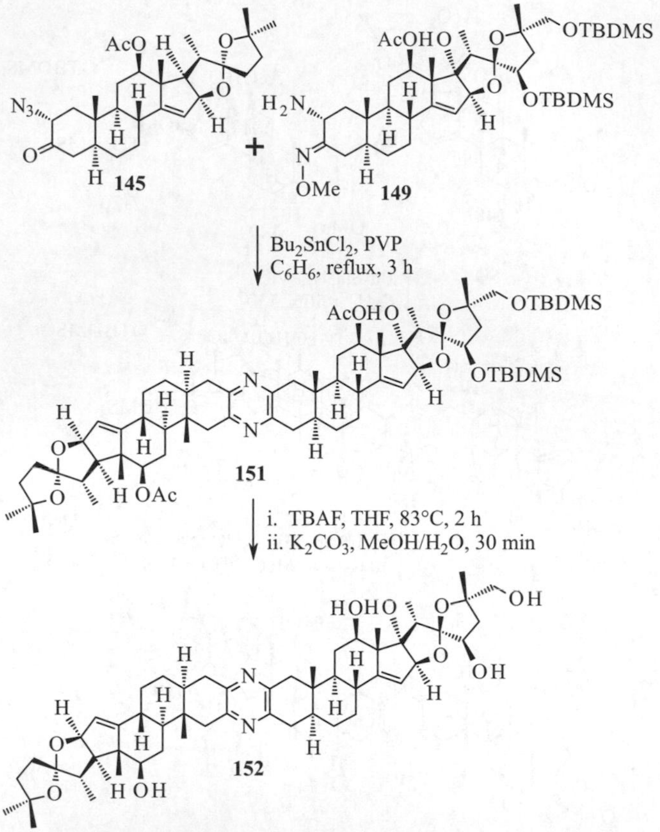

*Scheme 5.2.58 Synthesis of ritterostatin G_N1_N (**151**)*

1H, $J = 16.0$ Hz), 3.05 (d, 1H, $J = 17.0$ Hz), 1.65 (s, 3H), 1.47 (s, 3H), 1.47 (d, 3H, $J = 6.5$ Hz), 1.39 (s, 3H), 1.35 (d, 3H, $J = 7.0$ Hz), 1.33 (s, 3H), 0.75 (s, 3H), 0.72 (s, 3H). ^{13}C NMR (150 MHz, C_5D_5N): δ 8.99, 11.3, 11.7, 12.6, 15.5, 26.4, 28.0, 28.2, 28.7, 29.0, 29.5, 29.8, 32.4, 32.9, 33.8, 35.6, 35.7, 35.8, 36.3, 36.3, 38.8, 39.5, 41.2, 41.8, 44.2, 44.5, 45.8, 46.0, 47.3, 52.2, 53.2, 55.4, 61.8, 64.2, 69.3, 71.5, 75.6, 81.1, 81.5, 82.8, 91.7, 93.1, 110.9, 117.2, 122.3, 123.2, 148.3, 148.4, 148.7, 149.0, 149.5, 152.7, 211.8 (LaCour *et al.*, 1998).

Protected ritterostatin G_N1_N (**151**) was prepared from a mixture of 'North G' 2α-azidoketone **145** (4.7 mg, 0.0092 mmol), "North 1" 2α-aminomethoxime **149** (9.6 mg, 0.011 mmol), PVP (14 mg), freshly crushed 4Å molecular sieves (14 mg), Bu_2SnCl_2 and C_6H_6 (10 mL) under constant stirring (*Scheme 5.2.58*), according to the procedure as described for **150**, except that the distillate (7 mL) was collected (Dean–Stark trap set for azeotropic distillation) over 2 h without addition of any fresh solvent. Usual workup and purification by CC (mobile phase: C_6H_{14}:EtOAc = 2:1, 1.5:1 and EtOAc:MeOH:TEA = 100:10:1) provided **151** (5.9 mg, 49%). ^{1}H NMR (300 MHz, $CDCl_3$): δ 7.86 (m, 2H), 7.74 (dd, 2H, $J = 1.6, 7.9$ Hz), 7.46–7.28 (m, 6H), 5.56 (s, 1H), 5.45 (s, 1H), 5.06 (dd, 1H, $J = 5.1, 11.2$ Hz), 4.95 (s, 1H), 4.93 (dd, 1H, $J = 1.9, 8.6$ Hz), 4.40 (dd, 1H, $J = 4.5, 11.1$ Hz),

4.30 (dd, 1H, $J = 7.9$, 10.5 Hz), 3.97 (s, 1H), 3.10 (d, 1H, $J = 10.1$ Hz), 2.97 (d, 1H, $J = 10.1$ Hz), 2.87–2.82 (m, 4H), 2.78–2.44 (m, 5H), 2.38 (dd, 1H, $J = 8.7$, 8.8 Hz), 2.06 (s, 1H), 2.04 (s, 1H), 1.37 (s, 3H), 1.24 (s, 3H), 1.19 (s, 3H), 1.12 (d, 3H, $J = 6.0$ Hz), 1.11 (s, 6H), 1.00 (s, 9H), 0.99 (d, 1H, $J = 7$ Hz), 0.86 (s, 3H), 0.85 (s, 3H), 0.75 (s, 9H), -0.14 (s, 3H), -0.15 (s, 3H). FABMS: *m/z* 1134 $[M + H]^+$ (LaCour *et al.*, 1998).

Ritterostatin G_N1_N (**152**) was yielded from the protected ritterostatin G_N1_N (**151**) in the following fashion (*Scheme 5.2.58*). TBAF (1M THF solution, 11 μL, 3 equiv.) was added to a stirred solution of **151** (4.7 mg, 0.0035 mmol) in THF (1.5 mL) and refluxed for 2 h. The solution was cooled and MeOH (1 mL) and 10% KOH (100 μL) were added to the reaction mixture and refluxed for another 45 min. After cooling, brine (10 mL) was added and the mixture extracted with EtOAc (10 mL). The organic layer was washed with HCl (0.01% aqueous solution) and the aqueous layer was extracted with EtOAc (10 mL). The pooled organic extracts were washed with saturated aqueous $NaHCO_3$, dried (Na_2SO_4), concentrated under vaccum and the crude residue was subjected to CC, eluted with 3–7% MeOH in DCM, to obtain **152** (2.9 mg, 94%). ^{1}H NMR (500 MHz, C_5D_5N): δ 8.12 (d, 1H, $J = 7.2$ Hz), 6.59 (t, 1H, $J = 5.3$ Hz), 6.29 (d, 1H, $J = 4.8$ Hz), 6.24 (s, 1H), 5.63 (s, 1H), 5.54 (s, 1H), 5.27 (dd, 1H, $J = 1.7$, 8.4 Hz), 5.24 (s, 1H), 4.80 (m, 1H), 4.70 (d, 1H, $J = 1.5$ Hz), 4.04 (dd, 1H, $J = 4.7$, 10.6 Hz), 3.80 (dd, 1H, $J = 4.8$, 11.0 Hz), 3.71 (dd, 1H, $J = 4.8$, 11.0 Hz), 3.48 (m, 1H), 3.11–3.05 (m, 3H), 2.92–2.83 (m, 3H), 2.72 (dd, 1H, $J = 8.0$, 11.4 Hz), 2.67–2.61 (m, 4H), 2.32 (t, 1H, $J = 11.1$ Hz), 2.22 (dq, 1H, $J = 7.0$, 8.0 Hz), 2.16–2.00 (m, 7H), 1.92–1.72 (m, 4H), 1.64 (s, 3H), 1.46 (s, 3H), 1.35 (d, 3H, $J = 7.0$ Hz), 1.32 (s, 3H), 1.31 (s, 3H), 1.24 (d, 3H, $J = 6.7$ Hz), 1.19 (s, 3H), 0.94–0.83 (m, 2H), 0.75 (s, 3H), 0.73 (s, 3H). ^{13}C NMR (125 MHz, C_5D_5N): δ 9.0, 11.7, 11.8, 12.6, 13.9, 14.5, 26.4, 27.9, 28.4, 28.7, 28.9, 28.8, 29.7, 30.3, 30.9, 33.6, 33.8, 34.0, 35.8, 35.8, 36.2, 36.3, 37.8, 39.5, 41.8, 41.8, 42.1, 44.5, 46.0, 46.1, 52.6, 53.2, 55.4, 56.3, 69.3, 71.5, 75.6, 78.5, 81.5, 82.8, 85.0, 91.6, 93.2, 117.2, 117.8, 120.4, 122.3, 148.5, 148.6, 148.9, 150.2, 152.7, 157.1. FABMS: *m/z* 879 $[M + H]^+$. HRFABMS: *m/z* 879.5523 calculated for $C_{54}H_{75}N_2O_5Si$, found: 879.5449 (LaCour *et al.*, 1998).

The synthesis of protected ritterostatin G_N1_S (**153**) could be carried out in the following manner (*Scheme 5.2.57*). 'North G' 2α-aminomethoxime **146** (5.4 mg, 0.011 mmol), 'South 1' 2α-azido ketone **148** (6.0 mg, 0.011 mmol), Nafion H (11.5 mg), Bu_2SnCl_2 (catalyst) in C_6H_6 (10 mL) were refluxed (*Scheme 5.2.57*). After 8 h of heating, a total of 5 mL of content was removed (with a Dean–Stark trap set for azeotropic distillation), leaving the final reaction volume at 1.5 mL. The residue was cooled to room temperature, filtered off, and the crude solid was chromatographed using the same solvent system as described for **151**, to afford **153** (1.3 mg, 21%). ^{1}H NMR (500 MHz, $CDCl_3$): δ 5.48 (s, 1H), 5.44 (s, 1H), 5.16 (dd, 1H, $J = 4.6$, 6.7 Hz,), 4.93 (dd, 1H, $J = 1.8$, 8.2 Hz), 4.40 (dd, 1H, $J = 4.7$, 11.2 Hz), 3.85 (d, 1H, $J = 8.0$ Hz), 3.85 (d, 1H, $J = 8.0$ Hz), 2.91–2.78 (m, 4H), 2.71–2.28 (m, 6H), 2.08 (s, 3H), 2.05 (s, 3H), 1.37 (s, 3H), 1.30 (s, 3H), 1.28 (s, 3H), 1.18 (s, 3H), 1.11 (s, 3H), 1.07 (dd, 3H, $J = 6.2$ Hz), 1.00 (dd, 3H, $J = 6.7$ Hz), 0.85 (s, 3H), 0.84 (s, 3H). FABMS: *m/z* 947 $[M + H]^+$. HRFABMS: *m/z* 947.5786 calculated for $C_{58}H_{79}N_2O_9$, found: 947.5737 (LaCour *et al.*, 1998).

Ritterostatin G_N1_S (**154**) was accomplished from protected ritterostatin G_N1_S (**153**) in the following fashion (*Scheme 5.2.57*). K_2CO_3 (4 mg) was added to a stirred solution of **153** (5.1 mg, 0.0054 mmol) in 88% aqueous MeOH (2 mL) and refluxed for 30 min. After cooling, H_2O (10 mL) were added and the mixture was extracted with DCM. The solvent

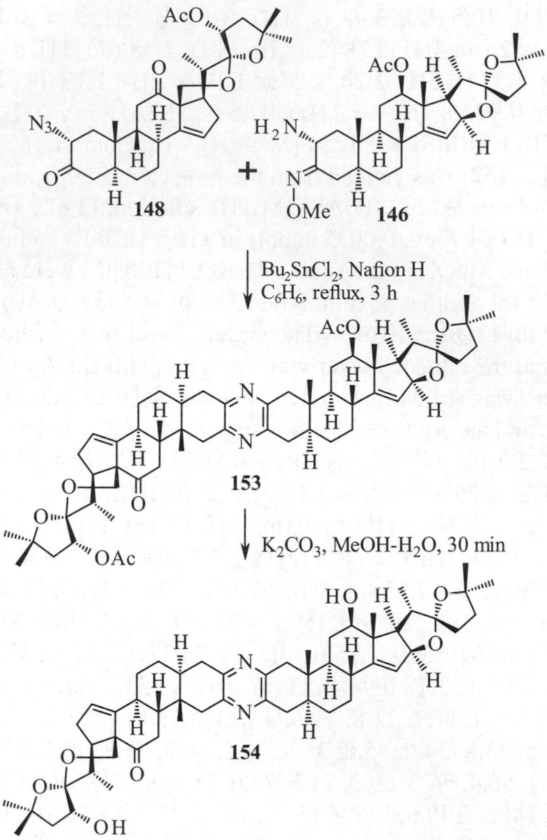

Bu_2SnCl_2, Nafion H
C_6H_6, reflux, 3 h

K_2CO_3, MeOH-H_2O, 30 min

Scheme 5.2.59 *Synthesis of ritterostatin G_N1_S (**154**)*

was dried (Na_2SO_4) and rotary evaporated to yield **154** (5.0 mg) in a quantitative yield (LaCour *et al.*, 1998).

^{1}H NMR: (300 MHz, C_5D_5N): δ 7.19 (br s, 1H), 5.55 (s, 1H), 5.54 (s, 1H), 5.28 (br d, 1H, J = 8.0 Hz), 4.80 (m, 1H), 4.07 (d, 1H, J = 11.9 Hz), 4.04 (d, 1H, J = 11.9 Hz), 3.49 (m, 1H), 3.18 (dq, 1H, J = 6.8 Hz), 3.10 (d, 1H, J = 17.5 Hz), 3.09 (q, 1H, J = 8.0 Hz), 3.08 (d, 1H, J = 16.9 Hz), 2.93 (dd, 1H, J = 5.0, 18.1 Hz), 2.91 (d, 1H, J = 5.0, 18.0 Hz), 2.85 (dt, 1H, J = 4.0, 12.0 Hz), 2.78 (t, 1H, J = 13.7 Hz), 2.81–2.76 (m, 1H), 2.71–2.61 (m, 3H), 2.64 (d, 1H, J = 16.1 Hz), 2.62 (dd, 1H, J = 3.2 and 13.5 Hz), 2.57 (d, 1H, J = 17.5 Hz), 2.35 (dd, 1H, J = 7.1, 12.3 Hz), 2.32 (m, 1H), 2.23 (dt, 1H, J = 6.9, 13.4 Hz), 2.17–1.80 (m, 10H), 1.70–1.53 (m, 6H), 1.47 (s, 3H), 1.47 (d, 3H, J = 6.8 Hz), 1.46 (s, 3H), 1.38 (s, 3H), 1.32 (s, 3H), 1.25 (d, 3H, J = 6.5 Hz), 1.19 (s, 3H), 0.77 (s, 3H), 0.73 (s, 3H). ^{13}C NMR (125 MHz, C_5D_5N): δ 11.3, 11.8, 13.9, 14.5, 15.5, 28.0, 28.4, 28.7, 29.4, 29.7, 29.7, 30.0, 30.3, 30.9, 32.4, 32.9, 33.6, 34.0, 35.6, 35.8, 36.2, 36.3, 37.8, 38.8, 41.2, 41.7, 42.1, 44.2, 45.8, 46.1, 47.3, 52.2, 52.6, 53.5, 56.4, 61.8, 64.2, 78.6, 81.1, 81.5, 81.5, 85.0, 110.9, 117.8, 120.4, 123.0, 148.3, 148.4, 148.7, 149.0, 149.4, 157.1. FABMS: m/z 863 $[M + H]^+$. HRFABMS: m/z 863.5574 calculated for $C_{54}H_{75}N_2O_7$, found: 863.5548 (LaCour *et al.*, 1998).

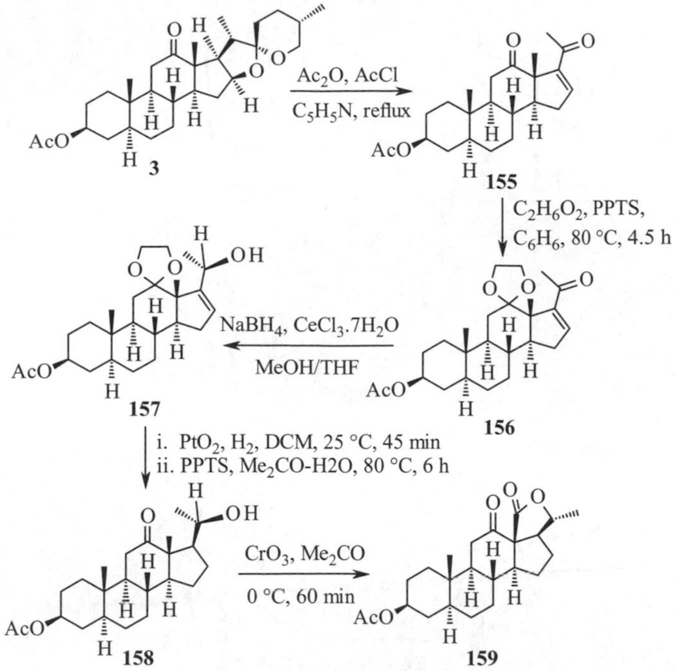

Scheme 5.2.60 *Synthesis of steroidal derivatives **155–159***

The protocol as described earlier for the synthesis of (+)-cephalostatin 1 (**1**) was successfully employed for the preparation of dihydrocephalostatin 1 (**182**) (Bhandaru and Fuchs, 1995; Guo *et al.*, 1996). In order to synthesize the protected dihydrocephalostatin 1 (**181**), a series of hecogenin derivatives **155–180** were synthesized from hecogenin acetate (**3**) in the following manner (Bhandaru and Fuchs, 1995a). Initially, hecogenin acetate (**3**, 80 g) was transformed by refluxing with Ac_2O and AcCl in C_5H_5N to enone **155** (60%) which in turn was converted by reacting with PPTS and $C_2H_6O_2$ to a protected enone **156** (*Scheme 5.2.60*). The treatment of **156** with $NaBH_4$ in the presence of $CeCl_3.7H_2O$ in a mixture of MeOH:THF (1:2) yielded an allylic alcohol **157** (86%) (*Scheme 5.2.60*). Hydrogenation of alcohol **157** using PtO_2 in DCM, followed by the treatment with PPTS, $Me_2CO–H_2O$ gave saturated alcohol **158** (77%) in two steps and CrO_3 oxidation of **158** (7.0 g) afforded lactone **159** (*Scheme 5.2.60*).

Alkaline hydrolysis of lactone **159** provided steroidal alcohol **160** (94%). Silylation of the hydroxyl group of **160** followed by $LiAlH_4$ reduction of the lactone moiety gave triol **161** (71%) in two steps. The overall yield from hecogenin acetate (**3**) to triol **161** was 13%. Syringe-drive addition of ethyldiazophosphonate to a 0.01M solution of **161** in C_6H_6 was treated with $Rh_2(OAc)_4$ (3 mol%) to produce a diastereomeric mixture of neopentyl α-alkoxyphosphono acetates **162** (96%). Bis-oxidation of **162**, following the procedure described in the literature (Brown *et al.*, 1971), yielded another diastereomeric mixture of phosphonate esters **163** (*Scheme 5.2.61*) (Bhandaru and Fuchs, 1995a).

Treatment of **163** with NaH in THF resulted in the intramolecular Wadsworth–Emmons reaction exclusively producing dihydropyran ester **164** (86%) in two steps. Reduction of

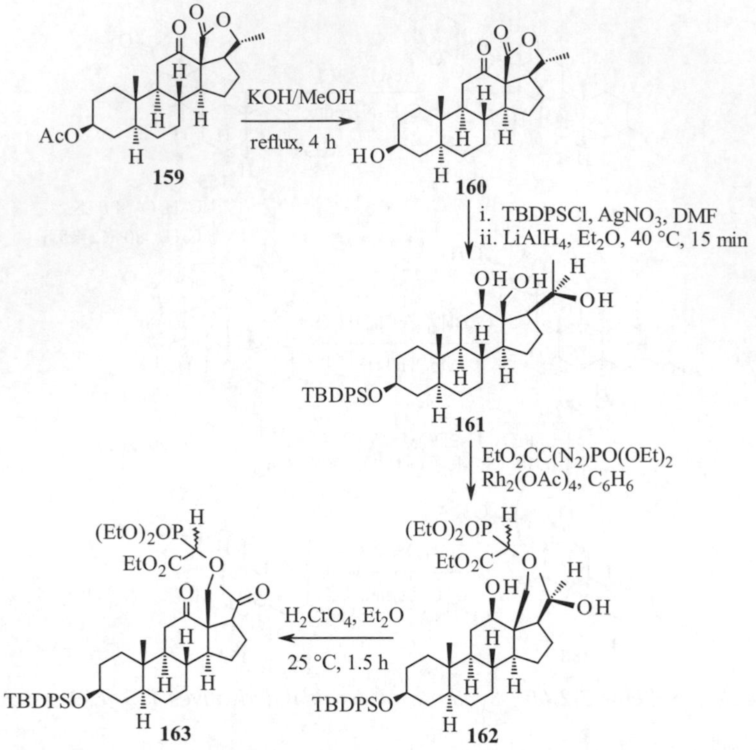

Scheme 5.2.61 *Synthesis of diastereomeric mixture of phosphonate esters* ***163***

ester **164** using $LiAlH_4$ yielded a mixture of diol **165**, which was then directly subjected to the Swern oxidation, giving the key pentacyclic keto-aldehyde **166** (87%) in two steps (*Scheme 5.2.62*) (Bhandaru and Fuchs, 1995a).

The keto-aldehyde **166** was reacted with methallyl stannane in DCM in the presence of $BF_3.Et_2O$ to achieve a readily separable mixture (1:3) of homoallyl alcohols, which upon Mitsunobu inversion of the undesired isomer gave alcohol **167** (79%). Benzylation of **167** gave benzyl ether **168** (90%), which was reduced to diastereomeric alcohols **169** (α:β = 1:9 at C-12) using $LiAlH_4$ in dry Et_2O (*Scheme 5.2.63*) (Bhandaru and Fuchs, 1995b).

The osmylation of **169** using OsO_4 in C_5H_5N followed by the treatment with $Pb(AcO)_4$ in C_6H_6 resulted a mixture (α:β = 1:9 at C-12) of diastereomeric keto-alcohols **170**. Grignard reaction of **170** using MeMgBr in Et_2O yielded a mixture of diastereomeric diols **171** that were converted to a readily separable mixture of three spiroketals **172**, **173**, and **174** (ratio 1:15:1), upon treatment with CSA at room temperature in two steps with a 78% afford (*Scheme 5.2.64*) (Bhandaru and Fuchs, 1995b).

The spiroketal diol **175** was accomplished by treating spiroketals **173** and **174** with TBAF in THF. Bis-oxidation of **175** using the procedure described by Brown *et al.* (1971) yielded diketone **176** which upon heating with CSA in DCE yielded quantitative conversion of 'South 1' diketone **177** (*Scheme 5.2.65*) (Bhandaru and Fuchs, 1995b).

Debenzylation of "South 1" diketone **177** was reduced with Pd-C (10% on carbon) under H_2 (1 atm) yielded alcohol which was instantly converted using DMAP, Ac_2O, Et_3N, DCM

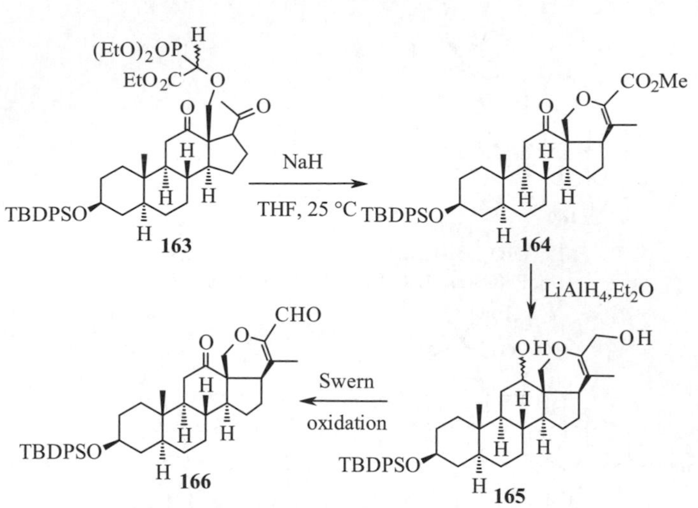

Scheme 5.2.62 *Synthesis of the key pentacyclic keto-aldehyde* **166**

to "South 1" acetate **178** (99%), followed by the treatment with PTAB (1.1 equiv.) in THF (6 min, 25 °C) to furnish "South 1" 2α-bromoketone **179** (70%). Subsequent reaction of **179** with TMGA in CH_3NO_2 yielded "South 1" 2α-azidoketone **180** (78%) (*Scheme 5.2.66*) (Guo *et al.*, 1996).

Protected dihydrocephalostatin 1 (**181**) was synthesized by employing a coupling reaction involving heating an equivalent molar mixture of "South 1" 2α-azidoketone **180** (12 mg) and "North 1" 2α-aminomethoxime **149** (20 mg) in C_6H_6 in the presence of Bu_2SnCl_2 (10 mol%) and PVP (32 mg, 0.02M solution in C_6H_6) in a flask equipped with a Dean–Stark trap set for azeotropic distillation. The reaction was stopped after 5 h to obtain

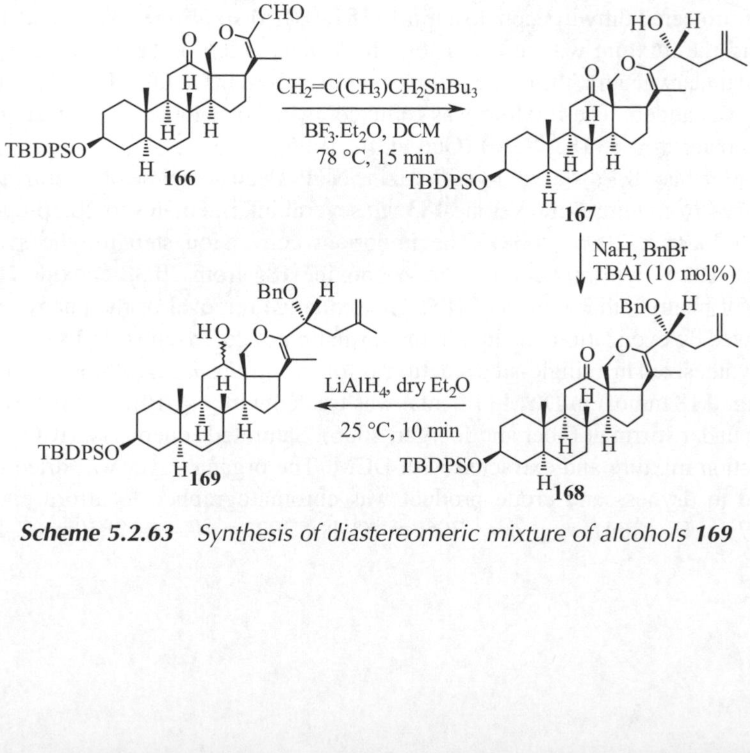

Scheme 5.2.63 *Synthesis of diastereomeric mixture of alcohols* **169**

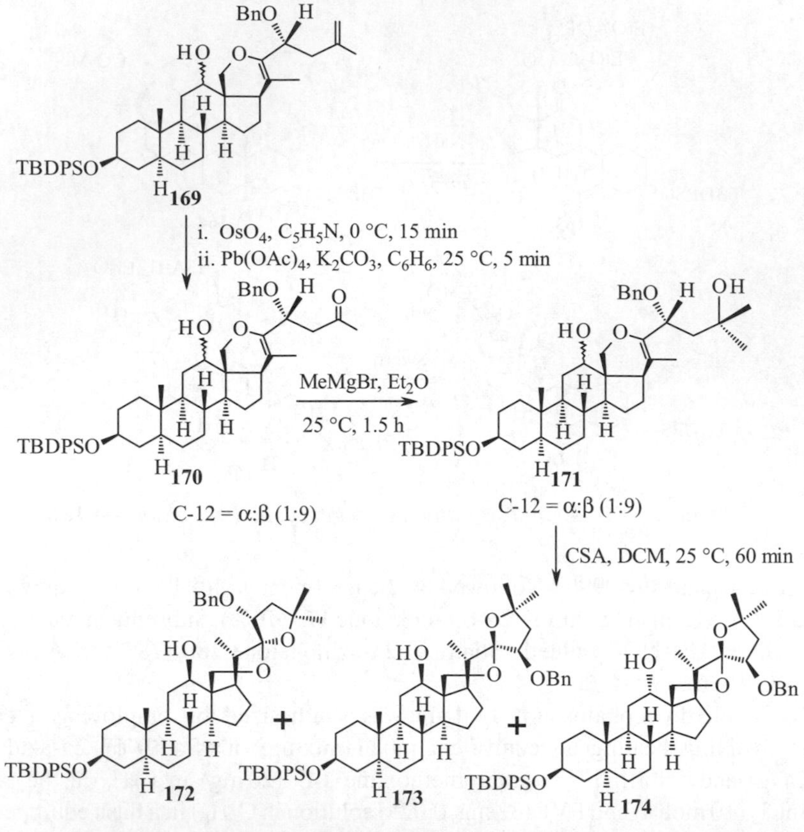

Scheme 5.2.64 *Synthesis of diastereomeric mixture of spiroketals 172–174*

protected **181** (51%). Finally, dihydrocephalostatin 1 (**182**) was prepared from a stirred mixture of protected dihydrocephalostatin 1 (**181**, 0.01M solution in THF) and TBAF (2.3 equiv.), and the mixture was refluxed for 2 h (*Scheme 5.2.67*). The organic solvent was evaporated under vacuum, the crude residue was dissolved in MeOH:H_2O (8:1) and K_2CO_3 (4 equiv.) was added. The mixture was refluxed again for another 30 min to obtain pure dihydrocephalostatin 1 (**182**, 83%) (Guo *et al.*, 1996).

Enantiopure bis-18,18′-desmethyl ritterazine N (**191**), an analogue of ritterazine N (**192**), was produced from tetracyclic ketone **183** *via* several intermediates to 2β,3β-epoxide **184** (Taber and Joerger, 2006, 2008). The important conversion step for the synthesis of enantiopure **191** was to produce the 2α,3α-epoxide **186** from 2β,3β-epoxide **184** *via* the formation of phenyl ether mesylate **185**. The oxidative removal of the phenyl ether from **185**, followed by cyclization resulted in the formation of 2α,3α-epoxide **186**. The **186** also could be synthesized in a single-step reaction as follows (*Scheme 5.2.68*). Tetracyclic ketone **183** (37 mg, 0.18 mmol) in DCM (1.2 mL) was treated with *m*-CPBA (43 mg, 0.25 mmol) for 30 min under stirring (Taber and Joerger, 2006). Saturated aqueous $NaHCO_3$ was added to the reaction mixture and extracted with DCM. The organic layer was dried (Na_2SO_4), evaporated to dryness and crude product was chromatographed to afford epoxide **186** (34 mg, 87%). IR (film): ν_{max} cm^{-1} 3053m (C–H), 2921m (C–H), 2850m (C–H), 1741s

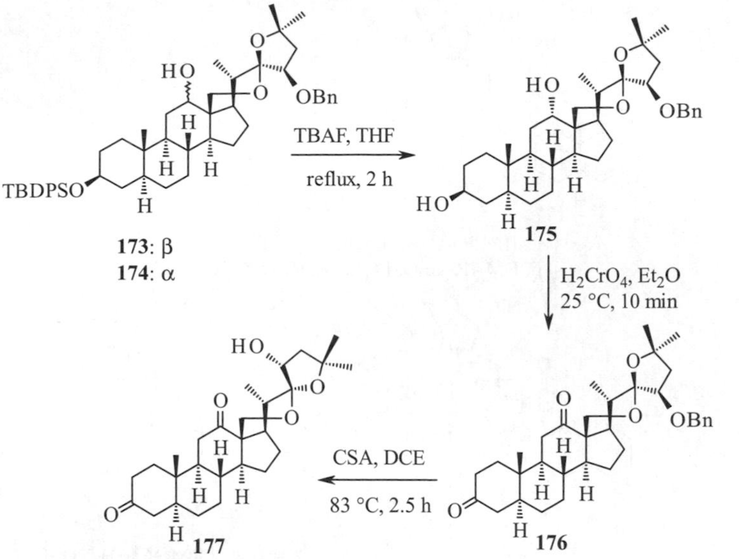

Scheme 5.2.65 *Synthesis of "South 1" diketone* ***177***

(C=O), 1604s (epoxide), 1448m (C–H), 1406m (C–H), 1380m (C–H), 1320 (C–H), 1265 (C–O), 1169 (epoxide C–O), 1144 (epoxide C–O), 984m, 898w, 801w, 737w. ^{1}H NMR (400 MHz, $CDCl_3$): δ 3.23 (s, 1H), 3.14–3.16 (m, 1H,), 2.36–2.39 (m, 1H), 2.17–2.24 (dd, $J=7.0$, 17.0 Hz), 1.90–2.02 (m, 3H), 1.77–1.83 (m, 3H), 1.48–1.59 (m, 5H), 1.18–1.31 (m, 2H), 0.83 (s, 3H). ^{13}C NMR (100 MHz, $CDCl_3$): δ 12.5, 28.6, 31.9, 33.3, 36.2, 37.5, 38.8, 39.2, 46.1, 50.6, 52.5, 54.2, 217.9. LRMS: *m/z* 220 $[M]^+$. HRMS: *m/z* 220.146330 calculated for $C_{14}H_{20}O_2$, found: 220.145760 (Taber and Joerger, 2006).

Alkylated epoxy ketone **188** was synthesized from 2α,3α-epoxide **186** and spiroketal triflate **187** as follows (*Scheme 5.2.68*). To a stirred solution of **186** (234 mg, 1.06 mmol) in azeotropically dried with toluene, KH (128 mg of 50% w/w mixture of KH/paraffin, 1.59 mmol of KH) and THF (8 mL) were added at room temperature. After 60 min, spiroketal triflate **187** (190 mg, 0.53 mmol) in THF (8 mL) was added, and the reaction was allowed to continue for another 30 min. Saturated aqueous NH_4Cl was added and the mixture was extracted with EtOAc, the organic layer was dried (Na_2SO_4) and rotary evaporated to dryness. The solid was suspended in $CDCl_3$ (8 mL), 0.1 M HCl (3 drops) was added and stirred for 1.5 h. The mixture was extracted with EtOAc, and washed with saturated aqueous $NaHCO_3$ followed by brine. The solvent was dried (Na_2SO_4), evaporated *in vacuo* and the crude residue was subjected to CC to yield **188** (20 mg, 9%). IR (KBr): ν_{max} cm^{-1} 2967m (C–H), 2921m (C–H), 1739s (C=O), 999m. ^{1}H NMR (400 MHz, $CDCl_3$): δ 5.65 (dt, 1H, $J=9.8$, 16.9 Hz), 5.03–5.13 (m, 2H), 4.37 (m, 1H), 3.17–3.21 (m, 1H, CH), 3.10 (dd, 1H, $J=4.1$, 5.8 Hz), 2.73 (m, 1H, CH), 1.42–2.12 (m, 18H), 1.10–1.36 (m, 8H), 1.31 (s, 3H), 1.14 (s, 3H), 0.86 (d, 3H, $J=6.7$ Hz), 0.80 (s, 3H). ^{13}C NMR (100 MHz, $CDCl_3$): δ 11.2, 12.8, 28.3, 28.8, 29.0, 30.1, 32.0, 32.1, 33.2, 34.6, 36.0, 36.4, 37.8, 39.0, 43.4, 44.7, 47.5, 50.5, 52.0, 52.4, 53.3, 75.0, 81.6, 115.2, 117.6, 136.5, 219.7. FABMS: *m/z* 451 $[M + Na]^+$. HRFABMS: *m/z* 451.2824 calculated for $C_{27}H_{40}O_4Na$, found: 451.2815 (Taber and Joerger, 2008).

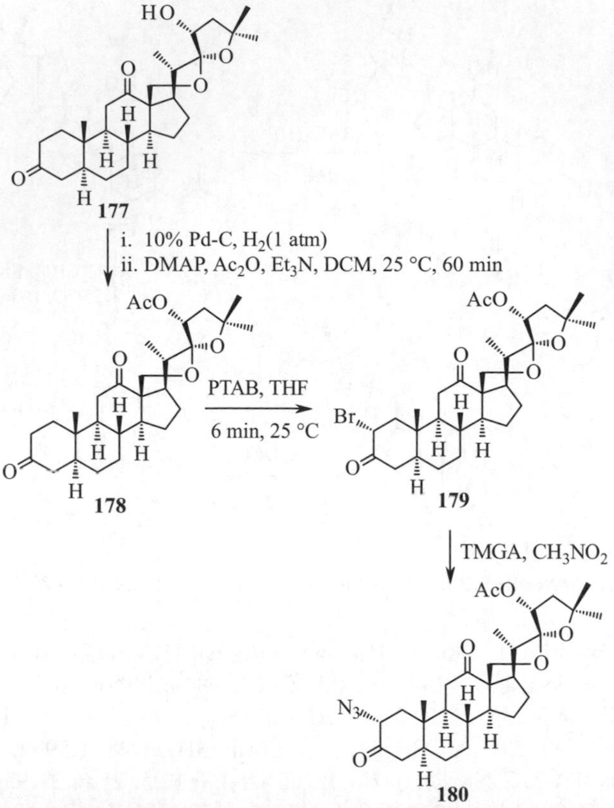

Scheme 5.2.66 Synthesis of "South 1" 2α-azidoketone **180**

A mixture of alkylated epoxy ketone **188** (20 mg, 46.6 μmol), NaN_3 (33 mg, 500 μmol), and $MeOH:H_2O$ (8:1, 1 mL) was sealed and heated at 102 °C for 5 h. The reaction mixture was cooled and extracted with EtOAc and the organic layer was washed with H_2O followed by brine (*Scheme 5.2.69*). The solvent was dried (Na_2SO_4), concentrated under vacuum and purified by CC to provide azido alcohol **189** (19.6 mg, 89%). IR (KBr): ν_{max} cm^{-1} 3439br (O–H), 2923m (C–H), 2102m, 1734s (C=O), 1000m. ^1H NMR (400 MHz, $CDCl_3$): δ 5.65 (m, 1H), 5.05–5.13 (m, 2H), 4.38 (m, 1H), 3.92 (m, 1H), 3.77 (m, 1H), 2.72 (m, 1H), 2.20–2.28 (m, 1H), 1.47–2.14 (m, 17H), 1.19–1.44 (m, 7H), 1.33 (s, 3H), 1.15 (s, 3H), 1.03 (s, 3H), 0.87 (d, 3H, $J = 6.7$ Hz). ^{13}C NMR (100 MHz, $CDCl_3$): δ 11.2, 12.3, 28.3, 28.5, 30.1, 31.6, 32.1, 32.3, 33.2, 35.4, 36.7, 36.8, 37.8, 38.7, 43.5, 44.7, 47.6, 52.0, 54.3, 61.3, 68.4, 75.1, 81.7, 115.3, 117.7, 136.4, 219.5. FABMS: *m/z* 494 $[M + Na]^+$. HRFABMS: *m/z* 494.2995 calculated for $C_{27}H_{41}N_3O_4Na$, found: 494.2985 (Taber and Joerger, 2008).

The synthesis of pyrazine bisalkene **190** could be performed utilizing azido alcohol **189** (*Scheme 5.2.69*). Dess–Martin periodinane (85 mg, 200 μmol) was added to a solution of **189** (14.0 mg, 39.7 μmol) in DCM (1 mL) at room temperature for 2 h. Saturated aqueous $NaHCO_3$, saturated aqueous $Na_2S_2O_3$, and H_2O mixture (11:1, 2 mL) were added to the reaction mixture and stirred for another 5 min, then partitioned between MTBE and brine.

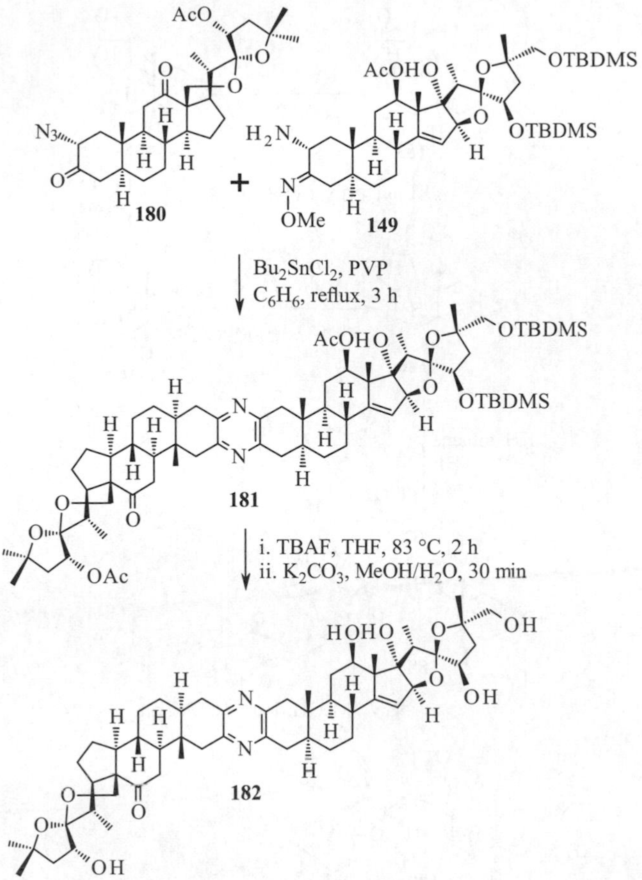

Scheme 5.2.67 *Synthesis of dehydrocephalostain 1 (**182**)*

The organic layer was dried (Na_2SO_4) and evaporated under pressure to provide crude azido ketone. In the next step, freshly prepared solution of NaTeH (0.033M, 2 mL, 66 μmol) was added to the crude azido ketone and stirred for 60 min at room temperature under N_2, and stirring was continued overnight at room temperature under O_2. DCM was added to the reaction and the organic layer was washed with brine. The organic layer was dried (Na_2SO_4), rotary evaporated and the crude product was subjected to CC to afford pyrazine bisalkene **190** (3.4 mg, 27%, mp: 244–246 °C). IR (KBr): v_{max} cm^{-1} 2985m (C–H), 2967m (C–H), 2924m (C–H), 1739s (C=O), 1002m. ^{1}H NMR (400 MHz, $CDCl_3$): δ 5.70 (m, 2H), 5.06–5.15 (m, 4H), 4.42 (m, 2H), 2.66–2.93 (m, 8H), 2.58 (dd, 2H, J = 12.3, 17.7 Hz), 2.31 (dd, 2H, J = 6.5, 16.4 Hz), 1.64–2.25 (m, 28H), 1.33 (s, 6H), 1.23–1.51 (m, 10H), 1.14 (s, 6H), 0.87 (d, 6H, J = 6.7 Hz), 0.83 (s, 6H). ^{13}C NMR (100 MHz, $CDCl_3$): δ 11.2, 11.7, 28.3, 29.1, 30.1, 31.7, 32.0, 33.1, 35.3, 36.0, 36.4, 37.8, 41.9, 43.5, 44.6, 46.1, 47.7, 52.0, 53.2, 74.9, 81.6, 115.3, 117.8, 136.2, 148.4, 148.5, 219.4. HRMS: *m/z* 871.5601 calculated for $C_{54}H_{76}N_2O_6Na$, found: 871.5596 (Taber and Joerger, 2008).

Finally, the pyrazine bisalkene **190** was employed for the synthesis of bis-18,18′-desmethylritterazine N (**191**), an analogue of ritterazine N (**192**), in the following fashion

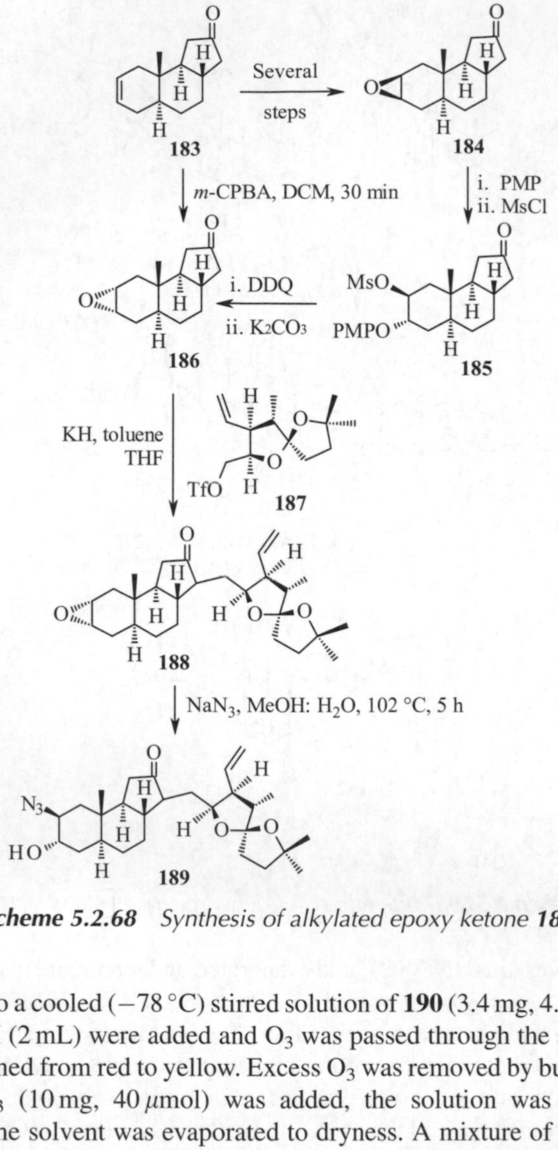

Scheme 5.2.68 *Synthesis of alkylated epoxy ketone* ***189***

(*Scheme 5.2.69*). To a cooled (−78 °C) stirred solution of **190** (3.4 mg, 4.0 μmol), Sudan-III (0.1 mg) and DCM (2 mL) were added and O_3 was passed through the solution at −78 °C until the colour turned from red to yellow. Excess O_3 was removed by bubbling N_2 through the solution. PPh_3 (10 mg, 40 μmol) was added, the solution was warmed to room temperature, and the solvent was evaporated to dryness. A mixture of THF:EtOH:NaOH (3:3:1, 2 mL) was added to the solid at room temperature and stirred for 1.5 h. EtOAc was added and the organic layer was washed with half-saturated aqueous NH_4Cl followed by brine. The organic layer was dried (Na_2SO_4), evaporated under pressure and the crude solid was chromatographed to furnish **191** (3.3 mg, 97%). IR (KBr): ν_{max} cm^{-1} 3411br (O–H), 2966m (C–H), 2925m (C–H), 1731s (C=O), 1400s (pyrazine), 999m. ^1H NMR (400 MHz, $CDCl_3$): δ 4.72 (m, 2H), 4.19 (m, 2H), 3.06 (d, 2H, J = 16.2 Hz), 2.96 (d, 2H, J = 16.1 Hz), 2.86 (dd, 2H, J = 5.5, 18.0 Hz), 2.42–2.70 (m, 6H), 1.66–2.17 (m, 26H), 1.52–1.63 (m, 2H), 1.38–1.52 (m, 2H), 1.34 (s, 6H), 1.22–1.38 (m, 8H), 1.18 (s, 6H), 1.10 (d, 6H, J = 6.7 Hz), 0.99 (s, 6H). ^{13}C NMR (100 MHz, $CDCl_3$): δ 13.4, 13.6, 28.5, 28.7, 30.0, 32.8, 33.3, 33.6, 34.9, 35.4, 37.3, 37.5, 42.9, 45.7, 45.7, 47.2, 55.5, 56.4, 68.2, 77.9, 82.0, 84.3, 120.0, 148.3,

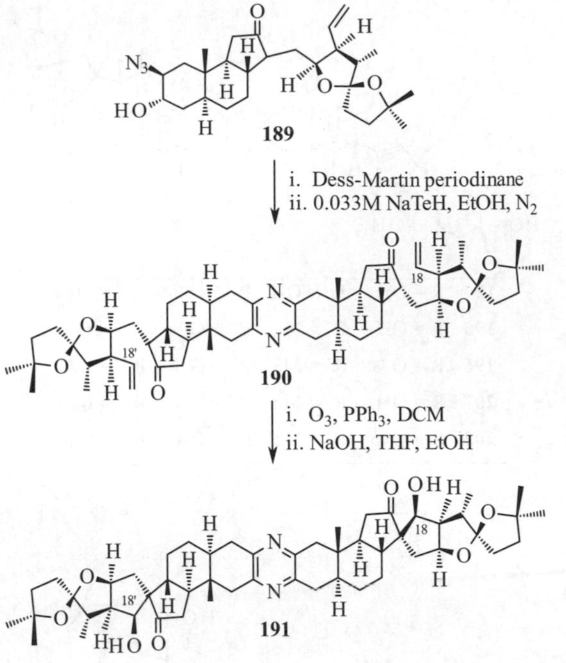

189

i. Dess-Martin periodinane
ii. 0.033M NaTeH, EtOH, N_2

190

i. O_3, PPh_3, DCM
ii. NaOH, THF, EtOH

191

Scheme 5.2.69 *Synthesis of bis-18,18′-desmethyl ritterazine N (**191**)*

148.3, 220.6. HRMS: *m/z* 875.5186 calculated for $C_{52}H_{72}N_2O_8Na$, found: 875.5186 (Taber and Joerger, 2008).

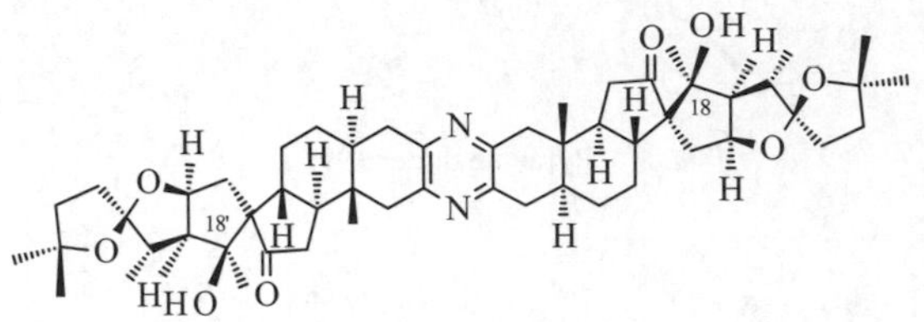

Ritterazine N (**192**)

Simple oxidation, reduction, methanolysis and acetylation of ritterazine B (**193**) could produce dimers **194–201** (Nahar *et al.*, 2007).

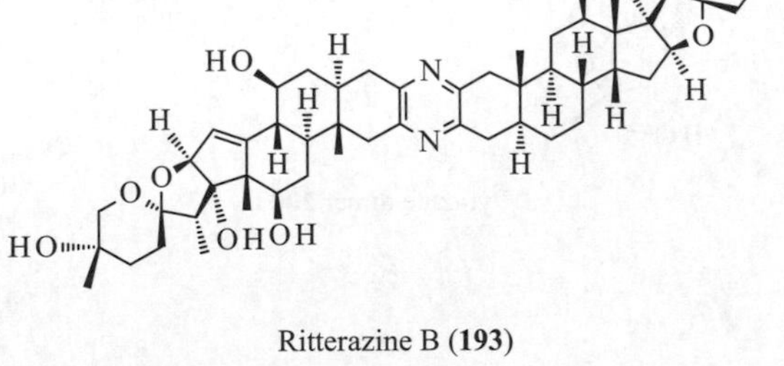

Ritterazine B (**193**)

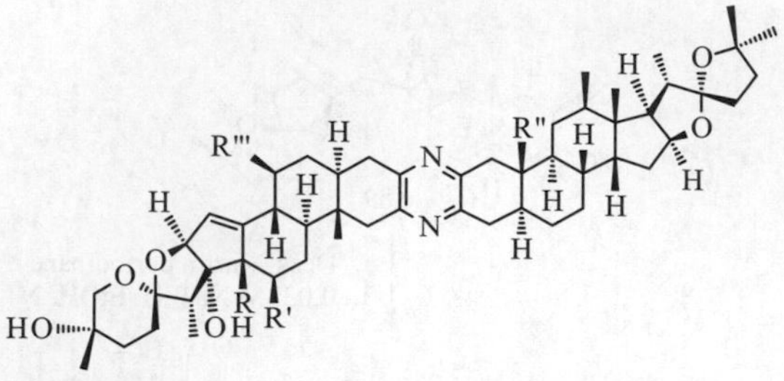

194	R = OH	R' = OH	R" = Oxo	R'" = Oxo
195	R = OH	R' = Me	R" = OAc	R'" = OAc
196	R = OAc	R' = Me	R" = OAc	R'" = OAc
207	R = OH	R' = Me	R" = OH	R'" = OAc
208	R = Oxo	R' = Me	R" = OAc	R'" = Oxo

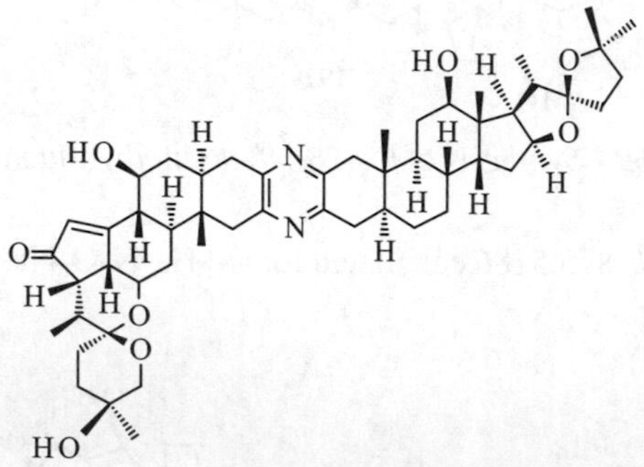

Pyrazine dimer **199**

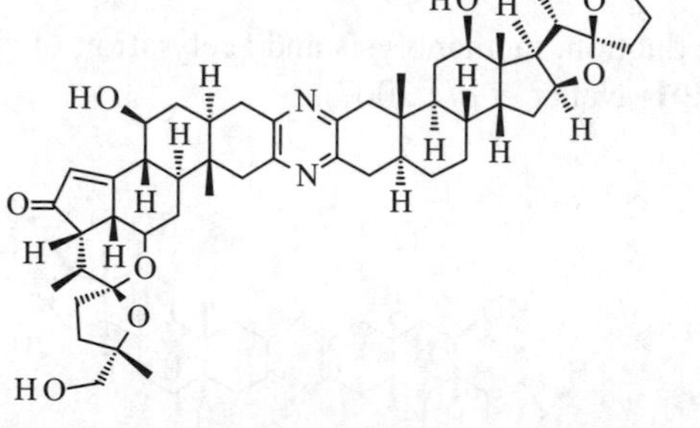

Pyrazine dimer **200**

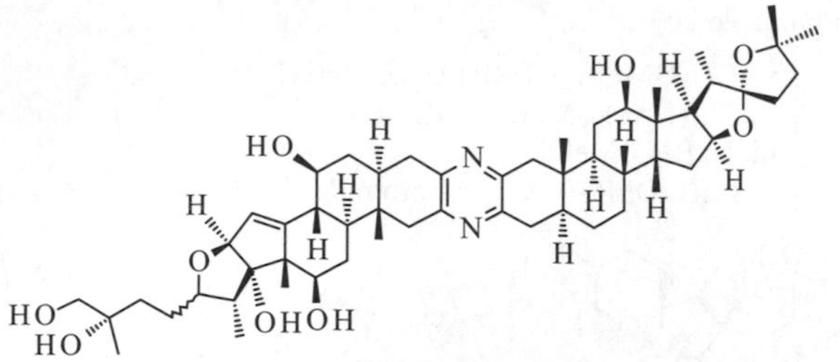

Pyrazine dimer **201**

5.3 Total Synthesis of Naturally Occurring Cephalostatin 1

Over a decade ago, LaCour *et al.* (1998) reported the convergent total syntheses of the extremely potent cell-growth inhibitor (+)-cephalostatin 1 (**1**) along with two hybrid analogues, ritterostatin G_N1_N (**151**) and ritterostatin G_N1_S (**153**). Comprehensive protocols for this multistep synthesis have been discussed earlier in *Section 5.2*. Since then, several attempts have been made to improve the synthetic protocol or to introduce new methodologies for the total synthesis of cephalostatin 1 (**1**). One of such protocols was reported by Fortner *et al.* (2010), who synthesized the enantioselective synthesis of (+)-cephalostatin 1 (**1**). The major steps of this synthesis were a unique methyl group selective allylic oxidation, directed C–H hydroxylation of a sterol at C-12, Au(I)-catalyzed 5-endo-dig cyclization, and a kinetic spiroketalization. Photochemical reaction of hecogenin acetate (**3**) gave an aldehyde **202**. Reaction of the aldehyde **202** led to the formation of the allylic *N*-Ph urazole hecogenin acetate **203** (*Scheme 5.3.1*) (Fortner *et al.*, 2010).

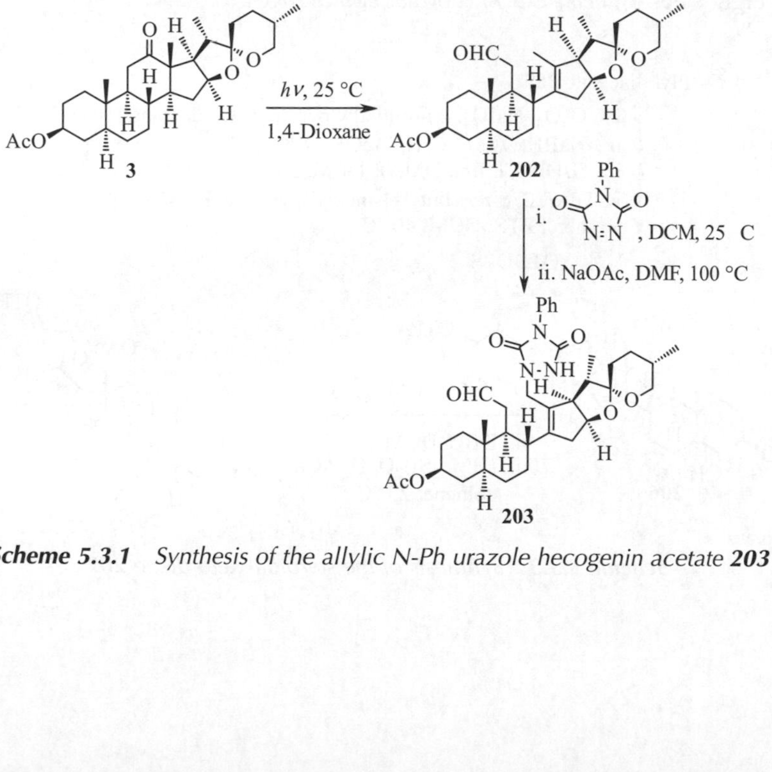

Scheme 5.3.1 *Synthesis of the allylic N-Ph urazole hecogenin acetate* ***203***

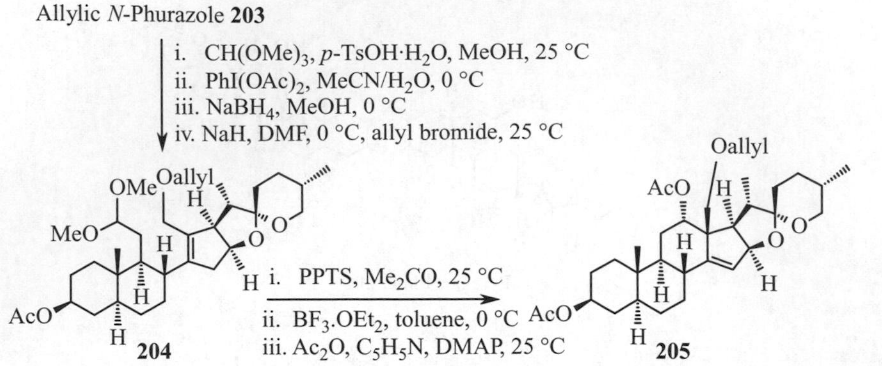

Allylic *N*-Phurazole **203**

i. $CH(OMe)_3$, *p*-TsOH·H_2O, MeOH, 25 °C
ii. $PhI(OAc)_2$, $MeCN/H_2O$, 0 °C
iii. $NaBH_4$, MeOH, 0 °C
iv. NaH, DMF, 0 °C, allyl bromide, 25 °C

204

i. PPTS, Me_2CO, 25 °C
ii. $BF_3.OEt_2$, toluene, 0 °C
iii. Ac_2O, C_5H_5N, DMAP, 25 °C

205

Scheme 5.3.2 *Synthesis of the O-allyl diacetate* ***205***

The allylic *N*-Ph urazole hecogenin acetate **203** was converted to steroidal monomer **204** in four steps (*Scheme 5.3.2*). An *O*-allyl diacetate **205** was conveniently achieved from **204** in three steps (*Scheme 5.3.2*) (Fortner *et al.*, 2010).

The steroidal keto diene **208** was carried out in seven steps using from **205** as the starting material as follows (*Scheme 5.3.3*). Finally, cephalostatin 1 western half (**211**) was prepared in 12 steps from **208** (*Scheme 5.3.4*) (Fortner *et al.*, 2010).

Androsterone enyne (**214**) was synthesized in five steps starting from *trans*-androsterone (**212**, 3β-hydroxy-5α-androstan-17-one) (*Scheme 5.3.5*) (Fortner *et al.*, 2010).

An easily separable mixture of bromomethylene spiroketals favouring the desired C_{22}-(*S*) isomer **216** by a 5:1 ratio was accomplished from androsterone enyne **214** in seven steps (*Scheme 5.3.6*).

Finally, bromomethylene spiroketal **216** was converted to cephalostatin 1 eastern half (**218**) in eight steps (*Scheme 5.3.7*) (Fortner *et al.*, 2010).

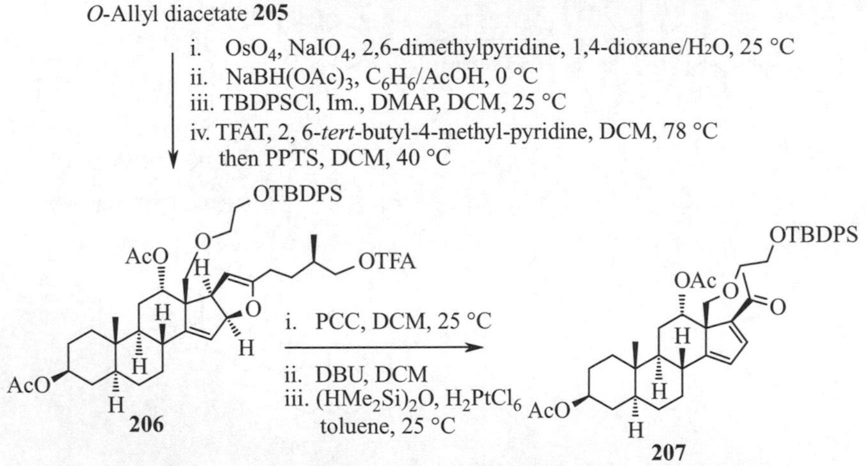

O-Allyl diacetate **205**

i. OsO_4, $NaIO_4$, 2,6-dimethylpyridine, 1,4-dioxane/H_2O, 25 °C
ii. $NaBH(OAc)_3$, C_6H_6/AcOH, 0 °C
iii. TBDPSCl, Im., DMAP, DCM, 25 °C
iv. TFAT, 2, 6-*tert*-butyl-4-methyl-pyridine, DCM, 78 °C then PPTS, DCM, 40 °C

206

i. PCC, DCM, 25 °C
ii. DBU, DCM
iii. $(HMe_2Si)_2O$, H_2PtCl_6 toluene, 25 °C

207

Scheme 5.3.3 *Synthesis of the steroidal keto diene* ***207***

Keto diene **207**

i. TBAF, AcOH, THF, 25 °C
ii. DMSO, *i*-Pr_2NEt, SO_3, C_5H_5N, DCM, 25 °C
iii. Piperidine, AcOH, 25 °C
iv. 1-methoxy-1-*tert*-butyldimethylsilyloxyethene $LiClO_4$, DCM

TBSO O OMe AcO O H H AcO H **208**

i. TBAF, THF, 25 °C
ii. Ph_3P, DIAD, $ClCH_2CO_2H$, THF, 25 °C
iii. HDTC, 2,6-dimethylpyridine, AcOH, 25 °C
iv. TBDPSCl, DMAP, DCM, 25 °C
v. MeMgBr, Et_2O, 25 °C
vi. TPAP, NMO, DCM, 25 °C

TBDPSO HO AcO O H H AcO H **209**

i. PhSeBr, C_5H_5N, DCM, −78 °C to 0 °C
ii. AIBN, Bu_3SnH, toluene, 100 °C
iii. CSA, DCE, 83 °C

TBDPSO O O AcO H H AcO H **210**

i. PTAB, THF, 0 °C
ii. TMGA, $EtNO_2$

TBDPSO O O AcO H N_3 H AcO H **211**

Scheme 5.3.4 *Synthesis of cephalostatin 1 western half (**211**)*

The reaction between cephalostatin 1 western half (**211**) and cephalostatin 1 eastern half (**218**) yielded (+)-cephalostatin 1 (**1**) in 47% yield in two steps (*Scheme 5.3.8*) (Fortner *et al.*, 2010).

Most recently, Shi *et al.* (2011) have reported a new practical synthetic strategy for the total synthesis of cephalostatin 1 (**1**). The main features of their strategy were: the use of pregnan-(3*S*,12*R*,16*S*,20*S*)-tetraol and a steroidal-16(*S*),22-lactone instead of the conventional pregnenolone or epiandrostenone as a starting point; the construction of chiral centres

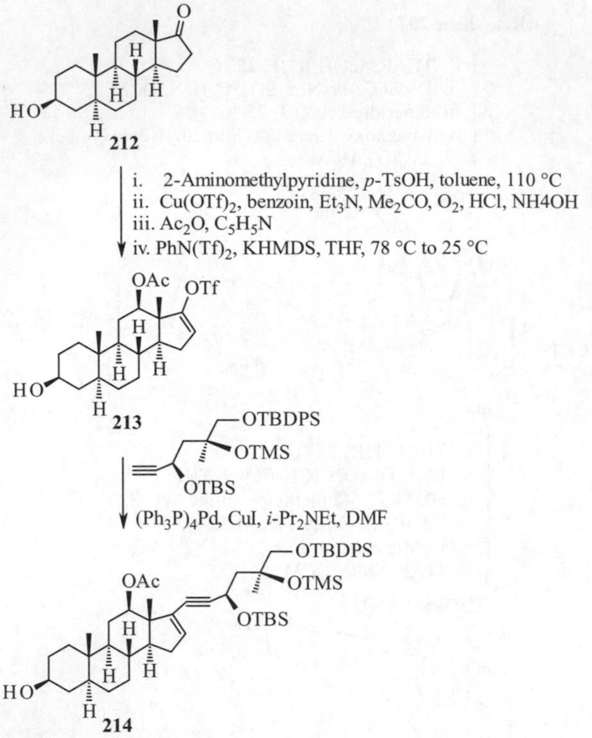

O
H
H H
HO
H
212

i. 2-Aminomethylpyridine, p-TsOH, toluene, 110 °C
ii. $Cu(OTf)_2$, benzoin, Et_3N, Me_2CO, O_2, HCl, NH4OH
iii. Ac_2O, C_5H_5N
iv. $PhN(Tf)_2$, KHMDS, THF, 78 °C to 25 °C

OAc OTf
H
H H
HO
H
213

OTBDPS
OTMS
OTBS
$(Ph_3P)_4Pd$, CuI, i-Pr_2NEt, DMF

OTBDPS
OAc OTMS
OTBS
H
H H
HO
H
214

Scheme 5.3.5 *Synthesis of the androsterone enyne **214***

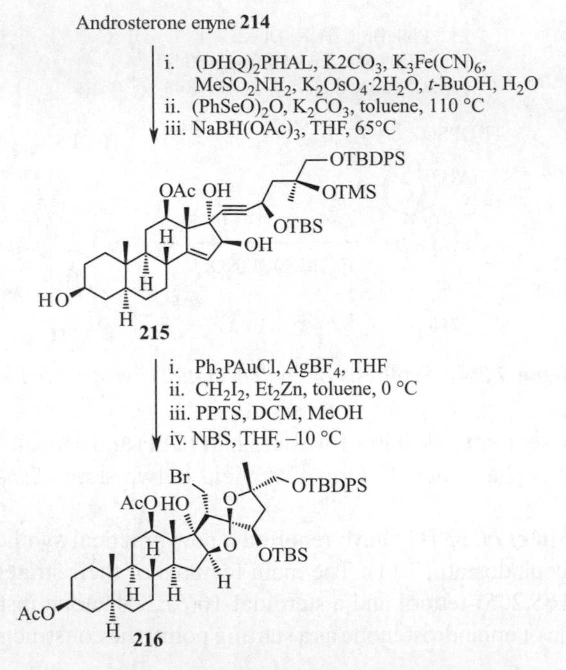

Scheme 5.3.6 *Synthesis of bromomethylene spiroketal C_{22}-(S) isomer **216***

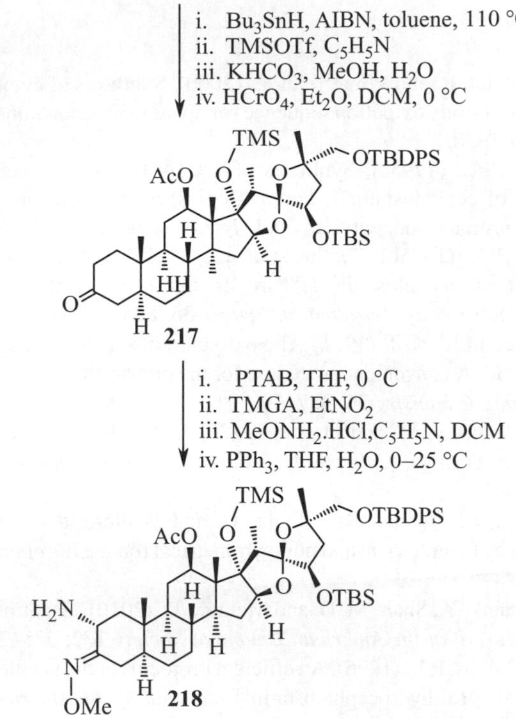

i. Bu3SnH, AIBN, toluene, 110 °C
ii. TMSOTf, C5H5N
iii. KHCO3, MeOH, H2O
iv. HCrO4, Et2O, DCM, 0 °C
TMS
OTBDPS
AcO
OTBS
217
i. PTAB, THF, 0 °C
ii. TMGA, EtNO2
iii. MeONH2.HCl,C5H5N, DCM
iv. PPh3, THF, H2O, 0–25 °C
H2N
OMe
218

Scheme 5.3.7 *Synthesis of cephalostatin 1 eastern half (**218**)*

of target molecules by employing either substrate control or chiral starting materials (particular emphasis was given to excluding expensive reagents and difficult operations); and the application of cascade reactions (the construction of spiroketals) and one-pot reactions. Their synthetic strategy is also an excellent demonstration of the rational use of readily available resource compounds.

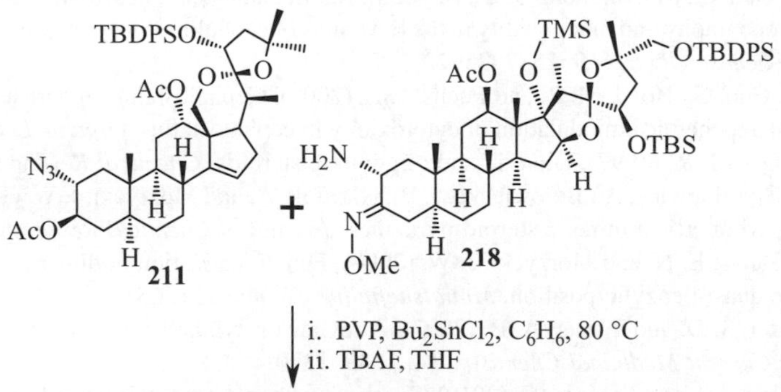

TBDPSO
AcO
N3
211
TMS
OTBDPS
OTBS
H2N
OMe
218
i. PVP, Bu2SnCl2, C6H6, 80 °C
ii. TBAF, THF
Cephalostatin 1 (1)

Scheme 5.3.8 *Synthesis of cephalostatin 1 (**1**)*

References

Bäsler, S., Brunck, A., Jautelat, R. and Winterfeldt, E. (2000). Synthesis of cytostatic tetradecacyclic pyrazines and a novel reduction - oxidation sequence for opening in sapogenins spiroketa. *Helvetica Chimica Acta* **83**, 1854–1880.

Bhandaru, S. and Fuchs, P. L. (1995a). Synthesis of a C14′,15′-dihydro derivative of the south hexacyclic steroid unit of cephalostatin 1. Part I: Regiospecific Rh[ll]-mediated intermolecular oxygen alkylation of a primary neopentyl alcohol. *Tetrahedron Letters* **36**, 8347–8350.

Bhandaru, S. and Fuchs, P. L. (1995b). Synthesis of a C-14′,15′. Dihydro derivative of the south hexacyclic steroid unit of cephalostatin 1. Part II: spiroketal synthesis and stereochemical assignment by NMR spectroscopy. *Tetrahedron Letters* **36**, 8351–8354.

Brown, H. C., Garg, C. P. and Liu, K. T. (1971). The oxidation of secondary alcohols in diethyl ether with aqueous chromic acid. A convenient procedure for the preparation of ketones in high epimeric purity. *Journal of Organic Chemistry* **36**, 387–390.

Černý, I., Pouzar, V., Budesinsky, M. and Drasar, P. (2000). Synthesis of symmetrical bis-steroid pyrazines connected *via* D-rings. *Collection of Czechoslovak Chemical Communications* **65**, 1597–1608.

Drögemüller, M., Flessner, T., Jautelat, R., Scholz, U. and Winterfeldt, E. (1998). Synthesis of cephalostatin analogues by symmetrical and non-symmetrical routes. *European Journal of Organic Chemistry*, 2811–2831.

Fortner, K. C., Kato, D., Tanak, Y., Shair, M. D. and Mathew, D. (2010). Enantioselective synthesis of (+)-cephalostatin 1. *Journal of the American Chemical Society* **132**, 275–280.

Guo, C., Bhandaru, S. and Fuchs, P. L. (1996). An efficient protocol for the synthesis of unsymmetrical pyrazines. Total synthesis of dihydrocephalostatin 1. *Journal of the American Chemical Society* **118**, 10672–10673.

Heathcock, C. H. and Smith, S. C. (1994). Synthesis and Biological Activity of Unsymmetrical Bis-Steroidal Pyrazines Related to the Cytotoxic Marine Natural Product Cephalostatin 1. *Journal of Organic Chemistry* **59**, 6828–6839.

Jautelat, R., Müller-Fahrnow, A. and Winterfeldt, E. (1999). A novel oxidative cleavage of the steroidal skeleton. *Chemistry-A European Journal* **5**, 1226–1233.

LaCour, T. G., Guo, C., Bhandaru, S., Boyd, M. R. and Fuchs, P. L. (1998). Interphylal product splicing: The first total syntheses of cephalostatin 1, the north hemisphere of ritterazine G, and the highly active hybrid analogue, ritterostatin G_N1_N. *Journal of the American Chemical Society* **120**, 692–697.

LaCour, T. G., Guo, C., Ma, S., Jeong, J. U., Boyd, M. R., Matsunaga, S., Fusetani, N. and Fuchs, P. L. (1999). On topography and functionality in the B-D rings of cephalostatin cytotoxins. *Bioorganic & Medicinal Chemistry Letters* **9**, 2587–2592.

LaCour, T. G., Guo, C., Boyd, M. R. and Fuchs, P. L. (2000). Cephalostatin support studies. Part 19. Outer-ring stereochemical modulation of cytotoxicity in cephalostatins. *Organic Letters* **2**, 33–36.

Li, Y. X. and Dias, J. R. (1997). Dimeric and oligomeric steroids. *Chemical Review* **97**, 283–304.

Łotowski, Z., Gryszkiewicz, A., Borowiecka, J. B., Nikitiuk, A. and Morzycki, J. W. (1999). A facile synthesis of symmetrical dimeric steroid-pyrazines. *Journal of Chemical Research (S)* 662–663.

Łotowski, Z., Dubis, E. N. and Morzycki, J. W. (2000). Functionalization of dimeric cholestanopyrazines at the quasi-benzylic position. *Monatshefte fur Chemie* **131**, 65–71.

Nahar, L., Sarker, S. D. and Turner, A. B. (2007). A review on synthetic and natural steroid dimers: 1997–2006. *Current Medicinal Chemistry* **14**, 1349–1370.

Poza, J. J., Rodrígue, J. and Jiménez, C. (2010). Synthesis of a new cytotoxic cephalostatin/ritterazine analogue from hecogenin and 22-epi-hippuristanol. *Bioorganic and Medicinal Chemistry* **18**, 58–63.

Shawakfeh, K. Q., Al-Said, N. H. and Al-Zoubi, R. M. (2008). Synthesis of bisdiosgenin pyrazine dimers: new cephalostatin analogs. *Steroids* **73**, 579–584.

Shawakfeh, K. Q. and Al-Said, N. H. (2011). Synthesis of new symmetrical bis steroidal pyrazine analogues from diosgenin. *Steroids* **76**, 273–277.

Shi, Y., Jia, L., Xiao, Q., Lan, Q., Tang, X., Wang, D., Li, M., Ji, Y., Zhou, T. and Tian, W. (2011). A practical synthesis of cephalostatin 1. *Chemistry - An Asian Journal* **6**, 786–790.

Smith, H. E. and Hicks, A. A. (1971). Optically active amines.12. synthesis and spectral properties of some optically active alpha-oximino ketones and alpha-amino ketone hydrochlorides-dimerization of alpha-amino ketones. *Journal of Organic Chemistry* **36**, 3659–3668.

Taber, D. F. and Taluskie, K. V. (2006). Computationally guided organometallic chemistry: preparation of the heptacyclic pyrazine core of ritterazine N. *Journal of Organic Chemistry* **71**, 2797–2801.

Taber, D. F. and Joerger, J-M. (2008). Synthesis of bis-18,18′-desmethyl ritterazine N. *Journal of Organic Chemistry* **73**, 4155–4159.

Tanaka, K., Itagaki, Y., Satake, M., Naoki, H., Yasumoto, T., Nakanishi, K. and Berova, N. (2005). Three challenges toward the assignment of absolute configuration of Gymnocin-B. *Journal of the American Chemical Society* **127**, 9561–9570.

6

Applications of Steroid Dimers

Dimerization of a steroid skeleton, with or without spacers, offers some valuable characteristics that can be exploited to different areas of applications including drug discovery, design and delivery. Dimeric steroids show micellar, detergent, and liquid-crystal properties, and can be used as catalysts for different types of organic reactions. A number of dimeric steroids, *e.g.*, cytotoxic cephalostatins and their analogues, can be utilized for potential anticancer drugs discovery and development. Steroid dimers have also found their applications as *'molecular umbrella'* for drug delivery, and cholaphanes and cyclocholates as potential artificial receptors. This chapter will look into various applications of steroid dimers, with particular emphasis on drug discovery, design and delivery.

6.1 Application of Steroid Dimers as '*Molecular Umbrellas*': Drug Delivery

Lipid bilayer, a universal and an essential component of cell membranes, typically about five nanometers thick, is composed of two layers of phospholipid molecules organized in two sheets. There are approximately 5×10^6 lipid molecules in a 1 μm^2 area of lipid bilayer, or about 10^9 lipid molecules in the plasma membrane of a small animal cell. The lipid bilayer forms the core structure of all cell membranes, and surrounds all cells.

In lipid bilayers, the two layers of lipid molecules are packed with their hydrophobic tails pointing inward and their hydrophilic heads outward, exposed to water. Lipid bilayers (*Figure 6.1.1*) play a pivotal role in cells by serving as barriers for transport; they only permit the transport of those molecules and ions that are necessary for maintaining the essential living state. Their hydrophobic core restricts the entry of a number of important classes of strongly hydrophilic bioactive molecules or drugs, *e.g.*, oligonucleotides.

To find ways of promoting the passive transport of polar drugs across hydrophobic barriers (lipid bilayers) to improve their efficacy appears to be one of the main challenges in medicinal chemistry today. For example, antisense oligonucleotides, which are being considered for the treatment of a wide range of diseases, *e.g.*, nonsmall-cell lung cancer, Crohn's disease, malignant melanoma, chronic lymphocytic leukaemia, multiple myeloma

Steroid Dimers: Chemistry and Applications in Drug Design and Delivery, First Edition. Lutfun Nahar and Satyajit D. Sarker. Published 2012 by John Wiley & Sons, Ltd.

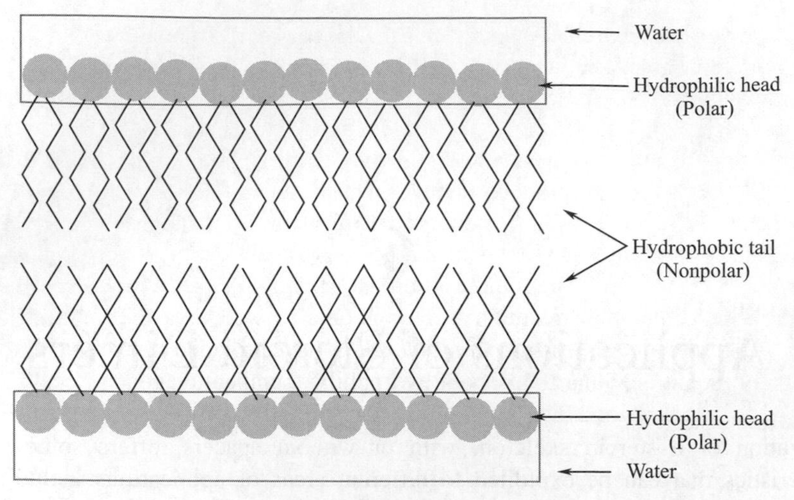

Figure 6.1.1 *Lipid bilayer*

and cytomegalovirus retinitis, are limited by their inefficient delivery to mRNA in the cytoplasm of cells (Kondo *et al.*, 2008). Antisense oligonucleotides are single strands of DNA or RNA, which are complementary to a chosen sequence. Antisense RNA prevents protein translation of certain messenger RNA strands by binding to them, and antisense DNA can be used to target a specific, complementary (coding or noncoding) RNA.

In theory, the passive transport of oligonucleotides (short nucleic acid polymers) across plasma membranes should significantly improve their efficacy. It is not only the oligonucleotides, but also the enhanced transport of smaller polar drugs, *e.g.*, peptides and nucleotides, across hydrophobic barriers that may produce improved therapeutic efficacy.

To meet the challenges of finding a way for promoting the passive transport of polar drugs across lipid bilayers, a new class of transporter molecules, *'molecular umbrellas'* (*Figure 6.1.2*) was introduced (Janout *et al.*, 1996). This new concept in surfactant chemistry is based on molecules that mimic the structure and function of umbrellas, *i.e.*, molecules that can cover an attached agent and shield it from an incompatible environment. *'Molecular umbrellas'* have provided a promising avenue in drug design and delivery, particularly, for the design of the intracellular delivery of hydrophilic therapeutic agents. These molecules

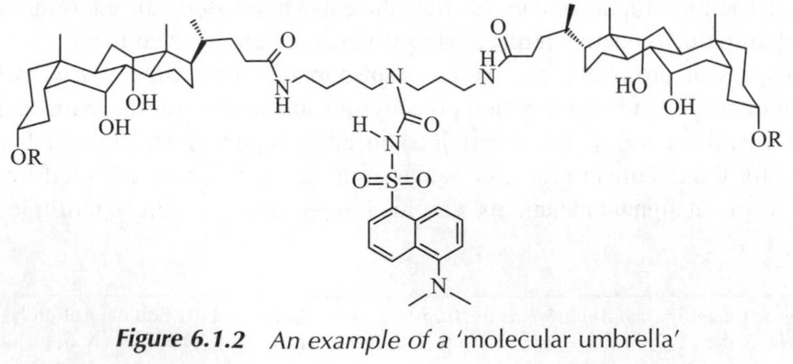

Figure 6.1.2 *An example of a* 'molecular umbrella'

have been proven to be extremely powerful in the delivery of molecules across incompatible barriers (*e.g.*, lipid bilayers), a process highly important to drug delivery.

'Molecular umbrellas' are a unique class of amphiphiles. Since the first introduction of *'molecular umbrellas'* almost a couple of decades ago (Janout *et al.*, 1996), the suitability of such molecules for application as vectors in drug-delivery systems has been extensively studied. They are 'amphomorphic' compounds, which can produce a hydrophobic or hydrophilic exterior when exposed to a hydrophobic or hydrophilic microenvironment, respectively.

Several such molecules have been synthesized involving two steroidal molecules (*e.g.*, cholic acid) to date. The synthesis of *'molecular umbrellas'* involves the use of amphiphilic molecules that maintain a hydrophobic as well as a hydrophilic face (see *Chapter 2* for further details). *'Molecular umbrellas'* have provided an exciting platform for the design of the intracellular delivery of hydrophilic therapeutic agents. Because of their low or no toxicity and good biocompatibility, the suitability of *'molecular umbrellas'* for application as carriers in drug-delivery systems looks promising. Apart from promoting the transport of polar agents across cellular membranes, *'molecular umbrellas'* may also improve water solubility and stability of hydrophobic drugs of therapeutic interest.

A *'molecular umbrella'* consists of two or more amphiphilic walls. The amphiphilic walls are composed of rigid steroid units having polar and nonpolar faces. They are coupled to carry the active agent. It mimics the structure and function of an umbrella, *i.e.*, a molecule that can cover an attached agent and shield it from an incompatible environment. Two or more such umbrella walls (amphiphiles) are coupled to a suitable scaffold either before or after a desired agent is linked to a central location.

The very first *'molecular umbrella'* was prepared by using cholic acid as 'wall material', spermidine as the scaffold, and an environmentally sensitive fluorescent probe, 5-dimethylamino-1-naphthalenesulfonyl (dansyl), as the agent (Janout *et al.*, 1996; Nahar *et al.*, 2007). Because of the relative hydrophobicity of the dansyl moiety, it was assumed that such an umbrella would favour exposed or fully exposed conformations in solvents of low polarity and a shielded conformation in water. Cholic acid was chosen as it possesses the requisite amphilicity and can be readily conjugated through its carboxylic acid group. Also, umbrella frameworks derived from cholic acid and spermidine were thought to be potentially biocompatible, since both compounds occur naturally in mammalian cells. It was also proposed that this *'molecular umbrella'* could be utilized for the sequestration of cytotoxic, nonpolar crosslinking agents within an umbrella, which would otherwise decompose in water or react with water-soluble nucleophiles, and for their release into target membranes by "flipping" from a shielded to a fully exposed state.

'Molecular umbrellas', composed of bile acids as umbrella "wall" polyamines such as spermidine and spermine as scaffold material, and L-lysine as "branches" are capable of transporting certain hydrophilic peptides, nucleotides, and oligonucleotides across liposomal membranes by passive diffusion (Janout and Regen, 2009). It has been proposed that a *'molecular umbrella'* permeates across a lipid membrane *via* the following sequence of events (Nahar *et al.*, 2007).

i. Diffusion of the conjugate to the biomembrane surface in a fully exposed state.
ii. Insertion into the outer monolayer leaflet by flipping into a shielded state.
iii. Diffusion to the inner monolayer leaflet, and
iv. Entry into the cytoplasm via the sequential reversal of steps (ii) and (i).

It was found that the *'molecular umbrellas'* could really assist in the transport of a hydrophilic peptide across a phospholipid membrane. It was suggested that these vectors could transport polar peptides like glutathione across 1-palmitoyl-2-oleoyl-*sn*-glycero-3-phosphocholine (POPC) bilayer membranes (Janout *et al.*, 2002). The feasibility of conjugating thiolated forms of ATP and AMP to *'molecular umbrellas'* and transporting these nucleosides across liposomal membranes was evaluated. The facile cleavage of such conjugates by entrapped glutathione, together with the fact that mammalian cells contain millimolar concentration of glutathione in their cytoplasm, led to the assumption that this chemistry might be extended to the development of prodrugs, *e.g.*, antisense oligonucleotides, which could be delivered to the cells. Janout *et al.* (2002) were able to provide convincing evidence for *'molecular-umbrella-assisted'* transport of glutathione (GSH) across lipid bilayers composed of 1-palmitoyl-2-oleyol*sn*-glycero-3-phosphocholine (POPC). It was suggested that this *'molecular-umbrella-assisted'* transport strategy might equally be applicable to practical prodrug delivery devices.

In a recent study aiming at gaining a better understanding of the cellular uptake mechanism for *'molecular umbrellas'*, it was observed that a fluorescently labelled *di*-walled *'molecular umbrella'*, synthesized from cholic acid, spermine, and 5-carboxy-fluorescein could greatly facilitate cellular uptake (into Hela cells) of hydrophilic agent, 5-carboxyfluorescein. *In vitro* experiments with diffuse marker, endocytic marker, and inhibitors indicated the involvement of several distinct uptake pathways, *e.g.*, passive diffuse, clathrin-mediated endocytosis, and caveolae/lipid-raft-dependent endocytosis, in the internalization of diwalled *'molecular umbrella'* (Ge *et al.*, 2009).

The persulphated *'molecular umbrella'* derived from one spermine, four lysine, and eight deoxycholic acid molecules, could exhibit ionophoric activity as shown by pH discharge and Na(+) and Cl(−) transport experiments (Chen *et al.*, 2009). On the other hand, a moderately more hydrophilic analogue derived from cholic acid displayed no such ionophoric property. Both *'molecular umbrellas'* crossed liposomal membranes by passive transport. It was concluded that the interactions between such amphomorphic molecules and phospholipid bilayers were a sensitive function of the molecular umbrella's hydrophilic/lipophilic balance.

6.2 Biological and Pharmacological Functions of Steroid Dimers: Drug Discovery and Design

Steroids are biologically active molecules and show various pharmacological actions. As a consequence, steroidal dimers are generally expected to retain, enhance or reduce those bioactive properties. In addition to any inherent biological and pharmacological properties offered by the monomeric steroid, dimerization tends to render new bioactive potency, reduced toxicity, or improved therapeutic value to the steroid dimers. Various steroid dimers exhibit several biological and pharmacological properties, namely, anti-malarial property, cytotoxicity (anticancer), serum cholesterol lowering effect, and insect moulting. Some of these properties could potentially be exploited for new drug discovery and design.

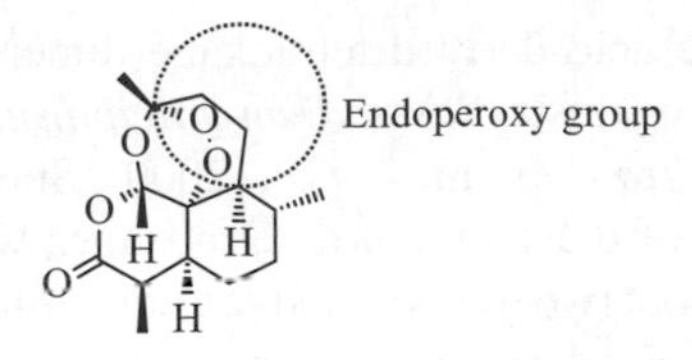

Figure 6.2.1 *Artemisinin, an antimalarial drug that contains an endoperoxy bridge*

6.2.1 Antimalarial Activity

Endemic malaria continues to be a major threat globally, affecting more than 40% of the global population, causing an estimated annual mortality of 1.5–2.7 million people. According to an estimate of The World Health Organization (WHO), about 90% of these deaths take place in sub-Saharan Africa among infants under the age of five. Because of the emergence of chloroquine-resistant strains of *Plasmodium falciparum,* it has become paramount to develop newer and more effective antimalarial drugs.

Since the discovery of the antimalarial molecule artemisinin (*Figure 6.2.1*) from the Chinese medicinal plant *Artemisia annua* over 25 years ago (Klayman, 1985), much efforts have been given to the synthesis of molecules bearing peroxy bridge within a ring system (endoperoxy functionality), as seen in the artemisinin molecule, for the potential development of further novel antimalarial agents.

With ever-increasing expansion in chemistry of antimalarials possessing ether peroxide moieties, *e.g.*, 1,2,4-trioxane rings, the 1,2,3,4-tetraoxane (1,2,4,5-tetraoxacyclohexane) systems have also been considered as possible oxygen donors in fighting against malaria. In fact, a trioxane or tetraoxane moiety has been considered to be an interesting pharmacophore for potential antimalarial activity since its antimalarial activity was found to be similar to that of 1,2,4-trioxanes, such as naturally occurring artemisinin, its semisynthetic derivatives, and related compounds. Theoretically, the design of an effective 1,2,4,5-tetraoxane-based antimalarial drug appears to address several problems, notably, the hydrophilic:lipophilic ratio of the carrier molecule, stability under physiological conditions, and the cytotoxic: antimalarial activity ratio of the investigated tetraoxane and of possible metabolites.

Bearing the above facts in mind, 5α- and 5β-cholestane-derived *gem*-dihydroperoxides and tetraoxanes were synthesized (see *Chapter 2*) and their potential antimalarial activity was studied (Todorović *et al.*, 1996). One of the these molecules, tetraoxane dimer **1**, displayed modest antimalarial activity towards mefloquine resistant *Plasmodium falciparum* clone designated Sierra Leone (D6) with an IC_{50} value of 155 nM (IC_{50} of artemisinin was 5.3 nM).

Tetraoxane dimer **1**

Later, several other cholic acid-derived tetraoxane dimers **2–9** were synthesized and evaluated for antimalarial (against *Plasmodium falciparum* D6 and W2 clones) and antiproliferative activity *in vitro* (Opsenica *et al*., 2000). Starting with chiral ketones, *cis* (**2–5**) and *trans* series (**6–9**) of diastereomeric tetraoxanes were synthesized, and the *cis* series (**2–5**) appeared to be about two times as active as the *trans* (**6–9**) against both clones of *P. falciparum*. In all those dimers, the position of the tetraoxane ring was exactly the same as in **1**. In addition to the central tetraoxane pharmacophore, the presence of the amide functionality appeared to be significant in enhancing antimalarial potency *in vitro* as compared to methyl esters and carboxylic acid moieties within the series of the congeners. For example, the antimalarial activity of bis(*N*-propylamide) (**2**) on the *P. falciparum* D6 clone was as potent as artemisinin, and the IC_{50} value of **2** was 9.29 nM as opposed to 8.6 nm for artemisinin.

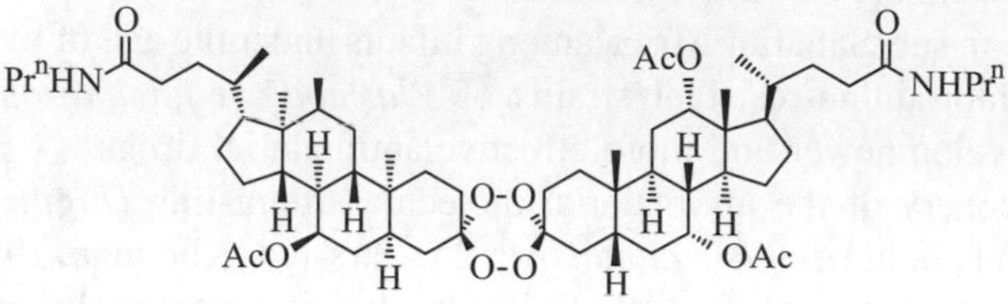

Tetraoxane dimer **2**

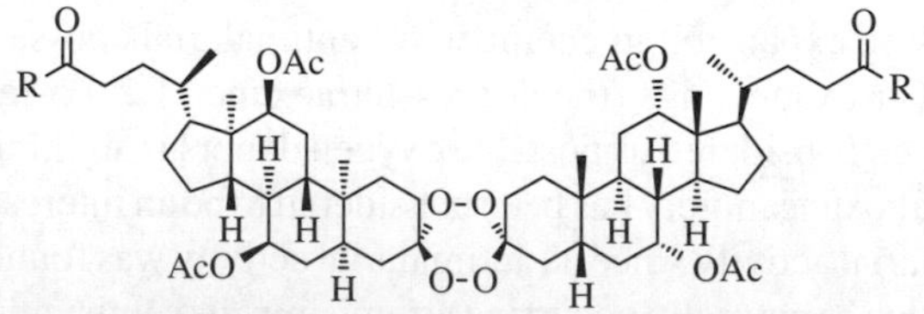

3 R = OMe **4** R = NH_2 **5** R = OH

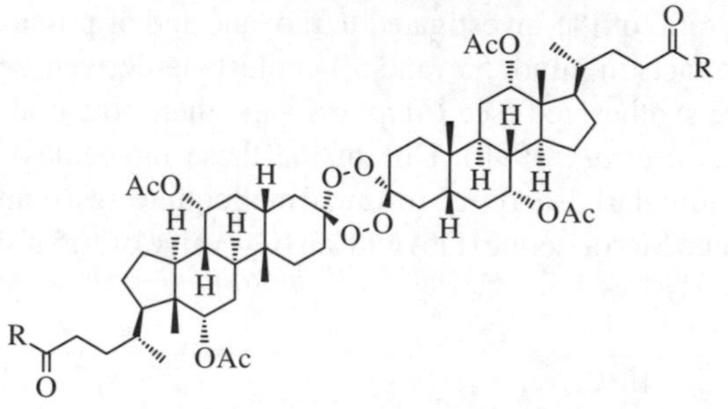

6 R = $NHPr^n$ **7** R = OMe **8** R = NH **9** R = OH

The tetraoxane dimer esters **3** and **7**, as well as the corresponding acids **5** and **9**, did not display any activity against *P. falciparum* D6 and W2 clones at test concentrations. The amide moiety in compounds **2**, **4**, **6** and **8**, introduced as the auxiliary functional group in a tetraoxane molecule, increased the antimalarial potency of tetraoxanes (Opsenica *et al.*, 2000). While the tetraoxanes with primary amides (**4** and **8**) were somewhat more active against the *P. falciparum* chloroquine-resistant W2 clone, *N*-propylamides (**2** and **6**) exhibited more antimalarial potency against the chloroquine-susceptible D6 clone.

Among these tetraoxanes (**2–9**), dimer **2** exhibited a low level of cytotoxicity [IC_{50} = > 100 μM against normal human peripheral blood mononuclear cells (PBMC)], but remarkable antimalarial potency. Thus, it is reasonable to assume that this compound (**2**) could be used as a template for the design and synthesis of newer and better antimalarial drugs.

The potential antimalarial activity of two other similar tetraoxane steroid dimers (**10** and **11**), together with dimers **3–5** and **7–9**, were also assessed against two different strains of *P. falciparum*, D6 and W2 (Bhattacharjee *et al.*, 2005) (*Table 6.2.1*).

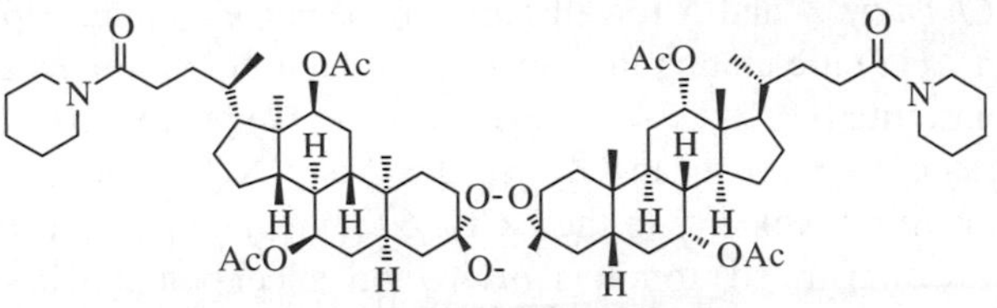

Tetraoxane **10**

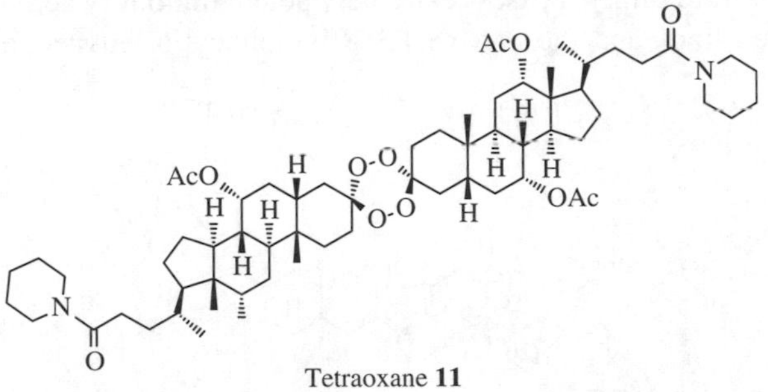

Tetraoxane **11**

The activity of steroidal tetraoxanes appeared to be primarily shape dependant. A clear difference between the activity of the *cis* (**3–5** and **10**) and the *trans* (**7–9** and **11**) isomers

Table 6.2.1 In vitro *antimalarial activity of tetraoxane steroid dimers* ***3–5*** *and* ***7–11*** *against* P. falciparum *D6 (mefloquine-resistant) and W2 (chloroquine-resistant) strains (Bhattacharjee et al.,* 2005)

Dimers	IC_{50} in ng/mL		Dimers	IC_{50} in ng/mL	
	D6	W2		D6	W2
3	>99.97	>96	**8**	128.58	59.35
4	23.74	18.79	**9**	> 100.31	>99
5	> 100.31	>99	**10**	885.26	417.94
7	>99.97	>96	**11**	254.05	124.62

was observed with these bis-steroidal tetraoxanes. From the studies with the tetraoxane steroidal dimers, it was noted that complex molecules like steroids could be excellent carriers of the tetraoxane pharmacophore. In fact, the introduction of cholic acid derived tetraoxane carrier resulted in enhanced antimalarial activity, as compared to simple tetraoxanes, and very low toxicity. It was also observed that replacing cholestane with derivatives of cholic acid could significantly improve the antimalarial potential of such compounds.

6.2.2 Cytotoxicity and Anticancer Potential

Most of the naturally occurring steroid dimers, *e.g.*, cephalostatins (*Figure 6.2.2*), crellastatins and ritterazines (*Figure 6.2.3*), and their structural analogues are well-known cytotoxins (see *Chapters 4* and *5* for all structures). For example, cephalostatin 1 (**12**) and cephalostatin 7 (**13**) are among the most potent natural cytotoxins ever tested by the National Cancer Institute (NCI), USA. Cephalostatin 1 (**12**) proved to be the most active of the 19 cephalostatins (~1 nM mean GI_{50} in the 2-day NCI-60 screen). In addition to subnanomolar *in vitro* cytotoxicity in the NCI's 60 human cancer cell line panel, cephalostatins have also demonstrated prominent *in vivo* anticancer potential in several xenografts including melanoma, sarcoma, brain tumour, in leukaemia and in a human mammary carcinoma model (Muller *et al.*, 2005).

These natural steroid dimers possess extremely potent inhibitory activity against several human cancer cell lines and the murine P388 lymphocytic leukaemia cell line (Nahar

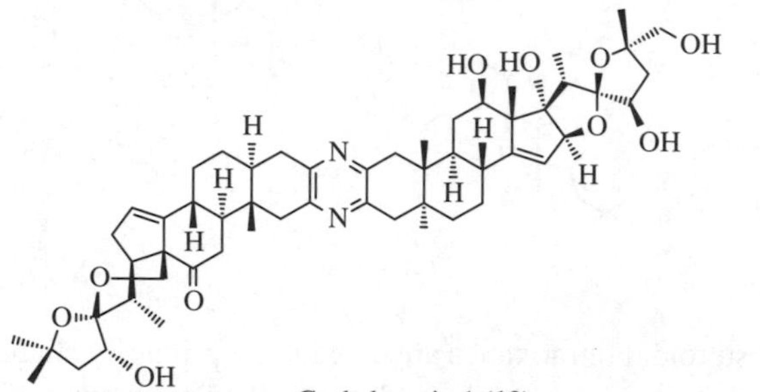

Cephalostatin 1 (**12**)

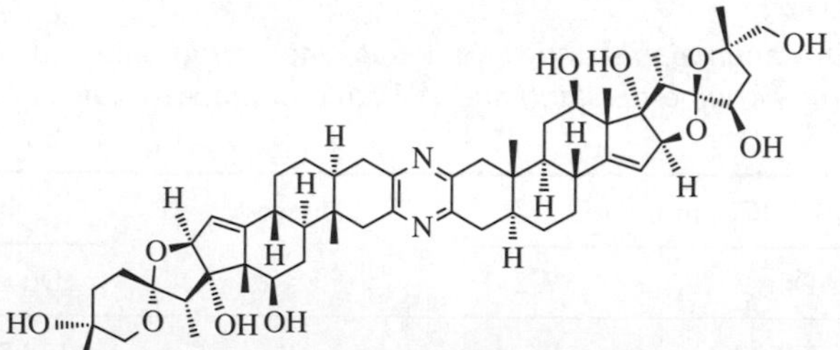

Cephalostatin 7 (**13**)

Figure 6.2.2 *Cephalostatins 1 (**12**) and 7 (**13**)*

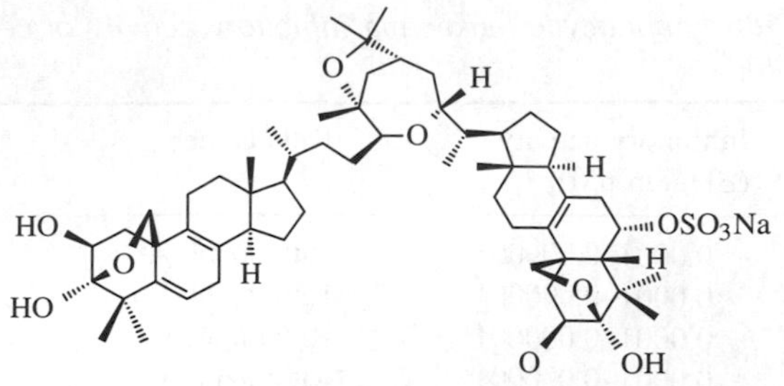

Crellastatin A (**14**)

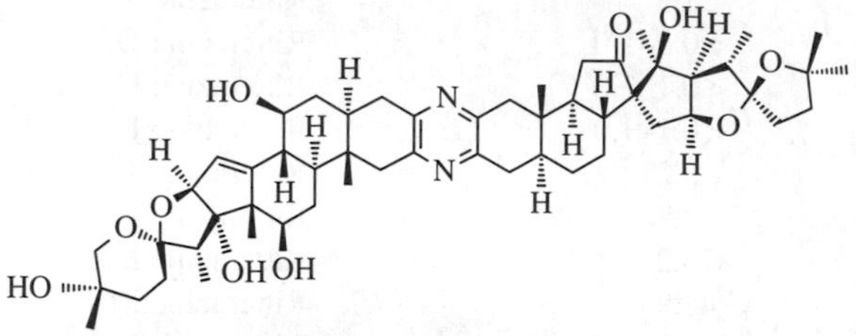

Ritterazine A (**5**)

Figure 6.2.3 *Crellastatin A (**14**) and ritterazine A (**15**)*

et al., 2007). The murine P388 lymphocytic leukaemia inhibitory activities of naturally occurring cephalostatins and ritterazines are presented in *Table 6.2.2* (Moser, 2008). Because of prominent cytotoxicity towards cancer cell lines, these compounds have the potential of being utilized as templates for antineoplastic drug discovery and design.

Cephalostatins comprise an important series of cancer cell growth inhibitors. The cytotoxic mechanism (s) of all 19 cephalostatins is thought to be the same and possibly involves novel molecular target (s). Cephalostatin 1 (**12**) is the most potent inhibitor of the family against the P388 cell line and 400 times more cytotoxic than the well known nonsteroidal anticancer agent Taxol® (Nahar *et al.*, 2007). Thorough investigation of the apoptotic pathways induced by **12** revealed some unusual features (von Schwarzenberg and Vollmar, 2010).

In leukaemic cells, cephalostatin 1 (**12**) could induce a unique apoptotic signalling pathway that activates caspase-9 independently of an apoptosome, because neither the release of cytochrome *c* from the mitochondrial intermembrane space nor an interaction of apoptotic protease-activating factor 1 with caspase-9 was detected (Muller *et al.*, 2005). The protein Smac/DIABLO was selectively and significantly released from mitochondria in response to cephalostatin 1 (**12**) treatment. This interesting finding was not specific to leukemic cells, but other cell lines, *e.g.*, SKMel-5 (melanoma) and MCF-7 (mammary carcinoma) cells were found to respond in a similar way to chephalostatin 1 (**12**) treatment (von Schwarzenberg and Vollmar, 2010). It was suggested that **12** may be able to trigger a signal directly influencing the mitochondrial pore formation.

Smac is known to promote apoptosis in response to various apoptosis inducers, *e.g.*, cephalostatin 1 (**12**), by antagonizing intracisternal A-particle ***(IAP)-mediated*** inhibition of caspases (von Schwarzenberg and Vollmar, 2010), and Smac agonists sensitized

Table 6.2.2 *Murine P388 lymphocytic leukaemia inhibitory activity of cephallostatins and ritterazines (Moser, 2008)*

Cephalostatins	Inhibitory activity (ED_{50} in nM)	Ritterazines	Inhibitory activity (ED_{50} in nM)
Cephalostatin 1	0.0001–0.000001	Ritterazine A	14.2
Cephalostatin 2	0.0001–0.000001	Ritterazine B	0.17
Cephalostatin 3	0.0001–0.000001	Ritterazine C	102.3
Cephalostatin 4	0.0001–0.000001	Ritterazine D	17.5
Cephalostatin 5	42.5	Ritterazine E	3.8
Cephalostatin 6	2.3	Ritterazine F	0.81
Cephalostatin 7	<0.1–1.0	Ritterazine G	0.81
Cephalostatin 8	<0.1–1.0	Ritterazine H	17.8
Cephalostatin 9	<0.1–1.0	Ritterazine I	15.3
Cephalostatin 10	3.2	Ritterazine J	14.0
Cephalostatin 11	2.7	Ritterazine K	10.4
Cephalostatin 12	76.2	Ritterazine L	11.1
Cephalostatin 13	47.9	Ritterazine M	16.7
Cephalostatin 14	4.4	Ritterazine N	522
Cephalostatin 15	26.2	Ritterazine O	2383
Cephalostatin 16	<1.1	Ritterazine P	819
Cephalostatin 17	4.4	Ritterazine Q	657
Cephalostatin 18	4.6	Ritterazine R	2461
Cephalostatin 19	7.9	Ritterazine S	539
		Ritterazine T	522
		Ritterazine U	2341
		Ritterazine V	2341
		Ritterazine W	3631
		Ritterazine X	3404
		Ritterazine Y	4.0
		Ritterazine Z	2200

various tumour cells for cell death. Cephalostatin 1 (**12**) treatment of Smac deleted cells was found to generate a strong reduction in caspase-9 and caspase-3 activation. However, cephalostatin 1 (**12**) did not seem to induce a binding between caspase-9 and Apaf-1, which means that caspase-9 activation in response to cephalostatin 1 treatment takes place independently of the formation of an apoptosome.

It was observed that apoptosis, induced by **12**, was almost completely inhibited in caspase-9-deficient cells, whereas cells stably retransfected with full-length caspase-9 died normally when treated with cephalostatin 1 (**12**) (Lopez-Anton *et al.*, 2006). Cephalostatin 1 (**12**) was shown to induce endoplasmatic reticulum (ER)-stress as well as caspase-4 activation, suggesting an activation of caspase-9 independent from apoptosome formation. The significant reduction in caspase-2 activity triggered by **12** in Smac-depleted cells indicated that caspase-2 as a further caspase involved in the apoptosome-independent cell death induced by cephalostatin 1 (**12**). As caspase-2 is activated in caspase-9-deficient cells similarly to parental Jurkat cells, it can be assumed that caspase-2 activation could be an upstream event.

Thus, cephalostatin 1 (**12**) has been proven to be a valuable tool to discover novel aspects in apoptotic signalling pathways including apoptosome-independent activation of caspase-9, and induction of ER-stress and caspase-4 activation as well as recruitment of a PIDDosome responsible for caspase-2 activation (von Schwarzenberg and Vollmar, 2010).

Cephalostatins 2–9 have one common steroidal unit (see *Chapter 4*), but differ in the other steroidal moiety. It appeared that the differences in the second steroidal unit did not affect the powerful cytotoxicity of this group of compounds in most cases. However, cephalostatins 5 and 6 showed significantly diminished cytotoxic potency. It was also found that the presence of an aromatic C′-ring (left-side) in cephalostatins 5 and 6 significantly reduced (P388 ED_{50} ~10-2 mg/mL) the cytostatic activity (Pettit *et al.*, 1992). Cephalostatins 5 and 6 were modestly cytotoxic only against renal SNl2Kl and CNS U-251 human cell lines.

Cephalostatins 7–9 exhibited significant cytotoxicity against nonsmall-cell lung HOP 62, small cell lung DMS-273, renal RXF-393, brain U-251 and SF-295, breast MCF-7 and leukaemia CCRF-CEM, HL-60, and RPM1-8226 cell lines (Pettit *et al.*, 1992) with ED_{50} values ranging between 10^{-8} to 10^{-10} μM (Pettit *et al.*, 1992). It was suggested that the pyridizine right-side unit was essential for cytotoxicity, and minor changes in configuration and substitution pattern (including an additional methyl in cephalostatin 8) in the left-side E′- and F′-rings had little or no influence on the cytotoxic activity.

The cancer cell growth inhibitory activity and cytotoxicity of cephalostatins 7–11 against diverse solid tumours in the NCI's *in vitro* disease-oriented antitumour screen were also established (Li and Dias, 1997; Nahar *et al.*, 2007). Cephalostatins 12 and 13 were comparatively evaluated alongside cephalostatin 1 (**12**) in the same antitumour screen (Pettit *et al.*, 1994a). While these compounds showed substantial growth inhibitory activity against many of the cell lines, the potency was much lower than cephalostatin 1, and the average GI_{50} values for cephalostatins 12 and 13 were 400 and >1000 nM, respectively. It was postulated that the somewhat diminished cytotoxicity of these compounds, compared to that of cephalostatin 1 (**12**), might be due to increased level of hydroxylation, *i.e.*, increased polarity, in cephalostatins 12 and 13. Thus, relative hydrophobicity may play a pivotal role in the cytotoxicity of cephalostatin molecules.

Similarly, cephalostatins 14 and 15, when tested against the same panel of cancer cell lines as outlined above, the overall panel-averaged cytotoxic potencies were found to be somewhat less (*e.g.*, 100 nM and 68 nM, respectively) than that of the benchmark compound, cephalostatin 1 (**12**) (Pettit *et al.*, 1994b). The results of this study established that the modification on the 'left-side' moiety in cephalostatins by introduction of the 8β-hydroxy-11-ene-12-one and (or) the 14α,15α-epoxy system (α-orientation) could substantially reduce *in vitro* cyctotoxicity of these dimers.

Cephalostatins 16 and 17 were also assessed for *in vitro* anticancer properties using the NCI's human cancer cell line panel, and it was found that both cephalostatins had the characteristic cytotoxicity profile as confirmed by 'Compare' pattern recognition analyses (Pettit *et al.*, 1995). The overall panel-averaged cytotoxic potencies of cephalostatins 16 and 17 were quite promising with the GI_{50} values of 4.0 and 1.0 nM, respectively, and were comparable to that of the benchmark compound **12**.

Cephalostatins 18 and 19 exhibited remarkable cytotoxicity against the murine P388 lymphocytic leukaemia cell line (ED_{50}*ca.* 10^{-3} μg/mL), a minipanel of human cancer cell lines including OVCAR-3, SF-295, A-498, NCI-H460, KM-20L2 and SK-MEL-5 ($GI_{50} <$ 10^{-3} μg/mL) and the NCI's 60 human cancer cell line panel (mean panel GI_{50} *ca.* 10^{-9} M) (Pettit *et al.*, 1998; Nahar *et al.*, 2007). The overall cytotoxic potency of Cephalostatins 18 and 19 was about 8 to 10 fold lower than that of cephalostatin 6.

Synthetic cephalostatin analogues (see *Chapter 5*) that lack in $\Delta^{14,15}$, for example, dimer **16**, appeared not to have any cytostatic activity in preliminary tests. It is assumed that the unusual double bond $\Delta^{14,15}$, which is present in naturally occurring cephalostatins, and the polar spiroketal functionality might be essential for cytostatic properties of cephalostatins and their synthetic analogues.

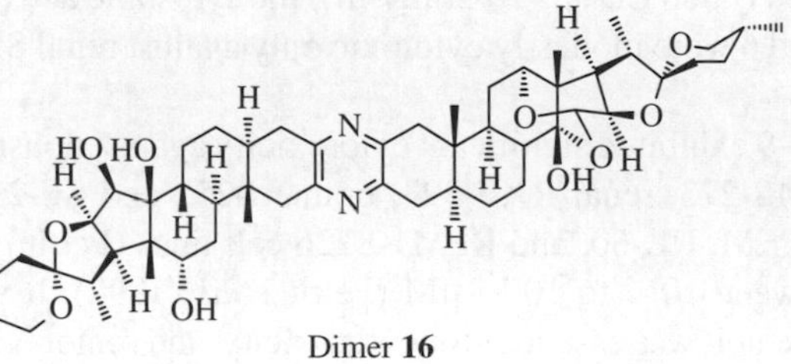

Dimer **16**

It has also been established that unsymmetric $\Delta^{14,15}$-cephalostatin analogues, natural or synthetic, generally possess a more prominent cytotoxic/cytostatic property than that of the symmetric $\Delta^{14,15}$ analogues. Therefore, in designing cytotoxic cephalostatin analogues for future anticancer drug development, it is important to introduce unsymmetry into the analogues.

Bearing this observation in mind, Poza *et al.* (2010) have recently synthesized a new unsymmetrical bis-steroidal pyrazine analogue (**17**) from the commercially available hecogenin acetate and 22-*epi*-hippuristanol, and evaluated the cytotoxic potential *in vitro* against MDA-MB-231, A-549 and HT-29 tumour cells. An increase in cytotoxicity in **17** compared to that of 22-*epi*-hippuristanol was observed. Dimer **17** shares a similar structural framework to ritterazines B, F-I, Y and cephalostatin 7 (see *Chapter 4*), differing in the presence of an additional methyl group at C-24, the absence of $\Delta^{14,15}$ in the left-hand part of the molecule, and the oxidation positions. Although dimer **17** shows molecular unsymmetry, which was considered to be a prerequisite for tumour inhibition, its moderate cytotoxicity (at μM level) in relation to the former ritterazines (at nM level) indicated that the appropriate oxidation positions might be essential for exceptional potency.

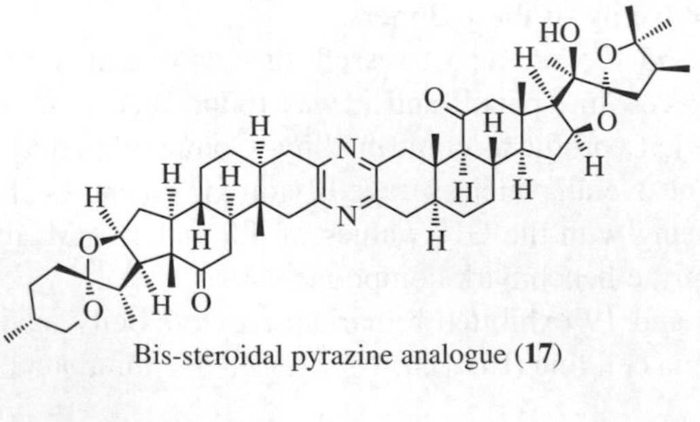

Bis-steroidal pyrazine analogue (**17**)

The structure–activity relationships (SARs) data available from the growing family of cephalostatins have shown the importance of a 17-OH group for strong antitumour activity (Nahar *et al*., 2007). For example, cephalostatins 1 (**12**) and 7 (**13**), both having 17-OH group, possess the most potent antitumour activity among all cephalostatins. It has also been shown that any additional methoxylations or hydroxylations on the steroidal A ring core structure seem to decrease cytotoxicity.

The ritterazines, which are structurally closely related to the cephalostatins, were reported by Fukuzawa *et al.* during 1994–1997 from *Ritterela tokioka* (Nahar *et al.*,2007). Ritterazines A-Z (see *Chapter 4*) show *in vitro* cytotoxicity, however, less potent than that of cephalostatins 1–19, against several human cancer or tumour cell lines including lung A-549, breast 1–3 type, colon HT-29, renal A-498, prostate PC-3 and bladder PACA-2. Like cephalostatins and their analogues, naturally occurring ritterazines and their synthetic analogues were also screened in the NCI's 60 human cancer cell line panel. The cytotoxic potentials of ritterazines A-Z against P388 murine leukaemia cells are summarized in *Table 6.2.2*. Reitterazine A (**15**), the first member of this group of cytotoxic steroidal dimers, exhibited cytotoxicity against the P388 murine leukaemia cell line with an IC_{50} value of 3.8×10^{-3} μg/mL (Fukuzawa *et al*., 1994).

Ritterazine B (**18**), which is structurally closely related to ritterazines F and G, was found to be the most potent cytotoxic agent among the natural ritterazines (for ritterazine structures see Chapters *4* and *5*) (Fukuzawa *et al*., 1995a). The IC_{50} value of ritterazine B (**18**) against the P388 murine leukaemia cell line was 0.018 ng/mL. It is assumed that both ritterazines and cephalostatins follow same or an extremely similar mechanism of action to exert their cytotoxicity.

Cephalostatin 1 (**12**), ritterazine B (**18**) and some of their derivatives are more cytotoxic than the synthetic analogue ritterostatin G_N7_S (7′-deoxyritterazine G, **19**), which itself has better activity than that of cephalostatin 7 (**13**) and ritterazine K.

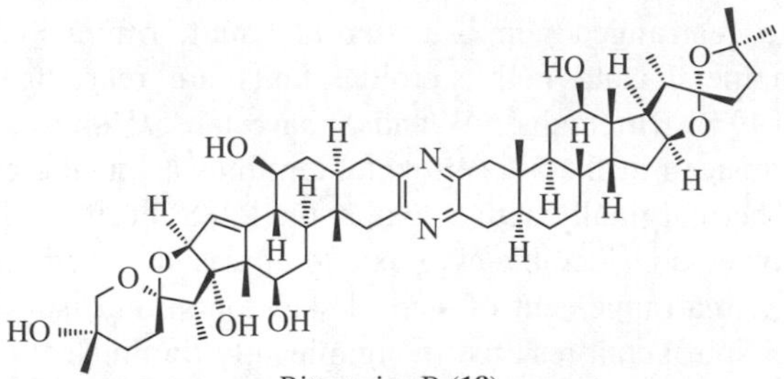

Ritterazine B (**18**)

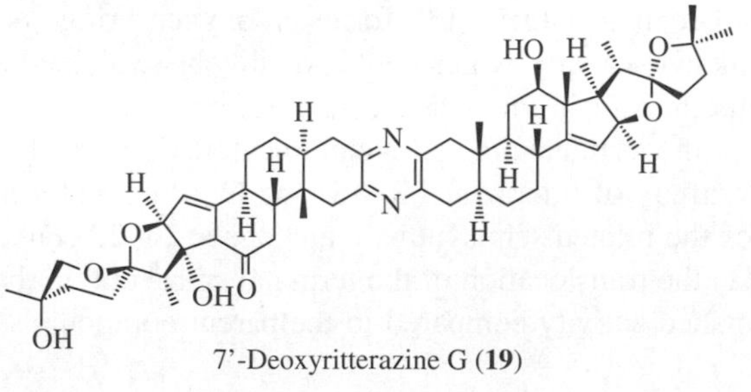

7'-Deoxyritterazine G (**19**)

The ritterazines and cephalostatins share many common structural features in which two highly oxygenated C_{27} steroidal units are fused *via* a pyrazine ring at C-2 and C-3 and both chains of the steroidal units usually form either 5/5 or 5/6 spiroketals (Fukuzawa *et al.*, 1995a). The cephalostatins in general are more oxygenated on the right side, and the ritterazines have more oxygenations on the left side. Hydroxyl groups are generally seen at C-12, C-17, C-23, C-26, C-12′, and C-23′ in the cephalostatins, whereas C-12, C-7′, C-12′, C-17′, and C-25′ are hydroxylated in the ritterazines. Despite notable structural similarities, the cephalostatins in general displayed more potent cancer cell growth inhibitory activity (P388) than the ritterazines (*Table 6.2.2*).

Ritterazines B (**18**), F and G (see *Chapter 4*), which have 5/5 and 5/6 spiroketal rings were most cytotoxic irrespective of C-22 stereochemistry or presence of Δ^{14} olefin, while oxidation of the 12-OH to a ketone (ritterazines H and I) rendered a considerable decrease in cytotoxic activity (Fukuzawa *et al*, 1995b). Ritterazines A (**15**), D, and E, which have rearranged steroid skeletons, exhibited similar activity as the 12-keto derivatives. In those dimers, stereochemistry at C-22 or methylation at C-24 did not affect cytotoxic potency (*Table 6.2.2*). Ritterazine J, where C-26 is oxidized and accordingly possesses two 5/6 spiroketal groups, was found to be less active than ritterazine F. Since ritterazines J–M displayed similar activity, the hydroxyl groups at C-7 and C-17 seemed to have no effect their cytotoxicity.

After the discovery of ritterazines N-Z (Fukuzawa *et al.*, 1997) and their cytotoxic behaviour, a clearer and much extended picture of the structure activity relationships of ritterazines began to emerge. Ritterazines N-S (see *Chapter 4*), with two nonpolar steroidal units, were much less potent cytotoxic agents than ritterazine B (**18**). Ritterazines T–Y are structurally related to ritterazines A (**15**) and B (**18**), having polar and nonpolar steroidal units that lack 7′-OH and 17′-OH functionalities. Ritterazine Y, on the other hand, differs from ritterazine B (**18**) only in the absence of the two hydroxyl groups. Ritterazines T–X are actually the derivatives of ritterazine Y with some modifications. Ritterazine T has a rearranged nonpolar steroidal unit, ritterazine U is an oxidized analogue of ritterazine T, and both steroidal units are rearranged in ritterazine V (Fukuzawa *et al.*, 1997). Ritterazines W and X have the 5/5 spiroketal terminus in the polar steroidal unit instead of the 5/6 spiroketal terminus as in ritterazine T. Ritterazines T–X were found to be marginally active against the P388 murine cell line (*Table 6.2.2*), but ritterazine Y showed considerable cytotoxicity. It was observed that the modifications of ritterazine Y, *e.g.*, rearrangement of steroid skeleton(s) and isomerization of the 5/6 spiroketal to the 5/5 spiroketal, resulted in significantly diminished cytotoxicity. Similar modifications of ritterazine B (**18**), providing ritterazine A (**15**) and ritterazines C-M (see *Chapter 4*), also caused notable reduction in the cytotoxic potency. Ritterazine Z, which is apparently related to cephalostatin 1 (**12**), forms an oxygen bridge between C-18 and C-22, and showed weak cytotoxicity, which could possibly be explained from the presence of the rearranged nonpolar unit in the 'Eastern hemisphere'.

The importance of the terminal 5/6 spiroketal to cytotoxicity could be evident from the comparison of the activity of ritterazines B (**18**) and C, which is further supported by the cytotoxic activity of the related semisynthetic analogues **20**–**22** (Fukuzawa *et al.*, 1997), where (in **20** and **21**) the translocation of the terminal spiroketal in the polar steroid units considerably diminished activity compared to the parent compounds, ritterazines A (**15**) and B (**18**).

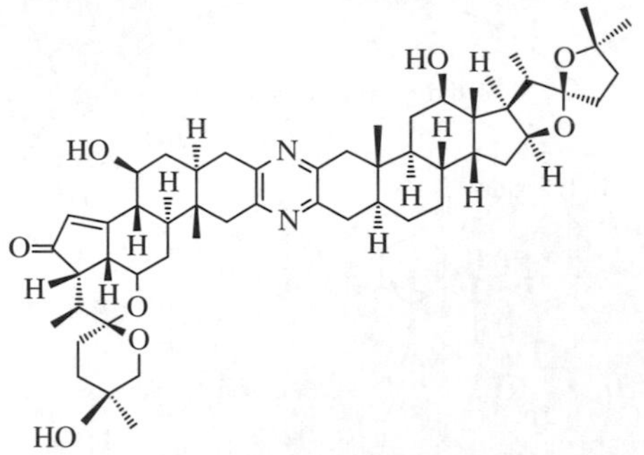

Ritterazine analogue **20**

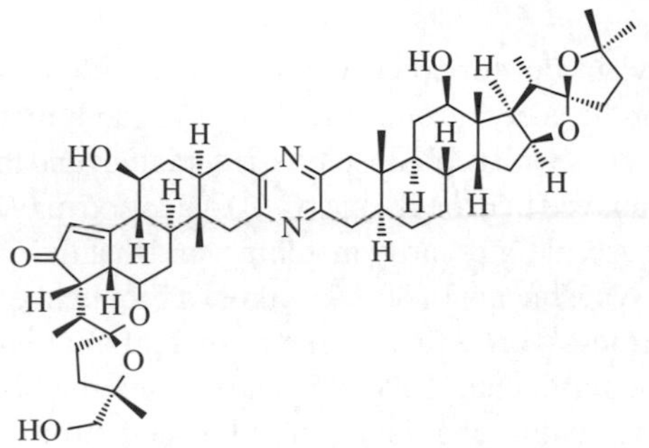

Ritterazine analogue **21**

It appeared that the spatial arrangement of the 5/6 spiroketal with respect to the rest of the skeleton was particularly important for the potent cytotoxic activity of ritterazines and their analogues. Another analogue **22**, which retains the ring E′, was weakly active, yet 10 times more potent than analogues **20** and **21**. The presence of the 5/6 spiroketal at the right position in the polar steroid unit is an essential feature, but alone it is not adequate for providing potent cytotoxic property. Symmetric or nearly symmetric ritterazines (ritterazines J, K, L, and M) having two polar steroidal units with 5/6 spiroketal were found to be 100 times less active than ritterazine B (Fukuzawa *et al.*, 1997).

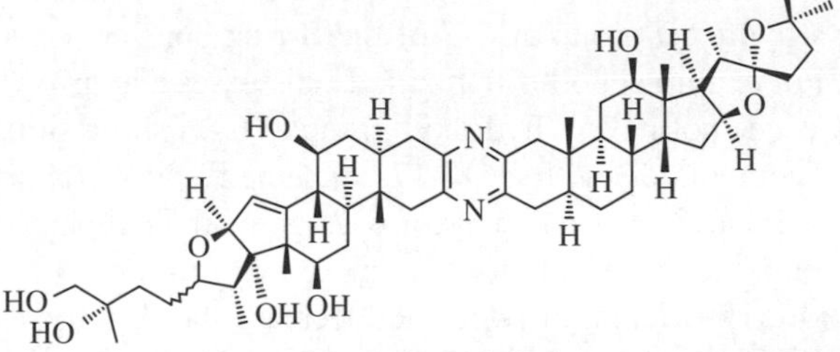

Ritterazine analogue **22**

The oxidation of the C-12 alcohol to a ketone (ritterazine B to H) was found to decrease the cytotoxicity, and similar was the case for the oxidation of C-7′. Introduction of one or more acetyl groups also reduced the cytotoxic potential, indicating the importance of all

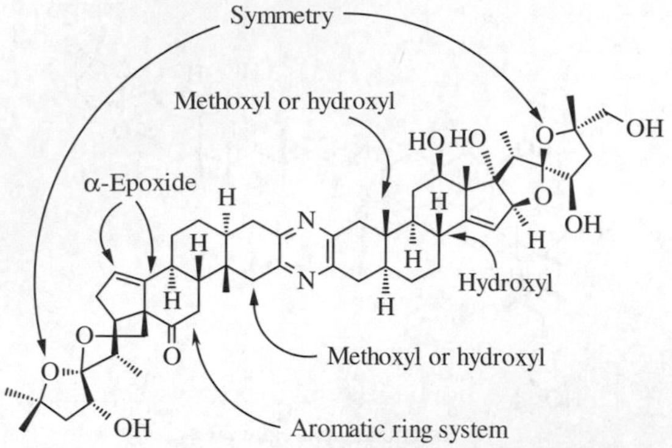

Figure 6.2.4 *Key features that reduce the cytotoxic potency of cephalostatins*

three secondary hydroxyl groups for potent cytotoxicity of ritterazines. It was observed that the higher the number of the acetyl groups the weaker the activity (Fukuzawa *et al.*, 1997).

Based on the SARs studies with cephalostatins, ritterazines and their synthetic analogues, carried out to date, it is believed that thc B-ring (C-7′) oxidation may not be necessary for high *in vitro* cytotoxicity, but useful for polarity matching, and that the C-12′ oxidation state may vary in a unique subunit type, but not in others without affecting the cytotoxic potency. It has also been postulated that the Δ^{14} or *cis* fusion in some subunits may enhance reactivity of some functionalities such as the spiroketals. Like in cephalostatins, the 17-OH group in ritterazines is also a crucial structural feature that is required for their strong antitumour activity. For example, ritterazine A (**15**), ritterazine B (**18**), ritterazine T and ritterazine Y which do not have any 17-OH group, generally display a 25- to 330-fold drop of tumour inhibitory property.

From the data available on various SARs studies on cephalostatins, ritterazines and their synthetic analogues, it is possible to comment on the structural features that generally tend to decrease the cytotoxicity of these compounds. These features are summarized below (*Figure 6.2.4*).

i. Additional methoxyl or hydroxyl functionality at C-1 on the steroidal A ring core structure. For example, the cytotoxicity of cephalostatins 10, 11, 18 and 19, which possess a hydroxyl/methoxyl group at C-1, are weaker than cephalostatin 1.
ii. Additional hydroxylation at C-9 on the B-ring of the steroidal unit. For example, cephalostatin 4, which has a hydroxyl group at C-9, is less cytotoxic than cephalostatin 1. However, a hydroxyl group at C-7 of the B-ring does not have any effect on the cytotoxicity. For example, both ritterazine J (possesses a hydroxyl group at C-7) and ritterazine K (does not have a hydroxyl group at C-7) show similar cytotoxicity.
iii. Aryl C′-ring compounds with a 12,17-connected spiroketal area. For example, cephalostatins 5 and 6 are weaker cytotoxic agents than cephalostatin 1.
iv. An 14α,15α-epoxide (α-orientation) as in cephalostatins 14 and 15. However, a 14β,15β-epoxide (β-orientation) does not decrease the activity, *e.g.* cephalostatin 4.
v. Loss of both 17-OH groups, and the 7-OH group. Example: ritterazines A and T, and ritterazines B and Y.
vi. Symmetry or almost symmetry. Cephalostatins and ritterazines that are symmetric, *e.g.*, cephalostatin 12, ritterazine K, ritterazine N and ritterazine R, and that are almost symmetric, *e.g.*, cephalostatin 13, ritterazine J, ritterazine L, ritterazine M,

ritterazine O and ritterazine S show diminished cytotoxicity. However, some of the symmetrical compounds, *e.g.*, ritterazine K (average $ED_{50} = 96$ nM in the NCI's 60 cell line panel), still show strong cytostatic properties.

The following structural features in cephalostatins and ritterazines are essential for potent cytotoxic activity.

i. Among the cephalostatins, the most potent compounds possess a 12-keto-12'-ol function, while in the ritterazine series the direct comparison of ritterazine B and ritterazine H favours the 12,12'-diol feature.
ii. The free hydroxyl group at C-12 in ritterazines.
iii. At least one $\Delta^{14,15}$ double bond. Ritterazines that lack this feature tend to show extremely low levels of cytotoxicity.
iv. At least one 17-OH group. Loss of one out of two 17-OH does not decrease activity (ritterazine K *vs.* ritterazine L), but of the second 17-OH (along with the 7-OH) as seen in the ritterazine series leads to a significant decrease in activity.
v. Asymmetry. All highly active cephalostatins and ritterazines are substantially asymmetric.
vi. Polarity match. One steroidal unit must be substantially more polar than the other, *e.g.*, cephalostatin 1 and ritterazine B.

Crellastatin A (**14**), an asymmetric bis-steroid sulphate, showed *in vitro* cytotoxic activity against NSCLC–N6 (human bronchopulmonary nonsmall-cell lung carcinoma) cells ($IC_{50} = 1.5$ μg/mL) as well as on its clones, C65 (IC_{50} 4.4 μg/mL), C92 (IC_{50} 6.3 μg/mL) and C98 (IC_{50} 9.2 μg/mL) (D'Auria *et al.*, 1998) (*Table 6.2.3*). The weaker activities exhibited by the corresponding 2,2′-diacetylated (IC_{50} 9.1 μg/mL) and desulphated (IC_{50} 9.3 μg/mL) derivatives suggested that the 2,2′-hydroxyl functionality as well as the sulfate groups are essential for cytotoxic activity of crellastatin A (**14**).

Although the hydroxyl and the sulphate groups appeared to be essential for the cytotoxic property of crellastatin A (**14**), when evaluated the cytotoxicities of crellastatins A–H (see *Chapter 4*) against various clones of NSCLC tumour cells, *e.g.*, C15, C65, C92 and C98, neither the sulphate groups nor the hydroxylation pattern at ring A, appeared to have any noticeable effect on the cytotoxicity of these compounds (Zampella *et al.*, 1999). It was concluded that the cytotoxicity of crellastatins could not be attributed to any specific functionalities present, but to the steroidal and dimeric nature of these compounds. The IC_{50} values of crellastatins A–H against the NSCLC were 1.52, 1.20, 3.40, 5.94, 9.87, 2.70, 2.50 and 2.30 μg/mL (*Table 6.2.3*).

Similarly, crellastatins I–M exhibited cytotoxicity against the above cell lines with IC_{50} values ranging from 1.0 to 8.0 μg/mL (Gianini *et al.*, 1999). Among these compounds,

Table 6.2.3 *Cytotoxic activity of crellastatins A–M against the NSCLC cell line*

Crellastatins	IC_{50} (μg/mL)	Crellastatins	IC_{50} (μg/mL)	Crellastatins	IC_{50} (μg/mL)
A	1.52	F	2.70	K	3.70
B	1.20	G	2.50	L	2.90
C	3.40	H	2.30	M	1.10
D	5.94	I	1.90		
E	9.87	J	7.60		

crellastatin M was the most active one ($IC_{50} = 1.10$ μg/mL), whereas crellastatin E (**23**) was the least cytotoxic ($IC_{50} = 9.87$ μg/mL) (*Table 6.2.3*).

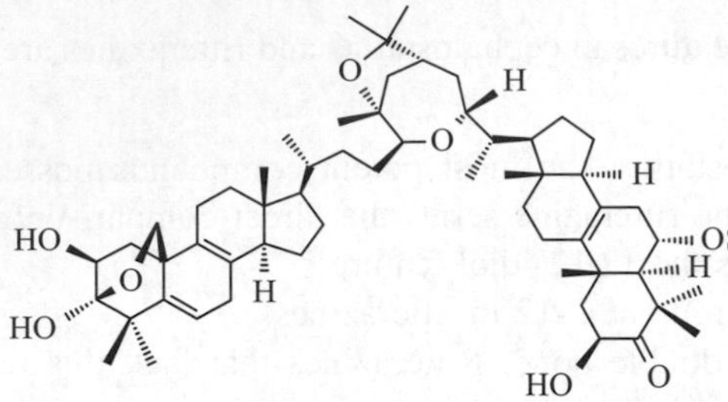

Crellastatin E (**23**)

Some cholic acid-based tetraoxanes, *e.g.*, **1–11**, initially synthesized as potential antimalarial agents, as discussed earlier, were also found to possess significant antiproliferative activity at micromolar concentrations against human melanoma Fem-X and human cervix carcinoma HeLa cell lines (Opsenica *et al.*, 2000; Nahar *et al.*, 2007). The activity of these tetraoxanes (**1–10**) seems to depend on the types of functionalities present on the cholic acid side chain.

Tetraoxanes with primary amide functionality on the cholic acid side chain, *e.g.*, in **4** and **8**, have more potent antiproliferative activity than the *N*-propylamide congeners (**2** and **6**) (Opsenica *et al.*, 2000. All these tetraoxanes were also screened in the NCI's 60 human cancer cell line panel for growth inhibitory activity. The antiproliferative activities of some of these tetraoxanes were quite similar to that of *cis*-platinum. Tetraoxane **4** seems to be active only against the renal cancer cell line UO-31 out of 60 cell lines tested, with a GI_{50} value of 0.10 μM.

A series of C_2-symmetric 17β-estradiol homodimers **25–30**, which are linked at position 17α of 17β-estradiol (**24**) with either an alkyl chain or a polyethylene glycol chain, were evaluated for potential cytotoxicities against two human breast tumour cell lines using the Sulforhodamine B (SRB) colourimetric assay (Rabouin *et al.*, 2003; Berube *et al.*, 2006).

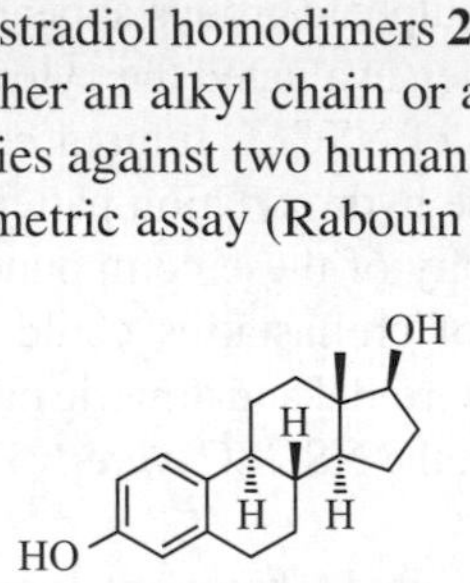

17β-Estradiol (**24**)

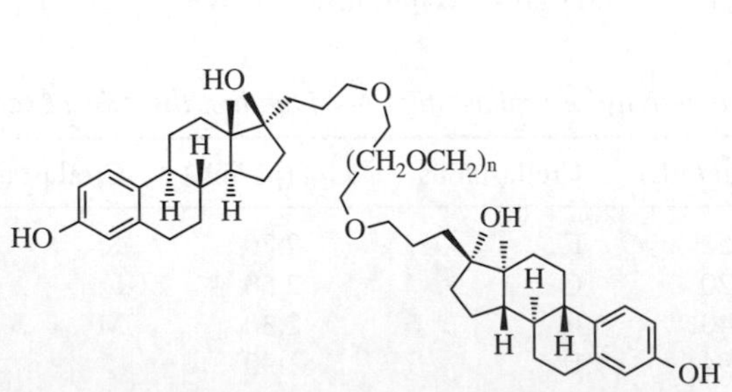

25 n = 1 **26** n = 2 **27** n = 3 **28** n = 4 **29** n = 5

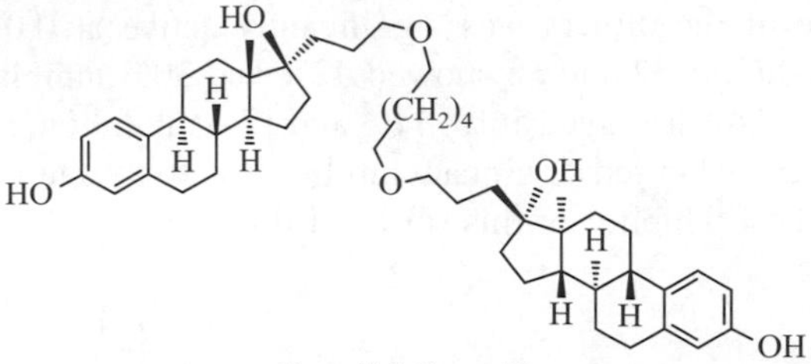

Estradiol dimer **30**

The cytotoxicity of these dimers, along with controls (17β-estradiol and tamoxifen), was tested *in vitro* against estrogen-dependent and -independent (ER^+ and ER^-) human breast tumour cell lines: MCF-7 and MDA-MB-231. As displayed by the SRB assay, dimers **26** and **27** had selective cytotoxicity against the ER^+ cell line with IC_{50} values 62 and 63 μM, respectively. The observed activity was generally less than that of the antiestrogen, tamoxifen (IC_{50} 11 μM). The lower homologue **25** and the higher homologues **28** and **29** appeared not to have any significant cytotoxicity ($IC_{50} > 100$ μM). All these dimers were generally less toxic towards breast cancer cells as compared to the cognate hormone 17β-estradiol (**24**) and the antiestrogen, tamoxifen (**31**).

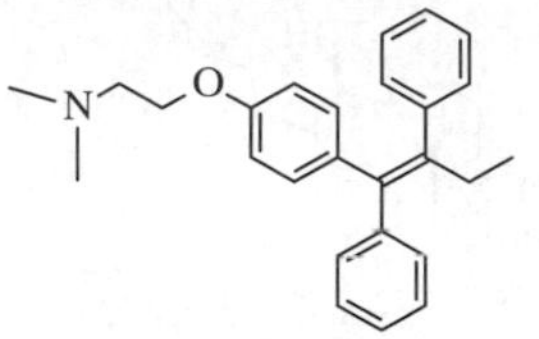

Tamoxifen (**31**)

These dimers were also tested on human intestinal cancer (HT-29) and on murine skin cancer (B16-F10) cell lines for additional comparison of activities The dimers **25–30** were essentially inactive towards intestinal HT29 cancer cell line with IC_{50} values ranging from 71 to >200 μM. However, surprisingly, all these dimers were found to be cytotoxic to the murine skin cancer cell line but inactive towards the intestinal cancer cell line with IC_{50} values ranging from 4.7 to 30 μM (Berube *et al.*, 2006). This appeared to be the first example of selective anticancer activity of 17β-estradiol dimers towards skin cancer cells. It was suggested that his unique property, if observed on human skin cancers, might in future provide an alternate treatment for skin cancer that is known to afflict numerous people globally.

Steroid sulphatase (STS) transforms inactive steroid sulphates, estrone sulphate (E1S) and dehydro-*epi*-androsterone sulphate (DHEAS) into active hydroxysteroids estrone (E1) and dehydro-*epi*-androsterone (DHEA). Because of their superior water solubility, steroid sulphates represent a transport form for steroids that can then be used in the local synthesis of active steroids. STS thus constitutes an important key enzyme in the control of intratumoural levels of active steroids. Therefore, inhibitors of steroid sulphatase can potentially lead to the development of anticancer agents.

Several estradiol dimers (**32–40**) have recently been screened for possible steroid sulphatase inhibitory activity in the transformation of estrone sulphate into estrone using homogenized HEK-293 cells overexpressing STS as the source of enzyme (Fournier and

Poirier, 2009). None of the dimers was significantly active at 0.01 μM concentration. However, at 0.1 μM, dimers **37** and **38** showed 42% and 30% inhibition, respectively. At 1 μM, dimers **37**, **38** and **40** displayed 56%, 54% and 32% inhibition, respectively. Only the C17–C17 linked dimers exhibited significant inhibitory activity against STS, and dimer **37** was found to be the best inhibitor of this series of dimers.

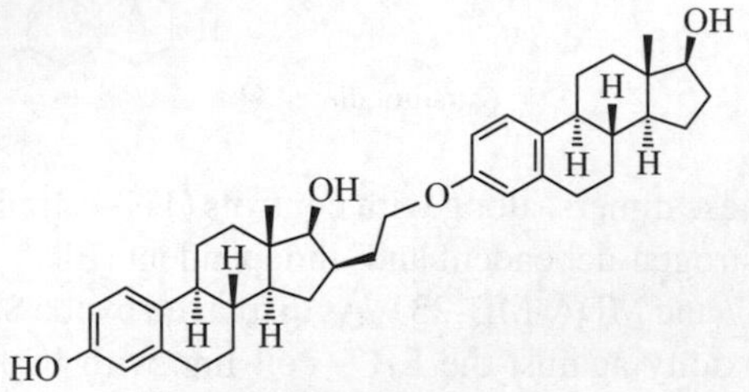

Estradiol dimer **32**

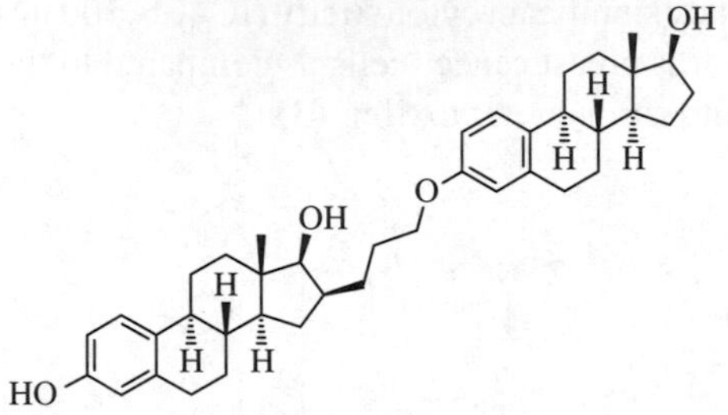

Estradiol dimer **33**

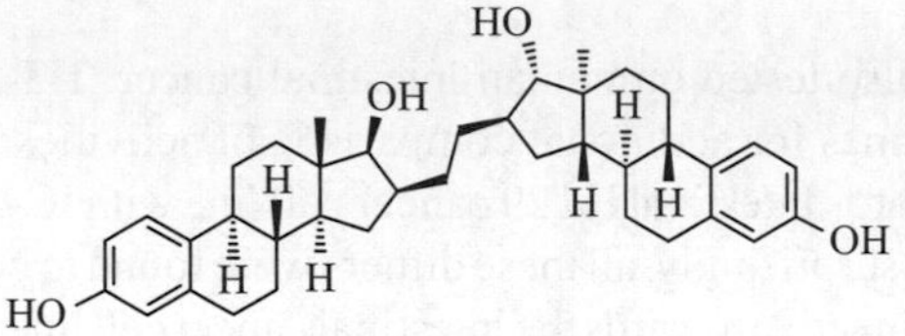

Estradiol dimer **34**

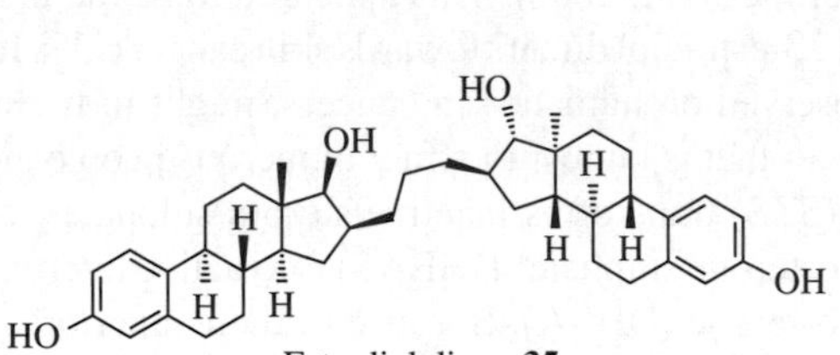

Estradiol dimer **35**

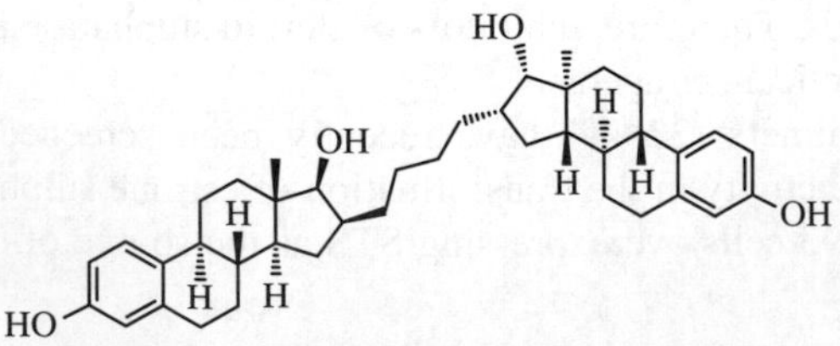

Estradiol dimer **36**

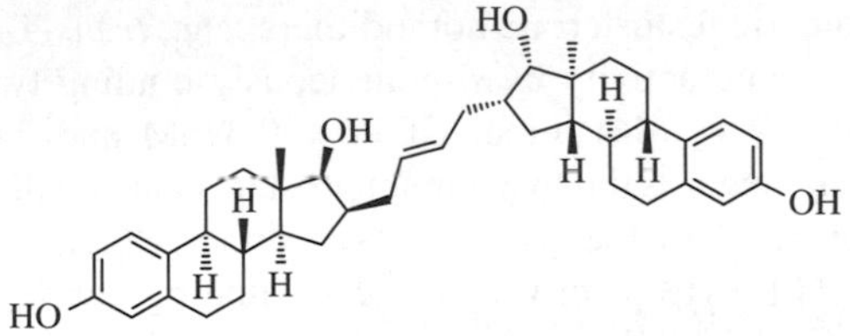

Estradiol dimer **37**

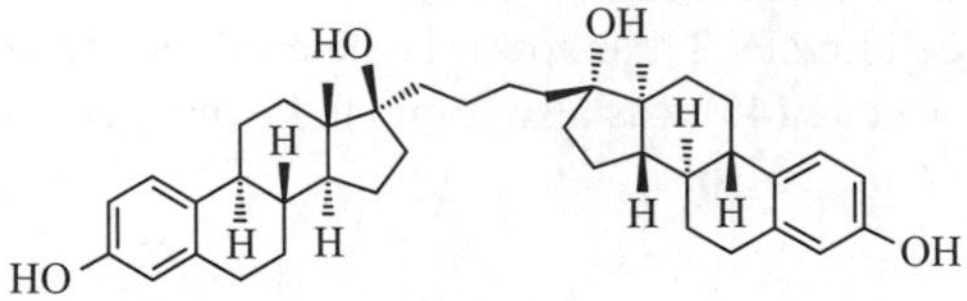

Estradiol dimer **38**

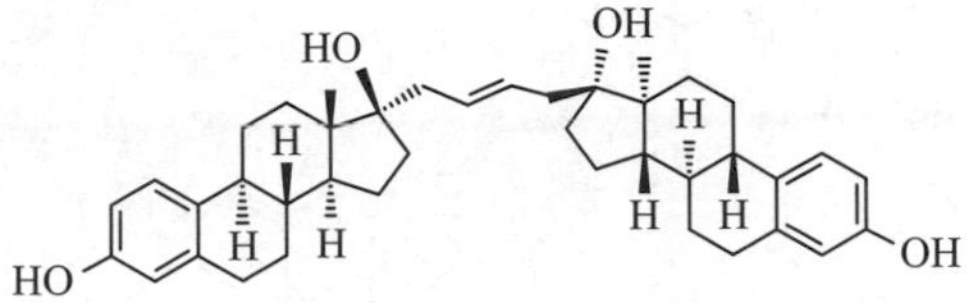

Estradiol dimer **39**

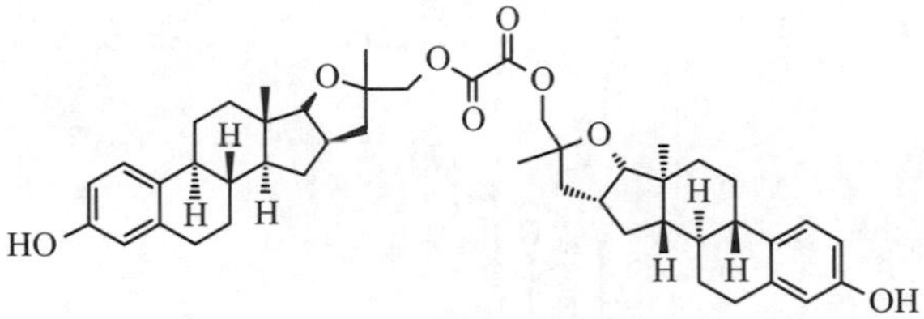

Estradiol dimer **40**

Androgens, *e.g.*, testosterone (**41**) and dihydrotestosterone (**42**) are important in the development and normal functions of prostate cells, as well as male sexual organ growth and sexual function. Testosterone (**41**) is the principal androgen in the blood while dihydrotestosterone (**42**) is the most potent androgen in the cells (Bastien *et al.*, 2010). Upon androgen stimulation, the proliferation of prostate cells is increased leading to the possible formation of a malignant tumour. The androgen receptor level is higher in prostate cancer cells compared to normal cells. Thus, androgens are implicated not only to prostate tumourigenesis, but also to hormone-dependent cancer progression, dictating the application of androgen-deprivation therapy in patients suffering from prostate cancer.

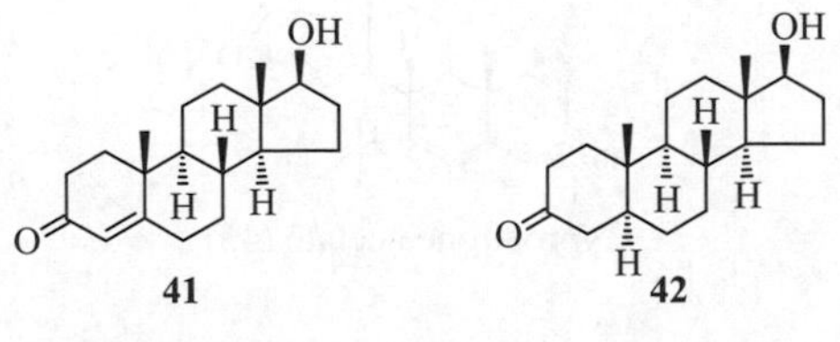

41 **42**

Testosterone (**41**) and dihydrotestosterone (**42**)

The synthesized isomeric testosterone homodimers, *e.g.*, *trans*-T_2 (**43**) and *cis*-T_2 (**44**), could exhibit antiandrogenic activity by simultaneously binding two androgen receptors (Bastien *et al.*, 2010). *cis*-T_2 (**44**) dimer (IC_{50} = 30.3 mM and 24.7 mM, respectively) displayed higher toxicity towards the two human prostate cancer cell lines, LNCaP (AR +) and PC3 (AR −), compared to the *trans*-T_2 (**43**) dimer (IC_{50} = 80 mM and 35.7 mM, respectively) in the MTT cytotoxicity assay. This finding supported the idea that the double-bond geometry of the dimer could influence its biological activity.

Dimer **44** was slightly more cytotoxic than cyproterone acetate (**45**), a clinically used steroid-based antiandrogen. Both dimers **43** and **44** were found to be more active against the hormone-independent cell line PC3 than towards the hormone-dependent cell line LNCaP. Similarly, cyproterone acetate (**45**) was also found to be more active on the PC3 cells than LNCap cells (Bastien *et al.*, 2010).

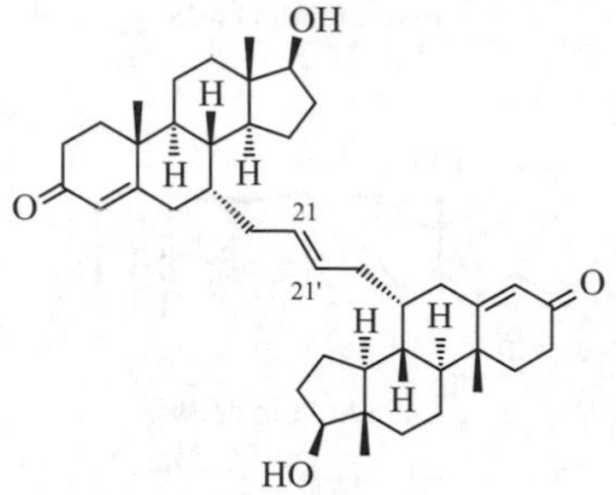

OH H H H 21 21' H H H O O HO

trans-T_2 (**43**)

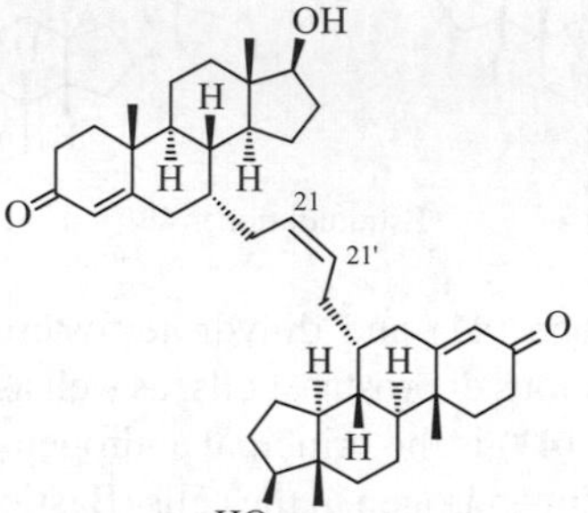

OH H H H 21 21' O H H H O HO

cis-T_2 (**44**)

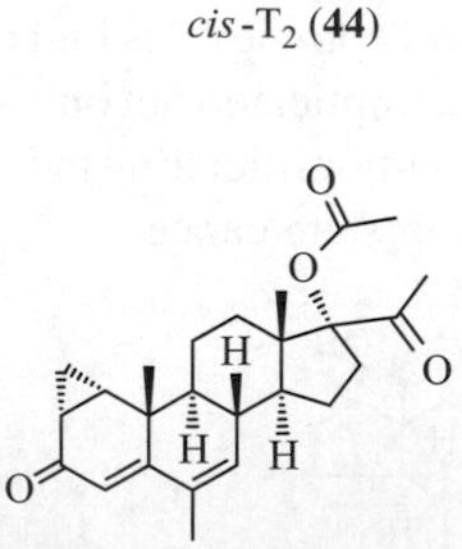

O O H O H H O

Cyproterone acetate (**45**)

The bile acid-based steroidal dimers, *N,N*-diethylenetriaminebis[cholic acid amide] (**46**) and *N,N*-ethylenediaminebis[deoxycholic acid amide] (**47**), were synthesized and their cytotoxic potential was evaluated (Salunke *et al.*, 2004). While *N,N*-diethylenetriamine bis [cholic acid amide] did not show any antiproliferative property, *N,N*-ethylenediamine bis [deoxycholic acid amide] totally inhibited the growth of human oral cancer (HEp-2) and human breast cancer (MCF-7) cells at nanomolar concentration.

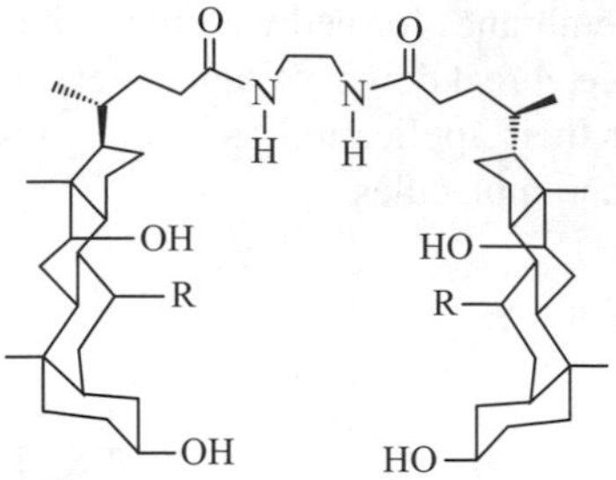

46: R = OH
47: R = H

Dimers **46** and **47**

6.2.3 Effect on Micellar Concentrations of Bile Salts and Serum Cholesterol Level

Bile salts are naturally occurring amphiphilic compounds that are stored in the gallbladder, and they play a pivotal role in the emulsification and transport of dietary fat and lipids in food for all mammals by the formation of micelles. The dimerization of bile salts is considered to be the critical step in the formation of bile salts micelles. The critical micellar concentration (CMC) of the bile salts are significantly lowered by the presence of a sodium cholate dimer (**48**), synthesized by linking two cholic acid molecules *via* a spacer. This indicates that the sodium cholate dimer can facilitate the micellization of bile salts (Gouin and Zhu, 1998). It can be assumed that this dimer may act as a nucleation agent, thereby promoting the formation of larger aggregates that is micelles.

Because modified bile acids, *e.g.*, bile acid dimers, show different levels of the ileal reabsorption and tend to aggregate with other bile acids, it is possible to use such compounds to reduce or prevent the reabsorption of bile acid in the intestinal tract. This should result in the lower blood cholesterol level. Several bile acid dimers containing ether or ester linkages between the C-3 positions of two bile acid molecules were synthesized to test this theory (Gouin and Zhu, 1996). These bile acid dimers are thought to reduce the reabsorption of bile salts *via* active transport, leading to an increased excretion of bile salts in faeces, which, in turn, would promote the hepatic conversion of cholesterol to bile acids and may eventually reduce the serum cholesterol level. Thus, bile acid dimers may be explored further for their possible therapeutic potential as blood-cholesterol-level reducing agent.

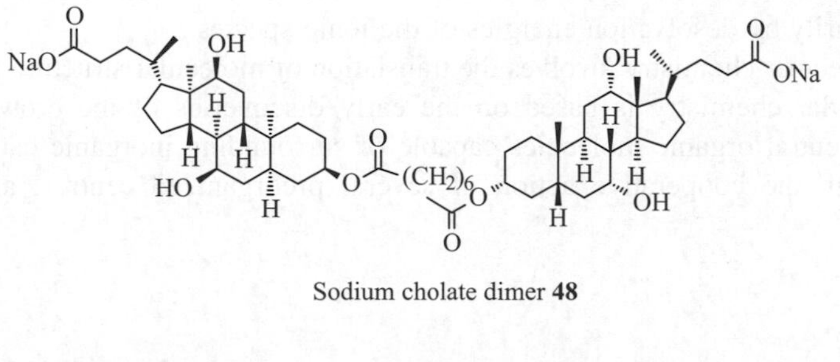

Sodium cholate dimer **48**

6.2.4 Effect on Bilayer Lipid Membranes

Some steroid dimers appear to have significant influence on the formation and structure of bilayer lipid membrane. For example, in presence of dimer **49** the membrane formation was found to be much faster and it was more stable than those formed in the absence of this dimer (Kalinowski *et al.*, 2000). This dimer was also able to significantly decrease the electrocompressibility. The presence of this dimer in the membrane limited the mobility of solvent inside the membrane. The membranes formed with the dimer possess such properties, like thickness, stability, resistance, breakdown voltage, electrocompressibility, and time of formation, more suitable for their application as a biomembrane model and support for sensors based on biomembrane molecules.

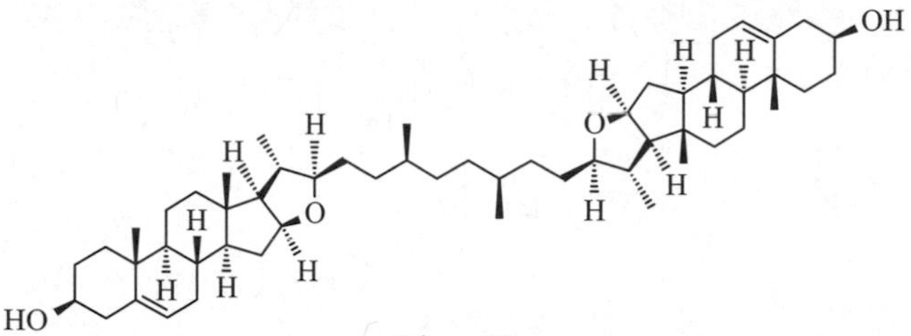

Dimer **49**

6.2.5 Supramolecular Transmembrane Ion Channels, and Artificial Receptors and Ionophores

A new group of supramolecular transmembrane ion channels using construction units possessing molecular amphiphilicity, *e.g.*, steroidal dimers **50** and **51**, has recently been synthesized (Kobuke and Nagatani, 2001). Linking of two amphiphilic cholic acid methyl ethers (as the membrane-insertible unit) through biscarbamate bonds afforded bis(7,12-dimethyl-24-carboxy-3-cholanyl)-*N,N*′-xylylene dicarbamate (**50**) and bis(7,12-dimethyl-24-hydroxy-3-cholanyl)-*N,N*′-xylylene dicarbamate (**51**). These dimers form stable ion channels characterized by relatively small conductance and long-lasting open states. It was shown that when incorporated into a planar bilayer membrane, dimers **50** and **51** exhibited stable (lasting 10 ms to 10 s) single ion channel current. Both these channels were cation selective, and the permeability ratios of K^+ to Cl^- were 17 and 7.9, respectively. These channels also displayed considerable K^+ selectivity over Na^+ by a factor of 3:1 and 3.2, respectively. This metal-ion selectivity appeared to be almost independent of the different external charges of the head-group. High cation/anion selectivities were observed and could be modulated by changing the ionic nature of the head-group (*e.g.*, a carboxylate or ammonium external headgroup) of channels. No Li+ current was observable, and the permeability preferences between different alkali metals and halides seemed to be determined primarily by desolvation energies of the ionic species.

Supramolecular chemistry involves the translation of molecular structure into function. Supramolecular chemistry is based on the early discoveries of the crown ethers and cryptands, neutral organic molecules capable of surrounding inorganic cations, binding them through the cooperative action of several preorganized centres, and providing lipophilic exteriors that promote solubility in nonpolar media. This was then extended to anion recognition.

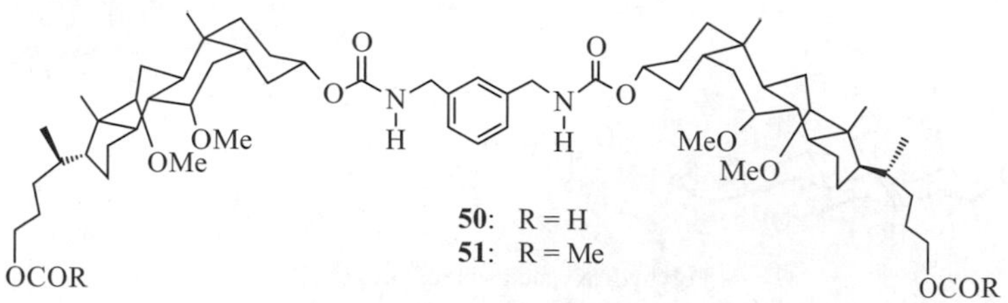

50: R = H
51: R = Me

Dimers **50** and **51**

A neutral macrocyclic organic receptor was synthesized for halide anions, in which there were a lipophilic exterior featuring flexible alkyl chains to maintain solubility in nonpolar organic media and a rigid framework to maintain a binding cavity and to limit the possibilities for intramolecular hydrogen-bond formation (Davies *et al*., 1996). This new host **52** was related to cholaphanes (see *Chapter 3*), and was a macrodilactum derived from two molecules of cholic acid. The hydroxy groups in this dimer were to create an environment that could mimic the aqueous solvation of a spherical anion, but preorganized for binding and surrounded by a lipophilic envelope. The prospects of anion recognition by **52**, were evaluated by molecular modelling and LSI-MS. It was observed that the macrocycle had an ovoid cavity about 3.3 Å long by 2.2 Å broad, sufficient to accommodate a fluoride, and possibly a chloride or bromide anion, but too small for significant penetration by $CHCl_3$ solvent molecules.

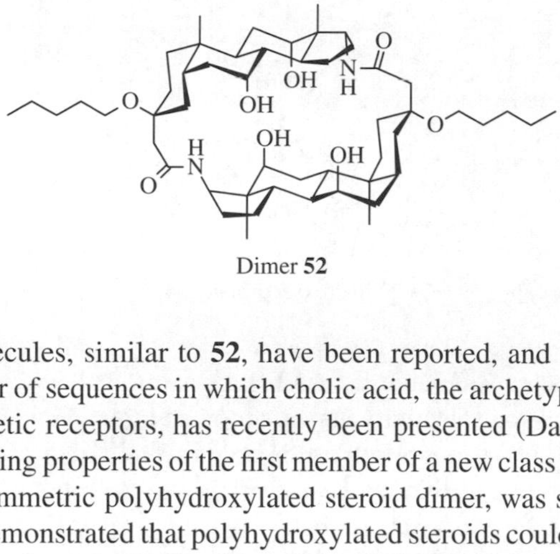

Dimer **52**

Several other molecules, similar to **52**, have been reported, and an excellent account, highlighting a number of sequences in which cholic acid, the archetypal bile acid, has been 'sculpted' into synthetic receptors, has recently been presented (Davies, 2007).

The Na^+-transporting properties of the first member of a new class of artificial ionophore **53**, based on a C_2-symmetric polyhydroxylated steroid dimer, was studied (De Riccardis *et al*., 2002). It was demonstrated that polyhydroxylated steroids could be versatile building blocks for the construction of effective ionophores. With respect to other related systems (*e.g*., cholic acid derivatives), polyhydroxylated steroids appear to have the advantage of the flat liphophilic steroid nucleus that optimizes the interaction with the lipid bilayer and possibly with other steroids naturally present in the cellular membrane. The activity of compound **53** was comparable with that of a tetrameric derivative based on cholic acid, despite the fact that only two steroid moieties are present in its structure. To investigate the ionophoric properties, the ability of **53** to promote the transport of Na^+ across a lipid bilayer using a $^{23}Na^+$-NMR-based methodology was applied.

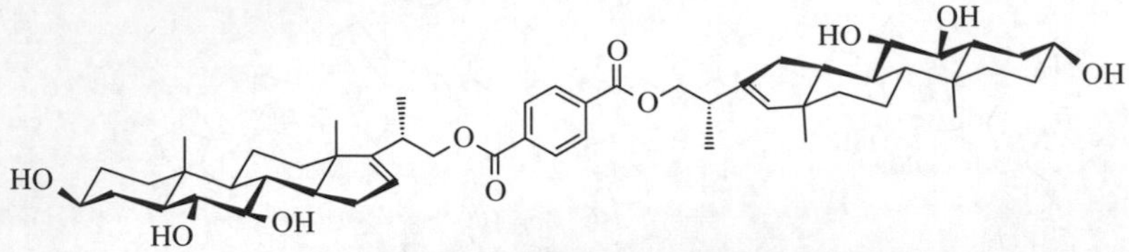

Polyhydroxylated stereoid dimer **53**

Interest in the design of membrane-spanning artificial ionophores (*e.g.*, **53**), inspired by the naturally occurring antifungal macrolide amphotericin B, has gained momentum in recent years because they may lead to new classes of antibiotics that are less susceptible toward resistance (De Riccardis *et al.*, 2002).

Several cholaphane molecules based on lithocholic and deoxycholic acids were synthesized by a sequence of reactions involving Cs-salt methodology of macrocyclisation (see *Chapter 3*), and it was demonstrated that these cholaphanes could act as artificial receptors for flavin analogues, *i.e.,* they could bind flavin molecules through hydrogen bonds (Chattopadhyay and Pandey, 2006; 2007).

6.2.6 Other Properties

The dimer (**55**) of 20-hydroxyecdysone (**54**), the insect-moulting hormone, seems to retain the high agonistic activity, similar to the parent monomer, on the ecdysteroid receptor in the B_{II} bioassay.

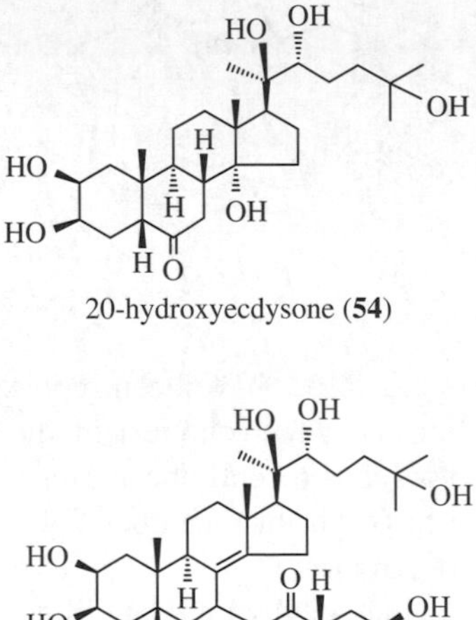

20-hydroxyecdysone (**54**)

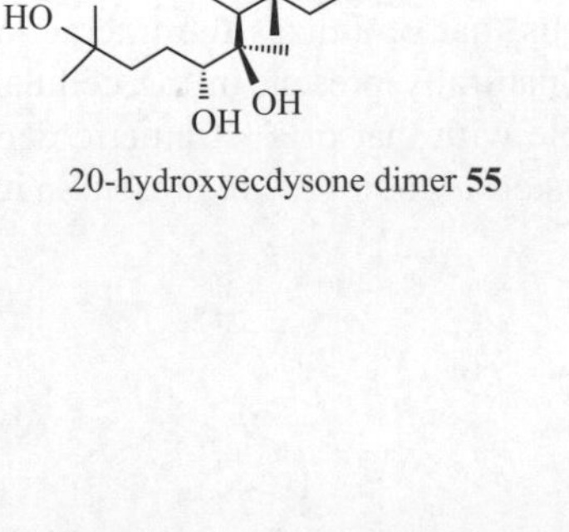

20-hydroxyecdysone dimer **55**

Bile acids derived dimers, *e.g.*, cholic acid methyl ester dimer and its 7,12-diacetyl derivative, possess complexing abilities towards alkaline metal cations (Nahar *et al.*, 2007). However, the lithocholic acid methyl ester dimer does not possess this property.

Bile acid-based novel amphiphilic topology in the form of steroidal dimers, N^1,N^3-diethylenetriaminebis[cholic acid amide] (**56**) and N^1,N^3-diethylenetriaminebis [deoxycholic acid amide] (**57**), was designed and synthesized (Salunke *et al.*, 2004). These dimers were tested for antifungal and antiproliferative activity *in vitro*. N^1,N^3-Diethylenetriaminebis[cholic acid amide] was found to be active against *Candida albicans*, *Yarrowia lipolytica*, and *Benjaminiella poitrassi* at nanomolar concentration and did not show any effect on cell proliferation.

56: R = OH
57: R = H

Dimers **56** and **57**

References

Bastien, D., Leblanc, V., Asselin, E. and Bérubé, G. (2010). First synthesis of separable isomeric testosterone dimers showing differential activities on prostate cancer cells. *Bioorganic and Medicinal Chemistry Letters* **20**, 2078–2081.

Berube, G., Rabouina, D., Perron, V., N'Zemba, B., Gaudreault, R-C., Parenta, S. and Asselin, E. (2006). Synthesis of unique 17-beta-estradiol homo-dimers, estrogen receptors binding affinity evaluation and cytocidal activity on breast, intestinal and skin cancer cell lines. *Steroids* **71**, 911–921.

Bhattacharjee, A. K., Carvalho, K. A., Opsenica, D. and Solaja, B. A. (2005). Structure-activity relationship study of steroidal 1,2,4,5-tetraoxane antimalarials using computational procedures. *Journal of Serbian Chemical Society* **70**, 329–345.

Chattopadhyay, P and Pandey, P. S. (2006). Synthesis and binding ability of bile acid-based receptors for recognition of flavin analogues. *Tetrahedron* **62**, 8620–8624.

Chattopadhyay, P and Pandey, P. S. (2007). Synthesis and binding ability of bile acid-based receptors for recognition of flavin analogues. *Bioorganic and Medicinal Chemistry Letters* **17**, 1553–1557.

Chen, W. H., Janout, V., Kondo, M., Mosoian, A., Mosoyan, G., Petroy, R. R., Klotman, M. E. and Regen, S. L. (2009). A fine line between molecular umbrella transport and ionophoric activity. *Bioconjugate Chemistry* **20**, 1711–1715.

D'Auria, M. V., Giannini, C., Zampella, A., Minale, L., Debitus, C. and Roussakis, C. (1998). Crellastatin A: a cytotoxic bis-steroid sulfate from the Vanuatu marine sponge *Crella* sp. *Journal of Organic Chemistry* **63**, 7382–7388.

Davis, A. P. (2007). Bile acid scaffolds in supramolecular chemistry: the interplay of design and synthesis. *Molecules* **12**, 2106–2122.

Davis, A. P., Gilmer, J. F. and Perry, J. J. (1996). A steroid-based cryptand for halide anions. *Angew. Chem. Int. Ed, Engl.* **35**, 1312–1314.

De Riccardis, F., Di Filippo, M., Garrisi, D., Izzo, I., Mancin, F., Pasquato, L., Scrimin, P. and Tecilla (2002). An artificial ionophore based on polyhydroxylated steroid dimer. *Chemical Communication*, 3066–3067.

Fournier, D. and Poirier, D. (2009). Estradiol dimers as a new class of steroid sulfatase reversible inhibitors. *Bioorganic and Medicinal Chemistry Letters* **19**, 693–696.

Fuzukawa, S., Matsunaga, S. and Fusetani, N. (1994). Ritterazine A, a highly cytotoxic dimeric steroidal alkaloid, from the Tunicate *Ritterella tokioka*. *Journal of Organic Chemistry* **59**, 6164–6166.

Fuzukawa, S., Matsunaga, S. and Fusetani, N. (1995a). Isolation and structure elucidation of ritterazines B and C, highly cytotoxic dimeric steroidal alkaloids, from the tunicate *Ritterella tokioka*. *Journal of Organic Chemistry* **60**, 608–614.

Fuzukawa, S., Matsunaga, S. and Fusetani, N. (1995b). Ten more ritterazines, cytotoxic steroidal alkaloids from the turnicate *Ritterella tokioka*. *Tetrahedron* **51**, 6707–6716.

Fuzukawa, S., Matsunaga, S. and Fusetani, N. (1997). Isolation of 13 new ritterazines from the tunicate *Ritterella tokioka* and chemical transformation of ritterazine B. *Journal of Organic Chemistry* **62**, 4484–4491.

Ge, D., Wu, D., Wang, Z., Shi, W., Wu, T., Zhang, A., Hong, S., Wang, J., Zhang, Y. and Ren, L. (2009). Cellular uptake mechanism of molecular umbrella. *Bioconjugate Chemistry* **20**, 2311–2316.

Giannini, C., Zampella, A., Debitus, C., Menou, J. L., Roussakis, C. and D'Auria, M. V. (1999). Isolation and structural elucidation of the crellastatins I-M: cytotoxic bis-steroid derivatives from the vanuatu marine sponge *Crella* sp. *Tetrahedron* **55**, 13749–13756.

Gouin, S. and Zhu, X. X. (1996). Synthesis of 3α- and 3β-dimers from selected bile acids. *Steroids* **61**, 664–669.

Gouin, S. and Zhu, X. X. (1998). Fluorescence and NMR Studies of the effect of a bile acid dimer on the micellization of bile salts. *Langmuir* **14**, 4025–4029.

Janout, V., Lanier, M. and Regen, S. L. (1996). Molecular umbrellas. *Journal of American Chemical Society* **118**, 1573–1574.

Janout, V., Jing, B. W. and Regen, S. L. (2002). Molecular umbrella-assisted transport of thiolated AMP and ATP across phospholipid bilayers. *Bioconjugate Chemistry* **13**, 351–356.

Janout, V. and Regen, S. L. (2009). Bioconjugate-based molecular umbrellas. *Bioconjugate Chemistry* **20**, 183–192.

Kalinowski, S., Łotowski, Z. and Morzycki, J. W. (2000). The influence of bolaamphiphilic steroid dimer on the formation and structure of bilayer lipid membranes. *Cellular and Molecular Biology Letters* **5**, 107–128.

Klayman, D. L. (1985). Qinghaosu (artemisinin): an antimalarial drug from China. *Science* **228**, 1049–1055.

Kobuke, Y. and Nagatani, T. (2001). Transmembrane ion channels constructed of cholic acid derivatives. *Journal of Organic Chemistry* **66**, 5094–5101.

Kondo, M., Mehiri, M. and Regen, S. L. (2008). Viewing membrane-bound molecular umbrellas by Parallax analyses. *Journal of American Chemical Society* **130**, 13771–13777.

Li, Y. X. and Dias, J. R. (1997). Dimeric and oligomeric steroids. *Chem. Rev.* **97**, 283–304.

Lopez-Anton, N., Rudy, A., Barth, N., Schmitz, M. L., Pettit, G. R., Schulze-Osthoff, K., Dirsch, V. M. and Vollmar, A. M. (2006). The marine product cephalostatin 1 activates an endoplasmic reticulum stress-specific and apoptosome-independent apoptotic signaling pathway. *Journal of Biological Chemistry* **281**, 33078–33086.

Moser, B. R. (2008). Review of cytotoxic cephalostatins and ritterazines: isolation and synthesis. *Journal of Natural Products* **71**, 487–491.

Muller, I. M., Dirsch, V. M., Rudy, A., Lopez-Anton, N., Pettit, G. R. and Vollmar, A. M. (2005). Cephalostatin 1 inactivates Bcl-2 by hyperphosphorylation independent of M-phase arrest and DNA damage. *Molecular Pharmacology* **67**, 1684–1689.

Nahar, L., Sarker, S. D. and Turner, A. B. (2007). A review on synthetic and natural steroid dimers: 1997–2006. *Current Medicinal Chemistry* **14**, 1349–1370.

Opsenica, D., Pocsfalvi, G., Juranic, Z., Tinant, B., Declercq, J-P., Kyle, D. E., Milhous, W. K., Solaja, B. A. (2000). Cholic acid derivatives as 1,2,4,5-tetraoxane carriers: Structure and antimalarial and antiproliferative activity. *Journal of Medicinal Chemistry* **43**, 3274–3282.

Pettit, G. R., Kamano, Y., Inoue, M., Dufresne, C., Boyd, M. R., Herald, C. L., Schmidt, J. M., Doubek, D. L. and Christie, N. (1992). Antineoplastic agents. 214. isolation and structure of cephalostatins 7–9. *Journal of Organic Chemistry* **57**, 429–431.

Pettit, G. R., Ichihara, Y., Xu, J-P., Boyd, M. R. and Williams, M. D. (1994a). Isolation and structure of the symmetrical disteroidal alkaloids cephalostatins 12 an 13. *Bioorganic and Medicinal Chemistry Letters* **4**, 1507–1512.

Pettit, G. R., Xu, J.-P., Ichihara, Y., Williams, M. D. and Boyd, M. R. (1994b) Antineoplastic agents. 285. Isolation and structures of cephalostatin-14 and cephalostatin-15. *Canadian Journal of Chemistry* **72**, 2260–2267.

Pettit, G. R., Xu, J-P. and Schmidt, J. M. (1995). Isolation and structure of the exceptional pterobranchia human cancer inhibitors cephalostatins 16 and 17. *Bioorganic and Medicinal Chemistry* **5**, 2027–2032.

Pettit, G. R., Tan, R., Xu, J-P., Ichihara, Y., Williams, M. D. and Boyd, M. R. (1998). Antineoplastic agents. 398. Isolation and structure elucidation of cephalostatins 18 and 19. *Journal of Natural Products* **61**, 955–958.

Poza, J. J., Rodriguez, J. and Jimenez, C. (2010). Synthesis of a new cytotoxic cephalostatin/ritterazine analogue from hecogenin and 22-*epi*-hippuristanol. *Bioorganic and Medicinal Chemistry* **18**, 58–63.

Rabouin, D., Perron, V., N'Zemba, B., C-Gaudreault, R. and Berube, G. (2003). A facile synthesis of C_2-symmetric 17β-estradiol dimers. *Bioorganic and Medicinal Chemistry Letters* **12**, 557–560.

Salunke, D. B., Hazra, B. G., Pore, V. S., Bhat, M. K., Nahar, P. B. and Despande, M. V. (2004). Newt steroidal dimers with antifungal and antiproliferative activity. *Journal of Medicinal Chemistry* **47**, 1591–1594.

Todorović, N. M., Stefanović, M., Tinant, B., Declercq, J-P., Makler, M. T. and Šolaja, B. A. (1996). Steroidal geminal dihydroperoxides and 1,2,4,5-tetraoxanes: structure determination and their antimalarial activity. *Steroids* **61**, 688–696.

von Schwarzenberg, K. and Vollmar, A. M. (2010). Targeting apoptosis pathways by natural compounds in cancer: marine compounds as lead structures and chemical tools for cancer therapy. *Cancer Letters* (in press) DOI: 10.1016/j.canlet.2010.07.004.

Zampella, A., Giannini, C., Debitus, C., Roussakis, C. and D'Auria, M. V. (1999). Isolation and structural elucidation of crellastatins B-H: cytotoxic bis(steroid) derivatives from the Vanuatu marine sponge *Crella* sp. *European Journal of Organic Chemistry* 949–953.

Index

Acid-catalyzed addition, 31
Acetylation, 73, 91, 124, 174, 328, 369
2-Acetyloxybenzoyl chloride, 115
3-Acetyloxybenzoyl chloride, 116
4-Acetyloxybenzoyl chloride, 117
(17β-Acetyloxy-5α-androstano)-(5α-cholestano)pyrazine, 321, 323, 324
3α-Acetyloxy-5α-androstan-17-one, 96, 97
17β-Acetyloxy-5α-androstan-3-one, 17, 18
3α-Acetyloxy-5α-androstan-17-one azine, 96, 97
17β-Acetyloxyandrost-4-en-3-one, 11
2-[17′β-Acetyloxy-2′-(5α-androsten)-3′-yl]-17β-acetyloxy-3-methoxy-2-(5α-androstane), 18
17β-Acetyloxy-4-chloroandrost-4-en-3-one, 11
3β-Acetyloxy-5α-cholestan-2-one, 316
3β-Acetyloxycholest-5-en-7-one, 174
3β-Acetyloxy-5α,6α-epoxycholestane, 76, 77
17β-Acetyloxyestra-4,9-dien-3-one, 174
(22β,25*R*)-3β-Acetyloxyfurost-5-en-26-al, 139, 140
(22β,25*R*)-3β-Acetyloxyfurost-5-en-26-ol, 139
3β-Acetyloxypregna-5,16-dien-20-one, 87, 92
3β-Acetyloxypregna-5,16-dien-20-one oxime, 92
20β-Acetyloxy-5α-pregnan-6-one, 78
3β-Acetyloxypregn-5-en-20-one, 77, 91, 92
20β-Acetyloxy-5α-pregn-2-en-6-one, 77, 78
3β-Acetyloxypregn-5-en-20-one oxime, 91, 92
3β-Acetyloxy-22-isospirosta-5,7-diene, 69
(25*R*)-3β-Acetyloxy-5α-spirostan-12-one, 85, 288
(25*R*)-3β-Acetyloxy-5α-spirostan-12α-propionate, 295
(25*R*)-3β-Acetyloxy-5α-spirost-14-en-12α-ol, 288
Acetylthioethoxy dimer, 16
Activated ALOX B, 296, 297
Active metal reduction, 7
Active transport, 401
Acyclic dilithocholate, 233, 234
Acyclic dimers, 2, 7, 215, 217, 221, 225, 227
Acyclic steroid dimers, 2–182
Adams' catalyst, 97
Agonistic activity, 404
Air oxidation, 312, 318
seco-Aldehyde, 352
Aldol reaction, 88
Alkaline hydrolysis, 34, 36, 127, 169, 222, 224, 288, 323, 324, 326, 361
Alkaloid-detecting reagents, 5
Alkylation reduction, 144
Alkyl glycol chain, 396
Allyl acetate, 82
7α-Allyl-androst-4-en-3-on-17β-yl acetate, 82
Allylation, 211, 214
Allyl bromide, 129, 211, 372
Allyl derivatives, 210
3α-Allyl ether-5β-cholan-7α,12α,24-triol, 214
24-Allyl ether 5β-cholan-3α,7α,12α-triol, 210–212
Allylmagnesium chloride, 78
Amaroxocane A, 279, 280
Amaroxocane B, 279, 280
Amberlyst 15, 300, 301
Amide linkages, 40
Amine-catalyzed dimerization, 174
24-Amino-5β-cholan-3α,7α,12α-yl triacetate, 166

Steroid Dimers: Chemistry and Applications in Drug Design and Delivery, First Edition. Lutfun Nahar and Satyajit D. Sarker. © 2012 John Wiley & Sons, Ltd. Published 2012 by John Wiley & Sons, Ltd.

2-Amino-22-*epi*-hippurist-1-ene-3,11-dione, 335, 338, 339
2α-Aminoketone, 302, 339
3α-Aminolithocholic acid, 235
2α-Amino-3-hydroxyimino-cholestane, 314
2α-Amino-3-hydroxyimino-5α-cholestane, 314, 315
2α-Amino-3-methoxyimino-cholestane, 316
2α-Amino-3-methoxyimino-5α-cholestane, 316, 323
23-Amino-24-nor-5β-cholan-3α,7α,12α-yl triacetate, 111
(25*R*)-2-Amino-5α-spirost-1-ene-3,12-dione, 337, 338
Amphipathic lipids, 4
Amphiphiles, 170, 381
Amphiphilic molecules, 381
Amphiphilicity, 170, 171, 187, 402
Amphiphilic walls, 381
Amphomorphic, 170, 381, 382
Androgen receptors, 400
Androgens, 399
Androstane, 18, 323, 328, 329, 332, 333
Androstadienedione, 14, 15
E/Z Androstadienedione dimers, 14, 15
Androsta-1,4-diene-3,17-dione, 14
Androsta-3,5-diene-7,17-dione, 19, 20
Androsta-4,6-dien-3-on-17β-yl acetate, 81, 82
Androsta-3,5-dien-3,17β-yl diacetate, 81
5α-Androstanolone, 66
5β-Androstanolone, 65, 66
5α-Androstanolone diacetate, 66
5α-Androstanolone diol dimer, 66
5β-Androstanolone diol dimer, 66
Androstenedione, 11, 151, 152
Androst-4-ene-3,17-dione, 11, 12, 151
Androst-4-ene-3,17-dion-19-oic acid, 12
5α-Androst-2-en-3,17β-yl diacetate, 321
5α-Androst-2-en-17β-ol, 89, 90
Androstanolone pyrazine dimer, 313
Androsterone, 96
Androsterone acetate, 96, 97
Androsterone derivatives, 96
trans-Androsterone, 372
Androsterone enyne, 372, 374
Androsterone hydrazone, 96
17,17′-Anhydro-3-*O*-ethoxymethyl-17α-hydroxymethyl-17β-estradiol, 95
Antiandrogen, 400
Antiandrogenic activity, 400
Anticancer drug, 241, 379, 390
Antiestrogen, 397
Antifungal macrolide amphotericin, 404
Antileukemic steroid dimers, 65
Antimalarial activity, 383, 384, 386
Antimalarial drugs, 383, 385
Antimalarial property, 32, 382
Antineoplastic drug, 387
Antiproliferative activity, 384, 396, 405
Antisense oligonucleotides, 379, 380, 382
Antitumour activity, 262, 391, 394
Antiviral activity, 173
Apocynaceae, 241
Apoptotic pathways, 387
Apoptotic signalling pathway, 387, 389
Archetypal bile acid, 403
Aromatic spacer groups, 188
Aroyl chloride, 115, 117
Artemisia annua, 383
Artemisinin, 383, 384
Artificial enzymes, 187
Artificial ionophore, 403, 404
Artificial lipid bilayers, 141
3-Arylpropionate ester
Asymmetric bis-steroid sulphate, 395
(25*R*)-3α-Azido-2β-chloro-5α-spirostan-12α-propionate, 298, 299
2α-Azido-5α-cholestan-3-one, 313–315
2α-Azido-17β-hydroxy-5α-androstan-3-one, 327
16α-Azido-3β-hydroxy-5α-androstan-l7-one, 331, 332
2α-Azido-3-hydroxyimino-5α-cholestane, 314
2α-Azidoketone, 313, 355, 356, 363, 366
23-Azido-24-nor-5β-cholan-3α,7α,12α-yl triacetate, 111
24-Azido-5β-cholan-3α,7α,12α-yl triacetate, 165
(25*R*)-3-Azido-5α-spirost-2-en-12α-propionate, 299

Backbone rearranged dimer, 77
Baeyer–Villiger oxidation, 99, 229
Barbier–Wieland procedure, 212
B_{II} bioassay, 404
Benzylated estrone, 129
Benjaminiella poitrassi, 405

[N-1,2,3-Benzotriazin-4(3H)one-yl]-3-(2-pyridyldithio) propionate, 173
Benzyl 3α-*tert*-butyldimethylsilyloxy-5β-cholan-24-oate, 155
Benzyl 7α,12α-diacetoxy-3α-hydroxy-24-nor-5β-cholan-23-oate, 207, 208
3-*O*-Benzyl-17α-(3-hydroxypropyl)-1,3,5(10)-estratrien-17β-ol, 130
Benzyl 3α-hydroxy-5β-cholan-24-oate, 49
Benzyl lithocholate, 155
3-*O*-Benzyl-17α-(prop-2′-enyl)-1,3,5(10)-estratrien-17β-ol, 129, 130
3-Benzyloxy-1,3,5(10)-estratrien-17-one, 129
BF_3 catalyzed rearrangement, 77
Bicholesta-3,5-dienyl, 9, 10
Bicholestane, 7, 8, 10, 72
6β,6′β-Bicholestane, 72
Bicholestadienedione, 71
6β,6β′-Bicholesta-4,4′-diene-3,3′-dione, 70–72
Bi(cholest-5-ene), 9
6β,6β′-Bicholesta-4,4′-diene-3,3′-dione, 70–72
Bicholesteryl, 9, 10
Bile acid, 1, 22, 24
Bile acid-based dimers, 45, 46, 102, 103, 106, 108, 109, 401, 405
Bile acid methyl cholate dimers, 43, 45
Bile acid-piperazine diamides, 118–120
Binding cavity, 403
Binding properties, 188
Bioassay-guided fractionation, 282
Bioassay-guided purification, 242
Bioassay-monitored isolation, 243
Biocompatibility, 381
Biomimetic chemistry, 188
Biosynthetic route, 262
2,2′-Bipyridine-4,4′-dicarboxylic acid, 45–47
Bis(17β-acetyloxy-4-chloro-3β-hydroxyandrost-4-en-3′α-yl), 11, 12
1,3-Bis[(22β,25*R*)-3β-acetyloxyfurost-5-en-26-amino], 140, 141
1,4-Bis[(22β,25*R*)-3β-acetyloxyfurost-5-en-26-amino]butane, 140
1,6-Bis[(22β,25*R*)-3β-acetyloxyfurost-5-en-26-amino]hexane, 140, 141
1,8-Bis[(22β,25*R*)-3β-acetyloxyfurost-5-en-26-amino]4-azaoctane, 141
Bis(3β-acetyloxypregna-5,16-dien-20-one oximinyl)oxalate, 93
Bis(3β-acetyloxypregn-5-en-20-one oximinyl) oxalate, 91–93
Bis-acryloylated derivative, 312, 313
Bis(5α-androst-2-en-17β-yl)oxalate, 90
Bis(androst-4-en-3-on-17β-yl)oxalate, 89
Bis-(20*S*)-5α-23,24-bisnorchol-16-en-3β,6α,7β-triol-22-terephthaloate, 122, 128, 129
3α,3α′-*O*-Bis(bromoacetyl)cholyl(deoxycholyl) ethylenediamine, 104
N,N′-Bis(3α-*O*-bromoacetylcholyl)ethylenediamine, 103, 215
N,N′-bis(3α-*O*-bromoacetylcholyl)-*m*-xylylenediamine, 105, 217
N,N′-Bis(3α-*O*-bromoacetyldeoxycholyl) ethylenediamine, 104
N,N′-Bis(3α-*O*-bromoacetyldeoxycholyl) pyridine-2,6-diamine, 109, 227
N,N′-Bis(3α-*O*-bromoacetyldeoxycholyl)-*m*-xylylenediamine, 105
N,N′-Bis(3α-*O*-bromoacetyllithocholyl)-pyridine-2,6-diamine, 109
Biscarbamate, 40, 56, 57
Biscarbamate bonds, 40, 402
Bischolamide, 102, 103
Bis(5β-cholan-24-oic acid 3β-yl)diethylene glycol, 26, 27
Bis(β-cholan-24-oic acid 3α-yl)sebacate, 24
Bis(5β-cholan-24-oic acid 3α-yl)suberate, 24
Bis(cholesta-3,5-diene), 14, 15
Bis(5α-cholestan-3β-yl)oxalate, 30
Bis(cholest-5-en-3β-yl)oxalate, 30
1,4-Bis(4-cholesten-6-on-3β-oxy)-2*E*-butene, 64
1,6-Bis(4-cholesten-6-on-3β-oxy)-3*E*-hexene, 65
Bischolesterienyl, 14
Bisconicasterone, 277, 278
Bisdeoxycholamide, 102, 104
N,N′-Bisdeoxycholylethylenediamine, 103
N,N′-Bisdeoxycholyl-pyridine-2,6-diamine, 108, 109
N,N′-Bisdeoxycholyl-*m*-xylylenediamine, 104–106
Bis-18,18′-desmethyl ritterazine N, 364, 369
(*E*)-1,2-Bis-(3α,12α-diacetoxy-5β-pregnan-20-yl)ethene, 146
Bis(3β,11α-diacetoxy-5α-pregn-16-en-20-on e oximinyl)oxalate, 92, 93
(*E*)-1,2-Bis-(3β,20β-diacetoxy-5-pregnen-19-yl)ethene, 182
N,N′-Bis(3α,12α-*O*-diformyldeoxycholyl)-2,6-diaminopyridine, 107

N,N′-Bis(3α,12α-*O*-diformyldeoxycholyl)-pyridine-2,6-diamine, 107
Bis(7α,12α-dihydroxy-5β-cholan-24-oic acid 3α-yl)sebacate, 25
Bis(7α,12α-dihydroxy-5β-cholan-24-oic acid 3α-yl)suberate, 25
Bis(7,12-dimethyl-24-carboxy-3-cholanyl)-*N,N′*-xylylene dicarbamate, 41, 402
1,4-Bis-(3,12-dioxo-5β-pregnan-20-yl)-2*E*-butene, 146–148
Bis(3-dioxy-5α-cholestane), 32
Bis(3-dioxy-7α,12α-diacetoxy-5β-cholan-24-amide), 37
Bis(3-dioxy-7α,12α-diacetoxy-5β-cholan-24-oic acid), 36, 37
Bisdinitrophenylhydrazone, 71
3β,3′β-Bisdinitrophenylhydrazone-6β,6′β-bicholesta-4,4′-diene, 70, 71
Bishydrazone, 84, 85
Bisergostatrienol, 2, 4
7α,7′α-Bisergostatrienol, 69, 70
7α,7′β-Bisergostatrienol, 69, 70
7β,7′β-Bisergostatrienol, 69, 70
Bis[estra-1,3,5(10)trien-17-on-3-yl]oxalate, 27
N,N′-Bis(3α-*O*-formylcholyl)-pyridine-2,6-diamine, 109
N,N′-Bis(3α-*O*-formyllithocholyl)-pyridine-2,6-diamine, 107
7α,12α-Bis(formyloxy)-24-nor-5β-chol-22-en-3α-ol, 201
7α,12α-Bis(formyloxy)-24-nor-5β-chol-22-en-3-one, 201
Bis(3-hydroperoxy-5β-cholestan-3-yl) peroxide, 31, 32
Bis(3β-hydroxycholest-4-en-3′α-yl), 8, 9, 11
1,4-Bis(6*E*-hydroximino-4-cholesten-3β-oxy)-2*E*-butene, 65
1,6-Bis(6*E*-hydroximino-4-cholesten-3β-oxy)-3*E*-hexene, 65
1,4-Bis-(6β-hydroxy-20-oxo-5α-pregnan-6-yl)-2*E*-butene, 80, 81
Bis(3-hydroperoxy-5β-cholestan-3-yl)-peroxide, 31, 32
N,N′-Bislithocholyl-pyridine-2,6-diamine, 107, 109
Bis(methyl 3β-acetyloxybisnorchola-5,7-dien-22-oate, 69, 70
Bis(methyl 5β-cholan-24-oate-3β-yl)diethylene glycol, 26
Bis(methyl 5β-cholan-24-oate-3α-yl)oxalate, 29
Bis(methyl 5β-cholan-24-oate-3α-yl)sebacate, 23
Bis(methyl 5β-cholan-24-oate-3α-yl)suberate, 22
Bis(methyl 5β-cholan-24-oate-3α-yl) terephthalate, 43
Bis(methyl 7α,12α-diacetoxy-5β-cholan-24-oate-3α-yl)terephthalate, 44
Bis(methyl 7α,12α-dihydroxy-5β-cholan-24-oate-3α-yl)sebacate, 24
Bis(methyl 7α,12α-dihydroxy-5β-cholan-24-oate-3α-yl)suberate, 23
Bis(methyl 7α,12α-dihydroxy-5β-cholan-24-oate-3α-yl)terephthalate, 44
Bis[(methyl 7,12-dimethyl-24-carboxylate)-3-cholanyl]-*N,N′*-xylylene dicarbamate, 41
Bis(methyl 3-dioxy-7α,12α-diacetoxy-5β-cholan-24-oate), 35
Bis(methyl-3α-hydroxyandrost-4-en-17-on-19-oate-3′α-yl), 11, 13
Bis(methyl 3α-trimethylsilyloxyandrost-4-en-17-on-19-oate-3′α-yl), 13, 14
Bis(methyl 3α-trimethylsilyloxyandrost-4-en-17-on-19-oate-3′β-yl), 12–14
Bis-oxidation, 361, 362
(*E*)-1,2-Bis-(3-oxo-4-pregnen-20-yl)ethene, 149
Bis(pentafluorophenyl)3α,3′α-(terephthaloyloxy)-bis(5β-cholan-24-oate), 50, 226
Bis(pregn-5-en-20-on-3β-yl) oxalate, 28
Bis(*N*-propylamide), 384
Bis[*N*-(*n*-propyl)-3-dioxy-7α,12α-diacetoxy-5β-cholan-24-amide], 38
3β,3′β-bis(pyridine-n-carboxy)-5β-cholan-24-oic acid ethane-1,2-diol diesters, 117
3α,3′α-Bis(pyridine-n-carboxy)-5β-cholan-24-oic acid piperazine diamides, 120
Bis-steroidal derivatives, 258
Bis-steroidal tetraoxanes, 386
Bis-steroid pyrazines, 320
Bistheonellasterone, 277
Bis(stigmasta-5,22*t*-dien-3β-yl)oxalate, 31
Bis(24-triphenylmethoxy)-3α,3′α-(isophthaloyloxy)-bis(5β-cholane), 48
Brain tumour, 386
Breast tumour cell lines, 396, 397
Brominated, 174

2α-Bromo-5α-cholestan-3-one, 317
trans-1α-Bromo-di(5α-cholestano)pyrazine, 319
trans-4α-Bromo-di(5α-cholestano) pyrazine, 318
(25*R*)-2α-Bromo-5α,6β-dihydroxyspirostan-3-one, 343, 347
(25*R*)-2α-Bromo-5α-spirost-14-ene-3,12-dione, 290
(25*R*)-2α-Bromo-4α,5α-epoxyspirostan-3, 6-dione, 344, 347
2α-Bromo-22-*epi*-hippuristan-3,11-dione, 335
2α-Bromo-17β-hydroxy-5α-androstan-3-one, 327
16α-Bromo-3β-hydroxy-5α-androstan-l7-one, 331
(25*R*)-2α-Bromo-5α-hydroxyspirostan-3, 6-dione, 348
Bromomethylene spiroketals, 372
(25*R*)-2α-Bromo-5α-spirostan-3,6-dione, 342, 350
(25*R*)-2α-Bromo-5α-spirostan-3,12-dione, 317, 337
(25*R*)-2α-Bromo-5α-spirostan-3-one, 341, 346
(25*R*)-2α-Bromospirost-4-en-3,6-dione, 342
(25*R*)-2α-Bromospirost-5-en-3-one, 342, 345
2-Butanone, 61
Butenandt acid, 21
3β-*tert*-Butyldiphenylsilyloxyandrost-5-en-17-one, 122
3α-*tert*-Butyldimethylsilyloxy-5β-cholan-24-oic acid, 156
3α-*tert*-Butyldimethylsilyloxy-5β-cholan-3α,7α,12α-triol, 211
3β-*tert*-Butyldimethylsiloxy-5β-cholan-24-yl ethane-1,2-diol tosylate, 136
3β-(*tert*-Butyldiphenylsilyloxy)-17,17-(ethylenedioxy)-androst-5-ene, 123
3α-*tert*-Butyldimethylsilyloxy-24-hydroxy-5β-cholan-7α,12α-yl diacetate, 164
tert-Butylisocyanide, 236
3-Butynyl alcohol, 61
3-Butynyl cholesteryl ether, 61
3β-Butynyloxycholest-4-en-6-one, 63
3β-Butynyloxy-5,6-epoxycholestane, 62
3β-Butynyloxy-5α-hydroxycholestan-6-one, 62

Cancer cell growth inhibitors, 241, 253, 387
Candida albicans, 405
5-Carboxyfluorescein, 382
Catalytic hydrogenation, 17, 97, 230, 290
Catalytic reduction, 17, 97, 230, 290
Cell proliferation, 405
Cephalodiscidae, 242, 287
Cephalodiscus gilchristi, 242, 243, 287
Cephalostatin analogues, 288, 291, 303, 304, 306, 324, 328, 329, 331, 332, 334, 339, 390
Cephalostatin 1, 5, 6, 241, 242, 246, 254, 287, 351, 356, 357, 371–373, 375, 386–389, 392, 394, 395
Cephalostatins 1–19, 242, 391
Cephalostatin/ritterazine analogues, 334, 338
Chenodeoxycholic acid, 162–164, 232, 235–237
Chenodeoxycholic acid-based dimer, 162–164
Chlorination, 7, 9, 11
2-Chlorobenzoyl chloride, 234
N-2-Chlorocarbonylmethyl-*N,N,N*-trimethylammonium chloride, 42
3β-Chloro-5α-cholestane, 7, 8
3β-Chlorocholest-5-ene, 7, 8
3β-Chlorocholest-5-en-7-one, 174
3β-Chlorocholest-5-en-7-spiro-(4′α,5′-oxa)-3′β-chloro-5′α-cholestan-7′-one, 176
Chlorotestosterone acetate, 11
(25*R*)-2β-Chloro-3α-hydroxy-5α-spirostan-12α-propionate, 297, 298
Chloroquine-resistant strains, 383, 385
4-Chlorotestosterone acetate, 11
5β-Cholane-3α,24-diol, 47, 134
5β-Cholan-3α,7α,12α,24-tetraol, 210, 213, 214
5β-Cholane-3α,12α,24-triol, 147
Cholaphane, 2, 3, 187, 188, 199, 205, 210, 211, 214, 215, 217, 219, 221, 223–225, 228, 404
Cholaphane cyclodimer, 188
5α-Cholestane, 7, 8, 314–317, 323, 326
Cholestane-based ether dimers, 76, 77
Cholestane-based oxime dimers, 74, 75
5α-Cholestan-2,3-dione, 317
5α-Cholestane-2α,3α-diamine, 316
Cholestane homodimers, 169
Cholestanol, 28, 30
Cholestanone, 15–17
Cholestanone-based dimers, 15
5α-Cholestan-3β-ol, 28
5α-Cholestan-3-one, 31, 32
5β-Cholestan-3-one, 31
Cholestane 3-ethylene hemithioketals, 16
Cholest-4,6-dien-3-one, 19

Cholestene dimer, 282
Cholest-5-ene-3β,19-diol-3-monoacetate
Cholestene monomers, 282
Cholest-5-en-3β-ol, 7
Cholestenone, 8, 9, 11
Cholest-4-en-3-one, 8, 9
Cholest-5-en-3-one, 58, 174
Cholest-5-en-7-one, 174
Cholestenone pinacol, 8, 9, 11
Cholest-5-en-3-spiro-(6′α,5′-oxa)-5′α-cholest-3′-one, 175
Cholest-5-en-7-spiro-(4′α,5′-oxa)-5′α-cholestan-7′-one, 176
2-(2′-Cholesten-3′-yl)-3-acetylthioethoxy-2-cholestene, 16, 17
2α-(2′-Cholesten-3′-yl)-3-cholestanone, 15–17
Cholestanyl chloride, 7, 8
Cholestenyl magnesium chloride, 7, 8
Cholesterol, 1, 7, 14, 60, 70, 138, 401
Cholesterol-based dimers, 174
Cholesterol 1,2-diol dimer, 60
Cholesteryl benzoate, 177
Cholesteryl benzoate aldol, 177
Cholesteryl chloride, 7, 9
Cholesteryl magnesium chloride, 9
Cholic acid, 21, 22, 109, 120, 121, 192, 202, 211, 382, 403
Cholic acid based cyclopeptides, 229
Cholic acid based dimers, 22, 36, 39, 41, 54, 84, 112, 164, 166, 168
Cholic acid based *gemini* surfactants, 120, 122
Cholic acid based glycine conjugate, 160
Cholic acid biscarbamate, 56, 57
Cholic acid peptide dimer, 161
Cholic acid derived tetraoxane dimers, 384
Cholic acid triformate, 200
Cholyldeoxycholamide, 104
N-Cholyl-*N*′-deoxycholylethylenediamine, 103
N-Cholylethylenedimine, 101–103
Chonemorpha macrophylla, 241
Chronic lymphocytic leukaemia, 379
Classical condensation, 339
Cognate hormone, 397
Conversion of ergosterol, 2
Corey–Schmidt oxidation, 336
Corey–Schmidt reagent, 334
Crella (Yvesia) spinulata, 278
Crellastatin A, 254–256, 387, 395
Crellastatins A–M, 395
Crellidae, 254
Critical micellar concentration, 401
Crohn's disease, 379
Crosscoupling reaction, 60
Cross metathesis reaction, 60
Cs-salt methodology, 215, 227, 404
Cupric acetate-pyridine oxidation
Cyclic cholapeptide, 207, 209
Cyclic ethylene hemithioketals, 16
Cyclic oligomeric cholate derivatives, 154
Cyclobutane, 20, 174
Cyclocholate dilactone, 232, 233
Cyclocholate lactone, 232, 233
Cyclocholates, 2, 187, 232, 233, 379
Cyclodichenodeoxycholate, 235–237
Cyclodichenodeoxycholate diacetate, 236, 237
Cyclodichenodeoxycholate diol, 236, 237
Cyclodimerization, 206, 233
Cyclodilithocholate, 233–235
1,2-Cyclopentano-perhydrophenanthrene, 1
Cyclopentane ring, 1
Cyclopentaphenanthrene, 1, 2
Cyclotetramerization, 233
Cyproterone acetate, 400
Cytomegalovirus retinitis, 380
Cytotoxic activity, 263, 334, 389, 392, 393, 395
Cytotoxic cephalostatins, 379
Cytotoxic steroid dimers, 5, 242
Cytotoxin, 6, 242, 386

Dansyl moiety, 170, 381
Deacetylated, 334
Deacetyl-*cis*-T_2, 81, 83, 84
Deacetyl-*trans*-T_2, 81, 83, 84
Dean–Stark trap, 18, 200, 356, 358, 359, 363
Debenzylation, 362
Dehydro-*epi*-androsterone, 28, 397
Dehydro-*epi*-androsterone sulphate, 397
Dehydrobrominated, 174
Dehydrocholic acid
16,17-Dehydrodigitoxigenin-3β-yl acetate, 87, 88
16-Dehydropregnenolone acetate, 87, 92
16-Dehydropregnenolone acetate dimer, 87
Deoxycholaphane, 228
Deoxycholic acid, 101, 106, 227, 229, 232, 382, 401, 405
N-Deoxycholylethylenedimine, 101, 102
Deoxygenation, 144
14-*epi*-7′-Deoxyritterazine B, 350

7′-Deoxyritterazine G, 350, 391
17-Deoxyritterazine K, 270
Dess–Martin periodinane, 366
Desulphated, 395
DHEA, 28, 122, 123, 125, 126, 397
DHT, 89
DIABLO, 387
2β,17β-Diacetoxy-5α-androstan-3-one
3β,3′β-Diacetoxy-6β,6′β-bicholesta-4, 4′-diene, 71
3α,7α-Diacetoxy-5β-cholanic acid, 164
3α,7α-Diacetoxy-5β-cholan-24-oic acid, 162
3β,5β-Diacetoxy-cholestan-6-one oxime, 74
3β,20β-Diacetoxy-19-ethenyl-pregn-5-ene, 181, 182
7α,12α-Diacetoxy-3α-hydroxy-24-nor-5β-cholan-23-oic acid, 207
7α,12α-Diacetoxy-3-oxo-5β-cholan-24-amide, 34
3α,7α-Diacetoxy-12-oxo-5β-cholan-24-oate, 85
7α,12α-Diacetoxy-3-oxo-5β-cholan-24-oic acid, 34, 35
7α,12α-Diacetoxy-3-oxo-5β-cholan-24-oyl chloride, 35
3β,20β-Diacetoxy-pregn-5-en-19-al, 181
3β,11α-Diacetoxy-5α-pregn-16-en-20-one, 92, 93
3β,11α-Diacetoxy-5α-pregn-16-en-20-one oxime, 92
(25*R*)-2α,12β-Diacetoxy-5α,14α-spirostan-3-one, 325, 326
Diacetylated, 395
Diacid, 21, 121, 122
3α,24-Diallyl ether-5β-cholan-7α,12α-diol, 210
Diamine, 56, 58, 107–109, 138, 140, 168, 227, 228
1,8-Diamino-4-azaoctane, 141
1,4-Diaminobutane, 58, 59, 140
2α,3α-Diamino-5α-cholestane, 316, 317
N,N′-Di(24-Amino-7α,12α-diacetoxy-5β-cholan)3α,3′α-oxamide, 221, 222
1,6-Diaminohexane, 58, 59, 138, 140
1,3-Diaminopropane, 58, 59, 138, 140
2,6-Diaminopyridine, 106, 107, 110
Diastereomeric keto-alcohols, 362
N,N′-Di(24-Azido-7α,12α-diacetoxy-5β-cholan)3α,3′α-oxamide, 55, 56
16-Diazo-3β-hydroxyandrost-5-en-17-one, 97
Dibenzyl-3α,3′α-(terephthaloyloxy)-bis(5β-cholan-24-oate), 49, 50
1,6-Dibromohexane, 133
Dibutyl 2,2-dibromomalonate, 191, 192
Di-*n*-butylether, 17
Dicarbonyl acid chloride, 22
N,N′-Di(5β-cholane-7α,12α,24-triol)3α,3′α-oxamide, 53
cis-Di(5α-cholestano)pyrazine, 313, 316–318, 320
trans-Di(5α-cholestano)pyrazine, 313, 314, 317, 318
2,6-Dichlorobenzoyl chloride, 161, 235, 236
N,N′-Dicholylethylenedimine, 102
N,N′-Dicholyl-*m*-xylylenediamine, 104, 105
N,N′-Di(7α,12α-diacetoxy-5β-cholan-24-ioc acid)3α,3′α-oxamide, 53
N,N′-Di(7α,12α-diacetoxy-24-iodo-5β-cholan) 3α,3′α-oxamide, 54, 55
Dihydrotestosterone, 89, 90, 399
5α-Dihydrotestosterone acetate, 17–19
5α-Dihydrotestosterone acetate acetylthioethyl enol ether, 18
5α-Dihydrotestosterone acetate dimers, 17, 19
5α-Dihydrotestosterone acetate 3-ethylene hemithioketal, 18
3β,3′β-Dihydroxy-6β,6′β-bicholesta-4, 4′-diene, 71, 72
Di[(25*R*)-5α,6β-dihydroxy-5α-spirostano] pyrazine, 347, 348
Di[(25*R*)-3,6-dioxo-5α-hydroxy-5α-spirostano] pyrazine, 348
Diels–Alder adducts, 277
Diels–Alder reaction, 87
Diels–Alder condensation, 87
Diels–Alder-type cycloaddition, 277
Diels–Alder-type dimerization, 277
N,N-Diethylenetriamine-bis[cholic acid amide], 401
Dienol ether formation, 87
Di[(25*R*)-5-en-spirostano]pyrazine, 345, 346
3α,12α-*O*-Diformyldeoxycholic acid, 106
Dihydrocephalostatin 1, 361, 363, 364
gem-Dihydroperoxide dimer, 31
Dihydroperoxy-peroxide, 31
Di[(25*R*)-4α,5α-epoxy-6-oxo-spirostano] pyrazine, 347, 348
Di(17β-hydroxy-5α-androstano)pyrazine, 311, 312, 327, 328

3α,7α-Dihydroxy-5β-cholanic acid, 162
3α,12α-Dihydroxy-5β-cholanic acid, 101
3α,7α-Dihydroxy-5β-cholan-24-oic acid, 232
3α,12α-Dihydroxy-5β-cholan-24-oic acid, 101
3α,3′α-Dihydroxy-5β-cholan-24-oic acid piperazine diamide, 118
Di(3β-hydroxyfurost-5-en-26-yl), 143
Dihydro-ornithostatin, 349–351
Dihydro-ornithostatin $O_1 1_N$, 349
(25*R*)-5α,6β-Dihydroxy-spirostan-3-one, 342
1,14-Diiodotetradecane, 144
N,N′-Di(methyl 7α,12α-diacetoxy-5β-cholan-24-oate)3α,3′α-oxamide, 53
Dimethyl ester of EDTA, 121
N,N′-Dimethylhydrazine, 65, 66
20-Dimethylketal, 87
2,2-Dimethoxypropane, 125
Dimeric cholaphanes, 187
Dimeric steroidal alkaloids, 250, 262
Dimerization, 2, 6, 15, 19, 87, 174, 277, 313, 348, 379, 382, 401
N,N′-Di(3α,7α,12α-triacetoxy-24-nor-5β-cholan)23,23′-oxamide, 111
N,N′-Di(3α,7α,12α-trihydroxy-24-nor-5β-cholan)23,23′-oxamide, 111
Diosgenin, 138–143, 339–341, 343–345
Diosgenin acetate, 139
Diosgenin derivatives, 142, 339–341, 343–345
Diosgenin polyamine dimers, 140, 141
Diosgenin pyrazine dimers, 339
Diosgenin TBDMS ether, 141, 142
3,8-Dioxabicyclo-[4.2.1]nonane, 279
Dioxane, 27, 31, 32, 40, 41, 71, 85, 86, 96, 120, 145, 203, 302, 320, 352
3,12-Dioxo-5β-cholan-24-al, 147
3,12-Dioxo-5β-chol-24-ene, 147
Di[(25*R*)-6-oxo-4-en-spirostano]pyrazine, 346, 347
Di[(25*R*)-6-oxo-5α-spirostano]pyrazine, 346, 347
cis-Di-[(25*R*)-12-oxo-5α-spirostano]pyrazine, 317, 319
trans-Di-[(25*R*)-12-oxo-5α-spirostano] pyrazine, 317, 319
Di-[(25*R*)-12-oxo-5α-spirostano]pyrazines, 317, 319
Di[(25*R*)-5α-spirostano]pyrazine, 346
Direct self-esterification, 152
Disteroidal alkaloids, 246
3,3′-Disteroidal *ortho*-phthalate, 210, 213
24,24′-Disteroidal *ortho*-phthalate, 214
Diwalled *'molecular umbrella'*, 173, 382
Dragendorff's reagent, 5

Ecdysteroid receptor, 404
ED_{50}, 241, 246, 388, 389, 395
EDTA spacer, 121
Electrolytic reduction, 8
Emulsification, 401
(25R)-2-Enamino-5α-spirost-14-ene-3,12-dione, 290, 299
Endoperoxy bridge, 383
Eosin, 69
2α,3α-Epoxy-5α-androstan-3β,17β-yl diacetate, 321, 322
(25*R*)-4α,5α-Epoxyspirostan-3,6-dione, 343, 344
(25*R*)-5α,6α-Epoxyspirostan-3β-ol, 344, 345
(25*R*)-5α,6α-Epoxyspirostan-3-one, 342
(25*R*)-2α,3α-Epoxy-5α-spirostan-12α-propionate, 297
(22*E*)-Ergosta-5,7,22-trien-3β-ol, 69
Ergosterol, 1, 2, 69
Ester linkage, 40, 91, 401
Eastern hemisphere, 255–263, 266, 268–270, 273–275, 392
17β-Estradiol, 129–134, 396, 397
17β-Estradiol dimers, 397
Estradiol-based ring D–ring D dimers, 149
17β-Estradiol homodimers, 129–134, 396
Estra-1,3,5(10)-triene-3,17β-diol, 151
Estrone, 27, 28, 43, 88, 89, 91, 92, 94, 95, 129, 174, 397
Estrone 3-methyl ether, 88
Estrone oxalate dimer, 27, 28, 89, 91, 92
Estrone sulphate, 397
Estrone terephthaloate dimer, 43
Ethane-1,2-diol diester, 113–118, 218
Ethane-1,2-diol mono *p*-tosylate THP ether, 134, 136
Ether bond formation, 25
Ether linkage, 22
21*R*-Ethoxy-16,17-dehydrodigitoxigenin-3β-yl acetate dimer, 88
21*S*-Ethoxy-16,17-dehydrodigitoxigenin-3β-yl acetate dimer, 88
O-Ethoxymethylestrone, 94, 95
Ethyl 7α,12α-diacetoxy-3α-hydroxy-5β-glycocholate, 160

Ethyldiazophosphonate, 361
1-Ethyl-3-(3-dimethyl aminopropyl)-carbodiimide hydrochloride, 68
Ethylenediamne dihydrochloride, 68
Ethyl glycinate hydrochloride, 160
Ethane-1,2-diol bridges, 134, 221
10β-Ethyl-5α,6α-epoxycholestane, 77
20-Ethenyl-pregn-4-en-3-one, 148, 149
Ethyl 3α,7α,12α-trihydroxy-5β-glycocholate, 160
Etienic acid, 152–154
Etienic acid nitrate, 152, 154

Fibrosterols A–C, 282
Flavin analogues, 227, 404
5-Fluorouracil, 65
p-Formaldehyde, 127, 236
Formylation, 106, 109, 189
3α-*O*-Formylcholic acid, 109
3α-*O*-Formyllithocholic acid, 106
Furostanone, 351, 352
(3β,22β,25*R*)-Furost-5-ene-3,26-diol, 139
(25*R*)-Furost-5-en-26-ol-3β-TBDMS ether, 142
trans-Fused rings, 2

Gemini surfactant, 120–122
GI_{50}, 386, 389, 390, 396
Glutathione, 171, 382
Grignard reaction, 192, 362
Grubbs 1st-generation catalyst, 215
Grubbs 2nd-generation catalyst, 82, 146

Hamigera hamigera, 278
Hamigerol A, 278
Hamigerol B, 279
'Head-to-head' cholaphane, 214–216, 218–223, 228
'Head-to-tail' cholaphane, 189, 224, 225, 229, 230
Hecogenin, 85–87, 288, 289, 296–298, 303, 324, 334, 336, 351, 361, 371, 372, 390
Hecogenin acetate, 85, 86, 288, 361, 371, 372, 390
Hecogenin derivatives, 289, 296–298, 303, 324, 336, 361
Hecogenine acetate dimers, 86, 87
Hemithioketal, 16, 18
22-*epi*-Hippuristan-3,11-dione, 334, 335
22-*epi*-Hippuristanol, 334, 335, 390
Homoallylic alcohol, 288
Hydrazide spacers, 51, 110
Hydrazone spacer, 84
Hydrolysis, 24, 34, 36, 74, 75, 127, 169, 179, 193, 222, 224, 229, 233, 288, 323, 324, 326, 361
Hydrophilic, 6, 170, 171, 173, 379–383
Hydrophilicity, 170
Hydrophilic peptides, 173, 381, 382
Hydrophobic, 4, 6, 170, 171, 173, 379–381
Hydrophobicity, 170, 381, 389
6*E*-Hydroximinosteroid homodimers, 60, 64
(17β-Hydroxy-5α-androstano)-(5α-cholestano) pyrazine, 323, 324
3β-Hydroxy-5α-androstan-17-one, 372
17β-Hydroxy-5β-androstan-3-one, 65
3α-Hydroxy-5α-androstan-17-one azine, 96
19-Hydroxyandrost-4-ene-3,17-dione, 11, 12
3′β-[(3β-Hydroxyandrost-5-ene-17β-carbonyl) oxy]-androst-5′-en-17′β-oate, 152–154
3β-Hydroxyandrost-5-en-17β-oic acid, 152
3β-Hydroxyandrost-5-en-17-one, 28, 97
17β-Hydroxyandrost-4-en-3-one, 81
12β-Hydroxycephalostatin, 1, 349
3α-Hydroxy-5β-cholanic acid, 21
3α-Hydroxy-5β-cholan-24-oic acid, 113, 159, 218, 227
3α-Hydroxy-5β-cholan-24-oic acid (cholan-24-oic acid methyl ester)-3α-yl ester, 159
3α-Hydroxy-5β-cholan-24-oic acid ethane-1, 2-diol diester, 113, 218
3α-Hydroxy-5β-cholan-24-ol ethane-1,2-diol mono THP ether, 134
3α-Hydroxy-5β-cholan-24-yl ethane-1, 2-diol, 134, 135
3α-Hydroxy-5β-cholan-24-yl ethane-1,2-diol tosylate, 135
19-Hydroxy-cholest-5-en-3β-yl acetate, 169
20-Hydroxyecdysone, 72, 73, 404
20-Hydroxyecdysone-based dimers, 72, 73
17β-Hydroxyestra-4,9-dien-3-one, 174
17β-Hydroxy-5(10)-estren-3-one, 174
3α-Hydroxy-24-iodo-5β-cholan-7α,12α-yl diacetate, 164, 165
Hydroxylation, 256, 371, 389, 391, 394, 395
3β-Hydroxypregn-5-en-20-one, 28
6α-Hydroxy-6β-(1-propenyl)-5α-pregnan-20-one, 80
6β-Hydroxy-6α-(1-propenyl)-5α-pregnan-20-one, 80

(25*R*)-5α-Hydroxyspirostan-3,6-dione, 344, 345
(25*R*)-2α-Hydroxy-5α-spirostan-3, 12-dione, 338, 339
(25*R*)-3β-Hydroxy-5α-spirostan-12-one, 334
(25*R*)-12α-Hydroxy-5α,14α-spirostan-3-one, 324, 325
(25*R*)-3β-Hydroxy-5α-spirostan-12α-propionate, 295, 296
17β-Hydroxysteroids, 89
N-Hydroxysuccinimide, 101, 170

IC_{50}, 254, 262, 383–385, 391, 395–397, 400
Ileal reabsorption, 401
Insect-moulting hormone, 72, 404
Intramolecular Wadsworth–Emmons reaction, 361
17-Iodoandrosta-5,16-dien-3β-ol, 88
23-Iodo-24-nor-5β-cholan-3α,7α,12α-yl triacetate, 110, 111
Ionophoric activity, 382
Ionophoric property, 382
Isophthaloyl dichloride, 48
3α,3′α-(Isophthaloyloxy)-bis(5β-cholan-24-ol), 48
Izu Peninsula, 262
Japanese marine sponge, 278
Japanese tunicate, 262
Japindine, 3, 241
Jones' oxidation, 341, 344, 352, 354
Jones' reagent, 12, 34, 97, 157, 341

Ketobisnoaldehyde, 138
Keto cholesterol, 58, 59, 138
3-Keto-4-methylene sterols, 277
Keto stigmasterol, 58, 59, 138

$LiAlH_4$ reduction, 47, 77, 167, 210, 212, 361
Liebermann–Burchard reagent, 5
Linear dimeric oligomeric cholate derivatives, 154
Linear dimers, 2, 7, 154
Lipid bilayer, 141, 379–382, 403, 404
Lipid molecule, 1, 379
Liposomal membranes, 173, 381, 382
Lipophilic, 1, 2, 382, 383, 403
Lipophilic envelope, 403
Lipophilic exterior, 403
Lipophilicity, 173
Lipophilic molecules, 1
Lissodendoryx (Acanthodoryx) fibrosa, 282
Lithocholaphane, 224, 226, 227
Lithocholic acid, 21–25, 47–50, 106, 113, 115, 117–119, 134, 135, 137, 154–159, 218, 220, 221, 224, 225, 227, 233–235, 405
Lithocholic acid-based dimers, 22–25, 47–50, 113, 115, 117–119, 134, 137, 154–159
3α-Lithocholic acid dimer, 22–24
3β-Lithocholic acid dimers, 25
Lithocholic acid derivatives, 47, 135
Lithocholic acid-based ethane-1,2-diol diester
Liquid crystal, 6
L-Lysine, 172, 381
Lyophilized sponge, 254

Macrocyclic cholaphane, 100, 189
Macrocyclic steroid dimers, 3, 187
Macrocyclization, 227
Macrocyles, 232
Macrodilactum, 403
Macrolactamization, 210
Macrolactonization, 210, 233
3β-Magnesium chloro-5α-cholestane, 7
β-Marcaptoethanol, 16
Marine organisms, 241
Marine sponges, 2, 254, 256, 277, 278, 287
Marine worms, 242, 287
Mediterranean sponge, 278
Mefloquine-resistant, 385
Methallyl stannane, 362
Methanolysis, 369
trans-1α-Methoxy-di(5α-cholestano) pyrazine, 320, 321
trans-4β-Methoxy-di(5α-cholestano) pyrazine, 320
Methoxylation, 391
Methyl 3β-acetyloxy-bisnorchola-5,7-dien-22-oate, 69
Methyl 3α-amino-7α,12α-diacetoxy-5β-cholan-24-oate, 52, 166
Methyl 2-amino-3,12-dioxo-5β-chol-1,14-dien-24-oate, 302
Methyl androst-4-ene-3,17-dion-19-oate, 12, 13
17α-Methyl-androst-4-en-3-on-17β-ol, 20
Methyl 3α-azido-7α,12α-diacetoxy-5β-cholan-24-oate, 51, 52
Methyl 2β-bromo-3,12-dioxo-5β-chol-14-en-24-oate, 301, 302

Methyl 3α-*tert*-butyldimethylsilyloxy-7α, 12α-diacetoxy-5β-cholan-24-oate, 164
Methyl chenodeoxycholate, 45, 46
Methyl cholate, 22, 32, 40, 43–45, 56, 84, 101, 211
Methyl 3α-chlorocarbonyloxy-7α,12α-diacetoxy-5β-cholan-24-oate, 56
Methyl deoxycholate, 45, 47, 101, 102
Methyl 3α,7α-diacetoxy-12α-hydroxy-5β-cholan-24-oate, 84
Methyl 7α,12α-diacetoxy-3α-hydroxy-5β-cholan-24-oate, 34, 44, 51, 56, 164, 221
Methyl 7α,12α-diacetoxy-3α-iodo-5β-cholan-24-oate, 51
Methyl 7α,12α-diacetoxy-3-oxo-5β-cholan-24-oate, 34
Methyl 3α,7α-diacetoxy-12-oxo-5β-cholan-24-oate, 85
Methyl 7α,12α-diacetoxy-3-oxo-5β-cholan-24-oate, 34
Methyl 3α,7α-dihydroxy-5β-cholanate, 45
Methyl 3α,12α-dihydroxy-5β-cholanate, 45
Methyl 3α,7α-dihydroxy-5β-cholan-24-oate, 45
Methyl 3α,12α-dihydroxy-5β-cholan-24-oate 45
Methyl 3α,12α-dihydroxy-5β-chol-14-en-24-oate, 300
Methyl 7,12-dimethylcholate, 40, 41
Methyl 3,12-dioxo-5β-chol-14-en-24-oate, 300, 301
Methyl ethyl ketone, 62
Methylenation, 169
Methyl etienic carboxylate, 152–154
Methyl 3α-hydroxy-5β-cholanate, 22
Methyl 3α-hydroxy-5β-cholan-24-oate, 22
Methyl 7α-acetyloxy-3α-hydroxy-5β-cholan-24-oate, 163, 164
Methyl 3β-hydroxyandrost-5-en-17β-oate, 152
Methyl 3β-hydroxy-D-norandrost-3-en-16β-oate, 97
Methyl 3β-(2-hydroxyethoxy)-5β-cholan-24-oate, 25
Methyl lithocholate, 22, 25, 28, 29, 43, 45, 157
N-Methylmorpholine *N*-oxide, 203
Methyl 3′β-[(3β-nitratoandrost-5-ene-17β-carbonyl)oxy]-androst-5′-en-17′β-oate, 152, 153
N-Methylpyrrolidone, 51, 55
Methyl 3-oxo-D-nor-5α-androstan-16β-oate, 97, 98
17α-Methyltestosterone, 20
17α-Methyltestosterone dimers, 20
Methyl 3α-(2′-tetrahydropyranyl)cholate, 40
Methyl 3α,7α,12α-triacetoxy-5β-cholan-24-oate, 33
Methyl 3α,7α,12α-trihydroxy-5β-cholan-24-oate, 22
Microwave-assisted metathesis, 80
Miscellaneous dimers, 174
Mitsunobu inversion, 362
Montmorillonite K10, 60
MTT cytotoxicity assay, 400
Molecular cavities, 187, 188
Molecular recognition, 187
Molecular umbrella, 6, 7, 170–174, 379–382

Natural cytotoxins, 6, 386
Naturally occurring steroid dimers, 241–284, 386
Nematic mesophase, 180
Neopentyl α-alkoxyphosphono acetates, 361
3β-Nitratoandrost-5-en-17β-oic acid, 152, 153
Nonsymmetric pyrazine dimer, 324, 326, 327
Nonsymmetric pyrazine analogue, 324, 326, 327
Nonsymmetric trisdecacyclic pyrazines, 324, 326, 327
D-Nor-5α-androstan-16β-oic acid, 98
D-Nor-5α-androstan-16β-ol, 99
D-Nor-5α-androstan-16-one, 97, 99, 100
D-Nor-5α-androstan-16-one azine, 97, 100
D-Nor-5α-androstan-16β-yl acetate, 99
Norcholic acid, 189, 190, 192, 194, 196, 202, 204, 207
24-Nor-5β-cholan-3α,7α,12α, 23-tetraol, 214
24-Nor-5β-chol-22-en-3α,12α-yl diacetate, 146
D-Norpregnan-20-methyl ether, 98, 99

2-Octyl nitrite, 311, 312
Okinawan marine sponge, 277
Olefin metathesis, 77, 146
Oligonucleotides, 173, 379, 380, 381, 382
Organolithium intermediate, 94
Oxalate dimers, 28, 89
Oxalate ester linkage, 91
Oxalate spacer group, 27, 28, 89, 91
Oxalic acid, 21
Oxamide spacers, 51, 221

Oxidation, 34, 59, 70, 97, 139, 191, 329, 339, 363, 392, 393
3-Oxo-5α-androstan-17β-ol, 328
3-Oxo-5α-androstan-2β,17β-yl diacetate, 322
3-Oxo-5β-cholan-24-oic acid, 34, 35, 157, 158
3-Oxo-5β-cholan-24-oic acid (cholan-24-oic acid methyl ester)-3α-yl ester, 158–160
3-Oxo-2(3′)-en-bis(6,7-*seco*-cholest-4-en-6, 7-dioic)anhydride, 21
3-Oxo-pregn-4-en-20β-al, 148
3-Oxopregn-4-ene-20β-carboxaldehyde, 138
2,2′-Oxybis(ethylamine), 226, 227
2,2′-Oxydiacetylchloride, 225
Oxygenated steroidal spiroketal, 241
2-Oximino-5α-androstan-l7β-ol-3-one
Oxyminyl oxalate dimers, 91, 93

Passive diffusion, 173, 381
Passive transport, 379, 380, 382
PEG chain, 129
Pentacyclic keto-aldehyde, 362, 363
Phenanthrene, 1
Phorbas amaranthus, 279
Phospholipid bilayers, 382
Phosphonate esters, 361, 362
Photochemical condensation, 7
Photochemical reaction, 19, 371
Photochemical transformation, 72
anti-photodimer, 174, 175
Photohecogenine acetate, 86, 87
Phytosterols, 1
Piperazine, 118–120, 220
PKCζ inhibitors, 282
Plant sterols, 1
Plasmodium falciparum, 383
Poecilosclerida, 254
Polyamine dimeric steroid, 6
Polyamine dimers, 138
Polyamines, 172, 381
Polyclinidae, 262, 287
Polyethylene glycol chain, 396
Polyhydroxylated dimer, 122
Polyoxygenated steroidal units, 242
Polysulphate sterol dimers, 278
Potent cytotoxins, 242
Pregnane dimers, 144, 145
Pregnene dimers, 138
Pregnene 1,3-diaminopropane dimer, 138
Pregnene 1,6-diaminohexane dimer, 138
Pregn-5-en-3β-ol-20-one, 28
Pregnenolone, 28, 66, 91, 373
Pregnenolone acetate, 77, 91
Pregnenolone derivatives, 68, 79, 91
Pregnenolone dimer, 66, 67
Pregnenolone hemisuccinate, 66, 67
Pregnenolone homodimers, 77
Pregnenolone-3β-yl-*N*-(2-amino-ethyl)-succinamic acid, 66
Pregnanoic ester, 144
Pregnanoic ester dimers, 144
Prodrugs, 382
Propargyl alcohol, 60, 61
3β-Propargyloxycholest-4-en-6-one, 63
Propargyl cholesteryl ether, 60, 61
3β-Propargyloxy-5,6-epoxycholestane, 62
3β-Propargyloxy-5α-hydroxycholestan-6-one, 62, 63
6α-(1-propenyl)-5α-pregnane-20β,6β-diol, 78, 80
6β-(1-Propenyl)-5α-pregnane-20β,6α-diol, 78, 79
Propionic anhydride, 295
N-Propylamide congeners, 396
N-(*n*-Propyl)-7α,12α-diacetoxy-3-oxo-5β-cholan-24-amide, 34
Protected cephalostatin 1, 356, 357
Protected dihydrocephalostatin 1, 361, 363, 364
Protected ritterostatin G_N1_N, 358, 359
Protected ritterostatin G_N1_S, 359
ortho-Phthalic anhydride, 210, 211
Pyrazine core, 242, 262
Pyrazine dimers, 339
Pyrazine-linked steroid dimers, 288
Pyridinium bromide perbromide, 174, 175

Reduction, 7, 11, 14, 47, 72, 98, 172, 203, 211, 293, 356, 361, 388, 392
Reductive amination, 58, 138
Reductive condensation, 11
Reductive process, 68
Reductive ring F fission, 142
Remote oxidation, 303
Ring-closing metathesis, 211
Ritterazine A, 5, 262–264, 266, 287, 288, 387, 388, 392, 394
Ritterazine B, 264, 267, 268, 369, 391–393, 395
Ritterazines A–Z, 262, 391
Ritterazine N, 271, 364, 367, 369, 388
Ritterella tokioka, 243, 262, 263, 271, 287
Ritterostatin G_N1_N, 351, 358, 359, 371

Ritterostatin G_N1_S, 351, 359, 360, 371
Ritterostatin G_N7_S, 349, 350, 391

Scaffold material, 172, 381
Selective tosylation, 136, 229
L-Selectride, 291–293
Shallow coral reefs, 279
Shishicrellastain A, 278
Shishicrellastain B, 279
Silica gel, 5, 81
Silylated dimers, 13
Silyl-modified Sakurai reaction, 229
Simmons–Smith methylenation, 169
Sodium cholate dimer, 401
Spacer group, 2, 4, 21, 27, 28, 43, 74, 84, 85, 100, 106, 149, 187, 188
Sprays reagents, 5
Spermidine, 170–172, 381
N^1,N^3-Spermidinebis(cholic acid amide), 173, 174, 405
Spermine, 172, 381, 382
Spiroketal diol, 362
Spiroketalization, 371
Spiroketals, 242, 362, 364, 372, 375, 392, 394
Spiroketal pyrazine dimers, 242, 262
Spiroketal triflate, 365
(25*R*)-5α-Spirostan-3,6-dione, 342
(25*R*)-5α-Spirostan-2,14-diene-12α-propionate, 297
(25*R*)-5α,14α-Spirostan-3α,12α-diol, 324
(25*R*)-5α-Spirostan-3,12-dione, 336, 337
(25*R*)-5α-Spirostan-3-one, 340, 341
(25*R*)-5α-Spirost-14-ene-3α,12α-diol, 288, 289
(25*R*)-5α-Spirost-14-ene-3,12-dione, 289, 290
(25*R*)-Spirost-4-en-3,6-dione, 341, 342
(25*R*)-Spirost-5-en-3β-ol, 138
(25*R*)-Spirost-5-en-3-one, 339, 340
(25*R*)-Spirost-5-en-3β-yl acetate, 139
Stereospecific transformation, 169
Steroid, 1, 2
Steroid alcohol, 4
Steroidal amine, 52, 121, 166, 170, 174, 230
Steroid dimer, 1–5, 21, 59, 68, 277, 379
Steroid sulphatase, 151, 397
Sterol, 1, 4, 278, 279, 371
Stigmasta-5,22*t*-dien-3β-ol, 28
Stigmasta-5,22*t*-dien-3-one, 58
Stigmasterol, 28, 31, 58, 138
Stigmasterol 1,2-diol dimer, 60
Structure–activity relationships, 391
Sudan-III, 368
Succinic anhydride, 66, 67
Succinimido cholate, 101, 102, 104
Succinimido deoxycholate, 101–103, 105
Sulforhodamine B, 396
Sulphur-containing dimeric steroid, 241
Supramolecular chemistry, 187, 188, 402
Supramolecular hosts, 100
Supramolecular transmembrane ion channels, 40, 402
Surfactant molecules, 170
Swern oxidation, 362, 363
Symmetrical dimers, 4
Symmetrical pyrizine dimer, 313
C_2-Symmetric dimers, 94, 95
C_2-Symmetric 17β-estradiol homodimers, 129, 130
C_2-Symmetric estrone diol dimer, 94, 95
C_2-Symmetric polyhydroxylated dimer, 122
Synthesis of cephalostatin 1, 371–375
Synthetic dimers, 2
Synthetic molecular receptors, 188

cis-T_2, 81–84, 400
trans-T_2, 81–84, 400
Tamoxifen, 397
Terephthalic acid, 43, 215, 219, 220
Terephthalic acid diester, 220
Terephthalate dimers, 43, 44
Terephthaloyl chloride, 43, 44, 128
Terephthaloyl dichloride, 49
3α,3′α-(Terephthaloyloxy)-bis(5β-cholan-24-oic acid), 50
Testosterone, 11, 81–83, 89, 399, 400
Testosterone acetate, 11
Testosterone homodimers, 81–83, 400
Tetracyclic cores, 256
Tetradecacyclic cephalostatin analogues, 303
Tetrahydroxy cholaphane, 198, 200, 206
Tetrakis(formyloxy) cholaphane, 198, 200, 206
1,2,4,5-Tetraoxacyclohexane, 383
Tetraoxane dimers, 31, 32, 383, 384
Tetraoxane moiety, 383
Tetraoxane pharmacophore, 384, 386
1,2,4,5-Tetraoxane system, 31
Theonellasterone, 277
Theonella swinhoei, 277, 278
Therapeutic efficacy, 380
Thin-layer chromatography, 5
Thiolate disulphide, 172

3α-Tosyloxy-5β-cholan-24-oate, 25
(25*R*)-3β-Tosyloxy-5α-spirostan-12α-propionate, 296, 297
Transbilayer movement, 173
Transient Diels–Alder condensation, 87
3α,7α,12α-Triacetoxy-5β-glycocholic acid, 160
3α,7α,12α-Triacetoxy-24-nor-5β-cholan-23-oic acid, 212
Triamine spacers, 138
3α-Trifluoroacetoxy-5β-cholan-24-oic acid, 113, 114
3α-Trifluoroacetoxy-5β-cholan-24-oic acid ethane-1,2-diol diester, 113
3α-Trifluoroacetoxy-5β-cholan-24-oyl chloride, 114, 118
Trifluoroacetoxy piperazine diamide, 118
3α,7α,12α-Trihydroxy-5β-cholan-24-amide, 120
3α,7α,12α-Trihydroxy-5β-cholan-24-amine, 120
3α,7α,12α-Trihydroxy-5β-cholan-24-oate, 22
3α,7α,12α-Trihydroxy-5β-cholan-24-oic acid, 22
24-Triisopropylsilyl ether-5β-cholan-3α,7α,12α-triol, 214
Trioxabicyclononanes, 177, 179, 180
Trioxane moiety, 383
1,2,4-Trioxane rings, 383
24-Triphenylmethoxy-5β-cholan-3β-ol, 47
Trisdecacyclic pyrazine dimers, 324, 326, 327
3α,7α,12α-Tris(formyloxy)-24-nor-5β-chol-22-ene, 200, 201
Tumour cell lines, 250, 263, 334, 391, 396, 397
Tumour-inhibiting properties, 303
Tumour xenografts, 386

Umbrella walls, 172, 381
Unidirectional Ugi-MiB approach, 235
Unsaturated B-rings, 69
Unsymmetrical pyrizine dimers, 349

Vanillin-sulphuric acid, 5
Vanuatu Island marine sponge, 254, 256

Western hemisphere, 255–263, 266–270, 274, 276
Wittig reaction, 147, 148, 181
Wolff–Kishner reduction, 98

Xenografts, 386
Xylene, 174, 175, 177
m-Xylenediamine, 217

Yamaguchi method, 236
Yarrowia lipolytica, 405